Integrated Management of Insect Pests on Canola and Other *Brassica* Oilseed Crops

Integrated Management of Insect Pests on Canola and Other *Brassica* Oilseed Crops

Edited by

Gadi V.P. Reddy

CABI

CABI is a trading name of CAB International

CABI
Nosworthy Way
Wallingford
Oxfordshire OX10 8DE
UK

Tel: +44 (0)1491 832111
Fax: +44 (0)1491 833508
E-mail: info@cabi.org
Website: www.cabi.org

CABI
745 Atlantic Avenue
8th Floor
Boston, MA 02111
USA

Tel: +1 (617)682-9015
E-mail: cabi-nao@cabi.org

A catalogue record for this book is available from the British Library, London, UK.

Library of Congress Cataloging-in-Publication Data

Names: Reddy, Gadi V. P., editor.
Title: Integrated management of insect pests on canola and other Brassica
oilseed crops / edited by Gadi V.P. Reddy.
Description: Boston, MA : CABI, [2017] | Includes bibliographical references and index.
Identifiers: LCCN 2016045651| ISBN 9781780648200 (hbk : alk. paper) | ISBN
9781780648224 (epub)
Subjects: LCSH: Rape (Plant)--Diseases and pests--Integrated control. |
Canola--Diseases and pests--Integrated control. | Brassica--Diseases and pests--Integrated control.
Classification: LCC SB608.R2 I58 2017 | DDC 633.8/53--dc23 LC record available at
https://lccn.loc.gov/2016045651

ISBN-13: 978 1 78064 820 0

Commissioning editor: Ward Cooper
Editorial assistant: Emma McCann
Production editor: Tim Kapp

Typeset by SPi, Pondicherry, India
Printed and bound in the UK by CPI Group (UK) Ltd, Croydon, CR0 4YY

Contents

About the Editor

Dr Gadi V.P. Reddy is working as an Entomologist at Montana State University Western Triangle Ag Research Center in Conrad, Montana, USA. He has 26 years of entomological research, teaching and extension experience and has worked in various countries and programmes. Dr Reddy has a strong background in pest management, biocontrol, behavioural and chemical ecology and multitrophic interactions. His research achievements, which include 150 publications in international journals in the key areas of pest management, are a testimony to the high calibre of his research, productivity and expertise.

Contributors

Francesc Rubén Badenes-Pérez Instituto de Ciencias Agrarias, 28006-Madrid, Spain. E-mail: frbadenes@ica.csic.es

Harit K. Bal Michigan State University, Department of Entomology, 578 Wilson Road, 205 CIPS, East Lansing, Michigan 48824, USA. E-mail: bal@msu.edu

Ronald E. Batallas Department of Biological Sciences, CW 405 Biological Sciences Building, University of Alberta T6G 2E9, Canada. E-mail: batallas@ualberta.ca

Tulsi Bhardwaj Division of Agricultural Extension, Indian Agricultural Research Institute, Pusa, New Delhi-110012, India. E-mail: tbhardwaj2003@yahoo.com

James D. Blande Department of Environmental and Biological Sciences, University of Eastern Finland, PO Box 1627, FIN- 70211, Kuopio, Finland. E-mail: james.blande@uef.fi

Dhana Raj Boina Plant Health Management Division, National Institute of Plant Health Management, Hyderabad-500 030 Telangana State, India. E-mail: boinaraj@gmail.com

Randall Brandt Agriculture and Agri-Food Canada, 5403 - 1 Ave S. Lethbridge, Alberta T1J 4B1, Canada. E-mail: randall.brandt@agr.gc.ca

G. David Buntin Department of Entomology, University of Georgia – Griffin Campus, Griffin, Georgia 30223, USA. E-mail: gbuntin@uga.edu

Héctor A. Cárcamo Agriculture and Agri-Food Canada, 5403 - 1 Ave S. Lethbridge, Alberta T1J 4B1, Canada. E-mail: hector.carcamo@agr.gc.ca

Surendra Dara Strawberry and Vegetable Crops Advisor and Affiliated IPM Advisor, San Luis Obispo, Santa Barbara, and Ventura Counties, University of California Cooperative Extension, 2156 Sierra Way, Ste. C, San Luis Obispo, California 93401, USA. E-mail: skdara@ucdavis.edu

Tim J. Dumonceaux Agriculture and Agri-Food Canada, 107 Science Place, Saskatoon, Saskatchewan S7N 0X2, Canada. E-mail: tim.dumonceaux@agr.gc.ca

Robert (Bob) H. Elliott Agriculture and Agri-Food Canada, 107 Science Place, Saskatoon, Saskatchewan S7N 0X2, Canada. E-mail: bob.elliott@agr.gc.ca

Maya Evenden Department of Biological Sciences, CW 405 Biological Sciences Building, University of Alberta T6G 2E9, Canada. E-mail: mevenden@ualberta.ca

Yaghoub Fathipour Department of Entomology, Faculty of Agriculture, Tarbiat Modares University, PO Box 14115-336, Tehran, Iran. E-mail: fathi@modares.ac.ir

Kevin D. Floate Lethbridge Research and Development Centre, Agriculture and Agri-Food Canada, PO Box 3000, Lethbridge, Alberta T1J 4B1, Canada. E-mail: kevin.floate@agr.gc.ca

John Gavloski Manitoba Agriculture, Crop Industry Branch, 65-3rd Ave NE, Box 1149, Carman, Manitoba R0G 0J0, Canada. E-mail: john.gavloski@gov.mb.ca

Kristopher L. Giles Department of Entomology and Plant Pathology, 127 NRC, Oklahoma State University, Stillwater, Oklahoma 74078-3033, USA. E-mail: kris.giles@okstate.edu

Larry Grenkow Agriculture and Agri-Food Canada, Saskatoon Research Centre, 107 Science Place, Saskatoon, Saskatchewan S7N 0X2, Canada. E-mail: larry.grenkow@agr.gc.ca

Parwinder S. Grewal College of Sciences, University of Texas Rio Grande Valley, Edinburg, Texas 78539, USA. E-mail: parwinder.grewal@utrgv.edu

Rebecca H. Hallett School of Environmental Sciences, University of Guelph, Guelph, Ontario N1G 2W1, Canada. E-mail: rhallett@uoguelph.ca

Vincent A. Hervet Alberta Agriculture and Forestry, J.G. O'Donoghue Building, 307, 7000 - 113 Street, Edmonton, Alberta T6H 5T6, Canada. E-mail: vincent.hervet@gmail.com

Matthew P. Hill CSIRO, Canberra ACT 2601, Australia. E-mail: matt.hill@csiro.au

Sari J. Himanen Ecological Crop Science, Natural Resources and Bioproduction, Natural Resources Institute Finland (Luke), Lönnrotinkatu 5, Mikkeli FI-50100, Finland. E-mail: sari.himanen@luke.fi

Heikki M.T. Hokkanen Department of Agricultural Sciences, Box 27 [Latokartanonkaari 5], FIN-00014 University of Helsinki, Finland. E-mail: heikki.hokkanen@helsinki.fi

Jarmo K. Holopainen Department of Environmental and Biological Sciences, University of Eastern Finland, PO Box 1627, FIN- 70211, Kuopio, Finland. E-mail: jarmo.holopainen@uef.fi

Prashant Jha Southern Agricultural Research Center, Montana State University, 748 Railroad Hwy., Huntley, Montana 59037, USA. E-mail: pjha@montana.edu

Janet J. Knodel Department of Plant Pathology, North Dakota State University, Walster Hall, Room 210, NDSU Dept 7660, PO Box 6050, Fargo, North Dakota 58108-6050, USA. E-mail: janet.knodel@ndsu.edu

Tao Li Terrestrial Ecology Section, Department of Biology, University of Copenhagen, Universitetsparken 15, DK-2100 Copenhagen Ø, Denmark. E-mail: tao.li@bio.ku.dk

Sarina Macfadyen CSIRO, Canberra ACT 2601, Australia. E-mail: sarina.macfadyen@csiro.au

Peter Mason Agriculture and Agri-Food Canada, 960 Carling Ave, Ottawa, Ontario K1A 0C6, Canada. E-mail: masonp@agr.gc.ca

Ingeborg Menzler-Hokkanen Department of Agricultural Sciences, Box 27 [Latokartanonkaari 5], FIN-00014 University of Helsinki, Finland. E-mail: ingeborg.menzler-hokkanen@helsinki.fi

Mohammad Ali Mirhosseini Department of Entomology, Faculty of Agriculture, Tarbiat Modares University, PO Box 14115-336, Tehran, Iran. E-mail: mohammad.mirhosseini@modares.ac.ir

Christine Noronha Agriculture and Agri-Food Canada, 440 University Avenue, PO Box 1210, Charlottetown, Prince Edward Island C1A 7M8, Canada. E-mail: christine.noronha@agr.gc.ca

Chrystel Olivier Science and Technology Branch, Agriculture and Agri-Food Canada, 107 Science Place, Saskatoon, Saskatchewan S7N 0X2, Canada. E-mail: drchrystelolivier@gmail.com

Edel Pérez-López Fraccionamiento La Pradera, Retorno Fucsia, No 10-B, Emiliano Zapata, Veracruz, Mexico. E-mail: edellopez1987@gmail.com

S. Jesu Rajan Plant Health Management Division, National Institute of Plant Health Management, Hyderabad-500 030 Telangana State, India. E-mail: sjrajan83@gmail.com

Gadi V.P. Reddy Western Triangle Agricultural Research Center, Montana State University, 9546 Old Shelby Road, PO Box 656, Conrad, Montana 59425, USA. E-mail: reddy@montana.edu

Tom A. Royer Department of Entomology and Plant Pathology, 127 NRC, Oklahoma State University, Stillwater, Oklahoma 74078-3033, USA. E-mail: tom.royer@okstate.edu

Muhammad Sarwar Pakistan Atomic Energy Commission, National Institute for Biotechnology & Genetic Engineering (NIBGE), Faisalabad 44000, Punjab, Pakistan. E-mail: drmsarwar64@yahoo.com

Barbara J. Sharanowski Department of Biology, University of Central Florida, Orlando, Florida, USA. E-mail: barb.sharanowski@gmail.com

Govinda Shrestha Western Triangle Agricultural Research Center, Montana State University, 9546 Old Shelby Road, PO Box 656, Conrad, Montana 59425, USA. E-mail: govinda.shrestha@montana.edu

Juliana J. Soroka Science and Technology Branch, Agriculture and Agri-Food Canada, 107 Science Place, Saskatoon, Saskatchewan S7N 0X2, Canada. E-mail: julie.soroka@agr.gc.ca

Rajagopalbabu Srinivasan University of Georgia, Department of Entomology, 2360 Rainwater Road, Tifton, Georgia 31793, USA. E-mail: babusri@uga.edu

Raj Kumar Thakur Project coordinator, AICRP (Honeybees & Pollinators), Division of Entomology, Indian Agricultural Research Institute, New Delhi-110012, India. E-mail: rkt_apic@rediffmail.com

Sally Vail Agriculture and Agri-Food Canada, 107 Science Place, Room SSW 317, Saskatoon, Saskatchewan S7N 0X2, Canada. E-mail: sally.vail@agr.gc.ca

Leyun Wang Department of Entomology, College of Plant Protection, China Agricultural University, Beijing 100193, China. E-mail: wangleyun@yahoo.com

R.W.M. Udari M. Wanigasekara Department of Entomology, 214 Animal Science Bldg., University of Manitoba, Winnipeg R3T 2N2, Canada. E-mail: udari_madu@yahoo.com

Chaminda De Silva Weeraddana Department of Biological Sciences, CW 405 Biological Sciences Building, University of Alberta, Edmonton T6G 2E9, Canada. E-mail: weeradda@ualberta.ca

Tyler J. Wist Agriculture and Agri-Food Canada, Saskatoon Research Centre, 107 Science Place, Saskatoon, Saskatchewan S7N 0X2, Canada. E-mail: tyler.wist@agr.gc.ca
Zi-Hua Zhao Department of Entomology, College of Plant Protection, China Agricultural University, Beijing 100193, China. E-mail: zhzhao@cau.edu.cn

Acknowledgements

The following reviewers are gratefully acknowledged for taking time in reviewing the manuscripts.

Francesc Rubén Badenes-Pérez
Rayappa Balikai
Héctor A. Cárcamo
Xuedong Chen
Sam M. Cook
William Rodney Cooper
John Gavloski
Rieta Gols
Tim Haye
Murray B. Isman
Vivek Kumar
Geneviève Labrie
Robert J. Lamb
Peter Mason
Ross H. Miller
Boyd A. Mori
Joseph E. Munyaneza
Robert S. Nofemela
Christine Noronha
Scott L. Portman
Marisol Quintanilla
Anantanarayanan Raman
Srinivasan Ramasamy
Muhammad Riaz
Jhalendra Rijal
Tom Royer
Muhammad Salman
Ian M. Scott
David I. Shapiro-Ilan
Tony Shelton
Hugh A. Smith
Junji Takabayshi
Matthew D. Thom
Charles Vincent
Robert Wright
Shaohui Wu

1 Flea Beetles (*Phyllotreta* spp.) and Their Management

Janet J. Knodel*

North Dakota State University, Fargo, North Dakota, USA

1.1 Introduction

Flea beetles (Coleoptera: Chrysomelidae), in the genus *Phyllotreta*, are important economic pests of canola production worldwide. The crucifer flea beetle, *Phyllotreta cruciferae* (Goeze), and the striped flea beetle, *Phyllotreta striolata* (Fabricius), are the two most common pest species in canola production in North America (Lamb, 1984; Weiss *et al.*, 1991; Palaniswamy and Lamb, 1992). Although there are five species of flea beetle that infest *Brassica* spp. crops, these other species are generally not a major threat to oilseed crop production (Burgess, 1977a; Wylie, 1979). *P. cruciferae* is widespread in Europe, Asia and Africa (Brown, 1967; Wylie, 1979). It was introduced into North America in the early 1920s in British Columbia and is now found across southern Canada and the USA (Wylie, 1979). *P. striolata* is native to Eurasia and was first discovered in North America in sediment dating before 1668 in Québec, Canada (Rousseau and LeSage, 2016). Its current distribution includes Canada, the eastern and western USA, Mexico, South America and South Africa (Chittenden, 1923; Wylie, 1979). Yield losses from *Phyllotreta* species in canola have been estimated at tens of millions of US dollars annually (Burgess, 1977a; Lamb and Turnock, 1982; Madder and Stemeroff, 1988). Lamb and Turnock (1982) reported that yield losses of 8–10% occur from flea beetle feeding injury even when the crop is protected with an insecticide. Control costs for *Phyllotreta* flea beetles have been estimated to be about US$300 million annually in canola production areas of the northern Great Plains (Knodel and Olson, 2002).

1.2 Biology

Phyllotreta cruciferae and *P. striolata* have a single generation in their northern distribution and two or more generations in their southern distribution (Burgess, 1977a; Wylie, 1979; Lamb, 1983; Knodel and Olson, 2002; Andersen *et al.*, 2005). Adult *P. cruciferae* are about 2–3 mm long and black with an iridescent blue sheen on their elytra (Fig. 1.1). Adult *P. striolata* are black with two yellow stripes on their elytra and are about 1.5–2.5 mm long (Fig. 1.2). Adults of *Phyllotreta* flea beetles are often observed 'jumping' when they are disturbed, which is due to their enlarged femurs on their hind legs.

P. cruciferae prefer to overwinter in the leaf litter beneath shrubs and brush or in wooded areas (Andersen *et al.*, 2005; Ulmer and Dosdall, 2006). Turnock *et al.* (1987) found that the overwintering survival of *Phyllotreta* flea beetles was about 70% in temperate areas. It is suggested that females have a higher success rate of overwintering, since spring-emerging *P. cruciferae* have a sex ratio of 1.5 females to one male (Weiss *et al.*, 1994; Ulmer and Dosdall, 2006). The hibernating adults begin to emerge from their overwintering sites in early spring as the mean ground temperature rises to 10–12°C and emergence peaks when ground temperatures reach 14–15°C (Ulmer and Dosdall, 2006). Depending on the fluctuations in spring temperature, it may take up to 3 weeks or more for the adults to leave the overwintering sites (Westdal and Romanow, 1972; Burgess, 1977a; Wylie, 1979). Beetles typically feed on volunteer canola, mustard and weeds of Brassicaceae before moving to spring-planted canola

*E-mail: janet.knodel@ndsu.edu

(Westdal and Romanow, 1972; Burgess, 1977a, 1981; Wylie, 1979). *Phyllotreta* flea beetles are strong flyers and can easily disperse to find spring-planted canola fields, regardless of the proximity of fields to their overwintering sites (Lamb, 1983; Burgess and Spurr, 1984; Ulmer and Dosdall, 2006).

Adults of *Phyllotreta* species feed and reproduce for about 10–12 weeks (Burgess, 1977a; Wylie, 1979; Ulmer and Dosdall, 2006). Females oviposit in the soil during June and deposit up to 25 eggs per female; eggs are deposited singly or in groups of three or four on the roots of host plants (Westdal and Romanow, 1972). Eggs are oval, yellow and about 0.38–0.46 mm long by 0.18–0.25 mm wide. Larvae hatch from the eggs in about 12 days and feed on the secondary roots of the plant. Larvae are small (approximately 3 mm), whitish, slender, cylindrical worms with tiny legs and a brown head and anal plate. Larvae progress through three instars over a period of 25–34 days and then form an earthen puparium. Pupae are white except for the black eyes. The pupal stage lasts for about 7–9 days.

The new generation of adult *Phyllotreta* flea beetles emerges from puparia beginning in August and emergence continues through September in northern North America (Ulmer and Dosdall, 2006). Ulmer and Dosdall (2006) found that the sex ratio of *P. cruciferae* for the new generation was 1.2 females to one male. Beetles feed on the epidermis of green foliage and pods (Fig. 1.3); however,

Fig. 1.1. Adult crucifer flea beetle (*Phyllotreta cruciferae* (Goeze)). Photograph by P. Beauzay, NDSU.

Fig. 1.2. Adult striped flea beetle (*Phyllotreta striolata* (Fabricius)). Photograph by P. Beauzay, NDSU.

Fig. 1.3. Feeding on seed pods of canola in late summer by new generation of *Phyllotreta* flea beetles. Photograph by P. Beauzay, NDSU.

due to the advanced stage of the crop, feeding injury is usually negligible (Knodel and Olson, 2002). In fall, adults move back to overwintering sites and repeat their life cycle (Burgess, 1977a, 1981; Wylie, 1979).

1.3 Hosts

Phyllotreta flea beetles are selective oligophagous herbivores that feed primarily on host plants in the family Brassicaceae (Feeny *et al.*, 1970). Many of the Brassicaceae plants produce a mustard oil called allyl isothiocyanate, a glucosinolate breakdown product, which is attractive to *Phyllotreta* flea beetles in the field (Vincent and Stewart, 1984; Chew, 1988; Pivnick *et al.*, 1992; Hopkins *et al.*, 2009). The most preferred agricultural hosts attacked by *Phyllotreta* species include oil rapeseed or Argentine canola (*Brassica napus* L.), Polish canola (*Brassica rapa* L.) and oriental or brown mustard (*Brassica juncea* (L.) Czern.) (Palaniswamy and Lamb, 1992; Palaniswamy *et al.*, 1992; Soroka and Grenkow, 2013). Mustard species (*Sinapis alba* L. and *B. juncea*) typically have less damage from feeding *Phyllotreta* flea beetles and higher yield compared with canola *Brassica* species (Brandt and Lamb, 1991, 1993; Hopkins *et al.*, 1998; Brown *et al.*, 2004). Henderson *et al.* (2004) found that *S. alba* has a volatile deterrent phytochemical that inhibits feeding by *Phyllotreta* flea beetles. Crambe (*Crambe abyssinica* Hochst. and *Crambe hispanica* L.) also had lower feeding injury by *Phyllotreta* flea beetles compared with canola *Brassica* species due to a non-volatile deterrent phytochemical (Chengwang *et al.*, 1992; Henderson *et al.*, 2004; Soroka and Grenkow, 2013). False flax (*Camelina sativa* (L.) Crantz) was found to act as a non-host of *Phyllotreta* flea beetles since it lacked the cues to initiate feeding (Pachagounder *et al.*, 1998; Henderson *et al.*, 2004; Soroka and Grenkow, 2013). In the garden, *Phyllotreta* flea beetles will feed on other *Brassica* species, such as broccoli (*Brassica oleracea* var. *italica*), cabbage (*B. oleracea* var. *capitata*), cauliflower (*B. oleracea* var. *botrytis*), kale (*B. oleracea* var. *acephala*), Brussels sprouts (*B. oleracea* var. *gemmifera*), turnip (*Brassica rapa* subsp. *rapifera*), horseradish (*Armoracia rusticana* Gaertn., Mey. and Scherb.) and radish (*Raphanus sativus* L.) (Westdal and Romanow, 1972). Weeds infested by *Phyllotreta* species include flixweed (*Descurainia sophia* (L.)), field pennycress/stinkweed (*Thlaspi arvense* L.), peppergrass (*Lepidium virginicum* L.) and wild mustard (*Sinapis arvensis arvensis* L.) (Westdal and Romanow, 1972). Laboratory feeding studies showed that *P. cruciferae* also fed on plants in the caper family (Capparidaceae), the nasturtium family (Tropaeolaceae) and the marshflower family (Limnanthaceae) (Feeny *et al.*, 1970).

1.4 Crop Damage

The greatest crop loss from this pest occurs in the seedling stage within the first 2 weeks after plant emergence (Turnock and Lamb, 1982; Lamb, 1984; Bracken and Bucher, 1986). Adult feeding on cotyledons causes the tissue to die around the feeding sites, creating a shot-hole appearance and necrosis on seedlings (Fig. 1.4). Feeding injury is often concentrated on one cotyledon only, since *P. cruciferae* tend to aggregate during feeding (Anderson *et al.*, 1992). As a result of herbivory by *Phyllotreta*, the plant's ability to conduct photosynthesis is negatively affected, often causing wilting and seedling death (Westdal and Romanow, 1972). Damage from *Phyllotreta* herbivory on seedlings results in reduced crop stands, causing lower seed yield and quality, and uneven plant growth, causing delayed maturity (Putnam, 1977; Lamb and Turnock, 1982; Lamb, 1984; Weiss *et al.*, 1991). Fields may need to be reseeded when canola stands are below 43 plants/m^2 (Kandel and Knodel, 2011). Gavloski and Lamb (2000) found that compensation by canola seedlings from

Fig. 1.4. Canola seedling damaged by *Phyllotreta* flea beetles feeding on cotyledons (note pitting and shot holing). Photograph by P. Beauzay, NDSU.

flea beetle herbivory was a function of whether the insects destroyed the apical meristem and the degree of defoliation. *Phyllotreta* flea beetles have been recorded attacking the growing point (meristem tissue) of four- to six-leaf seedlings, killing the plant when populations are significant and environmental conditions are hot and dry (Burgess, 1977a; Lamb, 1984; Knodel and Olson, 2002). Warm, dry weather promotes flea beetle flight and increases feeding activity (Burgess, 1977a; Turnock and Lamb, 1982; Lamb, 1984). In contrast, cool, wet and windy conditions cause flea beetles to creep slowly just into field edges, where feeding damage is often lower (Westdal and Romanow, 1972).

During the summer, larval feeding injury on the secondary root hairs causes only a negligible effect on canola yield. Bracken and Bucher (1986) reported a yield loss of 5% from larval densities of 0.16/cm^2 in Manitoba, Canada. In the summer, a new generation of *Phyllotreta* flea beetles feeds on the epidermis of green foliage and pods of mature canola (Feeny *et al.*, 1970). This feeding damage results in poor seed fill, premature pod drying, shrivelled seeds or pod shattering, and provides an entry point for fungal growth within pods in damp weather (Knodel and Olson, 2002). When populations of the new generation are significant, reduced seed production and quality can occur, especially on the upper or younger pods or late-seeded crop (Knodel and Olson, 2002).

1.5 Integrated Pest Management

Implementing an integrated pest management (IPM) programme provides the best pest management strategies for *Phyllotreta* flea beetles in canola. IPM uses pest monitoring in conjunction with economic thresholds and multiple strategies to promote the judicious use of insecticides and conservation of natural enemies of pests (Pedigo and Rice, 2009).

1.6 Monitoring

The most critical time to monitor for *Phyllotreta* species is when spring-planted canola is emerging and ground temperatures are above 14–15°C (Knodel and Olson, 2002; Ulmer and Dosdall, 2006). Canola fields should be scouted by walking a 'W' pattern in the field and inspecting ten plants randomly at five sites per field for a total of 50 plants/field. The amount of defoliation is used as a guide to determine the need for management action (Knodel and Olson, 2002). The percentage defoliation is estimated for each plant selected. Scouting should continue until plants have reached the four- to six-leaf stage (Knodel and Olson, 2002).

Traps also are used for monitoring populations of *Phyllotreta* flea beetles in canola. Yellow and white are preferred by *Phyllotreta* species over other colours, such as red or blue (Vincent and Stewart, 1985; Adams and Los, 1986; Láska *et al.*, 1986). Yellow sticky traps (Fig. 1.5) are available commercially from insect trap suppliers and can be used to detect first emergence and population peaks (Knodel

Fig. 1.5. Yellow sticky trap used for monitoring *Phyllotreta* flea beetles. Photograph by J. Knodel, NDSU.

and Olson, 2002). Lamb (1983) reported that neither sticky traps nor suction traps (baited or unbaited) are a reliable system for predicting densities of *Phyllotreta* flea beetles. However, Andersen *et al.* (2005) found that adult *P. cruciferae* catch on yellow sticky card was highly correlated with subsequent feeding injury on *Brassica* plants.

Male *Phyllotreta* flea beetles produce an aggregation pheromone, sesquiterpenes, which attracts both sexes (Chengwang and Weiss, 1992; Chengwang *et al.*, 1999; Toth *et al.*, 2005; Bartelt *et al.*, 2011). Using modified yellow plastic boll weevil traps, Soroka *et al.* (2005) found that combinations of the aggregation pheromone and allyl isothiocyanate (a crucifer-specific volatile) generally attracted higher numbers of flea beetles than either component by itself in the field. Vincent and Stewart (1984) reported that *P. cruciferae* was more attracted to allyl isothiocyanate than *P. striolata*; however, *Psylliodes punctulata* Melsheimer was not affected by the presence of allyl isothiocyanate. Gruber *et al.* (2009) found that allyl isothiocyanate was attractive to *Phyllotreta* flea beetles only in the spring and early fall, but it was inhibitory in the late fall. Currently, no aggregation pheromone is sold commercially by insect trap suppliers for monitoring *Phyllotreta* flea beetles.

The narrow time frame between spring emergence of *Phyllotreta* flea beetles and crop infestation limits the use of traps as a decision-making tool for scouts, crop consultants and producers (Ulmer and Dosdall, 2006).

1.7 Chemical Control

Systemic insecticides in the neonicotinoid group of insecticides (IRAC 4A) that are applied as a seed treatment to canola seeds are the primary means of *Phyllotreta* pest management (Lamb and Turnock, 1982; Weiss *et al.*, 1991, 1994; Knodel *et al.*, 2008). Active ingredients in registered neonicotinoid insecticides for flea beetle control in canola in the USA include imidacloprid, thiamethoxam and clothianidin (Knodel *et al.*, 2015). Neonicotinoid seed treatments are the most widely used management strategy for control of *Phyllotreta* flea beetles. More than 90% of canola acreage in Canada and 95% in the USA are planted with neonicotinoid-treated seed (Soroka *et al.*, 2008). Cyantraniliprole (IRAC 28), a newer active ingredient, is also available as an insecticide seed treatment in canola for flea beetle control. Insecticide seed treatments generally have an advantage over foliar spray applications of insecticides, because of their convenience during planting and extended protection against flea beetles when producers are busy planting other crops. Research has shown that neonicotinoid seed treatments are effective for reducing flea beetle damage to seedling canola, and seedling protection typically lasts for about 21 days (Antwi *et al.*, 2007a, b; Knodel *et al.*, 2008). Soroka *et al.* (2008) found that decreasing the ratio of neonicotinoid (acetamiprid or clothianidin) treated seed to two-thirds of the seed coated with insecticide was comparable to those in 1× treatments for flea beetle feeding injury, plant stands and seed yield except when feeding pressure was significant; however, there was no cost advantage to decreasing the volume of insecticide-treated seed. As result of this study, canola producers are discouraged from mixing untreated seed with insecticide-coated seed to reduce costs or insecticide use, or to use up stock of on-farm bin-run seed (Soroka *et al.*, 2008). The prophylactic use of insecticide seed treatments may be unnecessary in some situations where *Phyllotreta* populations are below economically damaging populations. However, producers must decide before planting whether an insecticide seed treatment will be used and before the population levels of *Phyllotreta* spp. are known. Currently, there is no forecasting model that predicts outbreaks of *Phyllotreta* flea beetles in the spring and their potential for damage to the canola crop (Thomas, 2003).

Since canola is a preferred crop for honey production by honey bees, *Apis mellifera* (L.), there are many concerns about the adverse risks of systemic neonicotinoid seed treatments in canola to bee health. Although systemic insecticide seed treatments are generally considered more ecologically sound than foliar-applied insecticides, systemic insecticides may translocate toxins to pollen or nectar during crop development, which could negatively impact foraging pollinators and cause pesticide poisoning. However, several studies indicate that long-term exposure of honey bees to neonicotinoid insecticide seed treatments in canola poses a minor risk to bee health (Schmuck *et al.*, 2001; Maus *et al.*, 2003; Schmuck, 2004; Faucon *et al.*, 2005; Cutler and Scott-Dupree, 2007; Nguyen *et al.*, 2009). Cutler and Scott-Dupree (2007) placed bee hives in the middle of 1 ha clothianidin-treated canola and untreated control fields during bloom for 3 weeks and found no differences in bee mortality, worker longevity or brood development between control and clothianidin-treated canola. Maus *et al.* (2003) reviewed the safety of imidacloprid to honey bees and reported that the majority of studies found no acute or chronic toxicity of imidacloprid at ≤ 20 ppb. However, honey bees rejected imidacloprid-contaminated food at 20 ppb in the laboratory (Kirchner, 1999). Decourtye *et al.* (2001) reported compromised learning in honey bees after exposure to low rates of imidacloprid-contaminated food at 12–48 ppb in the laboratory. Suchail *et al.* (2001) observed honey bees fed low concentrations of imidacloprid had high chronic toxicity. Additional research is needed to improve knowledge on the interactions of pollinators and pesticides; both are key components of modern agriculture.

Foliar-applied insecticides are often necessary when the peak emergence of *Phyllotreta* flea beetles is delayed beyond the 21-day window of protection from insecticide seed treatments due to cool temperature (Weiss *et al.*, 1991; Ulmer and Dosdall, 2006; Knodel *et al.*, 2008). In addition, foliar-applied insecticidal sprays are often necessary to protect the canola crop when cool, wet weather slows the growth and uptake of insecticide seed treatment in the canola plant and *Phyllotreta* populations are significant (Knodel *et al.*, 2008). Some active ingredients of pyrethroid (IRAC 3A) insecticides registered for flea beetle control in canola include bifenthrin, deltamethrin, gamma-cyhalothrin, lambda-cyhalothrin and zeta-cypermethrin (Knodel *et al.*, 2015).

Reduced-risk insecticides have been tested for efficacy against *Phyllotreta* flea beetles. Elliott *et al.* (2007) found that spinosad was more toxic by ingestion than topical contact and that flea beetle mortality was increased with longer exposure times (more than 120 h) and toxicity was increased when an ionic surfactant was added. For *P. cruciferae* management in canola, spinosad was found to be the most effective ecorational insecticide compared with neem (azadirachtin), pyrethrin, kaolin (a clay) and the fungal entomopathogen *Beauvaria bassiana* (Andersen *et al.*, 2006; Antwi *et al.*, 2007a, b). Reddy *et al.* (2014) reported that the combination of two entomopathogenic fungi, *Metarhizium brunneum* and *B. bassiana*, applied twice were as effective as conventional insecticides and presented a more ecologically sound alternative to conventional insecticide applications.

To protect canola from yield loss, foliar insecticides are often applied at a nominal threshold level of 25% defoliation to cotyledon and true leaves (Weiss *et al.*, 1991; Anon., 1997; Knodel and Olson, 2002). Most recently, research by Tangtrakulwanich *et al.* (2014) found that foliar treatments must be made quickly if damage exceeds a lower action threshold of 15–20% defoliation. Knodel and Olson (2002) recommended applying insecticides during the sunny, warm part of the day when flea beetles are actively feeding on the plants. Canola plants that have reached the four- to six-leaf vegetative growth stage or beyond can tolerate more feeding damage, unless flea beetles are damaging the growing point (Knodel and Olson, 2002).

Phyllotreta flea beetles have been effectively controlled by neonicotinoid-treated canola seed for more than a decade. Insecticide resistance occurs more rapidly when there is widespread adoption of one insecticide or insecticide class used year after year against an abundant pest. In Canada, Tansey *et al.* (2008, 2009) found that the two species, *P. cruciferae* and *P. striolata*, had different susceptibilities when exposed to canola neonicotinoid seed treatment with active ingredients clothianidin, imidacloprid and thiamethoxam. *P. cruciferae* had a higher mortality and exhibited less feeding when exposed to thiamethoxam and clothianidin compared with *P. striolata* (Tansey *et al.*, 2008, 2009). These effects were even more apparent when the two species were subjected to intraspecific crowding and stresses from overwintering, such as depletion of lipid and glycogen reserves (Tansey *et al.*, 2008). Differences in efficacy could cause a shift in the prevalence of flea beetle species from *P. cruciferae* to *P. striolata* where these species occur sympatrically and where seed-treated canola is grown in large acreage.

Pesticide resistance has costly consequences. Pimentel (2005) estimated that US$1.5 billion of agronomic losses occurred each year in the USA due to the development of pesticide resistance in insect pests. Alternative IPM strategies for *Phyllotreta* flea beetle control are needed in canola. For example, Zhao *et al.* (2008) cloned and characterized the arginine kinase (AK) gene from *P. striolata* and constructed dsRNA to impair the beetle's development and to enhance mortality of adults. In addition, RNAi targeting the AK gene reduced fecundity and fertility of *P. striolata*, suggesting that this is a potential new strategy to help delay insecticide resistance (Zhao *et al.*, 2008).

1.8 Cultural Control

Cultural control studies have demonstrated that different cropping systems (described below) have the ability to decrease the extent of feeding injury by *Phyllotreta* flea beetles in canola. Although the use of cultural strategies may not completely eliminate the need for insecticidal control, they offer the possibility for managing *Phyllotreta* flea beetles and reducing insecticide use in canola production.

1.8.1 Seeding dates

Peak flea beetle emergence often coincides with the germination of early-seeded canola in April or early May resulting in higher flea beetle feeding injury due to most of the canola being in the susceptible seedling stage (Ulmer and Dosdall, 2006; Knodel *et al.*, 2008). In contrast, less flea beetle feeding injury was observed on late-seeded canola from late May to early June (Lamb, 1984, 1988; Knodel *et al.*, 2008). However, yield of late-seeded canola is often lower than early-seeded canola, due to heat stress during flowering (Angadi *et al.*, 2000; Knodel *et al.*, 2008). As a result, agronomists recommend seeding canola early for optimal yields due to heat sensitivity during flowering, in spite of the higher risks of flea beetle damage (Angadi *et al.*, 2000). Another seeding date strategy is dormant seeded canola in the fall. Dosdall and Stevenson (2005) found that fall-seeded canola emerged and developed past the susceptible seedling stage before significant numbers of spring *Phyllotreta* flea beetles emerged from overwintering sites. However, fall-seeded canola is considered a high-risk practice by producers, especially in growing areas with extreme temperatures and dry soils in the winter (Kandel and Knodel, 2011).

1.8.2 Seed size, seeding rates and row spacing

Increased seeding rates have been shown to reduce the mean flea beetle damage per plant because there are more plants per unit area (Dosdall and Stevenson, 2005). Dosdall *et al.* (1999) found that increasing the seeding rate also reduced flea beetle injury to canola, since leaf biomass was greater at high seeding rates. Planting large-seeded rather than small-seeded varieties of canola has been shown to lower flea beetle damage and result in higher plant establishment, shoot weight and yield (Bodnaryk and Lamb, 1991; Elliott *et al.*, 2008). Canola planted in wider row spacing of 20–30 cm resulted in decreased feeding injury per plant than narrower row spacing of 10 cm (Dosdall *et al.*, 1999).

1.8.3 Tillage systems

Lower population densities of flea beetles have been observed in no-till fields compared with conventionally tilled fields, probably due to the cooler microenvironment which is less preferred by flea beetles (Dosdall *et al.*, 1999; Milbrath *et al.*, 1995).

1.9 Plant Resistance

Research has found that species and cultivars of Brassicaceae can vary in their levels of resistance to feeding injury by *Phyllotreta* flea beetles (Lamb, 1980, 1984, 1988; Lamb and Palaniswamy, 1990; Bodnaryk and Lamb, 1991; Palaniswamy *et al.*, 1992; Pachagounder and Lamb, 1998; Gavloski *et al.*, 2000). Bodnaryk and Lamb (1991) found that larger seed size in *B. napus* and *S. alba* increased seedling survival due to a lower proportion of cotyledon area damaged compared with smaller seeds, and that this could be a 'desirable' trait for host plant resistance against *Phyllotreta* flea beetles. Gavloski *et al.* (2000) found resistance to *Phyllotreta* flea beetles in yellow mustard (*S. alba* and some *S. alba* × *B. napus* crosses), which was due to the deterrent effect of p–hydroxybenzyl glucosinolate (HOBE or glucosinalbin) (Bodnaryk, 1991). Other researchers found that yellow mustard also has tolerance to flea beetle feeding injury (Bodnaryk and Lamb, 1991; Elliott and Rakow, 1999).

Trichomes on plant leaves, stems and pods in some species of Brassicaceae reduce feeding injury of *Phyllotreta* flea beetles and could be used for a host plant resistance trait in canola breeding. For example, the pods of white mustard, *S. alba*, are covered with large numbers of trichomes and these trichomes reduce feeding by *Phyllotreta* flea beetles compared with the glabrous pods of *B. napus* (Lamb, 1980). Leaves of a wild species, *Brassica villosa* Biv., are hirsute and were resistant to *P. cruciferae* feeding compared with the glabrous leaves of *B. rupestris* Raf., *B. macrocarpa* Guss. and *B. napus* (Palaniswamy and Bodnaryk, 1994). Soroka *et al.* (2011) found that the dense pubescence of leaves deterred feeding of *P. cruciferae* and doubled the time to reach satiation compared with glabrous leaves of *B. napus*. Henderson *et al.* (2004) found that the pre-feeding behaviour of *P. cruciferae* involved use of the antennae, tarsi and mouthparts to determine the suitability of a potential host. Any interruptions, such as dense pubescence preventing the pre-feeding behaviour, could negatively impact feeding time and the amount of feeding, or both (Henderson *et al.*, 2004).

1.10 Biological Control

Most biological control with predators and parasitoids has not been successful in reducing populations of *Phyllotreta* flea beetles in canola. Parasitic wasps, such as *Microtonus* species (Hymenoptera: Braconidae), are known to attack both *P. cruciferae* and *P. striolata;* however, the rate of parasitization is low and/or establishment of introduced parasitoids is unsuccessful (Wylie, 1982, 1983, 1984, 1988; Wylie and Loan, 1984). Only a few predators have been recorded as predaceous on *Phyllotreta* flea beetles: field crickets (Orthoptera: Gryllidae: *Gryllus pennsylvanicus* Burmeister), big-eyed bugs (Hemiptera: Geocoridae: *Geocoris bullatus* (Say)), damsel bugs (Hemiptera: Nabidae: *Nabis alternatus* Parshley and *Nabicula americolimbata* Carayon), lacewing larvae (Neuroptera: Chrysopidae: *Chrysopa* spp.) and two-lined collops (Coleoptera: Melyridae: *Collops vittatus* (Say)) (Burgess, 1977b; Thomas, 1986; Burgess and Hinks, 1987). Since adult beetles emerge during a narrow period in the spring, this makes it difficult for natural enemies to have a negative impact on populations of *Phyllotreta* species; and other life stages (egg, larva and pupa) are protected in the soil.

Entomopathogenic nematodes (Rahbditida: Steinernematidae and Heterorhabditidae) have been evaluated for management of *Phyllotreta* species on crucifer vegetables and canola in the laboratory and field (Xu *et al.*, 2010; Yan *et al.*, 2013; Antwi and Reddy, 2016). Yan *et al.* (2013) found that *Steinernema carpocapsae* All and *Heterorhabditis indica* LN2 controlled the soil-dwelling life stages of *P. striolata* in the field and resulted in lower adult populations and subsequent leaf shot-hole damage and higher yields in cabbage grown in China. Xu *et al.* (2010) discovered that nematodes with higher pathogenicity, greater heat tolerance and reproduction potential increased the potential for biological control of *P. striolata.* Reddy *et al.* (2014) also found that foliar applications of the entomopathogenic nematode *S. carpocapsae* reduced feeding injury and provided control of *P. cruciferae* in canola. Antwi and Reddy (2016) tested a sprayable polymer gel formulation of two entomopathogenic nematodes, *Steinernema* spp. and *Heterorhabditis* spp., and found that these formulations had the highest yield under significant *P. cruciferae* densities and higher yield than conventional neonicotinoid seed treatments. Nematodes offer a viable alternative to conventional insecticides for pest management of *Phyllotreta* flea beetles.

References

Adams, R.G. and Los, L. (1986) Monitoring adult corn flea beetles (Coleoptera: Chrysomelidae) in sweet corn fields with color sticky traps. *Environmental Entomology* 15, 867–873.

Andersen, C.L., Hazzard, R., Van Driesche, R. and Mangan, F.X. (2005) Overwintering and seasonal patterns of feeding and reproduction in *Phyllotreta cruciferae* (Coleoptera: Chrysomelidae) in the Northeast United States. *Environmental Entomology* 34, 749–800.

Andersen, C.L., Hazzard, R., Van Driesche, R. and Mangan, F.X. (2006) Alternative management tactics for control of *Phyllotreta cruciferae* and *Phyllotreta striolata* (Coleoptera: Chrysomelidae) on *Brassica rapa* in Massachusetts. *Journal of Economic Entomology* 99, 803–810.

Anderson, M.D., Peng, C. and Weiss, M.J. (1992) Crambe, *Crambe abyssinica* Hochst., as a flea beetle resistant crop (Coleoptera: Chrysomelidae). *Journal of Economic Entomology* 85, 594–600.

Angadi, S.V., Cutforth, H.W., Miller, P.R., McConkey, B.G., Entz, M.H., Brandt, S.A. and Volkmar, K.M. (2000) Response of three *Brassica* species to high temperature stress during reproductive growth. *Canadian Journal of Plant Science* 80, 693–701.

Anon. (1997) *Flea beetle management for canola, rapeseed and mustard in the northern Great Plains. Sustainable Agriculture Facts. Growing for Tomorrow*. Agriculture and Agri-Food Canada, Ottawa, Canada. Available at: http://www.saskatchewan.ca/

Antwi, F.B. and Reddy, G.V.P. (2016) Efficacy of entomopathogenic nematodes and sprayable polymer gel against crucifer flea beetle (Coleoptera: Chrysomelidae) on canola. *Journal of Economic Entomology* 109, 1706–1712. doi: 10.1093/jee/tow140.

Antwi, F.B., Knodel, J.J. and Olson, D.L. (2007a) Comparative evaluation and economic potential of ecorational versus chemical insecticides for crucifer flea beetle (Coleoptera: Chrysomelidae) management in canola. *Journal of Economic Entomology* 100, 710–716.

Antwi, F.B., Olson, D.L. and Carey, D.R. (2007b) Comparative of ecorational and chemical insecticides against crucifer flea beetle (Coleoptera: Chrysomelidae) on canola. *Journal of Economic Entomology* 100, 1201–1209.

Bartelt, R.J., Zilkowski, B.W., Cossé, A.A., Schnupf, U., Vermillion, K. and Momany, F.A. (2011) Male-specific sesquiterpenes from *Phyllotreta* flea beetles. *Journal of Natural Products* 74, 585–595.

Bodnaryk, R.P. (1991) Developmental profile of sinalbin (p-hydroxybenzylglucosinolate) in mustard seedlings, *Sinapis alba* L., and its relationship to insect resistance. *Journal of Chemical Ecology* 17, 1543–1556.

Bodnaryk, R.P. and Lamb, R.J. (1991) Influence of seed size in canola, *Brassica napus* L., and mustard, *Sinapis alba* L., on seedling resistance against flea beetles, *Phyllotreta cruciferae* (Goeze). *Canadian Journal of Plant Science* 71, 397–404.

Bracken, G.K. and Bucher, G.E. (1986) Yield losses in canola caused by adult and larval flea beetles, *Phylotreta cruciferae* (Coleoptera: Chrysomelidae). *The Canadian Entomologist* 118, 319–324.

Brandt, R.N. and Lamb, R.J. (1991) Mechanisms of resistance to the flea beetle, *Phyllotreta cruciferae* (Goeze) (Coleoptera: Chrysomelidae), in mustard seedlings, *Sinapis alba* L. *Canadian Journal of Plant Science* 71, 13–20.

Brandt, R.N. and Lamb, R.J. (1993) Distribution of feeding damage by *Phyllotreta cruciferae* (Goeze) (Coleoptera: Chrysomelidae), on oilseed rape and mustard seedlings in relation to crop resistance. *The Canadian Entomologist* 125, 1011–1021.

Brown, W.J. (1967) Notes on the extralimital distribution of some species of Coleoptera. *The Canadian Entomologist* 99, 85–93.

Brown, J., McCaffrey, J.P., Brown, D.A., Harmon, B.L. and Davis, J.B. (2004) Yield reduction in *Brassica napus*, *B. rapa*, *B. juncea*, and *Sinapis alba* caused by flea beetle (*Phyllotreta cruciferae* (Goeze) (Coleoptera: Chrysomelidae)) infestation in northern Idaho. *Journal of Economic Entomology* 97, 1642–1647.

Burgess, L. (1977a) Flea beetles (Coleoptera: Chrysomelidae) attacking rape crop in the Canadian Prairie Provinces. *The Canadian Entomologist* 109, 21–32.

Burgess, L. (1977b) *Geocoris bullatus*, an occasional predator on flea beetles (Hemiptera: Lygaeidae). *The Canadian Entomologist* 109, 1519–1520.

Burgess, L. (1981) Winter sampling to determine overwintering sites and estimate density of adult flea beetles of rape (Coleoptera: Chrysomelidae). *The Canadian Entomologist* 113, 441–447.

Burgess, L. and Hinks, C.F. (1987) Predation on adult of the crucifer flea beetle by the northern fall field cricket. *The Canadian Entomologist* 119, 495–496.

Burgess, L. and Spurr, D.T. (1984) Distribution and abundance of overwintering flea beetles (Coleoptera: Chrysomelidae) in a grove of trees. *Environmental Entomology* 13, 941–944.

Chengwang, P. and Weiss, M.J. (1992) Evidence of an aggregation pheromone in the flea beetle, *Phyllotreta cruciferae* (Goeze) (Coleoptera: Chrysomelidae). *Journal of Chemical Ecology* 18, 875–884.

Chengwang, P., Weiss, M.J. and Anderson, M.D. (1992) Flea beetle (Coleoptera: Chrysomelidae) response, feeding, and longevity on oilseed rape and crambe. *Environmental Entomology* 32, 604–609.

Chengwang, P., Bartelt, R.J. and Weiss, M.J. (1999) Male crucifer flea beetle produces an aggregation pheromone. *Physiological Entomology* 24, 98–99.

Chew, F.S. (1988) Biological effects of glucosinolates. In: Cutler, H.G. (ed.) *Biologically Active Natural Products: Potential Uses in Agriculture*. American Chemical Society, Washington, DC, pp. 155–181.

Chittenden, F.H. (1923) Notes on the distribution and habits of North American *Phyllotreta* (Coleop.). *Proceedings of the Entomological Society of Washington* 25, 131–139.

Cutler, G. and Scott-Dupree, C.D. (2007) Exposure to clothianidin seed-treated canola has no long-term impact on honey bees. *Journal of Economic Entomology* 100, 765–772.

Decourtye, A., Le Matayer, M., Pottiau, H., Tisseur, M., Odoux, J.F. and Pham-Delegue, M.H. (2001) Impairment of olfactory learning performances in the honey bee after long term ingestion of imidacloprid. In: Belzunces, L.P., Pelissier, C. and Lewis, G.B. (eds) *Hazard of Pesticides to Bees*. Institut National de la Recherche Agronomique, Paris, pp. 113–117.

Dosdall, L.M. and Stevenson, F.C. (2005) Managing flea beetles (*Phyllotreta* spp.) (Coleoptera: Chrysomelidae) in canola with seeding date, planting density, and seed treatment. *Agronomy Journal* 97, 1570–1578.

Dosdall, L.M., Dolinski, M.G., Cowle, N.T. and Conway, P.M. (1999) The effect of tillage regime, row spacing, and seeding rate of feeding injury by flea beetles, *Phyllotreta* spp. (Coleoptera: Chrysomelidae) in canola in central Alberta, Canada. *Crop Protection* 18, 217–224.

Elliott, R.H. and Rakow, G.F.W. (1999) Resistance of *Brassica* and *Sinapis* species to flea beetles, *Phyllotreta cruciferae*. In: Wratten, N. and Salisbury, P.A. (eds) *New Horizons for an Old Crop: Proceedings of 10th International Rapeseed Congress, Canberra, Australia, 26–29 September 1999*. Canberra GCIRC, Canberra, Australia.

Elliott, R.H., Benjamin, M.C. and Gillott, G. (2007) Laboratory studies on the toxicity of spinosad and deltamethrin to *Phyllotreta cruciferae* (Coleoptera: Chrysomelidae). *The Canadian Entomologist* 139, 534–544.

Elliott, R.H., Franke, C. and Rakow, G.F.W. (2008) Effects of seed size and seed weight on seedling establishment, vigour and tolerance of Argentine canola (*Brassica napus*) to flea beetles, *Phyllotreta* spp. *Canadian Journal of Plant Science* 88, 207–217.

Faucon, J.P., Aurieres, C., Drajnudel, P., Ribiere, M., Martel, A.C., Zeggane, S., Chauzat, M.P. and Aubert, M.F.A. (2005) Experimental study on the toxicity of imidacloprid given in syrup to honey bee (*Apis mellifera*). *Pest Management Science* 61, 111–125.

Feeny, P., Paauwe, K. and Demong, N. (1970) Flea beetles and mustard oils: host plant specificity of *Phyllotreta cruciferae* and *P. striolata* adults (Coleoptera:

Chrysomelidae). *Annals of the Entomological Society of America* 63, 832–841.

Gavloski, J.E. and Lamb, R.J. (2000) Compensation by cruciferous plants is specific to the type of simulated herbivory. *Environmental Entomology* 29, 1273–1282.

Gavloski, J.E., Ekuere, U., Keddie, A., Dosdall, L., Kott, L. and Good, A.G. (2000) Identification and evaluation of flea beetle, *Phyllotreta cruciferae*, resistance within Brassicaceae. *Canadian Journal of Plant Science* 80, 881–887.

Gruber, M.Y., Xu, N., Grenkow, L., Li, X., Onyilagha, J., Soroka, J.J., Westcott, N.D. and Hegedus, D.D. (2009) Responses of the crucifer flea beetle to *Brassica* volatiles in an olfactometer. *Environmental Entomology* 38, 1467–1479.

Henderson, A.E., Hallett, R.H. and Soroka, J.J. (2004) Prefeeding behavior of the crucifer flea beetle, *Phyllotreta cruciferae*, on host and nonhost crucifers. *The Journal of Insect Behavior* 17, 17–39.

Hopkins, R.J., Ekbom, B. and Henkow, L. (1998) Glucosinolate content and susceptibility for insect attack of three populations of *Sinapis alba*. *Journal of Chemical Ecology* 24, 1203–1216.

Hopkins, R.J., van Dam, N.M. and van Loon, J.J.A. (2009) Role of glucosinolate in insect-plant relationships and multitrophic interactions. *Annual Review of Entomology* 54, 57–83.

Kandel, H. and Knodel, J.J. (eds) (2011) *Canola Production Field Guide.* North Dakota State University Cooperative Extension Service Publication A1280, Fargo, North Dakota.

Kirchner, W.H. (1999) Mad-bee disease? Sublethal effects of imidacloprid (Gaucho) on the behavior of honey bees. *Apidologie* 30, 422–423.

Knodel, J.J. and Olson, D.L. (2002) *Biology and Integrated Pest Management of the Crucifer Flea Beetle in Canola*. North Dakota State University Cooperative Extension Service Publication E-1234, Fargo, North Dakota.

Knodel, J.J., Olson, D.L., Hanson, B.K. and Henson, R.A. (2008) Impact of planting dates and insecticide strategies for managing crucifer flea beetle (Coleoptera: Chrysomelidae) in spring-planted canola. *Journal of Economic Entomology* 101, 810–821.

Knodel, J.J., Beauzay, P. and Boetel, M. (2015) *2016 North Dakota Field Crop Insect Management Guide*. North Dakota State University Cooperative Extension Service Publication E-1143, Fargo, North Dakota.

Lamb, R.J. (1980) Hairs protect pods of mustard (*Brassica hirta* 'Gisilba') from flea beetle feeding damage. *Canadian Journal of Plant Science* 60, 1439–1440.

Lamb, R.J. (1983) Phenology of flea beetle (Coleoptera: Chrysomelidae) flight in relation to their invasion of canola fields in Manitoba. *The Canadian Entomologist* 115, 1493–1502.

Lamb, R.J. (1984) Effects of flea beetle *Phyllotreta* spp. (Coleoptera: Chrysomelidae) on survival, growth, seed yield and quality of canola, rape and yellow mustard. *The Canadian Entomologist* 116, 269–280.

Lamb, R.J. (1988) Assessing the susceptibility of crucifer seedlings to flea beetle (*Phyllotreta* spp.) damage. *Canadian Journal of Plant Science* 68, 85–93.

Lamb, R.J. and Palaniswamy, P. (1990) Host discrimination by a crucifer-feeding flea beetle, *Phyllotreta striolata* (Coleoptera: Chrysomelidae). *The Canadian Entomologist* 122, 817–824.

Lamb, R.J. and Turnock, W.J. (1982) Economics of insecticidal control of flea beetles (Coleoptera: Chrysomelidae) attacking rape in Canada. *The Canadian Entomologist* 114, 827–840.

Láska, P., Zelenkova, I. and Bičík, V. (1986) Colour attraction in species of the genera: *Delia* (Diptera: Anthomyidae), *Ceutorhychus*, *Meligethes*, and *Phyllotreta* (Coleoptera: Curculionidae, Nitudulidae, and Chrysomelidae). *Acta Entomologica Bohemoslovaca* 83, 418–424.

Madder, D.J. and Stemeroff, M. (1988) The economics of insect control on wheat, corn, and canola, 1980–1985. *Bulletin of Entomological Society of Canada Supplement* 2.

Maus, C., Cure, G. and Schmuck, R. (2003) Safety of imidacloprid seed dressings to honey bees: a comprehensive overview and compilation of the current state of knowledge. *Bulletin of Insectology* 56, 51–57.

Milbrath, L.R., Weiss, M.J. and Schatz, B.G. (1995) Influence of tillage strategies, planting dates, and oilseed crucifers on flea beetle populations (Coleoptera: Chrysomelidae). *The Canadian Entomologist* 127, 289–293.

Nguyen, B.K., Saegerman, C., Pirard, C., Mignon, J., Widart, J., Thirionet, B., Verheggen, F.J., Berkvens, D., De Pauw, E. and Haubruge, E. (2009) Does imidacloprid seed-treated maize have an impact on honey bee mortality? *Journal of Economic Entomology* 102, 616–623.

Pachagounder, P. and Lamb, R.J. (1998) Feeding preference of a flea beetle, *Phyllotreta cruciferae* (Coleoptera: Chrysomelidae) among wild crucifers. *The Canadian Entomologist* 130, 241–242.

Pachagounder, P., Lamb, R.J. and Bodnaryk, R.P. (1998) Resistance to the flea beetle, *Phyllotreta cruciferae* (Coleoptera: Chrysomelidae), in false flax *Camelina sativa* (Brassicaceae). *The Canadian Entomologist* 130, 235–240.

Palaniswamy, P. and Bodnaryk, R.P. (1994) A wild *Brassica* from Italy provides trichome resistance against flea beetles. *The Canadian Entomologist* 126, 1119–1120.

Palaniswamy, P. and Lamb, R.J. (1992) Host preference of the flea beetles, *Phyllotreta cruciferae* and *P. striolata* (Coleoptera: Chrysomelidae), for crucifer seedlings. *Journal of Economic Entomology* 85, 743–752.

Palaniswamy, P., Lamb, R.J. and McVetty, P.B.E. (1992) Screening for antixenosis resistance to flea beetles, *Phyllotreta cruciferae* (Goeze) (Coleoptera: Chrysomelidae), in rapeseed and related crucifers. *The Canadian Entomologist* 124, 895–906.

Pedigo, L.P. and Rice, M.E. (2009) *Entomology and Pest Management*, 6th edn. Waveland Press, Inc., Long Grove, Illinois.

Pimentel, D. (2005) Environmental and economic costs of the application of pesticides primarily in the United States. *Environment, Development and Sustainability* 7, 229–252.

Pivnick, K.A., Lamb, R.J. and Reed, D. (1992) Response of flea beetles, *Phyllotreta* spp., to mustard oils and nitriles in field trapping experiments. *Journal of Chemical Ecology* 18, 863–873.

Putnam, L.G. (1977) Response of four *Brassica* seed crop species to attack by the crucifer flea beetle, *Phyllotreta cruciferae*. *Canadian Journal of Plant Science* 57, 987–989.

Reddy, G.V.P., Tangtrakulwanich, K., Wu, S., Miller, J.H., Ophus, V.L. and Prewett, J. (2014) Sustainable management tactics for control of *Phyllotreta cruciferae* (Coleoptera: Chrysomelidae) on canola in Montana. *Journal of Economic Entomology* 107, 661–666.

Rousseau, M. and LeSage, L. (2016) Earliest North American occurrence of *Phyllotreta striolata* (Coleoptera: Chrysomelidae) from Québec, Canada. *The Canadian Entomologist* 148, 476–478.

Schmuck, R.O. (2004) Effects of a chronic dietary exposure of the honey bees *Apis mellifera* (Hymenoptera: Apidae) to imidacloprid. *Archives of Environmental Contamination and Toxicology* 47, 471–478.

Schmuck, R., Schoening, R., Stork, A. and Schramel, O. (2001) Risk posed to honey bees *Apis mellifera* (Hymenoptera: Apidae) by an imidacloprid seed dressing of sunflowers. *Pest Management Science* 57, 225–238.

Soroka, J. and Grenkow, L. (2013) Susceptibility of brassicaceous plants to feeding by flea beetles, *Phyllotreta* spp. (Coleoptera: Chrysomelidae). *Journal of Economic Entomology* 106, 2557–2567.

Soroka, J.J., Bartelt, R.J., Zilkowksi, B.W. and Cossé, A.A. (2005) Response of flea beetles, *Phyllotreta cruciferae* in the field to synthetic aggregation pheromone components and plant host volatiles. *Journal of Chemical Ecology* 31, 1829–1843.

Soroka, J.J., Grenkow, L.F. and Irvine, R.B. (2008) Impact of decreasing ratios of insecticide-treated seed on flea beetle (Coleoptera: Chrysomelidae, *Phyllotreta* spp.) feeding levels and canola seed yields. *Journal of Economic Entomology* 101, 1811–1820.

Soroka, J.J., Holowachuk, J.M., Gruber, M.Y. and Grenkow, L.F. (2011) Feeding by flea beetles (Coleoptera: Chrysomelidae, *Phyllotreta* spp.) is decreased on canola (*Brassica napus*) seedlings with increased trichome density. *Journal of Economic Entomology* 104, 125–136.

Suchail, S., Guez, D. and Belzunces, L.P. (2001) Discrepancy between acute and chronic toxicity induced by imidacloprid and its metabolites in *Apis mellifera*. *Environmental Toxicology and Chemistry* 20, 2482–2486.

Tangtrakulwanich, K., Reddy, G.V.P., Wu, S., Miller, J.H., Ophus, V.L. and Prewett, J. (2014) Developing nominal threshold for *Phyllotreta cruciferae* (Coleoptera: Chrysomelidae) damage on canola in Montana, USA. *Crop Protection* 66, 8–13.

Tansey, J.A., Dosdall, L.M., Keddie, B.A. and Sarfraz, R.M. (2008) Differences in *Phyllotreta cruciferae* and *Phyllotreta striolata* (Coleoptera: Chrysomelidae) response to neonicotinoid seed treatments. *Journal of Economic Entomology* 10, 159–167

Tansey, J.A., Dosdall, L.M. and Keddie, B.A. (2009) *Phyllotreta cruciferae* and *Phyllotreta striolata* response to insecticidal seed treatments with different modes of action. *Journal of Applied Entomology* 133, 201–209.

Toth, M., Csonka, E., Bartelt, R.J., Cosse, A.A., Zilkoski, B.W., Muto, S.-E. and Mori, K. (2005) Pheromonal activity of compounds identified from male *Phyllotreta cruciferae*: field tests of racemic mixtures, pure enantiomers, and combination with allyl isothiocyanate. *Journal of Chemical Ecology* 31, 2705–2720.

Thomas, P. (2003) *Canola Growers Manual*. Canola Council of Canada, Winnipeg, Manitoba, Canada.

Thomas, W.C. (1986) Predation on adult *Phyllotreta striolata* (F.) flea beetles by *Podisus maculiventris* (Heteroptera: Pentatomidae) and *Nabicula americolimbata* (Hemiptera: Nabidae). *The Canadian Entomologist* 118, 731–732.

Turnock, W.J. and Lamb, R.J. (1982) Protecting canola from flea beetle injury. *Canadian Agriculture* 27, 4–8.

Turnock, W.J., Lamb, R.J. and Bilodeau, R.J. (1987) Abundance, winter survival, and spring emergence of flea beetles (Coleoptera: Chrysomelidae) in a Manitoba grove. *The Canadian Entomologist* 119, 419–426.

Ulmer, B.J. and Dosdall, L.M. (2006) Emergence of overwintered and new generation adults of the crucifer flea beetle, *Phyllotreta cruciferae* (Goeze) (Coleoptera: Chrysomelidae). *Crop Protection* 25, 23–30.

Vincent, C. and Stewart, R.K. (1984) Effect of allyl isothiocyanate on field behavior of crucifer-feeding flea beetles (Coleoptera: Chrysomelidae). *Journal of Chemical Ecology* 10, 33–39.

Vincent, C. and Stewart, R.K. (1985) Influence of trap color on captures of adult crucifer-feeding flea beetles. *Journal of Agricultural Entomology* 3, 120–124.

Weiss, M.J., McLeod, P., Schatz, B.G. and Hanson, B.K. (1991) Potential for insecticidal management of flea beetle (Coleoptera: Chrysomelidae) on canola. *Journal of Economic Entomology* 84 1597–1603.

Weiss, M.J., Schatz, B.G., Gardner, J.C. and Nead, B.A. (1994) Flea beetle (Coleoptera: Chrysomelidae) populations and crop yield in field pea and oilseed rape intercrops. *Environmental Entomology* 23, 654–658.

Westdal, P.H. and Romanow, W. (1972) Observation on distribution, seasonal life history, and abundance of flea beetles (Coleoptera: Chrysomelidae). *Manitoba Entomologist* 6, 35–45.

Wylie, H.G. (1979) Observations on distribution, seasonal life history, and abundance of flea beetles (Coleoptera: Chrysomelidae) that infest rape crops in Manitoba. *The Canadian Entomologist* 111, 1345–1353.

Wylie, H.G. (1982) An effect of parasitism by *Microctonus vittatae* (Hymenoptera: Braconidae) on emergence of *Phyllotreta cruciferae* and *Phyllotreta striolata* (Coleoptera: Chrysomelidae) from overwintering sites. *The Canadian Entomologist* 114, 727–732.

Wylie, H.G. (1983) Oviposition and survival of the European parasite *Microctonus bicolor* (Hymenoptera: Braconidae) in crucifer-infesting flea beetles (Coleoptera: Chrysomelidae) in Manitoba. *The Canadian Entomologist* 115, 55–58.

Wylie, H.G. (1984) Oviposition and survival of three Nearctic euphorine braconids in crucifer-infesting flea beetle (Coleoptera: Chrysomelidae). *The Canadian Entomologist* 116, 1–4.

Wylie, H.G. (1988) Release in Manitoba, Canada of *Townesilitus bicolor* [Hym.: Braconidae], an European parasite of *Phyllotreta* spp. [Col.: Chrysomelidae]. *Entomophaga* 33, 25–32.

Wylie, H.G. and Loan, C. (1984) Five Nearctic and one introduced euphorine species (Hymenoptera: Braconidae) that parasitize adults of crucifer-infesting flea beetles (Coleoptera: Chrysomelidae). *The Canadian Entomologist* 116, 235–246.

Xu, C., de Clercq, P., Moens, M., Chen, S. and Han, R. (2010) Efficacy of entomopathogenic nematodes (Rhabditida: Steinernematidae and Heterorhabditidae) against striped flea beetle, *Phyllotreta striolata* (Coleoptera: Chrysomelidae). *BioControl* 55, 789–797.

Yan, X., Han, R., Moens, M., Chen, S. and de Clercq, P. (2013) Field evaluation of entomopathogenic nematodes for biological control of striped flea beetle, *Phyllotreta striolata* (Coleoptera: Chrysomelidae). *BioControl* 58, 247–256.

Zhao, Y., Yang, G., Wang-Pruski, G. and You, M. (2008) *Phyllotreta striolata* (Coleoptera: Chrysomelidae): arginine kinase cloning and RNAi-based pest control. *European Journal of Entomology* 105, 815–822.

2 Diamondback Moth (*Plutella xylostella*) Management

Yaghoub Fathipour* and Mohammad Ali Mirhosseini

Tarbiat Modares University, Tehran, Iran

2.1 Introduction

Brassicaceae (formerly Cruciferae) includes 375 genera and 3200 species of plants. *Brassica* consists of approximately 100 species, including *Brassica napus* L., commonly known as oilseed rape, rapeseed or canola. Between 1990 and 2013 the area planted to *Brassica* crops worldwide increased by more than 48% (FAOSTAT, 2015). New canola varieties have also been developed that confer better oil quality and easier oil extraction (Anon., 1997). Because these varieties have more economic value, the management of yield-reducing pests has become more important.

Several pests, especially insects, damage canola crops (Talekar and Shelton, 1993; Liu *et al.*, 1994; Karimi *et al.*, 2012; Goodarzi *et al.*, 2015) but among these diamondback moth (DBM) *Plutella xylostella* (L.) (Lepidoptera: Plutellidae) is the most destructive (Talekar and Griggs, 1986; Shelton *et al.*, 2000; Furlong *et al.*, 2013). DBM is important as a canola pest because: (i) it is the most widely distributed of all the lepidopteran pests of canola (Talekar and Shelton, 1993); (ii) it has a high reproductive rate (more than 20 generations per year in the tropics); (iii) its natural enemies are either absent in some areas or disrupted by pesticides in others (Hui *et al.*, 2010); and (iv) its high genetic variability facilitates rapid development of insecticide resistance (Sayyed *et al.*, 2004; Sarfraz *et al.*, 2006; Ahmad *et al.*, 2012). The estimated annual cost of controlling DBM worldwide is US$1.4 billion (Zalucki *et al.*, 2012); if the yield losses caused by DBM are added, costs are US$5 billion (Furlong *et al.*, 2013).

Chemical insecticides are the primary means of controlling DBM (Grzywacz *et al.*, 2010; Ahmad *et al.*, 2012) but chemical control can be unreliable due to insecticide resistance (Li *et al.*, 2012) and outbreaks of DBM caused by the destruction of natural enemy populations. According to the Arthropod Pesticide Resistance Database (APRD), 110 references from 235 locations worldwide document DBM resistance to 91 active ingredients (APRD, 2015). Furthermore, the use of insecticides is economically inefficient, with only 1% of pesticides applied reaching the target pest. Thus the efficient, effective control of DBM requires an integrated pest management (IPM) programme that includes all compatible methods of managing DBM populations, with the aim of reducing insecticide use while maintaining profitability, yield and crop quality. The absence of such an IPM package is, in many regions, a key gap in our knowledge of management of DBM populations. This chapter examines the key aspects of such a comprehensive IPM-based DBM management programme.

2.2 DBM Control Strategies

Although using only insecticides is currently the major tactic employed for DBM management, integration of additional measures (cultural, biological and behavioural controls) can improve control (Fig. 2.1) and these are discussed below.

2.2.1 Cultural control

For many years, farmers used cultural practices as the main control measure for many pests. Cultural control tactics can be highly effective components of pest management systems (Newsom, 1975). By

*Corresponding author. E-mail: fathi@modares.ac.ir

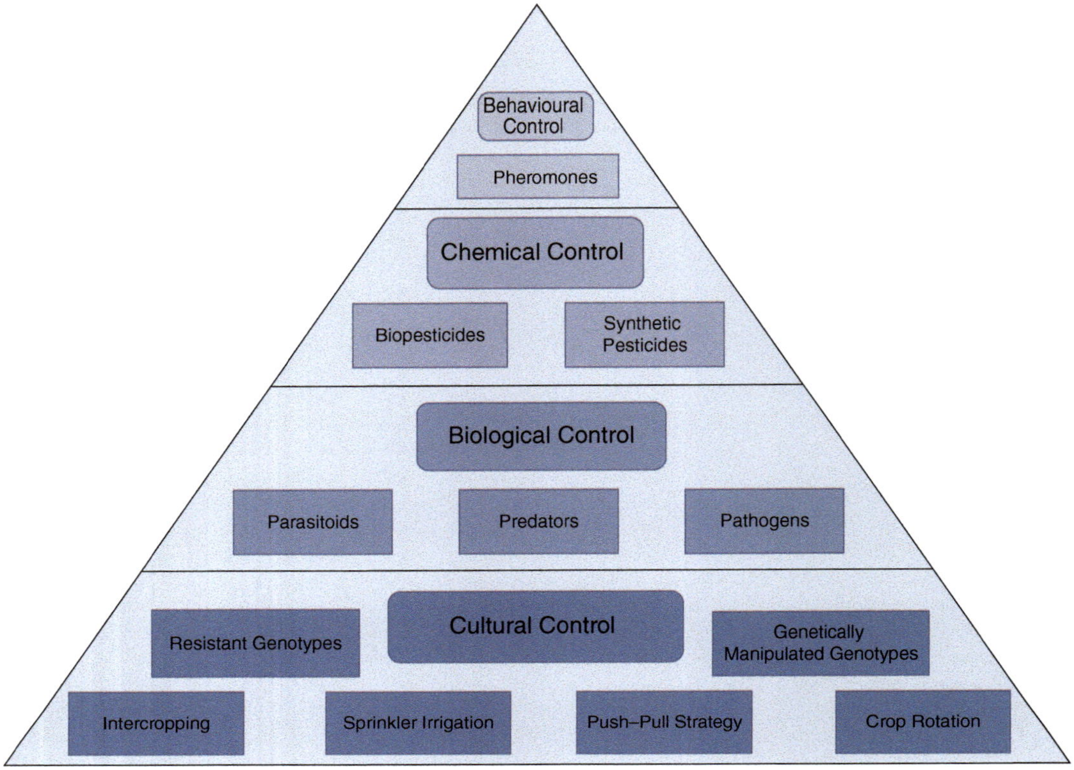

Fig. 2.1. Different control strategies and their relative importance in an idealized integrated DBM management programme.

definition, cultural control is the deliberate manipulation of cropping and soil system environments to make them less favourable for pests and more favourable for their natural enemies (Fathipour and Sedaratian, 2013). Almost all the tactics of cultural control are compatible with other IPM components and the environment and include such measures as use of host plant resistance, intercropping, crop rotation, the push–pull strategy and modification of fertilizer applications.

Host plant resistance

Host plant resistance can be an effective replacement for use of broad-spectrum insecticides. Potentially, deployment of resistant crop varieties is inexpensive, easy to introduce, specific to one or several pests, persistent, easy to integrate into other farm operations, compatible with other IPM control tactics and safe to the environment. While such resistance is based on heritable traits, some of these traits may fluctuate widely under different environmental conditions. Accordingly, host plant resistance may be classified either as genetic (traits that are under the primary control of genetic factors), which are very popular in IPM programmes, or as ecological (traits that are under the primary control of environmental factors).

Genetic resistance includes both induced resistance (biotic and abiotic environmental factors that lower insect fitness or negatively affect the pest's host selection processes) and constitutive resistance (inherited characteristics whose expression, although influenced by the environment, is not triggered by environmental factors) (Metcalf and Luckmann, 1994). Mechanisms of genetic resistance include antixenosis, antibiosis and tolerance.

Host plant resistance to pests can be inherited either vertically (controlled by a single gene) or horizontally (controlled by many genes). Because of the large number of genes involved, it is much more difficult to breed cultivars with horizontal resistance to insect pests (Smith, 1989). To determine which mechanism is active in particular cases, experiments can be designed that assess the pest's life table parameters among varieties (Chi, 1988; Carey, 1993; Fathipour and Maleknia, 2016) (Table 2.1), as well as their nutritional indices (Waldbauer, 1968) (Table 2.2), and the pest's digestive proteolytic and amylolytic activities (Fathipour and Naseri, 2011).

Canola cultivars differ in their susceptibility to attack by DBM. To evaluate plant resistance to DBM,

Table 2.1. Equations and concepts of female-based and two-sex-based life table parameters.

Female-based life table	Two-sex-based life table
x (day) Age	x (day) Age
N_x Number of surviving individuals (only females in adult stage) entering the age x	N_x Number of surviving individuals (females and males in adult stage) entering the age x
M_x Daily mean number of eggs produced per female of age x	f_{xj} Age-stage specific fecundity (daily number of eggs produced per female of age x)
$l_x = \frac{N_x}{N_0}$ The age-specific survival; N_0 = number of individuals at the age $x = 0$	s_{xj} ; $l_x = \sum_{j=1}^{k} s_{xj}$ Age-stage specific survival rate (x = age; j = stage); k = number of stages
m_x Daily mean number of female eggs produced per female of age x	$m_x = \frac{\sum_{j=1}^{k} s_{xj} f_{xj}}{\sum_{j=1}^{k} s_{xj}}$ Age specific fecundity [daily number of eggs produced per individual, i.e. this number is divided by all individuals (males and females) of age x]; k = number of stages
$R_0 = \sum_{x=\alpha}^{\beta} l_x m_x$ The net reproductive rate (female eggs per female)	$R_0 = \sum_{x=0}^{\omega} \sum_{j=1}^{k} s_{xj} f_{xj}$ The net reproductive rate (eggs per individual)
$\sum_{x=0}^{\omega} e^{-r(x+1)} l_x m_x = 1$ Intrinsic rate of increase (r) [number of females added to the population per female per day, i.e. the intrinsic birth rate (b) minus the intrinsic death rate (d)] (day^{-1})	$\sum_{x=0}^{\omega} e^{-r(x-1)} l_x m_x = 1$ Intrinsic rate of increase (r) [number of individuals added to the population per individual per day, i.e. the intrinsic birth rate (b) minus the intrinsic death rate (d)] (day^{-1})
$\lambda = e^r$ Finite rate of increase [the rate at which the population (only females) increases from one day to the next day] (day^{-1})	$\lambda = e^r$ Finite rate of increase [the rate at which the population (females and males) increases from one day to the next day] (day^{-1})
$GRR = \sum_{x=\alpha}^{\beta} m_x$ The gross reproductive rate (female eggs per female)	$GRR = \sum_{x=\alpha}^{\beta} m_x$ The gross reproductive rate (eggs per individual)
$T = \frac{\ln R_0}{r}$ Mean generation time (day)	$T = \frac{\ln R_0}{r}$ Mean generation time (day)

Table 2.2. Nutritional indices for determining consumption and utilization of food by insects.

Nutritional index	Equation
Relative consumption rate (RCR) or Consumption index (CI) (unit: mg/mg/day)	$\frac{\text{Fresh or dry weight of food eaten (ingested) during feeding period (mg)}(E)}{\text{Mean fresh or dry weight of insect during feeding period (mg)}(A)\times\text{Duration of feeding period (day)}(T)}$ $\frac{E}{A\times T}$
Relative growth rate (RGR) (unit: mg/mg/day)	$\frac{\text{Fresh or dry weight gain (increase) of insect during feeding period (mg)}(G)}{\text{Mean fresh or dry weight of insect during feeding period (mg)}(A)\times\text{Duration of feeding period (day)}(T)}$ $\frac{G}{A\times T}$
Efficiency of conversion of ingested food (ECI) (unit: %)	$\left[\frac{\text{Fresh or dry weight gain (increase) of insect during feeding period (mg)}(G)}{\text{Fresh or dry weight of food eaten (ingested) during feeding period (mg)}(E)}\right]\times 100$ $\left(\frac{G}{E}\right)\times 100$
or	$\left(\frac{RGR}{RCR}\right)\times 100$
Efficiency of conversion of digested food (ECD) (unit: %)	$\left[\frac{\text{Fresh or dry weight gain (increase) of insect during feeding period (mg)}(G)}{\text{Fresh or dry weight of food eaten (ingested) during feeding period (mg)}(E) - \text{Fresh or dry weight of faeces produced during feeding period (mg)}(F)}\right]\times 100$ $\left(\frac{G}{E-F}\right)\times 100$
Approximate digestibility (AD) (unit: %)	$\left[\frac{\text{Fresh or dry weight of food eaten (ingested) during feeding period (mg)}(E) - \text{Fresh or dry weight of faeces produced during feeding period (mg)}(F)}{\text{Fresh or dry weight of food eaten (ingested) during feeding period (mg)}(E)}\right]\times 100$ $\left(\frac{E-F}{E}\right)\times 100$

several researchers have calculated the DBM's life table parameters on different host plants. Although canola may be an inferior host for DBM compared with host plants such as cauliflower, cabbage and kohlrabi (Golizadeh *et al.*, 2009b), DBM does cause economic damage to some canola cultivars and canola genotype can strongly affect DBM performance in terms of developmental time, adult longevity, survival rate, other life table parameters, nutritional indices and digestive enzyme activities (see Section 2.3 for details). For instance, the highest and lowest DBM population growth was observed on canola genotypes SLM_{046} and RGS_{003}, respectively (Soufbaf *et al.*, 2010a). Other research found RGS_{003} to have the lowest suitability for oviposition of DBM among different genotypes of canola tested (Ebrahimi *et al.*, 2008). Integrating these results with nutritional indices of DBM on SLM_{046} and RGS_{003} genotypes may show that SLM_{046} is a more suitable genotype for DBM. Larval weight, food consumption, efficiency of conversion of ingested food (ECI) and efficiency of conversion of digested food (ECD) of DBM on SLM_{046} were all found to be significantly higher than on RGS_{003} (Kianpour *et al.*, 2014). This may be the result of differences in primary metabolites (e.g. the amount of nitrogen) between canola genotypes and the effect of these differences on both second (DBM) and third (its natural enemies) trophic levels in canola food webs (Soufbaf *et al.*, 2012). The levels of nitrogen among canola genotypes and the reproductive performance of DBM seem to be linked (Soufbaf *et al.*, 2013). Optimal canola genotypes, however, would have to have genes leading to both higher yield and lower suitability for the development and reproduction of DBM. It must be noted that a genotype with resistance to DBM may not be resistant to other canola pests. Thus, resistance of canola genotypes should be tested against all economic pests in a given area, such as *Brevicoryne brassicae* (L.), *Spodoptera exigua* (Hübner) and *Helicoverpa armigera* (Hübner).

Genetically manipulated Brassica crops

The integration of genetic transformation technology with conventional plant breeding has great potential to improve crop performance. Genetically manipulated *Brassica* crops have the potential to control DBM populations. In one study, the effect of different *Brassica* plants including canola's progenitor (*Brassica rapa* L.), two cultivated canola cultivars (Opera and RGS_{003}), one hybrid ($Hyula_{401}$), one gamma-ray mutated (Mutant-RGS_{003}) and one transgenic (PF) genotype on the life table parameters of DBM were determined (Nikooei *et al.*, 2015a). It was concluded that these manipulations significantly affected performance of DBM and its parasitoid (see Section 2.3 for details). There is evidence that gamma-ray mutation of several canola cultivars can reduce the suitability of these cultivars for DBM (Akandeh *et al.*, 2016). Furthermore, transmission of the gene of *Bacillus thuringiensis* (Bt) to canola and other *Brassica* crops has been accomplished (Ramachandran *et al.*, 1998; Wang *et al.*, 2014) and these types of crop can produce Bt toxin, making them very resistant to DBM and other lepidopteran pests.

Intercropping

Intercropping is growing two or more crops in close proximity to enhance yield or pest control. The most common reason for intercropping is to enhance yield on limited land. In doing so, it is important to avoid crop competition for physical space, nutrients, water or sunlight. However, enhancing vegetation diversity through intercropping can also result in significant reductions in density and damage of some pests (Landis *et al.*, 2000). In one study, the survival of DBM was significantly lower in a Chinese cabbage (*Brassica chinensis*)/garlic (*Allium sativum*)/lettuce (*Lactuca sativa*) intercrop than in Chinese cabbage alone (Cai *et al.*, 2011). Intercropping cabbage with tomato, garlic, dill, pepper and clover reduces the density of DBM in comparison with cabbage monocultures (Dover, 1986; Talekar and Griggs, 1986; Asman *et al.*, 2001; Mohammed *et al.*, 2010). Furthermore, many wild flowering plants and non-cruciferous crops such as legumes support natural enemies by providing nectar and pollen and can be more attractive to natural enemies due to a variety of odours (Suso *et al.*, 2016).

Crop rotation

Crop rotation is the practice of growing different crops in the same area over crop seasons. Crop rotation is an environmentally friendly approach and a traditional production practice used to increase soil fertility and tilth, enhance crop vigour and reduce pest build-up. Crop rotation can reduce DBM population density because of the pest's narrow host range (crucifers only). Mandatory host-free periods for a region have been used to reduce DBM activity in

Mexico and Australia (Sayyed *et al.*, 2002). In large commercial cultivations, crop rotation of crucifers with rice, cucurbits, beans, peas, tomato or melons all strongly suppressed DBM populations (Li *et al.*, 2016). Because of the high mobility of DBM adults, crop rotation strategies need to be practised communally to restrict DBM over a wide area. However, this approach may be not be feasible in commercial vegetable-producing areas, because of the demand for and high price of crucifer vegetables.

Sprinkler irrigation

Intermittent sprinkler irrigation can reduce DBM populations by up to around 85% (Tabashnik and Mau, 1986; McHugh and Foster, 1995). The physical disruption of flying activity, oviposition and to some extent wash-off of larvae (especially young larvae) and adults were presumably the major causes of the observed effects. Applying sprinkler irrigation in combination with the push–pull strategy can push female DBM to trap crops for oviposition.

Push–pull strategy

The 'push–pull' approach to pest management was first introduced as an IPM tactic in Australia for the cotton bollworm by Pyke *et al.* (1987). They investigated the simultaneous use of repellent and attractive stimuli to manipulate the distribution of this pest with an aim to reduce the use of insecticides. This same strategy was then implemented for control of the onion maggot (*Delia antiqua* (Meigen)) and termed stimulo-deterrent diversion (Miller and Cowles, 1990). The push–pull strategy (Fig. 2.2) involves the behavioural manipulation of pests and their natural enemies by integrating stimuli that make the protected resource unattractive or unsuitable to the pests (push) while luring them toward an attractive source (pull) where the pests are subsequently removed (Fathipour and Sedaratian, 2013).

Trap plants are an important source of 'pull' stimuli. The concept of trap cropping fits into the ecological framework of habitat manipulation of an agroecosystem to achieve pest control. Recently, interest in trap cropping has increased considerably because of concerns about harm from pesticides to human health or the environment, as well as general economic considerations of agricultural production (Shelton and Badenes-Pérez, 2006). With respect to DBM, several studies have evaluated trap plants to divert DBM from *Brassica* crops. Although most of this work was done in other crops, such as cabbage, their findings should be applicable in canola fields as well. In one such study, both DBM larval density and resultant damage in cabbage were significantly reduced by planting a border of Indian mustard (*Brassica juncea* (L.) Vassiliĭ Matveievitch Czernajew) (Hasheela *et al.*, 2010). In another study, DBM was offered multiple trap plants including glossy (waxy) collards (*Brassica oleracea* L.), Indian mustard (*B. juncea*), and yellow rocket (*Barbarea vulgaris* R.Br.), both concurrently and non-concurrently in a cabbage field (Badenes-Pérez *et al.*, 2004). When these

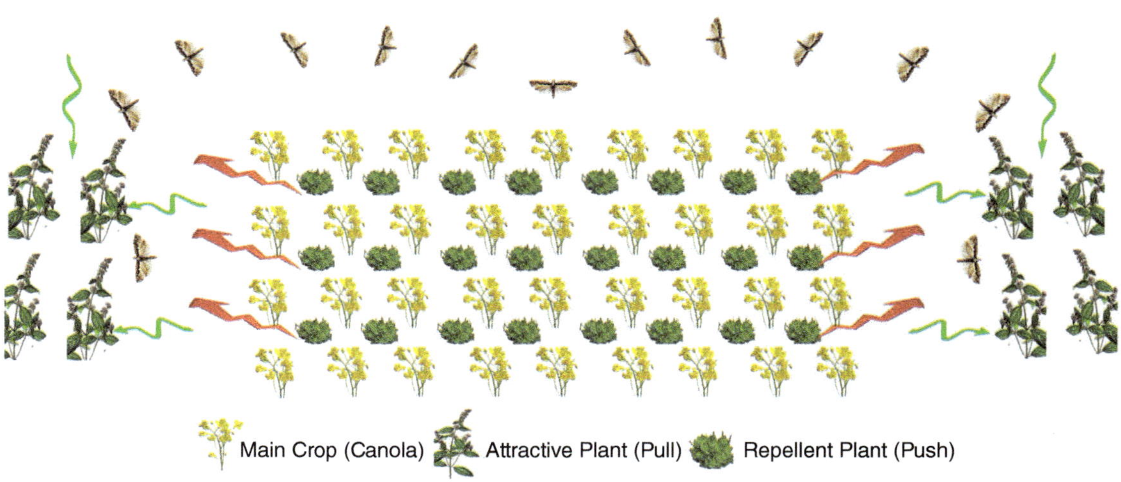

Fig. 2.2. Diagrammatic representation of the components of push–pull strategy.

trap plants were deployed concurrently with the crop (choice test), the number of eggs laid by DBM on glossy collards, Indian mustard and yellow rocket was three, 18 and 12 times greater than on cabbage, respectively, while for the non-concurrent offering (no-choice test), the number of eggs laid on these same crops was 300, 19 and 110 times greater than on cabbage, respectively. In addition, larval survival rates were significantly lower on glossy collards (6.7%) and yellow rocket (0%) than on cabbage (22.2%). A field survey in a subsequent study revealed that the population density of DBM larvae in cabbage plots without trap plants was 5.2–11.3 times higher than in cabbage plots with several rows of yellow rocket bordering the plot (Badenes-Pérez *et al.*, 2005). With an increasing planting area of yellow rocket, the percentage of eggs laid on cabbage was found to decrease significantly. Other field and laboratory studies have found that when yellow rocket was used as a trap plant in broccoli and cabbage plots, not only did DBM preferentially lay its eggs on this plant, but also the larvae did not survive on this plant (Shelton and Nault, 2004). This type of trap plant is called a dead-end trap plant since it is highly attractive for oviposition to an insect pest but offspring of the pest cannot survive on it. Another study also found that nearly all DBM larvae on yellow rocket died as neonates or early-instar larvae (Lu *et al.*, 2004). Yellow rocket can be made approximately 1.5 times more attractive to DBM females for oviposition if treated with sulfur fertilizer (Badenes-Pérez *et al.*, 2010).

Unfortunately, there are few studies of plants or materials that are either repellent or deterrent to DBM and could be used to push the pest away from the main crop. Applications of yeheb (*Cordeauxia edulis* Hemsl.) extract can strongly reduce adult oviposition and larval feeding of DBM (Egigu *et al.*, 2010). In addition, this extract may be more useful for attracting some DBM natural enemies, such as *Cotesia vestalis* (Haliday), when mixed with volatile organic compounds collected from the intact and DBM-damaged brassicaceous plants. Cultivating plants with strong, unpleasant volatiles in between rows of the main crop can also be effective in reinforcing the 'push' part of the push–pull strategy.

2.2.2 Biological control

There is a wide range of natural enemies, including parasitoids, predators and pathogens, that attack different developmental stages of DBM (Table 2.3).

Parasitoids of DBM

By definition, parasitoids are beneficial insects that lay their eggs inside or on the outside of any life stage of its host. All immature stages of DBM are attacked by many parasitoids (Table 2.3). Considerable basic and applied research has focused on these organisms, especially hymenopteran parasitoids (Sarfraz *et al.*, 2005).

Egg parasitoids of lepidopteran pests most commonly used in biological control belong to the polyphagous genus *Trichogramma* (Trichogrammatidae), which contribute little to natural control and require frequent mass releases to be effective. For instance, Tabone *et al.* (2010) assessed the parasitism rate of 12 *Trichogramma* species on DBM under greenhouse and laboratory conditions. This study revealed that three species, including *T. chilonis* Ishii, *T. evanescens* Westwood and *T. ostriniae* (Pang et Chen), are suitable agents in greenhouses, causing up to 70% parasitism of DBM eggs under some conditions. In an earlier study (Guo *et al.*, 1999), among 29 species or strains of *Trichogramma* wasps evaluated in the laboratory conditions, *T. chilonis* and *T. pretiosum* Riley were found to be suitable candidates to control DBM in fields in China. *Trichogramma* wasps are generalist egg parasitoids and in a multi-host situation they may preferentially parasitize a non-target host species to the detriment of the desired control programme, making their host preference an important issue as DBM may be only one of a number of species in the canola pest complex. In one study, the host preferences of *T. pretiosum* were determined when it was offered the eggs of DBM and *Pseudoplusia includens* (Walker) (Lep.: Noctuidae) (Pluke and Leibee, 2006) and it was found that, for both choice and no-choice tests, *T. pretiosum* parasitized significantly more DBM eggs than *P. includens* eggs. The results of this study revealed that in the no-choice experiments, *T. pretiosum* parasitized 80% of DBM eggs and 13% of *P. includens* eggs available, while in the choice experiments, *T. pretiosum* parasitized 76% of DBM eggs and 20% of *P. includens* eggs available in the environment. Furthermore, trichogrammatid wasps can be used in combination with other compatible agents (such as Bt) to reduce crop damage (Hwang *et al.*, 2010). For example, the survival rate of DBM was found to be 32% and 14% when exposed to egg parasitism or Bt separately, but only 1% when exposed to both *T. ostriniae* and Bt. The use of efficient species

Table 2.3(a). Major natural enemies of diamondback moth, arranged by development stage and geographical region: parasitoids.

Order	Family	Scientific name	Stage of DBM attacked[a]	Main region(s) reported
Hymenoptera	Ichneumonidae	*Diadegma semiclausum*	L	Asia, Australia, Africa, Europe
		Diadegma leontiniae	L	Africa, America
		Diadegma insulare	L	America, Asia, Europe
		Diadegma mollipla	L	Africa, Asia
		Diadegma fenestrale	L	Asia, Europe
		Diadegma novaezealandiae	L	Australasia
		Diadegma rapi	L	Australia
		Diadegma exareolator	L	Europe
		Diadegma armillata	L	Europe
		Diadegma chrysosticta	L	Europe
		Diadegma vestigialis	L	Europe
		Diadegma cerophaga	L	Asia, Europe
		Diadegma tibialis	L	Europe
		Diadegma trochanterata	L	Europe
		Diadegma gracilis	L	Europe
		Diadegma gibbula	L	Europe
		Diadegma holopyga	L	Europe
		Diadegma interrupta	L	Europe
		Diadegma monospila	L	Europe
		Diadegma majale	L	Asia
		Diadegma anurum	L	Asia
		Hyposoter ebeninus	L	Europe
		Diadromus ustulatus	P	Europe
		Diadromus subtilicornis	P	America, Asia, Europe
		Diadromus collaris	P	Asia, Africa, Europe, America, Australia
		Diadromus subtilis	P	Europe
		Diadromus varicolor	P	Asia
		Herpestomus brunnicornis	P	Asia
		Itoplectis viduata	P	Europe
		Itoplectis alternans	P	Europe
		Itoplectis naranyae	P	Asia
		Itoplectis tunetanus	P	Europe
	Braconidae	*Cotesia plutellae* (*C. vestalis*)	L	Asia, Africa, Europe, America, Australia
		Cotesia glomerata	L	Asia
		Bracon hebetor	L	Africa
		Apanteles piceotrichosus	L	Africa
		Apanteles litae	L	Africa
		Apanteles eriophyes	L	Africa
		Apanteles fuliginosus	L	Europe
		Apanteles ruficrus	L	Europe
		Apanteles rubecula	L	Europe
		Apanteles ippeus	L	Asia, Australia
		Microplitis plutellae	L	America
		Microplitis mediator	L	Europe
	Eulophidae	*Oomyzus sokolowskii*[b]	L-P	Africa, America, Asia
		Tetrastichus howardi	P	Africa
	Pteromalidae	*Dibrachys cavus*	P	Europe
	Chalcididae	*Conura pseudofulvovariegata*	P	Africa, Europe

Continued

Table 2.3(a). Continued.

Order	Family	Scientific name	Stage of DBM attacked[a]	Main region(s) reported
		Conura unimaculata	P	Africa
		Brachymeria excarinata	P	Asia
		Brachymeria phya	P	Asia
	Trichogrammatidae	*Trichogramma* spp.	E	Cosmopolitan
Diptera	Tachinidae	*Actia* sp.	L	Africa

[a]E, egg; L, larva; P, pupa
[b]Occasionally facultative hyperparasitoid

Table 2.3(b). Major natural enemies of diamondback moth, arranged by developmental stage and geographical region: predators and pathogens.

Order	Family	Scientific name	Stage of DBM attacked[a]	Main region(s) reported[b]
PREDATORS				
Hymenoptera	Formicidae	*Anomma nigricans*	E-L	Africa
		Pheidole sp.	E-L	Africa
Hemiptera	Pentatomidae	*Podisus maculiventris*	L	America
		Podisus nigrispinus	L	Africa
		Oechalia schellenbergii	L	Australia
	Nabidae	*Nabis kinbergii*	L	Australia
Araneae	Thomisidae	*Ebrechtella tricuspidata*	L	Asia
	Miturgidae	*Cheiracanthium inclusum*	L	Africa
	Lycosidae	*Pardosa milvina*	L	America
		Pardosa astrigena	L	Asia
		Pardosa laura	L	Asia
		Pardosa pseudoannulata	L	Asia
Neuroptera	Chrysopidae	*Chrysoperla carnea*	E-L	America, Asia, Europe
PATHOGENS				
Hypocreales	Cordycipitaceae	*Beauveria bassiana*	L-P	**
		Isaria fumosorosea	L-P	**
	Clavicipitaceae	*Metarhizium anisopliae*	L-P	**
Entomophthorales	Entomophthoraceae	*Zoophthora radicans*	L-P	**
		Pandora blunckii	L-P	America, Asia, Europe
Rhabditida	Steinernematidae	*Steinernema carpocapsae*	L	**
	Heterorhabditidae	*Heterorhabditis bacteriophora*	L	**
Bacillales	Bacillaceae	*Bacillus thuringiensis*	L	**
Enterobacteriales	Enterobacteriaceae	*Xenorhabdus nematophila*	L	**

[a]E, egg; L, larva; P, pupal stage
[b]**Commercial product

of these parasitoid wasps at the right time (when host egg densities are high), in sufficient density and in combination with other compatible methods can be a good fit for DBM management programmes.

Larval parasitoids have the greatest potential for control of DBM, especially species of *Microplitis* and *Cotesia* (both Braconidae) and of *Diadegma* (Ichneumonidae) (Lim, 1986; Sarfraz *et al.*, 2005).

The first classical biological control programme against DBM was carried out in 1936 when the larval parasitoid *Diadegma semiclausum* (Hellen) (Hymenoptera: Ichneumonidae) and the pupal parasitoid *Diadromus collaris* (Gravenhorst) (Hymenoptera: Ichneumonidae) were successfully introduced into New Zealand from the UK (Talekar and Shelton, 1993). After that, *D. semiclausum* was introduced to other countries where it now causes more than 60% parasitism to DBM larvae in some instances (Poelking, 1992; Talekar, 1992; Momanyi *et al.*, 2006). Wang *et al.* (2004) found that this species parasitized 72–94% of DBM larvae in a broccoli field in 1999 at the Gatton Research Station in southern Queensland, Australia. Some studies have found that the population growth rate and performance of this parasitoid wasp changes with the level of primary metabolites (such as nitrogen) in canola, as well as by genetic manipulation (e.g. transgenic, gamma-ray mutant and hybrid genotypes) of the host plant (Soufbaf *et al.*, 2012; Nikooei *et al.*, 2015b, 2016). A comparison of the host-searching efficiency of *D. semiclausum* (a specialist parasitoid) and *Cotesia plutellae* (Kurdjumov) (an oligophagous species) found that the host location strategies employed by *D. semiclausum* were better adapted to the host's defensive behaviour; thus this wasp is more effective at detecting and parasitizing the host than *C. plutellae* (Wang and Keller, 2002). Another field study found that parasitism by *D. semiclausum* was low shortly after the crop (cabbage) was planted but increased as the plants grew older, in contrast to parasitism by *C. plutellae*, which was higher after crop transplant and decreased as plants matured (Talekar and Yang, 1993). *C. plutellae* is a solitary endolarval parasitoid of DBM that tolerates hot and humid climates (Talekar and Shelton, 1993) and is the most abundant larval parasitoid of DBM in South Africa and Hawaii (Johnson *et al.*, 1988; Kfir, 1997). According to some studies, a single female of this wasp has the potential to parasitize 100 larvae effectively, with 80–86% parasitism (Choubey *et al.*, 2014). Although *C. plutellae* can parasitize all instars of DBM larvae, it prefers to do so in seconds and thirds and the parasitism rate reduced sharply with increasing host age in the fourth instar (Shi *et al.*, 2002). This study also showed that females had the highest fecundity if they developed from hosts stung as third instars. Spraying fields with some volatiles such as limonene and methyl jasmonate has been found to attract *C. plutellae* and enhance larval parasitism (Ibrahim *et al.*, 2005). Laboratory experiments on the life table parameters and thermal requirements of *Diadegma anurum* (Thomson) have demonstrated that this parasitoid has good potential for control of DBM larvae (Golizadeh *et al.*, 2008).

Prepupal and pupal parasitoid wasps of DBM are principally species of *Diadromus* (Ichneumonidae), which can contribute significantly to DBM control (Kfir, 1997; Liu *et al.*, 2000; Braun *et al.*, 2004; Kirk *et al.*, 2004). Among these, *D. collaris* (Gravenhorst) has been recorded in many parts of the world as a major solitary pupal endoparasitoid of DBM (Talekar and Shelton, 1993). Although this species can complete its development at temperatures from 15°C to 33.5°C, its best performance (number of pupae parasitized per female, number of progeny per female and percentage of female in progeny) is at 30°C (Liu *et al.*, 2002b), suggesting that this wasp may be more effective in warmer regions. Another study found the performance of *D. collaris* to be affected by host pupal age and so its survival from larva to adult, size and the parasitism capacity of the resultant female adults decreased significantly with increasing host pupal age (Wang and Liu, 2002). Liu *et al.* (2001) showed that females of *D. collaris* fed a honey solution showed significantly increased parasitism capacity and percentage female progeny.

Predators of DBM

Although most researchers agree that predatory arthropods are good agents to use in pest management programmes in agricultural ecosystems, these agents have received less attention than other agents such as parasitoids in controlling DBM populations. Some generalist predators such as ants, lacewings, hemipterans, beetles and spiders can prey on different stages of DBM (Goudegnon *et al.*, 2002; Furlong *et al.*, 2004; Reddy *et al.*, 2004; Wang *et al.*, 2004) (Table 2.3). None of these predators is specific to DBM and therefore their densities cannot be directly correlated with the pest population. Although Harvey and Eubanks (2005) indicated that field populations of *Solenopsis invicta* Buren (Hymenoptera: Formicidae) could be employed to control DBM populations, more studies are needed to determine predator efficiency in field conditions. Reddy *et al.* (2004) showed that the common green lacewing, *Chrysoperla carnea* Stephens, is attracted to some semiochemicals from DBM such as the adult sex and larval frass. Use of DNA markers showed that the damsel bug *Nabis kinbergii* Reuter and some spiders such as *Lycosa* sp., *Ebrechtella tricuspidata* (F.), *Pardosa astrigena* Koch, *Pardosa laura* Karsch and *P. pseudoannulata* (Boesenberg et Strand) could reduce DBM density in crops (Ma *et al.*, 2005; Quan *et al.*, 2011). However, the impact of native predators on DBM populations in many parts of the world is unknown.

Pathogens of DBM

Pathogens formulated as microbial insecticides are being increasingly used in crop protection, due to their specificity against target pests and their low environmental pollution and residue. DBM is attacked by fungi (e.g. *Zoophthora radicans* (Brefeld), *Beauveria bassiana* (Balsamo), *Metarhizium anisopliae* (Metchnikoff), *Isaria* spp.), bacteria (e.g. *Bacillus thuringiensis*), baculoviruses and nematodes (e.g. *Steinernema* spp. and *Heterorhabditis* spp.) (Wilding 1986; Schroer and Ehlers, 2005). Godonou *et al.* (2009) tested eight isolates of entomopathogenic fungi *B. bassiana* and *M. anisopliae* from Benin against DBM larvae. The highest mortality of DBM larvae (94%) was associated with a *B. bassiana* isolate (Bba5653). In addition, crop yields for plots treated with a water formulation of this isolate were higher than in plots treated with an emulsion of water and oil. One laboratory experiment showed that the fungus *Isaria fumosoroseus* Wize not only caused mortality to DBM larvae but also decreased longevity of adults, fecundity and the population rate of intrinsic increase (Huang *et al.*, 2010). Some environmental factors (soil moisture, humidity, rainfall, ultraviolet and solar radiation), however, affect fungi conidia (Furlong and Pell, 1997) and, if unfavourable, lower efficacy.

Schroer and Ehlers (2005) found that a surfactant-polymer formulation of the entomopathogenic nematode (EPN) *Steinernema carpocapsae* (Weiser) caused 80% mortality to DBM larvae, though nematode survival was affected by relative humidity. In general, the type of spray application system can affect the performance of EPNs against DBM (Mason *et al.*, 1999).

Baculoviruses have been reported that infect DBM. A granulovirus from Kenya (Nya-01) at 3 × 10^{13} occlusion bodies/ha caused 90% infection in second-instar DBM larvae (Grzywacz *et al.*, 2001). In another study, feeding of second-instar larvae of DBM cabbage leaf discs treated with *Helicoverpa armigera* nucleopolyhedrovirus (HaNPV) at 3.8 × 10^4 PIB/ml caused 50–75% mortality (Magholi *et al.*, 2014). Some entomopathogenic viruses that infect DBM larvae can be transmitted by parasitoids (Furlong and Asgari, 2010), assisting in its natural dispersal.

2.2.3 Chemical control

Pesticides play a major role in DBM control but pose risks to DMB parasitoids and other non-target organisms. Indiscriminate use of chemical pesticides against DBM in high-value *Brassica* crop production has repeatedly led to pest resistance and environmental pollution.

Firstly, it is important to consider when the use of insecticides is needed, remembering that crops can tolerate some level of pest injury without any loss in yield. Calendar-based applications, carried out without any evaluation of whether or not they will produce an economic gain for the farmer, are often an ineffective use of insecticides. In contrast, responsive (need-based) applications, made when an evaluation of the potential gain from the application has been determined, are a more appropriate tool. In practice, the responsive use of insecticides is often dependent on the availability of an appropriate pest monitoring and forecasting system, and the control measures normally involved are pesticides (Dent, 1995). Spraying or other chemical applications should be based on efficient decision-making tools, one of which is economic injury level (EIL). EIL is the smallest number of insects (amount of injury) that will cause yield losses equal to the pest management costs, and thus the pest population density or extent of crop damage at which the value of the crop destroyed exceeds the cost of controlling the pest. The pest density at which management action should be taken to prevent an increasing pest population from reaching the economic injury level is called the economic threshold (ET) or action threshold (AT).

Although there is insufficient research to define an ET and EIL values for DBM on canola, some researchers have suggested that control action should be taken in canola fields when larval populations exceed 100–150 individuals/m^2 in immature-to-flowering plants and 200–300 individuals/m^2 in plants with flowers and pods. For seedlings, control is recommended when 25–33% of the cotyledons or true leaves are injured (Dosdall *et al.*, 2011). Choosing selective insecticides is important to minimize side effects on non-target organisms. Broad-spectrum insecticides like synthetic pyrethroids are usually very toxic to beneficial natural enemies and should be avoided. Hence, it is important to determine the life cycle of dominant species of DBM parasitoids in each location to avoid spraying broad-spectrum pesticides at the peak of their adult population. To this end, thermal requirements data of parasitoids alongside those of DBM can be useful for estimating the appearance of a given life stage of DBM and its parasitoids (Table 2.4). Using these thermal requirements and local meteorological data, the emergence of pests and their natural enemies can be predicted (for more information see Campbell *et al.*, 1974; Nematollahi *et al.*, 2016). Another finer point of effective chemical control is spraying at the most susceptible stage of pest in order to achieve the highest possible pest mortality rate. For DBM, the most susceptible stage is first-instar larvae (Golizadeh *et al.*, 2009a, b; Soufbaf *et al.*, 2010b). The emergence of each stage of DBM is also predictable using temperature thresholds and meteorological data (Table 2.4).

Large numbers of pesticides are used against DBM around the world. These can be divided into two groups: biopesticides and synthetic pesticides.

Biopesticides

Biopesticides, also known as biological pesticides, are derived from natural materials such as animals, plants and bacteria. The advantages of most biopesticides over synthetic compounds include their lower toxicity to non-target organisms and their quick decomposition. Most biopesticides target specific pests and generally pose little or no risk to humans or the environment.

Many biopesticides exist that can reduce the density of DBM populations. *B. thuringiensis* is a ubiquitous bacterium and its proteinaceous crystal toxin can be used against pests from several different orders, but mainly Lepidoptera. However, its toxicity depends on the strain and the pest treated. For instance, Bt IBT-15 strain is more effective and faster acting against DBM larvae than Bt-serovar entomocidus strain (Eswarapriya *et al.*, 2010). In one field study of a variety of pesticides (including two Bt products, neem products, the entomopathogenic fungus *B. bassiana* and the synthetic insecticide dichlorvos), the Bt products were the most effective (Vanlaldiki *et al.*, 2013). In commercial collards in South Carolina, spinosad and emamectin benzoate provided control of DBM, while azadirachtin, *B. bassiana* and Bt delta endotoxins could not consistently maintain DBM populations below the economic injury level, especially when DBM larvae averaged more than three per plant (Khan *et al.*, 2008). Another study found spinosad to be more effective against DBM eggs than against larvae, suggesting that the use of lower doses of spinosad against DBM eggs could reduce pest populations before any damage had been done to the crop (Legwaila *et al.*, 2014). However, some studies have found this compound to be very damaging to adults of *Diadegma insulare* and *C. plutellae*, causing 100% and 50% mortality, respectively (Hill and Foster, 2000; Haseeb *et al.*, 2004), demonstrating the need for correctly timed use of this pesticide to mitigate these unintended effects. Another biopesticide whose effect on DBM has been studied is cantharidin, a natural toxin isolated from blister beetles of the family Meloidae (Zhang *et al.*, 2002).

Botanical pesticides are based on plant extracts or essential oils. Many studies have investigated the effects of such extracts on DBM but only a few of these are commercially available. Some plant extracts, including those of *Azadirachta indica* Juss, *Melia azedarach* L. and *Catunaregam spinosa* (Thunb.), show feeding deterrent activity for DBM larvae (Charleston *et al.*, 2005; Gao *et al.*, 2011). Field applications of extracts of *A. indica* seed and *Lantana camara* L. leaf on cabbage plants controlled DBM and increased yields by 37% and 26%, respectively (Baidoo and Adam, 2012). Feeding third-instar DBM larvae for 2 days on crop leaves treated with an ethanol extract of *Peganum harmala* L. seed at concentrations of 30 or 40 mg/ml caused 66% and 100% mortality, respectively. In addition, larval and pupal weights of DBM larvae on treated leaves were both significantly lower than those in the control treatment (Abbasipour *et al.*, 2010). Similarly, an ethyl acetate extract of *Veratrum nigrum* L. showed insecticidal activity against the second- and third-instar larvae of DBM, with LC_{50} values of 225 ppm and 335 ppm, respectively (Vanichpakorn *et al.*, 2010). Although using botanical insecticides can reduce environmental hazards, they may be harmful to natural enemies and so their effect (especially long-term) on dominant natural enemies should be tested before application.

Synthetic pesticides

Using synthetic insecticides has been the usual method for control of DBM in many parts of the world. Since most synthetic pesticides have broad-spectrum activity against non-target organisms, including natural enemies, their effects should be tested on local dominant species of natural enemies. In most cases, not only lethal doses but also

Table 2.4. The lower temperature thresholds and thermal constants of diamondback moth (DBM) and some of its important parasitoids.

Species	Host plant	Temp. range (°C)	Egg		Larva		Pupa		Total		Reference
			T_0[a]	K[b]	T_0	K	T_0	K	T_0	K	
DBM	Cauliflower	10–35	6.48	58.11	7.71	131.90	7.57	64.54	7.06	263.75	Golizadeh *et al.*, 2007
DBM	Cabbage	10–35	6.48	58.11	8.17	128.72	7.92	65.76	7.84	261.59	Golizadeh *et al.*, 2007
DBM	Cabbage	10–30	7.30	52.10	7.00–7.60	135.10	7.80–7.90	64.80–72.70	7.40	268.20	Liu *et al.*, 2002a
DBM	Broccoli	10–35	7.74	50.99	5.03	185.18	6.89	80.64	6.34	312.50	Marchioro and Foerster, 2011
DBM[c]	Canola	2–38	–	–	1.15	66.67	5.62	80.00	4.23	142.86	Bahar *et al.*, 2014
DBM	–	12–35	–	–	–	–	–	–	7.40–9.50	229.00–313.00	Umeya and Yamada, 1973
DBM	–	17–33	7.2	52.00	8.5	161.00	9.8	61.00	–	274.00	Yamada and Kawasaki, 1983
DBM	Kale	17–29	–	–	–	–	–	–	9.2–9.9	294.1	Sarnthoy *et al.*, 1989
DBM	Radish	15–32.5	–	–	–	–	–	–	6.10–8.80	232.56–312.50	Shirai, 2000
Diadegma insulare	Canola	2–34	–	–	–	–	2.33	208.333	2.57	285.71	Bahar *et al.*, 2014
D. anurum	Cabbage	15–35	–	–	–	–	–	–	7.00	282.30	Golizadeh *et al.*, 2008
D. anurum	Cauliflower	15–35	–	–	–	–	–	–	6.90	277.70	Golizadeh *et al.*, 2008
D. mollipla	Cabbage	21–33	–	–	–	–	–	–	10.24	238.10	Nofemela, 2004
Cotesia plutellae	–	15–35	–	–	–	–	–	–	10.60–11.80	167.60	Shi and Liu, 1998
C. plutellae	Cabbage	21–33	–	–	–	–	–	–	8.14	217.39	Nofemela, 2004
Oomyzus sokolowskii	Cabbage	15–33	–	–	–	–	–	–	11.60	211.80	Ferreira *et al.*, 2003
O. sokolowskii	Cabbage	20–35	–	–	–	–	–	–	10.70	240.00	Wang *et al.*, 1999
Trichogramma pretiosum	Cabbage	18–32	–	–	–	–	–	–	12.52	123.03	Pereira *et al.*, 2004
T. exiguum	Cabbage	18–32	–	–	–	–	–	–	13.13	120.00	Pereira *et al.*, 2004

[a]Lower temperature threshold (°C): the temperature below which development stops.
[b]Thermal constant (DD): the number of degree days above lower temperature threshold required by an insect to complete its development.
[c]Except eggs and first-instar larvae.

sublethal doses of synthetic pesticides may adversely affect population parameters of DBM (Mahmoudvand *et al.*, 2011a, b, 2015) as well as those of its natural enemies and this trade-off should be taken into consideration in any pesticide application programme.

2.2.4 DBM resistance to pesticides

DBM populations have developed resistance to various chemical insecticides and Cry toxins of Bt in the field (Talekar and Shelton, 1993; Heckel *et al.*, 2001) and DBM is ranked second in resistance to different groups of insecticides based on the Arthropod Pesticide Resistance Database (APRD). DBM's ability to develop resistance rapidly is due mainly to its short life cycle, continuous host availability in crops and high genetic variation and detoxification ability. Several mechanisms of resistance, including acetylcholinesterase insensitivity, reduced penetration, nerve insensitivity and detoxification of insecticides, may have a role in DBM resistance. Without insecticide resistance management (IRM), DBM will continue to overcome new insecticides as they are developed. IRM should be an integral part of any DBM control programme and a prerequisite for integrated DBM management. To prevent and manage DBM resistance to insecticides, identifying resistance mechanisms is an important area requiring further research. Four mechanisms have been suggested: (i) metabolic resistance; (ii) altered target-site resistance; (iii) behavioural resistance; and (iv) penetration resistance.

Metabolic resistance

In this mechanism, resistant insects may detoxify or destroy the toxin faster than susceptible insects, or prevent the toxin from reaching target sites by binding it to proteins or other compounds in their bodies. This kind of resistance is the most common mechanism and often presents the greatest challenge to its avoidance. This detoxification method involves three phases: (i) recognition and hydrolysis of the toxin; (ii) conversion of toxin into a non-toxic molecule; and (iii) excretion of the molecule from the insect's body. Previous studies have highlighted the role of the enzymes in resistance of DBM and other insects to insecticides (Sonoda and Tsumuki, 2005; Wu *et al.*, 2011). This kind of resistance may be prevented or circumvented by using insecticides with a different mode of action or certain synergist compounds. For instance, Nehare *et al.* (2010) showed that increasing esterase enzymes is the reason for the successful resistance of DBM to indoxacarb and that this resistance can be controlled by using diethyl-maleate (DEM) and triphenyl phosphate (TPP). Piperonyl butoxide (PBO) is a good synergist to overcome DMB resistance to abamectin (Qian *et al.*, 2008). Similarly, a synergistic effect can be achieved by mixing insecticides from different classes. In this context, the toxicity of bifenthrin against bifenthrin-resistant DBM populations increased significantly when bifenthrin was combined with spinosad, emamectin or indoxacarb (Attique *et al.*, 2006). These synergistic effects may be attributed to the different modes of action of these insecticide classes acting on different components of nerve impulse transmission. However, mixing insecticides can also stimulate the development of multiple resistance (resistance to more than one class of insecticides) that may extend across other chemical classes and thus make DBM populations more difficult to manage. In this case, alternative strategies such as mosaics (applying several insecticides with different modes of action in different locations within an area) or rotations (choosing at least two different insecticides, each with a different mode of action for the control programme) should be considered. The constant use of insecticides from one chemical group (with the same mode of action) will increase the risk of rapid build-up of resistance to that chemical group. On the other hand, alternating the use of chemical groups with different modes of action will slow down the process of selection for resistance.

Altered target-site resistance

The target site where insecticides usually bind in insects can become modified to reduce insecticide effects. This is the most common mechanism of resistance in fungi and weeds and it is also prevalent in insects. For instance, resistance of DBM to nereistoxin insecticides is due to insensitivity of the acetylcholine receptor (Cheng *et al.*, 2008). For this kind of resistance, using other insecticides with different sites of action can be useful.

Behavioural resistance

Resistant insects may avoid the negative effects of insecticides by changing their normal activity patterns. Insects may stop feeding or move to the underside of the sprayed leaf. Although this type of resistance has not been found in DBM yet, there is evidence that DBM resistant to some insecticides can change its sex pheromone communication system (Xu *et al.*, 2010). In one recent study, wind tunnel experiments indicated that changes associated with insecticide resistance in the abamectin-resistant strain of DBM significantly reduced female attractiveness to susceptible males. In addition, mating choice experiments confirmed that non-random mating occurred between two susceptible and resistant strains. Resistance can be accelerated by the deliberate mating of resistant males and females. Using different insecticides with different modes of action can help to reduce the selection pressure for resistance.

Penetration resistance

In this case, resistant insects may absorb the toxin more slowly than susceptible individuals. Penetration resistance occurs when the outer cuticle develops barriers that can slow absorption of insecticides. This kind of resistance is frequently present in combination with other types of resistance and can protect insects from a wide range of insecticides. Use of oral insecticides such as Bt may mitigate this resistance.

2.2.5 DBM resistance management strategies

Minimizing insecticide use

Minimizing insecticide use is fundamental to pesticide resistance management. From an insecticide resistance viewpoint, every time an insect population is treated with an insecticide, selection for resistance is occurring. Although this selection is not evident at first because so few individuals survive, larval survival and subsequent plant damage becomes obvious as the resistant population increases. Intensification of spray programmes can exacerbate the problem by putting even more resistance selection pressure on DBM populations. As already mentioned, an understanding of the economic injury level is important, because trying to produce a crop with a zero tolerance for damage may work in the short term but leads to increased insecticide use, insecticide resistance and, ultimately, loss of pest control and increased crop loss. The use of non-chemical strategies such as host-free periods, crop rotation, biological control and weed control can help to reduce the need for chemicals and consequently slow the development of insecticide resistance.

Mixtures of insecticides with different modes of action

Mixing insecticides with different modes of action or different mechanisms of resistance can be effective in managing resistance development. A typical combination is two insecticides and a suitable synergist. Insecticides with similar modes of action generally express cross-resistance (selection for resistance to one product selects for resistance to all products with the same mode of action). However, long-term use of mixtures of different classes of insecticide can also give rise to insecticide resistance, if resistance mechanisms to different insecticides arise together in some individuals. Continued use of the mixture will select for these multiple insecticide-resistant pests.

Using insecticides with short persistence

Insects with resistant genes will be selected over susceptible ones whenever insecticide concentrations kill only the susceptible pests. An ideal insecticide quickly disappears from the environment so that persistence of a 'selecting dose' does not occur. In addition, environmental pollution, damage to natural enemies and increasing risk for consumers are all intensified by using persistent insecticides.

Rotation of insecticides

The rotation or alternation of insecticides is another tactic used to manage resistance development. Longer use of a single insecticide class will enhance the chance of resistance developing (especially in multivoltine insects such as DBM) since the survivors of the first generation and the next will most likely be tolerant to that class. In this tactic, it is important that pest resistance to both insecticides be rare, that the alternating insecticides belong to unrelated chemical classes and that the time interval between applications of the rotating insecticides is long enough for the pest population to return to its original level of susceptibility. Some studies suggest that such an insecticide rotation plan for DBM resistance management is crucial for the success of future DMB management programmes (Riley and Sparks, 2009; Walker *et al.*, 2012).

2.2.6 Behavioural control

Pest management strategies can use the behaviour of pests to manipulate and reduce their densities. Most moths rely on chemical cues (semiochemicals or infochemicals) to communicate with each other for mating (using sex pheromone) or to find suitable hosts (using allelochemicals) for oviposition. Semiochemicals or infochemicals are chemical cues that mediate the relationship between two organisms by inducing behavioural and/or physiological responses to one or both of the organisms. Sex pheromones in Lepidoptera, especially moths, are produced by the female pheromone gland and released at specific times during the day to attract males and mediate sexual behavior (Roelofs and Rooney, 2003). Use of sex pheromones can be a powerful tool for mass trapping of males as well as mating disruption aimed at creating mating failure and reducing the number of fertile eggs. This is a common approach for the control of moths because of its specificity and non-toxicity to non-target organisms. In addition, monitoring established populations is the most widespread use of pheromones in pest management programmes. Although the female sex pheromone of DBM has been synthesized (Chisholm *et al.*, 1979) and its commercial product CheckMate (Suterra LLC, Bend, Oregon, USA) is available, differences among populations of DBM may limit the use of these compounds (Yang *et al.*, 2007). The latest research has found that (Z)-9-tetradecenyl acetate, which was previously reported as a component of the sex attractant of a Canadian DBM population, was not detected in the gland extracts of the Korean population. Some researchers have produced pheromone microcapsules using gelatin and gum arabic as wall-forming materials that can release DBM sex pheromone slowly (over 6 weeks in the field) aimed at mating disruption (Chen *et al.*, 2007). Results of the latter study showed that the mating disruption efficiency of these microcapsules ranged from 77% to 99% over the season.

Green leaf volatiles (GLVs) from *Brassica* crops are attractive to both male DBM (as a cue for increased mating opportunities) and female DBM (as an odour from a suitable host for oviposition) (Fig. 2.3). For instance, GLVs from *Brassica oleracea* subsp. *capitata* including 1-hexanol, (Z)-3-hexen-1-ol, 1-hexen-3-ol, hexanal, (E)-2-hexenal, hexyl acetate, and (Z)-3-hexenyl acetate were identified as synergists of DBM pheromone for its attraction (Reddy and Guerrero, 2000). Laboratory experiments showed that mixtures of (Z)-3-hexenyl acetate, (E)-2-hexenal and (Z)-3-hexen-1-ol with pheromone induced attractant behaviour in 80–100% of unmated males, significantly higher than the pheromone alone. A field study meanwhile revealed that when (Z)-3-hexenyl acetate was mixed with the pheromone in a 1:1 ratio, it enhanced by six to seven times the number of females and by 20–30% the number of males caught by traps compared with pheromone alone. These relatively inexpensive and environmentally safe compounds hold potential for blending with pheromones and increasing the performance of traps. Another study found that DBM pheromone mixed with (Z)-3-hexenyl acetate and (Z)-3-hexen-1-ol significantly increased male catches compared with pheromone alone (Dai *et al.*, 2008). Furthermore, a blend of brassicaceous volatiles, DBM synthetic pheromone and odour of larval frass can attract natural enemies such as *Trichogramma chilonis*, *Cotesia plutellae* and *Chrysoperla carnea* (Reddy *et al.*, 2002), and treating trap plants with these compounds may reduce DBM densities by increasing natural enemy activity.

Plants respond to herbivore damage through various morphological and biochemical means (Fig. 2.3). Herbivore-induced plant volatiles (HIPVs) are involved in plant communication with natural enemies of herbivorous insects and neighbouring plants. These compounds are released at the onset of pest damage and can help plants attract natural enemies of the pest. Since the central component of this pathway is jasmonic acid (JA), the endogenous concentration and exogenous treatment of plants with JA may result in induced responses similar to plant defences induced by herbivory (Farmer and Ryan, 1992) and may induce plants to produce volatiles that attract natural enemies of herbivores. Although some studies have found that composition, timing and duration of HIPV emission differ between herbivory of DBM larvae and JA treatments in some crucifer species (Zhang *et al.*, 2010), it seems that treatment of plants with JA can be effective in attracting natural enemies.

2.3 DBM Management Plan

Although the use of parasitoids and other IPM components such as cultural and physical control have been employed for DBM management (Grzywacz *et al.*, 2010; Li *et al.*, 2016), many studies

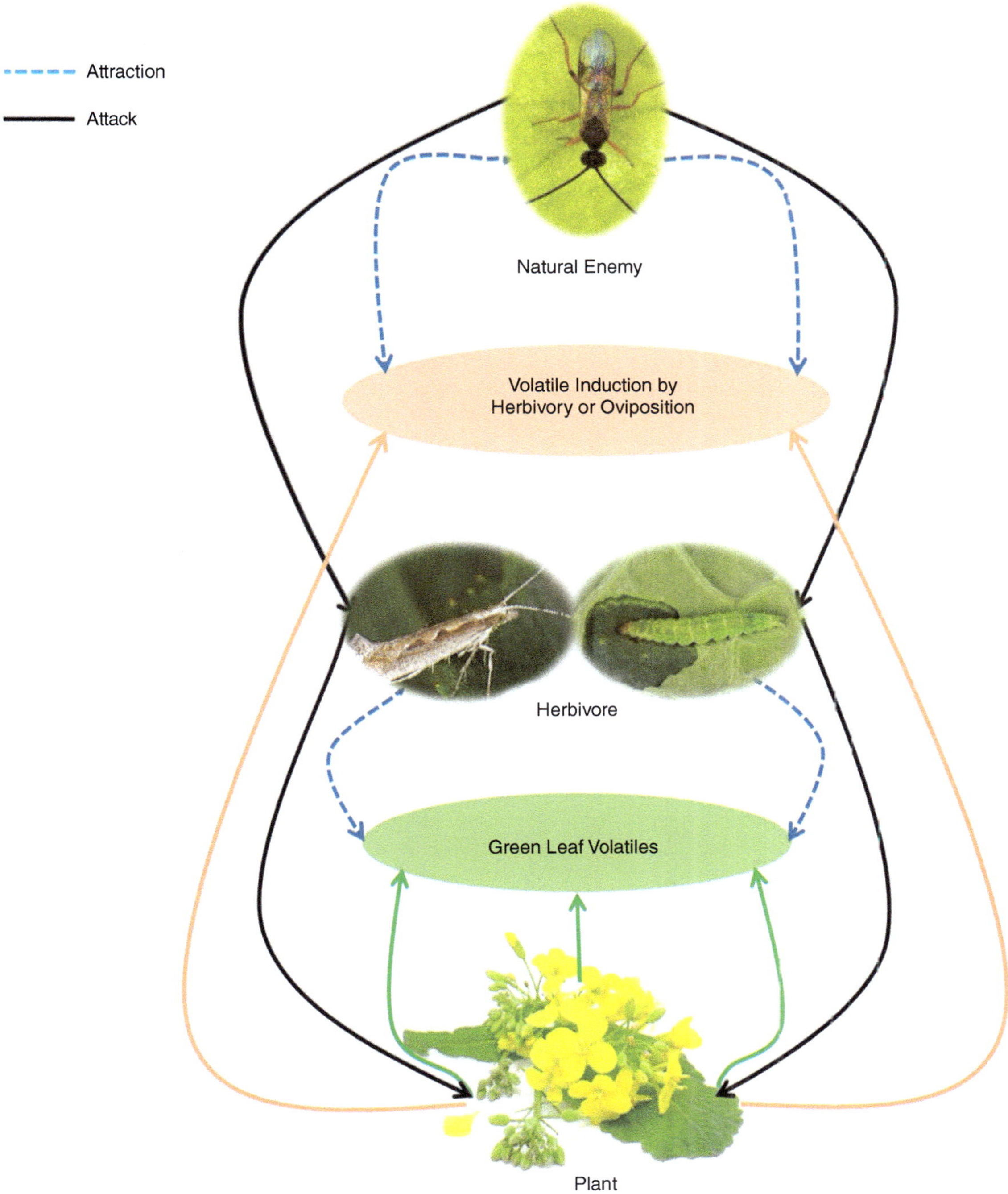

Fig. 2.3. Green leaf and herbivore-induced plant volatiles in a canola, DBM and natural enemy system.

indicate that farmers continue to spray broad-spectrum insecticides for control of DBM as a first, easily available and reliable option in most areas of the world. Early use of non-selective insecticides (such as pyrethroids) are an important initiating factor in subsequent DBM outbreaks by dramatically reducing parasitoid populations (Talekar and Shelton, 1993). Furthermore, calendar-based instead of need-based spraying, the use in some countries of unregistered or fraudulent insecticides of poor quality and the over-application of highly toxic insecticides not only increase DBM resistance to these chemical compounds but also cause high mortality of natural enemies of this economically important pest. Therefore, to reduce the risk of DBM resistance to insecticides, the threat to food safety, environmental pollution, human health problems, harmful side effects and deleterious effects to non-target beneficial organisms including natural enemies and pollinators, the following strategies are suggested regarding the use of insecticides: (i) the use of bio-insecticides instead of synthetic ones; (ii) the use of selective insecticides instead of broad-spectrum ones; (iii) minimizing insecticide use; (iv) mixing insecticides with different modes of action; (v) using insecticides with short persistence; and (vi) rotation of insecticides.

Comparing the developmental time, pre-adult mortality and life table parameters, including the net reproductive rate (R_0), intrinsic rate of increase (r), finite rate of increase (λ) and mean generation time (T) of DBM on different host plants, shows that canola is one of the most important host plants of DBM (Table 2.5). Although the literature suggests that the potential population growth of DBM on some genotypes of cabbage and cauliflower is higher than on canola (Table 2.5), there are some especially susceptible canola genotypes where this population growth is noticeable. The suitability of canola for DBM and the considerable potential for the population increase of this pest on canola compel growers to design a management programme to reduce crop losses caused by this noxious pest.

Since IPM is a holistic approach that integrates many components to maximize the advantages and minimize the disadvantages of the management plan (Fathipour and Sedaratian, 2013), all possible interactions among components of the IPM system should be taken into consideration. Furthermore, it is necessary to consider the different trophic interactions in both below-ground and above-ground locations (Fig. 2.4) in any combination of control measures, especially resistant genotypes and biological control agents. For success in this integration, the effect of resistant cultivars on the population growth and performance of the third trophic level (natural enemies) should be tested. Indeed, tritrophic interactions, and in particular the interactions of the bottom-up (force of plants) and top-down (force of insect natural enemies) forces, might be relevant in the biological control of pests (Singh, 2003) (Fig. 2.4).

Two important components of IPM – resistant genotypes and biological control – must be integrated successfully to reduce DBM population density. Host plant resistance can be a valuable component of an IPM system, compatible with other control measures such as chemical control and biocontrol agents and making beneficial natural enemies more effective. A long-term study on different aspects of canola pests such as DBM, *H. armigera*, *S. exigua* and *B. brassicae* and their natural enemies reveals interesting findings on integrated crop management (ICM) in a canola cropping system in which a part of these findings is summarized in Table 2.6 and some in Table 2.3, 2.4 and 2.5. As well as a comprehensive survey on DBM control strategies, the data presented in Table 2.6 suggest that canola genotype has a profound effect on the development, survival rate, reproduction, population growth and nutritional indices of DBM. As shown in Table 2.6, some genetically manipulated resistant or constitutively resistant genotypes reduced the potential population growth of DBM by up to 77% (e.g. Transgenic-PF genotype) (Nikooei *et al.*, 2015a), demonstrating the important role of resistant genotypes in integrated DBM management programmes. Since host plant resistance and biological control are used simultaneously in integrated control programmes, the effect of resistance (mainly secondary biochemical effects) on the performance of biological control agents should be taken into consideration when designing an IPM programme. In some experiments carried out on tritrophic interactions in a canola–DBM–natural enemy system, natural enemies (especially parasitoids) were adversely affected by resistant canola genotypes but not as much as DBM. For instance, the Transgenic-PF genotype that caused 77% reduction in the performance of DMB only caused 26% reduction in the performance of its parasitoid *Diadegma semiclausum* (Nikooei *et al.*, 2015b, 2016). Sometimes the resistant

Table 2.5. Developmental and life table parameters of diamondback moth (DBM) on different host plants.

Host plant	Temp. (°C)	Development time (day)	Pre-adult mortality (%)	R_0 (offspring)	r (day^{-1})	λ (day^{-1})	T (day)	Reference
Canola	25	15.00–16.60	20.80–48.30	30.60–57.30	0.241–0.304	1.270–1.350	13.20–14.60	Soufbaf *et al.*, 2010a, b
Canola	27	–	31.00–56.00	32.54–98.59	0.143–0.245	1.153–1.278	18.70–24.37	Akandeh *et al.*, 2015; Akandeh *et al.*, 2016
Canola	25	13.92–24.61	17.00–48.00	7.88–60.79	0.071–0.236	1.074–1.266	17.26–28.39	Nikooei *et al.*, 2015a
Canola	25	15.06	29.44	58.26	0.244	1.270	16.64	Golizadeh *et al.*, 2009b
Canola	25	15.05–16.87	47.65–65.75	37.32–124.37	0.214–0.287	1.230–1.330	13.94–17.76	Ebrahimi *et al.*, 2008; Ebrahimi *et al.*, 2009
Canola	23	14.90–17.38	32.10–43.00	31.41–52.26	0.178–0.230	1.194–1.258	17.23–19.97	Fathi *et al.*, 2011
Canola	25	10.20	6.00	81.80	0.290	1.336	15.19	Saeed *et al.*, 2010
Canola	25	12.96–14.7	36.00–62.00	4.70–26.93	0.080–0.169	1.087–1.185	18.06–21.36	Kianpour and Fathipour, 2013
Cabbage	20	22.8	8.05	141.44	0.190	1.210	26.62	Marchioro and Foerster, 2014
Cabbage	25	14.03–19.83	8.00–19.00	129.56–201.62	0.216–0.289	1.241–1.335	16.87–22.84	Pan *et al.*, 2014
Cabbage	25	13.64–15.26	19.00–31.00	26.70–68.58	0.180–0.250	1.200–1.290	16.66–18.26	Yin *et al.*, 2009
Cabbage	25	14.13–14.52	19.05–19.89	43.13–183.81	0.256–0.285	1.290–1.330	14.71–17.28	Golizadeh *et al.*, 2009b
Cabbage	10–30	85.13–11.84	13.79–41.73	21.39–183.81	0.033–0.315	1.030–1.370	13.98–91.33	Golizadeh *et al.*, 2007; Golizadeh *et al.*, 2009a
Cabbage	27	15.90	16.60	146.00	0.310	1.370	15.80	Ayalew *et al.*, 2006
Cabbage	25	11.40	22.00	32.29	0.200	1.221	17.37	Saeed *et al.*, 2010
Cabbage	25	16.29–17.4	54.00–66.00	4.65–9.49	0.069–0.113	1.070–1.119	19.61–22.18	Kianpour and Fathipour, 2013
Cauliflower	10–30	81.26–11.43	15.87–47.01	29.67–179.90	0.038–0.340	1.030–1.400	14.13–89.59	Golizadeh *et al.*, 2007; Golizadeh *et al.*, 2009a
Cauliflower	20	25.3	18.05	101.95	0.180	1.200	27.44	Marchioro and Foerster, 2014
Cauliflower	25	14.12	18.58	159.84	0.293	1.340	17.28	Golizadeh *et al.*, 2009b
Cauliflower	25	11.00	18.00	96.48	0.300	1.350	15.23	Saeed *et al.*, 2010
Cauliflower	25	15.79	24.00	7.96	0.107	1.113	19.15	Kianpour and Fathipour, 2013
Kohlrabi	25	13.76	18.94	51.62	0.261	1.300	15.100	Golizadeh *et al.*, 2009b
Mustard	25	10.80	10.00	65.86	0.260	1.297	16.11	Saeed *et al.*, 2010
Ethiopian mustard	27	16.38	20.70	91.00	0.290	1.340	15.50	Ayalew *et al.*, 2006
Black mustard	27	17.73–18.26	71.90–74.20	21.00–26.00	0.160–0.180	1.170–1.190	18.00–18.90	Ayalew *et al.*, 2006
Turnip	25	13.00	36.00	64.51	0.220	1.248	18.94	Saeed *et al.*, 2010
Turnipweed	20	25.00	23.20	73.88	0.150	1.160	28.46	Marchioro and Foerster, 2014
Radish	25	12.80	14.00	40.25	0.190	1.209	19.45	Saeed *et al.*, 2010
Wild radish	20	24.3	20.00	83.18	0.150	1.160	28.72	Marchioro and Foerster, 2014
Broccoli	20	24.70	6.75	138.71	0.210	1.230	23.17	Marchioro and Foerster, 2014
Erucastrum arabicum	27	16.65	47.00	87.00	0.260	1.300	17.00	Ayalew *et al.*, 2006

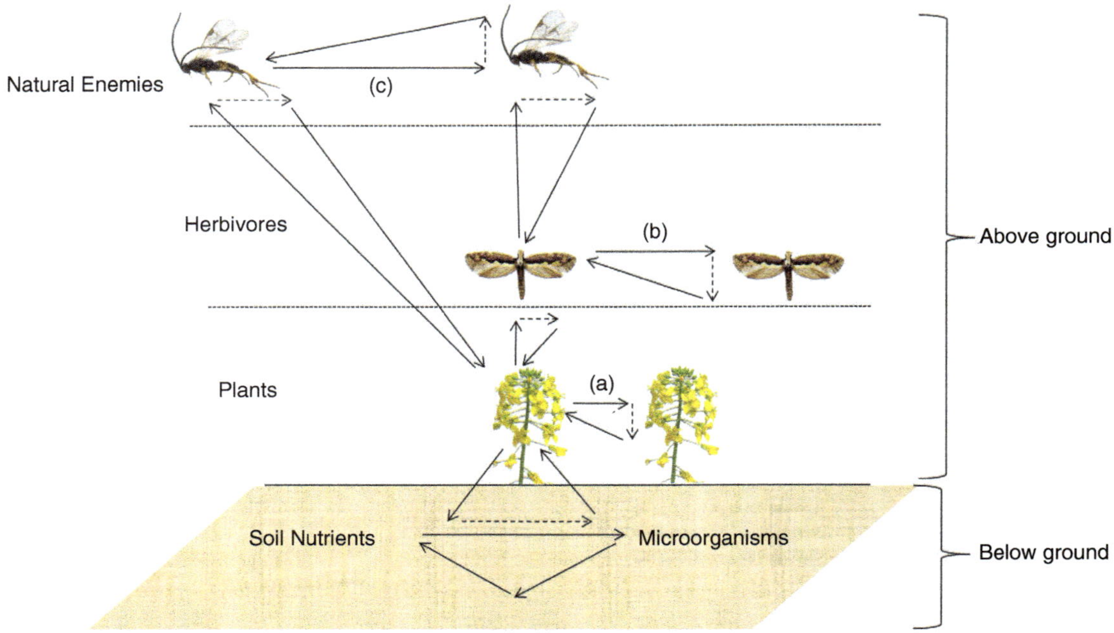

Fig. 2.4. Potential interactions among the plants, insect herbivores and insect natural enemies and between above-ground and below-ground components. (a), (b) and (c) are cross-forces acting inside the first, second and third trophic level, respectively. Dotted arrows are the path in which the incoming force changes qualitatively/quantitatively by receiver level and then sends back to the sender level.

canola genotype shows synergistic activity with a biological control agent. For instance, Soufbaf *et al.* (2012) found that the parasitoid wasp *D. semiclausum* performed better on RGS_{003}, which was the inferior (partially resistant) host to DBM compared with SLM_{046}, which was most susceptible to DBM. One field study revealed that, among canola genotypes tested, the lowest larval density of DBM was observed on the Opera genotype, a resistant cultivar, while the percentage of larvae parasitized by *Diadegma majale* (Gravenhorst) was significantly higher (88.7%) on this cultivar (Fathi *et al.*, 2012). Although in most cases the adverse effects of resistant canola genotypes on the performance of natural enemies of DBM is negligible, these negative effects should be taken into consideration when evaluating any combination of resistant canola genotypes and biological control agents (whether inoculation, augmentation or conservation).

When a crop is attacked by multiple pests, the resistance status of the genotype chosen for the management programme should be considered for all pests in the area. A genotype that is highly resistant to one pest might be susceptible to another pest in the same cropping system. The data in Table 2.6 reveal the susceptibility and resistance rate of different canola genotypes to four main pests and two parasitoid wasps. As these data show, a genotype resistant or partially resistant to a pest might be resistant to another pest, but it might be highly or moderately susceptible to other pests feeding on canola. For instance, the canola genotype RGS_{003} that is moderately susceptible or resistant to DBM, *H. armigera* and *S. exigua* shows high resistance to the aphid *B. brassicae*, a major pest of canola. Similarly, the Okapi genotype that is susceptible to DBM and *H. armigera* and highly susceptible to *S. exigua* shows high resistance to *B. brassicae* (Table 2.6). Such conflicts should be taken into account before designing an appropriate programme to control DBM.

Many cultural control methods are also capable of being integrated with biological control agents and some, such as intercropping, can improve the performance of natural enemies. The push–pull

Table 2.6. The development, life table parameters and nutritional indices of some important pests and natural enemies on different genotypes of canola. In each column and for each species, numbers in parentheses indicate percentage of increasing (development time, pre-adult mortality and *T*) or decreasing (other) parameters compared with the most susceptible genotype.

Canola genotypes	Temp. (°C)	Development time (day)	Pre-adult mortality (%)	R_0 (offspring)	r (day^{-1})	λ (day^{-1})	T (day)	ECI (%)	ECD (%)	AD (%)	References
(a) Diamondback moth (DBM)											
SLM_{046}	25	15.70 (21.14)	21.0 (4.0)	51.50 (10.12)	0.304 (0)	1.355 (0)	13.40 (1.52)	2.298 (0)	2.471 (0)	95.867 (1.69)	Soufbaf *et al.*, 2010a; Soufbaf *et al.*, 2010b; Kianpour *et al.*, 2014
Okapi	25	15.10 (16.51)	29.0 (12.0)	46.10 (19.55)	0.286 (5.92)	1.331 (1.77)	13.20 (0)	–	–	–	Soufbaf *et al.*, 2010a; Soufbaf *et al.*, 2010b
Medina	25	15.30 (18.06)	32.0 (15.0)	35.10 (38.74)	0.259 (14.80)	1.295 (4.43)	13.60 (3.03)	–	–	–	Soufbaf *et al.*, 2010a; Soufbaf *et al.*, 2010b
RGS_{003}	25	16.20 (25.00)	33.0 (16.0)	30.60 (46.60)	0.241 (20.72)	1.272 (6.13)	14.40 (9.09)	0.539 (1.76)	0.563 (1.91)	97.557 (0)	Soufbaf *et al.*, 2010a; Soufbaf *et al.*, 2010b; Kianpour *et al.*, 2014
Talaye	25	16.10 (24.23)	31.0 (14.0)	38.00 (33.68)	0.270 (11.18)	1.310 (3.32)	13.30 (0.76)	–	–	–	Soufbaf *et al.*, 2010a; Soufbaf *et al.*, 2010b
Zarfam	25	16.20 (25.00)	26.0 (9.0)	57.30 (0)	0.278 (8.55)	1.320 (2.58)	14.60 (10.61)	–	–	–	Soufbaf *et al.*, 2010a; Soufbaf *et al.*, 2010b
Opera	25	16.60 (28.09)	48.0 (31.0)	41.70 (27.23)	0.258 (15.13)	1.294 (4.50)	14.50 (9.85)	–	–	–	Soufbaf *et al.*, 2010a; Soufbaf *et al.*, 2010b
Sarigol	25	15.70 (21.14)	36.0 (19.0)	38.20 (33.33)	0.253 (16.78)	1.288 (4.94)	14.00 (6.06)	–	–	–	Soufbaf *et al.*, 2010a; Soufbaf *et al.*, 2010b
Licord	25	16.10 (24.23)	38.0 (21.0)	36.80 (35.78)	0.255 (16.12)	1.290 (4.80)	13.60 (3.03)	–	–	–	Soufbaf *et al.*, 2010a; Soufbaf *et al.*, 2010b
Hayula420	25	15.00 (15.74)	44.0 (27.0)	55.20 (3.66)	0.287 (5.59)	1.332 (1.70)	13.80 (4.55)	–	–	–	Soufbaf *et al.*, 2010a; Soufbaf *et al.*, 2010b
Brassica rapa	25	14.36 (10.80)	19.0 (2.0)	40.77 (28.85)	0.206 (32.24)	1.228 (9.37)	17.95 (35.98)	–	–	–	Nikooei *et al.*, 2015a
Hybrid- $Hyula_{401}$	25	14.24 (9.88)	17.0 (0)	39.04 (31.87)	0.207 (31.91)	1.231 (9.15)	17.57 (33.11)	–	–	–	Nikooei *et al.*, 2015a
Mutant-RGS_{003}	25	20.34 (56.94)	37.0 (20.0)	19.03 (66.79)	0.121 (60.20)	1.129 (16.68)	24.05 (82.20)	–	–	–	Nikooei *et al.*, 2015a
Transgenic- PF	25	24.61 (89.89)	48.0 (31.0)	7.88 (86.24)	0.071 (76.64)	1.074 (20.74)	28.39 (115.08)	–	–	–	Nikooei *et al.*, 2015a
Star	25	14.70 (13.42)	36.0 (19.0)	25.21 (56.00)	0.169 (44.41)	1.185 (12.55)	18.92 (43.33)	–	–	–	Kianpour and Fathipour, 2013

Continued

Table 2.6. Continued.

Canola genotypes	Temp. (°C)	Development time (day)	Pre-adult mortality (%)	R_0 (offspring)	r (day^{-1})	λ (day^{-1})	T (day)	ECI (%)	ECD (%)	AD (%)	References
Elite	25	12.96 (0)	35.0 (18.0)	20.73 (63.82)	0.159 (47.70)	1.173 (13.43)	18.99 (43.86)	–	–	–	Kianpour and Fathipour, 2013
NSA2	25	13.67 (5.48)	60.0 (43.0)	26.93 (53.00)	0.154 (49.34)	1.167 (13.87)	21.36 (61.82)	–	–	–	Kianpour and Fathipour, 2013
Ebonite	23	15.45 (19.21)	36.0 (19.0)	49.35 (13.87)	0.214 (29.61)	1.239 (8.56)	18.22 (38.03)	–	–	–	Fathi *et al.*, 2011
Adder	23	15.14 (16.82)	37.0 (20.0)	45.54 (20.52)	0.212 (30.26)	1.236 (8.78)	18.03 (36.59)	–	–	–	Fathi *et al.*, 2011
$Hyula_{401}$	23	17.38 (34.10)	41.0 (24.0)	39.06 (31.83)	0.183 (39.80)	1.201 (11.37)	19.97 (51.29)	–	–	–	Fathi *et al.*, 2011
Option500	23	16.94 (30.71)	42.0 (25.0)	33.46 (41.61)	0.188 (38.16)	1.206 (11.00)	18.72 (41.82)	–	–	–	Fathi *et al.*, 2011
(b) *Diadegma semiclausum*											
SLM_{046}	25	–	31.0 (7.0)	5.56 (72.31)	0.147 (22.63)	1.160 (4.05)	20.25 (34.55)	–	–	–	Soufbaf *et al.*, 2012
RGS_{003}	25	12.27 (0)	34.0 (10.0)	14.86 (26.00)	0.178 (6.32)	1.195 (1.16)	15.05 (0)	–	–	–	Nikooei *et al.*, 2016
B. rapa	25	12.29 (0.16)	37.0 (13.0)	14.87 (25.95)	0.173 (8.95)	1.189 (1.65)	15.05 (0)	–	–	–	
Opera	25	12.37 (0.81)	24.0 (0)	20.08 (0)	0.190 (0)	1.209 (0)	15.74 (4.74)	–	–	–	
Hybrid – $Hyula_{401}$	25	15.21 (23.96)	31.0 (7.0)	12.21 (39.19)	0.141 (25.79)	1.152 (4.71)	17.58 (16.81)	–	–	–	
Mutant – RGS_{003}	25	14.94 (21.76)	26.0 (2.0)	13.95 (30.53)	0.143 (24.74)	1.153 (4.63)	18.34 (21.86)	–	–	–	
Transgenic – PF	25	15.11 (23.15)	30.0 (6.0)	12.53 (37.60)	0.141 (25.79)	1.152 (4.71)	17.74 (17.87)	–	–	–	
(c) *Helicoverpa armigera*											
Talaye	25	36.60 (5.17)	41.0 (18.0)	157.40 (52.52)	0.159 (11.17)	1.172 (2.01)	31.10 (0)	12.323 (0)	32.357 (0)	44.364 (37.87)	Chegeni and Fathipour, 2011; Karimi *et al.*, 2012
Opera	25	35.10 (0.86)	23.0 (0)	320.80 (3.23)	0.179 (0)	1.196 (0)	32.10 (3.22)	7.616 (4.71)	12.956 (19.40)	62.839 (19.40)	
Licord	25	36.50 (4.89)	34.0 (11.0)	180.80 (45.46)	0.163 (8.94)	1.177 (1.59)	31.80 (2.25)	10.114 (2.21)	27.369 (4.99)	53.551 (28.69)	
Modena	25	36.40 (4.60)	27.0 (4.0)	235.10 (29.08)	0.160 (10.61)	1.174 (1.84)	34.20 (9.97)	11.095 (1.23)	28.895 (3.46)	40.057 (42.18)	
SLM_{046}	25	36.00 (3.45)	42.0 (19.0)	177.60 (46.43)	0.155 (13.41)	1.168 (2.34)	33.30 (7.07)	6.326 (6.0)	11.072 (21.29)	51.209 (31.03)	

Hayula420	25	36.80	41.0	297.70	0.158	1.171	35.20	9.940	14.291	71.141	
		(5.75)	(18.0)	(10.20)	(11.73)	(2.09)	(13.18)	(2.38)	(18.07)	(11.10)	
Zarfam	25	34.80	40.0	277.20	0.166	1.181	34.90	10.569	17.142	53.299	
		(0)	(17.0)	(16.38)	(7.26)	(1.25)	(12.22)	(1.75)	(15.22)	(28.94)	
Okapi	25	36.10	33.0	312.10	0.163	1.177	35.30	5.947	6.922	82.236	
		(3.74)	(10.0)	(5.85)	(8.94)	(1.59)	(13.50)	(6.38)	(25.44)	(0)	
RGS_{003}	25	35.00	37.0	331.50	0.168	1.183	34.70	10.288	13.635	73.933	
		(0.57)	(14.0)	(0)	(6.15)	(1.09)	(11.58)	(2.03)	(18.72)	(8.30)	
Sarigol	25	36.80	51.0	256.90	0.153	1.165	36.10	6.265	7.834	80.713	
		(5.75)	(28.0)	(22.5)	(14.53)	(2.59)	(16.08)	(6.06)	(24.52)	(1.52)	
(d) *Spodoptera exigua*											
Sarigol	25	26.70	48.0	164.52	0.169	1.185	29.54	11.063	15.970	69.423	Pourghasem and Fathipour, 2011; Goodarzi *et al.*, 2015
		(3.41)	(8.0)	(53.42)	(19.14)	(3.81)	(7.54)	(2.20)	(1.37)	(17.15	
SLM_{046}	25	27.09	58.0	209.99	0.176	1.192	29.88	7.997	11.345	70.61	
		(4.92)	(18.0)	(40.55)	(15.79)	(3.25)	(8.77)	(5.27)	(5.99)	(15.96)	
Hayula420	25	26.11	48.0	181.63	0.183	1.201	27.81	13.265	17.337	76.753	
		(1.12)	(8.0)	(48.58)	(12.44)	(2.52)	(1.24)	(0)	(0)	(9.82)	
RGS_{003}	25	26.44	52.0	227.81	0.179	1.196	29.66	10.183	14.130	72.327	
		(2.40)	(12.0)	(35.50)	(14.35)	(2.92)	(7.97)	(3.08)	(3.21)	(14.25)	
Opera	25	27.39	50.0	153.27	0.161	1.174	30.73	7.647	9.736	78.676	
		(6.08)	(10.0)	(56.61)	(22.97)	(4.71)	(11.87)	(5.62)	(7.60)	(7.90)	
Okapi	25	25.82	44.0	353.20	0.209	1.232	27.47	11.412	16.540	68.955	
		(0)	(4.0)	(0)	(0)	(0)	(0)	(1.85)	(0.80)	(17.62)	
Licord	25	26.69	48.0	283.24	0.187	1.205	29.58	4.201	4.877	86.573	
		(3.37)	(8.0)	(19.81)	(10.53)	(2.19)	(7.68)	(9.06)	(12.46)	(0)	
Modena	25	29.94	56.0	225.52	0.153	1.165	34.56	11.804	15.346	77.579	
		(15.96)	(16.0)	(36.15)	(26.79)	(5.44)	(25.81)	(1.46)	(1.99)	(8.99)	
Zarfam	25	29.09	47.0	233.77	0.169	1.185	31.17	6.646	8.154	81.815	
		(12.66)	(7.0)	(33.81)	(19.14)	(3.81)	(13.47)	(6.62)	(9.18)	(4.76)	
Talaye	25	29.96	40.0	161.29	0.147	1.158	33.38	4.798	5.705	83.814	
		(16.03)	(0)	(54.33)	(29.67)	(6.0)	(21.51)	(8.47)	(11.63)	(2.76)	
(e) *Brevicoryne brassicae*											
Karaj-1	25	7.17	78.8	5.813	0.140	1.151	12.421	–	–	–	Karami and Fathipour, 2016
		(21.53)	(46.20)	(85.49)	(57.70)	(17.31)	(11.79)				
Karaj-2	25	7.79	75.9	4.314	0.113	1.119	12.721	–	–	–	
		(32.03)	(43.30)	(89.23)	(65.86)	(19.61)	(14.49)				
Karaj-3	25	7.47	76.6	3.068	0.088	1.092	12.135	–	–	–	
		(26.61)	(44.00)	(92.34)	(73.41)	(21.55)	(9.22)				
Licord	25	6.86	57.6	18.644	0.247	1.281	11.738	–	–	–	
		(16.27)	(25.00)	(53.46)	(25.38)	(7.97)	(5.64)				

Continued

Table 2.6. Continued.

Canola genotypes	Temp. (°C)	Development time (day)	Pre-adult mortality (%)	R_0 (offspring)	r (day^{-1})	λ (day^{-1})	T (day)	ECI (%)	ECD (%)	AD (%)	References
Okapi	25	6.94 (17.63)	85.6 (53.00)	1.373 (96.57)	0.023 (93.05)	1.023 (26.51)	11.111 (0)	–	–	–	
Opera	25	5.90 (0)	32.6 (0)	40.057 (0)	0.331 (0)	1.392 (0)	11.143 (0.29)	–	–	–	
RGS_{003}	25	8.67 (46.95)	95.7 (63.10)	0.072 (99.82)	−0.242 (173.11)	0.774 (44.40)	11.217 (0.95)	–	–	–	
Sarigol	25	8.56 (45.08)	64.8 (32.20)	4.605 (88.50)	0.106 (67.98)	1.112 (20.11)	14.020 (26.18)	–	–	–	
Talaye	25	7.84 (32.88)	63.5 (30.90)	3.595 (91.03)	0.106 (67.98)	1.113 (20.04)	11.661 (4.95)	–	–	–	
Zarfam	25	7.67 (30.00)	82.0 (49.40)	2.947 (92.64)	0.084 (74.62)	1.088 (21.84)	12.359 (11.23)	–	–	–	
(f) *Diaeretiella rapae*											
Okapi	25	10.00 (12.74)	3.24 (3.23)	71.01 (44.56)	0.341 (24.93)	1.407 (8.81)	12.44 (9.70)	–	–	–	Karami and Fathipour, 2016
Opera	25	8.87 (0)	0.01 (0)	128.09 (0)	0.426 (0)	1.531 (0)	11.34 (0)	–	–	–	

AD, approximate digestibility; ECD, efficiency of conversion of digested food; ECI, efficiency of conversion of ingested food.

strategy and the use of sex pheromones are also compatible measures that can have additive effects in a DBM management programme. In this method, the application of insecticides on trap plants only can reduce DBM resistance and insecticidal residue in the crop.

Integrated pest management is a decision support system for the selection and use of pest management tactics, either alone or harmoniously coordinated into a management strategy, based on cost–benefit analyses that take into account the interests of and impacts on producers, society and the environment. Although IPM programmes for DBM population control can help to reduce the use of harmful chemical insecticides and promote food safety, they have received less attention than they deserve because of a key gap in our knowledge, which is the availability of comprehensive regional management programmes that consolidate all the available techniques in a unified programme to manage DBM populations in such a manner that economic damage is avoided and adverse side effects are minimized. This chapter has tried to introduce all the available and potential control measures of DBM that might form different components of an IPM programme and to discuss the factors determining their interactions.

References

Abbasipour, H., Mahmoudvand, M., Rastegar, F. and Basij, M. (2010) Insecticidal activity of *Peganum harmala* seed extract against the diamondback moth, *Plutella xylostella*. *Bulletin of Insectology* 63, 259–263.

Ahmad, N., Ansari, M.S. and Hasan, F. (2012) Effects of neem based insecticides on *Plutella xylostella* (Linn.). *Crop Protection* 34, 18–24.

Akandeh, M., Kocheili, F., Soufbaf, M. and Rasekh, A. (2015) The effect of physical mutation on the life history parameters and ovipositional preference of the *Plutella xylostella*. *Journal of Crop Protection* 4, 633–642.

Akandeh, M., Kocheili, F., Soufbaf, M. and Rasekh, A. (2016) Effect of canola's physical mutation on *Plutella xylostella* (L.) life table. *Journal of Agricultural Science and Technology* 18, 985–998.

Anon. (1997) Consensus document on the biology of *Brassica napus* L. (oilseed rape). Series on the harmonization of regulatory oversight in biotechnology. *Environmental Directorate, Organization for Economic Co-operation and Development*, Paris, No. 7.

APRD (2015) *Arthropod Pesticide Resistance Database*. Available at: www.pesticideresistance.org (accessed 20 December 2015).

Asman, K., Ekbom, B. and Ramert, B. (2001) Effect of intercropping on oviposition and emigration behavior of the leek moth (Lepidoptera: Acrolepiidae) and the diamondback moth (Lepidoptera: Plutellidae). *Environmental Entomology* 30, 288–294.

Attique, M., Khaliq, A. and Sayyed, A. (2006) Could resistance to insecticides in *Plutella xylostella* (Lep., Plutellidae) be overcome by insecticide mixtures? *Journal of Applied Entomology* 130, 122–127.

Ayalew, G., Lohr, B., Ogol, C. and Baumgartner, J. (2006) Suitability of cultivated and wild crucifers for the development of diamondback moth, *Plutella xylostella* L. (Lepidoptera: Plutellidae). *Journal of Entomology* 3, 82–88.

Badenes-Pérez, F.R., Shelton, A.M. and Nault, B.A. (2004) Evaluating trap crops for diamondback moth, *Plutella xylostella* (Lepidoptera: Plutellidae). *Journal of Economic Entomology* 97, 1365–1372.

Badenes-Pérez, F.R., Shelton, A.M. and Nault, B.A. (2005) Using yellow rocket as a trap crop for diamondback moth (Lepidoptera: Plutellidae). *Journal of Economic Entomology* 98, 884–890.

Badenes-Pérez, F.R., Reichelt, M. and Heckel, D.G. (2010) Can sulfur fertilization improve the effectiveness of trap crops for diamondback moth, *Plutella xylostella* (L.) (Lepidoptera: Plutellidae)? *Pest Management Science* 66, 832–838.

Bahar, M., Soroka, J., Grenkow, L. and Dosdall, L. (2014) New threshold temperatures for the development of a North American diamondback moth (Lepidoptera: Plutellidae) population and its larval parasitoid, *Diadegma insulare* (Hymenoptera: Ichneumonidae). *Environmental Entomology* 43, 1443–1452.

Baidoo, P. and Adam, J. (2012) The effects of extracts of *Lantana camara* (L.) and *Azadirachta indica* (A. Juss) on the population dynamics of *Plutella xylostella*, *Brevicoryne brassicae* and *Hellula undalis* on cabbage. *Sustainable Agriculture Research* 1, 229–235.

Braun, L., Olfert, O., Soroka, J., Mason, P. and Dosdall, L. (2004) Diamondback moth biocontrol activities in Canada. In: Kirk, A.A. and Bordat, D. (eds) *Improving Biocontrol of Plutella xylostella. Proceedings of the International Symposium, 21–24 October, 2002.* CIRAD, Montpellier, France, pp. 144–146.

Cai, H.J., Li, S.Y., Ryall, K., You, M.S. and Lin, S. (2011) Effects of intercropping of garlic or lettuce with Chinese cabbage on the development of larvae and pupae of diamondback moth (*Plutella xylostella*). *African Journal of Agricultural Research* 6, 3609–3615.

Campbell, A., Frazer, B., Gilbert, N., Gutierrez, A. and Mackauer, M. (1974) Temperature requirements of some aphids and their parasites. *Journal of Applied Ecology* 11, 431–438.

Carey, J.R. (1993) *Applied Demography for Biologists with Special Emphasis on Insects*. Oxford University Press, New York.

Charleston, D., Kfir, R., Vet, L. and Dicke, M. (2005) Behavioural responses of diamondback moth *Plutella xylostella* (Lepidoptera: Plutellidae) to extracts derived from *Melia azedarach* and *Azadirachta indica*. *Bulletin of Entomological Research* 95, 457–465.

Chegeni, E. and Fathipour, Y. (2011) Nutritional indices and host preference of *Helicoverpa armigera* (Lepidoptera: Noctuidae) on 10 canola cultivars. MSc thesis, Tarbiat Modares University, Tehran, Iran.

Chen, Z., Fang, Y. and Zhang, Z. (2007) Synthesis and assessment of attractiveness and mating disruption efficacy of sex pheromone microcapsules for the diamondback moth, *Plutella xylostella* (L.). *Chinese Science Bulletin* 52, 1365–1371.

Cheng, L.G., Guang, Y., Chen, Z.H. and Li, Z.Y. (2008) Insensitive acetylcholine receptor conferring resistance of *Plutella xylostella* to nereistoxin insecticides. *Agricultural Sciences in China* 7, 847–852.

Chi, H. (1988) Life-table analysis incorporating both sexes and variable development rates among individuals. *Environmental Entomology* 17, 26–34.

Chisholm, M., Underhill, E. and Steck, W. (1979) Field trapping of the diamondback moth *Plutella xylostella* using synthetic sex attractants. *Environmental Entomology* 8, 516–518.

Choubey, B.K., Srinivasa, M.K., Jalali, S.K. and Venkatesan, T. (2014) Determination of host parasitoid ratio for diamond back moth *Plutella xylostella* (Linnaeus) and its parasitoid *Cotesia vestalis* Haliday. *International Journal of Current Research* 6, 10042–10045.

Dai, J., Deng, J. and Du, J. (2008) Development of bisexual attractants for diamondback moth, *Plutella xylostella* (Lepidoptera: Plutellidae) based on sex pheromone and host volatiles. *Applied Entomology and Zoology* 43, 631–638.

Dent, D.R. (1995) Principles of integrated pest management. In: Dent, D.R. (ed.) *Integrated Pest Management*. Chapman & Hall, London, pp. 8–46.

Dosdall, L.M., Soroka, J.J. and Olfert, O. (2011) The diamondback moth in canola and mustard: current pest status and future prospects. *Prairie Soils and Crops* 4, 66–76.

Dover, J.W. (1986) The effect of labiate herbs and white clover on *Plutella xylostella* oviposition. *Entomologia Experimentalis et Applicata* 42, 243–247.

Ebrahimi, N., Talebi, A.A., Fathipour, Y. and Zamani, A.A. (2008) Host plants effect on preference, development and reproduction of *Plutella xylostella* (L.) (Lepidoptera: Plutellidae) under laboratory conditions. *Advances in Environmental Biology* 2, 108–114.

Ebrahimi, N., Talebi, A.A. and Fathipour, Y. (2009) Demographic parameters of the diamondback moth, *Plutella xylostella* (Lep: Plutellidae) on five rapeseed cultivars. *Journal of Entomological Society of Iran* 28, 49–59.

Egigu, M.C., Ibrahim, M.A., Yahya, A. and Holopainen, J.K. (2010) Yeheb (*Cordeauxia edulis*) extract deters feeding and oviposition of *Plutella xylostella* and attracts its natural enemy. *BioControl* 55, 613–624.

Eswarapriya, B., Gopalsamy, B., Kameswari, B., Meera, R. and Devi, P. (2010) Insecticidal activity of *Bacillus thuringiensis* IBT-15 strain against *Plutella xylostella*. *International Journal of PharmTech Research* 2, 2048–2053.

FAOSTAT (2015) *Production Statistics*. FAO, Rome. Available at: http://faostat.fao.org/site/567/default.aspx#ancor (accessed 12 November 2015).

Farmer, E.E. and Ryan, C.A. (1992) Octadecanoid precursors of jasmonic acid activate the synthesis of wound-inducible proteinase inhibitors. *The Plant Cell* 4, 129–134.

Fathi, S.A.A., Bozorg-Amirkalaee, M. and Sarfaraz, R.M. (2011) Preference and performance of *Plutella xylostella* (L.) (Lepidoptera: Plutellidae) on canola cultivars. *Journal of Pest Science* 84, 41–47.

Fathi, S.A.A., Bozorg-Amirkalaee, M., Sarfraz, R.M. and Rafiee-Dastjerdi, H. (2012) Parasitism and developmental parameters of the parasitoid *Diadegma majale* (Gravenhorst) in control of *Plutella xylostella* (L.) on selected cultivars of canola. *BioControl* 57, 49–59.

Fathipour, Y. and Maleknia, B. (2016) Mite predators. In: Omkar (ed.) *Ecofriendly Pest Management for Food Security*. Elsevier, San Diego, California, pp. 329–366.

Fathipour, Y. and Naseri, B. (2011) Soybean cultivars affecting performance of *Helicoverpa armigera* (Lepidoptera: Noctuidae). In: Tzi-Bun, N. (ed.) *Soybean Biochemistry, Chemistry and Physiology*. INTECH Open Access Publisher, Rijeka, Croatia, pp. 599–630.

Fathipour, Y. and Sedaratian, A. (2013) Integrated management of *Helicoverpa armigera* in soybean cropping systems. In: El-Shemy, H. (ed.) *Soybean-Pest Resistance*. INTECH Open Access Publisher, Rijeka, Croatia, pp. 231–280.

Ferreira, S.W., Barros, R. and Torres, J.B. (2003) Thermal requirements of *Oomyzus sokolowskii* (Kurdjumov) (Hymenoptera: Eulophidae) and estimation of the number of generations in crucifers in Pernambuco State, Brazil. *Neotropical Entomology* 32, 407–411.

Furlong, M.J. and Asgari, S. (2010) Effects of an ascovirus (HvAV-3e) on diamondback moth, *Plutella xylostella*, and evidence for virus transmission by a larval parasitoid. *Journal of Invertebrate Pathology* 103, 89–95.

Furlong, M.J. and Pell, J.K. (1997) The influence of environmental factors on the persistence of *Zoophthora radicans* conidia. *Journal of Invertebrate Pathology* 69, 223–233.

Furlong, M.J., Shi, Z.H., Liu, Y.Q., Guo, S.J., Lu, Y.B., Liu, S.S. and Zalucki, M.P. (2004) Experimental analysis of the influence of pest management practice on the efficacy of an endemic arthropod natural enemy complex of the diamondback moth. *Journal of Economic Entomology* 97, 1814–1827.

Furlong, M.J., Wright, D.J. and Dosdall, L.M. (2013) Diamondback moth ecology and management: problems, progress, and prospects. *Annual Review of Entomology* 58, 517–541.

Gao, G., Lu, Z., Tao, S., Zhang, S. and Wang, F. (2011) Triterpenoid saponins with antifeedant activities from stem bark of *Catunaregam spinosa* (Rubiaceae) against *Plutella xylostella* (Plutellidae). *Carbohydrate Research* 346, 2200–2205.

Godonou, I., James, B., Atcha-Ahowe, C., Vodouhe, S., Kooyman, C., Ahanchede, A. and Korie, S. (2009) Potential of *Beauveria bassiana* and *Metarhizium anisopliae* isolates from Benin to control *Plutella xylostella* L.(Lepidoptera: Plutellidae). *Crop Protection* 28, 220–224.

Golizadeh, A., Kamali, K., Fathipour, Y. and Abbasipour, H. (2007) Temperature-dependent development of diamondback moth, *Plutella xylostella* (Lepidoptera: Plutellidae) on two brassicaceous host plants. *Insect Science* 14, 309–316.

Golizadeh, A., Kamali, K., Fathipour, Y. and Abbasipour, H. (2008) Life table and temperature-dependent development of *Diadegma anurum* (Hymenoptera: Ichneumonidae) on its host *Plutella xylostella* (Lepidoptera: Plutellidae). *Environmental Entomology* 37, 38–44.

Golizadeh, A., Kamali, K., Fathipour, Y. and Abbasipour, H. (2009a) Effect of temperature on life table parameters of *Plutella xylostella* (Lepidoptera: Plutellidae) on two brassicaceous host plants. *Journal of Asia-Pacific Entomology* 12, 207–212.

Golizadeh, A., Kamali, K., Fathipour, Y. and Abbasipour, H. (2009b) Life table of the diamondback moth, *Plutella xylostella* (L.) (Lepidoptera: Plutellidae) on five cultivated brassicaceous host plants. *Journal of Agricultural Science and Technology* 11, 115–124.

Goodarzi, M., Fathipour, Y. and Talebi, A.A. (2015) Antibiotic resistance of canola cultivars affecting demography of *Spodoptera exigua* (Lepidoptera: Noctuidae). *Journal of Agricultural Science and Technology* 17, 23–33.

Goudegnon, A., Kirk, A., Arvanitakis, L. and Bordat, D. (2002) Status of the diamondback moth and *Cotesia plutellae*, its main parasitoid, in the Cotonou and Porto Novo periurban areas of Benin. In: Kirk, A. and Bordat, D. (eds) *Improving Biocontrol of Plutella xylostella, Proceedings of the International Symposium*. CIRAD, Montpellier, France, pp. 172–178.

Grzywacz, D., Parnell, M., Kibata, G., Oduor, G., Ogutu, W., Miano, D. and Winstanley, D. (2001) The development of endemic baculoviruses of *Plutella xylostella* (diamondback moth, DBM) for control of DBM in East Africa. In: Endersby, N.M. and Ridland, P.M. (eds) *The Management of Diamondback Moth and other Cruciferous Pests*. Department of Natural Resources and Environment, Melbourne, Australia, pp. 271–280.

Grzywacz, D., Rossbach, A., Rauf, A., Russell, D., Srinivasan, R. and Shelton, A. (2010) Current control methods for diamondback moth and other brassica insect pests and the prospects for improved management with lepidopteran-resistant Bt vegetable brassicas in Asia and Africa. *Crop Protection* 29, 68–79.

Guo, M., Zhu, D. and Li, L. (1999) Selection of *Trichogramma* species for controlling the diamondback moth *Plutella xylostella* (L.). *Entomologia Sinica* 6, 187–192.

Harvey, C.T. and Eubanks, M.D. (2005) Intraguild predation of parasitoids by *Solenopsis invicta*: a non-disruptive interaction. *Entomologia Experimentalis et Applicata* 114, 127–135.

Haseeb, M., Liu, T.X. and Jones, W. (2004) Effects of selected insecticides on *Cotesia plutellae*, endoparasitoid of *Plutella xylostella*. *BioControl* 49, 33–46.

Hasheela, E.B., Nderitu, J.H., Olubayo, F.M. and Kasina, M. (2010) Evaluation of border crops against infestation and damage of cabbage by diamondback moth (*Plutella xylostella*). *Tunisian Journal of Plant Protection* 5, 99–105.

Heckel, D.G., Tabashnik, B.E., Liu, Y.B., Gahan, L.J., Shelton, A.M., Zhao, J.Z and Baxter, S.W. (2001) Diamondback moth resistance to Bt: relevance of genetics and molecular biology to detection and management. In: Ridland, P.M. and Endersby, N. (eds) *The Management of Diamondback moth and Other Crucifer Pests: Proceedings of the 4th International Workshop*. Department of Natural Resources and Environment, Melbourne, Australia, pp. 27–36.

Hill, T.A. and Foster, R.E. (2000) Effect of insecticides on the diamondback moth (Lepidoptera: Plutellidae) and its parasitoid *Diadegma insulare* (Hymenoptera: Ichneumonidae). *Journal of Economic Entomology* 93, 763–768.

Huang, Z., Ali, S., Ren, S.X. and Wu, J.H. (2010) Effect of *Isaria fumosoroseus* on mortality and fecundity of *Bemisia tabaci* and *Plutella xylostella*. *Insect Science* 17, 140–148.

Hui, W., Juan, W., Li, H.S., Dai, H.G. and Gu, X.J. (2010) Sub-lethal effects of fenvalerate on the development, fecundity, and juvenile hormone esterase activity of diamondback moth, *Plutella xylostella* (L.). *Agricultural Sciences in China* 9, 1612–1622.

Hwang, I.C., Park, C., Kang, D.K., Jin, N.Y., Jung, S.Y., Seo, M.J., Kim, J.E., Youn, Y.N. and Yu, Y.M. (2010) Combined application of *Trichogramma ostriniae* and *Bacillus thuringiensis* for eco-friendly control of *Plutella xylostella*. *Journal of the Korean Society for Applied Biological Chemistry* 53, 316–322.

Ibrahim, M.A., Nissinen, A. and Holopainen, J.K. (2005) Response of *Plutella xylostella* and its parasitoid *Cotesia plutellae* to volatile compounds. *Journal of Chemical Ecology* 31, 1969–1984.

Johnson, M.W., Mitchell, W., Robin, M., Cushing, N. and Rethwisch, M. (1988) Parasitization of the diamondback moth, *Plutella xylostella* (L.) (Lepidoptera: Plutellidae), in Hawaii. *Hawaiian Entomological Society* 28, 197–204.

Karami, A. and Fathipour, Y. (2016) The effects of different canola cultivars on performance of *Brevicoryne brassicae* and its parasitoid wasp *Diaeretiella rapae*. MSc thesis, Tarbiat Modares University, Tehran.

Karimi, S., Fathipour, Y., Talebi, A.A. and Naseri, B. (2012) Evaluation of canola cultivars for resistance to *Helicoverpa armigera* (Lepidoptera: Noctuidae) using demographic parameters. *Journal of Economic Entomology* 105, 2172–2179.

Kfir, R. (1997) The diamondback moth with special reference to its parasitoids in South Africa. In: Sivapragasam, A., Loke, W.H., Hussan, A.K. and Lim, G.S. (eds) *The Management of Diamondback Moth and Other Crucifer Pests: Proceedings of the 3rd International Workshop*. Malaysian Agricultural Research Institute, Kuala Lumpur, pp. 54–60.

Khan, M.F.R., Griffin, R.P., Carner, G.R. and Gorsuch, C.S. (2008) Evaluating biological based insecticides for managing diamondback moth, *Plutella xylostella* (L.) (Lepidoptera: Plutellidae). *Journal of Entomology* 5, 252–261.

Kianpour, R. and Fathipour, Y. (2013) Tritrophic interactions in a system of host plant-diamondback moth-parasitoid wasp *Diadegma semiclausum*. PhD dissertation, Tarbiat Modares University, Tehran.

Kianpour, R., Fathipour, Y., Karimzadeh, J. and Hosseininaveh, V. (2014) Influence of different host plant cultivars on nutritional indices of *Plutella xylostella* (Lepidoptera: Plutellidae). *Journal of Crop Protection* 3, 43–49.

Kirk, A., Mercadier, G., Bordat, D., Delvare, G., Pichon, A., Arvanitakis, L., Goudegnon, A. and Rincon, C. (2004) Variability in *Plutella* and its natural enemies: implications for biological control. In: Endersby, N.M. and Ridland, P.M. (eds) *The Management of Diamondback Moth and Other Crucifer Pests: Proceedings of the 4th International Workshop*. Department of Natural Resources and Environment, Melbourne, Australia, pp. 71–77.

Landis, D.A., Wratten, S.D. and Gurr, G.M. (2000) Habitat management to conserve natural enemies of arthropod pests in agriculture. *Annual Review of Entomology* 45, 175–201.

Legwaila, M.M., Munthali, D.C., Obopile, M. and Kwerepe, B.C. (2014) Effectiveness of spinosad against diamondback moth (*Plutella xylostella* L.) eggs and larvae on cabbage under Botswana conditions. *International Journal of Insect Science* 6, 15–21.

Li, Z., Zalucki, M.P., Bao, H.L., Chen, H.Y., Hu, Z.D. and Feng, X. (2012) Population dynamics and 'outbreaks' of diamondback moth, *Plutella xylostella,* in Guangdong Province, China: climate or the failure of management? *Journal of Economic Entomology* 105, 739–52.

Li, Z., Feng, X., Liu, S.S., You, M. and Furlong, M.J. (2016) Biology, ecology, and management of the diamondback moth in China. *Annual Review of Entomology* 61, 277–296.

Lim, G.S. (1986) Biological control of diamondback moth. In: Talekar, N.S. and Griggs, T.D. (eds) *Diamondback Moth Management: Proceedings of the 1st International Workshop.* Asian Vegetable Research and Development Center, Tainan, Taiwan, pp. 159–171.

Liu, S.S., Hommes, M. and Hildenhagen, R. (1994) Damage to white cabbage by the aphid *Brevicoryne brassicae* (L.): influence of aphid density and stage of plant growth. *IOBC WPRS Bulletin* 17, 75–75.

Liu, S.S., Wang, X.G., Guo, S.J., He, J.H. and Shi, Z.H. (2000) Seasonal abundance of the parasitoid complex associated with the diamondback moth, *Plutella xylostella* (Lepidoptera: Plutellidae) in Hangzhou, China. *Bulletin of Entomological Research* 90, 221–231.

Liu, S.S., Wang, X.G., Shi, Z.G. and Gebremeskel, F. (2001) The biology of *Diadromus collaris* (Hymenoptera: Ichneumonidae), a pupal parasitoid of *Plutella xylostella* (Lepidoptera: Plutellidae), and its interactions with *Oomyzus sokolowskii* (Hymenoptera: Eulophidae). *Bulletin of Entomological Research* 91, 461–469.

Liu, S.S., Chen, F.Z. and Zalucki, M.P. (2002a) Development and survival of the diamondback moth (Lepidoptera: Plutellidae) at constant and alternating temperatures. *Environmental Entomology* 31, 221–231.

Liu, S.S., Gebremeskel, F. and Shi, Z.H. (2002b) Reproductive compatibility and variation in survival and sex ratio between two geographic populations of *Diadromus collaris*, a pupal parasitoid of the diamondback moth, *Plutella xylostella*. *BioControl* 47, 625–643.

Lu, J.H., Liu, S.S. and Shelton, A.M. (2004) Laboratory evaluations of a wild crucifer *Barbarea vulgaris* as a management tool for the diamondback moth *Plutella xylostella* (Lepidoptera: Plutellidae). *Bulletin of Entomological Research* 94, 509–516.

Ma, J., Li, D., Keller, M., Schmidt, O. and Feng, X. (2005) A DNA marker to identify predation of *Plutella xylostella* (Lep., Plutellidae) by *Nabis kinbergii* (Hem., Nabidae) and *Lycosa* sp. (Aranaea, Lycosidae). *Journal of Applied Entomology* 129, 330–335.

Magholi, Z., Abbasipour, H. and Marzban, R. (2014) Effects of *Helicoverpa armigera* nucleopolyhedrosis virus (HaNPV) on the larvae of the diamondback moth, *Plutella xylostella* (L.) (Lepidoptera: Plutellidae). *Plant Protection Science* 50, 184–189.

Mahmoudvand, M., Abbasipour, H., Garjan, A.S. and Bandani, A.R. (2011a) Sublethal effects of hexaflumuron on development and reproduction of the diamondback moth, *Plutella xylostella* (Lepidoptera: Yponomeutidae). *Insect Science* 18, 689–696.

Mahmoudvand, M., Abbasipour, H., Garjan, A.S. and Bandani, A.R. (2011b) Sublethal effects of indoxacarb on the diamondback moth, *Plutella xylostella* (L.) (Lepidoptera: Yponomeutidae). *Applied Entomology and Zoology* 46, 75–80.

Mahmoudvand, M., Moharramipour, S. and Iranshahi, M. (2015) Effects of pyriproxyfen on life table indices of *Plutella xylostella* in multigenerations. *Psyche* 2015, 1–7.

Marchioro, C.A. and Foerster, L.A. (2011) Development and survival of the diamondback moth, *Plutella xylostella* (L.) (Lepidoptera: Yponomeutidae) as a function of temperature: effect on the number of generations in tropical and subtropical regions. *Neotropical Entomology* 40, 533–541.

Marchioro, C.A. and Foerster, L.A. (2014) Preference–performance linkage in the diamondback moth, *Plutella xylostella*, and implications for its management. *Journal of Insect Science* 14, 85–98.

Mason, J.M., Matthews, G.A. and Wright, D.J. (1999) Evaluation of spinning disc technology for the application of entomopathogenic nematodes against a foliar pest. *Journal of Invertebrate Pathology* 73, 282–288.

McHugh, J.J. and Foster, R.E. (1995) Reduction of diamondback moth (Lepidoptera: Plutellidae) infestation in head cabbage by overhead irrigation. *Journal of Economic Entomology* 88, 162–168.

Metcalf, R.L. and Luckmann, W.H. (1994) *Introduction to Insect Pest Management*. Wiley, New York.

Miller, J.R. and Cowles, R.S. (1990) Stimulo-deterrent diversion: a concept and its possible application to onion maggot control. *Journal of Chemical Ecology* 16, 3197–3212.

Mohammed, A., Addo-Quaye, A. and Elvis, A. (2010) Control of diamondback moth (*Plutella xylostella*) on cabbage (*Brassica oleracea* var *capitata*) using intercropping with non-host crops. *American Journal of Food Technology* 5, 269–274.

Momanyi, C., Löhr, B. and Gitonga, L. (2006) Biological impact of the exotic parasitoid, *Diadegma semiclausum* (Hellen), of diamondback moth, *Plutella xylostella* L., in Kenya. *Biological Control* 38, 254–263.

Nehare, S., Moharil, M., Ghodki, B., Lande, G., Bisane, K., Thakare, A. and Barkhade, U. (2010) Biochemical analysis and synergistic suppression of indoxacarb resistance in *Plutella xylostella* L. *Journal of Asia-Pacific Entomology* 13, 91–95.

Nematollahi, M.R., Fathipour, Y., Talebi, A.A., Karimzadeh, J. and Zalucki, M.P. (2016) Comparison of degree-day distribution models for predicting emergence of the cabbage aphid on canola. *Crop Protection* 80, 138–143.

Newsom, L.D. (1975) Pest management: concept to practice. In: Pimental, D. (ed.) *Insects, Science and Society*. Elsevier, San Diego, California, pp. 257–277.

Nikooei, M., Fathipour, Y., Jalali Javaran, M. and Soufbaf, M. (2015a) How different genetically manipulated *Brassica* genotypes affect life table parameters of *Plutella xylostella* (Lepidoptera: Plutellidae). *Journal of Economic Entomology* 108, 515–524.

Nikooei, M., Fathipour, Y., Jalali Javaran, M. and Soufbaf, M. (2015b) Influence of genetically manipulated Brassica genotypes on parasitism capacity of *Diadegma semiclausum* parasitizing *Plutella xylostella*. *Journal of Agricultural Science and Technology* 17, 1743–1753.

Nikooei, M., Fathipour, Y., Jalali Javaran, M. and Soufbaf, M. (2016) Genetically manipulated *Brassica* genotypes affect demography and performance of *Diadegma semiclausum* parasitizing *Plutella xylostella*. *Journal of Applied Entomology*. doi: 10.1111/jen.12316.

Nofemela, S.R. (2004) Studies on parasitoids of the diamondback moth, *Plutella xylostella* (L.) (Lepidoptera: Plutellidae), in South Africa. MSc thesis, The University of Rhodes, Grahamstown, South Africa.

Pan, Q.J., Chen, L., Lin, X.L., Ridsdill-Smith, T.J. and Liu, T.X. (2014) Geographical variations in life histories of *Plutella xylostella* in China. *Journal of Pest Science* 87, 659–670.

Pereira, F.F., Barros, R., Pratissoli, D. and Parra, J.R. (2004) Biology and thermal requirements of *Trichogramma pretiosum* Riley and *T. exiguum* Pinto & Platner (Hymenoptera: Trichogrammatidae) reared on eggs of *Plutella xylostella* (L.) (Lepidoptera: Plutellidae). *Neotropical Entomology* 33, 231–236.

Pluke, R.W. and Leibee, G.L. (2006) Host preferences of *Trichogramma pretiosum* and the influence of prior ovipositional experience on the parasitism of *Plutella xylostella* and *Pseudoplusia includens* eggs. *BioControl* 51, 569–583.

Poelking, A. (1992) Diamondback moth in the Philippines and its control with *Diadegma semiclausum*. In: Talekar, N. (ed.) *Diamondback Moth and Other Crucifer Pests: Proceedings of the 2nd International Workshop.* Asian Vegetable Research and Development Center, Tainan, Taiwan, pp. 271–278.

Pourghasem, M. and Fathipour, Y. (2011) Nutritional indices and oviposition preference of *Spodoptera exigua* (Lepidoptera: Noctuidae) on different canola cultivars in laboratory condition. MSc thesis, Tarbiat Modares University, Tehran.

Pyke, B., Rice, M., Sabine, B. and Zalucki, M.P. (1987) The push-pull strategy – behavioral control of *Heliothis*. *Australian Cotton Grower* 9, 7–9

Qian, L., Cao, G., Song, J., Yin, Q. and Han, Z. (2008) Biochemical mechanisms conferring cross-resistance between tebufenozide and abamectin in *Plutella xylostella*. *Pesticide Biochemistry and Physiology* 91, 175–179.

Quan, X., Wu, L., Zhou, Q., Yun, Y., Peng, Y. and Chen, J. (2011) Identification of predation by spiders on the diamondback moth *Plutella xylostella*. *Bulletin of Insectology* 64, 223–227.

Ramachandran, S., Buntin, G.D., All, J.N., Raymer, P.L. and Stewart, C.N. (1998) Greenhouse and field evaluations of transgenic canola against diamondback moth, *Plutella xylostella*, and corn earworm, *Helicoverpa zea*. *Entomologia Experimentalis et Applicata* 88, 17–24.

Reddy, G.V.P. and Guerrero, A. (2000) Behavioral responses of the diamondback moth, *Plutella xylostella*, to green leaf volatiles of *Brassica oleracea* subsp. *capitata*.

Journal of Agricultural and Food Chemistry 48, 6025–6029.

Reddy, G.V.P., Holopainen, J. and Guerrero, A. (2002) Olfactory responses of *Plutella xylostella* natural enemies to host pheromone, larval frass, and green leaf cabbage volatiles. *Journal of Chemical Ecology* 28, 131–143.

Reddy, G.V.P., Tabone, E. and Smith, M. (2004) Mediation of host selection and oviposition behavior in the diamondback moth *Plutella xylostella* and its predator *Chrysoperla carnea* by chemical cues from cole crops. *Biological Control* 29, 270–277.

Riley, D.G. and Sparks, A.N. (2009) *Insecticide Resistance Management for Diamondback Moth in Georgia*. The University of Georgia, College of Agricultural and Environmental Sciences, Athens, Georgia.

Roelofs, W.L. and Rooney, A.P. (2003) Molecular genetics and evolution of pheromone biosynthesis in Lepidoptera. *Proceedings of the National Academy of Sciences of the USA* 100, 9179–9184.

Saeed, R., Sayyed, A.H., Shad, S.A. and Zaka, S.M. (2010) Effect of different host plants on the fitness of diamondback moth, *Plutella xylostella* (Lepidoptera: Plutellidae). *Crop Protection* 29, 178–182.

Sarfraz, M., Keddie, A.B. and Dosdall, L.M. (2005) Biological control of the diamondback moth, *Plutella xylostella*: a review. *Biocontrol Science and Technology* 15, 763–789.

Sarfraz, M., Dosdall, L.M. and Keddie, B.A. (2006) Diamondback moth–host plant interactions: implications for pest management. *Crop Protection* 25, 625–639.

Sarnthoy, O., Keinmeesuke, P., Sinchaisri, N. and Nakasuji, F. (1989) Development and reproductive rate of the diamondback moth *Plutella xylostella* from Thailand. *Applied Entomology and Zoology* 24, 202–208.

Sayyed, A.H., Rizvi, M.R. and Alvi, A.H. (2002) Management of diamondback moth, *Plutella xylostella* (Lepidoptera: Plutellidae): a lesson from South East Asia for sustainable integrated pest management. *Pakistan Journal of Biological Sciences* 5, 234–245.

Sayyed, A.H., Omar, D. and Wright, D.J. (2004) Genetics of spinosad resistance in a multi-resistant field-selected population of *Plutella xylostella*. *Pest Management Science* 60, 827–832.

Schroer, S. and Ehlers, R.U. (2005) Foliar application of the entomopathogenic nematode *Steinernema carpocapsae* for biological control of diamondback moth larvae (*Plutella xylostella*). *Biological Control* 33, 81–86.

Shelton, A.M. and Badenes-Pérez, F.R. (2006) Concepts and applications of trap cropping in pest management. *Annual Review of Entomology* 51, 285–308.

Shelton, A.M. and Nault, B.A. (2004) Dead-end trap cropping: a technique to improve management of the diamondback moth, *Plutella xylostella* (Lepidoptera: Plutellidae). *Crop Protection* 23, 497–503.

Shelton, A.M., Sances, F.V., Hawley, J., Tang, J.D., Boune, M., Jungers, D., Collins, H.L. and Farias, J. (2000) Assessment of insecticide resistance after the outbreak of diamondback moth (Lepidoptera: Plutellidae) in California in 1997. *Journal of Economic Entomology* 93, 931–936.

Shi, Z. and Liu, S.S. (1998) Influence of temperature on the development, survival and reproduction of *Cotesia plutellae*, a larval parasite of *Plutella xylostella*. *Acta Phytophylacica Sinica* 26, 142–146.

Shi, Z.H., Liu, S.S. and Li, Y.X. (2002) *Cotesia plutellae* parasitizing *Plutella xylostella*: host-age dependent parasitism and its effect on host development and food consumption. *BioControl* 47, 499–511.

Shirai, Y. (2000) Temperature tolerance of the diamondback moth, *Plutella xylostella* (Lepidoptera: Yponomeutidae) in tropical and temperate regions of Asia. *Bulletin of Entomological Research* 90, 357–364.

Singh, R. (2003) Tritrophic interactions with reference to biological control of insect pests. *Biological Memoirs* 29, 55–70.

Smith, C.M. (1989) *Plant Resistance to Insects: A Fundamental Approach*. Wiley, New York.

Sonoda, S. and Tsumuki, H. (2005) Studies on glutathione S-transferase gene involved in chlorfluazuron resistance of the diamondback moth, *Plutella xylostella* L.(Lepidoptera: Yponomeutidae). *Pesticide Biochemistry and Physiology* 82, 94–101.

Soufbaf, M., Fathipour, Y., Karimzadeh, J. and Zalucki, M.P. (2010a) Bottom-up effect of different host plants on *Plutella xylostella* (Lepidoptera: Plutellidae): a life-table study on canola. *Journal of Economic Entomology* 103, 2019–2027.

Soufbaf, M., Fathipour, Y., Karimzadeh, J. and Zalucki, M.P. (2010b) Development and age-specific mortality of diamondback moth on *Brassica* host plants: pattern and causes of mortality under laboratory conditions. *Annals of the Entomological Society of America* 103, 574–579.

Soufbaf, M., Fathipour, Y., Zalucki, M.P. and Hui, C. (2012) Importance of primary metabolites in canola in mediating interactions between a specialist leaf-feeding insect and its specialist solitary endoparasitoid. *Arthropod–Plant Interactions* 6, 241–250.

Soufbaf, M., Fathipour, Y., Zalucki, M.P. and Karimzadeh, J. (2013) Relevance of reproductive correlates in response of diamondback moth (Lepidoptera: Plutellidae) to plant quality. *Journal of Entomological and Acarological Research* 45, 52–57.

Suso, J.M., Bebeli, P.J., Christmann, S., Mateus, C., Negri, V., Carvalho, M.P., Torricelli, R. and Veloso, M.M. (2016) Enhancing legume ecosystem services through an understanding of plant–pollinator interplay. *Frontiers in Plant Science* 7, 333. doi: 10.3389/fpls.2016.00333.

Tabashnik, B.E. and Mau, R.F. (1986) Suppression of diamondback moth (Lepidoptera: Plutellidae) oviposition by overhead irrigation. *Journal of Economic Entomology* 79, 189–191.

Tabone, E., Bardon, C., Desneux, N. and Wajnberg, E. (2010) Parasitism of different *Trichogramma* species and strains on *Plutella xylostella* L. on greenhouse cauliflower. *Journal of Pest Science* 83, 251–256.

Talekar, N.S. (1992) Introduction of *Diadegma semiclausum* for the control of diamondback moth in Taiwan. In: Talekar, N.S. (ed.) *Diamondback Moth and Other Crucifer Pests: Proceedings of the 2nd International Workshop*. Asian Vegetable Research and Development Center, Tainan, Taiwan, pp. 263–270.

Talekar, N. and Shelton, A. (1993) Biology, ecology, and management of the diamondback moth. *Annual Review of Entomology* 38, 275–301.

Talekar, N.S. and Griggs, T.D. (1986) *Diamondback Moth Management: Proceedings of the 1st International Workshop*. Asian Vegetable Research and Development Center, Tainan, Taiwan.

Talekar, N.S. and Yang, J.C. (1993) Influence of crucifer cropping system on the parasitism of *Plutella xylostella* (Lep., Yponomeutidae) by *Cotesia plutellae* (Hym., Braconidae) and *Diadegma semiclausum* (Hym., Ichneumonidae). *Entomophaga* 38, 541–550.

Umeya, K. and Yamada, H. (1973) Threshold temperature and thermal constants for development of the diamond-back moth, *Plutella xylostella* L., with reference to their local differences. *Japanese Journal of Applied Entomology and Zoology* 17, 19–24.

Vanichpakorn, P., Ding, W. and Cen, X.X. (2010) Insecticidal activity of five Chinese medicinal plants against *Plutella xylostella* L. larvae. *Journal of Asia-Pacific Entomology* 13, 169–173.

Vanlaldiki, H., Singh, M.P. and Sarkar, P. (2013) Efficacy of eco-friendly insecticides on the management of diamondback moth (*Plutella xylostella* Linn.) on cabbage. *The Bioscan* 8, 1225–1230.

Waldbauer, G.P. (1968) The consumption and utilization of food by insects. *Advances in Insect Physiology* 5, 229–288.

Walker, G., Davis, S., MacDonald, E. and Herman, T. (2012) Update on diamondback moth (*Plutella xylostella*) insecticide resistance and the vegetable brassica insecticide management strategy. *New Zealand Plant Protection* 65, 114–119.

Wang, X.G. and Keller, M.A. (2002) A comparison of the host-searching efficiency of two larval parasitoids of *Plutella xylostella*. *Ecological Entomology* 27, 105–114.

Wang, X.G. and Liu, S.S. (2002) Effects of host age on the performance of *Diadromus collaris*, a pupal parasitoid of *Plutella xylostella*. *BioControl* 47, 293–307.

Wang, X.G., Liu, S.S., Guo, S.J. and Lin, W.C. (1999) Effects of host stages and temperature on population parameters of *Oomyzus sokolowskii*, a larval–pupal parasitoid of *Plutella xylostella*. *BioControl* 44, 391–403.

Wang, X.G., Duff, J., Keller, M.A., Zalucki, M.P., Liu, S.S. and Bailey, P. (2004) Role of *Diadegma semiclausum* (Hymenoptera: Ichneumonidae) in controlling *Plutella xylostella* (Lepidoptera: Plutellidae): cage exclusion experiments and direct observation. *Biocontrol Science and Technology* 14, 571–586.

Wang, Y., Zhang, Y., Wang, F., Liu, C. and Liu, K. (2014) Development of transgenic *Brassica napus* with an optimized cry1C* gene for resistance to diamondback moth (*Plutella xylostella*). *Canadian Journal of Plant Science* 94, 1501–1506.

Wilding, N. (1986) The pathogens of diamondback moth and their potential for its control – a review. In: Talekar, N.S. and Griggs, T.D. (eds) *Diamondback Moth Management. Proceedings of the First International Workshop, 11–15 March 1985, Tainan, Taiwan*. The Asian Vegetable Research and Development Center, Shanhua, Taiwan, AVRDC Publication no. 86-248, pp. 219–232.

Wu, S., Yang, Y., Yuan, G., Campbell, P.M., Teese, M.G., Russell, R.J., Oakeshott, J.G. and Wu, Y. (2011) Overexpressed esterases in a fenvalerate resistant strain of the cotton bollworm, *Helicoverpa armigera*. *Insect Biochemistry and Molecular Biology* 41, 14–21.

Xu, Z., Cao, G.C. and Dong, S.L. (2010) Changes of sex pheromone communication systems associated with tebufenozide and abamectin resistance in diamondback moth, *Plutella xylostella* (L.). *Journal of Chemical Ecology* 36, 526–534.

Yamada, H. and Kawasaki, K. (1983) The effect of temperature and humidity on the development, fecundity and multiplication of the diamondback moth, *Plutella xylostella* (L.). *Japanese Journal of Applied Entomology and Zoology* 27, 17–21.

Yang, C.Y., Lee, S., Choi, K.S., Jeon, H.Y. and Boo, K.S. (2007) Sex pheromone production and response in Korean populations of the diamondback moth, *Plutella xylostella*. *Entomologia Experimentalis et Applicata* 124, 293–298.

Yin, X.H., Wu, Q.J., Li, X.F., Zhang, Y.J. and Xu, B.Y. (2009) Demographic changes in multigeneration *Plutella xylostella* (Lepidoptera: Plutellidae) after exposure to sublethal concentrations of spinosad. *Journal of Economic Entomology* 102, 357–365.

Zalucki, M.P., Shabbir, A., Silva, R., Adamson, D., Shu-Sheng, L. and Furlong, M.J. (2012) Estimating the economic cost of one of the world's major insect pests, *Plutella xylostella* (Lepidoptera: Plutellidae): just how long is a piece of string? *Journal of Economic Entomology* 105, 1115–1129.

Zhang, Y., Zhou, Y. and Zhang, Z. (2002) Effect of cantharidin on the midgut of the orient armyworm (*Mythimna separata*) and diamondback moth (*Plutella xylostella*). *Acta Entomologica Sinica* 46, 272–276.

Zhang, P.J., Shu, J.P., Dicke, M. and Liu, S.S. (2010) Different headspace profiles in wild crucifer species in response to *Plutella xylostella* herbivory and exogenous jasmonic acid application. *Insect Science* 17, 29–37.

3 The Challenge of Swede Midge Management in Canola

Rebecca H. Hallett*

University of Guelph, Guelph, Ontario, Canada

3.1 Introduction

Since its discovery in Ontario, Canada, in 2000, the swede midge, *Contarinia nasturtii* (Kieffer) (Diptera: Cecidomyiidae) (Fig. 3.1), has become a serious pest of *Brassica* crops in its invaded North American range. Swede midge causes mild to severe economic damage to cruciferous vegetables and canola in Ontario, Quebec, Nova Scotia, Prince Edward Island and adjacent US states. It is an emergent pest of canola in Saskatchewan and Manitoba. The swede midge is of Nearctic origin and is an important and widespread pest in Europe (Readshaw, 1961). European reports of increasing problems with swede midge seemed to become more common in the late 1990s and early 2000s (R. Baur, Switzerland, 2000, personal communication; Gemmar and Koch, 2002a, b) and it has been described as an increasingly common pest of *Brassica* vegetables in Europe (Frey *et al.*, 2004; Wyss and Daniel, 2004). With its introduction to North America, swede midge was first a pest issue in the *Brassica oleracea* L. group of cruciferous vegetable crops, known as the cole crops, such as cabbage, broccoli and cauliflower. Over the past 5–10 years, swede midge has become a serious economic pest of spring canola, *Brassica napus* L., in Ontario and its range has continued to expand throughout North America to include regions where canola is a major crop.

Swede midge attacks all cruciferous crops, and a range of cruciferous weeds, in most developmental stages (Stokes, 1953; Hallett, 2007). Characteristic symptoms include leaf crumpling, swelling of buds and petioles and corky scarring (Barnes, 1946). The swede midge impacts canola yield by killing the meristem, preventing primary raceme bolting and killing developing flower buds (J. Williams and R.H. Hallett, 2016, unpublished results).

The swede midge is a highly successful invasive insect in its North American invaded range. Its success is due, at least in part, to its variable life history, which allows the midge to persist in uncertain environments. Because swede midge is a crucifer specialist, with multiple, overlapping generations each year, a short adult lifespan and larvae that are protected by plant tissues, multiple tactics are required for effective management. Crop rotation, cruciferous weed management and planting date considerations are critical in reducing damage and population growth. Damage prevention is difficult with high populations and insecticides are a necessary but often insufficient management tactic.

This chapter is intended to help to increase capacity to address this invasive species, which is increasingly problematic in canola, by summarizing its invasion history and patterns of damage, synthesizing the knowledge and experience gained over the past 16 years of swede midge research in North America, identifying characteristics of both the midge and canola that contribute to making the swede midge a challenging pest to control and discussing current management recommendations in this context. These challenges include the cryptic nature of larval feeding, multiple generations, variable patterns of diapause entry and termination, a wide host range and a lack of natural enemies. Characteristics of canola growth and production systems that contribute to the challenge of swede midge pest management include factors affecting compensatory growth in canola, as well as the economics of canola production, which make intensive and effective

*E-mail: rhallett@uoguelph.ca

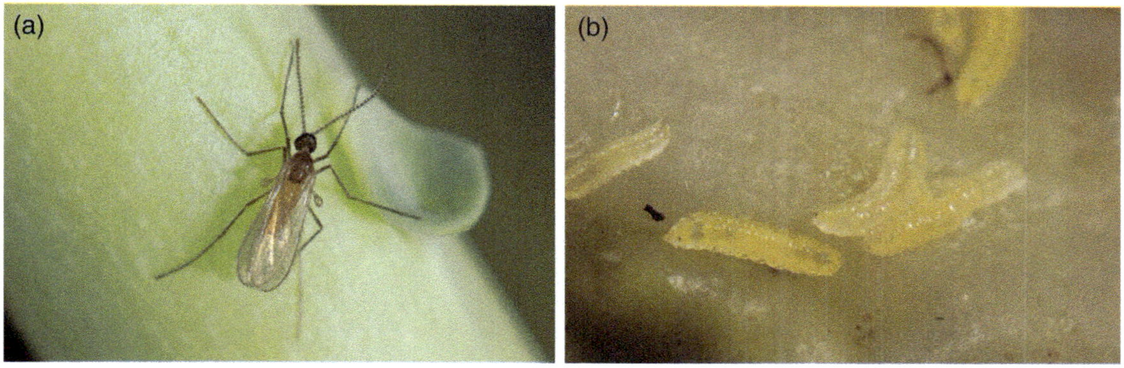

Fig. 3.1. Swede midge, *Contarinia nasturtii*, (a) adult male and (b) larvae. Photographs: D.K.B. Cheung.

management of the swede midge in canola difficult. Gaps in knowledge and research needs, as well as future approaches that could improve integrated pest management of the swede midge, are also discussed.

3.2 History of Infestation in North America

The first North American record of swede midge was made in southern Ontario in June 2000 based on captures of swede midge adults on yellow sticky cards at two vegetable farms in Markham and Stouffville, Ontario (Hallett and Heal, 2001). However, growers in the area first observed plants with damage symptoms typical of swede midge in the mid-1990s. In 2001, a yellow sticky card-based survey of cruciferous vegetable fields throughout Ontario found evidence that swede midge was present in nine counties in Ontario and one county in Quebec (R.H. Hallett, unpublished data). Swede midge was declared a regulated pest in Canada in 2002 by the Canadian Food Inspection Agency (CFIA), which began visual surveys of swede midge in Ontario and Quebec. Swede midge was first reported in Quebec in 2003 (Corlay and Boivin, 2008). From 2002 to 2004, swede midge was identified in 14 counties in southern Ontario and in four counties in Quebec where cole crops were grown (CFIA, 2004a, b). It was not until the synthetic sex pheromone lure became available experimentally to researchers in 2004 that the first US records were made, beginning with New York State, though symptomatic cole crop plants had been observed for several years prior (Kikkert *et al.*, 2006). Use of the pheromone trap by CFIA, beginning in 2005, led to a rapid expansion of counties positive for swede midge in Ontario and Quebec, as well as to the first CFIA records in canola fields (CFIA, 2009). The presence of swede midge was confirmed in Massachusetts and New Jersey in 2005, and in Connecticut and Vermont in 2006 (Chen *et al.*, 2007, 2009a; MIPOP, 2007; Brown, 2014). In 2007, both the most westerly and easterly records to date in North America were made by CFIA in Saskatchewan and Nova Scotia, followed by Manitoba and Prince Edward Island in 2008 (CFIA, 2009). Ohio was added to the list of swede midge positive states in 2009 (Chen *et al.*, 2011). In 2009, the CFIA and the USDA Animal and Plant Health Inspection Service deregulated the swede midge, due to its rapid spread despite the implementation of phytosanitary regulations and the limited risk posed to the cole crop industry (CFIA, 2009). In the intervening years, expansion of swede midge range within positive jurisdictions has been reported but there had been no new state or provincial records until the detection of swede midge in Michigan was announced in 2015 (MDARD, 2015). Thus, the current recorded range of swede midge in North America encompasses six Canadian provinces and seven US states.

3.2.1 Swede midge in eastern canola production regions

Ontario

A bioclimatic model determined that most of Canada is suitable for establishment of swede midge, but that southern Ontario, southern Quebec, New Brunswick, Nova Scotia and Prince Edward Island are most favourable eco-climatically for swede midge population growth (Olfert *et al.*, 2006). The greatest risk of pest outbreaks and of crop losses is predicted to occur in southern Ontario between the Great Lakes and along the St Lawrence River; however, favourable conditions, i.e. those that allow population growth in all years, are projected to occur as far north as Sudbury and Timiskaming districts in Ontario (Olfert *et al.*, 2006). When first discovered in Ontario, the swede midge was considered to be a serious pest with a medium overall risk rating but high potential for economic impact, notably because of its high potential to spread and ability to cause significant damage to crops (CFIA, 2002). Swede midge was initially a concern in cruciferous vegetable crops, where crop rotation is limited in Ontario and plants are susceptible to infestation through the entire season.

In Ontario, ~52% of canola production occurs in western Ontario (Grey, Wellington, Dufferin, Simcoe, Bruce, Huron, Perth, Waterloo and Peel counties) and ~34% in northern Ontario (Timiskaming, Nipissing, Cochrane, Manitoulin, Greater Sudbury and Sudbury) (OMAFRA, 2011a). All counties in the western Ontario region were identified as swede midge positive between 2002 and 2007 (CFIA, 2007). The first swede midge damage to spring canola fields was observed in the Grand Valley, Dufferin County, in July 2003, about 7 years after it was first observed in cole crops in Ontario (T. Baute, Ontario, 2003, personal communication). Discovery of swede midge in canola in Ontario coincided with a decline in canola production in Ontario (from 24,300 ha in 2002 to 6100 ha in 2006) (OMAFRA, 2014a) due to economic considerations. In 2004, about 60% of Ontario canola production occurred in counties positive for swede midge. Canola production in Ontario increased again in the years following to a peak of 35,600 ha in 2011 (OMAFRA, 2015). In western Ontario, swede midge has been a perennial pest since 2005 in Wellington, Bruce, Grey and Dufferin counties, particularly in late-planted fields (OMAFRA, 2006, 2008, 2009, 2010, 2011b, 2013, 2014b). Swede midge infestations in canola have been favoured by: (i) the lack of insecticides registered for swede midge management in canola (until 2012); (ii) a consequent lack of management efforts; (iii) weather conditions that contributed to suboptimal planting dates; and (iv) likely reductions in crop rotation due to advantageous canola prices. In years where planting is delayed due to poor weather in May, entire fields in these areas have been rendered unharvestable due to pervasive damage. Swede midge damage in western Ontario has been particularly prevalent near Shelburne, Dundalk and Grand Valley.

In northern Ontario, swede midge was detected in a cole crop field in the Greater Sudbury district in the 2001 sticky trap survey (R.H. Hallett, unpublished data) and confirmed in the region by the CFIA in 2006. In the northern region, only the Greater Sudbury district was recorded as swede midge positive prior to the deregulation of swede midge as a quarantine pest in 2009 and the consequent termination of CFIA survey activities.

In 2011, about 10 years after its first detection in northern Ontario, swede midge became a significant economic pest in Timiskaming district in the northern Ontario canola production region, causing damage to about 80% of canola near New Liskeard (B. Hall, Ontario, 2011, personal communication). Swede midge incidence rapidly expanded in this district and resulted in high levels of damage in each year from 2011 to 2014 (OMAFRA, 2011b, 2013, 2014b).

After growers in severely affected areas of the northern region experienced 63–81% yield losses to swede midge in 2014, the Ontario Canola Growers Association (OCGA) recommended a 3-year moratorium on canola production in Timiskaming in order to reduce populations to manageable levels (Phillips, 2015). Even prior to the moratorium, there was a significant decline in canola production area in Ontario, largely attributable to the swede midge, with the 2014 area (12,900 ha) being a 55% decline from the previous 5-year average canola area (i.e. 2009–2013, mean 28,480 ha/year) (OMAFRA, 2014a). In 2015, average pheromone trap captures in Timiskaming were ≥ 50 midges/trap/day from June until late July in non-canola fields, but where canola had been grown in 2014, suggesting that populations in this region will continue to impact canola production seriously for several years to come (Hallett, 2016).

Quebec

Swede midge appears to have been slower to spread into canola in Quebec than in Ontario but it is now a significant concern, particularly in the northern canola production region. A survey conducted in 2009 indicated that swede midge was present in seven areas, particularly in the three main canola production areas in Quebec: Temiscamingue, Saguenay-Lac-St-Jean and Bas St-Laurent (G. Labrie, Quebec, 2016, personal communication). The first damage to canola in the field was observed in the Temiscamingue area in 2013, and 50% of canola producers (representing 1991 ha of the total 2494 ha of canola) insured by La Financière Agricole declared yield losses due to swede midge. In 2014, there was a 50% reduction in the number of canola growers in the Temiscamingue area. In Saguenay-Lac-St-Jean, the first economic damage was observed in 2015 (G. Labrie, Quebec, 2016, personal communication).

3.2.2 Swede midge in western canola production regions

The first Prairie records of swede midge were made in August 2007 (CFIA, 2009) at three sites in Saskatchewan (Nipawin, Melfort, Canora-Yorkton). Each of these records consisted of a single male capture in a pheromone trap. Despite annual pheromone trap monitoring from 2007 to 2011, no further midges were captured (J. Soroka, Saskatchewan, 2011, personal communication). However, in 2012 the first reports of field damage in Saskatchewan were made near Nipawin. By 2014, swede midge was recorded from dozens of field sites spanning the breadth of Saskatchewan (AAFC, 2015). In 2015, swede midge was recorded from several sites on the Alberta–Saskatchewan border (J. Soroka, unpublished data).

The first report of swede midge in Manitoba was made in 2008 (CFIA, 2009). In 2014, swede midges were captured in pheromone-baited traps at three sites in Manitoba and larvae were found in the field at one additional site (Gavloski, 2014; AAFC, 2015). In 2015, no swede midges were captured in traps at any of the 36 survey sites in Manitoba; however, larvae were found in canola flower buds in northwest Manitoba in mid-August (Gavloski, 2015; J. Soroka, unpublished data).

The range of swede midge in the Prairie provinces continues to expand, leading to increased concerns about the potential for economic impacts. Bioclimatic modelling indicated that most of the Canadian Prairies were suitable for swede midge establishment and population growth, with conditions near Edmonton, Alberta, being more favourable than in surrounding areas (Olfert *et al.*, 2006). The latter study also showed that conditions would significantly improve for swede midge in years when rainfall in the Prairie Ecozone is well above normal, leading to increased risk of pest establishment (Olfert *et al.*, 2006). Although most records to date have occurred at sites quite removed from the US border, there is a risk that swede midge will also make its way into canola production areas in Minnesota (MDA, 2016), North and South Dakota and Montana. Mika *et al.* (2008) found that these states were considered suitable to very favourable for swede midge establishment and population growth, should it be introduced. In addition, future climate projections, particularly those utilizing the Canadian Centre for Climate Modelling and Analysis CGCM2 model, indicated further expansion of areas with favourable and very favourable climates for swede midge over the next 10–80 years (Mika *et al.*, 2008). Therefore western North America is likely to experience continued expansion of the range of swede midge and increased population densities, unless restricted deliberately by management efforts.

3.3 Implications of Ecology and Behaviour for Management

Invasive species typically have characteristics that help them to establish and thrive in diverse environments (Sakai *et al.*, 2001). Many aspects of the life history of the swede midge help to ensure that it survives in unpredictable environments and that a population persists once it enters an area. These qualities have a direct impact on our ability to manage the swede midge and the factors that need to be considered when developing an integrated pest management (IPM) programme for this pest.

3.3.1 Life cycle and damage

The life cycle of the swede midge is relatively rapid and highly variable – characteristics that appear to contribute to its success as an invasive species. Females are reproductively mature within 8 h of emerging as adults and may lay up to 100 eggs (in clusters of two to 50 eggs) on apical and floral meristematic tissues of host plants during the short 1–4-day adult lifespan (Barnes, 1946; Readshaw, 1961, 1966).

Larvae feed gregariously for 7–21 days (Readshaw, 1966), developing between the tightly appressed leaves and petioles around the central meristem and within buds and bud clusters at the central rosette and in leaf axils. Developing larvae are thus concealed and are protected from direct contact with insecticidal sprays. Larval feeding results in swelling, twisting and scarring of tissue (Fig. 3.2). Damage symptoms may not be visible for 5–10 days after feeding begins (Hallett, 2007; Hallett and Sears, 2013) and damaged tissue remains on plants for the remainder of the season. Therefore, the presence or

Fig. 3.2. Swede midge damage symptoms at (a) the vegetative stage and (b) the pod bouquet effect. Photographs: L. Des Marteaux.

absence of symptomatic tissue cannot be used as an indicator of the current presence of larvae or for assessing the efficacy of an insecticide treatment.

Mature larvae drop to the soil for pupation or for diapause entry. The pupal stage may last 10–48 days (Readshaw, 1966). Duration of the life cycle depends upon temperature and soil moisture (Readshaw, 1966; Hallett *et al.*, 2009a) but under Ontario summer conditions the swede midge can complete its entire life cycle in 3 weeks.

Larvae form spherical cocoons for overwintering as late-instar larvae or ovoid cocoons for pupation (Readshaw, 1961). Almost all cocoons are found within the top 1 cm of soil (Chen and Shelton, 2007). Consequently, cocoons can be picked up on the tyres of farm equipment and moved inadvertently from site to site. Avoiding the movement of farm equipment from infested to uninfested fields can help to prevent spread of swede midge. Restrictions on the movement of farm equipment from infested areas was thus included in CFIA guidelines when swede midge was considered a regulatory pest in Canada (CFIA, 2009).

3.3.2 Voltinism and reproductive potential

Swede midge has a high population potential due to its relatively rapid life cycle, reproductive output and multivoltinism throughout its range. Although each female can lay approximately 100 eggs, reports of egg and larval numbers in laboratory experiments indicate that typically 18.1–27.8 eggs are laid per female (Chen and Shelton, 2010; Williams, 2015). Investigations of the effect of female density on oviposition rates on canola plants at the early bud stage (i.e. inflorescence visible at centre of rosette, stage 3.1) (Harper and Berkenkamp, 1975) revealed a significant positive relationship between female density and total oviposition per plant but no effects on oviposition rates per female (Williams, 2015). No density-related decline in oviposition was observed at densities of up to 100 females/plant. At that density, total oviposition was 2492 ± 690 eggs/plant and the highest number exceeded 4000 eggs/plant (Williams, 2015), indicating an extraordinary potential for population growth on canola. The high reproductive rate of swede midge females and apparent lack of negative competitive effects, even at extremely high densities, have likely contributed to the rapid spread of swede midge in North America (Chen *et al.*, 2011). Rapid population growth is likely to lead to increasing issues with swede midge, particularly when the midge becomes established in regions where canola is a dominant crop in the landscape.

The number of generations varies geographically and with environmental conditions from two to five per year (Readshaw, 1961, 1966; Rygg and Braekke, 1980; Hallett and Heal, 2001). Pheromone-based monitoring and the MidgEmerge population dynamics model indicate that there are four overlapping generations per year in its eastern North American range (i.e. Ontario and Quebec) (Hallett *et al.*, 2009a), with the possibility of five generations under future climates (Mika *et al.*, 2008). The MidgEmerge model also indicated there are likely two to three generations per year in the Canadian Prairies; however, emergence trap data to date indicate the presence of only two generations (J. Soroka, Saskatchewan, 2016, personal communication).

There can be considerable variation in occurrence of swede midge population peaks among canola sites (Fig. 3.1), likely due to local differences in precipitation and soil moisture, which affects adult emergence (Hallett *et al.*, 2009a), as well as to differential rates of movement from overwintering sites into canola fields in June. Canola sites in western Ontario routinely experience very high populations (i.e. ≥ 30 males/trap/day; Hallett *et al.*, 2009b) from early June to September (Fig. 3.3).

3.3.3 Variable life history traits

In addition to variation in voltinism, the swede midge exhibits variability in diapause entry and termination. The MidgEmerge model revealed the presence of two, and possibly three, emergence phenotypes of the swede midge (Hallett *et al.*, 2009a). In southern Ontario, first emergence typically occurs in mid to late May, with emergence of the first phenotype peaking in early June, followed by the second phenotype in late June (Hallett *et al.*, 2009a). A third emergence phenotype that occurs in mid-August, and is a relatively small cohort compared with the spring emergence phenotypes, has now been observed in multiple years and locations in southern Ontario (Goodfellow, 2005; Des Marteaux *et al.*, 2014). Prolonged emergence of overwintering midges has also been described in Norway, where emergence occurs from late May to August (Rygg and Braekke, 1980). This strategy is assumed to ensure that at least some of the overwintered population will emerge into favourable environmental conditions where suitable host plants are present and, thus, the survival of the population in newly colonized areas or areas with unpredictable environments. The combination of

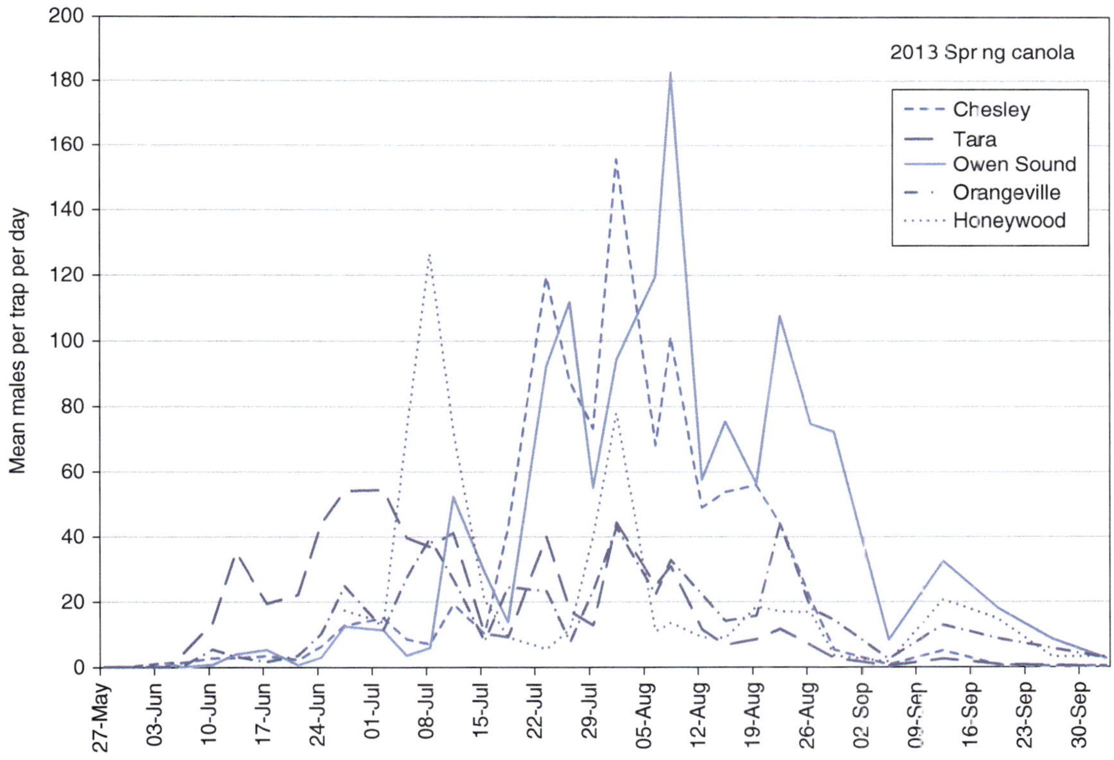

Fig. 3.3. Mean captures of swede midge males on four pheromone traps placed around the edges of each of five spring canola fields, southern Ontario, 2013.

prolonged emergence of overwintered midges and multiple generations leads to the overlap of successive generations throughout the season (Rogerson, 1963), so that there is almost continuous swede midge pressure from mid-June to September in southern Ontario. In addition, prolonged emergence ensures that any field contaminated with swede midge in the previous year continues to act as a source of new midges that disperse to new fields into the late summer (Fig. 3.4).

Diapause entry in the swede midge is inversely correlated with photoperiod and absolute maximum air temperature (Des Marteaux *et al.*, 2014). The critical photoperiod (i.e. photoperiod at which 50% of individuals enter diapause) varies from 13.27 h in UK to 13.73 h in Ontario; however, ~4% rates of diapause entry are observed at photoperiods as long as 15.4 h (Readshaw, 1961; Des Marteaux *et al.*, 2014). Therefore, some portion of the larval population from every generation produced from the early summer to the fall will enter diapause with 100% diapause rates occurring at photoperiods < 12 h (Readshaw, 1961; Des Marteaux *et al.*, 2014). This strategy helps to ensure the persistence of swede midge in a given site. As swede midge larvae are present nearly continually from early June to September, and foliar applications of insecticides at less than biweekly intervals are not likely to be employed in canola production systems, it is impossible to prevent there being some larvae in the soil that will persist until the next season. Thus, a canola field that was infested with swede midge should never be considered safe to plant to canola for a second consecutive year. There will always be some swede midge emergence from that field, so proper crop rotation and control of cruciferous weeds are essential.

Swede midge exhibit prolonged diapause, meaning that although most midges will diapause for one winter, some portion of the population remains

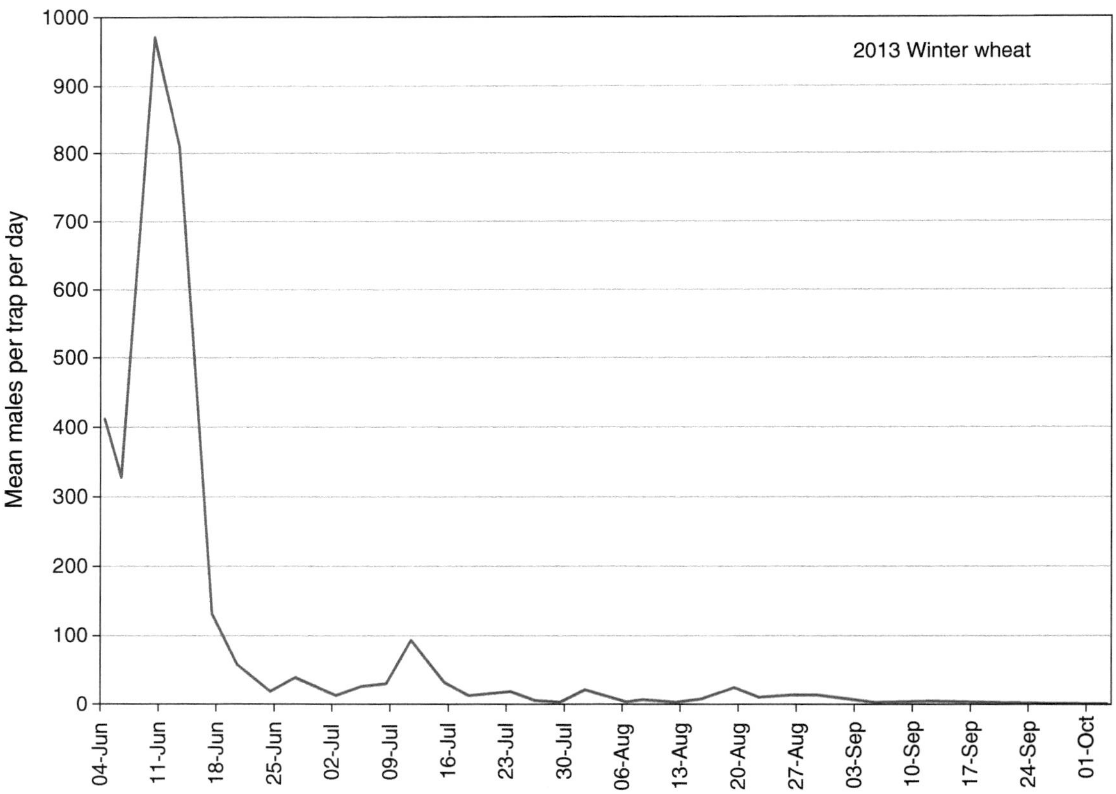

Fig. 3.4. Mean captures of swede midge males on four pheromone traps placed around the edges of a winter wheat field in 2013, where spring canola had been grown in 2012. Shelburne, Ontario, 2013.

in diapause for at least two winters. In an Ontario study, about 2% of the population remained in the soil for at least two winters (Des Marteaux *et al.*, 2014). Diapause for more than one winter has also been reported in Norway (Rygg and Braekke, 1980). Other *Contarinia* species, such as *C. tritici* (Kirby) and *C. vincetoxici* Kieffer, can remain in diapause for up to 3 and 13 years, respectively (Barnes, 1952; Solbreck and Widenfalk, 2012). Maximum duration of swede midge diapause has not been determined but as it is possible that some swede midge may stay in the soil for > 2 years, crop rotations of at least 3 years are recommended.

3.3.4 Crucifer specialists

Although there are 60 other species of *Contarinia* found in North America, swede midge is the only one that develops on plants in the Brassicaceae (Gagné, 1989). Within the Brassicaceae, swede midge has a large number of reported hosts, including the *B. oleracea* group of vegetables (e.g. cabbage, broccoli, cauliflower, Brussels sprouts), radish, rutabaga, turnips and the leafy cruciferous vegetables, such as mustard greens and bok choy (Hallett, 2007).

Cruciferous weed species also serve as hosts of swede midge and some species may sustain midge populations before or after the cultivation of cruciferous crops (Stokes, 1953; Hallett, 2007; Chen *et al.*, 2009a). Cruciferous weeds may play a temporal role as biological bridges for swede midge to reach later planted crops (Chen *et al.*, 2009a), as well as a spatial role in assisting the movement of swede midge through the landscape. Cruciferous weed management is therefore an important component of swede midge management. Movement of swede midge through the landscape may also be aided by close proximity of host crop fields, as well as by the movement of commercial cruciferous vegetable transplants from infested to non-infested areas (CFIA, 2009).

Although there have been limited attempts to date to identify resistant varieties among host crops, there has been no strong evidence seen for the occurrence of host plant resistance within the genus *Brassica* (Hallett, 2007; R.H. Hallett *et al.*, unpublished data). Among several *B. oleracea* varieties evaluated, differences in susceptibility to swede midge were found but not resistance (Hallett, 2007). Previous research comparing susceptibility of spring canola varieties indicated that where swede midge is a concern, *B. juncea* and/or *Sinapis alba* varieties of canola should be selected over *B. napus* varieties, which were more susceptible to damage (R.H. Hallett *et al.*, unpublished data). However, no varieties were found to be highly resistant to swede midge and these trials were conducted under relatively low population pressure (R.H. Hallett *et al.*, unpublished data). Consequently, there is no emphasis on varietal selection in current IPM recommendations for canola. Winter canola was found not to incur serious swede midge damage in the spring, presumably due to advanced plant development at the time of swede midge emergence (Table 3.1). However, the presence of winter canola in the spring may serve as a bridge host to later emerging or later planted crucifer crops. In the fall, winter canola could serve as a host to swede midge larvae developing in the late summer and early fall when other cruciferous crops have been harvested, and thus contribute to the overwintering midge population.

Table 3.1. Swede midge damage ratings (mean±SE) on primary and secondary racemes of winter canola, *Brassica napus* 'Kronos', treated with weekly foliar applications of acetamiprid insecticide compared with an untreated control. Planted 3 September 2005, Elora Research Station, Ontario. Means within a date followed by the same letter are not significantly different by Tukey–Kramer, $\alpha = 0.05$.

	Primary raceme[a]				Secondary raceme[a]
Treatment	7 Oct, 2005[b]	14 Oct, 2005[c]	21 Oct, 2005[d]	12 Jun, 2006[e]	12 Jun, 2006[f]
Insecticide	0.49±0.06 A	0.53±0.06 A	0.49±0.06 A	0±0 A	0.03±0.01 A
Control	0.10±0.03 B	0.08±0.03 B	0.14±0.04 B	0.03±0.03 A	0.02±0.01 A

[a]Primary raceme damage rating, 0 to 3 scale; Secondary raceme damage rating, 0 to 11 scale.
[b]$F = 36.2$, df = 1,158, $P < 0.0001$
[c]$F = 50.19$, df = 1,158, $P < 0.0001$
[d]$F = 26.27$, df = 1,158, $P < 0.0001$
[e]$F = 1$, df = 1,158, $P > 0.3$
[f]$F = 0.15$, df = 1,158, $P > 0.7$

3.3.5 Lack of natural enemies

The enemy release hypothesis ascribes the success of alien species in their invaded range to escape from the natural enemies present in their native range (Elton, 1958). In Europe and the UK, *Pirene eximia* Haliday (Hymenoptera: Pteromalidae), *Synopeas* sp. and *Platygaster* sp. (Hymenoptera: Platygastridae) have been reported as larval parasitoids of swede midge (Bovien and Knudsen, 1950; Rogerson, 1963; Readshaw, 1966, 1968). Four endoparasitoids of swede midge were found in a recent European survey to identify parasitoids of swede midge for potential introduction to Canada as classical biological control agents (Abram *et al.*, 2012). One species belonged to the Pteromalidae (*Macroglenes chalybeus* Haliday) and three to the Platygastridae (*Inostemma opacum* Thomson, *Synopeas myles* Walker and *S. osaces* Walker) (Abram *et al.*, 2012). *M. chalybeus* and *S. myles* caused the highest levels of parasitism observed and were found throughout both the full season and the surveyed range (Abram *et al.*, 2012). These results seemed promising; however, parasitism rates by these two species were generally very low (< 3% average parasitism) and only rarely reached 30–40% (Abram *et al.*, 2012). Ultimately, neither of these species was considered suitable for introduction to North America, primarily due to concerns about lack of host specificity (Abram *et al.*, 2013).

North American Cecidomyiidae are attacked by several natural enemies, including parasitoids in nine hymenopteran families that may be generalists or specialists on the gall midges (Gagné, 1989). However, no parasitoids of larval swede midge were found in surveys conducted in 2004 and 2005 in Quebec (Corlay *et al.*, 2007). In 2013 in Saskatchewan, a female *Gastrancistrus* sp. (Hymenoptera: Pteromalidae) was observed ovipositing into swede midge larvae developing within swollen, fused canola buds, and both this species and an *Inostemma* sp. were reared from field-collected swede midge larvae in 2013–2015 (L. Andreassen and J. Soroka, Saskatchewan, 2015, personal communication). The *Inostemma* sp. is currently undergoing further identification, but species identification of the *Gastrancistrus* sp. cannot be achieved due to the lack of North American keys for these genera (J. Soroka, Saskatchewan, 2015, personal communication).

A specimen in the genus *Synopeas* was reared from swede midge-infested canola in the Temiscamingue region, Quebec, in 2015, and identification to the species level has been sought (G. Labrie, Quebec, 2015, personal communication).

No surveys for natural enemies have been undertaken in Ontario to date but adult *Medetera* sp. (Diptera: Dolichopodidae) were observed preying on swede midge adults in the field (Goodfellow, 2005) and are occasional predators in swede midge colonies housed at the University of Guelph.

The combination of life history traits, a broad host range and a lack of natural enemies allows for persistence and spread of swede midge in the landscape and gradual population growth to economically damaging levels, which are then difficult to manage.

3.4 Impact of Swede Midge on Canola

3.4.1 Patterns of damage and infestation

In Ontario, the first field reports of damage to canola by swede midge were made in 2003 (T. Baute, Ontario, 2003, personal communication), about 7 years after symptoms were first noticed by cole crop growers and 2 years after the first official record in Ontario (Hallett and Heal, 2001). In all subsequent years for which reports are available, significant damage due to swede midge has been reported in Ontario, particularly to late-planted canola (i.e. seeded late May to June) (OMAFRA, 2006, 2008, 2009, 2010, 2011b, 2013, 2014b). Initially, spring canola plants were typically damaged during the early growth stages (seedling to early rosette) (OMAFRA, 2006). As swede midge spread and its prevalence increased, damage was also reported in fields with earlier planting dates and reports of multiple infestations throughout the season became more common (OMAFRA, 2013). By 2011, swede midge was widespread throughout canola growing regions of Ontario, including the northern production region, and significant damage was observed (OMAFRA, 2011b, 2013).

Swede midge damage is often first observed along field edges and near windbreaks, as swede midge moves into new fields. A similar spatial pattern has been observed in newly infested cole crop fields and is assumed to be due to swede midge being better able to descend to crop level near windbreaks, when they are being carried by winds. This pattern of invasion in canola is also attributable to the movement of swede midge into new fields after dispersal from overwintering sites (i.e. canola fields from the previous year). However, as the season progresses, and in fields with crucifer production in consecutive years, swede midge can be found throughout the field, though damage may continue to be most severe near field edges.

The impact of swede midge on canola depends upon the timing of infestation, whether a plant is subjected to single or multiple infestations, and environmental conditions that impact the ability of the plant to compensate for damage. Thus, while early season damage was problematic in both 2013 and 2014, plant recovery from damage was poor in 2013 due to weather-related crop stress, but better than expected in 2014 due to good growing conditions (OMAFRA, 2013, 2014b). Despite favourable conditions in 2014, swede midge caused yield losses of 20–50% in areas where pressure was highest (OMAFRA, 2014b). Canola infested by swede midge may branch extensively to compensate for damage to the main stem. Thus, swede midge infestation often leads to uneven maturation of plants in affected fields and consequent delays to harvest. In northern regions especially, delayed maturation may have severe economic consequences if it is not possible to harvest prior to snowfall. Thus, both direct and indirect yield impacts may occur as a result of swede midge infestation.

3.4.2 Damage symptoms and plant stage interactions

Swede midge damage symptoms depend upon plant stage at the time of infestation. The earliest that swede midge damage to canola has been observed in the field is at the two-leaf stage. Throughout the vegetative stage (i.e. rosette stage, or stage 2) (Harper and Berkenkamp, 1975), swede midge infestation will cause damage to the central meristem, potentially resulting in twisting of the stem, crumpling of leaves, swelling of leaf petioles and/or death of the central meristem. Damage at the early reproductive stages likely has the greatest impact on yields, as infestation of the floral meristem can result in lack of bud initiation, direct feeding on buds and the introduction of secondary rots. Depending upon the degree of damage, infestation at this stage may result in reduction or complete absence of flowering and pod set on the main raceme. Infestation of flower buds by swede midge larvae can lead to swollen, fused flower buds and reduce seed production (Barnes, 1946). Although relatively rare in Ontario, swollen buds are the primary symptom observed in the Prairies (J. Soroka, Saskatchewan, 2013, personal communication). Manifestation of damage symptoms from early infestation becomes more evident during later growth stages as injured plants may fail to complete main stem elongation (i.e. bolting) and bud and/or pod formation (CFIA, 2009). Incomplete stem elongation often leads to the bunching of pods at the top of a stunted main stem, a symptom that is known as the 'bouquet' or 'umbrella' effect (Fig. 3.2). Compensatory growth to replace damaged or killed tissue may result in additional branching, but the extent to and conditions under which this growth recoups lost yield still require elucidation.

Females do not appear to oviposit into buds that have begun to expand and open (Readshaw, 1961). Thus, after stem elongation and bud swelling on the main stem, females will lay eggs on new meristematic tissue developing in leaf axils for branching and production of secondary and tertiary buds (Williams, 2015). Such infestation may lead to the death of buds and/or branches. Once all buds have begun to open, swede midge no longer poses a threat to yield, but oviposition may still occur wherever young, rapidly growing tissue develops. Although there is no risk of economic damage by swede midge once all flower buds have begun to open, larvae may still develop on any new tissues produced. For example, swede midge larvae have been found in new leaf tissue growing on canola stems after harvest in late September in Ontario. Development of larvae on this tissue will contribute to the overwintering population, and thus pose a threat to nearby cruciferous crops in the following spring.

The first assessment of the impact of swede midge on canola in Ontario was made in 2003 at a vegetable farm with very high swede midge populations (i.e. ~30 males/trap/day; Hallett *et al.*, 2009b), and with canola planted at relatively low densities. Under these conditions, when not protected from swede midge with multiple foliar insecticide applications, *B. napus* 'Hyola 700RR' plants suffered a 10% reduction in height and produced 55% more branches than when protected by insecticides (Table 3.2). Despite increased branching, untreated plants produced 98% fewer pods and had only 2% of the seed yield of insecticide-treated plants.

Planting-date trials conducted at two locations near Guelph in 2006 confirmed that damage at all growth stages was more severe in plots with later planting dates (Table 3.3). At both locations, pod

Table 3.2. Impact of swede midge on mean (±SE) growth parameters (per plant) and seed yield (per plot) of spring canola, *Brassica napus* 'Hyola 700RR', planted 2 June, and treated with three biweekly applications of acetamiprid or left untreated. Stouffville, Ontario, 2003. Means in the same column and growth parameter followed by the same letter are not significantly different by ANOVA, $\alpha = 0.05$.

Growth parameter	Assessment date		
	8 Aug 2003[a]	26 Aug 2003[b]	24 Sep 2003[c]
Plant height (cm)			
Insecticide	63.76±1.98 A	68.99±2.66 A	76.00±1.73 A
Control	43.28±1.60 B	54.75±2.37 B	69.01±1.69 B
No. of branches			
Insecticide	10.46±0.60 A	11.81±1.04 A	14.99±0.69 B
Control	9.21±0.64 A	12.13±0.95 A	23.31±1.76 A
No. of pods			
Insecticide	18.13±2.49 A	32.84±3.80 A	22.89±1.70 A
Control	2.10±0.67 B	6.31±1.28 B	0.40±0.16 B
Seed yield (g)[d]			
Insecticide	94.73±12.16 A		
Control	1.55 ± 0.47 B		

[a]Height: $F = 64.79$, df = 1,158, $P < 0.0001$. Branches: $F = 2.03$, df = 1,158, $P > 0.15$. Pods: $F = 38.49$, df = 1,158, $P < 0.0001$.
[b]Height: $F = 16.02$, df = 1,158, $P < 0.0001$. Branches: $F = 0.05$, df = 1,158, $P > 0.82$. Pods: $F = 43.81$, df = 1,158, $P < 0.0001$.
[c]Height: $F = 8.34$, df = 1,158, $P < 0.005$. Branches: $F = 19.43$, df = 1,158, $P < 0.0001$. Pods: $F = 173.93$, df = 1,158, $P < 0.0001$.
[d]$F = 58.67$, df = 1,14, $P < 0.0001$.

Table 3.3. Effect of planting date on swede midge damage (mean±SE) to spring canola, Brassica napus 'Hyola 700RR', treated with alternating weekly foliar applications of lambda-cyhalothrin and acetamiprid. Elora and Arkell, Ontario, 2006. Means in the same column followed by the same letter are not significantly different by Tukey–Kramer, $\alpha = 0.05$.

Planting date	Treatment	Primary raceme rating (0 to 4)			Secondary raceme rating (0 to 11)	
		Vegetative	Flowering	Pod-filling	Flowering	Pod-filling
Elora						
24 May	Untreated	0.05 ± 0.04 B	0.03 ± 0.03 C	0.70 ± 0.17 C	–	2.83 ± 0.66 B
	Foliar	0.10 ± 0.05 AB	0.03 ± 0.03 C	0.80 ± 0.18 C	–	2.95 ± 0.68 B
7 June	Untreated	–	1.03 ± 0.19 B	1.08 ± 0.19 BC	3.78 ± 0.71 B	2.93 ± 0.64 B
	Foliar	0.03 ± 0.03 AB	1.05 ± 0.19 B	1.25 ± 0.20 BC	3.88 ± 0.70 B	4.18 ± 0.74 B
21 June	Untreated	0.53 ± 0.14 A	2.45 ± 0.19 A	2.70 ± 0.11 A	10.98 ± 0.03 A	9.78 ± 0.40 A
	Foliar	0.30 ± 0.10 AB	2.38 ± 0.17 A	1.75 ± 0.23 B	9.35 ± 0.49 A	4.95 ± 0.67 B
Arkell						
24 May	Untreated	0.08 ± 0.04 A	0 ± 0 C	0.08 ± 0.06 B	–	0.20 ± 0.20 B
	Foliar	0 ± 0 A	0.13 ± 0.05 BC	0.05 ± 0.05 B	–	0.05 ± 0.05 B
7 June	Untreated	– A	0.15 ± 0.09 BC	0.55 ± 0.14 B	0.28 ± 0.28 C	2.69 ± 0.57 AB
	Foliar	0 ± 0 A	0.05 ± 0.04 C	0.38 ± 0.13 B	0 ± 0 C	1.13 ± 0.44 B
21 June	Untreated	0 ± 0 A	1.58 ± 0.21 A	1.80 ± 0.22 A	5.80 ± 0.72 A	5.28 ± 0.60 A
	Foliar	0.10 ± 0.06 A	0.58 ± 0.16 B	0.58 ± 0.16 B	3.00 ± 0.64 B	1.75 ± 0.45 B

set was too low in the latest planted plots (21 June planting date) to allow mechanical harvest, essentially amounting to a 100% yield loss in the last planting-date treatment. However, results were inconsistent between sites with respect to swede midge effects on plant height and yield (Table 3.4). At Elora, plant height was reduced by > 30% between the first and third planting dates, and plants in untreated plots were 16% shorter than those in plots protected with insecticides (Table 3.4). At Arkell, no height differences were observed; however, yield for the second planting date was > 30% less than

Table 3.4. Effect of planting date and insecticide treatment on yield and plant height (mean±SE) of spring canola, Brassica napus 'Hyola 700RR', treated with alternating weekly foliar applications of lambda-cyhalothrin and acetamiprid. Elora and Arkell, Ontario, 2006. Means in the same column followed by the same letter are not significantly different by LSD, $\alpha = 0.05$.

	Elora[a]		Arkell[b]	
Treatment	Yield	Height (cm)	Yield	Height (cm)
Planting date				
24 May	991 A	113.0 A	1027 A	121.3 A
7 June	980 A	116.3 A	706 A	117.1 A
21 June	–[c]	77.4 B	–[c]	120.0 A
P	0.91	<0.0001	0.08	0.96
LSD	NS	14.6	362	NS
Insecticide treatment				
Untreated	1030 A	93.4 B	662 B	111.8 A
Foliar	941 A	111.0 A	1071 A	127.1 A
P	0.41	0.007	0.03	0.24
LSD	NS	11.9	362	NS

[a]A significant planting date by foliar treatment interaction was found for crop height ($P = 0.0007$).
[b]No significant planting date by foliar treatment interactions were found for any measured variable.
[c]Yield not determined, insufficient pods to allow mechanical harvest.

that of the first planting date, and yield in untreated plots was > 38% less than in insecticide-treated plots, regardless of planting date. The greater effect of swede midge on later plantings of canola is presumably due to interactions between swede midge numbers and plant stage at the time of infestation, as well as swede midge plant stage preferences.

When given a choice in the laboratory, swede midge preferred to oviposit on plants at the seven-leaf (i.e. seventh true leaf expanded, stage 2.7; Harper and Berkenkamp, 1975) and early bud stages, rather than on plants that had started to flower (i.e. first flower open, stage 4.1; Harper and Berkenkamp, 1975) (Williams, 2015). Most oviposition at the early bud stage was on secondary racemes rather than on the primary raceme, suggesting a preference for oviposition on secondary growth points in the early bud stage and older growth stages (Williams, 2015). These results confirm that, as for other hosts, swede midge prefer young, fast-growing tissues of canola for oviposition, with the highest oviposition on stages with the greatest meristematic tissue (Barnes, 1946; Readshaw, 1961; Hallett, 2007). Given no choice, females will oviposit on less favourable canola growth stages; however, there may be increased risk of larval mortality, due to desiccation or starvation, if oviposition occurs on plants with relatively little meristematic tissue (Williams, 2015).

A laboratory study tracking damage development after a single infestation at the early bud stage revealed a decline in the proportion of secondary and tertiary racemes exhibiting damage symptoms over time (J. Williams and R.H. Hallett, 2016, unpublished results). These results suggest compensation by plants through the production of additional branches following damage to the primary raceme and initial secondary and tertiary racemes. New racemes were produced from leaf axils where severely damaged secondary inflorescences were present previously. Whether a beneficial compensatory effect will occur in the field is complicated by the presence of multiple generations of swede midge and by the potential for compensatory racemes to have delayed maturity and thereby not contribute to final yield (J. Williams and R.H. Hallett, 2016, unpublished results).

3.4.3 Yield impacts

At high population levels, economic impacts by swede midge on yield are quite apparent, though actual yield losses in field-realistic conditions have been difficult to quantify (Williams, 2015). In laboratory studies examining density-dependent effects of swede midge following a single infestation with 0–100 females on plants at the early bud stage, the total number of pods produced and total seed weight produced on primary and secondary racemes significantly declined with increasing swede midge density (J. Williams and R.H. Hallett, 2016, unpublished results). Seed weight per pod on primary racemes and production of secondary racemes were also negatively related to swede midge density. No density-dependent effects were observed on these growth and yield parameters for tertiary racemes. However, contributions of tertiary racemes to seed production were not sufficient to negate the density-dependent decline in total seed weight on a whole-plant basis. Consequently, total pods produced and total seed weight per plant declined by approximately 30% and 35%, respectively (J. Williams and R.H. Hallett, 2016, unpublished results). Development of plants from the early bud stage to flowering took approximately 1 week longer on plants subjected to high densities of midges than for uninfested controls. Thus, swede midge infestation at the bud stage has significant negative effects on canola yield and development, with the primary raceme being most severely impacted (J. Williams and R.H. Hallett, 2016, unpublished results). Delays in maturation could lead to complications at harvest, if plant tissue and seed produced by compensatory racemes are still green or insufficiently dried; and their contribution to yield could be lost at harvest due to stunted height and underdeveloped seeds or pods (Canola Council of Canada, 2011). Variability in the number and timing of swede midge infestation events among plants in a given field can also cause in-field variability in maturation, which may interfere with harvest. Detailed studies on the impact of single infestations at other susceptible growth stages and of multiple infestations are needed to elucidate the impact of swede midge on canola yield more completely and the ability of canola to compensate with subsequent growth. This information is critical to determining the optimum timing of insecticide applications in conjunction with a pheromone-based action threshold.

In recent field trials, utilizing the currently registered insecticides for swede midge in canola (i.e. lambda-cyhalothrin and chlorantraniliprole), it has been difficult to assess impacts of swede midge on canola yields, primarily due to difficulties in maintaining an undamaged control for yield comparisons. However, comparisons between yields in trials conducted in heavily infested versus less infested areas suggest yield losses of up to 50% due to swede midge (R.H. Hallett *et al.*, unpublished data). The OCGA reported 63–81% yield losses in affected areas in 2014 (Phillips, 2015), demonstrating the devastating impact that swede midge can have on canola yields.

3.5 Current IPM Recommendations

3.5.1 Statement of recommendations

Current IPM recommendations for the swede midge in canola in Ontario (Hallett and Hall, 2014) are based on knowledge of swede midge biology and ecology, results of field trials, field observations and consultations with industry representatives, as well as practices used in managing swede midge in cruciferous vegetable crops. The rationale for, and further discussion of, the current IPM recommendations is provided in subsequent sections below. Development of a pheromone-based economic threshold in canola is in progress, so the action threshold guidelines for insecticide application timing given here are considered tentative and are a best estimate given our knowledge and experience to date. Current recommendations consist of the following key elements:

1. Crop rotation and host plant deprivation:
Canola production should be rotated with non-cruciferous crops, on a minimum 3-year canola rotation plan. Canola should not be planted in fields adjacent to the previous year's canola.
2. Planting dates:
Canola should be planted early in order to avoid the presence of swede midge during vulnerable plant stages. Depending on the area, it may be best to plant canola first, i.e. before other crops, in order to escape significant damage.
3. Weed management:
Cruciferous weeds in and around fields should be managed to reduce availability of host plants for swede midge.
4. Vulnerable plant stages:
The most vulnerable stages of canola to swede midge infestation are: during the vegetative (i.e. rosette) stage; when buds are developing in the centre of the rosette (i.e. early bud stage); and when buds are developing in secondary leaf axils. These are the key stages to protect from swede midge infestation.
5. Monitoring swede midge populations:
Swede midge populations must be monitored 3 times per week (or twice weekly at a minimum)

with pheromone traps to determine treatment timing. The action threshold (see below) is based on average male captures per trap per day, from four pheromone traps placed around each field.

6. Foliar insecticide application timing:
Insecticide applications must occur quickly after the adult swede midge population threshold numbers are reached and should be made before larvae or damage to canola buds are visible.

7. Tentative action threshold:
The first insecticide treatment should be based on a cumulative trap capture of 20 midges, beginning at the first true leaf stage. Subsequent insecticide treatments should be made when average captures in pheromone traps reach ≥ 5 midges/trap/day, with a minimum 7-day interval between insecticide applications.

8. Insecticide selection:
If multiple insecticide applications are needed, rotation between active ingredients should be practised. Currently in Canada, only lambda-cyhalothrin (Matador 120 EC, Syngenta) and chlorantraniliprole (Coragen, DuPont) are registered for control of swede midge in canola. Coragen cannot be applied within 60 days of planting, if the field is planted with Lumiderm-treated seed, as they are both Group 28 (Diamide) insecticides (PMRA registration no. 28982).

9. Spray coverage and surfactants:
Thorough spray coverage is critical, so use of a medium droplet size is recommended. On larger plants (e.g. full rosette to bud stage), the water volume should be increased (15–25 gallons/acre (168–281 l/ha)) to ensure good coverage. Use of an appropriate surfactant with Coragen can improve spray coverage and translaminar entry of the active ingredient into plant tissue (DuPont, 2015).

3.5.2 Crop rotation and host plant deprivation

As swede midge is a crucifer specialist, host deprivation through crop rotation is a key element of an IPM programme and will help to reduce population growth locally. The later occurrence of noticeable swede midge damage in canola compared with cole crops in Ontario is attributable, at least in part, to the longer crop rotations typically practised in field crop systems than in vegetable production systems. Differences in length of rotations out of crucifer crops have also been credited with the relatively low incidence of economic damage in cruciferous crops in New York State, where vegetable growers plant cabbage only every 3–5 years (Chen *et al.*, 2009b), compared with Ontario, where growers practise shortened rotations due to limited land and market accessibility concerns. Swede midge infestations were considerably reduced in the Netherlands after a 2-year period without crucifer cultivation (Theunissen *et al.*, 1997). However, prolonged diapause (Des Marteaux *et al.*, 2014) means that there may still be viable swede midge larvae in the soil for 2 years after cultivation of a cruciferous crop. In addition, swede midge has a wide range of alternative host plants that may serve as reservoir hosts in the absence of crop hosts. Thus, a minimum 3-year rotation is recommended, so that by the time canola is planted again, all of the pupae will be gone from the field. Attention should also be given to the spatial distribution of canola in the landscape over time, in order to maximize dispersal distances from overwintering sites to new fields. Planting in fields upwind from a previous crop and/or planting > 1.5 km away from a previous canola or cruciferous vegetable field may help to delay or minimize swede midge infestation.

Three-year crop rotation should substantially reduce swede midge populations and damage early in the season. However, where canola is dominant in the landscape, swede midge easily moves from overwintering sites to new canola fields and 3-year rotations are insufficient to maintain populations at manageable levels, as seen in Timiskaming. In the absence of highly effective insecticides and action thresholds for timing applications, canola growers may need to extend rotations to 4–6 years, which may well represent an economic hardship to growers in regions where canola is the main source of income.

In 2015, the OCGA recommended a 3-year moratorium on canola production in Timiskaming in order to deprive the midge of host plants and reduce populations back to manageable levels (Phillips, 2015). However, during the 2015 season, average trap captures at multiple non-canola sites were above 50 midges/trap/day from June until late July (Hallett, 2016). Thus, midges were present at numbers greatly exceeding the interim action threshold of 5 midges/trap/day for the duration of all vulnerable stages. These numbers show that swede midge was able to persist in the landscape, despite very little canola being grown, and indicate that populations may not have decreased sufficiently for planting of canola to be considered in this area in 2016. Thus, continuing with the moratorium in this region is advisable, in order to help to ensure that midge populations decline to manageable levels again. It is anticipated that populations in this region will continue to have a serious impact on canola production for several years to come, especially if the moratorium is not followed by all growers.

3.5.3 Planting dates

Given that swede midge prefers the young, fast-growing tissues of canola for oviposition, late-planted fields, in which plants at the vegetative to early bud stage coincide with typically high emergence peaks in June, are more susceptible to significant damage than early-planted fields, in which plants are already at a stage with less meristematic tissue when overwintered midges emerge. Thus, in areas where swede midge occurs, growers are advised to avoid late May to June plantings of canola. In areas with consistently high swede midge populations, it may be advisable to plant canola before planting other crops, if conditions allow. Planting or re-seeding of a failed crop is not recommended after mid to late June, as damage will be very high, the crop will likely be unharvestable and the resulting overwintering midge populations will present a risk to the following year's crop.

3.5.4 Weed management

Many cruciferous weeds are suitable hosts for the swede midge, which means that even in years when a host crop is not grown, the swede midge can persist in an area on weed hosts. Thus, management of weed hosts, as well as maximizing the distance between fields of host crops, is important for minimizing build-up of swede midge populations in an area.

3.5.5 Insecticides and action thresholds

Conventional cole crop growers in southern Ontario have been able to manage swede midge successfully through diversification of vegetable farms, pheromone trap monitoring, use of action thresholds and timely insecticide applications. However, the economics of canola production do not support the application of insecticides multiple times over the course of the season. Based on 2010 production costs and prices (OCGA, 2011; OMAFRA, 2014a), a single insecticide application can be assumed to cost CAN$69.19/ha. With an average canola price of CAN$455.00/tonne, swede midge damage would only have to result in 6% yield losses for an insecticide application to be economically worthwhile. However, with average production costs of CAN$903.55/ha and a gross return of CAN$1071.03/ha, an unnecessary insecticide application would reduce profits by 40%. Based on 2016 canola production estimates (OCGA, 2016), a single insecticide application in canola is economically worthwhile with 9% yield losses; however, an unnecessary insecticide application would reduce profits by 32%, and multiple applications may be needed against swede midge. Thus, it is critical to determine when midge populations warrant management efforts, and at what plant stage, in order to minimize both insecticide costs and yield losses.

Currently, growers are advised to protect the late vegetative stage when buds of the primary raceme are developing, and when the secondary and tertiary buds are forming in subsequent leaf axils, in order that insecticide applications coincide with growth stages with young meristematic and bud tissue. Applications prior to and/or at the seven-leaf stage could protect the primary growth point, while applications at or shortly after the early bud stage could protect secondary growth points.

Matador 120 EC (lambda-cyhalothrin) and Coragen (chlorantraniliprole) were registered for use against swede midge in canola in 2012 and 2013, respectively. Both registered products provide contact and residual activity. Chlorantraniliprole has translaminar movement, but growth that emerges after treatment is not protected (DuPont, 2015). Lambda-cyhalothrin has known efficacy against swede midge and was one of the first products registered against swede midge in cole crops in Canada (Hallett *et al.*, 2009b). In addition, there is no evidence of resistance to synthetic pyrethroids in Ontario swede midge populations (Hallett *et al.*, 2009b), making lambda-cyhalothrin a good candidate against swede midge in canola. Chlorantraniliprole is comparable in efficacy to lambda-cyhalothrin (Williams, 2015; R.H. Hallett *et al.*, unpublished data). Although early trials with acetamiprid showed effective reduction in swede midge damage with foliar insecticide treatments (e.g. Tables 3.1, 3.2 and 3.4), in recent field trials, it has been difficult to achieve sufficient protection of canola with the two currently registered products. Crop-stage timing trials (~three-leaf stage, ~eight-leaf stage to primary bud stage, early secondary bud stage) were conducted in grower fields in multiple years; however, the most effective application frequency and timing differed among sites, likely due to population differences and how well spray timing coincided with population peaks (Williams, 2015; R.H. Hallett and J. Williams, unpublished data). Neither active ingredient exhibited very high efficacy, especially when application timings were not optimal and swede midge numbers were high (Williams, 2015). The recommendation to leave at least 7 days between insecticide applications is based on residual efficacy trials that demonstrated ~7 days' residual activity of lambda-cyhalothrin against swede midge (Hallett and Sears, 2013), as well as registration labels for both insecticides, which specify minimum retreatment intervals of 5 and 7 days for chlorantraniliprole and lambda-cyhalothrin, respectively (PMRA registration nos. 24984 and 28982).

Inadequate and variable control of swede midge has been reported by Ontario canola growers (OMAFRA, 2013, 2014b). Suboptimal timing of insecticide applications is likely an important factor in control efficacy, as is the difficulty of controlling swede midge with only one to three insecticide applications when midges may be present constantly in the field from first emergence of seedlings through all vulnerable stages. Suboptimal insecticide application timing may result if growers make spray timing decisions on the basis of the presence of swede midge damage. As damage symptoms take some time to develop, larvae may well have left the plant for pupation before damage is observed and an insecticide application is made. As damage symptoms remain on the plant for the duration of the season, failure of insecticide control could be wrongly assumed if growers are not careful to distinguish damage on old tissues from that on new tissues when assessing the efficacy of control.

Pheromone-based action thresholds can be used to determine whether and when insecticide applications are needed during vulnerable crop stages. Swede midge populations can be effectively monitored with sex pheromone traps that capture male midges (Hillbur *et al.*, 2005; Boddum *et al.*, 2009) and pheromone-based action thresholds have been established to time insecticide applications effectively in conventional cabbage crops (Hallett and Sears, 2013). Appropriate pheromone-based action thresholds for swede midge in canola are under development (see Section 3.6, below).

Since adults live only 1–4 days, pheromone traps need to be checked every 2–4 days to keep track of populations so that insecticide applications can be appropriately timed against developing eggs and larvae. Insecticide applications must occur quickly after the adult swede midge population threshold numbers are reached, as they mate and lay eggs soon after emergence and larvae hatch ~3 days later.

The current recommendation for timing of the first insecticide application is that, with monitoring beginning at the first true leaf stage, the first application should be made when a cumulative total of 20 midges have been captured in four traps placed around the field edges. This recommendation is based on the observation in 2013 that fields with little economic damage later in the season had total captures of less than ~30 midges by the four-leaf stage, while fields with high levels of damage had total captures of several hundred midges by this time (R.H. Hallett and J. Williams, unpublished data). Thus, this is a conservative interim recommendation intended to prevent early application of insecticides to fields with low swede midge pressure, while at the same time ensuring that fields facing high numbers of midges are protected at an appropriate stage. A recommendation specific to the first insecticide application may be unnecessary when a pheromone-based economic action threshold is determined.

The current action threshold for subsequent insecticide applications is five males/trap/day, which is the same as the recommended action threshold for swede midge in cabbage (Hallett and Sears, 2013). This threshold may prove to be too conservative for use in canola, but high frequency and severity of swede midge damage, coupled with the difficulty of determining yields losses in field trials, have led to a cautious approach. An accurate pheromone-based economic action threshold would be a valuable tool for growers to optimize spray timing and efficacy. Careful examination of the effect of infestation density and timing on canola yield is needed to inform action threshold development.

3.6 Future Needs and Opportunities

The swede midge is a challenging pest to manage in canola due to its high potential for population growth in the crop, the cryptic nature of larval feeding that makes effective control with non-systemic insecticides difficult, and the limited control that can be achieved with few insecticide applications given the economic constraints of canola production. Canola that is at risk of swede midge infestation requires intensive management and monitoring. While horticultural crop producers are used to intensive crop and pest management practices, field crop producers typically do not have to monitor pests multiple times per week and thus the intensive management style needed for swede midge may be difficult for field crop producers to adopt. Based on the knowledge and experience gained to date with the management of swede midge in canola and cruciferous vegetables, several approaches are identified below that require further development and exploration and that hold the potential to improve our ability to manage swede midge effectively in canola: (i) a pheromone-based economic threshold; (ii) systemic insecticides; (iii) biological control agents; (iv) planting regimes; and (v) host plant resistance. These approaches have the greatest potential to enhance the efficacy of insecticide-based control tactics and/or will serve to increase the number of different tactics that growers can employ to reduce swede midge survival and limit population growth.

3.6.1 Pheromone-based action thresholds

As canola is vulnerable to swede midge from the time that the first true leaves are produced until all buds have begun to open, and as swede midge may be present throughout this entire period, it is necessary to establish a pheromone-based action threshold that can be used to time insecticide applications optimally in order to effectively protect against yield losses while being economically viable for growers. The documentation of density-dependent effects on damage and yield indicates that an action threshold approach should be effective for swede midge on canola and threshold development trials are underway (J. Williams and R.H. Hallett, in preparation). Laboratory and field trials to date suggest that the action threshold may need to be in the range of five to ten midges/trap/day (R.H. Hallett and J. Williams, unpublished data), but this needs further evaluation in the field.

Action threshold development should be informed by detailed laboratory studies examining relationships between midge density and yield impacts at different crop stages. Density-dependent effects on damage and yield have been documented after a single infestation event at the early primary bud stage (J. Williams and R.H. Hallett, 2016, unpublished results). Detailed studies on the impact of single infestations at other susceptible growth stages and of multiple infestations are needed to elucidate more fully the impact of swede midge on canola yield and the ability of canola to compensate with subsequent growth. This information is critical to determining the optimum timing of insecticide applications in conjunction with a pheromone-based action threshold.

Action thresholds may be affected by insecticide efficacy (Nault and Shelton, 2010; Hallett and Sears, 2013) and it may be necessary to re-evaluate action thresholds should new active ingredients receive registration in the future. Control failures have occurred under high swede midge populations (Hallett *et al.*, 2009b), thus the multiple insecticide applications that may be prescribed by the action threshold may not achieve adequate control nor be economically feasible when swede midge populations are high. The economics and success of employing an action threshold will thus be affected by insecticide efficacy as well and will need to be re-evaluated when any new active ingredients are registered. However, availability of a pheromone action threshold should improve pest management and ensure that insecticides are employed only when necessary, thereby reducing potential impacts on pollinators and other beneficial insects found in canola agroecosystems.

3.6.2 Systemic insecticides

Effective management of swede midge with insecticides is difficult due to the protected nature of larval feeding, which makes spray penetration and larval exposure to insecticides difficult. Cole crop growers in Canada have access to five different active ingredients registered against swede midge, including acetamiprid and spirotetramat, which have systemic activity (OMAF and MRA, 2014a). Canola growers have access to two insecticides, lambda-cyhalothrin and chlorantraniliprole (OMAF and MRA, 2014b), neither of which has systemic activity, although the latter is translaminar. Spirotetramat is particularly effective against swede midge in both canola and cole crops (R.H. Hallett, unpublished data). The two neonicotinoids, acetamiprid and imidacloprid, have known efficacy against swede midge in cole crops as both foliar and drench applications (Hallett *et al.*, 2009b; Heal *et al.*, 2011), but in several canola field trials the neonicotinoid seed treatments registered for flea beetle control in canola showed no evidence of protecting canola from swede midge damage early in the season (R.H. Hallett, unpublished data). Given the population potential in canola and observed populations in highly infested areas, the availability of a systemic insecticide seems to be critical to achieving a high level of control and effective population suppression. However, with current concerns about the effects of systemic insecticides on pollinators (Health Canada, 2014; Godfray *et al.*, 2015), it may be difficult to find a suitable systemic option in the near future for registration in canola. A systemic insecticide, applied as a foliar or seed treatment, could offer protection during the late rosette and early bud stages when buds on the primary and secondary racemes are being set. However, if a systemic insecticide is to be acceptable for use in canola, its residual titre in pollen and nectar must be within a safe level for pollinators prior to the start of flowering. Whether or not this will be achievable will depend on the difference between the concentration needed to control swede midge and that which has lethal or sublethal effects on bees.

3.6.3 Biological control agents

North American Cecidomyiidae are subjected to parasitism and predation by several natural enemies, primarily in the larval stage, as adults are very short-lived (Gagné, 1989). In terms of predators, there are several predaceous flies, including *Lestodiplosis* (Cecidomyiidae) and larval Phoridae, which prey on larval cecidomyiids, and adult Empididae that catch adult midges in flight (Gagné, 1989). Predation by a dipteran *Medetera* sp. on swede midge adults in Ontario (Goodfellow, 2005) has not been reported elsewhere and may warrant further investigation to determine abundance in the field and predation rates. Some hemipterans also feed on cecidomyiid larvae through gall tissue (Gagné, 1989). Several of these predators are known from US states where canola is grown and/or that are relatively close to Canadian canola production regions (Neiswander, 1962; Jones *et al.*, 1983), therefore surveys to determine their presence and prevalence in canola fields may be warranted, particularly those known to prey on cecidomyiids in simple galls.

Parasitoids of cecidomyiids may be generalists or specialists and belong to nine families of Hymenoptera, with the Platygasteridae, Pteromalidae and Torymidae being the most important (Gagné, 1989). The detection of parasitoids of swede midge in Canada in recent years is a positive development. Further evaluation of the species present, their ranges and parasitism rates, in canola-producing regions where swede midge is also present, are needed in order to determine the potential contribution that these natural enemies can make to swede midge population suppression and integrated pest management.

3.6.4 Planting density

Alterations to planting regimes, such as crop rotation, planting dates and planting density, can modify the suitability of an agroecosystem to a pest species and help to limit pest abundance and damage (Speight *et al.*, 2008). Since the plasticity of canola growth allows for changes in pod production and branching with plant density (Angadi *et al.*, 2003), and since swede midge females prefer meristematic tissue for oviposition, studies are needed to determine the optimum planting density of canola in areas infested with swede midge. There are two possible, and contrasting, ways in which planting density may be manipulated to limit losses to swede midge. On the one hand, a low seeding rate in canola fields with high swede midge pressure may have a positive effect on compensatory branching and yield, as at low density there would be additional space in which additional branches could be accommodated. Low planting density may be particularly beneficial in areas in which swede midge populations frequently lead to damage to the primary raceme, resulting in reduced pod set on the primary inflorescence (J. Williams and R.H. Hallett, 2016, unpublished results). A contrasting approach could be to utilize very high planting densities in order to limit branching and secondary and tertiary bud production. The reasoning to support this approach is that if less branching occurs, then there are fewer meristematic tissues present on the plant that could be infested by swede midge and thereby contribute to swede midge population growth (H.J. Earl, Ontario, 2013, personal communication). However, this approach only has the potential to be beneficial if the canola is planted early enough that damage to the primary racemes is avoided completely. If the primary racemes and primary buds are damaged, then plants may not be able to compensate through branching and secondary and tertiary pod set, which could further limit yield. Which of these approaches has the greatest potential to limit the impact of swede midge needs to be determined through field trials, and will likely vary according to field and environmental conditions affecting how early in the spring cultivation and planting of a field are possible, as well as plant stand establishment and growth rates.

3.6.5 Host plant resistance

The cultivation of insect-resistant plant varieties is a highly effective pest management tactic that can reduce herbivore damage, reproduction and/or survival, and provide significant economic returns on investment (Wiseman, 1999). Host plant resistance has been found in host plants of other Cecidomyiidae, most notably the Hessian fly, *Mayetolia disruptor* (Say), the rice gall midge, *Orseolia oryzae* (Wood-Mason) and the wheat midge, *Sitodiplosis mosellana* (Géhin) (Harris *et al.*, 2003). Host plant resistance to cecidomyiids occurs through mechanisms of both antixenosis and antibiosis. However, antibiotic resistance in wheat is frequently overcome by adaptation of *M. disruptor* biotypes (El Bouhssini *et al.*, 1998) and some *S. mosellana* larvae have been able to survive on wheat with antibiotic resistance (Smith *et al.*, 2007). Antixenotic resistance in wheat to *S. mosellana* deters oviposition through production of deterrent volatiles (Gharalari *et al.*, 2009, 2011), while antixenotic resistance to the sorghum midge, *Contarinia sorghicola* (Coquillett), in sorghum is related to glume and grain characteristics (Sharma *et al.*, 1996). Host plant resistance is considered an effective means of controlling sorghum midge populations and considerable genetic variability exists in sorghum for resistance to this insect (Sharma *et al.*, 1991). However, there are relatively simple genetics underlying resistance traits in host plants and resistance adaptation in a number of cecidomyiids (Harris *et al.*, 2003), therefore whether or not sources of durable resistance to swede midge exists in a cruciferous host remains to be investigated.

Preliminary evaluation of canola varieties for resistance to swede midge revealed differences in susceptibility, but did not reveal any sources of strong resistance (R.H. Hallett and H.J. Earl, unpublished data). Further screening for resistance in a broader range of accessions and genetic material may uncover resistance traits in the Brassicaceae. Given that swede midge is a crucifer specialist, has a high population potential in canola and can survive on weedy hosts in the landscape, identification or development of canola with antibiotic resistance seems to have greater potential to manage swede midge successfully than canola with an antixenotic mechanism of resistance, though use of susceptible refugia would likely be necessary (Cerda and Wright, 2004).

A strain of *Bacillus thuringiensis israeliensis* showed insecticidal activity against swede midge (Wu *et al.*, 2006); however, three Bt insecticidal proteins (Cry2Ab, Cry4Ba and Cry11Ba) did not cause mortality in swede midge larvae and thus were not suitable for expression in a crucifer host (Tian *et al.*, 2011). The investigation of other insecticidal proteins with activity against dipterans may reveal more promising candidates for development of swede midge-resistant transgenic canola. The use of genetic engineering may be more promising for the development of antibiotic resistance in canola and could be achieved through transgenic approaches or other genetic manipulation, such as the use of RNA interference (Huvenne and Smagghe, 2010; Terenius *et al.*, 2011).

3.7 Conclusion

The swede midge is a highly successful invasive insect in North America, capable of causing serious damage to yields in canola and other cruciferous crops. If recent trends continue, then further range expansion and increasing populations of swede midge in the Canadian Prairies and adjacent US states can be expected. Based on experiences in eastern North America, it is important to adopt measures, such as extended crop rotation regimes, in order to limit swede midge population growth early on. Once established, swede midge requires intensive management and can be very difficult to control with insecticides, particularly in canola where the current economics of production do not allow for multiple insecticide applications. A pheromone-based action threshold is needed in order to optimize the timing of insecticide applications during vulnerable crop stages. New technologies and approaches may offer solutions to achieve improved management of swede midge that will allow for continued canola production in heavily infested areas.

Acknowledgements

I am very grateful to my technicians, Jon Williams and James Heal, for their contributions to our swede midge and canola research programme, and for their insights and discussions of swede midge biology and management. I wish to thank Julie Soroka and Genevieve Labrie, who allowed me to include some of their observations on swede midge in their regions in this chapter, and Jon Williams and Hugh Earl, who have allowed me to include their insights on planting density considerations. I am also grateful to my many colleagues and collaborators who have engaged in discussions about swede midge and contributed to my research programme in various ways, including Mark Sears, Hugh Earl, Tony Shelton, Julie Soroka, Ross Weiss, Lars Andreasson, Owen Olfert, Jon Williams, Sheila Goodfellow, Lauren Des Marteaux, Braden Evans, Hannah Fraser, Brian Hall, Tracey Baute, Tim Haye and Laima Kott. I am particularly grateful for the longstanding support of Carrie James and the Ontario Canola Growers Association.

Trials examining effects of swede midge on winter canola, and planting date effects on swede midge damage (Tables 3.1, 3.3 and 3.4) were conducted in collaboration with Hugh J. Earl and Mark K. Sears, of the Department of Plant Agriculture and the former Department of Environmental Biology, respectively, University of Guelph.

My research programme on swede midge has been supported by the OMAFRA–University of Guelph Research Partnership, OMAFRA New Directions, the Growing Forward 2 programme in partnership with the Canola Council of Canada and the Eastern Canada Oilseeds Development Association, the Canola Council of Canada's Canola Agronomic Research Programme and the Ontario Canola Growers Association.

References

AAFC (Agriculture & Agri-Food Canada) (2015) 2015 Forecast and risk maps for insect pests of Prairie field crops. Available at: http://www.westernforum.org/IPMNMaps.html (accessed 5 March 2016).

Abram, P.K., Haye, T., Mason, P.G., Cappuccino, N., Boivin, G. and Kuhlmann, U. (2012) Identity, distribution, and seasonal phenology of parasitoids of the swede midge, *Contarinia nasturtii* (Kieffer) (Diptera: Cecidomyiidae) in Europe. *Biological Control* 62, 197–205.

Abram, P.K., Boivin, G., Haye, T. and Mason, P.G. (2013) *Contarinia nasturtii* Kieffer, swede midge (Diptera: Cecidomyiidae). In: Mason, P.G. and Gillespie, D.R. (eds) *Biological Control Programmes in Canada 2001–2012*. CAB International, Wallingford, Oxfordshire, UK, pp. 134–138.

Angadi, S.V., Cutforth, H.W., McConkey, B.G. and Gan, Y. (2003) Yield adjustment by canola grown at different plant populations under semiarid conditions. *Crop Science* 43, 1358–1366.

Barnes, H.F. (1946) *Gall Midges of Economic Importance. Vol. I: Gall Midges of Root and Vegetable Crops*. Crosby, Lockwood & Son, London.

Barnes, H.F. (1952) Studies of fluctuations in insect populations XII. Further evidence of prolonged larval life in the wheat-blossom midges. *Annals of Applied Biology* 39, 370–373.

Boddum, T., Skals, N., Wirén, M., Baur, R., Rasucher, S. and Hillbur, Y. (2009) Optimisation of the pheromone blend of the swede midge, *Contarinia nasturtii*, for monitoring. *Pest Management Science* 65, 851–856.

Bovien, P. and Knudsen, P. (1950) Krusesygegalmyggen (*Contarinia nasturtii* Kieff.), dens biologi og bekaempelse. *Tidsskrift for Planteavl* 53, 235–257.

Brown, J.E. (2014) The midge that eats more kale. Available at: http://phys.org/news/2014-11-midge-kale.html (accessed 4 March 2016).

Canola Council of Canada (2011) *Canola Grower's Manual*. Available at: http://www.canolacouncil.org/crop-production/canola-grower's-manual-contents/ (accessed 4 March 2016).

Cerda, H. and Wright, D.J. (2004) Modeling the spatial and temporal location of refugia to manage resistance in Bt transgenic crops. *Agriculture, Ecosystems and Environment* 102, 163–174.

CFIA (2002) Plant Health Risk Assessment: *Contarinia nasturtii* (Kieffer), the swede midge. Plant Health Risk Assessment Unit, Canadian Food Inspection Agency, Ottawa, Canada, PHPD Request: 2002-04.

CFIA (2004a) Swede midge / Cécidomyie du chou-fleur, *Contarinia nasturtii* (Keiffer), Ontario 2004. Regulated Area Map, Canadian Food Inspection Agency, Ottawa, Canada, 2004-08-23.

CFIA (2004b) Swede midge / Cécidomyie du chou-fleur, *Contarinia nasturtii* (Keiffer), Québec 2004. Regulated Area Map, Canadian Food Inspection Agency, Ottawa, Canada, 2004-09-16.

CFIA (2007) Swede midge / Cécidomyie du chou-fleur, *Contarinia nasturtii* (Keiffer), Ontario 2007. Regulated Area Map, Canadian Food Inspection Agency, Ottawa, Canada.

CFIA (2009) RMD-08-03: Review of the pest status of the swede midge (*Contarinia nasturtii*) in Canada. Canadian Food Inspection Agency, Ottawa, Canada. Available at: http://www.inspection.gc.ca/plants/plant-pests-invasive-species/directives/risk-management/rmd-08-03/eng/1304794114305/1304822057238 (accessed 4 March 2016).

Chen, M. and Shelton, A.M. (2007) Impact of soil type, moisture, and depth on swede midge (Diptera: Cecidomyiidae) pupation and emergence. *Environmental Entomology* 36, 1349–1355.

Chen, M. and Shelton, A.M. (2010) Effect of insect density, plant age, and residue duration on acetamiprid efficacy against swede midge. *Journal of Economic Entomology* 103, 2107–2111.

Chen, M., Zhao, J.Z. and Shelton, A.M. (2007) Control of *Contarinia nasturtii* Keiffer (Diptera: Cecidomyiidea) by foliar sprays of acetamiprid on cauliflower transplants. *Crop Protection* 126, 1574–1578.

Chen, M., Shelton, A.M., Wang, P., Hoepting, C.A., Kain, W.C. and Brainard, D.C. (2009a) Occurrence of the new invasive insect *Contarinia nasturtii* (Diptera: Cecidomyiidae) on cruciferous weeds. *Journal of Economic Entomology* 102, 115–120.

Chen, M., Weiwei, L. and Shelton, A.M. (2009b) Simulated crop rotation systems control swede midge, *Contarinia nasturtii. Entomologia Experimentalis et Applicata* 133, 84–91.

Chen, M., Shelton, A.M., Hallett, R.H., Hoepting, C.A., Kikkert, J.R. and Wang, P. (2011) Swede midge (Diptera: Cecidomyiidae), ten years of invasion of crucifer crops in North America. *Journal of Economic Entomology* 104, 709–716.

Corlay, F. and Boivin, G. (2008) Seasonal development of an invasive exotic species, *Contarinia nasturtii* (Diptera: Cecidomyiidae), in Quebec. *Environmental Entomology* 37, 907–913.

Corlay, F., Boivin, G. and Bélair, G. (2007) Efficiency of natural enemies against the swede midge *Contarinia nasturtii* (Diptera: Cecidomyiidae), a new invasive species in North America. *Biological Control* 43, 195–201.

Des Marteaux, L.E., Habash, M.B., Schmidt, J.M. and Hallett, R.H. (2014) Patterns of diapause frequency and emergence in swede midges of southern Ontario. *Agricultural & Forest Entomology* 17, 77–89. doi: 10.1111/afe.12083.

DuPont (2015) Coragen Canola and Cereal Technical Bulletin. Available at: http://www.dupont.ca/content/dam/dupont/products-and-services/crop-protection/documents/ca/cp_PSD-67_Coragen_Canola_Cereal_Technical_Bulletin_2015_E.pdf (accessed 17 April 2016).

El Bouhssini, M., Hatchett, J.H. and Wilde, G.E. (1998) Survival of Hessian fly (Diptera: Cecidomyiidae) larvae on wheat cultivars carrying different genes for antibiosis. *Journal of Agricultural Entomology* 15, 183–193.

Elton, C.S. (1958) *The Ecology of Invasions by Animals and Plants*. Methuen, London.

Frey, J.E., Frey, B. and Baur, R. (2004) Molecular identification of the swede midge (Diptera: Cecidomyiidae). *The Canadian Entomologist* 136, 771–780.

Gagné, R.J. (1989) *The Plant-Feeding Gall Midges of North America*. Cornell University Press, Ithaca, New York.

Gavloski, J. (2014) Summary of insects on crops in Manitoba in 2014. *Manitoba Agriculture, Food and Rural Development*, November 2014. Available at: http://www.gov.mb.ca/agriculture/crops/insects/pubs/2014summary.pdf (accessed 5 March 2016).

Gavloski, J. (2015) Summary of insects on crops in Manitoba in 2015. *Manitoba Agriculture, Food and Rural Development*, November 2015. Available at: http://www.gov.mb.ca/agriculture/crops/insects/pubs/2015summary.pdf (accessed 5 March 2016).

Gemmar, A. and Koch, I. (2002a) Incidence of the cabbage dry-heart midge in broccoli in the Palatinate. *Gemuse (Munchen)* 38, 12–13.

Gemmar, A. and Koch, I. (2002b) The swede midge can be monitored by heat sum models. *Gemuse (Munchen)* 38, 26–27.

Gharalari, A.H., Fox, S.L., Smith, M.A.H. and Lamb, R.J. (2009) Oviposition deterrence in spring wheat, *Triticum aestivum*, against orange wheat blossom midge, *Sitodiplosis mosellana*: implications for inheritance of deterrence. *Entomologia Experimentalis et Applicata* 133, 74–83.

Gharalari, A.H., Smith, M.A.H., Fox, S. and Lamb, R. (2011) Volatile compounds from non-preferred wheat spikes reduce oviposition by *Sitodiplosis mosellana*. *The Canadian Entomologist* 143, 388–391.

Godfray, H.C.J., Blacquière, T., Field, L.M., Hails, R.S., Potts, S.G., Raine, N.E., Vanbergen, A.J. and McLean, A.R. (2015) A restatement of recent advances in the natural science evidence base concerning neonicotinoid insecticides and insect pollinators. *Proceedings of the Royal Society B: Biological Sciences* 282, 20151821. Available at: http://dx.doi.org/10.1098/rspb.2015.1821.

Goodfellow, S.A. (2005) Population dynamics and predictive modeling of the swede midge, *Contarinia nasturtii* (Kieffer), in Ontario. MSc thesis, University of Guelph, Guelph, Ontario, Canada.

Hallett, R.H. (2007) Host plant susceptibility to the swede midge (Diptera: Cecidomyiidae). *Journal of Economic Entomology*, 100, 1335–1343.

Hallett, R.H. (2016) Northern swede midge populations and Ontario swede midge research. *Ontario Canola Growers Association Newsletter*, January 2016, p. 10. Available at: https://media.wix.com/ugd/241082_b17ba108041b43938cabd26197a7ecb6.pdf (accessed 5 March 2016).

Hallett, R.H. and Hall, B. (2014) Swede midge treatment recommendations. Available at: https://media.wix.com/ugd/241082_bfa51458b2fa476eb94e95b831dea6a9.pdf (accessed 5 March 2016).

Hallett, R.H. and Heal, J.D. (2001) First nearctic record of the swede midge, *Contarinia nasturtii* (Kieffer) (Diptera: Cecidomyiidae), a pest of cruciferous crops in Europe. *The Canadian Entomologist* 133, 713–715.

Hallett, R.H. and Sears, M.K. (2013) Pheromone-based action thresholds for control of the swede midge,

Contarinia nasturtii (Diptera: Cecidomyiidae), and residual insecticide efficacy in cole crops. *Journal of Economic Entomology* 106, 267–276.

Hallett, R.H., Goodfellow, S.A., Weiss, R.M. and Olfert, O. (2009a) MidgEmerge, a new predictive tool, indicates the presence of multiple emergence phenotypes of the overwintered generation of swede midge. *Entomologia Experimentalis et Applicata* 130, 81–97.

Hallett, R.H., Chen, M., Sears, M.K. and Shelton, A.M. (2009b) Field efficacy of insecticides for control of the swede midge (Diptera: Cecidomyiidae) on cole crops. *Journal of Economic Entomology* 102, 2241–2254.

Harper, F.R. and Berkenkamp, B. (1975) Revised growth stage key for *Brassica campestris* and *B. napus. Canadian Journal of Plant Science* 55, 657–658.

Harris, M.O., Stuart, J.J., Mohan, M., Nair, S., Lamb, R.J. and Rohfritsch, O. (2003) Grasses and gall midges: plant defense and insect adaptation. *Annual Review of Entomology* 48, 549–577.

Heal, J.D., Hallett, R.H., Shelton, A.M. and Olmstead, D. (2011) Control of swede midge on broccoli, 2010. *Arthropod Management Tests* 36, Report #E3. doi: 10.4182/amt.2011.E3.

Health Canada (2014) Update on neonicotinoid pesticides and bee health. Canadian Electronic Library, Health Canada, Ottawa, Ontario. Available at: http://www.hc-sc.gc.ca/cps-spc/pubs/pest/_fact-fiche/neonicotinoid/neonicotinoid-eng.php (accessed 5 March 2016).

Hillbur, Y., Celander, M., Baur, R., Rauscher, S., Haftmann, J., Franke, S. and Francke, W. (2005) Identification of the sex pheromone of the swede midge, *Contarinia nasturtii*. *Journal of Chemical Ecology* 31, 1807–1828. doi: 10.1007/s10886-005-5928-3.

Huvenne, H. and Smagghe, G. (2010) Mechanisms of dsRNA uptake in insects and potential of RNAi for pest control: a review. *Journal of Insect Physiology* 56, 227–235.

Jones, R.G., Gagné, R.J. and Barr, W.F. (1983) A systematic and biological study of the gall midges (Cecidomyiidae) of *Artemisia tridentata* Nuttall (Compositae) in Idaho. *Contributions of the American Entomological Institute* 81, 10–79.

Kikkert, J.R., Hoepting, C.A., Wu, Q.J., Wang, P., Baur, R. and Shelton, A.M. (2006) Detection of *Contarinia nasturtii* (Diptera; Cecidomyiidae) in New York, a new pest of cruciferous plants in the United States. *Journal of Economic Entomology* 99, 1310–1315.

MDA (2016) *Swede midge*. Minnesota Department of Agriculture, St Paul, Minnesota. Available at: http://www.mda.state.mn.us/plants/insects/swedemidge.aspx (accessed 5 March 2016).

MDARD (2015) *New invasive vegetable pest found in Michigan*. Michigan Department of Agriculture, Lansing, Michigan. Available at: http://www.michigan.gov/som/0,4669,7-192-47796-366419--,00.html (accessed 4 March 2016).

Mika, A.M., Weiss, R.M., Olfert, O., Hallett, R.H. and Newman, J.A. (2008) Will climate change be beneficial or detrimental to swede midge in North America? Contrasting predictions using climate projections from different general circulation models. *Global Change Biology* 14, 1721–1733.

MIPOP (2007) *Pest Alert: swede midge found in Connecticut* (June 6, 2007). Massachusetts Introduced Pests Outreach Project, Massachusetts Department of Agricultural Resources, Boston, Massachusetts. Available at: http://massnrc.org/pests/linkeddocuments/pestalerts/swedemidgeinCT.htm (accessed 4 March 2016).

Nault, B.A. and Shelton, A.M. (2010) Impact of insecticide efficacy on developing action thresholds for pest management: a case study of onion thrips on onions. *Journal of Economic Entomology* 103, 1315–1326.

Neiswander, C.R. (1962) The phlox midge, *Hyperdiplosis phlox* Greene, and its insect enemies. *Ohio Agricultural Experimental Station Research Bulletin* 915, 10–19.

OCGA (2011) Cost of production crop comparison chart. *Ontario Canola Growers Association Newsletter*, January 2011, p. 6.

OCGA (2016) Crop budgets comparison chart. *Ontario Canola Growers Association Newsletter*, January 2016, p. 9. Available at: https://media.wix.com/ugd/241082_b17ba108041b43938cabd26197a7ecb6.pdf (accessed 5 March 2016).

Olfert, O.O., Hallett, R.H., Weiss, R., Soroka, J.J. and Goodfellow, S. (2006) Potential distribution and relative abundance of swede midge, *Contarinia nasturtii* (Diptera: Cecidomyiidae), an invasive pest in Canada. *Entomologia Experimentalis et Applicata* 120, 221–228.

OMAF and MRA (2014a) *Vegetable Crop Protection Guide, 2014–2015*. Publication 838. Ontario Ministry of Agriculture & Food and Ministry of Rural Affairs, Guelph, Ontario, Canada.

OMAF and MRA (2014b) *Field Crop Protection Guide, 2014–2015*. Publication 812. Ontario Ministry of Agriculture & Food and Ministry of Rural Affairs, Guelph, Ontario, Canada..

OMAFRA (2006) *Ontario field crop report 2006 canola and edible bean seasonal summary*. Ontario Ministry of Agriculture, Food & Rural Affairs, Guelph, Ontario, Canada. Available at: http://www.omafra.gov.on.ca/english/crops/field/reports/2006summary-b.htm (accessed 4 March 2016).

OMAFRA (2008) *Ontario field crop report 2008 edible beans and canola seasonal summary*. Ontario Ministry of Agriculture, Food & Rural Affairs, Guelph, Ontario, Canada. Available at: http://www.omafra.gov.on.ca/english/crops/field/reports/2008summary-a.htm#canolasummary (accessed 4 March 2016).

OMAFRA (2009) *2009 Canola seasonal summary*. Ontario Ministry of Agriculture, Food & Rural Affairs, Guelph,

Ontario, Canada. Available at: http://www.omafra.gov.on.ca/english/crops/field/reports/2009summary-canola.htm (accessed 4 March 2016).

OMAFRA (2010) *2010 Canola seasonal summary*. Ontario Ministry of Agriculture, Food & Rural Affairs, Guelph, Ontario, Canada. Available at: http://www.omafra.gov.on.ca/english/crops/field/reports/2010summary-canola.htm (accessed 4 March 2016).

OMAFRA (2011a) *Canola: area and production, by county, 2011*. Ontario Ministry of Agriculture, Food & Rural Affairs, Guelph, Ontario, Canada. Available at: http://www.omafra.gov.on.ca/english/stats/crops/cty-canola11.htm (accessed 4 March 2016).

OMAFRA (2011b) *2011 Canola seasonal summary*. Ontario Ministry of Agriculture, Food & Rural Affairs, Guelph, Ontario, Canada. Available at: http://www.omafra.gov.on.ca/english/crops/field/reports/2011summary-canola.htm (accessed 4 March 2016).

OMAFRA (2013) *2013 Canola and edible bean seasonal summary*. Ontario Ministry of Agriculture, Food & Rural Affairs, Guelph, Ontario, Canada. Available at: http://www.omafra.gov.on.ca/english/crops/field/reports/2013summary-can&edbean.htm#pest (accessed 4 March 2016).

OMAFRA (2014a) *Historical provincial estimates by crop, 1981–2014 (Metric Units)*. Ontario Ministry of Agriculture, Food & Rural Affairs, Guelph, Ontario, Canada. Available at: http://www.omafra.gov.on.ca/english/stats/crops/estimate_hist_metric.htm (accessed 12 April 2016).

OMAFRA (2014b) *2014 Canola and edible bean seasonal summary*. Ontario Ministry of Agriculture, Food & Rural Affairs, Guelph, Ontario, Canada. Available at: http://www.omafra.gov.on.ca/english/crops/field/reports/2014summary-canola&ediblebean.htm#pest (accessed 4 March 2016).

OMAFRA (2015) *Estimated area, yield, production and farm value of specified field crops, Ontario, 2011–2015 (imperial and metric units)*. Ontario Ministry of Agriculture, Food & Rural Affairs, Guelph, Ontario, Canada. Available at: http://www.omafra.gov.on.ca/english/stats/crops/estimate_new.htm (accessed 12 April 2016).

Phillips, T. (2015) President's message. *Ontario Canola Growers Association Newsletter*, January 2015, p. 1. Available at: http://www.ontariocanolagrowers.ca/linked/jan_15_newsletter.pdf (accessed 5 March 2016).

Readshaw, J.L. (1961) The biology and ecology of the swede midge, *Contarinia nasturtii* (Kieffer) (Diptera: Cecidomyidae). PhD thesis, University of Durham, Durham, UK.

Readshaw, J.L. (1966) The ecology of the swede midge, *Contarinia nasturtii* (Kieff.) (Diptera, Cecidomyiidae). I. Life-history and influence of temperature and moisture on development. *Bulletin of Entomological Research* 56, 685–700.

Readshaw, J.L. (1968) Damage to swedes by the swede midge, *Contarinia nasturtii* (Kieff.), and a possible method of cultural control. *Bulletin of Entomological Research* 58, 25–29.

Rogerson, J.P. (1963) Swede midge on two Northumberland farms, 1959–1961. *Plant Pathology* 12, 161–171.

Rygg, T.D. and Braekke, H.P. (1980) Swede midge (*Contarinia nasturtii* Kieffer) (Diptera, Cecidomyiidae): investigations on biology, symptoms of attack and effects on yield. *Meldinger fra Norges landbrukshøgskole* 59, 1–9.

Sakai, A.K., Allendorf, F.W., Holt, J.S., Lodge, D.M., Molofsky, J., With, K.A., Baughman, S., Cabin, R.J., Cohen, J.E., Ellstrand, N.C., McCauley, D.E., O'Neil, P., Parker, I.M., Thompson, J.N. and Weller, S.G. (2001) The population biology of invasive species. *Annual Review of Ecology and Systematics* 32, 305–332.

Sharma, H.C., Vidyasagar, P. and Leuschner, K. (1991) Componential analysis of the factors influencing resistance to sorghum midge, *Contarinia sorghicola* Coq. *Insect Science and Its Application* 11, 889–898.

Sharma, H.C., Abraham, C.V., Vidyasagar, P. and Stenhous, J.W. (1996) Gene action for resistance in sorghum to midge, *Contarinia sorghicola*. *Crop Science* 36, 259–265.

Smith, M.A.H., Wise, I.L. and Lamb, R.J. (2007) Survival of *Sitodiplosis mosellana* (Diptera: Cecidomyiidae) on wheat (Poaceae) with antibiosis resistance: implication for the evolution of virulence. *The Canadian Entomologist* 139, 133–140.

Solbreck, C. and Widenfalk, O. (2012) Very long diapause and extreme resistance to population disturbance in a galling insect. *Ecological Entomology* 37, 51–55.

Speight, M.R., Hunter, M.D. and Watt, A.D. (2008) *Ecology of Insects*, 2nd edn. Wiley, Chichester, UK.

Stokes, B.M. (1953) The host plant range of the swede midge (*Contarinia nasturtii* Kieffer) with special reference to types of plant damage. *Tijdschrift over plantenziekten* 59, 82–90.

Terenius, O., Papanicolaou, A., Garbutt, J.S., Eleftherianos, I., Huvenne, H. *et al.* (2011) RNA interference in Lepidoptera: an overview of successful and unsuccessful studies and implications for experimental design. *Journal of Insect Physiology* 57, 231–245.

Theunissen, J., Ouden, H.D. and Schelling, G. (1997) Can the cabbage gall midge, *Contarinia nasturtii* (Diptera, Cecidomyiidae) be controlled by host plant deprivation? *Mededelingen van de Faculteit der Landbouwkundige en Toegepaste Biologische Wetenschappen Universiteit Gent* 62, 617–622.

Tian, J., Chen, M. and Shelton, A.M. (2011) Cry2Ab, Cry4Ba and Cry11Ba proteins are not toxic to swede midge. *Arthropod Management Tests* 36, L14.

Williams, J.L. (2015) Impact of crop phenology and infestation timing on damage and pest management of the swede midge, *Contarinia nasturtii* (Diptera: Cecidomyiidae), in canola. MSc thesis, University of Guelph, Guelph, Ontario, Canada.

Wiseman, B.R. (1999) Mechanisms of plant resistance against arthropod pests. In: Ruberson, J.R. (ed.) *Handbook of Pest Management*. Marcel Dekker Inc., New York, pp. 155–174.

Wu, Q.J., Zhao, J.Z., Taylor, A.G. and Shelton, A.M. (2006) Evaluation of insecticides and application methods against *Contarinia nasturtii* (Diptera: Cecidomyiidae), a new invasive insect pest in the United States. *Journal of Economic Entomology* 99, 117–122.

Wyss, E. and Daniel, C. (2004) The effect of exclusion fences on the colonization of broccoli and kohlrabi by the swede midge, *Contarinia nasturtii* (Diptera: Cecidomyiidae) [Die Wirksamkeit von Einflugbarrieren gegen die Besiedlung von Broccoli und Kohlrabi durch die Kohldrehherzgallmücke *Contarinia nasturtii* (Diptera: Cecidomyiidae)]. *Mitteilungen der Deutschen Gesellschaft für allgemeine und angewandte Entomologie* 14, 387–390.

4 Biology and Management of Sucking Pests of Canola

SURENDRA K. DARA*

University of California Cooperative Extension, San Luis Obispo, California, USA

4.1 Sucking Pests of Canola

A complex of insect pests attack canola worldwide, including species of Coleoptera, Diptera, Hemiptera, Hymenoptera, Lepidoptera, Orthoptera and Thysanoptera (Burgess and Weegar, 1988; Lamb, 1989; Schwartz and Foottit, 1992; Buntin and Raymer, 1994; Miles and McDonald, 1999; Cárcamo *et al.*, 2002; Weiss *et al.*, 2006). The sucking pests that feed on canola leaves, stems, buds, flowers, pods or seeds (Table 4.1) include several Hemiptera (such as aphids and plant bugs), with piercing and sucking mouthparts, and some thrips, with rasping and sucking mouthparts.

The importance of these different pests and their presence varies in different countries and even within local regions. Several aphids, mirids and thrips are pests of canola in different countries (Burgess and Weegar, 1988; Lamb 1989; Buntin and Raymer, 1994; Talpur, 2001; Fathi *et al.*, 2011) (Table 4.1). However, a few pests occur only in one location, such as the aster leafhopper (*Macrosteles quadrilineatus* Forbes) in Canada (Gavloski *et al.*, 2011), the painted bug (*Bagrada cruciferarum* Kirk.) in Pakistan (Saljoqi *et al.*, 2006) and two lygaeids, the Rutherglen bug (*Nysius vinitor* Bergroth) and the cluster bug (*N. clevelandensis* Evans), in Australia (Miles and McDonald, 1999; Baker, 2009).

4.2 Pest Damage

Pest damage from sucking pests can be significant in one or more of the following growth stages of canola (Canola Council of Canada, 2016): 0 – pre-emergence; 1 – seedling; 2 – rosette; 3 – bud; 4 – flower; and 5 – ripening. Each stage is further divided from 0 to 9. Temperature and other environmental factors influence the duration of each growth stage and the occurrence of pests and their damage during those stages.

Aphids feed on both inter- and intracellular plant sap to obtain nitrogenous compounds (in the form of amino acids and amides) and sugars (Auclair, 1963). Lygaeid bugs suck plant sap from various parts of the plant for their amino acid, lipid and carbohydrate requirements (Cohen, 2015). Thrips rasp on plant tissues and suck plant sap from ruptured cells (Kirk, 1995). Thrips also feed on pollen and seed to meet their high nitrogen needs. Therefore, canola, with its rich lipid and amino acid content, offers a good source of food for hemipteran and thysanopteran pests.

4.2.1 Aphids

Aphids mainly infest leaves, stems, flowers and pods of canola. Their feeding stunts plant height, delays development and causes withering of buds and flowers, flower abortion, low pod set and pod deformation (Buntin and Raymer, 1994; Harries *et al.*, 2016). Aphid feeding also leads to reduced seed size, quality and oil content along with associated yield losses. Some earlier studies in Australia (Berlandier and Cartwright, 1998; Berlandier and Valentine, 2003) suggested that aphids often do not cause significant yield losses, as sufficient soil moisture or rainfall allows plants to compensate for aphid damage by producing more flowers. However, other studies in Australia (Jenkins *et al.*, 2011; Micic and Wahlsten, 2015) and the USA (Buntin and Raymer, 1994; Brown *et al.*, 1999) have found that aphid infestations can cause significant yield losses and not

*E-mail: skdara@ucdavis.edu

Table 4.1. Major sucking pests of canola[a].

Common name	Scientific name	Family	Order
Aphids			
Cabbage aphid	*Brevicoryne brassicae* (L.)	Aphididae	Hemiptera
Cowpea aphid	*Aphis craccivora* Koch		
Green peach aphid	*Myzus persicae* (Sulzer)		
Turnip aphid	*Lipaphis erysimi* (Kaltenbach)		
Mirid and lygaeid bugs			
Lygus bug	*Lygus borealis* (Kelton)	Miridae	Hemiptera
Pale legume bug	*Lygus elisus* Van Duzee		
Tarnished plant bug	*Lygus lineolaris* (Palisot de Beauvois)		
N/A	*Lygus keltoni* Schwartz and Foottit		
N/A	*Lygus rubrosignatus* Knight		
N/A	*Lygus shulli* Knight		
N/A	*Lygus solidaginis* (Kelton)		
Grey cluster bug	*Nysius clevelandensis* Evans	Lygaeidae	
Rutherglen bug	*Nysius vinitor* Bergroth		
Other sucking insects			
Aster leafhopper	*Macrosteles quadrilineatus* Forbes	Cicadellidae	
N/A	*Balclutha* sp. Kirkaldy		
Painted bug	*Bagrada cruciferarum* Kirk.	Pentatomidae	
Thrips			
Banded thrips	*Aeolothrips fasciatus* (L.)	Aeolothripidae	Thysanoptera
Barley thrips	*Limothrips denticornis* Haliday	Thripidae	
Flower thrips	*Frankliniella tritici* (Fitch)		
Grass thrips	*Anaphothrips obscurus* (Müller)		
Lupin flower thrips	*Odontothrips loti* (Haliday)		
Onion thrips	*Thrips tabaci* Linderman		
Plague thrips	*Thrips imaginis* Bagnall		
Rose thrips	*Thrips* nr. *fuscipennis* Haliday		
Western flower thrips	*Frankliniella occidentalis* (Pergande)		
White flower thrips	*Thrips vulgatissimus* Haliday		

[a]**References:** Burgess and Weegar, 1988; Lamb, 1989; Schwartz and Foottit, 1992; Buntin and Raymer, 1994; Miles and McDonald, 1999; Cárcamo *et al.*, 2002; Saljoqi *et al.*, 2006; Weiss *et al.*, 2006; Baker, 2009; Gavloski *et al.*, 2011

all canola cultivars can compensate for aphid damage. A recent study in Australia reported that heavy infestations of the cabbage aphid, *Brevicoryne brassicae* (L.), caused significant yield losses even in a high rainfall environment (Harries *et al.*, 2016). Plants are especially vulnerable to aphid damage during the seedling and rosette stages. Depending on the crop stage and the level of infestation, yield loss can be as high as 33% (Micic and Mangano, 2016). However, when the green peach aphid, *Myzus persicae* (Sulzer), was involved in the transmission of *Beet western yellows virus* (BWYV), up to 46% of yield loss in seed was seen from the combination of aphid feeding and disease (Jones et al., 2007). Viral infection causes a reduction in oil and an increase in erucic acid content in the seed and yield loss is estimated to be 6–12 kg/ha for each 1% increase in BWYV infection (Coutts, 2015).

BWYV is not seed-borne, but as aphids migrate to canola from wild radish (*Raphanus raphinistrum* L.) or other hosts in the vicinity they can transmit the virus to canola (Coutts and Jones, 2000). Several cultivated and wild hosts also serve as a source of BWYV. While *M. persicae* is the primary vector of BWYV, *B. brassicae* and the cowpea aphid, *Aphis craccivora* Koch, can also contribute to late-season virus infections (Baker, 2009; Coutts, 2015). Both species of these aphids as well as the turnip aphid, *Lipaphis erysimi* (Kaltenbach), vector *Cauliflower mosaic virus* (CaMV) and *Turnip mosaic virus* (TuMV) (Baker, 2009) in canola.

4.2.2 Mirid and lygaeid bugs

Mirid bugs (*Lygus* spp.) mainly feed on buds, flowers, pods or developing seeds. Damage results in bud or flower blasting (where buds and flowers become white and fail to develop), delay in flowering, flower or pod drop, collapsed or non-viable seed and shrunken or shrivelled pods. *Lygus* spp. can also feed on the growing tips of seedlings and affect plant growth and development. Flowering to early pod stage is the period most vulnerable to damage by *Lygus* spp. when older nymphs and adults are most common (Butts and Lamb, 1991; Weiss *et al.*, 2006). Collapsed seeds are more frequently found in small pods in the upper part of the canola canopy (Turnock *et al.*, 1995). In field studies conducted in Canada, *Lygus* spp. caused 6–20% yield loss (Butts and Lamb, 1991; Wise and Lamb, 1998).

Lygaeid bugs reported from Australia include *N. vinitor* and *N. clevelandensis*, which feed on vegetative parts, flowers, developing pods and seeds (Baker, 2009; National Invertebrate Pest Initiative, 2014). This feeding results in aborted flowers and decreased pod set, reduced seed development and lower seed viability. It can also affect the quality and quantity of oil. Between the two *Nysius* species, *N. vinitor* is considered the more important one in Australia (Baker, 2009) and heavy infestations can lead to plant death during the seedling stage. Nymphs migrating from weeds can cause significant damage to canola on the field margins.

4.2.3 Thrips

Thrips feed on various plant parts and this reduces photosynthesis and pollination, eventually reducing yield (Fathi et al., 2011). Thrips feeding on developing pods results in twisted or curled pods and can cause premature pod drop. Earlier Canadian studies conducted in Alberta and Saskatchewan listed *Frankliniella tritici* (Fitch) as an abundant species of thrips in canola fields (Burgess and Weegar, 1988; Olfert *et al.*, 2002a), while the onion thrips, *Thrips tabaci* Linderman, is an important canola pest in Iran (Fathi *et al.*, 2011). While species such as *Thrips vulgatissimus* Haliday might contribute to pollination, thrips damage can cause up to 40% yield loss (Olfert *et al.*, 2002a).

4.2.4 Leafhoppers

Several species of leafhoppers also feed on canola and other crops in Saskatchewan (Olivier *et al.*, 2007; Soroka *et al.*, 2015) and some of these are vectors of plant diseases such as aster yellows, caused by '*Candidatus* Phytoplasma asteris'. While most of the leafhopper species that are found in canola occurred only at low densities, the aster leafhopper, *M. quadrilineatus* Forbes, and *Balclutha* sp. Kirkaldy were abundant. Information on their feeding damage is, however, not available.

In general, it takes 3–5 weeks for the symptoms of aster yellows infection to appear in canola and some plants can remain symptomless (Knodel, 2012). Symptoms of infection appear as witches' broom with sterile and distorted flowers and excessive branching. Pods become blue-green bladder-like structures with little or no seed. Olivier *et al.* (2007) reported that a large proportion of asymptomatic canola plants carried aster yellows phytoplasma, indicating the prevalence of the disease in Canada. With the periodic increase in leafhopper populations, the risk of disease transmission and associated yield loss can also increase.

4.3 Pest Biology, Monitoring and Economic Thresholds

4.3.1 Aphids

Aphid infestations occur on actively growing parts in the top 10–15 cm area of the plants and are more common during the flowering and pod-set stages (Weiss *et al.*, 2006; Canola Council of Canada, 2016). While *B. brassicae* and *L. erysimi* are found in dense colonies on racemes, *M. persicae* is found on the underside of the lower, older leaves (Micic and Mangano, 2016). *Myzus persicae* can be recognized by its yellowish green body colour, but both *B. brassicae* and *L. erysimi* are yellowish green, grey-green or olive green with a waxy coating on the body.

Aphids overwinter on cruciferous weed hosts and alate (winged) forms migrate to canola during spring or fall. Depending on the geographical region and environmental conditions, aphid species composition is variable and their infestations can be sporadic in patches of fields or widespread outbreaks. Aphid infestations occur during the crop establishment phase and again during flowering and pod-forming stages (Jenkins *et al.*, 2011). Populations generally build up under warm and dry conditions (Baker, 2009). Monitoring twice a week by checking five well scattered, random locations in a field and sampling at least 20 plants at each location is recommended (Baker, 2009). There is no economic threshold for aphids in Canada, but in Australia control measures are recommended when > 20% of plants have aphid infestations. The treatment threshold for *B. brassicae* is infestation of 20–50% of racemes with a length of at least 2 cm infested with aphids (Harries *et al.*, 2016).

4.3.2 Mirid and lygaeid bugs

In Canada, the seasonal abundance of *Lygus* spp. varies among different species and is also dependent on geographical region, host plant composition in that area and other factors (Otani and Cárcamo, 2011). Adults are 6–8 mm long with greenish or brownish bodies and wings. There are five nymphal instars, which are greenish and develop wing pads in late stages. Life history parameters vary among species and depend on environmental conditions and food sources, but in a laboratory study (at 25–27°C, 50–70% relative humidity and a 18L:6D photoperiod), *Lygus keltoni* Schwartz and Foottit males survived for about 34 days and females for about 48 days (Cárcamo *et al.*, 2006). The average number of nymphs that developed from the eggs of each female was 133. For *Lygus lineolaris* (Palisot de Beauvois), at 24–27°C and a 14L:10D photoperiod, mean longevity was 38 days for males and 40 for females (Slaymaker and Tugwell, 1982). Mean fecundity was 135 eggs/female.

While *L. lineolaris* is the predominant species in Saskatchewan (Braun *et al.*, 2001) in canola, *L. lineolaris* and *Lygus elisus* Van Duzee are more abundant in Manitoba (Timlick *et al.*, 1993). In Alberta, *L. lineolaris* and *L. keltoni* are the dominant species in the northern region, while *L. elisus* is dominant in the southern region (Otani and Cárcamo, 2011). Regardless of the species composition, all *Lygus* species are equally important in terms of their damage potential. They typically overwinter on alternative hosts or plant debris and move to canola to complete one or more generations, depending on the environmental conditions in the region (Weiss *et al.*, 2006; Otani and Cárcamo, 2011).

In Canada, *Lygus* damage is important from bud (stage 3) to pod ripening (5) stages, but economic thresholds are available only for flowering to early pod (4.4–5.1) and late-pod ripening (5.2) stages (Otani and Cárcamo, 2011; Canola Council of Canada, 2016). One Manitoba study reported that each *Lygus* found per sweep contributed to a 1.5% increase in collapsed seed (Turnock *et al.*, 1995). Wise and Lamb (1998) calculated that each *Lygus* in a ten-sweep sample could reduce yield by 6.35 kg/ha. Monitoring should occur from bolting to the early or mid-pod stages to assess *Lygus* populations and make treatment decisions. To monitor *Lygus* populations, ten walking sweeps with a 180° arc should be taken from each of ten sampling locations within a field. Sweeps should be taken under sunny, warm (> 15°C), calm conditions between late morning and mid-afternoon. The cost of insecticide applications and estimated returns from protected crop value can be determined based on the thresholds (Table 4.2).

Nysius vinitor is an important pest of canola in many parts of Australia, sometimes occurring in large numbers, while *N. clevelandensis* occurs only in some regions within Australia (Baker, 2009; National Invertebrate Pest Initiative, 2014). Adults of both species are about 5 mm long, greyish brown with slender bodies, dark markings, silver-grey wings and conspicuous black eyes. Nymphs are reddish brown and have swollen bodies compared with adults (Bailey, 2007). The forewings of *N. clevelandensis* have short hairs while those of *N. vinitor* are smooth. Females lay up to 400 eggs and the number of generations is variable. Both species migrate from weed hosts to canola and other crop plants in spring. Summer crops also serve as a source of infestation for winter canola. Monitoring should be conducted by examining 20 racemes at each of five to ten locations within a field. The economic threshold for *N. vinitor* is 10 adults or 20 nymphs per raceme (inflorescence) during flowering to early pod stages.

Rainfall and sufficient soil moisture during sensitive stages can compensate for losses caused by mirid and lygaeid bugs in canola. Plants under water stress are more vulnerable to pest damage and may not recover.

4.3.3 Thrips

Thrips are slender-bodied insects with fringed wings and are typically 1–2 mm long. Seasonal abundance of thrips varies with species, canola variety, local weather conditions, crop development stage and other factors (Talpur, 2001; Olfert *et al.*, 2002b; Fathi *et al.*, 2011). In Iran (at 25°C, 55% relative humidity and 16L:8D photoperiod), the generation time of *T. tabaci* varied from 21.5 to 26.5 days, female longevity from 15.1 to 25.3 days and fecundity from 18.9 to 44.1 eggs/female on different canola varieties (Fathi *et al.*, 2011). Under field conditions, many generations may occur during the cropping season. According to Olfert *et al.* (2002a), counting the number of thrips on a 20 cm

Table 4.2. Economic thresholds for *Lygus* spp. adapted from Otani and Cárcamo (2011) based on 0.715 g/m² loss per *Lygus* caught in ten sweeps during flowering to early pod stages and 0.512 g/m² during late pod ripening stage.

Control costs (CAN$)/ha	Number of *Lygus* per 10 sweeps						
	During late flowering to early pod (4.4 to 5.1) stages						
$19.77	8	6	5	4	4	3	3
$24.71	10	8	7	6	5	4	4
$29.65	12	9	8	7	6	5	5
$34.59	14	11	9	8	7	6	5
$39.54	16	13	10	9	8	7	6
$44.48	18	14	12	10	9	8	7
$49.42	20	16	13	11	10	9	8
	During late pod-ripening (5.2) stage						
$19.77	11	9	7	6	5	5	4
$24.71	14	11	9	8	7	6	5
$29.65	16	13	11	9	8	7	7
$34.59	19	15	13	11	10	9	8
$39.54	22	18	15	13	11	10	9
$44.48	25	20	16	14	12	11	10
$49.42	27	22	18	16	14	12	11
Protected crop value (CAN$)/ha	$352.42	$440.53	$528.63	$616.74	$704.85	$792.95	$881.06

section of plant stalk or raceme is a better way to monitor than using sweep nets. Demirel and Cranshaw (2005) suggested neon yellow, orange, neon orange and neon pink coloured sticky traps as effective tools for monitoring thrips in canola.

4.3.4 Leafhoppers

Increasing numbers of *M. quadrilineatus* in some areas pose the risk of aster yellows transmission (Canola Council of Canada, 2013). Adult *M. quadrilineatus* are yellowish and have four distinct black lines on the head. Recently, surveys in Canada have found an increase from 1–4% to 12% in phytoplasma-carrying *M. quadrilineatus* (Canola Council of Canada, 2013). Young plants or canola with plentiful rainfall might be attractive to *M. quadrilineatus*, but leafhoppers typically prefer wheat, barley, vegetables and other hosts for developmental and reproductive needs and leave canola as the crop matures (Knodel, 2012). No thresholds are available for *M. quadrilineatus* and information on other species of leafhoppers is also lacking because of their minor pest status.

4.4 Pest Management

Successful pest management depends on several factors, including a good understanding of the seasonal dynamics of the pest, vulnerable life stages for effective suppression, the influence of various agronomic and environmental factors and the availability of chemical and non-chemical control options. Sometimes, treatments targeting major pests can also control minor pests. Similarly, certain treatment decisions such as pesticide application can harm natural enemy populations and cause outbreaks of minor pests. An integrated pest management (IPM) approach provides a good balance of cultural practices, the application of chemical pesticides or non-chemical alternatives and the promotion of ecological conditions that conserve natural enemies, enhance microbial control, or reduce pest populations. This approach will ensure the most environmentally safe pest management. In canola, it is not economically viable to have several frequent applications of insecticides and developing an IPM strategy is more practical.

4.4.1 Aphids

When aphid populations are high, management is important both to reduce direct feeding damage and to minimize transmission of BWYV. The following management practices can be used for controlling aphids (Jones *et al.*, 2007; Baker, 2009; Jenkins *et al.*, 2011; Coutts, 2015; Harries *et al.*, 2016).

Biological control

Lacewings, ladybird beetles, parasitic wasps and syrphid flies are some of the natural enemies associated with aphid control. While information on their specific role in canola is not available, selecting chemicals that are less harmful to natural enemy populations is a general management recommendation.

Chemical control

Insecticides from different mode of action groups are effective against aphids in canola and require one or more applications. However, due to the resistance of *M. persicae* to organophosphate and synthetic pyrethroid insecticides in Australia, and potential cross-resistance to carbamates, care should be taken in selecting appropriate chemical pesticides. Jones *et al.* (2007) reported that an imidacloprid seed treatment (585 g AI/100 kg seed) provided aphid control and reduced the spread of BWYV for 2.5 months, resulting in an increase in seed yield of up to 88%. In Australia, sulfoxaflor effectively controls aphid in canola (Harries *et al.*, 2016). Timing of the insecticide treatment is important to reduce yield loss from aphids and controlling aphids at seedling and rosette stage is critical (Buntin and Raymer, 1994). Insecticide applications at first bloom can have some positive effect, but treatments at the podfilling stage are not beneficial.

Cultural control

Recommended cultural practices include maintaining good stubble cover to prevent aphids from landing on canola, managing weeds that serve as a source of aphids and BWYV, seeding at high rates to reduce the spread of virus and aphid landing, selection of canola varieties that are less susceptible to BWYV and delayed sowing. Delayed sowing is generally recommended to avoid the chance of an aphid migration from alternative hosts during a sensitive canola growth stage. However, Jenkins *et al.* (2011) reported that a 3-week delay in sowing affected seed yield and reduced protein and oil contents as a result of aphid infestations and water stress on plants. This suggests that cultural practices should be adjusted based on local environmental conditions and seasonal patterns of aphid migration.

4.4.2 Mirid and lygaeid bugs

Biological control

Egg parasitoids (*Anaphes iole* Girault, *Telenomus* sp. and *Polynema* spp.), nymphal parasitoids (*Peristenus* spp.) and several predators such as crab spider (*Misumena vatia* (Clerck)), damsel bugs (*Nabis* spp.), lacewing larva (*Meleoma emuncta* (Fitch)), ladybird beetles (*Coccinella* spp.) and larvae of a tachinid fly (*Alophorella* sp.) have been identified as natural enemies of *Lygus* spp. in Canada (Braun *et al.*, 2001; Cárcamo *et al.*, 2007a, b). Parasitoids can cause 9–70% mortality in eggs or nymphal stages. Egg parasitoids (unknown species) are also important in controlling *N. vinitor* (Bailey, 2007).

Chemical control

Several insecticides are available for managing *Lygus* spp. and good control can be achieved by one or more applications at an appropriate time, especially starting at the early pod stage (Otani and Cárcamo, 2011; see Table 4.2 for economic thresholds). However, managing *N. vinitor* requires multiple insecticide treatments, because they frequently migrate to canola (Hangartner and McDonald, 2015). Several organophosphate and synthetic pyrethroids are available for *N. vinitor* control but their application might promote resistant aphid populations and affect natural enemy populations.

Cultural control

Managing *Lygus* spp. on crops such as seed alfalfa, where they are major pests, was previously thought to be important to reduce the risk of infestations in canola but Cárcamo *et al.* (2003) demonstrated no such impact. However, management of weeds, volunteer canola and other hosts to remove oviposition sites for *Lygus* spp. is beneficial. Weed control and digging deep furrows around the field are recommended cultural practices to prevent the migration of *N. vinitor* to canola (Hangartner and McDonald, 2015).

Multi-year IPM studies conducted in California strawberries, where *Lygus hesperus* Knight is a major pest, have demonstrated the potential of botanical and microbial pesticides and the importance of combining and rotating chemical and non-chemical alternatives for effective control with a reduced risk of resistance development (Dara 2016a, b). Azadirachtin has been used as a botanical alternative to the chemical insect growth regulator novaluron, which is targeted towards immature stages of *L. hesperus*. Entomopathogenic fungi – *Beauveria bassiana* (Bals.) Vuill., *Isaria fumosorosea* Wize and *Metarhizium brunneum* Petch – were used with azadirachtin or lower label rates of chemical pesticides in the California studies. This approach of integrating microbial and botanical pesticides with chemicals could be used for managing *Lygus* spp. and other pests in canola.

4.4.3 Thrips

Biological control

Information on biological control of thrips in canola is not available but predators such as lacewing larvae and ladybird beetles attack thrips.

Chemical control

Synthetic pyrethroid insecticides can be applied at late bud or early flowering stages for managing thrips but they have been found to re-infest 2–7 days after such treatment (Olfert *et al.*, 2002b).

Cultural control

Selection of cultivars less susceptible to thrips is a strategy to reduce thrips damage (Fathi *et al.*, 2011). Egg-hatching, developmental time, longevity and fecundity of thrips can be affected on certain canola cultivars.

4.4.4 Leafhoppers

Biological control

Information on natural enemies of leafhoppers on canola is not available but several predatory species found in canola can feed on leafhoppers.

Chemical control

Since leafhoppers migrate to canola over several weeks, multiple chemical treatments may be necessary if infestations are high. Carbamate, neonicotinoid, organophosphate and synthetic pyrethroid insecticides used against other canola pests are also effective against leafhoppers. Their application for managing major pests could control coincidental leafhopper infestations.

Cultural control

It is potentially useful to monitor potential alternate hosts in the vicinity of canola and take measures to prevent their migration to canola. Early planting may avoid the attractiveness of migrating *M. quadrilineatus* and the associated risk of aster yellows infection (Knodel, 2012).

4.5 Conclusions

Canola is an important commercial crop in different parts of the world and a host of multiple sucking pests. The pest status of different species may vary in different canola-producing regions but the majority are hemipteran pests, along with some thysanopterans. The feeding damage of both reduces the quality and quantity of yield. Some aphids and leafhoppers are also important vectors of viral and phytoplasma diseases, respectively. Understanding the biology of the pest, sources of infestation, migratory behaviour, economic thresholds and treatment options are important for effective pest management.

In some regions, naturally occurring biological control agents have provided fair control of some canola pests, emphasizing the importance of natural enemy conservation. While there are some cultural control options that help to avoid the occurrence of peak pest infestations during vulnerable stages of canola, most growers rely on chemical control, with treatment decisions made based on regular monitoring. Since some chemical pesticides used in canola pose the risk of development of resistance in target and other pest populations, sustainable control relies on the development of appropriate IPM strategies based on multiple control options.

References

Auclair, J.L. (1963) Aphid feeding and nutrition. *Annual Review of Entomology* 8, 439–490.

Baker, G. (2009) Pests of canola and their management. In: *Canola: Best Practice Management Guide for South-Eastern Australia*. Grains Research Development Corporation, Australia, pp. 47–58.

Bailey, P.T. (2007) *Pests of Field Crops and Pastures: Identification and Control*. CSIRO Publishing, Australia.

Berlandier, F. and Cartwright, L. (1998) Effect of aphid feeding damage on canola yields in 1998. In: *Highlights of Canola Research and Development in Western*

Australia 1998. WA Agribusiness Crop Updates, Department of Agriculture, Western Australia, Perth, WA, Australia, pp. 25–27.

Berlandier, F. and Valentine, C. (2003) Aphid damage to canola – not all cultivars are equal. In: Zaicou-Kunesch, C. (ed.) *2003 Oilseeds Updates*. Department of Agriculture, Western Australia, Perth, WA, Australia, pp. 25–27.

Braun, L., Erlandson, M., Baldwin, D., Soroka, J., Mason, P., Foottit, R. and Hegedus, D. (2001) Seasonal occurrence, species composition, and parasitism of *Lygus* spp. in alfalfa, canola, and mustard. *The Canadian Entomologist* 133, 565–577.

Brown, J., McCaffrey, J.P., Harmon, B.L., Davis, J.B., Brown, A.P. and Erickson, A. (1999) Effect of late season insect infestation on yield, yield components and soil quality of *Brassica napus*, *B. rapa*, *B. juncea* and *Sinapis alba* in the Pacific Northwest region of the United States. *Journal of Agricultural Science* 132, 281–288.

Buntin, D.G. and Raymer, P.L. (1994) Pest status of aphids and other insects in winter canola in Georgia. *Journal of Economic Entomology* 87, 1094–1104. doi: 10.1093/jee/87.4.1097.

Burgess, L. and Weegar, H.H. (1988) Thrips (Thysanoptera) in canola crops in Saskatchewan. *The Canadian Entomologist* 120, 815–819.

Butts, R.A. and Lamb, R.J. (1991) Pest status of lygus bugs on oilseed *Brassica* crops. *Journal of Economic Entomology* 84, 1591–1596.

Canola Council of Canada (2013) Leafhoppers building. Available at: http://www.canolawatch.org/2013/07/04/leafhoppers-building/ (accessed 17 May 2016).

Canola Council of Canada (2016) Canola grower's manual. Available at: http://www.canolacouncil.org/crop-production/canola-grower's-manual-contents/ (accessed 5 May 2016).

Cárcamo, H., Otani, J., Herle, C., Dolinski, M., Dosdall, L., Mason, P., Butts, R., Kaminski, L. and Olfert, O. (2002) Variation of *Lygus* species assemblages in canola agroecosystems in relation to ecoregion and crop stage. *The Canadian Entomologist* 134, 97–111.

Cárcamo, H.A., Otani, J., Gavloski, J., Dolinski, M. and Soroka, J. (2003) Abundance of *Lygus* spp. (Heteroptera: Miridae) in canola adjacent to forage and seed alfalfa. *Journal of the Entomological Society of British Columbia* 100, 55–63.

Cárcamo, H.A., Larson, T.R. and Herle, C.E. (2006) Life history of *Lygus keltoni* (Hemiptera: Miridae) in the laboratory. *The Canadian Entomologist* 138, 871–874.

Cárcamo, H., Goulet, H., Herle, C. and Otani, J. (2007a) Parasitism of *Lygus* bugs by native *Peristenus* species in southern Alberta, Canada. *Journal of Insect Science* 8, 5–6. Available at: http://144.92.199.79/8.49/i1536-2442-8-49.pdf.

Cárcamo, H., Huber, J., Larson, T. and Bourchier, R. (2007b) Egg parasitoids of lygus bugs in southern Alberta. *Journal of Insect Science* 8, 5. Available at: http://144.92.199.79/8.49/i1536-2442-8-49.pdf.

Cohen, A.C. (2015) *Insect Diets: Science and Technology*. CRC Press, Boca Rotan, Florida.

Coutts, B. (2015) Diagnosing beet western yellow virus in canola. Available at: https://www.agric.wa.gov.au/mycrop/diagnosing-beet-western-yellow-virus-canola (accessed 19 May 2016).

Coutts, B.A. and Jones, R.A.C. (2000) Viruses infecting canola (*Brassica napus*) in south-west Australia: incidence, distribution, spread, and infection reservoir in wild radish (*Raphanus raphinistrum*). *Australian Journal of Agricultural Research* 51, 925–936.

Dara, S.K. (2016a) IPM solutions for insect pests in California strawberries: efficacy of botanical, chemical, mechanical, and microbial options. *CAPCA Adviser* 19(2), 40–46.

Dara, S.K. (2016b) Managing strawberry pests with chemical pesticides and non-chemical alternatives. *International Journal of Fruit Science*, published online 20 July 2016. Available at: http://dx.doi.org/10.1080/15538362.2016.1195311.

Demirel, N. and Cranshaw, W. (2005) Attraction of color traps to thrips species (Thysanoptera: Thripidae) on Brassica crops in Colorado. *Pakistan Journal of Biological Science* 8, 1247–1249.

Fathi, S.A.A., Gholami, F., Nouri-Ganbalani, G. and Mohiseni, A. (2011) Life history parameters of *Thrips tabaci* (Thysanoptera: Thripidae) on six commercial cultivars of canola. *Applied Entomology and Zoology* 46, 505–510.

Gavloski, J., Cárcamo, H. and Dosdall, L. (2011) Insects of canola, mustard, and flax in Canadian grasslands. In: Floate K.D. (ed.) *Arthropods of Canadian Grasslands (Volume 2): Inhabitants of a Changing Landscape*. Biological Survey of Canada, Ottawa, Canada, pp. 181–214.

Hangartner, S. and McDonald, G. (2015) *Rutherglen bug*. PestNotes Southern, Cesar, Parkville, Victoria, Australia. Available at: http://www.cesaraustralia.com/sustainable-agriculture/pestnotes/insect/Rutherglen-bug (accessed 20 May 2016).

Harries, M., Micic, S. and Seymour, M. (2016) Reassessing canola yield loss to aphids. Grains Research & Development Corporation Update, Perth, 29 Feburary–1 March, 2016. Available at: http://www.giwa.org.au/_literature_209579/Micic,_Svetlana_et_al_-_Reassessing_canola_yield_loss_to_aphids (accessed 6 May 2016).

Jenkins, L., Brill, R. and McCaffery, D. (2011) Managing aphids in flowering canola in central west NSW. *Proceedings of the 17th Australian Research Assembly on Brassicas. Wagga Wagga. New South Wales, August 2011*. Australian Oilseeds Federation, Sydney, Australia, pp. 82–88.

Jones, R.A.C., Coutts, B.A. and Hawkes, J. (2007) Yield-limiting potential of *Beet western yellows virus* in

Brassica napus. *Australian Journal of Agricultural Research* 58, 788–801.

Kirk, W.D.J. (1995) Feeding behavior and nutritional requirements. In: Parker, B.L., Skinner, M. and Lewis, T (eds) *Thrips Biology and Management*. Springer, New York, pp. 21–29. doi: 10.1007/978-1-4899-1409-5_2.

Knodel, J.J. (2012) Aster leafhoppers threat to canola. North Dakota State University Crop & Pest Report, 24 May 2012. Available at: https://www.ag.ndsu.edu/cpr/entomology/aster-leafhopper-threat-to-canola-5-24-12 (accessed 17 May 2016)

Lamb, R.J. (1989) Entomology of oilseed *Brassica* crops. *Annual Review of Entomology* 34, 11–29.

Micic, S. and Mangano, P. (2016) Aphid management in canola crops. Available at https://www.agric.wa.gov.au/pest-insects/aphid-management-canola-crops?nopaging=1 (accessed 5 May 2016).

Micic, S. and Wahlsten, L. (2015) Does aphid feeding cause yield loss in unstressed determinate canola? 2015 WA Agribusiness Crop Updates, Perth, 24–25 February, 2015. Available at: http://www.giwa.org.au/pdfs/CR_2015/Micic_Svetlana_Does_aphid_feeding_cause_yield_loss_in_unstressed_determinate_canola_FINAL.pdf (accessed 5 May 2016)

Miles, M. and McDonald, G. (1999) Insect pests of canola. In: *Canola in Australia – the First Thirty Years*. Produced for the 10th International Rapeseed Congress, Canberra, Australia, pp. 53–58. Available at: http://www.australianoilseeds.com/__data/assets/pdf_file/0011/2711/Chapter_11_-_Insect_Pests_of_Canola.pdf

National Invertebrate Pest Initiative (2014) Rutherglen bug in canola. Available at: http://ipmguidelinesforgrains.com.au/pests/rutherglen-bug-rgb/rutherglen-bug-in-canola/ (accessed 4 May 2016).

Olfert, O., Elliott, B., Braun, M. (2002a) Impact of thrip populations in canola. Available at: http://www.saskcanola.com/quadrant/System/research/reports/report-Olfert-thrip-short.pdf (accessed 3 May 2016).

Olfert, O., Elliott, B., Braun, M. (2002b) Impact of thrip populations in canola. Final report to Saskatchewan Canola Development Commission. Available at: http://www.saskcanola.com/quadrant/System/research/reports/report-Olfert-thrip-long.pdf (accessed 3 May 2016).

Olivier, C.Y., Murza, G., Galk, B., McAleese, D., Hegedus, D.D., Barasubiye, T., Séguin-Swartz, G.T., Olfert, O.O. and Boudon-Padieu, E. (2007) Survey of aster yellows disease in canola crops in Saskatchewan, Canada, 2001–2005. *Cruciferae Newsletter* 26, 40–42.

Otani, J. and Cárcamo, H. (2011) Biology and management of *Lygus* in canola. *Prairie Soils & Crops Journal* 4, 42–53.

Saljoqi, A.U.R., Sadur-Rehman, Hussain, N. and Khan, S.A. (2006) Insect pests of canola crop (other than aphid). *Journal of Agricultural and Biological Science* 1, 19–21.

Schwartz, M.D. and Foottit, R.G. (1992) *Lygus* species on oilseed rape, mustard, and weeds: a survey across the prairie provinces of Canada. *The Canadian Entomologist* 124, 151–158.

Slaymaker, P.H. and Tugwell, N.P. (1982) Low-labor method for rearing the tarnished plant bug (Hemiptera: Miridae). *Journal of Economic Entomology* 75, 487–488.

Soroka, J., Olivier, C., Grenkow, L. and Séguin-Swartz, G. (2015) Interactions between *Camelina sativa* (Brassicaceae) and insect pests of canola. *The Canadian Entomologist* 147, 193–214.

Talpur, S. (2001) Incidence and abundance of thrips, *Thrips* spp. on varieties of brassica. Available at: http://agris.fao.org/agris-search/search.do?recordID=PK2003000789 (accessed 3 May 2016).

Timlick, B.H., Turnock, W.J. and Wise, I. (1993) Abundance of *Lygus* spp. (Heteroptera: Miridae) on alfalfa and canola in Manitoba. *The Canadian Entomologist* 125, 1033–1041.

Turnock, W.J., Gerber, G.H., Timlick, B.H. and Lamb, R.J. (1995) Losses of canola seeds from feeding by *Lygus* species (Heteroptera: Miridae) in Manitoba. *Canadian Journal of Plant Science* 75, 731–736.

Weiss, M.J., Knodel, J.J., and Olson, D. (2006) Insect pests of canola. In: Radcliffe, E.B., Hutchison, D. and Cancelado, R.E. (eds) *Radcliffe's IPM World Textbook*. University of Minnesota, St Paul, Minnesota. Available at: http://ipmworld.umn.edu.

Wise, I.L. and Lamb, R.J. (1998) Economic threshold for plant bugs, *Lygus* spp. (Heteroptera: Miridae), in canola. *The Canadian Entomologist* 130, 825–828.

5 Cabbage Seedpod Weevil Management

Héctor A. Cárcamo* and Randall Brandt

Agriculture and Agri-Food Canada, Lethbridge Research and Development Centre, Lethbridge, Alberta, Canada

5.1 Introduction

5.1.1 Distribution

The cabbage seedpod weevil (CSW), *Ceutorhynchus obstrictus* (Marsham) (Coleoptera: Curculionidae), is a European pest of several brassicaceous seed crops from the Mediterranean region to Scandinavia. In North America it was first reported from the Vancouver area in British Columbia in 1931 (McLeod, 1962). Since then it has been reported in several jurisdictions of North America: Pacific North West and California (Hagen, 1946), Georgia (Buntin and Raymer, 1994) and the Canadian Prairies (Butts and Byers, 1996). Laffin *et al.* (2005) demonstrated that the population in Quebec (Canada) stemmed from a separate accidental introduction. With the increase in global trade and transportation, this pest will likely occur in most regions where its hosts are cultivated.

5.1.2 Biology

The biology of CSW is well studied and discussed in detail elsewhere (Haye *et al.*, 2010; Dosdall and Cárcamo, 2011), hence, only a summary is presented here. CSW has only one generation per year regardless of latitude and requires a minimum chill period of 16 weeks at 4°C to complete its obligatory diapause (Ni *et al.*, 1990). Increasing cumulative subfreezing temperatures, on the other hand, will reduce its survivorship as its supercooling point is only −7°C (Cárcamo *et al.*, 2009). Peak emergence from its overwintering sites (field margins and treed areas) occurs in the spring when temperatures reach 15°C (Ulmer and Dosdall, 2006a). Adults (Fig. 5.1) feed on a wide range of Brassicaceae, including several early winter annual weeds such as flixweed (*Descurania sophia*). Adult invasion of host crops such as canola (*Brassica napus* or *B. rapa*) peaks during the flowering stage in spring-planted cultivars, being more abundant along the field edges during the early immigration phase (Free and Williams, 1979; Dosdall *et al.* 2002). Both sexes of the overwintered generation respond to pheromones released by unmated females (Evans and Bergeron, 1994). Oviposition occurs on pods 45–60 mm long (Dosdall and Moisey, 2004); females were found to prefer plants with high levels of sulfur and low levels of nitrogen in southern Alberta in Canada (Blake *et al.*, 2010). However, the results were inconsistent in a similar European study assessing only nitrogen (Veromann *et al.*, 2013). Larvae (Fig. 5.2) will only complete their life cycle in certain *Brassica* species hosts that produce large enough pods (Dmoch, 1965; Fox and Dosdall, 2003; Dosdall and Moisey, 2004). The three larval instars take about 50 days in southern Alberta (Dosdall and McFarlane, 2004). Each larva can consume up to six seeds (Dmoch, 1965), before chewing a pin-sized hole on the pod to exit and pupate in the ground. The new adults emerge in about 2 weeks, depending on temperature, and feed on any green portion of Brassicaceae plants.

*Corresponding author. E-mail: hector.carcamo@agr.gc.ca

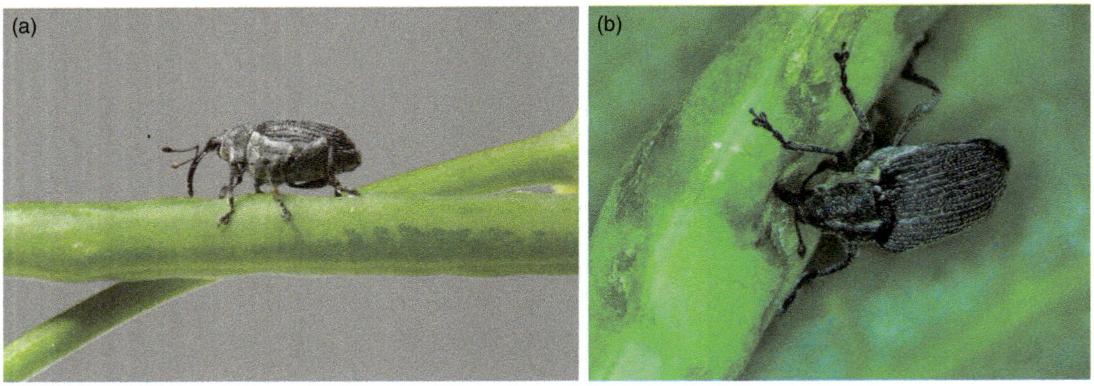

Fig. 5.1. Cabbage seedpod weevil (a) walking and (b) feeding on a canola pod. Photos by M. Dolinski.

Fig. 5.2. (a) Cabbage seedpod weevil larvae feeding on canola seeds inside pod and (b) exit holes made by larvae as they exit to pupate in the soil. Photos by M. Dolinski.

5.2 Crop Damage, Thresholds and Sampling

5.2.1 Crop damage

CSW has the potential to damage crops and reduce yield during three periods of plant growth. During the bud and flowering stage, reproductive weevils will aggregate to mate and feed on these structures, which can cause their abortion (Cárcamo, 2012). However, under adequate growing conditions the plants can compensate for up to 60% of abortion of flowers and buds (Free *et al.*, 1983). The main damage that reduces yield occurs during the larval stage, when about 20–30% of the seeds inside pods are consumed (Dmoch, 1965). A number of studies have demonstrated that when the percentage of pods with CSW exit holes at harvest time exceeds 25%, yield reduction is economically significant and warrants investment in a control action (Lerin, 1984; Buntin, 1999; Cárcamo, 2012). In addition to the direct seed consumption by larvae, the small holes made by the adults during feeding and oviposition can facilitate feeding by other insects or invasion by pathogens. In England, the pod midge *Dasineura brassicae* Winnertz (Diptera: Cecidomyiidae) uses these holes to oviposit and gain access to the seeds and further decrease yield (Free *et al.*, 1983). Exit holes made by mature larvae likely cause premature drying of the pod and may result in high levels of shattering; however, this aspect of yield loss remains to be quantified. The third period when crop damage may occur is when the new adults emerge and encounter hosts such as late-seeded canola crops with ripening seeds on which they feed. Buntin *et al.* (1995) found yield losses of only 2% in winter canola in northern Georgia from this late damage but damaged seeds had up to 40% lower germination, which would be a concern for the certified seed industry.

5.2.2 Threshold

Thresholds for seedpod weevils in oilseed rape (canola) vary with the region. In England a threshold of one adult per plant at flowering is used when the pod midge is present (Free *et al.*, 1983) but two per plant, similar to other parts of Europe (Sylven and Svenson, 1975; Tulisalo *et al.*, 1976), is followed when the midge is absent. In the Pacific North West of the USA, a range of three to six weevils per sweep at early flowering is used as a guideline to initiate control action (McCaffrey, 1992). In Alberta only two to three weevils per sweep are used as the nominal threshold (Dosdall *et al.*, 2001; Dosdall and Cárcamo, 2011). The threshold recommendations in Europe and Canada were derived from cage studies (Tulisalo *et al.*, 1976; Lerin, 1984; Cárcamo, 2012). Ongoing farm-level studies in southern Alberta are validating these thresholds (H.A. Cárcamo and S. Meers, unpublished data).

5.2.3 Monitoring

Adult CSW are sampled mainly with a sweep net. Dmoch (1965) recommended four samples of 25 sweeps to estimate abundances. Even smaller sample sizes (five to ten sweeps per plot) have been used in insecticide spray trials (Buntin, 1999; Cárcamo *et al.*, 2005). In regions where the crop is at a more advanced stage, other methods such as dislodging weevils into a bucket or tray have been used (Brown *et al.*, 1999). A number of studies have been conducted on the use of semiochemicals and plant volatiles to monitor adult populations (Smart and Blight, 1997). Yellow pan traps raised periodically to canopy level have been used successfully to determine the CSW's spatial distribution and that of its parasitoids in canola (Dosdall *et al.*, 2006; Blake *et al.*, 2010). These traps can be baited with isothiocyanates to improve weevil captures during the summer dispersal phase (Smart *et al.*, 1997). Sampling at least 50 m into a field is necessary to avoid overestimating abundances when only the border is sampled because of the pattern of high aggregation along field edges, especially during early flowering (Dosdall *et al.*, 2006). Populations of CSW larvae that successfully reach maturity at the end of the summer can be quantified at harvest time by counting exit holes in a subsample of the pods from the lowest, middle and main raceme of a plant, as recommended by Cárcamo *et al.* (2004). Infestation of pods below 25% is considered to cause insignificant yield losses in canola (Lerin, 1984; Buntin, 1999).

5.3 Management

5.3.1 Biological control

Biological control has received considerable attention as a potential strategy to reduce insecticide use. A large number of parasitoids attack CSW throughout its native and invasive range (Table 5.1) (Mason *et al.*, 2011; Haye *et al.*, 2013). In western Canada at least 15 parasitoids have expanded their host range to attack CSW, but the rates of attack are generally too low (<15%) to control the weevil (Dosdall *et al.*, 2006). In contrast, in Switzerland, larval ectoparasitoids in the family Pteromalidae can reduce populations by up to 52% (Haye *et al.*, 2010). In Canada, one of these parasitoids (*Trichomalus perfectus*), though not introduced intentionally, is established in Quebec and Ontario and appears to be spreading in the region (Mason *et al.*, 2011; Haye *et al.*, 2013); host range studies suggest that it attacks non-target species within the *Ceutorhynchus* genus but mainly those that feed on Brassicaceae (Haye *et al.*, 2015). This species and *Mesopolobus morys* (Walker) both deserve further research in terms of their potential efficacy to reduce CSW in other canola-growing regions and their potential non-target impacts on native weevils and *Ceutorhynchus* species released for biological control of weeds (Kuhlmann *et al.*, 2006).

5.3.2 Host plant resistance

Another strategy that holds promise to manage CSW sustainably is host plant resistance. A number of studies have quantified the relative susceptibilities of various oilseed brassicaceous genotypes

Table 5.1. *Ceutorhynchus obstrictus* parasitoids and distribution.

Parasitoid	Distribution[a]	Reference corresponding to distribution for species[a]
Chalcidae		
Conura albifrons (Walsh)[b]	Alberta[1,2], British Columbia[3], Quebec[4], Saskatchewan[2]	[1]Dosdall *et al*., 2006; [2]Dosdall *et al*., 2009; [3]Gillespie *et al*., 2006; [4]Mason *et al*., 2011
Conura torvina (Cresson)[b]	Alberta[1,2], Saskatchewan[2], Georgia[3]	[1]Dosdall *et al*., 2006; [2]Dosdall *et al*., 2009; [3]Gibson *et al*., 2006a
Eulophidae		
Euderus albitarsis (Zetterstedt)[b]	Alberta[3,5,6], Ontario[7], Quebec[3,7], Saskatchewan[3,6], Finland[2], India[1], Europe[3], England[3], Nova Scotia[3], Manitoba[3], Virginia[3], West Virginia[3], Israel[4]	[5]Dosdall *et al*., 2006; [6]Dosdall *et al*., 2009; [1]Gupta and Poorani, 2008; [7]Mason *et al*., 2011; [2]Pulkkinen and Yang, 1984; [4]Yefremova, 2015; [3]Yoshimoto, 1971
Euderus glaucus Yoshimoto	Ontario[1,3], Quebec[1,3], Florida[1], Texas[1], Georgia[2]	[2]Gibson *et al*., 2006a; [3]Mason *et al*., 2011; [1]Yoshimoto, 1971
Euderus sp.	British Columbia[1], Ontario[2]	[1]Gibson *et al*., 2006b; [2]Mason *et al*., 2011
Necremnus tidius (Walker)	Alberta[2,3,6], British Columbia[2,4,5], Ontario[7], Quebec[7], Saskatchewan[2,6], Georgia[1,2], Idaho[2], Illinois[2], Washington[2]	[3]Dosdall *et al*., 2006; [6]Dosdall *et al*., 2009; [2]Gibson *et al*., 2005; [1]Gibson *et al*., 2006a; [4]Gibson *et al*., 2006b; [5]Gillespie *et al*., 2006; [7]Mason *et al*., 2011
Eupelmidae		
Brasema allynii (French)[b]	Georgia	Gibson *et al*., 2006a
Eupelmus (Eupelmus) cyaniceps (Ashmead)[a]	Georgia	Gibson *et al*., 2006a
Eupelmus vesicularis (Retzius)	British Columbia[1,2], Ontario[3]	[1]Gibson *et al*., 2006b; [2]Gillespie *et al*., 2006; [3]Mason *et al*., 2011
Eurytomidae		
Eurytoma curculionum Mayr	Hungary[1], Switzerland[2], Germany[2], Romania[2]	[2]Haye *et al*., 2015; [1]Muller *et al*., 2011
Eurytoma sp.	British Columbia[1], Ontario[2], Quebec[2]	[1]Gibson *et al*., 2006b; [2]Mason *et al*., 2011
Eurytoma tylodermatis Ashmead[b]	Alberta[3], British Columbia[2] (probable), Saskatchewan[3], Georgia[1]	[3]Dosdall *et al*., 2009, [1]Gibson *et al*., 2006a; [2]Gillespie *et al*., 2006
Pteromalidae		
Catolaccus aenoviridis (Girault)[b]	Alberta, Saskatchewan	Dosdall *et al*., 2009
Chlorocytus sp.	Alberta[2,4], British Columbia[3], Ontario[5], Quebec[5], Saskatchewan[4], Germany[1]	[2]Dosdall *et al*., 2006; [4]Dosdall *et al*., 2009; [3]Gibson *et al*., 2006b; [1]Haye *et al*., 2015; [5]Mason *et al*., 2011
Lyrcus incertus (Ashmead)	Alberta[2], Ontario[3], Saskatchewan[2], Georgia[1]	[2]Dosdall *et al*., 2009; [1]Gibson *et al*., 2006a; [3]Mason *et al*., 2011
Lyrcus maculatus (Gahan)[b]	Alberta[2,4], British Columbia[3], Ontario[5], Saskatchewan[4], Georgia[1]	[2]Dosdall *et al*., 2006; [4]Dosdall *et al*., 2009; [1]Gibson *et al*., 2006a; [3]Gibson *et al*., 2006b; [5]Mason *et al*., 2011
Lyrcus perdubius (Girault)	Alberta[2], Ontario[3], Quebec[3], Georgia[1]	[1]Gibson *et al*., 2006a; [2]Dosdall *et al*., 2006; [3]Mason *et al*., 2011
Mesopolobus bruchophagi Gahan	Alberta	Dosdall *et al*., 2006
Mesopolobus gemellus Baur and Muller	Ontario[2], Quebec[2], Switzerland[1], Czech Republic[1], England[1], Germany[1] France[1]	[1]Baur *et al*., 2007; [2]Mason *et al*., 2011
Mesopolobus mediterraneus Mayr	Switzerland[1], Germany[1]	[1]Haye *et al*., 2015
Mesopolobus moryoides Gibson	Alberta[3,6], British Columbia[2,4,5], Ontario[7], Quebec[7], Saskatchewan[6], Georgia[1,2], California[2], Idaho[2], Illinois[2], Oregon[2], Washington[2]	[3]Dosdall *et al*., 2006; [6]Dosdall *et al*., 2009; [2]Gibson *et al*., 2005; [1]Gibson *et al*., 2006a; [4]Gibson *et al*., 2006b; [5]Gillespie *et al*., 2006; [7]Mason *et al*., 2011

Continued

Table 5.1. Continued.

Parasitoid	Distribution[a]	Reference corresponding to distribution for species[a]
Mesopolobus morys (Walker)	Hungary[1,2], Switzerland[1,2,3], France[2], Germany[2,3,5], Romania[3], Estonia[4,5], United Kingdom[5], Sweden[5], Poland[5]	[2]Baur *et al.*, 2007; [3]Haye *et al.*, 2015; [1]Muller *et al.*, 2011; [5]Ulber *et al.*, 2010; [4]Veromann *et al.*, 2010
Neocatolaccus tylodermae (Ashmead)	Ontario[2], Georgia[1]	[1]Gibson *et al.*, 2006a; [2]Mason *et al.*, 2011
Pteromalus spp.	Alberta[1,4], British Columba[3], Ontario[5], Quebec[5], Saskatchewan[4], Georgia[2]	[1]Dosdall *et al.*, 2006; [4]Dosdall *et al.*, 2009; [2]Gibson *et al.*, 2006a; [3]Gibson *et al.*, 2006b; [5]Mason *et al.*, 2011
Pteromalus cerealellae (Ashmead)	Georgia	Gibson *et al.*, 2006a
Stenomalina gracilis (Walker)	British Columbia[6,7], Poland[1,4], Switzerland[2], Germany[2,4], Romania[2], Estonia[3,4], United Kingdom[4], Sweden[4], Ireland[5]	[6]Gibson *et al.*, 2006b; [7]Gillespie *et al.*, 2006; [2]Haye *et al.*, 2015; [1]Klukowski and Kelm, 2000; [5]O Connor, 1996; [4]Ulber *et al.*, 2010; [3]Veromann *et al.*, 2010
Trichomalus lucidus (Walker)	Alberta[2,4,7], British Columbia[2,5,6], Ontario[8], Quebec[8], Saskatchewan[7], Germany[1,3], Switzerland[1,3], Idaho[2], Oregon[2], Washington[2], Romania[3]	[4]Dosdall *et al.*, 2006; [7]Dosdall *et al.*, 2009; [2]Gibson *et al.*, 2005; [5]Gibson *et al.*, 2006b; [6]Gillespie *et al.*, 2006; [3]Haye *et al.*, 2015; [8]Mason *et al.*, 2011; [1]Muller *et al.*, 2007
Trichomalus perfectus (Walker)	Ontario[5], Quebec[5], France[2], Germany[2,4,6], Switzerland[2,6], Estonia[3,4], United Kingdom[4], Sweden[4], Poland[4], Romania[6]	[6]Haye *et al.*, 2015; [5]Mason *et al.*, 2011; [2]Muller *et al.*, 2007; [4]Ulber *et al.*, 2010; [3]Veromann *et al.*, 2010

[a]Superscript numbers relate distribution range to appropriate reference.
[b]Possible hyperparasitoid (Gibson *et al.*, 2006a; Dosdall *et al.*, 2009)

and determined that *B. rapa* is the most susceptible whereas *B. napus* and various *Brassica juncea* mustards are similar in susceptibility (McCaffrey *et al.*, 1999; Kalischuk and Dosdall, 2004; Cárcamo *et al.*, 2007b). Likewise, all these and other studies (Ulmer and Dosdall, 2006b; Ross *et al.*, 2008) concur with early reports (Free and Williams, 1978) that *Sinapis alba* is immune to weevil attack. Thus, development of germplasm with resistance to CSW has focused on introgression of *Sinapis alba* traits into *B. napus* (McCaffrey *et al.*, 1999). These authors evaluated hybrid plants with introgressed traits and noted that the weevils attacked the plants at lower levels and fewer larvae completed development. Dosdall and Kott (2006) used embryo rescue methods to transfer the resistant traits to *B. napus* and noted similar results with the progeny. In a series of studies conducted by Tansey *et al.* (2010) (see review in Dosdall and Cárcamo, 2011), the authors demonstrated that the resistant germplasm possesses both antixenotic and antibiotic traits against the weevil and differed from susceptible germplasm in terms of olfactory and visual cues. A related strategy to manage CSW in semi-arid mustard-growing regions, consists of developing canola-quality *S. alba*. As shown by Cárcamo *et al.* (2007b), genotypes of *S. alba* with near canola quality attributes appear to keep their resistance to CSW.

5.3.3 Trap crops

Among cultural strategies, trap cropping (Fig. 5.3) has received the most attention. Buechi (1990) tested turnip rape (*B. rapa*) as a potential trap crop for a number of insect pests of oilseed rape (*B. napus*). The author concluded that turnip rape was not a suitable trap crop because CSW readily oviposited on the oilseed crop. Buechi, however, did not attempt to control pests in the trap crop with an insecticide. Buntin (1998) planted a spring cultivar of canola around the borders of winter canola in large plots (0.25 ha). Both cultivars were planted in the fall and the spring cultivar flowered earlier. The trap crop concentrated high numbers of weevils where they were sprayed with an insecticide. Although weevils were

Fig. 5.3. Trap crop of fall-planted *Brassica napus* planted around the perimeter of a main crop of *Brassica napus* planted in the spring. Weevils are effectively concentrated in the trap crop and sprayed with insecticides before they migrate to the main crop.

reduced in the main crop relative to control plots without a trap crop, there was insufficient yield protection. The author speculated that such trap crop systems may work in larger fields or under moderate weevil densities. This speculation turned out to be correct, as Cárcamo *et al.* (2007a) demonstrated that earlier flowering borders of canola in large square fields (e.g. 1.6 km square) in southern Alberta concentrated very large numbers of CSW. Growers were able to control the weevils with an insecticide to prevent their invasion and damage to the main crop. Smaller or narrow fields under heavy attack by CSW could not be protected with the trap crop. The authors concluded that this cultural method could result in significant financial savings by only spraying insecticide on 10% of the field and also protecting natural enemies and pollinators in the majority of the field.

5.3.4 Insecticide control

Despite the large research effort on developing alternative pest control strategies, insecticides remain the main form of CSW management. Several insecticides are registered in various countries to target adults during early flowering (Cárcamo *et al.*, 2001, 2005; Heimbach and Müller, 2013; Zamojska and Węgorek, 2014). In North America, insecticides used include bifenthrin, chlorantraniliprole, deltamethrin, gamma cyhalothrin, lambda cyhalothrin and zeta-cypermethrin (Buntin, 2015; Syngenta Canada Inc., 2015). Pyrethroid insecticides have higher efficacy than other classes and are routinely used in Europe (Alford *et al.*, 1996), USA (Buntin, 1999) and Canada (Cárcamo *et al.*, 2005). However, populations resistant to cyhalothrin lambda have been reported in Germany (Heimbach and Müller, 2013). Seed treatments in the form of neonicotinoid seed coatings have been investigated and produced mixed results. Bragg (1998) in Washington state (USA) and Dosdall (2009) in Alberta (Canada) reported a significant reduction in the proportion of canola pods with exit holes. However, Cárcamo *et al.* (2005) reported that in two of the three years of their field study, there was a numerically higher proportion of damaged pods in plots where plants were grown from seed treated with imidacloprid compared with those treated with lindane or the controls without insecticide-treated seed. Plants grown from imidacloprid-treated seed flowered a few days earlier than the controls and attracted more weevils, which would explain the slightly higher damage (Cárcamo *et al.*, 2005).

The spatial pattern of aggregation of CSW during early flowering can allow for reduced use of insecticide to protect parasitoids. A number of studies have reported that at moderate densities weevils are more abundant along the edges of the fields, particularly during early flowering (Free and Williams, 1979; Ferguson *et al.*, 2000; Dosdall *et al.*, 2006). As part of an integrated pest management programme in France for CSW, Jourdheuil *et al.* (1974) reported that spraying the borders provided adequate crop protection and conserved the parasitoids in the rest of the field. In some areas of Europe, the ectoparasitoid *T. perfectus* forages at the early pod stage when weevil larvae are at the third larval instar (Alford *et al.*, 1996). Therefore, they recommend spraying for CSW only up to early flowering to protect parasitoid populations (Murchie *et al.*, 1999).

5.4 Future Research Needs

A short-term priority for most canola-growing regions is to validate or refine economic thresholds. In North America, nominal thresholds are still in use, despite more than two decades of research. Results from cage and small plot studies should be validated at the field scale using sweeping as a practical sampling method. This will also require calibration to relate sweeping to actual weevil densities for a better understanding of the results from cage studies and to reconcile them with commercial field data. Also, none of the thresholds are dynamic – ideally they should include natural enemies in the decision.

Another ongoing effort in canola-growing regions is to maintain surveillance programmes to monitor the spread of the weevil into new areas. In the Canadian Prairies to date, the weevil has spread mostly from southern Alberta towards the east into southern Saskatchewan; however, low densities have been reported in the Parkland Ecoregion around Lacombe in south central Alberta – the main canola-growing region of the world (K. Gabert, Canola Council of Canada, personal communication). Establishment of CSW in the Parkland Ecoregion would pose another serious threat to crop production in this area, which currently is prone to *Lygus* bug damage as well.

In the long term, research efforts should continue to develop host plant resistance, biological control, cultural methods and their integration as long-term sustainable solutions to reduce insecticide use. It is unlikely that a cultivar with resistance to CSW will provide complete protection; therefore, integration with biological control should be a key strategy. Border trap crops that flower early effectively concentrate the weevil in a small area where they could be managed with an alternative bioinsecticide such as an entomopathogenic fungus or a strain of *Bacillus thuringiensis*. It might even be possible to deliver the biocontrol microbe using bee-vector technology currently used in other systems (Al-Mazra'Awi *et al.*, 2007). These strategies will require a high degree of management that may hinder adoption; furthermore they will require extensive and expensive research investment. A more practical and longer-term solution would be to have an effective parasitoid such as *Trichomalus perfectus* established in western North America and combine it with a cultivar that has some resistance to CSW. As recommended elsewhere (Haye *et al.*, 2013), more work should continue to assess this parasitoid for its potential to establish in North America. The accidental introduction and establishment of *T. perfectus* in eastern Canada will provide a unique opportunity to study its impact on the target and indigenous non-target *Ceutorhynchus* spp. Depending on the outcome, a redistribution of *T. perfectus* into other canola-growing areas in North America may be justified (Haye *et al.*, 2015).

Acknowledgements

This review would not have been possible without the key contributions made by many authors, but in particular by the late Dr Lloyd Dosdall who devoted a major part of his career to improving the knowledge on several aspects of the basic and applied biology of *C. obstrictus*. The authors relied heavily on two former reviews co-authored with Dr Dosdall. We dedicate this chapter to his memory.

References

Alford, D.V., Walters, K.F.A., Williams, J.H. and Murchie, A.K. (1996) A commercially viable low-cost strategy for the management of seed weevil populations on winter oilseed rape in the UK. In: *Brighton Crop Protection Conference – Pests and Diseases, Vol. 2, 18–21 November 1996, Brighton, UK*. British Crop Protection Council, Farnham, UK, pp. 609–614.

Al-Mazra'Awi, M.S., Kevan, P.G. and Shipp, L. (2007) Development of *Beauveria bassiana* dry formulation for vectoring by honey bees *Apis mellifera* (Hymenoptera: Apidae) to the flowers of crops for pest control. *Biocontrol Science and Technology* 17, 733–741.

Baur, H., Muller, F.J., Gibson, G.A.P., Mason, P.G. and Kuhlmann, U. (2007) A review of the species of *Mesopolobus* (Chalcidoidea: Pteromalidae) associated with *Ceutorhynchus* (Coleoptera: Curculionidae) host-species of European origin. *Bulletin of Entomological Research* 97, 387–397.

Blake, A.J., Dosdall, L.M. and Keddie, B.A. (2010) Plant nutrients and the spatiotemporal distribution dynamics of *Ceutorhynchus obstrictus* (Coleoptera: Curculionidae) and its parasitoids. *Environmental Entomology* 39, 1195–1205.

Bragg, D.E. (1998) *Control of insect pests on spring canola, 1998*. Technical Report, Washington State University, p. 5.

Brown, J., McCaffrey, J.P., Harmon, B.L., Davis, J.B., Brown, A.P. and Erickson, D.A. (1999) Effect of late season insect infestation on yield, yield components and oil quality of *Brassica napus*, *B. rapa*, *B. juncea* and *Sinapis alba* in the Pacific Northwest region of the United States. *Journal of Agricultural Science* 132, 281–288.

Buechi, R. (1990) Investigations on the use of turnip rape as trap plant to control oilseed rape pests. *Bulletin of the International Union of Biological Sciences, West Palearctic Region* 13, 32–39.

Buntin, G.D. (1998) Cabbage seedpod weevil (*Ceutorhynchus assimilis* Paykull) management by trap cropping and its effect on parasitism by *Trichomalus perfectus* (Walker) in oilseed rape. *Crop Protection* 17, 299–305.

Buntin, G.D. (1999) Damage loss assessment and control of the cabbage seedpod weevil (Coleoptera: Curculionidae) in winter canola using insecticides. *Journal of Economic Entomology* 92, 220–227.

Buntin, D. (2015) Canola: Canola insect control, 2015. In: *UGA Extension Special Bulletin 28, Georgia Pest Management Handbook* (2015 Commercial Edition). University of Georgia Extension, Athens, Georgia, pp. 140–141.

Buntin, G.D. and Raymer, P.L. (1994) Pest status of aphids and other insects in winter canola in Georgia. *Journal of Economic Entomology* 87, 1097–1104.

Buntin, G.D., McCaffrey, J.P., Raymer, P.L. and Romero, J. (1995) Quality and germination of rapeseed and canola seed damaged by adult cabbage seedpod weevil, *Ceutorhynchus assimilis* (Paykull) [Coleoptera: Curculionidae]. *Canadian Journal of Plant Sciences* 75, 539–541.

Butts, R.A. and Byers, J.R. (1996) Cabbage seedpod weevil: a potential new pest of canola in southern Alberta. *Pest Management News* 8, 5.

Cárcamo, H. (2012) Effect of insect herbivory on the architecture and seed production of canola, *Brassica napus* L. *Journal of Entomological Science* 47, 44–55.

Cárcamo, H.A., Dosdall, L., Dolinski, M., Olfert, O. and Byers, J.R. (2001) The cabbage seedpod weevil, *Ceutorhynchus obstrictus* (Coleoptera: Curculionidae) – a review. *Journal of the Entomological Society of British Columbia* 98, 201–210.

Cárcamo, H.A., Entz, T. and Blackshaw, R.E. (2004) Subsampling canola plants (*Brassica napus* L.) to estimate cabbage seedpod weevil (*Ceutorhynchus obstrictus* Marsham) damage. *Journal of Entomological Science* 39, 122–124.

Cárcamo, H.A., Dosdall, L.M., Johnson, D. and Olfert, O. (2005) Evaluation of foliar and seed treatments for control of the cabbage seedpod weevil (Coleoptera: Curculionidae) in canola. *The Canadian Entomologist* 137, 476–487.

Cárcamo, H.A., Dunn, R., Dosdall, L.M. and Olfert, O. (2007a) Managing cabbage seedpod weevil in canola using a trap crop – a commercial field-scale study in western Canada. *Crop Protection* 26, 1325–1334.

Cárcamo, H., Olfert, O., Dosdall, L., Herle, C., Beres, B. and Soroka, J. (2007b) Resistance to cabbage seedpod weevil among selected Brassicaceae germplasm. *The Canadian Entomologist* 139, 658–669.

Cárcamo, H.A., Herle, C.E., Otani, J. and McGinn, S.M. (2009) Cold hardiness and overwintering survival of the cabbage seedpod weevil, *Ceutorhynchus obstrictus*. *Entomologia Experimentalis et Applicata* 133, 223–231.

Dmoch, J. (1965) The dynamics of a population of the cabbage seedpod weevil (*Ceutorhynchus assimilis* Payk.) and the development of winter rape. Part I. *Ekologia Polska Seria A* 13, 249–287.

Dosdall, L.M. (2009) Responses of the cabbage seedpod weevil, *Ceutorhynchus obstrictus* (Marsham) (Coleoptera: Curculionidae), to seed treatments of canola (*Brassica napus* L.) with the neonicotinoid compounds clothianidin and imidacloprid. *Pest Management Science* 65, 1329–1336.

Dosdall, L.M. and Cárcamo, H. (2011) The cabbage seedpod weevil in prairie crops: its biology and management. *Prairie Soils and Crops ejournal* 4, 14–23. Available at: www.prairiesoilsandcrops.ca

Dosdall, L.M. and Kott, L.S. (2006) Introgression of resistance to cabbage seedpod weevil to canola from yellow mustard. *Crop Science* 46, 2437–2445.

Dosdall, L.M. and McFarlane, M.A. (2004) Morphology of the pre-imaginal life stages of the cabbage seedpod weevil, *Ceutorhynchus obstrictus* (Marsham) (Coleoptera: Curculionidae). *Coleopterists Bulletin* 58, 45–52.

Dosdall, L.M. and Moisey, D.W.A. (2004) Developmental biology of the cabbage seedpod weevil, *Ceutorhynchus obstrictus* (Coleoptera: Curculionidae), in spring canola, *Brassica napus*, in western Canada. *Annals of the Entomological Society of America* 97, 458–465.

Dosdall, L.M., Moisey, D., Cárcamo, H. and Dunn, R. (2001) *Cabbage Seedpod Weevil Fact Sheet.* Alberta Agriculture, Food and Rural Development, Alberta, Canada, Agdex 622-21.

Dosdall, L.M., Weiss, R.M., Olfert, O. and Cárcamo, H.A. (2002) Temporal and geographical distribution patterns of cabbage seedpod weevil (Coleoptera: Curculionidae) in canola. *The Canadian Entomologist* 134, 403–418.

Dosdall, L.M., Ulmer, B.J., Gibson, G.A.P. and Cárcamo, H.A. (2006) The spatio-temporal distribution dynamics of the cabbage seedpod weevil, *Ceutorhynchus obstrictus* (Coleoptera: Curculionidae), and its larval parasitoids in canola in western Canada. *Biocontrol Science and Technology* 16, 987–1006.

Dosdall, L.M., Gibson, G.A.P., Olfert, O.O. and Mason, P.G. (2009) Responses of Chalcidoidea (Hymenoptera) parasitoids to invasion of the cabbage seedpod weevil (Coleoptera: Curculionidae) in western Canada. *Biological Invasions* 11, 109–125.

Evans, K.A. and Bergeron, J. (1994) Behavioral and electrophysiological response of cabbage seed weevils (*Ceutorhynchus assimilis*) to conspecific odor. *Journal of Chemical Ecology* 20, 979–989.

Ferguson, A.W., Klukowski, Z., Walczak, B., Perry, J.N., Mugglestone, M.A., Clark, S.J. and Williams, I.H. (2000) The spatio-temporal distribution of adult *Ceutorhynchus assimilis* in a crop of winter oilseed rape in relation to the distribution of their larvae and that of the parasitoid *Trichomalus perfectus*. *Entomologia Experimentalis et Applicata* 95, 161–171.

Fox, A.S. and Dosdall, L.M. (2003) Reproductive biology of *Ceutorhynchus obstrictus* (Coleoptera: Curculionidae) on wild and cultivated Brassicaceae in southern Alberta. *Journal of Entomological Science* 38, 365–376.

Free, J.B. and Williams, I.H. (1978) The responses of the pollen beetle, *Meligethes aeneus*, and the seed weevil, *Ceutorhynchus assimilis*, to the oil-seed rape, *Brassica napus*, and other plants. *Journal of Applied Ecology* 15, 761–774.

Free, J.B. and Williams, I.H. (1979) The distribution of insect pests of oil-seed rape (*Brassica napus* L.) and the damage they cause. *Journal of Agricultural Science, Cambridge* 92, 139–149.

Free, J.B., Ferguson, A.W. and Winfield, S. (1983) Effect of various levels of infestation by the seed weevil (*Ceutorhynchus assimilis* Payk.) on the seed yield of oil-seed rape (*Brassica napus* L.). *Journal of Agricultural Science, Cambridge* 101, 589–596.

Gibson, G.A.P., Baur, H., Ulmer, B., Dosdall, L. and Muller, F. (2005) On the misidentification of chalcid (Hymenoptera: Chalcidoidea) parasitoids of the cabbage seedpod weevil (Coleoptera: Curculionidae) in North America. *The Canadian Entomologist* 137, 381–403.

Gibson, G.A.P., Gates, M.W. and Buntin, G.D. (2006a) Parasitoids (Hymenoptera: Chalcidoidea) of the cabbage seedpod weevil (Coleoptera: Curculionidae) in Georgia, USA. *Journal of Hymenoptera Research* 15(2), 187–207.

Gibson, G.A.P., Gillespie, D.R. and Dosdall, L. (2006b) The species of Chalcidoidea (Hymenoptera) introduced to North America for biological control of the cabbage seedpod weevil, and the first recovery of *Stenomalina gracilis* (Chalcidoidea: Pteromalidae). *The Canadian Entomologist* 138, 285–291.

Gillespie, D.R., Mason, P.G., Dosdall, L.M., Bouchard, P. and Gibson, G.A.P. (2006) Importance of long-term research in classical biological control: an analytical review of a release against the cabbage seedpod weevil in North America. *Journal of Applied Entomology* 130, 401–409.

Gupta, A. and Poorani, J. (2008) New distribution and host records of Chalcidoidea (Hymenoptera) from India. *Journal of Biological Control* 22, 261–265.

Hagen, K.S. (1946) The occurrence of *Ceutorhynchus assimilis* (Coleoptera: Curculionidae) in California. *Pan Pacific Entomologist* 22, 73.

Haye, T., Mason, P.G., Dosdall, L.M. and Kuhlmann, U. (2010) Mortality factors affecting the cabbage seedpod weevil, *Ceutorhynchus obstrictus* (Marsham), in its area of origin: a life table analysis. *Biological Control* 54, 331–341.

Haye, T., Mason, P.G., Dosdall, L.M., Gillespie, D.R., Gibson, G.A.P. and Kuhlmann, U. (2013) *Ceutorhynchus obstrictus* (Marsham), cabbage seedpod weevil (Coleoptera: Curculionidae). In: Mason, P.G. and Gillespie, D.R. (eds) *Biological Control Programs in Canada 2001–2012.* CAB International, Wallingford, UK, pp. 119–129.

Haye, T., Mason, P.G., Gillespie, D.R., Miall, J.H., Gibson, G.A.P., Diaconu, A., Brauner, A.M. and Kuhlmann, U. (2015) Determining the host specificity of the biological control agent *Trichomalus perfectus* (Hymenoptera: Pteromalidae): the importance of ecological host

range. *Biocontrol Science and Technology* 25(1), 21–47.
Heimbach, U. and Müller, A. (2013) Incidence of pyrethroid-resistant oilseed rape pests in Germany. *Pest Management Science* 69, 209–216.
Jourdheuil, P., Coutin, R. and Lacote, J.P. (1974) Evolution du niveau des populations de *Ceutorhynchus assimilis* Payk. dans le Lauragais (Aude, France). *Proceedings of the 4th International Rape Congress, Giessen, 4–8 June 1974*, pp. 511–518.
Kalischuk, A.R. and Dosdall, L.M. (2004) Susceptibilities of seven Brassicaceae species to infestation by the cabbage seedpod weevil (Coleoptera: Curculionidae). *The Canadian Entomologist* 136, 265–276.
Klukowski, Z. and Kelm, M. (2000) *Stenomalina gracilis* (Walker), a new parasitoid reared from *Ceutorhynchus napi* Gyll. in Poland. In: Paul, V.H., Foller, I., Evans, N. and Williams, I. (eds) *Working Group 'Integrated Control in Oilseed Crops': Proceedings of the Meeting at Prague (Czech Republic) May 31–June 2, 1999*. International Organization for Biological Control of Noxious Animals and Plants (OIBC/OILB), West Palearctic Regional Section (WPRD/SROP), Dijon, France, pp. 135–138.
Kuhlmann, U., Mason, P.G., Hinz, H.L., Blossey, B., De Clerck-Floate, R.A., Dosdall, L.M., McCaffrey, J.P., Schwarzlaender, M., Olfert, O., Brodeur, J., Gassmann, A., McClay, A.S. and Wiedenmann, R.N. (2006) Avoiding conflicts between insect and weed biological control: selection of non-target species to assess host specificity of cabbage seedpod weevil parasitoids. *Journal of Applied Entomology* 130, 129–141.
Laffin, R.D., Dosdall, L.M. and Sperling, F.A.H. (2005) Population structure of the cabbage seedpod weevil, *Ceutorhynchus obstrictus* (Marsham) (Coleoptera: Curculionidae): origins of North American introductions. *Environmental Entomology* 34, 504–510.
Lerin, J. (1984) Assessment of yield loss in winter rape due to seedpod weevil (*Ceutorhynchus assimilis* Payk.). II. Yield loss in a cage experiment. *Agronomy Journal* 4, 147–154.
Mason, P.G., Miall, J.H., Bouchard, P., Gillespie, D.R., Broadbent, A.B. and Gibson, G.A.P. (2011) The parasitoid communities associated with an invasive canola pest, *Ceutorhynchus obstrictus* (Coleoptera: Curculionidae) in Ontario and Quebec, Canada. *The Canadian Entomologist* 143, 524–537.
McCaffrey, J.P. (1992) Review of the US canola pest complex: cabbage seedpod weevil. In: *Proceedings, 1992 US Canola Conference, 5–6 March 1992, Washington DC*. American Pedigreed Seed Company, Memphis, Tennessee, pp. 140–143.
McCaffrey, J.P., Harmon, B.L., Brown, J., Brown, A.P. and Davis, J.B. (1999) Assessment of *Sinapis alba*, *Brassica napus* and *S. alba* × *B. napus* hybrids for resistance to cabbage seedpod weevil, *Ceutorhynchus assimilis* (Coleoptera: Curculionidae). *Journal of Agricultural Science* 132, 289–295.
McLeod, J.H. (1962) Cabbage seedpod weevil – *Ceutorhynchus assimilis* (Payk.) Curculionidae. In: McLeod, J.H., McGugan, B.M. and Coppel, H.C. (eds) *A Review of the Biological Control Attempts against Insects and Weeds in Canada.* CAB International, Wallingford, UK, pp. 5–6.
Muller, F.J., Baur, H., Gibson, G.A.P., Mason, P.G. and Kuhlmann, U. (2007) Review of the species of *Trichomalus* (Chalcidoidea: Pteromalidae) associated with *Ceutorhynchus* (Coleoptera: Curculionidae) host species of European origin. *The Canadian Entomologist* 139(5), 643–657.
Muller, F.J., Dosdall, L.M., Mason, P.G. and Kuhlmann, U. (2011) Larval phenologies and parasitoids of two seed-feeding weevils associated with hoary cress and shepherd's purse (Brassicaceae) in Europe. *The Canadian Entomologist* 143, 399–410.
Murchie, A.K., Williams, I.H. and Perry, J.N. (1999) Edge distributions of *Ceutorhynchus assimilis* and its parasitoid *Trichomalus perfectus* in a crop of winter oilseed rape (*Brassica napus*). *Biocontrol* 44, 379–390.
Ni, X., McCaffrey, J.P., Stoltz, R.L. and Harmon, B.L. (1990) Effects of postdiapause adult diet and temperature on oogenesis of the cabbage seedpod weevil (Coleoptera; Curculionidae). *Journal of Economic Entomology* 83, 2246–2251.
O'Connor, J.P. (1996) Further records of Irish Pteromalidae (Hymenoptera: Chalcidoidea), including ten species new to Ireland. *Irish Naturalists' Journal* 25(7), 254–257.
Pulkkinen, M. and Yang, Z.Q. (1984) The parasitoids and predators of *Saperda populnea* (Coleoptera: Cerambycidae) in Finland. *Annales Entomologici Fennici* 50(1), 7–12.
Ross, D.W., Brown, J., McCaffrey, J.P., Harmon, B.L. and Davis, J.B. (2008) Cabbage seedpod weevil resistance in canola (*Brassica napus* L.), yellow mustard (*Sinapis alba* L.) and canola × yellow mustard hybrids. *Canadian Journal of Plant Science* 88, 239–245.
Smart, L.E. and Blight, M.M. (1997) Field discrimination of oilseed rape, *Brassica napus* volatiles by cabbage seed weevil, *Ceutorhynchus assimilis*. *Journal of Chemical Ecology* 23(11), 2555–2567.
Smart, L.E., Blight, M.M. and Hick, A.J. (1997) Effect of visual cues and a mixture of isothiocyanates on trap capture of cabbage seed weevil, *Ceutorhynchus assimilis*. *Journal of Chemical Ecology* 23(4), 889–901.
Syngenta Canada Inc. (2015) *Approved pamphlet Voliam Xpress 30325*. Syngenta Canada Inc., Guelph, Ontario.
Sylven, E., and Svenson, G. (1975) Relationship between density of *Ceutorhynchus assimilis* Payk. (Col.) and damage by *Dasyneura brassicae* Winn. (Cec.) in a

cage experiment in summer turnip rape. *Statens Vaxtskyddsanstalt, Meddelanden* 16, 53–60.

Tansey, J.A., Dosdall, L.M., Keddie, A., Fletcher, R.S. and Kott, L.S. (2010) Antixenosis and antibiosis resistance to *Ceutorhynchus obstrictus* in novel germplasm derived from *Sinapis alba* × *Brassica napus*. *The Canadian Entomologist* 142, 212–221.

Tulisalo, U., Korpela, S. and Pohto, A. (1976) The yield loss caused by the seedpod weevil *Ceutorhynchus assimilis* Payk. (Col., Curculionidae) on summer turnip rape in cage experiments. *Annales Entomologici Fennici* 42, 98–102.

Ulber, B., Williams, I.H., Klukowski, Z., Luik, A. and Nilsson, C. (2010) Parasitoids of oilseed rape pests in Europe: key species for conservation biocontrol. In: Williams, I.H. (ed.) *Biocontrol-Based Integrated Management of Oil-Seed Rape Pests*. Springer, Dordrecht, The Netherlands, pp. 45–76.

Ulmer, B.J. and Dosdall, L.M. (2006a) Spring emergence biology of the cabbage seedpod weevil (Coleoptera: Curculionidae). *Annals of the Entomological Society of America* 99, 64–69.

Ulmer, B.J. and Dosdall, L.M. (2006b) Glucosinolate profile and oviposition behavior in relation to susceptibilities of Brassicaceae to the cabbage seedpod weevil. *Entomologia Experimentalis et Applicata* 121, 203–213.

Veromann, E., Williams, I.H., Kaasik, R. and Luik, A. (2010) Potential of parasitoids to control populations of the weevil *Ceutorhynchus obstrictus* (Marsham) on winter oilseed rape. *International Journal of Pest Management* 57(1), 85–92.

Veromann, E., Toome, M., Kännaste, A., Kaasik, R., Copolovici, L., Flink, J., Kovács, G., Narits, L., Luik, A. and Niinemets, Ü. (2013) Effects of nitrogen fertilization on insect pests, their parasitoids, plant diseases and volatile organic compounds in *Brassica napus*. *Crop Protection* 43, 79–88.

Yefremova, Z.A. (2015) An annotated checklist of the Eulophidae (excl Tetrastichinae) (Hymenoptera: Chalcidoidea) of Israel. *Zootaxa* 3957(1), 1–36.

Yoshimoto, C.M. (1971) Revision of the genus *Euderus* of America north of Mexico (Hymenoptera: Eulophidae). *The Canadian Entomologist* 103 541–578.

Zamojska, J. and Węgorek, P. (2014) Preliminary studies on the susceptibility level of *Ceutorhynchus assimilis* (Coleoptera: Curculionidae) to acetamiprid and chlorpyrifos in Poland and resistance mechanisms of the pest to acetamiprid. *Journal of Insect Science* 14(265), 1–6.

6 Biology, Ecology and Management of Pollen Beetle *Brassicogethes viridescens* (Coleoptera: Nitidulidae)

Christine Noronha[1]* and Peter G. Mason[2]

[1]*Agriculture and Agri-Food Canada, Charlottetown, Prince Edward Island, Canada;* [2]*Agriculture and Agri-Food Canada, Ottawa, Ontario, Canada*

6.1 Introduction

Audisio *et al.* (2009) reviewed the status of species in the subfamily Meligethinae (Coleoptera: Nitidulidae) using morphological and molecular data. Thirty-eight species of *Meligethes*, whose larval development is strictly associated with flowers of Brassicaceae, were re-assigned to the new genus *Brassicogethes*. Among the North American species are *Brassicogethes aeneus* (Fabricius), designated as the type species for the genus, *B. cleominis* (Easton), *B. simplipes* (Easton) and *B. viridescens* (Fabricius), which Hoebeke and Wheeler (1996) documented as being adventive. There is some doubt that the *B. aeneus* in North America is the same species as the *B. aeneus* in Europe, which is a major pest of oilseed rape crops, *Brassica napus* L., *B. rapa* L. (Brassicaceae) in that region (P.G. Mason *et al.*, unpublished).

6.2 Description

The pollen beetle (often referred to as bronzed blossom beetle in Europe) *Brassicogethes viridescens* is an important pest of canola/oilseed rape crops such as *Brassica napus*, *Brassica rapa* and *Brassica juncea* L. (Brassicaceae) in North America and Europe. Adults are shiny, bronze-coloured, with an elongate to oval convex body and yellowish legs and mouthparts. Pubescence is fine, short and sparse and uniformly distributed. Dorsal punctations are coarse and conspicuous, being denser and finer on the head and coarser and scattered on the elytra. Body size ranges from 2.0 mm to 2.9 mm. The distinct subapical tooth on the posterior edge of the middle femur is diagnostic and distinguishes *B. viridescens* from other species within this genus (Fig. 6.1) (Hoebeke and Wheeler, 1996). Morphology of the immature stages for both *B. aeneus* and *B. viridescens* can be found in Osborne (1965), though the distinguishing characters are difficult to see.

6.3 Life Cycle

In Europe, both *B. viridescens* and its sister species *B. aeneus* are univoltine and overwinter as adults in leaf litter debris along field borders and hedgerows beside rape fields (Nilsson, 1989a; Marczali and Nádasy, 2006). Adults of both species emerge from overwintering sites in early summer when ambient temperatures exceed 10°C throughout the day; they fly in search of flowers and pollen to feed on from a variety of different plant species (Free and Williams, 1978; Nilsson, 1989a; Williams, 2010). It is during this period that the reproductive organs of the female beetles begin to develop (Hoebeke and Wheeler, 1996). As temperatures reach a consistent 12–15°C throughout the day, the beetles seek out a larval host plant, i.e. plants from the family Brassicacae, to lay their eggs and to begin feeding on the pollen of these plants (Fig. 6.2) (Free and Williams, 1978; Alford *et al.*, 2003; Metspalu *et al.*, 2011). Females chew holes into the side of developing cruciferous buds and deposit one to two eggs inside the developing flower buds (Fig. 6.3). The eggs hatch

*Corresponding author. E-mail: christine.noronha@agr.gc.ca

Fig. 6.1. Middle femur of *Brassicogethes viridescens* showing the distinctive subapical tooth on the posterior edge.

Fig. 6.2. *B. viridescens* adults searching for oviposition sites on newly formed green buds.

into first-instar larvae 4–9 days later and moult to second instar soon after (Osborne, 1965; Nilsson, 1989b; Hoebeke and Wheeler, 1996). There are only two larval instars. First instars feed on the pollen within the bud they hatched in, while the second instars move out of the bud as it opens and migrate to feed on the pollen of other newly opened flowers (Williams and Free, 1979). On completion of the second instar development, the larvae drop to the soil to pupate. The new generation adults emerge and move in search of pollen

Fig. 6.3. A clutch of *B. viridescens* eggs within a green bud.

and flowers of other plant species to feed on before searching for overwintering sites within hedgerows and field borders (Nilsson, 1989 a, b, c; Marczali and Nádasy, 2006; Marczali *et al.*, 2006).

6.4 Ecology

6.4.1 Host plant and damage

Adults of pollen beetles feed on the pollen on several plant species during the post- and prehibernation period; however, oviposition only takes place in the buds of plants belonging to the family Brassicaceae. Thus, the pollen beetles are specialists when searching for oviposition sites and for larval development but generalists when searching for adult food. Because larvae feed and survive only on Brassicaceae, they can cause serious damage and yield reduction to oilseed crops such as *B. napus*, *B. rapa*, *B. juncea* and *Brassica carinata* (Ekbom and Borg, 1996). Adults use both olfactory and visual cues when searching for oviposition sites. Once sites are found, chemical cues from the bud surface are used to determine the quality of the oviposition site (Evans and Allen-Williams, 1994; Giamoustaris and Mithen, 1996; Hopkins and Ekbom, 1999; Cook *et al.*, 2002, 2004; Marczali *et al.*, 2006; Jönsson *et al.*, 2007; Hervé *et al.*, 2015). Once a suitable oviposition host is found the pollen beetle female will first chew a hole in the green buds before depositing one to two eggs per bud, with an average of 4.2 eggs per day (Ekbom and Borg, 1996). The presence of pollen plays an important role in oviposition site selection. Cook *et al.* (2004) demonstrated improved survival of the larvae and higher pupal and adult weight when larvae were fed pollen.

In Europe, immigration of adults into oilseed rape fields is synchronized with the development of the green buds and was found to occur when approximately half of the plants had formed their first buds (Nilsson, 1989a, b). However, temperature and precipitation influence the movement of pollen beetles into oilseed rape fields, causing the start of immigration to vary from year to year (Nilsson, 1989a; Finch *et al.*, 1990; Collier and Finch, 2001). *B. aeneus* is known to be active at lower temperatures: it moves into fall-planted (winter) oilseed rape fields early during bud formation and can seriously damage the crop; later in the season it will move into spring-planted (spring) oilseed rape fields but causes less damage in this crop. *B. viridescens*, on the other hand, is not as cold tolerant and emerges later in the season in Europe and mainly causes damage to the spring oilseed rape crop (Cook *et al.*, 2004). In North America, where the crop consists primarily of spring-planted canola, *B. viridescens* poses a significant risk to the industry (Dosdall and Mason, 2010).

Colonization and egg laying occur at the early bud stage, when the crop is the most susceptible. Larval feeding within the bud at the early bud stage causes the bud to drop off, resulting in yield reduction (Hansen, 2004). However, rape plants are known to compensate by producing auxiliary shoots with buds to compensate for the natural loss of some buds (Tatchell, 1983). This growth compensation is adequate when pollen beetle infestation is low but yield can be severely compromised under high pressure. Nilsson (1987) reported a 70% reduction in yield due to damage caused by *B. aeneus*. In addition to bud abortion, yield is compromised by adults and second-instar larvae feeding on pollen from open flowers, thus reducing pollen availability for fertilization (Fig. 6.4) (Hansen, 2004; Slater *et al.*, 2011). The economic threshold for pollen beetle, mainly *B. aeneus*, varies in different regions of Europe and is based on the number of plants/m^2. It can range from 25 pollen beetles per plant if less than 30 plants/m^2 to seven pollen beetles per plant if more than 70 plants/m^2 (Cook *et al.*, 2013). A threshold of three pollen beetles per plant was established for *B. aeneus* in spring oilseed rape in Denmark (Hansen, 2004). In North America, the economic threshold for *B. viridescens* has yet to be established.

Fig. 6.4. *B. viridescens* adult feeding on the pollen inside an open canola flower.

6.4.2 Distribution

Brassicogethes viridescens is widely distributed throughout Europe, the Near East, western Siberia (Audisio and De Biase, 1999; Audisio *et al.*, 2000, 2009) and north-west China (Lin *et al.*, 2015) and was introduced into North America through various ports along the Atlantic coast (Hoebeke and Wheeler, 1996). In North America, the first record of this Palaearctic species was from collections made in Nova Scotia (Brown, 1967). However, it was not until 1996 that its presence in Maine (USA) and the Canadian provinces of Nova Scotia, Prince Edward Island and Quebec was confirmed (Hoebeke and Wheeler, 1996; Mason *et al.*, 2003; Majka *et al.*, 2006, 2008). While *B. aeneus* is the dominant species in Europe, attacking mainly winter oilseed rape and to a lesser extent spring oilseed rape, *B. viridescens* feeds mainly on spring oilseed rape. It is usually found in smaller numbers but occasionally populations can be high enough to warrant control measures (Nilsson, 1989a). In North America, *B. viridescens* is found attacking spring canola, which is the primary crop grown in this region. Mason *et al.* (2003), using the Climex™ model, showed that the actual distribution of *B. viridescens* in eastern Canada matched the predicted model and projected the inevitable westward movement of this species into the principal canola-producing regions of Canada. The risk of *B. viridescens* populations establishing in western Canada and the potential to cause severe damage are heightened by the impact of climate change (Olfert and Weiss, 2006).

6.5 Management

6.5.1 Monitoring

Monitoring of *B. viridescens* populations to determine if control thresholds have been reached within a crop is essential to initiate an effective pest management programme and make decisions on which control measures should be applied. Because most pest species show an aggregated distribution or populations may be higher along the edge of the field, it is important to conduct an unbiased monitoring protocol to obtain an accurate estimation of the population size. Monitoring for pollen beetles can include: direct plant counts, i.e. number of beetles/plant when walking in a transect through the field; beat cloth sampling; yellow pan traps; and baited yellow sticky-card traps (Blight and Smart, 1999; Kaasik *et al.*, 2013; Metspalu *et al.*, 2015).

6.5.2 Trap crop

Trap crops have been used for many years as part of an integrated pest management (IPM) strategy for pest insects. This strategy includes either planting a crop that is more attractive to the pest or planting a strip earlier than the main crop to lure the pest away and reduce damage. Surrounding a field with a strip of an early-flowering trap crop would result in beetles being attracted to and congregating on the early crop. This strip could then be sprayed with an insecticide to reduce the population before the beetles enter the main crop (Hokkanen *et al.*, 1986). For pollen beetles, turnip rape (*B. rapa*) was found to be very attractive, making it a good potential trap crop for use in an IPM strategy (Cook *et al.*, 2006). However, the attractiveness of these trap crops to pollen beetles is dependent on the growth stage of the plant, which should be taken into consideration when using this technique (Veromann *et al.*, 2012; Kaasik *et al.*, 2014). Frearson *et al.* (2005) found that the trap crop should be in the early flowering stage in order to protect the vulnerable green bud stage of the main crop. The role of wild cruciferous plants around the field, which may be used as oviposition sites and as a food source, should not be discounted (Metspalu *et al.*, 2011). A good understanding of pollen beetle phenology and the most attractive growth stage of the trap crop is crucial when using this strategy.

6.5.3 Chemical control

Application of insecticides is the primary method used to control pollen beetle adults in Europe but resistance to several pyrethroids used against *B. aeneus* is fast developing. Richardson (2008a) found a significant correlation between the number of insecticide applications and the time for resistance to develop: the more spray applications, the earlier was the development of resistance. Because *B. aeneus* emerges early and attacks both winter and spring oilseed rape it is exposed to a higher number of insecticide applications, which has sped up resistance development. Resistance of *B. aeneus* to pyrethroids has been documented in several areas in Europe (Richardson, 2008a, b; Slater *et al.*, 2011). Increasing an insecticide's duration of control by mixing it with an adjuvant could decrease the number of spray applications and delay the development of resistance. Palagacheva *et al.* (2014) found that an application of indoxicarb (Avaunt®) at the buttoning stage (the appearance of yellow buds on the crop) gave good control of *B. aeneus*. Mixing it with a vegetable oil adjuvant (Codacide) increased the control duration.

To date, *B. viridescens* has not shown resistance to any insecticide used in Europe. In North America, currently there are no insecticides specifically registered for control of this pest species.

6.5.4 Biological control

Natural enemy communities that are associated with pollen beetles can suppress pest populations and their conservation and manipulation should be the basis of any IPM programme. Natural enemies that show impact in the area of origin (Europe) are good candidates for introduction as biological control agents into the invaded region (Canada) where there are vacant niches. To date, no parasitoids have been reared from *B. viridescens* in North America and predator communities are being studied (G. Labrie *et al.*, unpublished).

Among the nine parasitoid species associated with pollen beetles in Europe, the univoltine *Phradis interstitialis* (Thomson), *Phradis morionellus* (Holmgren) and *Tersilochus heterocerus* Thomson (Hymenoptera: Ichneumonidae) are the most abundant in fall-planted (winter) crops and the latter two species are most prevalent in spring-planted crops (Nilsson, 2003; Ulber *et al.*, 2010).

Thus, *P. morionellus* and *T. heterocerus* are good candidates for biological control of *B. viridescens* in spring-planted canola in eastern Canada. *P. morionellus* may be the superior of the two because it is more abundant in northern climates (Ulber *et al.*, 2010) and increased levels of parasitism by this species have been associated with significant declines in pollen beetle (*B. aeneus*) populations (Hokkanen, 2008). Although *Brachyserphus parvulus* (Nees) (Hymenoptera: Proctotrupidae), *Blacus nigricornis* Haeselbarth and *Diospilus capito* Nees (Hymenoptera: Braconidae) may also be common in some spring-planted crops, they are multivoltine and appear to attack pollen beetles in fall-planted crops during their first generation, followed by pollen beetles attacking spring-planted crops in their second generation (Nilsson, 2003). These species are unlikely to be of value as biological control agents of *B. viridescens* in Canada, because canola is almost exclusively spring-planted.

Nilsson (2003) provided details of the biology of *P. morionellus* and *T. heterocerus* in Europe and these are summarized here. Adults of both species usually appear in the crop at the beginning of flowering and females prefer to oviposit in large, second-instar larvae. Parasitoid larvae hatch from eggs just before host larvae are fully developed and ready to pupate in the soil. Larval growth occurs when the host is within the pupal chamber. After about 1 month as a pupa, the parasitoid adult has developed but stays within the cocoon, in diapause, until the following spring, coinciding with flowering of the crop. Upon emergence (280–330 degree-days above 5°C) the females are able to oviposit and live for about 1 week at summer temperatures. Males emerge before females and are shorter lived. Multi-parasitism is common and sometimes five or more eggs are found in a single host, but only a single parasitoid survives. Even so, parasitism levels of > 50% have been reported. Population studies suggest that to attain greatest biological control, several parasitoid species should be present.

Predator communities have some impact on pollen beetles in Europe. Although most studies have focused on *B. aeneus* the communities tend to be made up of generalist species that would also attack adults, larvae and pupae of *B. viridescens*. Adults may be prey to spiders, particularly web-making *Theridion* spp. (Theridiidae), orb-web-forming *Araneus* spp. (Araneidae) and *Tetragnatha* spp. (Tetragnathidae) and hunters such as the crab spider *Misumena vatia* (Hentz) (Thomisidae) (Alford, 2011). Species of all of these genera are common in North America (Bradley, 2013). Greatest impact is by ground-dwelling generalist predators, particularly ground beetles *Amara similata* (Gyllenhal), *Anchomenus dorsalis* (Pontoppidan), *Carabus* spp., *Clivina fossor* (L.), *Poecilus* spp., *Pterostichus melanarius* (Illiger) and *Trechus quadristriatus* (Schrank) (Coleoptera: Carabidae) that feed on mature pollen beetle larvae, prepupae and pupae (Büchs, 2003; Alford, 2011). Estimates of pollen beetle larval and pupal mortality caused by these predators are highly variable, ranging from 3% to 80%. Predator activity cycles, soil properties and agronomic practices appear to have significant influences (Büchs, 2003; Hokkanen, 2008). At least three of the European ground beetle species (*C. fossor*, *P. melanarius* and *T. quadristriatus*) are adventive in eastern North America and each of the ground beetle genera associated with pollen beetle habitats in Europe is well represented (Bousquet *et al.*, 2013).

Entomopathogens are present in natural ecosystems mainly in soil but their interactions with pests such as pollen beetle are poorly understood. Several pathogenic fungi, *Beauveria bassiana* (Balsamo) Vuillemin (Cordycipitaceae), *Metarhizium anisopliae* (Metschnikoff) Sorokīn (Clavicipitaceae) and *Paecilomyces fumosoroseus* (Wize) A.H.S. Br. & G. Sm. (Trichomaceae), the entomopathogenic nematode *Steinernema feltia* (Filipjev) (Steinernematidae) and *Nosema meligethi* I. & R. (Microsporida: Nosematidae) have been studied in Europe (Hokkanen, 2003). However, it is unlikely that entomopathogens will be useful in managing pollen beetle populations (Ekbom, 2010), including *B. viridescens*.

6.5.5 Land management

Crop management and landscape types have been shown to have an influence on pollen beetle damage and abundance. For example, Rusch *et al.* (2011) demonstrated that plants with a higher nitrogen index are better able to produce new buds to compensate for pollen beetle damage. They also determined that semi-natural habitats (overwintering sites) play a role in infestations through complex interactions that include such factors as distance from oilseed rape fields and wind direction. Thus, favourable winds and close proximity to oilseed rape fields may lead to increased pollen beetle populations. However, semi-natural habitats also have a positive effect on the abundance of pollen beetle parasitoids (Rusch *et al.*, 2011; Scheid *et al.*, 2011).

Maintaining natural enemy populations can be achieved through the use of conservation strips and no-till practices. Büchi (2002) showed that parasitism of pollen beetles was higher in fields with adjacent wildflower strips than in those fields adjacent to managed meadows, though the mechanism to explain this was not clear. The main parasitoid species overwinter in the soil in pupal chambers formed by the host. Tillage practices also influence parasitoid populations, with twice as many parasitoids emerging in spring from no-till (fallow or direct seeding) management systems (Nilsson, 2010). As Ekbom (2010) proposed, sowing some areas with a host crop where pollen beetles can lay eggs and provide hosts for parasitoids (and predators) would help to conserve natural enemy populations. Kaasik *et al.* (2014) found that parasitism of pollen beetles on *B. juncea* was higher compared with *B. napus*, *B. rapa* and *Sinapis alba* L. (Brassicaceae) and suggested that plant species that are close relatives of canola/oilseed rape might be used not only to trap pollen beetle adults but also to support natural enemy populations that could decrease the beetle numbers.

6.6 Outlook for North America

B. viridescens is well established in eastern North America and there is real potential for it to spread to the major canola-growing regions of western Canada. Climate change will likely result in an increase in the range and relative abundance of *B. viridescens* for temperature increases between 1°C and 7°C. Risks will likely become more intense, both in terms of severity where it presently occurs and its ability to become established in new areas (Olfert and Weiss, 2006). As the European experience shows, the widespread use of insecticides to control pollen beetle will only exacerbate the problem, leading to ongoing outbreak populations. Furthermore, there is a real threat that *B. viridescens* in North America, like *B. aeneus* in Europe, will develop resistance to pesticides. Thus, alternative management strategies must be considered. Of these, biological control is a key option but, for it to be effective, conservation of natural enemies must be a priority. Successful biological control of pollen beetle is unlikely if insecticide use is ongoing (Ekbom, 2010). Monitoring, trap cropping, the use of thresholds and conservation biological control have the potential to significantly reduce insecticide use against pollen beetles (Cook and Denholm, 2008). A proactive biologically based integrated pest management strategy would enable the threat posed by *B. viridescens* to be minimized. Waiting for the crisis will result in severe consequences for the North American canola industry.

References

Alford, D.V. (2011) *Plant Pests*. Harper Collins, London.

Alford, D.V., Nilsson, C. and Ulber, B. (2003) Insect pests of oilseed rape crops. In: Alford, D.V. (ed.) *Biocontrol of Oilseed Rape Pests*. Blackwell Publishing, Oxford, UK, pp. 9–41.

Audisio, P. and De Biase, A. (1999) Bionomic and morphological evidence of a new southern European species of the *Meligethes viridescens* complex (Coleoptera: Nitidulidae). *Folia Heyrovskyana* 7, 99–113.

Audisio, P., De Biase, A., Romanelli, P., Angelici, M.C., Ketmaiser, V. and De Matthaeis, E. (2000) Molecular re-examination of the taxonomy of the *Meligethes viridescens* species complex (Coleoptera: Nitidulidae). *Biochemical Systematics and Ecology* 28, 1–13.

Audisio, P., Cline, A.R., De Biase, A., Antonini, G., Mancini, E., Trizzino, M., Costantini, L., Strika, S., Lamanna, F. and Cerretti, P. (2009) Preliminary re-examination of genus-level taxonomy of the pollen beetle subfamily Meligethinae (Coleoptera: Nitidulidae). *Acta Entomologica Musei Nationalis Pragae* 49, 341–504.

Blight, M.M. and Smart, L.E. (1999) Influence of visual cues and isothiocynate lures on capture of the pollen beetle, *Meligethes aenus* in field traps. *Journal of Chemical Ecology*, 25, 1051–1515.

Bousquet, Y., Bouchard, P., Davies, A.E. and Sikes, D.S. (2013) *Checklist of Beetles (Coleoptera) of Canada and Alaska*, 2nd edn. Pensoft, Sofia-Moscow, Russia.

Bradley, R.A. (2013) *Common Spiders of North America*. University of California Press, Berkeley, California.

Brown, W.J. (1967) Notes on the extralimital distribution of some species of Coleoptera. *The Canadian Entomologist*, 99, 85–93.

Büchi, R. (2002) Mortality of pollen beetle (*Meligethes* spp.) larvae due to predators and parasitoids in rape fields and the effect of conservation strips. *Agriculture, Ecosystems and Environment* 90, 255–263.

Büchs, W. (2003) Predators as biocontrol agents of oilseed rape pests. In: Alford, D.V. (ed.) *Biocontrol of Oilseed Rape Pests*. Blackwell Publishing, Oxford, UK, pp. 279–298.

Collier, R.H. and Finch, S. (2001) Forecasting attacks by pest insects of cruciferous crops. In: Endersby, N.M. and Ridland, P.M. (eds) *The Management of Diamondback Moth and Other Crucifer Pests: Proceedings of the 4th International Workshop, 26–29 November 2001*,

Melbourne, Australia. The Regional Institute, Gosford, Australia, pp. 163–168.
Cook, S.M. and Denholm, I. (2008) Ecological approaches to the control of pollen beetles in oilseed rape. *EPPO Bulletin* 38, 110–113.
Cook, S.M., Bartlet, E., Murray, D.A and Williams, I.H. (2002) The role of pollen order in the attraction of pollen beetles to oilseed rape flowers. *Entomologia Experimentalis et Applicata* 104, 43–50.
Cook, S.M., Murray, D.A. and Williams, I.H. (2004) Do pollen beetles need pollen? The effect of pollen on oviposition, survival, and development of a flower-feeding herbivore. *Ecological Entomology* 29, 164–173.
Cook, S.M., Smart, L.E., Martin, J.L., Murray, D.A., Watts, N.P. and Williams, I.H. (2006) Exploitation of host plant preferences in pest management strategies for oilseed rape (*Brassica napus*). *Entomologia Experimentalis et Applicata* 119, 221–229.
Cook, S., Ellis, S. and Nicholls, C. (2013) *Monitoring and Control of Pollen Beetle in Oilseed Rape*. Information sheet 18.T0845 245 0009. HGCA Publications, London.
Dosdall, L.M. and Mason, P.G. (2010) Key pests and parasitoids of oilseed rape or canola in North America. In: Williams, I.H. (ed.) *Biocontrol-based Integrated Management of Oilseed Rape Pests*. Springer, Dordrecht, The Netherlands, pp. 167–213.
Ekbom, B. (2010) Pests and their enemies in spring oilseed rape in Europe and challenges to integrated pest management. In: Williams, I.H. (ed.) *Biocontrol-based Integrated Management of Oilseed Rape Pests*. Springer, Dordrecht, The Netherlands, pp. 151–165.
Ekbom, B. and Borg, A. (1996) Pollen beetle (*Meligethes aeneus*) oviposition preference of different host plants. *Entomologia experimentalis et Applicata* 78, 291–199.
Evans, K.A. and Allen-Williams, L.J. (1994) Laboratory and field response of the pollen beetle *Meligethes aeneus*, to the odour of oilseed rape. *Physiological Entomology* 19, 285–290.
Finch, S., Collier, R.H. and Elliott, M.S. (1990) Seasonal variations in the timing of bronzed-blossom beetle (*Meligethes aeneus/Meligethes viridescens*) on horticultural brassicas. In: *Proceedings British Crop Protection Conference – Pests and Diseases,* Brighton. British Crop Protection Council, Farnham, UK, pp. 349–354.
Frearson, D.J.T., Ferguson, A.W., Campbell, J.M. and Williams, I.H. (2005) The spatial dynamics of pollen beetles in relation to inflorescence growth stage of oilseed rape: implications for trap crop strategies. *Entomologia Experimentalis et Applicata* 116, 21–29.
Free, J.B. and Williams, I.H. (1978) The response of the pollen beetle, *Meligethes aeneus*, and the seed weevil, *Ceuthorhynchus assimilis*, to oil-seed rape, *Brassica napus* and other plants. *Journal of Applied Ecology* 15,761–764.
Giamoustaris, A. and Mithen, R. (1996) The effects of flower colour and glucosinolates on the interaction between oilseed rape and pollen beetles. *Entomologia Experimentalis et Applicata* 80, 206–208.
Hansen, L.M. (2004) Economic damage threshold model for pollen beetles (*Meligethes aeneus* F.) in spring oilseed rape (*Brassica napus* L.) crops. *Crop Protection* 23, 43–46.
Hervé, M.R., Garcia, N., Trabalon, M., Le Ralec, A., Delourme, R. and Cortesero, A.M. (2015) Oviposition behavior of the pollen beetle (*Meligethes aeneus*): a functional study. *Journal of Insect Behavior* 28, 107–119.
Hoebeke, E.R. and Wheeler, A.G. Jr (1996) *Meligethes viridescens* (F.) (Coleoptera: Nitidulidae) in Maine, Nova Scotia, and Prince Edward Island: diagnosis, distribution and bionomics of a Palaearctic species new to North America. *Proceedings of the Entomological Society of Washington* 98, 221–227.
Hokkanen, H.M.T. (2003) Pathogens of oilseed rape pests. In: Alford, D.V. (ed.) *Biocontrol of Oilseed Rape Pests*. Blackwell Publishing, Oxford, UK, pp. 299–322.
Hokkanen, H.M.T. (2008) Biological control methods of pest insects in oilseed rape. *EPPO Bulletin* 38, 104–109.
Hokkanen, H., Granlund, H., Husberg, G.B. and Markkula, I.M. (1986) Trap crops used successfully to control *Meligethes aeneus* (the rape blossom beetle). *Annales Entomologici Fennici* 52, 115–120.
Hopkins, R.J. and Ekbom, B. (1999) The pollen beetle *Meligethes aeneus,* changes egg production rate to match host quality. *Oecologia* 120, 274–278.
Jönsson, M., Rosdahl, K. and Anderson, P. (2007) Responses to olfactory and visual cues by overwintered and summer generations of pollen beetle *Meligethes aeneus. Physiological Entomology* 32, 188–193.
Kaasik R., Watts, N.P., Murray, D.A., Veromann, E. and Cook, S.M. (2013) Effects of monitoring position and time of day on pollen beetle numbers in crops of oilseed. *Integrated Control of Oilseed Crops IOBC/WPRS Bulletin* 96, 123–131.
Kaasik R., Kovács, G., Toome, M., Metspalu, L. and Veromann, E. (2014) The relative attractiveness of *Brassica napus*, *B. rapa*, *B. juncea* and *Sinapis alba* to pollen beetles. *BioControl* 59, 19–28.
Lin, X-L., Chen, Y., Huang, M. and Yang, X-K. (2015) A new species of the genus *Meligethes* Stephens (Coleoptera: Nitidulidae: Meligethinae) from China. *Zoological Systematics* 40, 268–289.
Majka, C.G., Webster, R. and Cline, A.R. (2006) Nitidulidae and Kateretidae (Coleoptera: Cucujoidea) of the Maritime Provinces of Canada. I. New records from Nova Scotia and Prince Edward Island. *The Canadian Entomologist* 138, 314–332.
Majka, C.G., Webster, R. and Cline, A.R. (2008) New record of Nitidulidae and Kateretidae (Coleoptera) from New Brunswick, Canada. *ZooKeys* 2, 337–356.
Marczali, Z. and Nádasy, M. (2006) Wintering characteristic of the *Meligethes* species in Hungary. *Central European of Agriculture* 7, 283–288.

Marczali, Z., Nádasy, M. and Keszthelyi, S. (2006) A study on swarming phenology and food plant selection of *Meligethes* species in four consecutive years. *Opatija* 34, 41–44.

Mason, P.G., Olfert, O., Sluchinski, L., Weiss, R.M., Boudreault, C., Grossrieder, M. and Kuhlmann, U. (2003) Actual and potential distribution of an invasive canola pest, *Meligethes viridescens* (Coleoptera: Nitidulidae), in Canada. *The Canadian Entomologist* 135, 405–413.

Metspalu, L., Williams, I.H., Jõgar, K., Ploomi, A., Hiiesaar, K., Lääniste, P., Švilponis, E., Mänd, M. and Luik, A. (2011) Distribution of *Meligethes aeneus* (F.) and *M. viridescens* (F.) on cruciferous plants. *Žemdirbystė Agriculture* 98, 27–34.

Metspalu, L., Veromann, E., Kaasik, R., Kovac, G., Williams, I.H., Mänd, M. (2015) Comparison of sampling methods for estimating the abundance of *Meligethes aeneus* on oilseed crops. *International Journal of Pest Management* 61, 312–319.

Nilsson, C. (1987) Yield losses in summer rape caused by pollen beetles *Meligethes* spp. *Swedish Journal of Agricultural Research* 17, 105–111.

Nilsson, C. (1989a) The pollen beetle (*Meligethes aeneus* F.) in winter and spring rape at Alnarp 1976–1978. I. Migration and sex ratio. *Växtskyddsnotiser* 52, 134–138.

Nilsson, C. (1989b) The pollen beetle (*Meligethes aeneus* F.) in winter and spring rape at Alnarp 1976–1978. II. Oviposition. *Växtskyddsnotiser* 52, 139–144.

Nilsson, C. (1989c) The pollen beetle (*Meligethes aeneus* F.) in winter and spring rape at Alnarp 1976–1978. III. Mortality factors. *Växtskyddsnotiser* 52, 145–150.

Nilsson, C. (2003) Parasitoids of pollen beetles. In: Alford, D.V. (ed.) *Biocontrol of Oilseed Rape Pests*. Blackwell Publishing, Oxford, UK, pp. 73–85.

Nilsson, C. (2010) Impact of soil tillage on parasitoids of oileed rape pests. In: Williams, I.H. (ed.) *Biocontrol-Based Integrated Management of Oilseed Rape Pests*. Springer, Dordrecht, The Netherlands, pp. 305–311.

Olfert, O. and Weiss R.M. (2006) Impact of climate change on potential distributions and relative abundance of *Oulema melanopus*, *Meligethes viridescens* and *Ceutorhynchus obstrictus* in Canada. *Agriculture, Ecosystems and Environment* 113, 295–301.

Osborne, P. (1965) Morphology of the immature stages of *Meligethes aeneus* (F.) and *M. viridescens* (F.) (Coleoptera, Nitidulidae). *Bulletin of Entomological Research* 55, 747–759.

Palagacheva, N., Dospatliev, L. and Dimitrov, Y. (2014) Efficient pest control of pollen beetle (*Meligethes aeneus* F.) and possibilities for protecting the pollinators in oilseed rape agrocenosis. *International Journal of Scientific and Research Publications* 4, 1–4.

Richardson, D.M. (2008a) Summary of findings from a participant country pollen beetle questionnaire. *OEPP/EPPO Bulletin* 38, 68–72.

Richardson, D.M. (2008b) Pollen beetle in the UK: the start of a resistance problem? *OEPP/EPPO Bulletin* 38, 73–74.

Rusch, A., Valantin-Morison, M., Sarthou, J.-P. and Roger-Estrade, J. (2011) Multi-scale effects of landscape complexity and crop management on pollen beetle parasitism rate. *Landscape Ecology* 26, 473–486.

Scheid, B.E., Thies, C. and Tscharntke, T. (2011) Enhancing rape pollen beetle parasitism within sown flower fields along a landscape complexity gradient. *Agriculture and Forest Entomology* 13, 173–179.

Slater, R., Ellis, S., Genay, J.P., Heimbach, U., Huart, G., Sarazin, M., Longhurst, C., Müller, A., Nauen, R., Rison, J.L. and Robin, F. (2011) Pyrethroid resistance monitoring in European populations of pollen beetle (*Meligethes* spp.) – a coordinated approach through the insecticide resistance committee IRAC. *Pest Management Science* 67, 633–638.

Tatchell, G.M. (1983) Compensation in spring-sown oilseed rape (*Brassica napus* L.) plants in response to injury to their flower buds and pods. *Journal of Agricultural Science* 101, 565–573.

Ulber, B., Williams, I.H., Klukowski, Z., Luik, A. and Nilsson, C. (2010) Parasitoids of oilseed rape pests in Europe: key species for conservation biological control. In: Williams, I.H. (ed.) *Biocontrol-based Integrated Management of Oilseed Rape Pests*. Springer, Dordrecht, The Netherlands, pp. 45–76.

Veromann, E., Metspalu, L., Williams, I.H., Hiiesaar, K., Mand, M., Kaasik, R., Kovács, G., Jogar, K., Svilponis, E., Kivimagi, I., Ploomi, A. and Luik, A. (2012) Relative attractiveness of *Brassica napus*, *Brassica nigra*, *Eruca sativa* and *Raphanus sativus* for pollen beetle (*Meligethes aeneus*) and their potential for use in trap cropping. *Arthropod–Plant Interactions* 6, 385–394.

Williams, I.H. (ed.) (2010) *Biocontrol-based Integrated Pest Management of Oilseed Rape Pests*. Springer, Dortrecht, The Netherlands.

Williams, I.H. and Free, J.B. (1979) Compensation of oilseed rape (*Brassica napus* L.) plants after damage to their buds and pods. *Journal of Agricultural Science*, 92, 53–59.

7 Noctuid (Lepidoptera: Noctuidae) Pests of Canola in North America

Kevin D. Floate[1]* and Vincent A. Hervet[2]

[1]*Agriculture and Agri-Food Canada, Lethbridge, Alberta, Canada;*
[2]*Alberta Agriculture and Forestry, Edmonton, Alberta, Canada*

7.1 The Cropping System

Canola is the dominant oilseed brassica crop in North America. Its name is a contraction of 'Canada' and 'ola' (meaning oil) and refers to specially bred varieties of *Brassica napus* L., *Brassica rapa* L. or *Brassica juncea* (L.). These varieties have low levels of erucic acid and low levels of glucosinolates (CCC, 2014). The oil extracted from canola seed is consequently safe for human consumption and has increased palatability.

Canola production is centred on the Northern Plains, primarily in Canada. In the 1940s, *B. napus* and *B. rapa* were grown as rapeseed. The development of canola in the 1970s has since led to an extraordinary increase in the annual acreage seeded to these species. Fields seeded to canola in Canada totalled 8.1 million hectares in 2015. Essentially all canola is grown in southern regions of the Prairie Provinces, i.e. Saskatchewan (53%), Manitoba (16%) and Alberta (30%) (Statistics Canada, 2015). In Alberta, a large acreage also is seeded in the north-west or Peace River region. Fields seeded to canola in the USA in 2015 totalled 0.6 million hectares with production centred in North Dakota (78%) (USDA, 2015).

Agronomic practices for growing canola are dictated by abiotic and biotic factors. The Northern Plains are characterized by a short growing season, dry summers and cold winters. For example, the City of Regina in southern Saskatchewan has an average growing season of 111 days (TOFA, 2016); annual precipitation averages 370 mm with average temperatures in the coldest and warmest months of −17°C (January) and 18°C (July), respectively (Weatherbase, 2016). Harsh winters preclude overwintering survival of canola seeded in autumn. Hence, all canola in western Canada is seeded in the spring. To reduce the build-up of pathogens and pests, canola is normally planted in a 3- or 4-year rotation cycle. Crops grown in rotation with canola will vary with a number of factors but may include cereals (e.g. barley, corn, wheat), broadleaf crops (e.g. flax, peas, soybean), or the field may be left fallow (CCC, 2013).

Canola is attacked by a complex of species in the family Noctuidae (Lepidoptera). These species are highly variable in terms of feeding behaviour, region of occurrence and timing of different life stages. This chapter summarizes general information on these pests, their natural enemies and general methods of control.

7.2 Noctuid Pests and Their General Biology

Canola is attacked by a complex of noctuid (Lepidoptera: Noctuidae) species that include cutworms, armyworms and semi-loopers (Table 7.1). Cutworms clip foliage or sever the stems of young plants and can be classified as subterranean, above-ground or climbers (Walkden, 1950). Subterranean cutworms feed almost exclusively underground; larvae cut the main stem of young plants but otherwise are not usually seen. Above-ground cutworms feed on foliage but older larvae may cut the main stem of young plants at or near the soil surface. They normally feed at night and hide in the soil during the day. Climbing cutworms climb up on to

*Corresponding author. E-mail: kevin.floate@agr.gc.ca

Table 7.1. Distribution, biology and ecology of selected noctuid pests affecting canola in North America[a].

Distribution	Biology	Additional comments
***Agrotis ipsilon* (Hufnagel) – Black cutworm**		
Throughout the USA and southern Canada	Above-ground cutworm; overwinters in southern locations as a pupa and migrates north in spring; 1–2 generations/year in Canada; 2–4 generations/year in the USA.	Eggs usually laid on live or dead foliage of cereals. Typically a pest in moister regions. Feeds on grasses and broadleaf species including corn, alfalfa, clover, sunflower and various weeds.
***Agrotis orthogonia* Morrison – Pale western cutworm**		
British Columbia to Manitoba; south to New Mexico; west to southern California	Subterranean cutworm; overwinters as an egg; 1 generation/year.	Serious pest during outbreak years. Eggs are laid in loose dusty soil in August–September. Preferentially feeds on cereals when present, particularly wheat, oats and barley. Other hosts include canola, mustard, flax, corn, sugar beets, legumes, and various weeds.
***Anarta trifolii* (Hufnagel) – Clover cutworm**		
Mexico to southern Alaska, excluding south-eastern USA	Climbing cutworm; overwinters as a pupa; 2–3 generations/year.	Minor occasional pest. Feeds on broadleaf species including alfalfa, tobacco, beets, turnips, cabbage and various weeds (e.g. *Chenopodium*, *Salsola*).
***Autographa californica* (Speyer) – Alfalfa looper**		
Alaska to northern Mexico, east to New Mexico, north to Manitoba	Climbing semi-looper; overwinters as a pupa; at least 2 generations/year in the southern regions of the prairie provinces, more in the southern USA.	Common but rarely a pest. Larvae feed on foliage but also clip flowers and young seedpods. Populations usually controlled by natural enemies. Feeds on a wide variety of broadleaf species.
***Euxoa auxiliaris* (Grote) – Army cutworm**		
Western North America, east to Michigan and north into the Northwest Territories	Surface-feeding cutworm; overwinters as a larva; 1 generation/year.	Serious pest during outbreak years. Eggs laid in loose, sandy soil in August–October. Availability of live plants and above-average rainfall in late summer and autumn increase likelihood of outbreaks in following spring. Because adults migrate from the Rocky Mountains in late summer, regions closer to the mountains are more susceptible to outbreaks. Host plants include wheat, oats, barley, canola, mustard, flax, alfalfa, sweet clover, peas, cabbage, sugar beet, various weeds and grasses.
***Euxoa messoria* (Harris) – Darksided cutworm**		
From Atlantic to Pacific coasts; north from the southern USA into southern Canada	Surface-feeding cutworm; overwinters as an egg; 1 generation/year.	Eggs are laid in loose sandy soils in autumn. Easily confused with other *Euxoa* spp. Broad range of herbaceous and woody hosts including vegetables, cereals, canola, corn, tobacco, flax, sunflower, vine, berry and tree fruits.
***Euxoa ochrogaster* (Guenée) – Redbacked cutworm**		
Throughout most of the Northern Hemisphere	Surface-feeding cutworm; overwinters as an egg; 1 generation/year.	Serious pest during outbreak years. Eggs are laid in loose sandy soil in autumn. Often co-occurs with other cutworm species with similar oviposition patterns. Will climb on to plants to feed if suitable plant material not available at ground level. Associated with dry, open habitats. Host plants include cereals, flax, canola, corn, mustard, sunflower, sugar beets, forage legumes, vegetables and other crops.

Continued

Table 7.1. Continued.

Distribution	Biology	Additional comments
Feltia herilis **(Grote),** ***Feltia jaculifera*** **(Guenée),** ***Feltia subgothica*** **(Haworth) – Dingy cutworm**		
Feltia jaculifera and *F. herilis* from Atlantic to Pacific coasts, north from the southern USA into southern Canada; *F. subgothica* restricted east of the Rocky Mountains but otherwise similar to *F. jaculifera* and *F. herilis*	Climbing cutworm; overwinters as a larva; 1 generation/year.	Serious pest during outbreak years. Eggs are laid on vegetation and the soil surface in late summer. Host plants include sunflowers, alfalfa, corn, flax, canola, oats, barley, rye and wheat, many other vegetables, grass, ornamental and various weeds.
Mamestra configurata **Walker – Bertha armyworm**		
Western North America, from British Columbia and Alberta to Manitoba, south to Mexico	Climbing armyworm, overwinters as a pupa; 1 generation/year.	Serious pest during outbreak years. Reaches economic levels when feeds on seedpods. Feeds on a wide variety of broadleaf species, including canola, flax, potatoes, cabbage, peas and alfalfa.
Melanchra picta **(Harris) – Zebra caterpillar**		
Throughout southern Canada; western USA and northern half of eastern USA	Climbing caterpillar; overwinters as a larva (west) or a pupa (east); 1 generation/year (west); 2 or more generations (east).	Not a reported pest of canola. However, because larvae are large, colourful and active during the day, sightings may cause concern. Generalist feeder, particularly on garden crops, and is reported to be an important pest in the eastern part of its range.
Mythimna unipuncta **(Haworth) – True armyworm**		
Parts of Europe, Asia, Africa, Central America, and South America; in North America, throughout the USA and southern Canada	Climbing armyworm; overwinters as a pupa in southern USA; 2–3 generations/year.	Sporadic pest of canola. Outbreaks occur every 5–20 years, usually during unusually wet years that follow unusually dry years. Economic damage mainly due to larval feeding on seedpods. Overwinters in the southern USA; spring migrations of adults cause outbreaks on the Northern Plains. Preferred hosts include native grasses, oats, wheat, fall rye, corn, barley and forage grasses; secondary hosts include alfalfa, beans, cabbage, onion, pea, sugar beet, turnips and other species.
Noctua pronuba **(Linnaeus) – Winter cutworm**		
Broad distribution in Eurasia; in North America, throughout most of the USA and southern Canada	Climbing/subterranean cutworm; overwinters as a larva; 1–2 generations/year.	First reported in North America in Nova Scotia in 1979; now present in all provinces and often common. Has a broad host range that includes alfalfa, oats, rye, wheat, grass hay, carrot, onion, potato, spinach, rhubarb, sugar beet, tomato, strawberry, hawkweed, dandelion, plantain, grape, marigold and chrysanthemum. No reports of outbreaks in canola, but recent outbreaks in other crops combined with its generalist feeding habit identify it as a possible future pest of canola.
Trichoplusia ni **(Hübner) – Cabbage looper**		
Throughout the USA and southern Canada; southern Palaearctic, Australasia and Hawaii	Climbing, semi-looper; overwinters as a pupa in southern USA; 2–3 generations/year in Canada, 5–7 generations/year in California.	Not a reported pest of canola, but is a destructive pest of vegetable crops with a preference for cruciferous species. Populations re-established each year at the northern extents of its distribution from adults that migrate from the southern USA.

[a]Information for most of these species is summarized from Walkden (1950), Crumb (1956), Beirne (1971) and Capinera (2001). Information for *N. pronuba* is from Bechinski *et al.* (2009). Additional sources of information are referenced in the text.

the plant to feed on foliage without necessarily damaging the main stem. Armyworms, particularly when present in large numbers, will aggregate into bands and march as an 'army' in the same direction in search of suitable host plants. Semi-loopers (Lepidoptera: Noctuidae: Plusiinae), more often termed loopers, move by drawing their hind end towards their front end to form a partial loop or arch with their body.

The noctuid pest complex in a given field can be highly variable, reflecting differences in regional climate, life cycles, the previous crop, soil type, agricultural practices and other factors. Species of particular importance include pale western cutworm (*Agrotis orthogonia* Morrison), redbacked cutworm (*Euxoa ochrogaster* (Guenée)) and bertha armyworm (*Mamestra configurata* Walker) (Strickland, 1923; Beirne, 1971; Ayre and Lamb, 1990). The preference and performance of this latter species increases on food plants with low glucosinolate levels (McCloskey and Isman, 1993; Ulmer *et al.*, 2001), which places canola at particular risk. Other pest species may include black cutworm (*Agrotis ipsilon* (Hufnagel)), clover cutworm (*Anarta trifolii* (Hufnagel)), army cutworm (*Euxoa auxiliaris* (Grote)), darksided cutworm (*Euxoa messoria* (Harris)), dingy cutworm (*Feltia jaculifera* (Guenée)), alfalfa looper (*Autographa californica* (Speyer)) and true armyworm (*Mythimna unipuncta* (Haworth)) (Beirne, 1971). Pale western cutworm is associated with drier regions and tends to be more dominant in the southern portions of Alberta and Saskatchewan. Redbacked cutworm is less adapted for dry conditions and more often is a pest in the northern parkland regions of these provinces. Excessive soil moisture favours the black cutworm, which is more common in Manitoba and further east. Redbacked cutworm and pale western cutworm overwinter on the Northern Plains, whereas infestations of true armyworm are triggered by flights of adults from the southern USA (McNeil, 1987). Pale western cutworm prefers cereal crops, but overwinters in the egg stage. Thus, outbreaks of this species can occur in canola fields that were planted to cereals in the previous year (Pruess and Roselle, 1958). Redbacked cutworm (King, 1926) and pale western cutworm (Sorenson and Thornley, 1941) are reported to oviposit preferentially in light-textured, dry, loose soils, whereas other noctuids usually oviposit on foliage. To add further complexity, two or more species may co-occur in the same field (Ayre and Lamb, 1990).

Noctuid larvae typically complete five or six instars (Guppy, 1961) but development may span seven to nine instars if larvae are parasitized, starved or exposed to low winter temperatures (Breeland, 1957; Byers *et al.*, 1993; Esperk *et al.*, 2007). Larvae generally appear hairless, curl up when disturbed and are roughly 25–50 mm long when mature, depending upon the species. Upon completion of the final instar, larvae of cutworms and armyworms stop feeding and form an earthen cell a few centimetres underground in which they pupate. Adults emerge from these pupal cells to feed on flower nectar and to mate; females typically mate only once. Different life stages of redbacked cutworm are illustrated in Fig. 7.1. A female may lay from several hundred to several thousand eggs, depending upon the species (Jacobson, 1965; Cheng, 1972; Turnock, 1985). In contrast to cutworms and armyworms, the larvae of semi-loopers pupate within cocoons that they spin and attach to foliage or in plant debris on the soil. Detailed descriptions and life history information of individual species is provided in Crumb (1956) and Capinera (2001).

The duration of the life stage varies with the species. Eggs of some species may hatch within a few days of oviposition or they may overwinter in the ground. Eggs laid in autumn may begin developing prior to winter and then hatch the following spring with the onset of warmer temperatures and after the soil has thawed (Jacobson, 1962). For species that overwinter as eggs (e.g. redbacked cutworm, pale western cutworm), a period of cold may be required before the eggs hatch (Jacobson, 1962). This requirement prevents eggs from hatching in autumn, when there is no food for the larvae. In spring, newly hatched larvae of some species, such as pale western cutworm, may survive without feeding for 2–3 weeks, especially at cooler temperatures, e.g. 10°C (Seamans and Rock, 1945; Jacobson, 1952). Some species may complete larval or pupal development in 4–6 weeks, whereas species that overwinter as larvae (e.g. army cutworm) or pupae (e.g. bertha armyworm, clover cutworm) may require 6–7 months to complete these life stages. Species with an obligate diapause necessarily have one generation per year, whereas other species (clover cutworm) may complete two or more generations each year (Guppy, 1961; Ayre and Lamb, 1990).

Most of the noctuid pests that affect canola on the Northern Plains overwinter locally. Other species re-establish populations in the Northern Plains from populations that have overwintered further

Fig. 7.1. Redbacked cutworm, *Euxoa ochrogaster*: (a) larval-feeding damage to young plant; (b) damage to cereal crop; (c) mature larva (ca. 3 cm in length); (d) pupa; (e) adult moth; (f) eggs; (g–j) four different adult colour morphs. Reproduced with permission from Agriculture Canada Insect Identification Sheet No. 6, 1976, Cat. No. A44-9/6.

south, e.g. true armyworm (Guppy, 1961) and black cutworm (Clement *et al.*, 1985). Still other species may overwinter in small numbers on the Northern Plains but outbreaks are normally triggered by large flights of adults from more southern populations. The risk for outbreaks of diamondback moth, *Plutella xylostella* (L.) (Lepidoptera: Plutellidae), in canola can be assessed by trapping adult moths and mapping wind trajectories in the spring, as adult moths are moved northwards from Mexico and the southern USA (Dosdall *et al.*, 2004).

7.3 Outbreaks and the Extent of Damage

Most of the noctuid pests that affect canola on the Northern Plains are native to North America (Beirne, 1971). They tend to be polyphagous, with preferences for either grasses or forbs. Replacement of what was originally an undisturbed and diverse plant community with an annual cropping system dominated by a few select cultivars of cereals and oilseeds has allowed some of these noctuid species to periodically reach high, economically damaging densities. Bertha armyworm has a general preference for plants in families Brassicaceae and Chenopodiaceae (King, 1928) and would have fed on native species in these families prior to European settlement. Subsequent conversion of native prairie to large monocultures of canola now provides it with an unlimited food supply. Annual tillage ensures loosely compacted soils that enhance the mobility of subterranean cutworms and their ability to locate suitable food plants.

Crop losses during noctuid outbreaks are highly variable. Damage may be limited to a few small patches in a field and not require control measures. Elsewhere, the entire crop may be lost. In the worst case, outbreaks reduce crop yields across large geographical areas in successive years. Damage by pale western cutworm was first reported in 1911 and 1912, in Alberta (Mabee, 1929). Widespread damage to cereal crops was subsequently observed in Montana from 1918 to 1921 where yield losses were estimated at that time to exceed US$4.1 million (Mabee, 1929). Between 1927 and 1933, pale western cutworm damaged cereal crops from eastern Saskatchewan to western Alberta and northwards up to 500 km from the Canada/USA border (Seamans, 1935). Total crop losses in Canada at that time were estimated at about CAN$19 million (McMillan, 1935). Outbreaks spanning this period also were reported in Utah (Sorenson and Thornley, 1941). Pale western cutworm outbreaks affected an estimated 100,000 ha in southern Alberta in 1985. Based on insecticide sales at that time, total losses were estimated at CAN$8 million (J.R. Byers, Lethbridge, Alberta, 2016, personal communication). Army cutworm outbreaks in 1990 affected more than 10,000 ha in southern Alberta, of which about 6000 ha were sprayed with insecticide and the remainder reseeded without spraying (Jones *et al.*, 1990, cited in Byers *et al.*, 1993). The outbreak history of bertha armyworm is reviewed by Mason *et al.* (1998). It was initially reported as a pest of flax, *Linum usitatissimum* L. (Linaceae), starting in the early 1920s. Outbreaks were then reported on rapeseed starting in the 1940s and then subsequently on canola starting in the 1970s. From 1994 to 1996, an estimated 1 million hectares of canola were sprayed in Saskatchewan to reduce bertha armyworm damage (Mason *et al.*, 1998). In recent years, localized outbreaks affecting canola in the Prairie Provinces have been reported for army cutworm, dingy cutworm, pale western cutworm, redbacked cutworm and true armyworm (WCCP, 2012, 2013). Many small outbreaks go unreported.

The occurrence, extent and persistence of outbreaks reflect a combination of abiotic and biotic factors. Outbreaks may be local or regional and often persist for 1–3 years before populations decline to pre-outbreak levels. Climate is the most important factor, given its ability to affect populations over large areas. Cold conditions essentially prevent some species, e.g. true armyworm (McNeil, 1987) and diamondback moth (Dosdall *et al.*, 2004), from overwintering on the Northern Plains. Successive dry years increase the potential for outbreaks of pale western cutworm, whereas successive wet years increase the potential for outbreaks of black cutworm (Walkden, 1950). These climatic cycles are likely the underlying cause for periods of outbreaks, which occur sporadically every 5–30 years (Cook, 1929; Beirne, 1971; Mason *et al.*, 1998). Shorter-term weather events are also important. A harsh winter can reduce pest populations in autumn to non-economically damaging levels by the following spring. Overwintering mortality is a key factor that moderates populations of bertha armyworm (King, 1928). However, levels of mortality are affected by snow depth, which influences soil temperatures (Lamb *et al.*, 1985). Biotic factors further reduce pest populations and often may work in concert with climate and weather events (Cook, 1929). Excessive rainfall has been reported as perhaps the key mortality factor reducing populations of pale western cutworm (Beirne, 1971). It is a subterranean-feeding species but wet soils force larvae to the soil surface, where they are vulnerable to parasitoids and predators. Wet spring conditions in combination with high larval densities also can promote epizootics (King, 1928; Beirne, 1971).

7.4 Natural Enemies

The collapse of outbreaks is often attributed to the actions of parasitoids, pathogens and predators (Walkden, 1950; Beirne, 1971). During a redbacked cutworm outbreak near Saskatoon, Saskatchewan, estimated parasitoidism averaged from 38% to 61% (King and Atkinson, 1928). Similarly high levels of parasitism for this species were reported in central Alberta (Schaaf, 1972). Average parasitoidism was estimated to be 32% during an outbreak of true armyworm in Tennessee (Breeland, 1957) and from 20% to 60% during an outbreak of army cutworm in southern Alberta (Byers *et al.*, 1993). High levels of parasitoidism have also been reported for bertha armyworm (Wylie and Bucher, 1977). A study on darksided cutworm in Ontario showed average parasitoidism to be 21% with a further 35% mortality caused by pathogens (Bucher and Cheng, 1971). Epizootics have also been associated with the collapse of bertha armyworm outbreaks (King, 1928; Beirne, 1971; Mason *et al.*, 1998) and with clover cutworm (Federici, 1978). Many noctuids are killed by predators but this type of mortality is difficult to quantify. With the possible exception of birds in localized fields, predators are unlikely to prevent outbreaks or significantly reduce levels of crop damage in the current year. Pest populations normally have to escalate for 1–2 years before densities of natural enemies increase sufficiently to be effective.

Wasps (Hymenoptera: Braconidae, Ichneumonidae, Encyrtidae) and flies (Diptera: Bombyliidae, Tachinidae) are common parasitoids of noctuids (see Table 7.2 below). Braconids and ichneumonids oviposit into early-instar larvae but generally kill late-instar larvae (Braconidae) or pre-pupae and pupal stages (Ichneumonidae). Encyrtids lay an egg into the egg of the host. The wasp egg undergoes a series of divisions that result in multiple larvae. These larvae delay development until the host has finished feeding as a late-instar larva and then emerge as adults. One *Copidosoma bakeri* (Howard) egg may produce up to 2500 adult offspring from an army cutworm (Byers *et al.*, 1993). Bombyliid eggs are laid on soil or vegetation where host species are likely to occur. First-instar larvae actively locate, penetrate and then complete development and pupate inside the host (Yeates and Greathead, 1997). Tachinid eggs may be injected directly into a host, laid on or near the host, or on vegetation with the eggs hatching only if they are consumed by the host (O'Hara, 2008). Mature larvae typically exit the host to pupate in soil or ground litter. Species of Sarcophagidae and Muscidae (Diptera) have been reported as parasitoids of cutworms (Crumb, 1929; Cheng, 1977) but these reports are rare and possibly suspect (Dahlem and Downes, 1996).

Table 7.2. Common parasitoids of noctuid pests affecting canola.

Parasitoid species	Pale western cutworm *Agrotis orthogonia*	Army cutworm *Euxoa auxiliaris*	Darksided cutworm *Euxoa messoria*	Redbacked cutworm *Euxoa ochrogaster*	Bertha armyworm *Mamestra configurata*
DIPTERA					
Tachinidae					
Athrycia cinerea (Coquillett)			Cheng (1977)		Wylie and Bucher (1977)
Bonnetia comta (Fallen)	Strickland (1923)				
Gonia sp. (incl. *G. aldrichi* Tothill)	Strickland (1923)			King and Atkinson (1928), Schaaf (1972)	
HYMENOPTERA					
Braconidae					
Apanteles laeviceps (Ashmead)		Strickland (1923)			
Meteorus dimidiatus (Cresson)	Strickland (1923)				
Meteorus rubens (Nees)	Walkden (1950)	Walkden (1950)	Walkden (1950)	King and Atkinson (1928), Schaaf (1972)	
Encyrtidae					
Copidosoma bakeri (Howard)		Walkden (1950), Byers *et al.* (1993)	Walkden (1950), Bucher and Cheng (1972), Cheng (1977),	King and Atkinson (1928), Schaaf (1972)	
Ichneumonidae					
Banchus flavescens Cresson					Wylie and Bucher (1977)
Campoletis atkinsoni (Viereck)				King and Atkinson (1928)	
Eutanyacra suturalis (Say)			Cheng (1977)	Schaaf (1972)	
Spilichneumon superbus (Provancher)				Schaaf (1972)	

Pathogens of noctuids include viruses, fungi, bacteria and nematodes. Once ingested, virus particles enter and repeatedly replicate in the cells of the insect. The cell walls break down and often give the dead or dying insect a liquefied appearance. Fungal spores attach to the outer surface of the host, germinate and then develop threadlike hyphae that penetrate the insect's cuticle to colonize the body cavity and kill the host. Spore germination and formation requires optimal temperatures and high humidity. Given the generally dry conditions of the Northern Plains, fungal epizootics are not normally a major source of mortality for pest species affecting canola (King and Atkinson, 1928; Walkden, 1950). Entomopathogenic bacteria produce toxins that kill the host by attacking nervous tissue or by attacking the cells that line the inside wall of the insect gut to cause starvation (Castagnola and Stock, 2014). Some groups of bacteria (e.g. *Bacillus*) form spores that allow them to persist outside the host in the soil and on vegetation. Other bacteria that cannot survive in the environment are symbionts of nematodes that infect insects. Bacteria pathogens isolated from pest species of cutworms include species of *Achromobacter*, *Bacillus*, *Enterobacter*, *Klebsiella*, *Pseudomonas*, *Serratia* and *Streptococcus* (Steinhaus and Marsh, 1962; Cheng, 1984). Juvenile entomopathogenic nematodes (e.g. *Steinernema* and *Heterorhabditis* spp.) infect insects by penetrating into their body cavity, usually through an existing opening (e.g. mouth, anus) or through thin regions of cuticle. Once established in the body cavity, the infective juveniles release bacteria (e.g. *Photorhabdus*, *Xenorhabdus*) that produce a toxin to kill the host.

Noctuid predators include a range of insect and vertebrate species. Ground beetles (Coleoptera: Carabidae) are among the most important insect predators in agroecosystems and will attack all immature noctuid stages (Frank, 1971; Brust *et al*., 1986a; Frank and Shrewsbury, 2004) (Fig. 7.2). There are about 400 carabid species on the Canadian Prairies and upwards of 80 species present in a given field (reviewed in Holliday *et al*., 2014); normal densities range from about ten to 50 beetles/m^2 (Thiele, 1977; Brust *et al*., 1986b) but can be much higher (Holliday and Hagley, 1979; Floate and Spence, 2015). One adult *Calosoma lugubre* Leconte reportedly will consume 16 final-instar larvae of the corn earworm (*Helicoverpa zea* (Boddie)) (Lepidoptera: Noctuidae) before losing interest (Walkden, 1950). In the absence of cutworms, carabids will feed on other pest species that occur in a typical canola crop rotation. These species include wheat midge (*Sitodiplosis mosellana* (Géhin); Diptera: Cecidomyiidae) (Floate *et al*., 1990), wireworms (Coleoptera: Elateridae) (Fox and MacLellan, 1956), aphids (Hemiptera: Aphididae), root maggot (*Delia radicum* L.; Diptera: Anthomyiidae) (Finch, 1996) and grasshoppers (Orthoptera: Acrididae) (Songa and Holliday, 1997). Other insect predators include ants (Hymenoptera: Formicidae), rove beetles (Coleoptera: Staphylinidae) (King and Atkinson, 1928; Frank, 1971) and the larvae of stiletto flies (Diptera: Therevidae) (King and Atkinson, 1928). Spiders feed on low numbers of adults (Cheng, 1973b) but will feed on even the largest cutworm larvae and may at times be important predators (Schaaf, 1972).

Fig. 7.2. *Carabus nemoralis* Müller (Coleoptera: Carabidae) feeding on the larva of a bertha armyworm, *Mamestra configurata* (Lepidoptera: Noctuidae). Photo: V.A. Hervet.

Vertebrate predators of noctuids include birds and mammals. Large numbers of larvae are consumed by crows, grackles, starlings, seagulls and hawks (King and Atkinson, 1928; Walkden, 1950; Cheng, 1973b) and can reduce potential crop losses (King, 1928). Flocks of seagulls following tractors during field operations can be a useful indicator of cutworm outbreaks. Small rodents feed on larvae and pupae in crops and a number of animals feed on army cutworm moths. Each summer, moths of this species migrate in massive numbers from the prairies to the mountains. At night, they feed on alpine and subalpine flower nectar where owls and bats can find and consume them. During the day, they aggregate under rock formations where they are at risk to predation by black bears, grizzly bears, coyotes, mountain bluebirds, ravens and American pipets (French *et al*., 1994; White *et al*., 1998).

A recent review listed the natural enemies of redbacked, darksided and army cutworms (Gavloski and Hervet, 2013). Other detailed lists of natural enemies affecting noctuid pests of canola are provided by Strickland (1923), Crumb (1929), Walkden (1950), Breeland (1957) and Schaaf (1972). A list of common parasitoids associated with selected noctuid pests is provided in Table 7.2.

7.5 Control Strategies

Elements of an integrated pest management (IPM) programme include forecasting to assess the risk of a pest outbreak, scouting to identify crop damage in early stages and application of methods either to prevent the build-up of pest populations (cultural, biological) or to control their numbers when densities exceed economic thresholds (insecticidal).

7.5.1 Forecasting

Sex pheromone traps are used to provide annual forecasts of bertha armyworm on the Northern Plains. Adults emerge from overwintered pupae early in the growing season. Numbers of adults recovered in traps can therefore be used to assess the likelihood of crop damage by their larval offspring later in the summer. Since 1995, Alberta Agriculture and Rural Development has organized and maintained a network of these traps throughout the province. Results are disseminated to farmers of the Prairie Provinces by the Prairie Pest Monitoring Network (PPMN, 2016), which advises farmers on when to scout their fields to determine if larval numbers exceed economic thresholds. Techniques for pheromone-based monitoring of true armyworm have also been researched (Turgeon *et al.*, 1983; Hendrix and Showers, 1990) and could identify the arrival of moths in a region.

Efforts to forecast outbreaks of other noctuid pests based on adult numbers have been unsuccessful (Ayre and Lamb, 1990; Turnock *et al.*, 1993). Sex pheromones have been developed and validated for a number of cutworm species (Byers and Struble, 1987). However, for most species, the larvae that damage crops in the current year have developed from eggs laid in the previous year (Table 7.1). Thus, depending upon overwintering mortality and spring precipitation, catches of adults in autumn often do not reflect numbers of larvae present in the following year (Ayre and Lamb, 1990). In addition, maintaining a network of traps to develop regional forecasts can be expensive and regional forecasts may not apply to individual fields, where factors such as soil type, crop history and stage of vegetation in the field when egg laying occurs can greatly influence the risk of noctuid damage.

Forecasting methods were developed in the 1920s and 1930s to predict outbreaks of pale western cutworm based on rainfall (Cook, 1926; Seamans, 1935). The underlying premise is that rain forces this subterranean cutworm species to the soil surface, where its numbers are reduced by parasitoids (Seamans, 1935) or by epizootics (Cook, 1926). The method was validated in each of 11 years with a high degree of accuracy across southern regions of Alberta and Saskatchewan (Seamans, 1935). It was not subsequently adopted, however, probably because of changes in farming practices and the subsequent availability of synthetic chemical insecticides.

7.5.2 Scouting

Frequent scouting of individual fields can detect developing outbreaks of noctuid pests. The recommended time to scout for cutworms is spring and early summer to detect species that have overwintered as eggs or larvae. Damage appears as thinning or bare patches surrounded by healthy young plants. The patches are areas of dead and dying plants with notched foliage and stems cut close to the soil surface and tend to be more common in lighter soils, on south-facing slopes, or on hill tops, i.e. warmer and drier locations within the field. Patch size increases as feeding continues along the periphery. Because feeding typically occurs at night, the presence of cutworms during daylight hours can be determined by searching for larvae in the top 2–5 cm of soil around the base of healthy plants along the periphery of patches. Economic injury levels and economic thresholds have not been developed specifically for cutworms in canola, thus nominal thresholds are currently used.

The recommended time to scout for bertha armyworm is the period immediately after peak flowering until it is too late in the season to apply insecticidal treatments prior to harvest. Sampling for bertha armyworm involves measuring a 0.25 m^2 sampling area, shaking the plants inside the sampling area, then examining the soil surface for larvae and moving earthen lumps and plant debris to expose hidden larvae (Turnock and Bilodeau, 1985; Wise *et al.*, 2009). Economic injury levels have been established for bertha armyworm in canola (Bracken and Bucher, 1977).

7.5.3 Cultural control

Cultural methods of control include crop rotation, tillage, seeding dates and use of resistant varieties. Annual rotation of cereal and broadleaf crops within fields is recommended for canola production to reduce the build-up of pathogens and crop-feeding insects, e.g. root maggot (CCC, 2013). For example, research in Manitoba showed that fields planted to broadleaf (versus cereal) crops had higher numbers of cutworms in the following year, albeit numbers in all fields were below economic thresholds (Turnock *et al.*, 1993). If this finding is generally valid, continuous cropping of broadleaf crops might reasonably be expected to favour populations of some species of cutworms.

Different tillage practices have been examined to control noctuid populations. Early in the last century, soil packing and use of a seed drill were shown to reduce the movement of subterranean species but did not provide satisfactory control (Parker *et al.*, 1921). The larvae of species that overwinter as eggs or larvae can be starved by frequent tillage of summer fallow fields in spring and summer to remove weeds and volunteer crop plants that provide a source of food (Parker *et al.*, 1921; Sorenson and Thornley, 1941; Walkden, 1950). Allowing soils to build up a crust later in the season can reduce oviposition by species that prefer to lay eggs in loose soils (King, 1926; Sorenson and Thornley, 1941). Soil tillage also mechanically kills immature stages and exposes them to natural enemies. However, frequent tillage reduces soil water storage capacity, organic matter and fertility and increases the risk of soil erosion. Furthermore, the advent of chemical herbicides in the late 1940s and early 1950s has alleviated the need to control weeds with tillage. Hence, frequent tillage has now been replaced with conservation tillage and zero tillage (Awada *et al.*, 2014). Turnock *et al.* (1993) examined the effect of this transition during a 7-year study in Manitoba. Their results indicated that conservation tillage favoured an increase in the number of cutworms and also the species diversity for cutworms and their parasitoids. They further detected reduced inter-field variation in cutworm numbers and concluded that this greater diversity enhanced population stability such that cutworm outbreaks might be less common with use of conservation tillage. Stinner and House (1990) review the literature on how adoption of conservation and zero tillage has affected different species of insect pests, including noctuids.

Cutworm populations can also be managed by delayed seeding. Crop damage by overwintered or newly hatched cutworms early in spring may require that fields be reseeded. In such cases, Salt and Seamans (1945) recommended that the field be tilled to remove living plants and starve the cutworms before reseeding the crop 10–14 days later. This method provided effective control of pale western cutworm but was not generally adopted, as it required an additional field operation and a greater risk that later seeded crops would be damaged by an early autumn frost (Seamans and Rock, 1945). With the availability of chemical insecticides, the requirement of a waiting period prior to reseeding is now moot.

No canola cultivars have been developed for resistance against noctuid pests, though work is ongoing (e.g. Mohammed, 2015). However, resistant cultivars are available for other pests associated with canola–cereal crop rotations on the Northern Plains. Solid-stemmed varieties of wheat have been bred to confer resistance to wheat stem sawfly, *Cephus cinctus* Norton (references cited in Cárcamo *et al.*, 2011). Wheat cultivars have been developed with genetic antibiotic resistance to wheat midge, *Sitodiplosis mosellana* (Géhin) (Vera *et al.*, 2013). Elsewhere, *Bacillus thuringiensis* (Bt) cultivars engineered to express insecticidal proteins are grown to control noctuid pests of corn (*Spodoptera frugiperda* (J.E. Smith)) and cotton (*Helicoverpa armigera* Hübner, *H. zea*) (Tabashnik *et al.*, 2013). However, studies show that insects routinely evolve mechanisms to overcome plant defences (Tabashnik *et al.*, 2013). Hence, if canola varieties resistant against noctuid pests become available in the future, steps will be needed to delay the adaptation by the pests to the resistance mechanism. For example, it is recommended or required that susceptible and resistant cultivars be grown as a blend (interspersed refuge) or in adjacent patches (separate refuge) to reduce selection pressures that favour development of virulent insect biotypes (Smith *et al.*, 2004; Carrière *et al.*, 2005).

7.5.4 Biological control

A number of studies have examined the efficacy of entomopathogens against noctuid pests of canola. For black cutworm, the pathogenicity of different nematode strains has been examined in laboratory and small-scale field trials (Capinera *et al.*, 1988; Ebssa and Koppenhöfer, 2012; Mahmoud, 2014). Different strains of viruses, bacteria, fungi and protozoa also have been screened for pathogenicity against this species (Ignoffo and Garcia, 1979; Johnson and Lewis, 1982). For darksided cutworm, baculoviruses (Cheng, 1984) and different preparations of Bt (Cheng, 1973a) have been tested in the laboratory for their efficacy. Morris *et al.* (1990) reported results of laboratory trials examining the pathogenicity of nematodes against bertha armyworm, black cutworm, variegated cutworm, *Peridroma saucia* (Hübner), true armyworm, *Mythimna unipuncta* and the redbacked cutworm. To our knowledge, the only microbial insecticide commercially available for use in canola is a Bt-based product registered in the USA for use against bertha armyworm (Knodel *et al.*, 2016). A Bt-based product registered for use in Canada in 2015 (PMRA, 2015) is not yet commercially available.

Adopting practices to maintain healthy populations of natural enemies (conservation biocontrol) can reduce the extent and duration of pest outbreaks (see Section 7.4). This can be most easily achieved by minimizing the use of insecticides; for example, spraying only when necessary, applying products with high specificity and limiting sprays to the affected area and a zone surrounding the affected area (see Section 7.5.5). Maintaining field boundaries (e.g. roadside margins, fence lines) in a semi-natural undisturbed state with a diversity of flowering plants also can be beneficial (see reviews by Landis *et al.*, 2000; Olfert *et al.*, 2005; Bianchi *et al.*, 2006). These boundary areas provide critical habitat to enhance the overwintering survival of natural enemies. Plant species flowering at different times provide a continuous source of nectar and pollen for adult parasitoids (and many species of pollinators). Field boundaries also provide habitat for prey and host species to sustain predators and parasitoids when noctuid pests are not present. However, adoption of conservation biocontrol is often incompatible with current farming methods on the Northern Plains (see Section 7.6).

A number of parasitoid species have been introduced or redistributed on to the Northern Plains as biocontrol agents of pests in canola crop rotation systems (De Clerck-Floate and Cárcamo, 2011). These include pests of alfalfa, wheat, corn and clover. In canola, a tachinid fly (*Eurithia consobrina* Meigen) (Turnock and Carl, 1995) and a braconid wasp (*Microplitis mediator* Haliday) (Mason and Youngs, 1994; Mason, 1999 – both cited in Mason *et al.*, 2002) have been released against bertha armyworm. The success of these releases is unknown. The tachinid fly *Bonnetia comta* (Fallen) was released into New Zealand as a biocontrol agent for black cutworm but apparently did not establish (Allan and Hill, 1984). *Cotesia vanessae* (Reinhard) is a braconid wasp that was reported recently for the first time in North America with a tentative report from southern Alberta (Hervet *et al.*, 2014). It is parthenogenetic, gregarious and multivoltine and can develop on a number of noctuid species in the laboratory, e.g. alfalfa looper, darksided cutworm, redbacked cutworm (Hervet *et al.*, 2012). Future monitoring is needed to determine its potential role as a mortality factor affecting noctuid pest populations.

7.5.5 Insecticides

Control of insect pests relies heavily on the use of synthetic chemical insecticides. Active ingredients in products targeting noctuid species in canola and currently registered in North America include carbamates (methomyl), organophosphates (chlorpyrifos), synthetic pyrethroids (bifenthrin, cypermethrin, deltamethrin, gamma-cyhalothrin, lambda-cyhalothrin, permethrin, zeta-cypermethrin) and diamides (chlorantraniliprole, cyantraniliprole) (Brook and Cutts, 2015). Most products contain one active ingredient, though some products contain two, e.g. chlorantraniprole + lambda-cyhalothrin. Registered products vary across regional jurisdictions and change annually. Bifenthrin and a microbial pesticide, listed for use in North Dakota, are not listed in Alberta. Conversely, methomyl and chlorpyrifos are listed for use in Alberta but not in North Dakota (Brook and Cutts, 2015; Knodel *et al.*, 2016). For this reason, specific products are not discussed here. Current information on products registered for use within a given jurisdiction can be obtained by contacting provincial or state authorities.

With one exception, these products are applied as sprays. The exception is a diamide (cyantraniliprole) based product that is registered as a seed treatment for early-season control of flea beetles and cutworms. Although we are unaware of any published studies that test the efficacy of this product in canola, its efficacy has been documented as a seed treatment for soybean against fall armyworm, *Spodoptera frugiperda* (Thrash *et al.*, 2013).

To conserve populations of parasitoids and predaceous insects (see Section 7.4), it is recommended that insecticides be applied only if necessary, according to label directions and only when most effective. Damage by cutworms and armyworms is typically isolated to one or a few patches in a field. Depending upon the size of the affected area(s) and potential for future damage, applications may not be justified. When treatments are required, sufficient control may be achieved by spot-spraying the affected area(s) plus a surrounding buffer zone, e.g. 10 m. Control is most effective for early-instar larvae. Last-instar larvae may be preparing to pupate and no longer be feeding, such that applications are not warranted. The larvae of many species feed nocturnally and shelter during daylight hours in the soil or under debris. Insecticide treatments for these species should be sprayed in early evening. For species whose larvae feed primarily underground (pale western cutworm), foliar sprays are ineffective. For cutworm and armyworm species that overwinter in egg or larval stages, insecticide-treated seeds may be an effective alternative and would avoid the inevitable non-target effects of insecticidal sprays on beneficial populations of parasitoids and predaceous insects.

Use of economic thresholds is recommended when deciding on whether insecticide treatments are warranted. Nominal thresholds for cutworms in canola are 3–4 larvae/m^2 (Brook and Cutts, 2015), whereas those for bertha armyworm may range from 10 to 34 larvae/m^2, varying with the cost of the seed and insecticide treatment (Anon., 2004). For some species (e.g. alfalfa looper, true armyworm), economic thresholds have not been established in canola.

7.6 Challenges

Farming methods on the Northern Plains are often inconsistent with recommended practices for integrated pest management programmes. Fields are generally large (e.g. 65–260 ha) to accommodate the use of large equipment and associated economies of scale. However, large fields reduce the amount of boundary areas in the landscape and, therefore, the undisturbed habitats that promote healthy populations of natural enemies. Essentially all of the canola grown in North America has been genetically modified for herbicide tolerance (CCC, 2010). Adoption of these varieties increases in-field control of annual flowering weedy species that provide sources of nectar for parasitic wasps. Planting these varieties also eliminates the need to till to control weeds. In previous years, tilling was recommended to control cutworms. Farmers opting to plant insecticide-treated seeds must make their purchase months in advance of spring seeding, i.e. the insecticide treatment will be applied without economic thresholds playing a part in the decision. These inconsistencies reflect an understandable desire by farmers to maximize their return on investment in an annual crop rotation system for which the control of noctuids in canola is only one consideration.

Reliance on insecticides to control these pests is expected to continue as long as products remain convenient, effective and relatively inexpensive. However, product efficacy predictably fails with the development of insecticide resistance by the target pest species (Mallet, 1989). Thus, there is an ongoing need for new products with novel modes of action to which pests have not yet developed resistance. Registrations of neonicotinoid- and diamide-based products are the latest examples of this phenomenon. Genetically modified cultivars that express insecticidal activity are not commercially available for canola but may become available in the future (e.g. Mohammed, 2015). However, pest species also develop resistance to insecticidal cultivars (Tabashnik *et al.*, 2013). Hence, insecticidal cultivars will need to incorporate new mechanisms to remain effective. Current research is examining the use of RNA interference (RNAi) to ‘switch off’ the production of selected proteins, such as proteins that protect insects from insecticide exposure (Kim *et al.*, 2015). The ability of insects to develop a counter-mechanism to RNAi is unknown.

The adoption and ongoing use of these new products and technologies continues to be assessed within a framework of national and international agreements and concerns regarding non-target effects. Lindane was deregistered as a seed treatment for canola in Canada as of 1 January 2005 in accordance with the Stockholm Convention on Persistent Organic Pollutants (ECCC, 2015). Clothianidin is a neonicotinoid-based product registered as a canola seed treatment in the USA (Knodel *et al.*, 2016). Adverse effects associated with adoption of neonicotinoid products to pollinators and other beneficial species (Pisa *et al.*, 2014) have led to their restricted use in certain jurisdictions (EU, 2013; OMAFRA, 2016).

In summary, we expect that farmers in North America will continue to rely on insecticides to control noctuid pests of canola; insecticidal cultivars may provide a possible future control option. We also expect that there will be ongoing discussions regarding the non-target effects of these products and whether their current level of use is justified. Efforts to reduce this reliance will continue to provide a rationale for adoption of integrated pest management programmes that include a greater reliance on cultural practices and maximizing the benefits of natural enemies.

Acknowledgements

Portions of this chapter have been extracted from a technology-transfer document prepared by K.D. Floate for the Canola Council of Canada (CARP Project No. 2012-1).

References

Allan, D.J. and Hill, M.G. (1984) Rearing and release of *Bonnetia comta* (Diptera: Tachinidae), a parasite of *Agrotis ipsilon* (Lepidoptera: Noctuidae). *New Zealand Entomologist* 8, 71–74. doi: 10.1080/00779962.1984.9722471.

Anon. (2004) *Economic Thresholds for Insects Attacking Oilseeds*. Alberta Agriculture and Forestry, Edmonton, Alberta, Canada. Available at: http://www1.agric.gov.ab.ca/$department/deptdocs.nsf/all/prm4848?opendocument (accessed 15 February 2016).

Awada, L., Lindwall, C.W. and Sonntag, B. (2014) The development and adoption of conservation tillage systems on the Canadian Prairies. *International Soil and Water Conservation Research* 2, 47–65. doi: 10.1016/S2095-6339(15)30013-7.

Ayre, G.L. and Lamb, R.J. (1990) Life histories, flight patterns, and relative abundance of nine cutworms (Lepidoptera: Noctuidae) in Manitoba. *The Canadian Entomologist* 122, 1059–1070. doi: 10.4039/Ent1221059-11.

Bechinski, E.J., Smith, L.J. and Merickel, F.W. (2009) *Large Yellow Underwing, a New Cutworm in Idaho*. CIS 1172. College of Agricultural and Life Sciences, University of Idaho, Moscow, Idaho. Available at: https://www.cals.uidaho.edu/edcomm/pdf/CIS/CIS1172.pdf (accessed 3 April 2016).

Beirne, B.P. (1971) Pest insects of annual crop plants in Canada: part I, Lepidoptera; II, Diptera; III, Coleoptera. *Memoirs of the Entomological Society of Canada* 103, 1–124. doi: 10.4039/entm10378fv.

Bianchi, F., Booij, C.J.H. and Tscharntke, T. (2006) Sustainable pest regulation in agricultural landscapes: a review on landscape composition, biodiversity and natural pest control. *Proceedings of the Royal Society B: Biological Sciences* 273, 1715–1727. doi: 10.1098/rspb.2006.3530.

Bracken, G.K. and Bucher, G.E. (1977) An estimate of the relation between density of bertha armyworm and yield loss on rapeseed, based on artificial infestations. *Journal of Economic Entomology* 70, 701–705. doi: 10.1093/jee/70.6.701.

Breeland, S.G. (1957) The armyworm, *Pseudaletia unipuncta* (Haworth), and its natural enemies. PhD thesis, University of Tennessee, Knoxville, Tennessee.

Brook, H. and Cutts, M.E. (eds) (2015) *Crop Protection 2015*. Alberta Agriculture and Rural Development, Edmonton, Alberta, Canada. Available at: https://open.alberta.ca/dataset/1be9589f-31f6-4841-a6b8-34b32e8e19b4/resource/fa9d2374-730d-477b-ac38-c377ee61a2f0/download/2204508-2015-Crop-Protection.pdf (accessed 15 February 2016).

Brust, G.E., Stinner, B.R. and McCartney, D.A. (1986a) Predation by soil inhabiting arthropods in intercropped and monoculture agroecosystems. *Agriculture, Ecosystems & Environment* 18, 145–154. doi: 10.1016/0167-8809(86)90137-4.

Brust, G.E., Stinner, B.R. and McCartney, D.A. (1986b) Predator activity and predation in corn agroecosystems. *Environmental Entomology* 15, 1017–1021. doi: 10.1093/ee/15.5.1017.

Bucher, G.E. and Cheng, H.H. (1971) Mortality in larvae of *Euxoa messoria* (Lepidoptera: Noctuidae) collected from the tobacco area of Ontario. *The Canadian Entomologist* 103, 888–892. doi: 10.4039/Ent103888-6.

Byers, J.R. and Struble, D.L. (1987) Monitoring population levels of eight species of noctuids with sex-attractant traps in southern Alberta, 1978–1983: specificity of attractants and effect of target species abundance. *The Canadian Entomologist* 119, 541–556. doi: 10.4039/Ent119541-6.

Byers, J.R., Yu, D.S. and Jones, W. (1993) Parasitism of the army cutworm, *Euxoa auxiliaris* (Grt.) (Lepidoptera: Noctuidae, by *Copidosoma bakeri* (Howard) (Hymenoptera: Encyrtidae) and effect on crop damage. *The Canadian Entomologist* 125, 329–335. doi: 10.4039/Ent125329-2.

Capinera, J.L. (2001) *Handbook of Vegetable Pests*. Academic Press, New York, NY. doi: 10.1016/B978-012158861-8/50011-9.

Capinera, J.L., Pelissier, D., Menout, G.S. and Epsky, N.D. (1988) Control of black cutworm, *Agrotis ipsilon* (Lepidoptera: Noctuidae), with entomogenous nematodes

(Nematoda: Steinernematidae, Heterorhabditidae). *Journal of Invertebrate Pathology* 52, 427–435. doi: 10.1016/0022-2011(88)90055-9.
Cárcamo, H.A., Beres, B.L., Herle, C.E., McLean, H. and McGinne, S. (2011) Solid-stemmed wheat does not affect overwintering mortality of the wheat stem sawfly, *Cephus cinctus*. *Journal of Insect Science* 11, 129. doi: 10.1673/031.011.12901.
Carrière, Y., Ellers-Kirk, C., Kumar, K., Heuberger, S., Whitlow, M., Antilla, L., Dennehy, T.J. and Tabashnik, B.E. (2005) Long-term evaluation of compliance with refuge requirements for Bt cotton. *Pest Management Science* 61, 327–330. doi: 10.1002/ps.1039.
Castagnola, A. and Stock, S.P. (2014) Common virulence factors and tissue targets of entomopathogenic bacteria for biological control of lepidopteran pests. *Insects* 5, 139–166. doi: 10.3390/insects5010139.
CCC (2010) *Estimated Acreage and Percentage*. Canola Council of Canada, Winnipeg, Manitoba, Canada. Available at: http://www.canolacouncil.org/markets-stats/statistics/estimated-acreage-and-percentage/ (accessed 25 February 2016).
CCC (2013) *Canola Encyclopedia, Field Characteristics, Crop Rotation*. Canola Council of Canada, Winnipeg, Manitoba, Canada. Available at: http://www.canola-council.org/canola-encyclopedia/field-characteristics/crop-rotation/#managing-your-rotation (accessed 27 February 2016).
CCC (2014) *Crop Grower's Manual. Chapter 2. Canola Varieties*. Canola Council of Canada, Winnipeg, Manitoba, Canada. Available at: http://www.canola-council.org/crop-production/canola-grower's-manual-contents/chapter-2-canola-varieties/ (accessed 26 January 2016).
Cheng, H.H. (1972) Oviposition and longevity of the dark-sided cutworm, *Euxoa messoria* (Lepidoptera: Noctuidae), in the laboratory. *The Canadian Entomologist* 104, 919–925. doi: 10.4039/Ent104919-6.
Cheng, H.H. (1973a) Laboratory and field tests with *Bacillus thuringiensis* against the dark-sided cutworm, *Euxoa messoria* (Lepidoptera: Noctuidae), on tobacco. *The Canadian Entomologist* 105, 941–945. doi: 10.4039/Ent105941-7.
Cheng, H.H. (1973b) Observations of the bionomics of the dark-sided cutworm, *Euxoa messoria* (Lepidoptera: Noctuidae), in Ontario. *The Canadian Entomologist* 105, 311–322. doi: 10.4039/Ent105311-2.
Cheng, H.H. (1977) Insect parasites of the darksided cutworm, *Euxoa messoria* (Lepidoptera: Noctuidae), in Ontario. *The Canadian Entomologist* 109, 137–142. doi: 10.4039/Ent109137-1.
Cheng, H.H. (1984) *Euxoa messoria* (Harris), darksided cutworm (Lepidoptera: Noctuidae). In: Kelleher, J.S. and Hulme, M.A. (eds) *Biological Control Programmes against Insects and Weeds in Canada 1969–1980.* CAB International, Wallingford, UK, pp. 33–37.
Clement, S.L., Kaster, L.V., Showers, W.B. and Schmidt, R.S. (1985) Seasonal changes in the reproductive condition of female black cutworm moths (Lepidoptera: Noctuidae). *Journal of the Kansas Entomological Society* 58, 62–68.
Cook, W.C. (1926) Some weather relations of the pale western cutworm (*Porosagrotis orthogonia* Morr.) a preliminary study. *Ecology* 7, 37–47.
Cook, W.C. (1929) A bioclimatic zonation for studying the economic distribution of injurious insects. *Ecology* 10, 282–293.
Crumb, S.E. (1929) *Tobacco Cutworms*. Technical Bulletin No. 88. United States Department of Agriculture, US Government Printing Office, Washington, DC.
Crumb, S.E. (1956) *The Larvae of the Phalaenidae*. Technical Bulletin No. 1135. United States Department of Agriculture, US Government Printing Office, Washington, DC.
Dahlem, G.A. and Downes, W.L. (1996) Revision of the genus *Boettcheria* in America north of Mexico (Diptera: Sarcophagidae). *Insecta Mundi* 10, 76–103.
De Clerck-Floate, R. and Cárcamo, H. (2011) Biocontrol arthropods: new denizens of Canada's grassland agroecosystems. In: Floate, K.D (ed.) *Arthropods of Canadian Grasslands (Volume 2): Inhabitants of a Changing Landscape*. Biological Survey of Canada, Ottawa, Ontario, Canada, pp. 291–321. doi:10.3752/9780968932155.ch12.
Dosdall, L.M., Mason, P.G., Olfert, O., Kaminski, L. and Keddie, B.A. (2004) The origins of infestations of diamondback moth, *Plutella xylostella* (L.), in canola in western Canada. In: Endersby, N.M. and Ridland, P.M. (eds) *Proceedings of the 4th International Workshop, 26–29, November 2001*. Department of Natural Resources and Environment, Melbourne, Australia, pp. 95–100.
Ebssa, L. and Koppenhöfer, A.M. (2012) Entomopathogenic nematodes for the management of *Agrotis ipsilon*: effect of instar, nematode species and nematode production method. *Pest Management Science* 68, 947–957. doi: 10.1002/ps.3259.
ECCC (2015) *Update to Canada's National Implementation Plan under the Stockholm Convention on Persistent Organic Pollutants*. Environment and Climate Change Canada, Gatineau, Quebec, Canada. Available at: https://www.ec.gc.ca/lcpe-cepa/default.asp?lang=En&n=E0F02793-1&offset=4&toc=show (accessed 1 March 2016).
Esperk, T., Tammaru, T. and Nylin, S. (2007) Intraspecific variability in number of larval instars in insects. *Journal of Economic Entomology* 100, 627–645. doi: 10.1093/jee/100.3.627.
EU (2013) Commission Implementing Regulation (EU) No. 485/2013 amending Implementing Regulation (EU) No. 540/2011, as regards the conditions of approval of the active substances clothianidin, thiamethoxam and imidacloprid, and prohibiting the use

and sale of seeds treated with plant protection products containing those active substances. *Official Journal of the European Union* L 139, pp. 12–26.

Federici, B.A. (1978) Baculovirus epizootic in a larval population of the clover cutworm, *Scotogramma trifolii,* in southern California. *Environmental Entomology* 7, 423–427. doi: 10.1093/ee/7.3.423.

Finch, S. (1996) Effect of beetle size on predation of cabbage root fly eggs by ground beetles. *Entomologia Experimentalis et Applicata* 81, 199–200. doi: 10.1111/j.1570-7458.1996.tb02032.x.

Floate, K.D. and Spence, J.R. (2015) 'Outbreaks' of *Amara* Stephens (Coleoptera: Carabidae) in Alberta, Canada. *The Coleopterists Bulletin* 69, 114–115. doi: 10.1649/0010-065X-69.1.114.

Floate, K.D., Doane, J.F. and Gillott, C. (1990) Carabid predators of the wheat midge (Diptera: Cecidomyiidae) in Saskatchewan. *Environmental Entomology* 19, 1503–1511. doi: 10.1093/ee/19.5.1503 1503-1511.

Fox, C.J.S. and MacLellan, C.R. (1956) Some Carabidae and Staphylinidae shown to feed on a wireworm, *Agriotes sputator* (L.), by the precipitin test. *The Canadian Entomologist* 88, 228–231. doi: 10.4039/Ent88228-5.

Frank, J.H. (1971) Carabidae (Coleoptera) as predators of the red-backed cutworm (Lepidoptera: Noctuidae) in central Alberta. *The Canadian Entomologist* 103, 1039–1044. doi: 10.4039/Ent1031039-7.

Frank, S.D. and Shrewsbury, P.M. (2004) Consumption of black cutworms, *Agrotis ipsilon* (Lepidoptera: Noctuidae), and alternative prey by common golf course predators. *Environmental Entomology* 33, 1681–1688. doi: 10.1603/0046-225X-33.6.1681.

French, S.P., French, M.G. and Knight, R.R. (1994) Grizzly bear use of army cutworm moths in the Yellowstone Ecosystem. *International Association for Bear Research & Management* 9, 389–399.

Gavloski, J. and Hervet, V. (2013) *Euxoa ochrogaster* (Guenée), redbacked cutworm, *Euxoa messoria* (Harris), darksided cutworm, and *Euxoa auxiliaris* (Grote), army cutworm (Lepidoptera: Noctuidae). In: Mason, P.G. and Gillespie, D.R. (eds) *Biological Control Programmes in Canada 2001–2012.* CAB International, Wallingford, UK, pp. 164–175. doi: 10.1079/9781780642574.0164.

Guppy, J.C. (1961) Life history and behaviour of the armyworm, *Pseudaletia unipuncta* (Haw.) (Lepidoptera: Noctuidae), in eastern Ontario. *The Canadian Entomologist* 93, 1141–1153. doi: 10.4039/Ent931141-12.

Hendrix, W.H. and Showers, W.B. (1990) Evaluation of differently colored bucket traps for black cutworm and armyworm (Lepidoptera: Noctuidae). *Journal of Economic Entomology* 83, 596–598. doi: 10.1093/jee/83.2.596.

Hervet, V.A.D., Laird, R.A. and Floate, K.D. (2012) *Cotesia 'vanessae'* (Hymenoptera: Braconidae): a potential biological control agent of cutworms (Lepidoptera: Noctuidae). *Joint Annual Meeting of the Entomological Societies of Canada and Alberta.* Coast Edmonton Plaza Hotel, Edmonton, Alberta, Canada. Available at: https://docs.google.com/viewer?a=v&pid=sites&srcid=ZGVmYXVsdGRvbWFpbnxiaW9sb2dpY2FsY29udHJvbGZvcnVtfGd4OjZkMDUyMzM3YzBhNGY2NTU (accessed 8 April 2016).

Hervet, V.A.D., Murillo, H., Fernández-Triana, J.L., Shaw, M.R., Laird, R.A. and Floate, K.D. (2014) First report of *Cotesia vanessae* (Hymenoptera: Braconidae) in North America. *The Canadian Entomologist* 146, 560–566. doi: 10.4039/tce.2014.9.

Holliday, N.J. and Hagley, E.A.C. (1979) Distribution and density of carabid beetles (Coleoptera) in a pest management apple orchard. *The Canadian Entomologist* 111, 759–770. doi: 10.4039/Ent111759-7.

Holliday, N.J., Floate, K.D., Cárcamo, H., Pollock, D.A., Stjernberg, A. and Roughley, R.E. (2014) Ground beetles (Coleoptera: Carabidae) of the prairie grasslands of Canada. In: Cárcamo, H. and Giberson, D. (eds) *Arthropods of Canadian Grasslands (Volume 4): Biodiversity and Systematics Part 2.* Biological Survey of Canada, Ottawa, Ontario, Canada, pp. 1–85. doi: 10.3752/9780968932179.ch1.

Ignoffo, C.M. and Garcia, C. (1979) Susceptibility of larvae of the black cutworm to species of entomopathogenic bacteria, fungi, protozoa, and viruses. *Journal of Economic Entomology* 72, 767–769. doi: 10.1093/jee/72.5.767.

Jacobson, L.A. (1952) Effects of starvation on larvae of the pale western cutworm, *Agrotis orthogonia* Morr. (Lepidoptera: Phalaenidae). *Canadian Journal of Zoology* 30, 194–200. doi: 10.1139/z52-018.

Jacobson, L.A. (1962) Diapause in eggs of the pale western cutworm *Agrotis orthogonia* Morr. (Lepidoptera: Noctuidae). *The Canadian Entomologist* 94, 515–522. doi: 10.4039/Ent94515-5.

Jacobson, L.A. (1965) Mating and oviposition of the pale western cutworm, *Agrotis orthogonia* Morrison (Lepidoptera: Noctuidae), in the laboratory. *The Canadian Entomologist* 97, 994–1000. doi: 10.4039/Ent97994-9.

Johnson, T.B. and Lewis, L.C. (1982) Pathogenicity of two nuclear polyhedrosis viruses in the black cutworm, *Agrotis ipsilon* (Lepidoptera: Noctuidae). *The Canadian Entomologist* 114, 311–316. doi: 10.4039/Ent114311-4.

Jones, H.W., Byers, J.R., Butts, R.A., Okuda, M. and Hanison, L. (1990) Insects and related pests of cereal crops – Alberta. *The Canadian Agricultural Insect Pest Review* 68, 13–14.

Kim, Y.H., Soumaila Issa, M., Cooper, A.M.W. and Zhu, K.Y. (2015) RNA interference: applications and advances in insect toxicology and insect pest management. *Pesticide Biochemistry and Physiology* 120, 109–117. doi: 10.1016/j.pestbp.2015.01.002.

King, K.M. (1926) *The Red-Backed Cutworm and Its Control in the Prairie Provinces*. Pamphlet 69 (new series), Canadian Department of Agriculture, Ottawa, Ontario, Canada.

King, K.M. (1928) *Barather configurata* Wlk., an army-worm with important potentialities on the Northern Prairies. *Journal of Economic Entomology* 21, 279–293. doi: 10.1093/jee/21.2.279.

King, K.M. and Atkinson, N.J. (1928) The biological control factors of the immature stages of *Euxoa ochrogaster* Gn. (Lepidoptera, Phalaenidae) in Saskatchewan. *Annals of the Entomological Society of America* 21, 167–188. doi: 10.1093/aesa/21.2.167.

Knodel, J.J., Beauzay, P. and Boetel, M. (2016) *2016 Field Crop Insect Management Guide*. E-1143 (Revised). North Dakota State University Extension Service, Fargo, North Dakota. Available at: https://www.ag.ndsu.edu/pubs/plantsci/pests/e1143.pdf (accessed 17 February 2016).

Lamb, R.J., Turnock, W.J. and Hayhoe, H.N. (1985) Winter survival and outbreaks of bertha armyworm, *Mamestra configurata* (Lepidoptera: Noctuidae), on canola. *The Canadian Entomologist* 117, 727–736. doi: 10.4039/Ent117727-6.

Landis, D.A., Wratten, S.D. and Gurr, G.M. (2000) Habitat management to conserve natural enemies of arthropod pests in agriculture. *Annual Review of Entomology* 45, 175–201. doi: 10.1146/annurev.ento.45.1.175.

Mabee, W.B. (1929) *How to Control The Pale Western Cutworm*. Bulletin No. 100. Montana State College of Agriculture and Mechanic Arts, Bozeman, Montana.

Mahmoud, M.F. (2014) Efficacy of entomopathogenic nematodes to certain insect pests infesting oilseed rape in the laboratory and greenhouse. *Egyptian Journal of Biological Pest Control* 24, 387–391.

Mallet, J. (1989) The evolution of insecticide resistance: Have the insects won? *Trends in Ecology and Evolution* 4, 336–340. doi: 10.1016/0169-5347(89)90088-8.

Mason, P.G. (1999) Release of the insect parasite *Microplitis mediator* for enhanced biological control of the bertha armyworm. Final Report, Saskatchewan Agriculture Development Fund Project 95000283.

Mason, P.G. and Youngs, B.J. (1994) Biological control of a Canadian canola pest, the bertha armyworm (*Mamestra configurata*), with the European parasitoid *Microplitis mediator*. *Norwegian Journal of Agricultural Sciences*, Supplement 16, 405–406.

Mason, P.G., Arthur, A.P., Olfert, O.O. and Erlandson, M.A. (1998) The bertha armyworm (*Mamestra configurata*) (Lepidoptera: Noctuidae) in western Canada. *The Canadian Entomologist* 130, 321–336. doi: 10.4039/Ent130321-3.

Mason, P.G., Turnock, W.J., Erlandson, M.A., Kuhlmann, U. and Braun, L. (2002) *Mamestra configurata* Walker, bertha armyworm (Lepidoptera: Noctuidae). In: Mason, P.G. and Huber, J.T. (eds) *Biological Control Programmes in Canada, 1981–2000*. CABI Publishing, Wallingford, UK, pp. 169–176.

McCloskey, C. and Isman, M.B. (1993) Influence of foliar glucosinolates in oilseed rape and mustard on feeding and growth of the bertha armyworm, *Mamestra configurata* Walker. *Journal of Chemical Ecology* 19, 249–266. doi: 10.1007/bf00993693.

McMillan, E. (1935) A survey of cutworm damage in a specimen locality in Saskatchewan. *Journal of Economic Entomology* 28, 428–431. doi: 10.1093/jee/28.2.428.

McNeil, J.N. (1987) The true armyworm, *Pseudaletia unipuncta*: a victim of the Pied Piper or a seasonal migrant? *International Journal of Tropical Insect Science* 8, 591–597. doi: 10.1017/S1742758400022657.

Mohammed, E.H.K. (2015) Production of transgenic canola plants resistant to black cutworm *"Agrotis ipsilon"*. PhD thesis, Cairo University, Cairo, Egypt. Abstract available at: http://erepository.cu.edu.eg/index.php/cutheses/article/view/5574 (accessed 8 April 2016).

Morris, O.N., Converse, V. and Harding, J. (1990) Virulence of entomopathogenic nematode-bacteria complexes for larvae of noctuids, a geometrid, and a pyralid. *The Canadian Entomologist* 122, 309–319. doi: 10.4039/Ent122309-3.

O'Hara, J.E. (2008) Tachinid flies (Diptera: Tachinidae). In: Capinera, J.L. (ed.) *Encyclopedia of Entomology*, 2nd edn. Springer Netherlands, Dordrecht, The Netherlands, pp. 3675–3686.

Olfert, O., Cárcamo, H. and Pepper, J. (2005) Insect pests and arthropod diversity in field margins of western Canada. In: Thomas, A.G. (ed.) *Topics in Canadian Weed Science*. Canadian Weed Science Society, Pinawa, Manitoba, Canada, pp. 135–163.

OMAFRA (2016) *Neonicotinoid Regulations*. Ontario Ministry of Agriculture, Food and Rural Affairs, Guelph, Ontario, Canada. Available at: https://www.ontario.ca/page/neonicotinoid-regulations (accessed 1 March 2016).

Parker, J.R., Strand, A.L. and Seamans, H.L. (1921) Pale western cutworm (*Porosagrotis orthogonia* Morr.). *Journal of Agricultural Research* 22, 289–321.

Pisa, L.W., Amaral-Rogers, V., Belzunces, L.P., Bonmatin, J.M., Downs, C.A. *et al.* (2014) Effects of neonicotinoids and fipronil on non-target invertebrates. *Environmental Science and Pollution Research* 22, 68–102. doi: 10.1007/s11356-014-3471-x.

PMRA (2015) Registration Decision RD2015-08, *Bacillus thuringiensis* subsp. *aizawai* strain ABTS-1857. Pest Management Regulatory Agency, Guelph, Ontario, Canada. Available at: http://www.hc-sc.gc.ca/cps-spc/alt_formats/pdf/pubs/pest/_decisions/rd2015-08/rd2015-08-eng.pdf (accessed 8 April 2016).

PPMN (2016) *Prairie Pest Monitoring Network Weekly Updates*. Prairie Pest Monitoring Network, Agriculture and Agri-Food Canada, Ottawa, Ontario,

Canada. Available at: http://www.westernforum.org/IPMNWeeklyUpdates.html (accessed 24 February 2016).

Pruess, K.P. and Roselle, R. (1958) *Pale Western Cutworm Control*. Extension Service, University of Nebraska, College of Agriculture and United States Department of Agriculture, Lincoln, Nebraska, E.C. 58-1580.

Salt, R.W. and Seamans, H.L. (1945) Experimental starvation of first-instar larvae of the pale western cutworm, *Agrotis orthogonia* Morr. *The Canadian Entomologist* 77, 150–155. doi: 10.4039/Ent77150-8.

Schaaf, A.C. (1972) The parasitoid complex of *Euxoa ochrogaster* (Guenee) (Lepidoptera: Noctuidae). *Quaestiones Entomologicae* 8, 81–120.

Seamans, H.L. (1935) Forecasting outbreaks of the pale western cutworm (*Agrotis orthogonia* Morr.) *Journal of Economic Entomology* 28, 425–428. doi: 10.1093/jee/28.2.425.

Seamans, H.L. and Rock, P.J.G. (1945) Starvation of the early instars of the pale western cutworm, *Agrotis orthogonia* Morr., and its use in the control of this pest. *The Canadian Entomologist* 77, 57–61. doi: 10.4039/Ent7757-4.

Smith, M.A.H., Lamb, R.J., Wise, I.L. and Olfert, O.O. (2004) An interspersed refuge for *Sitodiplosis mosellana* (Diptera: Cecidomyiidae) and a biocontrol agent *Macroglenes penetrans* (Hymenoptera: Pteromalidae) to manage crop resistance in wheat. *Bulletin of Entomological Research* 94, 179–188. doi: 10.1079/BER2004291.

Songa, J.M. and Holliday, N.J. (1997) Laboratory studies of predation of grasshopper eggs, *Melanoplus bivattatus* (Say), by adults of two species of *Pterostichus* Bonelli (Coleoptera: Carabidae). *The Canadian Entomologist* 129, 1151–1159. doi: 10.4039/Ent1291151-6.

Sorenson, C.J. and Thornley, H.F. (1941) *Pale Western Cutworm.* Bulletin No. 297. Agricultural Experiment Station, Utah State Agricultural College, Logan, Utah. *UAES Bulletins*, Paper 259. Available at: http://digitalcommons.usu.edu/uaes_bulletins/259 (accessed 17 February 2016).

Statistics Canada (2015) *Canola area, 2011*. Statistics Canada, Ottawa, Ontario, Canada. Available at: http://www5.statcan.gc.ca/cansim/a26?lang=eng&retrLang=eng&id=0010010&pattern=&csid= (accessed 26 January 2016).

Steinhaus, E. and Marsh, G. (1962) Report of diagnoses of diseased insects 1951–1961. *Hilgardia* 33, 349–490. doi: 10.3733/hilg.v33n09p349.

Stinner, B.R. and House, G.J. (1990) Arthropods and other invertebrates in conservation-tillage agriculture. *Annual Review of Entomology* 35, 299–318. doi: 10.1146/annurev.en.35.010190.001503.

Strickland, E.H. (1923) *Biological Notes on Parasites of Prairie Cutworms*. Bulletin 26. Canadian Department of Agriculture, Ottawa, Ontario, Canada.

Tabashnik, B.E., Brévault, T. and Carrière, Y. (2013) Insect resistance to Bt crops: lessons from the first billion acres. *Nature Biotechnology* 31, 510–521. doi: 10.1038/nbt.2597.

Thiele, H.U. (1977) *Carabid Beetles in their Environments. A Study on Habitat Selection by Adaptations in Physiology and Behavior*. Springer, New York, NY.

Thrash, B., Adamczyk, J.J., Lorenz, G., Scott, A.W., Armstrong, J.S., Pfannenstiel, R. and Taillon, N. (2013) Laboratory evaluations of lepidopteran-active soybean seed treatments on survivorship of fall armyworm (Lepidoptera: Noctuidae) larvae. *Florida Entomologist* 96, 724–728. doi: 10.1653/024.096.0304.

TOFA (2016) *Frost Chart for Canada*. The Old Farmers Almanac, Yankee Publishing, Dublin, New Hampshire. Available at: http://www.almanac.com/content/frost-chart-canada (accessed 27 February 2016).

Turgeon, J.J., McNeil, J.N. and Roelofs, W.L. (1983) Field testing of various parameters for the development of a pheromone-based monitoring system for the armyworm, *Pseudaletia unipuncta* (Haworth) (Lepidoptera: Noctuidae). *Environmental Entomology* 12, 891–894. doi: 10.1093/ee/12.3.891.

Turnock, W.J. (1985) Developmental, survival, and reproductive parameters of bertha armyworm, *Mamestra configurata* (Lepidoptera: Noctuidae) on four plant species. *The Canadian Entomologist* 117, 1267–1271. doi: 10.4039/Ent1171267-10.

Turnock, W.J. and Bilodeau, R.J. (1985) A comparison of three methods of examining the density of larvae of the bertha armyworm, *Mamestra configurata*, in fields of canola (*Brassica* spp.). *The Canadian Entomologist* 117, 1065–1066. doi: 10.4039/Ent1171065-8.

Turnock, W.J. and Carl, K.P. (1995) Evaluation of the Palaearctic *Eurithia consobrina* (Diptera: Tachinidae) as a potential biocontrol agent for *Mamestra configurata* (Lepidoptera: Noctuidae) in Canada. *Biocontrol Science and Technology* 5, 55–68. doi: 10.1080/09583159550040015.

Turnock, W.J., Timlick, B. and Palaniswamy, P. (1993) Species and abundance of cutworms (Noctuidae) and their parasitoids in conservation and conventional tillage fields. *Agriculture, Ecosystems & Environment* 45, 213–227. doi: 10.1016/0167-8809(93)90072-w.

Ulmer, B., Gillott, C. and Eerlandson, M. (2001) Feeding preferences, growth, and development of *Mamestra configurata* (Lepidoptera: Noctuidae) on Brassicaceae. *The Canadian Entomologist* 133, 509–519. doi: 10.4039/Ent133509-4.

USDA (2015) *Acreage (June 2015)*. National Agricultural Statistics Service, Agricultural Statistics Board, US Department of Agriculture, Washington, DC. Available at: http://www.usda.gov/nass/PUBS/TODAYRPT/acrg0615.pdf (accessed 12 March 2016).

Vera, C.L., Fox, S.L., DePauw, R.M., Smith, M.A.H., Wise, I.L., Clarke, F.R., Procunier, J.D. and Lukow, O.M. (2013) Relative performance of resistant wheat varietal blends and susceptible wheat cultivars exposed to wheat midge, *Sitodiplosis mosellana* (Géhin). *Canadian Journal of Plant Science* 93, 59–66. doi: 10.4141/cjps2012-019.

Walkden, H.H. (1950) *Cutworms, Armyworms, and Related Species Attacking Cereal and Forage Crops in the Central Great Plains*. Circular 849, US Department of Agriculture, Washington, DC.

WCCP (Western Committee on Crop Pests) (2012) Minutes of the Western Committee on Crop Pests, 52nd Annual Meeting, 2012, The Radisson Hotel, Saskatoon, Saskatchewan, pp. 1–80. Available at: http://www.westernforum.org/Documents/WCCP/WCCP%20Minutes/WCCP%202012%20minutes.pdf (accessed 8 April 2016).

WCCP (Western Committee on Crop Pests) (2013) Minutes of the Western Committee on Crop Pests, 53rd Annual Meeting, 2013, Canada Inns Fort Garry, Winnipeg, Manitoba, pp. 1–70. Available at: http://www.westernforum.org/Documents/WCCP/WCCP%20Minutes/Minutes%20of%20WCCP%202013.pdf (accessed 8 April 2016).

Weatherbase (2016) *Climate Records for Regina, Saskatchewan*. Canty & Associates LLC, Great Falls, Virginia. Available at: http://www.weatherbase.com/ (accessed 27 February 2016).

White, J.D., Kendall, K.C. and Picton, H.D. (1998) Grizzly bear feeding activity at alpine army cutworm moth aggregation sites in northwest Montana. *Canadian Journal of Zoology* 76, 221–227. doi: 10.1139/z97-185.

Wise, I.L., Turnock, W.J. and Gavloski, J. (2009) Sequential decision plan for controlling *Mamestra configurata* in spring canola. *The Canadian Entomologist* 141, 619–626. doi: 10.4039/n09-031.

Wylie, H.G. and Bucher, G.E. (1977) The bertha armyworm, *Mamestra configurata* (Lepidoptera: Noctuidae). Mortality of immature stages on the rape crop, 1972–1975. *The Canadian Entomologist* 109, 823–837. doi: 10.4039/Ent109823-6.

Yeates, D.K. and Greathead, D. (1997) The evolutionary pattern of host use in the Bombyliidae (Diptera): a diverse family of parasitoid flies. *Biological Journal of the Linnean Society* 60, 149–185. doi: 10.1006/bijl.1996.0097.

8 Biology and Management of the Generalist Herbivore, the Bertha Armyworm, *Mamestra configurata* (Lepidoptera: Noctuidae), on Canola in Western Canada

Maya L. Evenden*, Ronald E. Batallas and Chaminda Weeraddana

University of Alberta, Edmonton, Alberta, Canada

8.1 Introduction

The bertha armyworm, *Mamestra configurata* Walker (Lepidoptera: Noctuidae), is a native generalist herbivore in western Canada. It was first noted as a pest of flax, *Linum usitatissimum* L. (Linaceae), in the Canadian Prairie Provinces (Alberta, Saskatchewan and Manitoba) in the 1920s (King, 1928) but has been most notorious as a pest of rapeseed and canola crops, *Brassica napus* L. and *Brassica rapa* L. (Brassicaceae) (Mason *et al.*, 1998a). Populations of the bertha armyworm reach outbreak levels in canola at periodic intervals and cost producers millions of dollars in lost yield and control costs. Prairie-wide control costs have ranged from CAN$3.4 million in 1971 (Riegert, 1984) to CAN$16.5 million per year in 1994–1995 (Mason *et al.*, 1998a) and 2005–2007 (Erlandson, 2013). In addition, damage during bertha armyworm outbreaks (Fig. 8.1) resulted in lost yield worth CAN$14.2 million in 1971 (Riegert, 1984) and CAN$10–40 million per year in the outbreaks of 1994–1995 (Mason *et al.*, 1998a) and 2005–2007 (Erlandson, 2013). The occurrence of outbreak populations (Table 8.1) in the Prairie Provinces has increased in regularity with the increase in acreage of canola production in this region (Mason *et al.*, 1998a). Factors influencing the population dynamics of this insect have been the topic of intense study, primarily since the large outbreak of 1971–1972 in western Canada (Turnock, 1984a).

Two previously published extensive reviews focused on bertha armyworm biology and pest status. An early review (King, 1928) described the insect life stages, life history and bertha armyworm interactions with plant hosts and Prairie crops. Crow predators and disease of larvae were noted as important natural control agents. More recently Mason *et al.* (1998a) detailed research on the biology and control of bertha armyworm and described the periodicity of outbreaks in the Prairie Provinces. Population monitoring of bertha armyworm has become an important part of integrated pest management (IPM) of this pest and the development of a prairie-wide pheromone-based monitoring system is introduced in Mason *et al.* (1998a). There have been several reviews that target more specifically the natural control agents of the bertha armyworm and the potential for biological control of this pest in the Prairie Provinces (Turnock, 1984a; Mason *et al.*, 2002; Erlandson, 2013). This chapter reviews the biology and management of the bertha armyworm with an emphasis on newer literature published since Mason *et al.* (1998a). Future research needs

*Corresponding author. Email: mevenden@ualberta.ca

Fig. 8.1. Bertha armyworm feeding damage on (a) canola leaf and (b) canola pods. Photos: Michael Dolinski.

and potential new avenues for non-insecticidal control of the bertha armyworm are also discussed.

8.2 Biology

8.2.1 Phylogeny

The bertha armyworm is classified in the Noctuidae (Lafontaine and Schmidt, 2010), the most diverse family in the superfamily Noctuoidea, within the order Lepidoptera (Wagner, 2001). Within the Noctuidae, the bertha armyworm is part of the trifine Noctuidae moths, in which the cubital vein of the hindwing appears three-branched (Lafontaine, 1993). Agriculturally important noctuid moth pests assemble a monophyletic group, named the 'pest clade', which includes all cutworm and armyworm species (Mitchell *et al.*, 2006). The genus *Mamestra* was formerly in the subfamily Hadeninae; however, current taxonomic concepts place the genus in the tribe Hadenini, within the subfamily Noctuinae (Lafontaine and Schmidt, 2010). Moths in the Hadenini are characterized by stout apical setae on the foreleg tibia modified into 'claws', to aid adults in digging pupal cells out of the ground, and the presence of hairs on the surface of the compound eye (Fibiger and Lafontaine, 2005).

8.2.2 Distribution

The bertha armyworm is a Nearctic species (Mason *et al.*, 2002). Its distribution is limited to dry grassland habitats in western areas, from British Columbia eastwards to Manitoba and southwards to central California, Arizona, Texas and Mexico (Powell and Opler, 2009). A second species of *Mamestra* present in North America, *M. curialis* (Smith), occurs over a wider range from Quebec to British Columbia (Godfrey, 1972). Despite their overlapping distribution, *M. curialis* is prevalent in forested habitats in the north-west (King, 1928). Larvae of both species are similar in appearance but the moths are easily distinguished by wing patterns and genitalic characters (Godfrey, 1972).

8.2.3 Host range

The bertha armyworm is polyphagous and feeds on as many as 40 plant species (King, 1928; Bailey, 1976a; Turnock, 1985) in a variety of plant families: Brassicaceae, Compositae, Leguminosae, Chenopodiaceae and Linaceae (Dosdall and Ulmer, 2004). In a laboratory study comparing host suitability of plants in various families, bertha armyworm larvae developed fastest on *Brassica napus*, *Brassica rapa* and *Chenopodium album* L. (Chenopodiaceae) (Dosdall and Ulmer, 2004). Non-cultivated natural host plants of the bertha armyworm are not known but endemic species in the family Chenopodiaceae might be good candidates, based on the suitability of the invasive weed lamb's quarters, *Chenopodium album*, for growth and development of the bertha armyworm. Population growth on lamb's quarters may have promoted the first bertha armyworm outbreaks that affected flax production in western Canada in the 1920s, before widespread cultivation of *Brassica* crops (Dosdall and Ulmer, 2004). Similarly, bertha armyworm associated with lentil crops, *Lens culinaris* Medikus (Fabaceae) (King, 1928), probably initially developed on lamb's quarters in the vicinity rather than on lentil, which is not a suitable host (Turnock, 1985), and then moved to lentils in the sixth instar when 80% of

Table 8.1. Occurrence and area sprayed to control economically injurious populations of bertha armyworm in the Canadian Prairie Provinces.

Province	Years	Crop	Approximate sprayed area (ha)	Reference
Alberta	1927/28	Various	NA	King, 1928
	1932	Alfalfa	NA	Mason *et al.*, 1998a
	1943	Alfalfa	NA	Mason *et al.*, 1998a
	1971	Rapeseed	73,861	Turnock and Philip, 1977
	1972	Rapeseed	91,420	Turnock and Philip, 1977
	1973	Rapeseed	22,296	Turnock and Philip, 1977
	1974	Rapeseed	1161	Turnock and Philip, 1977
	1979	Canola	1200	Mason *et al.*, 1998a
	1980	Canola	6000	Mason *et al.*, 1998a
	1981	Canola	2700	Mason *et al.*, 1998a
	1983	Canola	1480	Mason *et al.*, 1998a
	1984	Canola	5000	Mason *et al.*, 1998a
	1989	Canola	> 40,000	Mason *et al.*, 1998a
	1990	Canola	> 60,000	Mason *et al.*, 1998a
	1991	Canola	2000	Mason *et al.*, 1998a
	1995	Canola	> 17,200	Mason *et al.*, 1998a
	1996	Canola	14,000	Mason *et al.*, 1998a
	1997	Canola	37,000	Mason *et al.*, 1998a
	2004	Canola	< 4000	WCCP, 2004
	2005	Canola	55,000	WCCP, 2005
	2006	Canola	12,140	WCCP, 2006
	2011	Canola	< 4000	WCCP, 2011
	2012	Canola	> 40,500	WCCP, 2012
	2013	Canola	40,500	WCCP, 2013
Saskatchewan	1922	Flax	NA	King, 1928
	1925	Flax, sweet clover, corn, garden peas	NA	Mason *et al.*, 1998a
	1943	Flax	NA	Mason *et al.*, 1998a
	1944	Flax, rapeseed	NA	Mason *et al.*, 1998a
	1947	Rapeseed	Unknown	Mason *et al.*, 1998a, Riegert, 1984
	1948	Rapeseed	Unknown	Mason *et al.*, 1998a
	1949	Rapeseed	NA	Mason *et al.*, 1998a
	1953	Flax, rapeseed	NA	Mason *et al.*, 1998a
	1954	Flax, rapeseed	NA	Mason *et al.*, 1998a
	1955	Flax, rapeseed	NA	Mason *et al.*, 1998a
	1956	Flax, rapeseed	6000	Mason *et al.*, 1998a
	1962	Flax, rapeseed	> 40	Mason *et al.*, 1998a
	1971	Rapeseed	> 800,000	Mason *et al.*, 1998a
	1972	Rapeseed	162,000	Mason *et al.*, 1998a
	1973	Rapeseed	22,000	Mason *et al.*, 1998a
	1981	Canola	3500	Mason *et al.*, 1998a
	1985	Canola	1200	Mason *et al.*, 1998a
	1987	Canola	6,000–8,000	Mason *et al.*, 1998a
	1990	Canola	10,000–15,000	Mason *et al.*, 1998a
	1994	Canola, flax	160,000	Mason *et al.*, 1998a
	1995	Canola, flax	652,500	Mason *et al.*, 1998a
	1996	Canola	250,000	Mason *et al.*, 1998a

Continued

Table 8.1. Continued.

Province	Years	Crop	Approximate sprayed area (ha)	Reference
	1997	Canola	32,000	Mason *et al.*, 1998a
	1998		358,000	WCCP, 1998
	2005	Canola	Unknown	WCCP, 2005
	2006	Canola	> 283,280	WCCP, 2006
	2007	Canola	Unknown	WCCP, 2007
	2008	Canola	5870	WCCP, 2008
	2011	Canola	Unknown	WCCP, 2011
	2012	Canola	Unknown	WCCP, 2012
	2013	Canola	Unknown	WCCP, 2013
Manitoba	1925	Sweet clover	NA	King, 1928
	1927	Sweet clover	NA	King, 1928
	1928	Unknown	NA	Mason *et al.*, 1998a
	1929	Unknown	NA	Mason *et al.*, 1998a
	1948	Flax	NA	Mason *et al.*, 1998a
	1949	Flax	NA	Mason *et al.*, 1998a
	1971	Canola	6180	Mason *et al.*, 1998a
	1972	Canola	> 20,000	Mason *et al.*, 1998a
	1980	Canola	2800	Mason *et al.*, 1998a
	1981	Canola	5680	Mason *et al.*, 1998a
	1993	Canola	1200	Mason *et al.*, 1998a
	1994	Canola, flax	500,000	Mason *et al.*, 1998a
	1995	Canola	364,000	Mason *et al.*, 1998a
	1996	Canola	< 50,000	Mason *et al.*, 1998a
	2005	Canola	5665	WCCP, 2005
	2006	Canola	Unknown	WCCP, 2006
	2007	Canola	Unknown	WCCP, 2007
	2011	Canola	Unknown	WCCP, 2011
	2012	Canola	Unknown	WCCP, 2012
	2013	Canola	Unknown	WCCP, 2013

larval food is eaten to obtain enough plant matter to complete development (Bailey, 1976a).

8.2.4 Life cycle

In the Canadian Prairie Provinces, the bertha armyworm has a univoltine life cycle in which pupae overwinter in facultative diapause. Museum collections of bertha armyworm adults indicate bivoltinism in more southern regions of its range (California, USA) (Wylie and Bucher, 1977).

Eggs

Female bertha armyworms oviposit an average of 2100 eggs over their lifetime (Howlader and Gerber, 1986) (Fig. 8.2). Mated females lay eggs in masses on the abaxial surface of leaves. Eggs are spherical and ridged and gradually darken in colour through embryonic development (Rempel, 1951). Several factors influence the oviposition decisions of female moths (Fig. 8.3). Egg development requires an accumulation of 82 degree days (DD) above a developmental threshold of 7°C for completion (Bailey, 1976b).

Bertha armyworm larvae develop through six instars in approximately 6 weeks under field conditions (Mason *et al.*, 1998a). Degree-day accumulation of 356 DD above 7°C is required for larval development. Larvae can develop at temperatures ranging from 8°C to 32°C but larvae do not complete development when held at 6°C or 36°C (Bailey, 1976b). Young larvae feed gregariously on foliage while the last two instars prefer to feed on the protein-rich pods (Bracken, 1984) (Fig. 8.4). Feeding occurs at night and the majority of damage is the result of feeding by sixth-instar larvae (L6), when 70–80% of the food is consumed (Bailey, 1976a). Larvae from L2 to L6 are sensitive to photoperiodic and temperature cues that induce pupal diapause (Hegdekar, 1977). Mature larvae drop from the plant and crawl into the soil, where they excavate cells for pupation at depths up to 15 cm below the soil surface (Wylie and Bucher, 1977).

Pupae

Facultative diapause of pupae is initiated by a photoperiod of 5–13 h at 20°C with a critical day length of 14.75 h. Diapause is induced regardless of photoperiod at a larval rearing temperature of 17°C, while continuous development without diapause occurs at 27°C (Hegdekar, 1977). The critical day length required for diapause induction varies with latitude and is generally longer for field populations (Hegdekar, 1983). The mechanism that initiates diapause is a low titre of ecdysone in bertha armyworm pupae. There is also a role for cyclic nucleotides in diapause control in bertha armyworm pupae in which cyclic adenosine monophosphate (cAMP) maintains diapause and cyclic guanosine monophosphate (cGMP) terminates it (Bodnaryk, 1987). Diapause proceeds through four distinct stages in which the response to injected ecdysone varies (Bodnaryk, 1977). There is a high transitory peak of ecdysone production during pupal–adult development in both sexes that is thought to serve as a negative feedback mechanism to terminate ecdysone production before adult eclosion (Bodnaryk, 1986).

Fig. 8.2. Bertha armyworm egg mass on abaxial leaf surface of canola plant. Photo: Chaminda Weeraddana.

Diapausing bertha armyworm pupae are freeze susceptible and survive winter through supercooling at temperatures below –18°C (Turnock *et al.*, 1983). As little as 5 cm of snow cover helps to protect overwintering pupae positioned 5 cm below the soil surface. Pupal survival in the field can be estimated by measurement of soil temperatures. Survival can be as low as 10% with exposure to

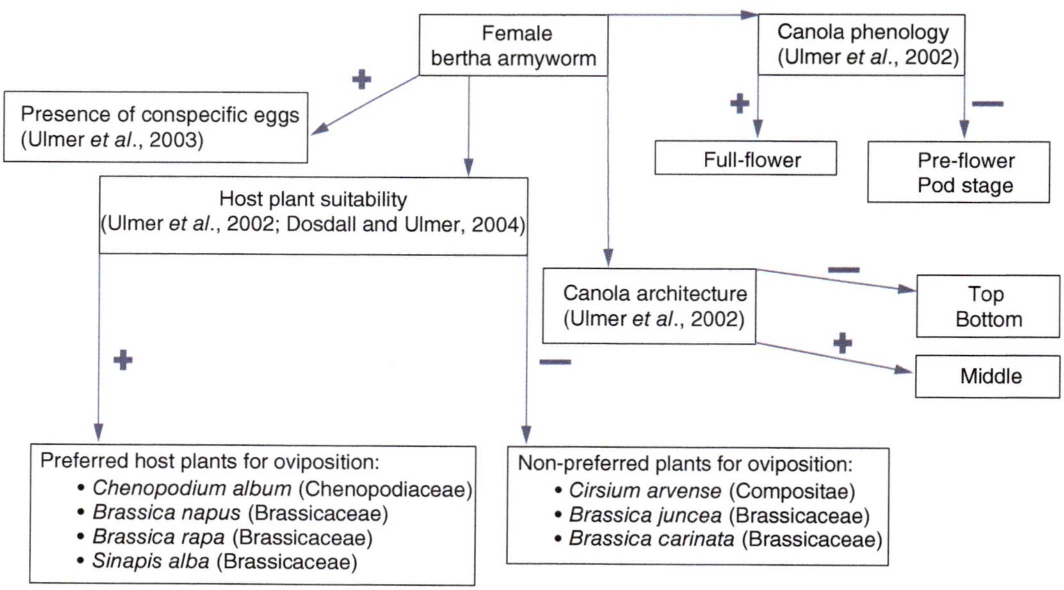

Fig. 8.3. Factors influencing oviposition decisions of bertha armyworm female moths.

Fig. 8.4. Bertha armyworm larva feeding on canola pod. Photo: Michael Dolinski.

Fig. 8.5. Bertha armyworm adult. Photo: Ronald Batallas.

soil temperature of –10°C or colder for 30 days (Lamb *et al.*, 1985). Diapausing pupae secrete more waxes and hydrocarbons for a longer period as compared with non-diapausing pupae, suggesting that a thicker wax layer will protect diapausing pupae from desiccation (Hegdekar, 1979).

Adults

The flight period of the bertha armyworm adult (Fig. 8.5) extends from the end of June to the beginning of August in the Canadian Prairie Provinces. Temperature has a significant effect on the longevity of bertha armyworm adults. Moths can live up to 30 days at 10°C but an increase in temperature results in reduced adult longevity. At 35°C, adult bertha armyworm longevity can be as short as 5.4 days (Gerber and Howlader, 1987).

Female bertha armyworm moths call mates by releasing a sex pheromone during the second or third scotophase after eclosion. Calling occurs after chorionated eggs are produced in the females' ovaries. The duration of calling behaviour is longer and occurs later in the scotophase under a photoperiod with relatively long nights. Females call earlier and for longer in the scotophase under cool (15–25°C) than warm (30–35°C) temperature regimes (Gerber and Howlader, 1987). Early studies revealed that the bertha armyworm pheromone signal comprises two major components, (Z)-11-hexadecenyl acetate (Z11-16:Ac) and (Z)-9 tetradecenyl acetate (Z9-14:Ac) (Chisholm *et al.*, 1975; Struble *et al.*, 1975), that are most attractive to males in trapping studies in a 19:1 ratio of Z11-16:Ac to Z9-14:Ac (Underhill *et al.*, 1977). Although several other minor components were later identified from female gland extracts, male bertha armyworm antennae only responded to Z11-16:Ac and Z9-14:Ac in electroantennogram studies. The addition of several of these minor components to the pheromone blend did not radically enhance male moth attraction but did improve the specificity of the lure (Struble *et al.*, 1984). Male bertha armyworm produce a courtship pheromone made up of 2-phenylethanol and benzaldehyde in a gland located on the ventral side of the abdomen (Clearwater, 1975).

Copulation occurs at dawn (Swailes *et al.*, 1975) and the first copulation lasts 17 ± 0.2 h with oviposition commencing shortly after (Howlader and Gerber, 1986). Seventy-five per cent of the total eggs are laid by the end of the seventh scotophase. Mated females have a refractory period of 2 days before they call for another mate and less time is spent calling by mated than by virgin females (Howlader and Gerber, 1986). Both calling behaviour and oogenesis are affected by high temperatures (35°C), when degeneration of oocytes occurs in the basal region of ovarioles (Gerber and Howlader, 1987).

8.3 Management: Monitoring

8.3.1 Adults

Prior to the bertha armyworm outbreak in the Canadian Prairie Provinces in 1971–1973, light traps were the most common tool for monitoring adult abundance of pestiferous noctuid moths in agricultural fields; however, they were not successful in predicting infestations (Mason *et al.*, 1998a). In addition, bertha armyworm moths are not strongly attracted to light traps (Bucher and Bracken,

1979). Identification of the components of the female sex pheromone (Chisholm *et al.*, 1975; Struble *et al.*, 1975) and the development of synthetic lures (Underhill *et al.*, 1977) provided an opportunity to establish an effective monitoring system for bertha armyworm adults. Sex pheromone-baited traps captured about 15 times more male moths than lights traps at multiple sites across the Prairie Provinces and thus delivered a more reliable estimate of adult populations (Steck *et al.*, 1979). The number of moths captured in sex pheromone-baited traps is affected by crop developmental stage, in part dependent on the degree of synchronization between the moth flight period and the attractive crop stage (early bloom) (Turnock, 1984b). Similarities between cumulative curves for male moth-trap catch and oviposition confirmed that male moth captures in sex pheromone-baited traps provide a valid index of abundance of female moths and their egg-laying capacity (Turnock, 1984b). The total number of male moths captured is positively related to the density of late-instar larvae in fields within a 5 km radius but field-to-field variability in larval density is too high to allow prediction of accurate estimates (Turnock, 1987). Currently, adult monitoring with sex pheromone-baited traps is an early warning system to indicate risk of damaging larval infestations. Guidelines for adult monitoring by the Prairie Pest Monitoring Network recommend monitoring sex pheromone-baited traps on canola fields weekly from the first week of June until the last week of July (Gavloski, 2015). Two all-green model Unitraps (Universal Moth Trap) are placed about 1 m above ground, 2 m from the field edge and 50 m apart (Gavloski, 2015). Cumulative bertha armyworm male moth counts in pheromone-baited traps (Table 8.2) throughout the monitoring season indicate the areas to emphasize for larval sampling.

A coordinated monitoring programme was implemented throughout the canola-growing region of the Prairie Provinces in 1995, led by the Western Forum for Pest Management, as an early warning system and forecast for potential risk of bertha armyworm larval infestation (Mason *et al.*, 1998a). A network of sex pheromone-baited traps distributed to participants is set up annually by the respective provincial agriculture departments from each Prairie Province as part of the Prairie Pest Monitoring Network. Weekly counts are reported to each provincial entomologist and data are gathered to generate a risk assessment map showing the spatial distribution and predicted density of moths across the Prairie Provinces (Mason *et al.*, 1998a). Risk assessment maps indicate regions where larval monitoring is required. The Prairie Pest Monitoring Network also generates updates on the average pupal development for the prairie region based on a degree-day model to anticipate maximum moth emergence to target with pheromone-baited traps (Mason *et al.*, 1998a).

Fermented sugar solutions can also be used as baits to attract moth pest species (Landolt and Mitchell, 1997). Characterization of the volatile profile from fermented sugar baits led to the development of lures to monitor noctuid moth pest species. Food bait lures have an advantage over sex pheromones, as these lures attract both sexes of moth. Mass capture of females may reduce female populations prior to oviposition, which can reduce subsequent larval damage. The combination of acetic acid and 3-methyl-1-butanol (1:1 by weight mixture) attracts bertha armyworm moths and captures both male and females at a ratio near 1:1 (Landolt, 2000; Landolt and Alfaro, 2001).

Table 8.2. Cumulative bertha armyworm moth counts in sex pheromone-baited traps in relation to risk of larval infestation. Source: Gavloski, 2015.

Cumulative number of moths per trap	Larval infestation risk level
0–300	**Low.** Infestations are unlikely to be widespread, but fields should be inspected for signs of insects or damage.
300–900	**Uncertain.** Infestations may not be widespread, but fields that were particularly attractive to egg-laying females could be infested. Check your fields.
900–1200	**Moderate.** Canola fields should be sampled regularly for larvae and for evidence of damage.
1200–1500+	**High.** Canola fields should be sampled frequently for larvae and for evidence of damage.

8.3.2 Larvae

Standard guidelines for larval sampling were developed initially during the course of the outbreak in 1971–1973. The 'early larval survey' method was based on scouting for occurrence of early-instar larvae in fields, and results were reported as the percentage of fields infested, with a larval density index (Turnock and Philip, 1977). Fields were surveyed for the presence of L1 and L2 instar larvae on plants and on the ground in three to six sample units of 0.5 m^2. This method was not effective, since timing of sampling was critical to detect infestations. Early surveys underestimated infestations whereas late surveys did not allow sufficient timing to alert producers (Turnock and Philip, 1977). Furthermore, damage from early-instar larvae is negligible, since they feed on primary leaves that drop as the crop matures (Bracken and Bucher, 1977). Yield loss occurs when L5 and L6 larvae feed on flowering shoots and pods, and thus later efforts focused on scouting for mature larvae when the crop was at half-full bloom to the mature pod stage (Bracken and Bucher, 1977). These sampling methods estimate the mean larval density/m^2 by counting the total number of larvae from ten sample units of 0.5 m^2 (Turnock, 1984b). Initial larval sampling protocols differed between the Prairie Provinces but all methods produced accurate estimates of larval density (Turnock and Bilodeau, 1985). Proper timing for larval sampling is based on information gathered from pheromone traps. Current guidelines recommend sampling 2 weeks after peak moth-trap catch when the crop is at peak flowering stage, and continuing until mean larval density exceeds the economic threshold or until the crop is swathed (Gavloski, 2015). For accurate estimates of larval densities, three samples of 1 m^2 or a minimum of five units of 0.25 m^2 should be taken, although 10–15 units of 0.25 m^2 results in more accurate estimates. At each sample unit, an open three-sided frame is placed on the ground surface beneath the plant canopy and plants are shaken by hand; then the larvae are counted on the ground surface (Turnock, 1984b). Bracken and Bucher (1977) estimated the relationship between larval density and yield loss under controlled field conditions. The larval density/m^2 at which insecticide treatment would be warranted is summarized in Table 8.3, based on an assumed yield loss of 0.058 bushel/acre for each larva/m^2 (Bracken and Bucher, 1977). The economic threshold for bertha armyworm in canola varies with the cost of insecticide application (CAN$/acre) and current crop value (CAN$/bushel) (Gavloski, 2015).

8.4 Control

8.4.1 Insecticides

Chemical insecticides

During the first reported outbreaks of bertha armyworm in the 1920s in the Canadian Prairie Provinces, producers had arsenic-based insecticidal dusts, poison bran baits and pyrethrum dusts at their disposal. Most success was achieved with arsenic dusts applied to infested fields of sweet clover (*Trifolium* spp. (Fabaceae)), alfalfa (*Medicago sativa* L.

Table 8.3. Economic threshold for bertha armyworm larvae infesting canola. Source: Gavloski, 2015.

Spraying cost (CAN$/acre)	Expected seed value (CAN$/bushel)										
	6	7	8	9	10	11	12	13	14	15	16
	Number of larvae/m^2										
7	20	17	15	13	12	11	10	9	9	8	8
8	23	20	17	15	14	13	11	11	10	9	9
9	26	22	19	17	16	14	13	12	11	10	10
10	29	25	22	19	17	16	14	13	12	11	11
11	32	27	24	21	19	17	16	15	14	13	12
12	34	30	26	23	21	19	17	16	15	14	13
13	37	32	28	25	22	20	19	17	16	15	14
14	40	35	31	27	24	22	20	19	17	16	15
15	43	37	32	29	26	23	22	20	19	17	16

(Fabaceae)) and flax (King, 1928). Little spraying was conducted at this time, however, because of the ephemeral nature of the outbreaks. High population densities in Saskatchewan in the late 1940s (Table 8.1) were effectively managed with applications of DDT over an unknown acreage (Riegert, 1984). At the time of the major bertha armyworm outbreak in 1971, no effective insecticides were registered for its control. Experimental insecticides were tested and the carbamate insecticide methomyl received an emergency registration. Fifty insecticides were screened for toxicity against the bertha armyworm but none were as toxic as methomyl (Harris and Turnbull, 1975). Stage-specific toxicity bioassays of methomyl and the organophosphate insecticides chlorpyrifos, leptophos and methidathion showed that eggs and young larvae (L1–L3) were susceptible, but later larval instars (L5–L6), pupae and adults were tolerant of insecticide applications (Harris and Turnbull, 1975). Insecticides are still the primary control tactic for bertha armyworm populations above economic threshold densities (Tables 8.1 and 8.3). The carbamate insecticide methomyl is still registered for use to control bertha armyworm on canola in Canada along with two organophosphate insecticides (chlorpyrifos, methamidophos) and three pyrethroid insecticides (deltamethrin, lambda-cyhalothrin, cypermethrin) (Brook and Cutts, 2016). A new insecticide registered for bertha armyworm control is chlorantraniliprole (Brook and Cutts, 2016), which belongs to a whole new class of insecticides, the anthranilamides. As this compound acts to disrupt insect muscles instead of the nervous system, it may remain effective against populations that have developed resistance to the older compounds.

Botanical insecticides

Studies have also examined the potential effect of botanical insecticides against the bertha armyworm. Larval growth is inhibited in a dose-dependent manner by consumption of azadirachtin, the active ingredient of neem. Topical application of azadirachtin to fourth-instar larvae reduces growth and relative consumption rates (Isman, 1993). Oral and topical applications of the liminoid cedrelone inhibit growth of larvae. High concentrations of cedrelone applied to leaf discs deter larval feeding (Koul and Isman, 1992). Although these botanical insecticides show growth inhibitor properties that could control young bertha armyworm larval instars, field experiments are required to demonstrate their potential.

Microbial insecticides

Early field applications of commercial formulations of the microbial insecticide *Bacillus thuringiensis* Berliner var. *kurstaki* (Bacillaceae) (Btk) targeting the bertha armyworm were of variable efficacy and promoted detailed study on the potential toxicity of Btk to the larvae (Morris, 1986). Feeding studies in the laboratory showed that very high doses of commercial Btk formulations were required to produce mortality of bertha armyworm larvae, with young larvae (L3–L4) more susceptible to poisoning than older larvae (L5–L6). Sublethal effects of larval weight loss did, however, reduce survival to the adult stage, suggesting that application of Btk may exert an effect on field populations (Morris, 1986). A combination of the δ-endotoxin in commercial Btk formulations and the β-exotoxin from naturally occurring Btk showed some synergistic toxicity to bertha armyworm but did not improve the sublethal effects on development of surviving larvae (Morris, 1988). Further studies compared numerous isolates of various *B. thuringiensis* (Bt) strains in laboratory feeding assays and revealed that several strains of *B. thuringiensis* var. *aizawai* were more toxic than the reference Btk strain against third-instar larvae (Trottier *et al.*, 1988). Closer examination of the toxicity of 51 different strains of Bt var. *aizawai* showed only seven of the tested *aizawai* strains were more toxic than the reference Btk strain, though mixing strains of different varieties with different toxic protein types was synergistic (Morris *et al.*, 1996). Other attempts to increase the toxicity of Btk to bertha armyworm include combining lethal concentrations of Btk with sublethal concentrations of a variety of compounds (Morris *et al.*, 1994, 1995). Incorporation of 0.1 % caffeine into a Btk spray applied to canola plants increased the toxicity of the formulation and reduced feeding activity as compared with Btk alone (Morris *et al.*, 1994). Other potential synergists identified in laboratory and greenhouse trials include boric acid, magnesium sulfate, acetamide, sodium carbonate, calcium carbonate and potassium carbonate (Morris *et al.*, 1995). The most recent work involved screening several δ-endotoxins against bertha armyworms using both *in vitro* and *in vivo* laboratory assays. The Bt δ-endotoxins Cry1Ac, Cry1Ab, Cry1Ca and the strain HD511 showed the highest activity in the *in vitro* assays and also had the most toxic effects on larvae (Erlandson *et al.*, 2002). It remains to be seen if a Bt transgenic canola plant would be effective against the bertha armyworm.

8.4.2 Biological control

The reliance on chemical insecticides and the poor efficacy of commercial formulations of Bt has promoted research into the potential use of natural enemies for biological control of bertha armyworm. There is a large number of natural mortality agents that play a role in the termination of bertha armyworm outbreaks but it is not known how natural enemies influence bertha armyworm population dynamics on the whole and what promotes the onset of outbreaks. Biological control against this insect has been reviewed extensively (Turnock, 1984a; Mason *et al.*, 2002; Dosdall and Mason 2010; Erlandson, 2013) and little research has occurred in this area since the publication of the most recent review.

Parasitoids

There is a large complex of parasitoids that have been reared from bertha armyworm under natural or laboratory conditions (Table 8.4). Based on their prevalence and distribution (Turnock, 1988), the most important native parasitoids of bertha armyworm in Canada are the ichneumonid wasp *Banchus flavescens* Cresson (Hymenoptera: Ichneumonidae) and the tachinid fly *Athrycia cinerea* (Coquillette) (Diptera: Tachinidae) (Mason *et al.*, 2002). Both of these parasitoids are univoltine, have relatively narrow host ranges and attack hosts in the larval stage. *Banchus flavescens* is a solitary endoparasitoid that exploits early larval instars (L1–L3) (Arthur and Mason, 1985) and *A. cinerea* is an exoparasitoid that lays eggs on the integument of larger larvae (L4–L6) (Wylie, 1977). Parasitism by *B. flavescens* can occur in populations of bertha armyworm at both low and high population densities but *A. cinerea* is found mostly associated with outbreak population densities of its host (Turnock, 1988). Egg parasitism of the bertha armyworm is rare but *Trichogramma inyoense* Pinot & Oatman (Hymenoptera: Trichogrammatidae) was reared from three egg masses collected in Saskatchewan in 1996 (Mason *et al.*, 1998b).

There have been several attempts to introduce classical biological control agents from Europe to control bertha armyworm. Two parasitoids of the closely related *Mamestra brassicae* L. (Lepidoptera: Noctuidae) in Europe, *Ernestia consobrina* (Meigen) (Diptera: Tachinidae) and *Microplitis mediator* Haliday (Hymenoptera: Braconidae), have been released in the Canadian Prairies. Although laboratory studies have shown that *E. consobrina* (Turnock and Carl, 1995) and *M. mediator* (Arthur and Mason, 1986; Pivnick, 1993) are efficient parasitoids of bertha armyworm that can survive the conditions on the Canadian Prairies, sampling efforts to determine whether introduced parasitoids have established populations in the new range have not yielded any parasitoids (Mason *et al.*, 2002).

Predators

Little information is known about the role of predators in population regulation of the bertha armyworm. Early observations (King, 1928) of bird predation on mature larvae at outbreak population densities in flax and sweet clover crops are corroborated by recent observations in canola (personal observation). Outbreaks of bertha armyworm that threaten complete crop destruction have been reduced due to consumption by flocks of crows to levels that resulted in less than 25% damage (King, 1928). Important invertebrate predators of bertha armyworm have not been identified under field conditions (Wylie and Bucher, 1977) but *Nabis americoferus* Carayon (Hemiptera: Nabidae) caused some mortality of young larvae in a greenhouse study conducted on sugar beet (Tamaki and Weeks, 1972) and an unreported species of cricket consumed bertha armyworm in field cage experiments (Wylie and Bucher, 1977).

Pathogens

Pathogens are among the most important natural mortality agents of bertha armyworm populations at high density (Wylie and Bucher, 1977). Fungal infection of bertha armyworm by *Entomophthora* sp. occurs in populations across the Prairie Provinces at low incidence levels (Turnock, 1988). Although infection with *Entomophthora* sp. was implicated in the collapse of the bertha armyworm outbreak in the Prairie Provinces in 1972 (Wylie

Table 8.4. Parasitoids of bertha armyworm.

Order	Family	Species	Reference
Diptera	Tachinidae	*Athrycia cinerea*	Wylie, 1977
		Blondelia sp.	O'Hara, 1999
		Chetogena claripennis sp. complex	Turnock, 1984a
		Chetogena tachinomoides	O'Hara, 1999
		Ernestia consobrina[a]	Turnock and Carl, 1995
		Exorista mella	Wylie and Bucher, 1977
		Lespesia archippivora	Turnock, 1984a
		Panzeria (Mericia) ampelus	Wylie and Bucher, 1977
		Phryxe pecosensis	Wylie and Bucher, 1977
		Phryxe vulgaris	Turnock, 1984a
		Spallanzania hebes	O'Hara, 1999
		Winthemia rufopicta	Turnock, 1984a
		Winthemia quadripustulata	Turnock, 1984a
Hymenoptera	Braconidae	*Apanteles xylinus*	Wylie and Bucher, 1977
		Cotesia laeviceps	Turnock, 1984a
		Microplitis mediator[a]	Arthur and Mason, 1986
	Eulophidae	*Eulophus* nr. *nebulosus*	Wylie and Bucher, 1977
		Euplectrus bicolor	Turnock, 1984a
	Ichneumonidae	*Banchus flavescens*	Wylie and Bucher, 1977
		Ichneumon canadensis	Wylie and Bracken, 1977
	Trichogrammatidae	*Trichogramma inyoense*	Mason *et al.*, 1998b

[a]Non-native species evaluated for classical biological control

and Bucher, 1977), environmental conditions of high humidity (Mason *et al.*, 2002) must coincide with high larval density for the development of widespread fungal epizootics. A microsporidian, presumed to be in the genus *Nosema*, was recovered from a small number of bertha armyworm larvae collected in surveys conducted in 1973 and 1974 in Manitoba but was not an important mortality factor (Wylie and Bucher, 1977). A high level of infection of early sixth-instar bertha armyworm larvae by several species of Steinernematidae and Heterorhabditidae nematodes was achieved in a soil-based laboratory bioassay (Morris and Converse, 1991). Further field research is required to determine if nematodes could provide a viable control option for bertha armyworm. As larvae are only in the soil for a brief period before pupation, field applications of nematodes would need to be timed precisely.

A large amount of research has been conducted on baculoviruses that cause disease in bertha armyworm (Erlandson, 2013). Recent genome sequencing of baculovirus strains from infected bertha armyworm revealed two species of nucleopolyviruses in the genus *Alphabaculovirus* (Baculoviridae). These two species, *Mamestra configurata nucleopolyherovirus-A* (MacoNPV-A) (Li *et al.*, 2002a) and *Mamestra configurata nucleopolyherovirus-B* (MacoNPV-B) (Li *et al.*, 2002b), differ in their virulence and prevalence of infection in bertha armyworm as well as their host range (Li *et al.*, 2002b). Although MacoNPV-A was more prevalent in infected bertha armyworm from populations across western North America than MacoNPV-B, the latter has a broader host range that includes several other pest species in the family Noctuidae (Li *et al.*, 2002b). Erlandson (2013) suggested that MacoNPV-B should be further developed as a biopesticide that could potentially target several noctuid pests that are common in prairie agroecosystems. Feeding studies to test formulations of various strains of MacoNPV indicate that bertha armyworm larvae fed virus-sprayed canola plants exhibit disease symptoms 10–21 days after feeding initiation in laboratory and field cage assays, respectively (Mason *et al.*, 2002). Access to the genome of these viruses also allows research into specific gene products that enhance the infectivity of the virus and can lead to the production of recombinant viruses with broader or more virulent activity (Erlandson, 2013).

8.4.3 Cultural control

The susceptibility of bertha armyworm pupae to winter temperatures affects the spatial distribution of populations but has not been linked to the initiation of outbreak population densities (Lamb *et al.*, 1985). As snow cover insulates pupae that are normally positioned approximately 5 cm under the soil surface (Lamb *et al.*, 1985), exposure of pupae to ambient temperatures can be achieved through tillage after larvae have pupated in the soil in the fall (King, 1928; Turnock and Bilodeau, 1984). A comparison of pupal mortality in untilled and tilled *B. rapa* and *B. napus* fields revealed greater overwinter survival in untilled fields, perhaps due to more snow cover in fields with crop stubble (Turnock and Bilodeau, 1984). Modern no-till agriculture techniques, however, are not compatible with this approach and tilling is not a control tactic that is commonly used against the bertha armyworm (Mason *et al.*, 1998a). Cultivation of fields in the spring may cause direct injury to pupae and expose them to bird predators (Wylie and Bucher, 1977).

Although the bertha armyworm is highly polyphagous, it maintains a feeding preference hierarchy among and within plant species (Turnock, 1985; Ulmer *et al.*, 2001; Dosdall and Ulmer, 2004) that might be exploited in plant breeding efforts. Within the Brassicaceae, antixenosis and antibiosis resistance to bertha armyworm feeding occurs in *Brassica juncea* L. (McCloskey and Isman, 1993; Bodnaryk, 1997; Ulmer *et al.*, 2001; Dosdall and Ulmer, 2004), *Thlaspi arvense* L. (Dosdall and Ulmer, 2004) and *Sinapis alba* L. (Bodnaryk, 1991; Ulmer *et al.*, 2001). The foliar glucosinolate sinigrin, allyl- or 2-propenyl glucosinolate, found in *B. juncea* is deterrent to bertha armyworm larvae (Ulmer *et al.*, 2001) and ovipositing female moths (Ulmer *et al.*, 2002). Larvae that do consume *B. juncea* foliage (McCloskey and Isman, 1993; Ulmer *et al.*, 2001) or artificial diet containing sinigrin (McCloskey and Isman, 1993) show slowed growth, low survival and reduced pupal weight (Dosdall and Ulmer, 2004) compared with bertha armyworm development on canola varieties. Sinalbin, *p*-hydroxybenzyl glucosinolate, found in the foliage of *S. alba*, appears to stimulate oviposition by females (Ulmer *et al.*, 2002) and feeding of neonate larvae (McCloskey and Isman, 1993). Larval weight gain, however, is slowed on *S. alba* compared with canola lines (Bodnaryk, 1991; Ulmer *et al.*, 2001) but this slow growth does not reduce the weight of the resulting pupae (Dosdall and Ulmer, 2004). Plant breeding efforts to increase certain foliar glucosinolates, especially sinigrin, may confer some level of resistance against bertha armyworm and potentially other generalist herbivores that include canola in their host range. Planting time of canola may also influence its susceptibility to feeding damage from bertha armyworm. Early-seeded canola varieties are more vulnerable to damage than late-seeded varieties probably because the attractive flowering stage (Fig. 8.3) better coincides with ovipositing female moths (Mason *et al.*, 1998a).

8.5 Conclusion

Despite the vast amount of research that has been conducted to understand and better manage the bertha armyworm in western Canada, our understanding of when and why populations outbreak is still lacking. Further population dynamics research may uncover important abiotic and biotic factors that contribute to synchronous population growth and collapse that occurs prairie-wide and determine how bertha armyworm may respond to a warming climate. An important piece of the puzzle would be the determination of endemic plant species that the bertha armyworm used as hosts prior to extensive cultivation of the Prairie Provinces. The continuation of the Prairie Pest Monitoring Network to provide yearly population estimates through the pheromone trapping network is crucial as the foundation for understanding population dynamics of this species in the long run and alerting producers to impending outbreaks (Mason *et al.*, 1998a).

There are also promising avenues of research on non-insecticidal management of the bertha armyworm that should be pursued. Development of a commercial biopesticide based on MacoNPV-B and gene products to enhance its infectivity and virulence is warranted (Erlandson, 2013). Plant breeding to produce canola varieties with increased foliar levels of sinigrin is worth testing, as it may reduce the severity of bertha armyworm larval feeding. Transgenic Bt canola should be tested against bertha armyworm even though *B. thuringiensis* applied to foliage does not produce much toxicity in this species. The manipulation of planting dates and choice of late-flowering canola varieties may provide the most cost-effective management strategy in the short term but its implementation would need to be closely linked with accurate population estimates obtained from the pheromone trapping network. A final area of research that has not received any attention to date is the development of pheromone-based control tactics such as mass trapping or pheromone-based mating disruption. Mass trapping would be most effective if an attractant could be found that draws in female moths in addition to males that are attracted to pheromone lures. The importance of field research that can integrate the effects of various management tactics on bertha armyworm and other herbivores and natural enemies present in canola agroecosystems is of paramount importance.

Acknowledgements

Recent research on bertha armyworm in the Evenden laboratory at the University of Alberta has been funded by the Canola Council of Canada (Canola Agronomic Research Program), Alberta Crop Industry Development Fund, Western Grain Research Foundation, Alberta Wheat Commission and the University of Alberta. Continuous funding from the Natural Science and Engineering Research Council (NSERC) of Canada to the Evenden laboratory has made contribution of this chapter possible.

References

Arthur, A.P. and Mason, P.G. (1985) The life history and immature stages of *Banchus flavescens* (Hymenoptera: Ichneumonidae), a parasitoid of the bertha armyworm, *Mamestra configurata* (Lepidoptera: Noctuidae) in western Canada. *The Canadian Entomologist* 117, 1249–1255.

Arthur, A.P. and Mason, P.G. (1986) Life history and immature stages of the parasitoid *Microplitis mediator* (Hymenoptera: Braconidae), reared on the bertha armyworm *Mamestra configurata* (Lepidoptera: Noctuidae). *The Canadian Entomologist* 118, 487–491.

Bailey, C.G. (1976a) A quantitative study of consumption and utilization of various diets in the bertha armyworm, *Mamestra configurata* (Lepidoptera: Noctuidae). *The Canadian Entomologist* 108, 1319–1326.

Bailey, C.H. (1976b) Temperature effects of non-diapause development in *Mamestra configurata* (Lepidoptera: Noctuidae). *The Canadian Entomologist* 108, 1339–1344.

Bodnaryk, R.P. (1977) Stages of diapause development in the pupa of *Mamestra configurata* based on the beta ecdysone sensitivity index. *Journal of Insect Physiology* 5, 537–542.

Bodnaryk, R.P. (1986) Ecdysteroid hormone levels in the pupa of the bertha armyworm, *Mamestra configurata* Wlk., during spermatogenesis. *Journal of Insect Physiology* 32, 931–935.

Bodnaryk, R.P. (1987) Dual control of pupal diapause by cyclic nucleotides in the bertha armyworm, *Mamestra configurata* Wlk. *Journal of Insect Physiology* 33, 33–37.

Bodnaryk, R.P. (1991) Developmental profile of sinalbin (p-hydroxybenzyl glucosinolate) in mustard seedlings, *Sinapis alba* L., and its relationship to insect resistance. *Journal of Chemical Ecology* 17, 1543–1556.

Bodnaryk, R.P. (1997) Will low-glucosinolate cultivars of the mustards *Brassica juncea* and *Sinapis alba* be vulnerable to insect pests? *Canadian Journal of Plant Science* 77, 283–287.

Bracken, G.K. (1984) Within plant preferences of larvae of *Mamestra configurata* (Lepidoptera: Noctuidae) feeding on oilseed rape. *The Canadian Entomologist* 116, 45–49.

Bracken, G.K. and Bucher, G.E. (1977) Estimate of relation between density of bertha armyworm (Lepidoptera: Noctuidae) and yield loss on rapeseed, based on artificial infestations. *Journal of Economic Entomology* 70, 701–705.

Brook, H. and Cutts, M.E. (2016) *Crop Protection 2016*. Alberta Agriculture and Forestry, Information Management Branch, Edmonton, Alberta, Canada.

Bucher, G.E. and Bracken, G.K. (1979) The bertha armyworm, *Mamestra configurata* (Lepidoptera: Noctuidae): an estimate of light and pheromone trap efficiency based on captures of newly emerged moths. *The Canadian Entomologist* 111, 977–984.

Chisholm, M.D., Underhill, E.W., Arthur, A.P. and Steck, W.F. (1975) Evidence for cis-11-hexadecen 1-ol acetate as a major component of the sex pheromone of the bertha armyworm, *Mamestra configurata* (Lepidoptera: Noctuidae). *The Canadian Entomologist* 107, 361–366.

Clearwater, J.R. (1975) Synthesis of a pheromone precursor in the male noctuid moth, *Mamestra configurata*. *Insect Biochemistry* 5, 737–746.

Dosdall, L.M. and Mason, P.G. (2010) Key pests and parasitoids of oilseed rape or canola in North America and the importance of parasitoids in integrated management. In: Williams, I.H. (ed.) *Biocontrol-based Integrated Management of Oilseed Rape Pests*. Springer, Dordrecht, The Netherlands, pp. 167–212. doi 10.1007/978-90-481-3983-5_6.

Dosdall, L.M. and Ulmer, B.J. (2004) Feeding, development, and oviposition of bertha armyworm (Lepidoptera: Noctuidae) on different host plant species. *Environmental Entomology* 33, 756–764.

Erlandson, M., Muller-Cohn, J., Bradfisch, G. and Braun, L. (2002) Screening *Bacillus thuringiensis* delta endotoxins for activity against the bertha armyworm, *Mamestra configurata* (Lepidoptera: Noctuidae). *Journal of Invertebrate Pathology* 80, 191–193.

Erlandson, M.A. (2013) *Mamestra configurata* Walker, bertha armyworm (Lepidoptera: Noctuidae). In: Mason, P.G. and Gillespie, D.R. (eds) *Biological Control Programmes in Canada 2001–2012.* CAB International, Wallingford, UK, pp. 228–232.

Fibiger, M. and Lafontaine, J.D. (2005) A review of the higher classification of the Noctuoidea (Lepidoptera) with special reference to the Holarctic fauna. *Esperiana* 11, 7–92.

Gavloski, J. (2015) *Bertha armyworm*. Manitoba Agriculture, Food and Rural Development, Winnipeg, Manitoba, Canada. Available at: http://www.gov.mb.ca/agriculture/crops/insects/bertha-armyworm.html.

Gerber, G.H. and Howlader, M.A. (1987) The effects of photoperiod and temperature on calling behaviour and egg development of the bertha armyworm, *Mamestra configurata* (Lepidoptera: Noctuidae). *Journal of Insect Physiology* 33, 429–436.

Godfrey, G.L. (1972) *A Review and Reclassification of Larvae of the Subfamily Hadeninae (Lepidoptera, Noctuidae) of America North of Mexico*. Agricultural Research Service, Department of Agriculture, Washington, DC.

Harris, C.R. and Turnbull, S.A. (1975) Laboratory studies on the toxicity of insecticides to the bertha armyworm (*Mamestra configurata*) (Lepidoptera: Noctuidae). *The Canadian Entomologist* 107, 865–872.

Hegdekar, B.M. (1977) Photoperiodic and temperature regulation of diapause induction in the bertha armyworm, *Mamestra configurata* Walker. *Manitoba Entomologist* 11, 56–60.

Hegdekar, B.M. (1979) Epicuticular wax secretion in diapause and non-diapause pupae of the bertha armyworm. *Annals of the Entomological Society of America* 72, 13–15.

Hegdekar, B.M. (1983) Effect of latitude on the critical photoperiod for diapause induction in the bertha armyworm, *Mamestra configurata* (Lepidoptera: Noctuidae). *The Canadian Entomologist* 115, 1039–1042.

Howlader, M.A. and Gerber, G.H. (1986) Effects of age, egg development, and mating on calling behavior of the bertha armyworm, *Mamestra configurata* Walker (Lepidoptera: Noctuidae). *The Canadian Entomologist* 118, 1221–1230.

Isman, M.B. (1993) Growth inhibitory and antifeedant effects of azadirachtin on six noctuids of regional economic importance. *Pesticide Science* 38, 57–63.

King, K. (1928) *Barathra configurata* Wlk, an armyworm with important potentialities on the northern prairies. *Journal of Economic Entomology* 21, 279–293.

Koul, O. and Isman, M.B. (1992) Toxicity of the limonoid allelochemical cedrelone to noctuid larvae. *Entomologia Experimentalis et Applicata* 64, 281–287.

Lafontaine, J.D. (1993) Cutworm systematics: confusions and solutions. *Memoirs of the Entomological Society of Canada* 165, 189–196.

Lafontaine, J.D. and Schmidt, B.C. (2010) Annotated check list of the Noctuoidea (Insecta, Lepidoptera) of North America north of Mexico. *Zookeys* 40, 1–239.

Lamb, R.J., Hayhoe, H.N. and Turnock, W.J. (1985) Winter survival and outbreaks of bertha armyworm, *Mamestra configurata* (Lepidoptera: Noctuidae), on canola. *The Canadian Entomologist* 117, 727–736.

Landolt, P.J. (2000) New chemical attractants for trapping *Lacanobia subjuncta*, *Mamestra configurata*, and *Xestia c-nigrum* (Lepidoptera: Noctuidae). *Journal of Economic Entomology* 93 101–106.

Landolt, P.J. and Mitchell, E.R. (1997) Attraction of tobacco budworm moths (Lepidoptera: Noctuidae) to jaggery, a palm sugar extract. *Florida Entomologist* 80, 402–407.

Landolt, P.J. and Alfaro, J.F. (2001) Trapping *Lacanobia subjuncta*, *Xestia c-nigrum*, and *Mamestra configurata* (Lepidoptera: Noctuidae) with acetic acid and 3-methyl-1-butanol in controlled release dispensers. *Environmental Entomology* 30, 656–662.

Li, Q., Donly, C., Li, L., Willis, L.G., Theilmann, D.A. and Erlandson, M. (2002a) Sequence and organization of the *Mamestra configurata* nucleopolyhedrovirus genome. *Virology* 294, 106–121. Available at: http://www.sciencedirect.com/science/journal/00426822.

Li, L., Donly, C., Li, Q., Willis, L.G., Keddie, B.A., Erlandson, M.A. and Theilmann, D.A. (2002b) Identification and genomic analysis of a second species of nucleopolyhedrovirus isolated from *Mamestra configurata*. *Virology* 297, 226–244. Available at: http://www.sciencedirect.com/science/journal/00426822.

Mason, P.G., Erlandson, M.A., Olfert, O.O. and Arthur, A.P. (1998a) The bertha armyworm (*Mamestra configurata*) (Lepidoptera: Noctuidae) in western Canada. *The Canadian Entomologist* 130, 321–336.

Mason, P.G., Harris, J.L., Long, Z.L. and Pinto, J.D. (1998b) First record of *Trichogramma inyoense* (Hymenoptera: Trichogrammatidae) attacking the bertha armyworm, *Mamestra configurata* (Lepidoptera: Noctuidae). *The Canadian Entomologist* 130, 105–106

Mason, P.G., Turnock, W.J., Erlandson, M.A., Kuhlmann, U. and Braun, L. (2002) *Mamestra configurata* Walker, bertha armyworm (Lepidoptera: Noctuidae). In: Mason, P.G. and Huber, J. (eds) *Biological Control Programmes Against Insects and Weeds in Canada, 1980–2000.* CAB International, Wallingford, UK, pp. 169–176.

McCloskey, C. and Isman, M.B. (1993) Influence of foliar glucosinolates in oilseed rape and mustard on feeding and growth of the bertha armyworm, *Mamestra configurata* Walker. *Journal of Chemical Ecology* 19, 249–266.

Mitchell, A. Mitter, C. and Regier, J. (2006) Systematics and evolution of the cutworm moths (Lepidoptera: Noctuidae): evidence from two protein-coding nuclear genes. *Systematic Entomology* 31, 21–46.

Morris, O.N. (1986) Susceptibility of the bertha armyworm, *Mamestra configurata* (Lepidoptera: Noctuidae), to commercial formulations of *Bacillus thuringiensis* var. *kurstaki*. *The Canadian Entomologist* 118, 473–478.

Morris, O.N. (1988) Comparative toxicity of delta-endotoxin and thuringiensin of *Bacillus thuringiensis* and mixtures of the two for the bertha armyworm (Lepidoptera: Noctuidae). *Journal of Economic Entomology* 81, 135–141.

Morris, O.N. and Converse, V. (1991) Effectiveness of steinernematid and heterorhabditid nematodes against noctuid, pyralid, geomatrid species in soil. *The Canadian Entomologist* 123, 55–61.

Morris, O.N., Converse, V., McLaughlin, N.B. and Trottier, M. (1994) Interaction of caffeine and related compounds with *Bacillus thuringiensis* ssp. *kurstaki* in bertha armyworm (Lepidoptera: Noctuidae). *Journal of Economic Entomology* 87, 610–617.

Morris, O.N., Kanagaratnam, P. and Converse, V. (1995) Chemical additive effects on the efficacy of *Bacillus thuringiensis* Berliner subsp. *kurstaki* against *Mamestra configurata* (Lepidoptera: Noctuidae). *Journal of Economic Entomology* 88, 815–824.

Morris, O.N., Kanagaratnam, P., Converse, V. and Trottier, M. (1996) Toxicity of *Bacillus thuringiensis* subsp. *aizawai* for *Mamestra configurata* (Lepidoptera: Noctuidae). *Journal of Economic Entomology* 89, 359–365.

O'Hara, J.E. (1999) Tachinidae (Diptera) parasitoids of bertha armyworm (Lepidoptera: Noctuidae). *The Canadian Entomologist* 131, 11–28.

Pivnick, K.A. (1993) Diapause initiation and pupation site selection of the braconid parasitoid *Microplitis mediator* (Haliday): a case of manipulation of host behaviour. *The Canadian Entomologist* 125, 825–830.

Powell, J.A. and Opler, P.A. (2009) *Moths of Western North America,* University of California Press, Berkeley, California.

Riegert, P.W. (1984) Insects on the Canadian plains. *Prairie Forum* 9, 327–344.

Rempel, J.G. (1951) A study of the embryology of *Mamestra configurata* (Walker) (Lepidoptera, Phalaenidae). *The Canadian Entomologist* 83, 1–19.

Steck, W., Philip, H.G., Arthur, A.P., Peters, C.C., Underhill, E.W. and Chisholm, M.D. (1979) Sex pheromone traps in population monitoring of adults of the bertha armyworm, *Mamestra configurata* (Lepidoptera: Noctuidae). *The Canadian Entomologist* 111, 91–95.

Struble, D.L., Warthen, J.D., Green, N. and Jacobson, M. (1975) Bertha armyworm (Lepidoptera: Noctuidae): detection of a sex pheromone and the stimulatory effect of some synthetic chemicals. *The Canadian Entomologist* 107, 355–359.

Struble, D.L., Byers, J.R. and Ayre, G.L. (1984) Minor sex-pheromone components of *Mamestra configurata* (Lepidoptera: Noctuidae) and improved blends for attraction of male moths. *The Canadian Entomologist* 116,103–105.

Swailes, G.E., Holmes, N.D. and Struble, D.L. (1975) Use of traps baited with virgin females for field observations on the bertha armyworm (Lepidoptera: Noctuidae). *The Canadian Entomologist* 107, 781–784.

Tamaki, G. and Weeks, R.E. (1972) Efficiency of three predators, *Geocoris bullatus*, *Nabis americoferus*, and *Coccinella transversoguttata*, used alone or in combination against three insect prey species, *Myzus persicae*, *Ceramica picta*, and *Mamestra configurata*, in a greenhouse study. *Environmental Entomology* 1, 258–263.

Trottier, M.R., Dulmage, H.T. and Morris, O.N. (1988) Susceptibility of the bertha armyworm, *Mamestra configurata* (Lepidoptera, Noctuidae), to sixty-one strains from ten varieties of *Bacillus thuringiensis*. *Journal of Invertebrate Pathology* 51, 242–249.

Turnock, W.J. (1984a) *Mamestra configurata* Walker, bertha armyworm (Lepidoptera: Noctuidae). In: Kelleher, J.S. and Hulme, M.A. (eds) *Biological Control Programmes against Insects and Weeds in Canada, 1969–1980.* CAB International, Wallingford, UK, pp. 49–55.

Turnock, W.J. (1984b) Effects of the stage of development of canola (*Brassica napus*) on the capture of moths in sex attractant traps and on larval density of *Mamestra configurata* (Lepidoptera, Noctuidae). *The Canadian Entomologist* 116, 579–590.

Turnock, W.J. (1985) Developmental, survival, and reproductive parameters of bertha armyworm, *Mamestra configurata* (Lepidoptera: Noctuidae) on four plant species. *The Canadian Entomologist* 117, 1267–1271.

Turnock, W.J. (1987) Predicting larval abundance of the bertha armyworm, *Mamestra configurata* Wlk., in Manitoba from catches of male moths in sex attractant traps. *The Canadian Entomologist* 119, 167–178.

Turnock, W.J. (1988) Density, parasitism, and disease incidence of larvae of the bertha armyworm, *Mamestra configurata* walker (Lepidoptera: Noctuidae), in Manitoba, 1973–1986. *The Canadian Entomologist* 120, 401–413.

Turnock, W.J. and Bilodeau, R.J. (1984) Survival of pupae of *Mamestra configurata* (Lepidoptera: Noctuidae) and two of its parasites in untilled and tilled soil. *The Canadian Entomologist* 116, 257–267.

Turnock, W.J. and Bilodeau, R.J. (1985) A comparison on three methods of examining the density of larvae of the bertha armyworm, *Mamestra configurata*, in fields of canola (*Brassica* spp.). *The Canadian Entomologist* 117, 1065–1066.

Turnock, W.J. and Carl, K.P. (1995) Evaluation of the Palaearctic *Eurithia consobrina* (Diptera: Tachinidae) as a potential biocontrol agent for *Mamestra configurata* (Lepidoptera: Noctuidae) in Canada. *Biocontrol Science and Technology* 5, 55–67.

Turnock, W.J. and Philip, H.G. (1977) The outbreak of bertha armyworm *Mamestra configurata* (Noctuidae: Lepidoptera) in Alberta, Canada 1971 to 1975. *Manitoba Entomologist* 11, 10–21.

Turnock, W.J., Lamb, R.J. and Bodnaryk, R.P. (1983) Effects of cold stress during pupal diapause on the survival and development of *Mamestra configurata* (Lepidoptera: Noctuidae). *Oecologia* 56, 185–192.

Ulmer, B., Erlandson, M. and Gillott, C. (2001) Feeding preferences, growth, and development of *Mamestra configurata* (Lepidoptera: Noctuidae) on Brassicaceae. *The Canadian Entomologist* 133, 509–519.

Ulmer, B., Erlandson, M. and Gillott, C. (2002) Oviposition preferences of bertha armyworm *Mamestra configurata* Walker (Lepidoptera: Noctuidae) on different crucifer cultivars and growth stages. *Environmental Entomology* 31, 1135–1141.

Ulmer, B., Erlandson, M. and Gillott, C. (2003) Conspecific eggs and bertha armyworm, *Mamestra configurata* (Lepidoptera: Noctuidae), oviposition site selection. *Environmental Entomology* 32, 529–534.

Underhill, E.W., Chisholm, M.D. and Steck, W.F. (1977) A sex pheromone mixture for the bertha armyworm moth, *Mamestra configurata*: (Z)-9-tetradecen-1-ol acetate and (Z)-11-hexadecen-1-ol acetate. *The Canadian Entomologist* 109, 1335–1340.

Wagner, D.L. (2001) Moths. In: Levin, S.A. (ed.) *Encyclopedia of Biodiversity*. Elsevier Inc., San Diego, California, pp. 249–270.

WCCP (Western Committee on Crop Pests) *Annual General Meeting Minutes* 1998, 2004–2008, 2011–2013. Available at: http://www.westernforum.org/WCCP%20Minutes.html.

Wylie, H.G. (1977) Observations of *Athrycia cinerea* (Diptera: Tachinidae), a parasite of *Mamestra configurata* (Lepidoptera: Noctuidae). *The Canadian Entomologist* 109, 747–754.

Wylie, H.G. and Bracken, G.K. (1977) Observations on *Ichneumon canadensis*, a parasite of noctuid pupae in southern Manitoba. *The Canadian Entomologist* 109, 1295–1296.

Wylie, H.G. and Bucher, G.E. (1977) The bertha armyworm, *Mamestra configurata* (Lepidoptera: Noctuidae) mortality of immature stages on the rape crop, 1972–1975. *The Canadian Entomologist* 109, 823–837.

9 Entomopathogenic Nematodes for Management of Insect Pests of Canola and Other Oilseed Crops

HARIT K. BAL[1*] AND PARWINDER S. GREWAL[2]

[1]Michigan State University, East Lansing, Michigan; [2]University of Texas Rio Grande Valley, Edinburg, Texas

9.1 Introduction

Insect pests cause significant problems to agricultural crops all over the world (Pedigo and Rice, 2009). To control these insect pests, chemical insecticides are being readily used with a worldwide expenditure of US$58.46 billion in 2015 (Mordor Intelligence, 2016). The excessive and indiscriminate use of chemical insecticides has a detrimental impact on the environment and human health (Pimental, 2005). In order to overcome such harmful side effects of insecticides, application of biological control agents, including microorganisms, predators and parasitoids, has been encouraged for pest management. Entomopathogenic nematodes (EPNs), belonging to families Heterorhabditidae and Steinernematidae, are soil-inhabiting lethal insect parasites that are ubiquitously distributed and have emerged as alternatives to insecticides that are now used commercially to control insect pests in many parts of the world (Grewal *et al.*, 2005). Infection is caused by the third-stage infective juveniles (IJs) which, after emerging from the host cadaver, start searching for a suitable insect host in the soil. Infection is aided by the symbiotic bacteria (*Xenorhabdus* in the case of *Steinernema* and *Photorhabdus* in the case of *Heterorhabditis*) carried by the IJs which are released into the insect haemocoel. The bacteria kill the insect within 24–72 h through septicaemia and the nematodes then feed on the multiplying bacteria, completing two to three generations in the host cadaver (Poinar, 1990) (Fig. 9.1). Currently, EPNs have been found to be effective against several key pests such as codling moth, fruit fly, European crane fly, sciarid flies, fungus gnats, citrus root weevil, black vine weevil, pecan weevil, grape root borer, white grubs, cutworms, etc. in high-value crops such as strawberry plantations, ornamental plants, mushrooms, orchards and turf (Grewal *et al.*, 2005; Grewal, 2012). Specifically, EPNs have provided the most effective control of citrus root weevil in Florida (Shapiro-Ilan *et al.*, 2005) and of the black vine weevil in ornamentals in Europe (Grewal *et al.*, 2005). This chapter reviews the potential of EPNs as biological control agents against major insect pests of canola and other *Brassica* oilseed crops, including mustard, camelina and crambe. Available data on EPN efficacy against different pests is summarized in Table 9.1 and discussed below by insect order. Future prospects to enhance biocontrol potential of EPNs against oilseed crop pests are also discussed.

9.2 Coleoptera

9.2.1 Flea beetle

Flea beetles, crucifer flea beetle *Phyllotreta cruciferae* (Goeze), striped flea beetle *Phyllotreta striolata* (Fabricius) and cabbage stem flea beetle *Psylliodes chrysocephala* Linnaeus (Coleoptera: Chrysomelidae) are the key insect pests of canola and mustard in the Northern Great Plains of USA and Canada (Soroka and Grenkow, 2013), southern China

*Corresponding author. E-mail: bal@msu.edu

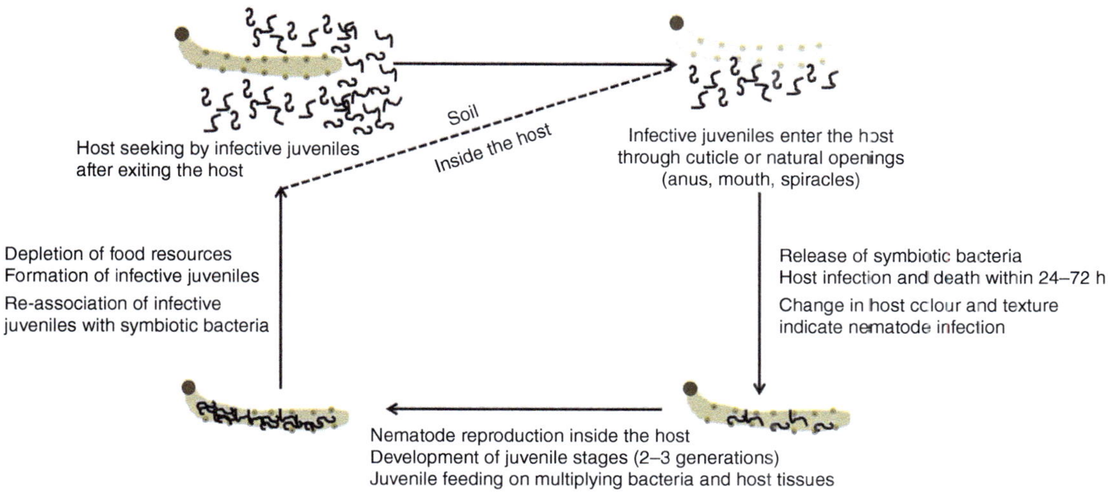

Fig. 9.1. Generalized life cycle of entomopathogenic nematodes belonging to families Heterorhabditidae and Steinernematidae. Dotted line separates the parts of the nematode life cycle occurring inside the host and in the soil.

(Yan *et al.*, 2013) and Europe (Williams, 2010). However, there is limited incidence of flea beetles on camelina (Henderson *et al.*, 2004; Vollmann and Eynck, 2015). Crambe has also been reported to be resistant to flea beetles (Anderson *et al.*, 1992; Soroka and Grenkow, 2013). The nematode *Howardula phyllotretae* Oldham has been reported to be a very common parasite of flea beetle adults in Europe (Sommer, 1981; Ekbom, 1991; Lipa and Ekbom, 2003) but it is not pathogenic and thus has limited value as a biological control agent for flea beetles. Although the EPN *Steinernema carpocapsae* Weiser has been reported to cause infection in the crucifer flea beetle (Morris, 1985), its application to the soil against the overwintering population of crucifer flea beetle on caged canola plants did not affect the subsequent generation of the beetle (Morris, 1987). Nematode infectivity declined after single treatments of *S. carpocapsae* before or after beetle colonization. A few studies in China have reported control of striped flea beetles with *Steinernema feltiae* Filipjev (Li and Wang, 1990; Wei *et al.*, 1992) and *S. carpocapsae* (Wei and Wang, 1993; Hou *et al.*, 2001; Kakizaki, 2004) in the field. However, further research deemed *S. carpocapsae* and *Heterorhabditis indica* Poinar as the most promising biological control agents against striped flea beetles in south China (Xu *et al.*, 2010; Yan *et al.*, 2013). Trdan *et al.* (2008) reported *S. feltiae* to be the most effective EPN species to control the adults of different species of flea beetles, causing up to 77% mortality in laboratory conditions, and suggested its potential for flea beetle control in *Brassica* crops in Slovenia. Further, field tests in Europe reported > 40% control of *Phyllotreta* spp. and 39% of *P. chrysocephala* in oilseed rape (Hokkanen *et al.*, 2006; Hokkanen, 2008), with similar level of control reported with a slow-release formulation of *S. feltiae* containing 15,000 IJs/m^2 and nematode spray at 450 million IJs/m^2 in canola (Menzler-Hokkanen and Hokkanen, 2003). A recent field study from the USA also reported a significant reduction in leaf injury in canola due to crucifer flea beetle in *S. carpocapsae* treated versus untreated plots (Reddy *et al.*, 2014).

9.2.2 Pollen beetle

Pollen beetles, *Meligethes* spp. (Coleoptera: Nitidulidae), are one of the major pests of oilseed rape (Alford, 2003). The larvae feed on pollen of oilseed rape and result in considerable yield losses when larvae are found in high numbers (Nilsson, 1987). Since the larvae enter the soil for pupation, they become vulnerable to EPN infection. The recovery of pollen beetle larvae after exposure to EPNs has been shown to decrease with an increase in the number of nematodes (*Steinernema bicornutum*

Table 9.1. Efficacy of entomopathogenic nematodes (EPNs) against major insect-pests of canola and oilseed crops (NT, not tested; ND, not determined).

Insect order	Insect pest	Life stage	EPN species	Level (%) of control: Laboratory	Level (%) of control: Field	Reference(s)
Coleoptera	*Ceutorhynchus assimilis*	Larvae	*Steinernema* sp., *Heterorhabditis* sp.	55–93	NT	Philipsen and Nielsen, 2003
			Steinernema feltiae	22–75	NT	Nielsen and Philipsen, 2004
			S. feltiae	NT	47.7	Hokkanen *et al.*, 2006
	C. napi	Larvae	*S. feltiae*	NT	72.4	Hokkanen *et al.*, 2006
	C. pallidactylus	Larvae	*S. feltiae*	50	NT	Nielsen and Philipsen, 2004
				NT	17.8	Hokkanen *et al.*, 2006
	Meligethes aeneus	Larvae	*S. feltiae*	NT	93.8	Menzler-Hokkanen and Hokkanen, 2005
				NT	60	Hokkanen *et al.*, 2006
	Meligethes spp.	Larvae	*Steinernema* sp., *Heterorhabditis* sp.	16–56	NT	Philipsen and Nielsen, 2003
			S. feltiae	5–13	NT	Nielsen and Philipsen, 2004
			Heterorhabditis bacteriophora	40	NT	
			Steinernema bicornutum, *Steinernema carpocapsae*, *S. feltiae*, *H. bacteriophora*	50–90	NT	Nielsen and Philipsen, 2005
	Phaedon cochleariae	Larvae	*Heterorhabditis indica*	60–97.5	NT	Mahar *et al.*, 2007
			H. bacteriophora	70–87.5	NT	
			S. carpocapsae	65–90	NT	
			S. feltiae	52.5–80	NT	
		Pupae	*H. indica*	100	NT	Mahar *et al.*, 2012
			H. bacteriophora	92	NT	
			S. carpocapsae	95	NT	
			S. feltiae	87.5	NT	
	Phyllotreta cruciferae	Adults	*S. carpocapsae*	31	NT	Morris, 1985
		Larvae		61–93	NT	Morris, 1987
				NT	42	Reddy *et al.*, 2014
	Phyllotreta striolata	Larvae	*S. feltiae*	86.6–100	77–94.2	Li and Wang, 1990
			S. carpocapsae	NT	38–84	Wei and Wang, 1993
				NT	71	Hou *et al.*, 2001
				NT	ND	Kakizaki, 2004
		Larvae, pupae	*Steinernema* sp., *S. carpocapsae*, *S. feltiae*, *S. glaseri*, *S. pakistanense*, *S. longicaudum*, *S. scapterisci*, *Heterorhabditis* sp., *H. bacteriophora*, *H. indica*	6.7–100	NT	Xu *et al.*, 2010
		Larvae, pupae	*H. indica*	NT	18	Yan *et al.*, 2013
			S. carpocapsae	NT	67	

	Phyllotreta nemorum, *P. undulata*, *P. nigripes*, *P. atra*	Adults	*S. feltiae*	6.7–100	NT	Trdan *et al.*, 2008
			S. carpocapsae	4.0–95.7	NT	
			H. megidis	4.0–66.7	NT	
			H. bacteriophora	1.8–100	NT	
	Phyllotreta spp.	Larvae	*S. feltiae*	NT	50.1	Menzler-Hokkanen and Hokkanen, 2003
				NT	41.5	Hokkanen *et al.*, 2006
	Psylliodes chrysocephala	Larvae	*S. feltiae*	NT	39	Hokkanen *et al.*, 2006
Lepidoptera	*Actebia fennica*	Larvae	*S. feltiae*, *S. carpocapsae*	70–80	70–100	West and Vrain, 1997
	Agrotis ipsilon	Larvae	*S. feltiae*	70–90	50	Capinera *et al.*, 1988
			S. carpocapsae	50–95	NT	Baur *et al.*, 1997b
			S. feltiae	50–100	NT	Mahmoud, 2014
			S. carpocapsae	75–100	NT	
			H. bacteriophora	35–85	NT	
			S. feltiae, *S. carpocapsae*, *H. bacteriophora*	NT	40–74	
	Mamestra brassicae	Larvae	*Steinernema* sp., *Heterorhabditis* sp.	81	NT	Philipsen and Nielsen, 2003
			S. carpocapsae	NT	ND	Beck *et al.*, 2014
	Phytometra gamma	Larvae	*S. feltiae*	15–100	NT	Mahmoud, 2014
			S. carpocapsae	35–75	NT	
			H. bacteriophora	5–40	NT	
			S. feltiae, *S. carpocapsae*, *H. bacteriophora*	NT	24–36	
	Pieris brassicae	Larvae	*S. feltiae*	75–97.5	NT	Wu and Chow, 1989
			H. tayserae	30–100	NT	Saleh, 1995
			P. luminescens (isolated from *H. indica*-IARI strain)	NT	100	Mohan *et al.*, 2003
			H. indica, *H. bacteriophora*, *S. carpocapsae*, *S. feltiae*	100	NT	Mahar *et al.*, 2005
			H. indica	12.5–100	NT	Lalramliana and Yadav, 2010
			S. thormophilum	37.5–100	NT	
			S. glaseri	37.5–100	NT	
			H. bacteriophora	43.3–100	NT	Ayoob and Zaki, 2014

Continued

Table 9.1. Continued.

Insect order	Insect pest	Life stage	EPN species	Level (%) of control		Reference(s)
				Laboratory	Field	
	Plutella xylostella	Pupae	*S. feltiae*, *H. bacteriophora*	60–66	NT	Morris, 1985
		Larvae, pupae	*H. bacteriophora*	20.3–76.3	NT	Baur *et al.*, 1995
			H. megidis	82.7	NT	
			Heterorhabditis sp.	15.2–75.7	NT	
			S. carpocapsae	84.8–95.9	NT	
			S. riobrave	89	NT	
			S. feltiae	66.7	NT	
			S. glaseri	37.8	NT	
			S. kushidai	6.9	NT	
		Larvae	*Steinernema* sp.	3.3–100	NT	Mason and Wright, 1997
			H. indica	86.7–100	NT	
		Larvae	*S. carpocapsae*	21.3–81	NT	Baur *et al.*, 1997b
			S. carpocapsae	NT	41	Baur *et al.*, 1998
		Larvae	*Steinernema* sp.	5–80	NT	Mason *et al.*, 1999
			Heterorhabditis sp.	10-85	NT	
		Pupae	*Photorhabdus luminescens* (isolated from *H. bacteriophora*)	60	NT	Abdel-Razek, 2003
			Xenorhabdus nematophilus (isolated from *S. carpocapsae*)	40	NT	
		Larvae	*S. carpocapsae*, *H. bacteriophora*	96–98	NT	Hussaini, 2003
			S. carpocapsae	60–80	NT	Schroer and Ehlers, 2005
			S. carpocapsae	12.1–81.9	NT	Schroer *et al.*, 2005a
			S. thermophilum	NT	35–46	Somvanshi *et al.*, 2006
			H. indica	84–96	NT	Nyasani *et al.*, 2007
			S. karii	70.7–93.3	NT	
			S. waiseri	74.7–92	NT	
			Steinernema sp.	74.7–88	NT	
			Heterorhabditis sp.	73.3–86.7	NT	
	Spodoptera exigua	Larvae	*S. feltiae*	40–100	NT	Mahmoud, 2014
			S. carpocapsae	45–100	NT	
			H. bacteriophora	25–85	NT	
			S. feltiae, *S. carpocapsae*, *H. bacteriophora*	NT	30–68	
	Spodoptera littoralis	Larvae	*S. carpocapsae*, *S. abbasi*, *S. riobrave*, *H. indica*, *H. bacteriophora*	100	NT	Salem *et al.*, 2007
Diptera	*Dasineura brassicae*	Larvae	*Steinernema* sp., *Heterorhabditis* sp.	0–2	NT	Philipsen and Nielsen, 2003
			S. feltiae	1	NT	Nielsen and Philipsen, 2004

		Pupae	*S. bicornutum*, *S. feltiae*, *H. bacteriophora*	≤ 50	NT	Nielsen and Philipsen, 2005
		Larvae	*S. feltiae*	NT	4.3	Hokkanen *et al*., 2006
	Delia radicum	Larvae	*Steinernema* spp.	70	NT	Welch and Briand, 1961
			Steinernema sp.	13–100	NT	Szczygiel, 1980
			Heterorhabditis sp.	33–100	NT	
			S. feltiae	37	NT	Georgis *et al*., 1983
			H. bacteriophora	80	NT	
			S. feltiae, *H. bacteriophora*	80	NT	Morris, 1985
			S. feltiae, *S. bibionis*, *H. bacteriophora*, *H. heliothidis*	50	81	Bracken, 1990
		Larvae, puparia	*H. zealandica*	100	NT	Lei *et al*., 1992
		Larvae	*S. carpocapsae*, *H. bacteriophora*	ND	NT	Simser, 1992
			S. feltiae	ND	NT	Vänninen *et al*., 1992, 1999
				ND	NT	Sulistyanto *et al*., 1994
			S. carpocapsae	0.5–50	NT	Royer *et al*., 1996
			S. feltiae	48–64	ND	Schroeder *et al*., 1996
				45	NT	Chen and Moens, 2003
			S. feltiae, *S. carpocapsae*, *S. arenarium*, *H. megidis*, *H. bacteriophora*	40–50	NT	Chen *et al*., 2003
			S. feltiae	NT	86	
			S. feltiae	77	NT	Nielsen, 2003
				100	NT	Nielsen and Philipsen, 2004
				60.87–85	NT	Susurluk, 2011
Hymenoptera	*Athalia lugens proxima*	Larvae	*S. feltiae*	NT	84	Narayanan and Gopalakrishnan, 2003
			H. indica	28–100	NT	Yadav and Lalramliana, 2012
			S. thermohilum	15–100	NT	
			S. glaseri	15–90	NT	
	Athalia rosae	Larvae	*S. carpocapsae*	NT	40–67	Sáringer *et al*., 1996

Tallosi, Peters & Ehlers, *S. carpocapsae*, *S. feltiae* and *Heterorhabditis bacteriophora* Poinar) in the laboratory tests (Philipsen and Nielsen, 2003; Nielsen and Philipsen, 2004, 2005). However, pollen beetle larvae parasitized by parasitoid wasps were found to be less affected by EPNs than non-parasitized larvae (Nielsen and Philipsen, 2005). A field study in Finland showed 93.8% reduction in pollen beetle emergence after the application of *S. feltiae* at 1 × 10^6 IJs/m^2 at the beginning of pollen beetle pupation (Menzler-Hokkanen and Hokkanen, 2005). Field trials to control key pests of oilseed rape using EPNs in six European countries also showed 60% reduction in pollen beetle population over a period of 2 years (Hokkanen *et al.*, 2006). In another study in Finland, simultaneous application of *S. feltiae* with the fungus *Isaria fumosorosea* strain IME-05 synergistically improved control of the larvae of *Meligethes aeneus*, although infection with *I. fumosorosea* resulted in a significant reduction of *S. feltiae* progeny (Zec-Vojinovic, 2009).

9.2.3 Mustard beetle

Mustard beetle, *Phaedon cochleariae* Fabricius (Coleoptera: Chrysomelidae), is a serious insect pest of cruciferous crops, particularly in Europe (Uddin *et al.*, 2008) with very limited research conducted on its biological control in oilseed rape with EPNs. There are two reports of susceptibility of mustard beetle larvae and pupae to EPNs *S. carpocapsae, S. feltiae, H. indica* and *H. bacteriophora* in laboratory-based sand bioassays (Mahar *et al.*, 2007, 2012). While the reproduction potential of *S. carpocapsae* was the highest at 25°C compared with the other EPN species, both heterorhabditid species caused more mortality in mustard beetle pupae than steinernematids at 30°C and at 12% moisture in sand bioassays.

9.2.4 Weevils

Larvae of cabbage seed weevil *Ceutorhynchus assimilis* (Paykull), cabbage seed pod weevil *Ceutorhynchus obstrictus* (Marsham) (syn. *C. assimilis* (Paykull)), and cabbage stem weevil *Ceutorhynchus pallidactylus* Marsham (Coleoptera: Curculionidae) cause considerable damage to the pods, stems and leaf stalks, respectively, of oilseed rape. The widespread occurrence of the cabbage seed weevil as a major pest of both winter and spring rape crops has been reported in Europe and North America (Dosdall *et al.*, 2002; Williams, 2010). However, cabbage seed pod weevil occurrence is limited in camelina and crambe (Cárcamo *et al.*, 2012; Lenssen *et al.*, 2012). The *Ceutorhynchus* species have been found to be susceptible to EPNs with up to 93% infection (Philipsen and Nielsen, 2003; Nielsen and Philipsen, 2004). Field trials in oilseed rape have also shown EPNs to reduce the population of *C. assimilis* by 47.7%, *Ceutorhynchus napi* Gyllenhal by 72.4% and *C. pallidactylus* by 17.8% (Hokkanen *et al.*, 2006).

9.3 Lepidoptera

9.3.1 Diamondback moth

Diamondback moth (DBM), *Plutella xylostella* (Linnaeus) (Lepidoptera: Plutellidae), is a major pest of *Brassica* oilseed crops with similar infestations on crambe and mustard, but with camelina being a non-preferred host (Lenssen *et al.*, 2012). DBM has developed resistance to chemical and biological insecticides, including *Bacillus thuringiensis* endotoxins (Furlong *et al.*, 2013). EPNs have emerged as one alternative to control DBM. Susceptibility of DBM larvae and pupae to two EPN species, *S. feltiae* and *H. bacteriophora*, was demonstrated in Petri dish assays by Morris (1985). While Baur *et al.* (1995) achieved > 95% control of DBM larvae with *S. carpocapsae*, Nyasani *et al.* (2007) reported 70–96% DBM mortality by five different *Steinernema* and *Heterorhabditis* isolates from Kenya in leaf disc assays. *Steinernema carpocapsae* also showed infectivity against both *B. thuringiensis* subsp. *kurstaki* (Btk) susceptible and resistant Hawaiian DBM strains in laboratory conditions; however, sufficient field control was not obtained (Baur *et al.*, 1997a). The bacterial symbionts *Xenorhabdus nematophilus* Thomas & Poinar and *Photorhabdus luminescens* Thomas & Poinar of *S. carpocapsae* and *H. bacteriophora*, respectively, have also shown virulence against DBM pupae in laboratory bioassays but did not assure field efficacy (Abdel-Razek, 2003). Environmental factors such as temperature, ultraviolet (UV) radiation and desiccation have been considered as the major reasons for inactivation of the EPNs and their failure in the foliar environment (Grewal *et al.*, 2005). Indigenous nematode species/strains that are better adapted to the local environment could withstand such extreme abiotic conditions and provide desired insect pest control. For example, Mason and Wright (1997) screened indigenous nematode strains from Malaysia and identified

Steinernema spp. that tolerated extreme environmental conditions and were suitable for foliar applications against DBM larvae. Alternatively, Baur *et al.* (1997b) suggested high humidity and rehydration for at least 48 h before application for effective foliar treatments of commercially formulated *S. carpocapsae* against DBM larvae. *S. carpocapsae* provided 41% control of DBM alone and 58% when combined with Btk in field tests conducted on *Nasturtium aquaticum* Hayek farms in Hawaii (Baur *et al.*, 1998). The addition of various antidesiccant or UV-protective adjuvants to the spray mixture has also been shown to improve EPN survival on exposed foliage and increase DBM control (Baur *et al.*, 1997a; Mason *et al.*, 1998; Vyas *et al.*, 2000; Schroer *et al.*, 2005a, b). Schroer and Ehlers (2005) developed a surfactant-polymer formulation of *S. carpocapsae* that enhanced nematode efficacy on the foliage to control DBM by improving conditions that enabled rapid host invasion. High field efficacy against DBM has also been achieved by increasing the nematode concentration in the spray mixture (Somvanshi *et al.*, 2006).

9.3.2 Cabbage moth

Cabbage moth, *Mamestra brassicae* Linnaeus (Lepidoptera: Noctuidae), is a minor pest of oilseed rape. There is limited information on cabbage moth control using EPNs. One study by Philipsen and Nielsen (2003) reported susceptibility of cabbage moth larvae collected from oilseed rape fields to EPNs and suggested it to be a potential host for EPN propagation and recycling given its large size (~40 mm). A recent study (Beck *et al.*, 2014) in cauliflower showed field efficacy of *S. carpocapsae* against cabbage moth larvae using a spray technique in combination with adjuvants that effectively lowered cabbage moth damage.

9.3.3 Cabbage butterfly

Cabbage butterfly, *Pieris brassicae* Linnaeus (Lepidoptera: Pieridae), a polyphagous pest of *Brassica* crops, has been shown to be susceptible to *S. feltiae* (Wu and Chow, 1989), *Heterorhabditis tayserae* (Saleh, 1995), *H. indica*, *H. bacteriophora*, *S. carpocapsae* and *S. feltiae* (Mahar *et al.*, 2005), *Steinernema thermophilum* Ganguly & Singh (Lalramliana and Yadav, 2010) and *H. bacteriophora* SKUASTK-EPN-Hr-1 (Ayoob and Zaki, 2014) in laboratory bioassays. Foliar spray application of the bacterial symbiont *Photorhabdus luminescens* isolated from *H. indica*-IARI strain has also been reported to cause 100% mortality of cabbage butterfly larvae within 24 h in ornamental nasturtium, *Tropaeolum majus* Linnaeus (Mohan *et al.*, 2003).

9.3.4 Armyworms, cutworms and other Lepidoptera

Beet armyworm *Spodoptera exigua* Hübner, black cutworm *Agrotis ipsilon* Hufnagel and silver Y moth *Phytometra gamma* Linnaeus (Lepidoptera: Noctuidae) are serious pests of *Brassica* oilseed crops in Egypt (Sayed and Teilep, 2013; Mahmoud and Shebl, 2014). EPN species *S. carpocapsae* and *S. feltiae* have shown greater virulence than *H. bacteriophora* against these three insect pests in both laboratory and greenhouse experiments (Salem *et al.*, 2007; Mahmoud, 2014). Some laboratory and field tests have shown greater potential of *S. carpocapsae* and *S. feltiae* against black cutworm (Capinera *et al.*, 1988; Baur *et al.*, 1997b; Ebssa and Koppenhöfer, 2011, 2012) and black army cutworm, *Actebia fennica* Tauscher (West and Vrain, 1997) as compared with other EPN species; however, appropriate temperature and humidity conditions govern their efficacy. These studies have suggested further investigation into the field efficacy of different EPN strains and appropriate application technology to control these lepidopteran pests.

9.4 Diptera

9.4.1 Cabbage maggot

In the past few years, cabbage maggot, *Delia radicum* Linnaeus (Diptera: Anthomyiidae), has become a major pest in oilseed rape in Europe and North America.

Several studies have documented the susceptibility of cabbage maggot to steinernematid nematodes (Welch and Briand, 1961; Szczygiel, 1980; Georgis *et al.*, 1983; Morris, 1985; Bracken, 1990; Simser, 1992; Lei *et al.*, 1992; Vänninen *et al.*, 1992; Jaworska, 1993; Sulistyanto *et al.*, 1994; Royer *et al.*, 1996; Schroeder *et al.*, 1996; Nielsen and Philipsen, 2004), particularly *S. feltiae* (Chen and Moens, 2003; Chen *et al.*, 2003; Nielsen, 2003; Nielsen and Philipsen, 2004; Susurluk, 2011). However, larval control has been variable (Schroeder *et al.*, 1996), with limited field control (Bélair *et al.*, 2005), generally lower than the insecticide treatments (Vänninen *et al.*, 1999) but comparable in some cases (Bracken, 1990). EPN dosage and environmental conditions have contributed to this variability in EPN efficacy against cabbage maggot. In addition, previous research has emphasized timing as an important factor. EPN application must coincide with the occurrence of the most destructive stage of the pest, that is, third-instar larva (Sulistyanto *et al.*, 1994), which is most vulnerable to EPN invasion for efficient control of cabbage maggot in oilseed rape (Chen *et al.*, 2003; Nielsen and Philipsen, 2004; Susurluk and Ehlers, 2008; Susurluk, 2011). Selection of an appropriate EPN species with the ability to withstand the temperature conditions that prevail at the time of the occurrence of the most destructive stage of the pest is also essential (Susurluk, 2011).

9.4.2 Brassica pod midge

Brassica pod midge, *Dasineura brassicae* Winnertz (Diptera: Cecidomyiidae), is a major pest of oilseed rape, particularly in Europe (Alford, 2003). *Brassica* pod midge larvae have been reported to be rarely infected by EPNs, presumably due to their small size (2 mm) that hinders nematode reproduction, making it an unsuitable host for EPNs (Philipsen and Nielsen, 2003; Nielsen and Philipsen, 2004). However, pupating brassica pod midges were reported to be susceptible to *S. feltiae* (Nielsen and Philipsen, 2005), which lowered the midge population by 4.3% in oilseed rape fields in Europe (Hokkanen *et al.*, 2006).

9.5 Hymenoptera

9.5.1 Sawfly

Turnip sawfly *Athalia rosae* Linnaeus and mustard sawfly *Athalia lugens proxima* Klug (Hymenoptera: Tenthredinidae) are insect pests of minor importance in oilseed rape (Alford, 2003). A free-living nematode, *Mermis albicans* Siebold, showed a significant amount of parasitism against the larvae of turnip sawfly in a study by Hoffmann and Schmutterer (1983). *S. carpocapsae* caused 40% mortality of larvae when sprayed on the foliage of oilseed rape and 67% when sprayed on the soil surface (Sáringer *et al.*, 1996); laboratory tests showed greater efficacy of *S. carpocapsae* than *S. feltiae*. The population of mustard sawfly, a serious pest of mustard and radish in India, was significantly reduced by 84% in radish plots treated with *S. feltiae* compared with the untreated control plots in a study from India (Narayanan and Gopalakrishnan, 2003). Consequently, a 74% increase in radish yield was observed in nematode-treated plots compared with the untreated control plots. Recently, the efficacy of three indigenous strains of EPNs, *H. indica*, *S. thermohilum*, and *S. glaseri* Steiner from Meghalaya, India, was tested against mustard sawfly larvae (Yadav and Lalramliana, 2012). Among the three strains, *H. indica* and *S. thermophilum* showed greater pathogenicity and reproduction in the last-instar larvae of mustard sawfly, thus indicating their potential as biological control agents against mustard sawfly in Indian mustard.

9.6 Future Prospects

Based on the literature reviewed above, EPNs appear to have potential for biological control of several insect pests of canola and other *Brassica* oilseed crops. However, most of the studies were based on laboratory bioassays of EPNs against the target insect pests, with a few studies focusing on their field efficacy. This, along with limited EPN efficacy against some insect pests both in the laboratory and field, suggests the need for further research aimed at identification of the most effective EPN species and strains, improvements in host finding and invasion, survival, dispersal and persistence of EPNs in the field following application for sustainable pest management in oilseed crops. The following sections discuss a systems approach (Barbercheck and Hoy, 2005; Shapiro-Ilan *et al.*, 2012; Shapiro-Ilan and Dolinski, 2015) to identifying, selecting and using EPNs to enhance their efficacy against oilseed crop pests that may help to guide future development of sustainable pest management practices.

9.6.1 Evaluation of new strains

EPNs are ubiquitous in the soil. The two families Heterorhabditidae and Steinernematidae contain three genera and > 80 species (Campos-Herrera *et al.*, 2012) and a rich diversity of strains with desired traits. Several studies that screened the genetic heterogeneity and variability in biological traits in local populations of EPNs and identified the species/strains with superior traits such as improved host finding and virulence, high infection rate, heat and desiccation tolerance have been documented (Jindal *et al.*, 2012; Campos-Herrera, 2015). In fact, differences in infectivity between species and isolates of the same species of EPNs have been reported against oilseed rape pests, DBM (Morris, 1985; Baur *et al.*, 1995; Mason and Wright, 1997), flea beetles (Trdan *et al.*, 2008; Xu *et al.*, 2010; Yan *et al.*, 2013), pollen beetle (Philipsen and Nielsen, 2003; Nielsen and Philipsen, 2005), mustard beetle (Mahar *et al.*, 2007, 2012), cabbage seed weevil (Nielsen and Philipsen, 2004, 2005), cabbage maggot (Chen *et al.*, 2003; Nielsen and Philipsen, 2004; Susurluk, 2011), cabbage butterfly (Mahar *et al.*, 2005), mustard sawfly (Yadav and Lalramliana, 2012) and armyworms and cutworms (Salem *et al.*, 2007; Mahmoud, 2014). Some of these studies have identified indigenous nematode strains that are more tolerant to the local environmental conditions than the imported commercial ones and are therefore more effective against oilseed pests in the region (Mason and Wright, 1997; Lalramliana and Yadav, 2010; Yadav and Lalramliana, 2012; Ayoob and Zaki, 2014). Thus, identifying the most promising indigenous species/strain against the target pest has the potential to utilize the biological control services of these beneficial nematodes efficiently.

9.6.2 Genetic improvement

Genetic improvement by classical (selection, hybridization, mutagenesis) and advanced (recombinant DNA methods) techniques have been shown to enhance EPN performance successfully by improving a desired trait such as host finding, virulence and tolerance of heat, cold and desiccation (Jindal *et al.*, 2012; Glazer, 2015). For example, genetic selection has improved the host-finding ability of EPNs against *Galleria mellonella* (Gaugler *et al.*, 1989, 1990, 1991; Gaugler and Campbell, 1991; Bal *et al.*, 2014a), *Acheta domesticus* (Gaugler *et al.*, 1991), *Popillia japonica* (Gaugler and Campbell, 1991; Selvan *et al.*, 1994; Grewal *et al.*, 2004), *Cyclocephala borealis* (Grewal *et al.*, 2004), *Cylas* sp. (Jansson *et al.*, 1990), *Lycoriella mali* (Grewal *et al.*, 1993), *Diaprepes abbreviatus* (Stuart *et al.*, 2004) and *Diabrotica virgifera virgifera* (Hiltpold *et al.*, 2010a, b). However, there has been no report on the genetic improvement of any EPN species in host finding or virulence against oilseed pests, thus necessitating future research on developing genetically improved nematode strains with high response to host cues and high virulence against insect pests of oilseed crops. Previous research has shown that herbivore-induced plant volatiles increase the host-finding ability and biological control activity of EPNs against root-feeding insect pests in *Thuja occidentalis* (van Tol *et al.*, 2001), citrus (Ali *et al.*, 2012) and maize (Hiltpold *et al.*, 2010a, b). The latter studies also found that artificial selection of *H. bacteriophora* for high responsiveness to a foraging cue, (*E*)-β-caryophyllene, doubled its host-finding ability and significantly reduced *D. v. virgifera* populations in an (*E*)-β-caryophyllene-emitting maize variety than the original strain. Oilseed *Brassica* plants produce glucosinolates that evolved as a defence mechanism against pests but became a foraging signal for insects that developed ways to detoxify these compounds (Bohinc *et al.*, 2012). Glucosinolates may contribute to resistance of camelina to insect pests such as DBM, mustard sawfly, cabbage maggot, cabbage seedpod weevil and cabbage stem weevil (Vollmann and Eynck, 2015). Their impact on EPNs is not known. Therefore, identification of glucosinolates that may attract EPNs, genetic improvement of nematode strains showing glucosinolate attraction and manipulation of these plant volatiles as insect feeding stimulants in other *Brassica* oilseed crops like canola could be useful for employing these environmentally friendly biological control agents in the management of economically important insect pests of oilseed crops. Apart from enhancing host-finding ability, genetic improvement to stress traits such as heat, cold, desiccation, UV radiation and pesticide tolerance would also increase performance of EPNs in the field (Jindal *et al.*, 2012; Glazer, 2015). Selective breeding or cross-hybridization of indigenous strains that are adapted to the local environment with strains that are known for high host-finding ability or high virulence could be one approach to be considered for the development of genetically improved EPNs to control oilseed crop pests. Given that few studies have investigated the efficiency of genetically improved nematode strains in the field (Grewal *et al.*, 1993, 2004; Selvan *et al.*, 1994; Gaugler *et al.*, 1997; Hiltpold *et al.*, 2010b), such research on developing genetically improved nematode strains against oilseed crop pests should focus on testing their efficacy in the field for sustainable pest management.

9.6.3 Improved application technology

Development of improved EPN application technology and formulation has made considerable progress in recent years to control below-ground as well as above-ground insect pests in different cropping systems (Shapiro-Ilan and Dolinski, 2015). Prior to improving their delivery to the target site, the choice of EPN species for a particular target pest and the timing of application are critical. One such example of using the right EPN species and appropriate time of application for cabbage maggot control in oilseed rape has been discussed earlier (Section 1.4.1).

One way to achieve superior biocontrol efficacy of EPNs is through improved formulation. Considerable progress has been made in developing EPN formulations and adjuvants to achieve efficient pest control, particularly for above-ground pests as discussed earlier in the case of DBM (Baur *et al.*, 1997a; Mason *et al.*, 1998; Vyas *et al.*, 2000; Schroer and Ehlers, 2005; Schroer *et al.*, 2005a, b). A shift from developing these formulations to enhance EPN survival to improve conditions for rapid host invasion is evident from these studies.

Improvement in the application equipment and equipment components is another way to improve EPN delivery and efficacy. Such improved efficacy of EPNs using improved spray application systems has been previously demonstrated against oilseed pests, DBM (Mason *et al.*, 1999) and cabbage moth (Beck *et al.*, 2014), suggesting its potential for controlling other above-ground insect pests of oilseed crops. In recent years, application technology has also been advanced by developing novel EPN delivery methods, particularly targeting below-ground pests that enable slow release of EPNs, ensuring their longer availability to attack insects as compared with aqueous nematode suspensions. In Finland, a controlled-release formulation of *S. feltiae* containing concentrated nematode suspension mixed with polyacrylate beads and sealed into permeable bags ('NemaBag') was found effective in controlling flea beetle (*Phyllotreta* sp.) and pollen beetle (*Meligethes aeneus*) in oilseed rape (Menzler-Hokkanen and Hokkanen, 2003). Recently, *H. bacteriophora* was encapsulated in alginate capsules that showed greater success than nematode spray application in controlling *D. v. virgifera* both in the laboratory and in field conditions (Hiltpold *et al.*, 2012). EPNs have also been formulated as insect cadavers, using different production and application techniques (reviewed by Dolinski *et al.*, 2015), as an alternative to aqueous application for the control of agricultural pests.

Although encapsulation of EPNs and their formulation as insect cadavers has good potential, as these approaches have shown greater EPN survival, dispersal, and infectivity and lower water consumption as compared with aqueous spray applications, they still need improvements in terms of storage, cost, transport and handling (Dolinski *et al.*, 2015; Hiltpold, 2015). Therefore, further investigations to develop cost-effective EPN formulation for efficient below-ground as well as above-ground applications to control oilseed pests are needed, given the limited amount of research in this area targeting insect pests of oilseed crops.

9.6.4 Conservation practices and integrated approaches

Although EPNs are generally used for short-term inundative or augmentative biological control (Grewal *et al.*, 2005), their natural widespread occurrence in diverse ecosystems (Alumai *et al.*, 2006) and persistence beyond a season in some managed ecosystems (Klein and Georgis, 1992; Bal *et al.*, 2014b) suggest that the conservation biological control approach may be practical and cost effective (Campos-Herrera, 2015). Habitat management practices such as conservation tillage, presence of crop residue, and soil amendments such as compost and mulches may enhance the survival and persistence of endemic and applied EPNs in various agroecosystems (Shapiro-Ilan and Dolinski, 2015) but it may not hold true for all the EPN species (Millar and Barbercheck, 2002). Nevertheless, tillage has been shown to have a negative impact on *H. bacteriophora* persistence in oilseed rape (Susurluk and Ehlers, 2008). Even insect pests respond differently to habitat management and conservation tillage (Chaplin-Kramer *et al.*, 2011). While zero tillage and high plant density have been shown to reduce flea beetle damage in oilseed crops such as canola (Dosdall *et al.*, 1999; Morison *et al.*, 2007), it resulted in greater populations of cabbage maggot in no-tillage or shallow tillage regimes as compared with conventional tillage regime in oilseed rape (Dosdall *et al.*, 1998; Morison *et al.*, 2007). Alford (2003) has also indicated greater abundance of predators and parasitoids of several insect pests such as pollen beetles, stem weevils and flea beetles in minimally cultivated canola and other oilseed crops. A crop management approach that includes appropriate tillage and cover-cropping practices that may not only enhance survival and persistence of endemic or applied EPNs but may also increase the incidence of other predators could improve the overall soil food web conditions for sustainable pesticide-free pest management of oilseed crops. Such an approach could also consider integration of EPNs with other pest control agents. There are a few examples of synergistic combinations of EPNs with *B. thuringiensis* (Baur *et al.*, 1998), entomopathogenic fungi (Zec-Vojinovic, 2009) and azadirachtin (Yan *et al.*, 2013) providing greater control against oilseed pests such as DBM, pollen beetle and striped flea beetle, respectively, than the EPN or the other pest control agents alone.

9.7 Conclusions

EPNs appear to be an attractive option for the control of several insect pests of canola and other oilseed crops. Studies from different parts of the world have documented laboratory and field-based assessments of different species/strains of EPNs against economically important coleopteran, lepidopteran, dipteran and hymenopteran insect pests of oilseed crops. However, literature is limited for some insect pests, such as cabbage moth, cabbage butterfly, mustard beetle and mustard sawfly. Specifically, more emphasis on evaluating field efficacy of EPNs is required. To enhance biocontrol potential of EPNs against insect pests of canola and other oilseed crops, this chapter has suggested a systems approach involving strain discovery and evaluation, genetic improvement for host finding, virulence and stress tolerance traits, appropriate application techniques, conservation practices and integration with other pest control measures for sustainable and integrated pest management of oilseed crop pests.

References

Abdel-Razek, A.S. (2003) Pathogenic effects of *Xenorhabdus nematophilus* and *Photorhabdus luminescens* (Enterobacteriaceae) against pupae of the diamondback moth, *Plutella xylostella* (L.). *Journal of Pest Science* 76, 108–111.

Alford, D.V. (2003) *Biocontrol of Oilseed Rape Pests*. Blackwell Science, Oxford, UK.

Ali, J.G., Alborn, H.T., Campos-Herrera, R., Kaplan, F., Duncan, L.W., Rodriguez-Saona, C., Koppenhöfer, A.M. and Stelinski, L.L. (2012) Subterranean, herbivore-induced plant volatile increases biological control activity of multiple beneficial nematode species in distinct habitats. *PLoS ONE* 7, e38146.

Alumai, A., Grewal, P.S., Hoy, C.W. and Willoughby, D.A. (2006) Factors affecting the natural occurrence of entomopathogenic nematodes in turfgrass. *Biological Control* 36(3), 368–374.

Anderson, M.D., Peng, C. and Weiss, M.J. (1992) Crambe, *Crambe abyssinica* Hochst., as a flea beetle resistant crop (Coleoptera: Chrysomelidae). *Journal of Economic Entomology* 85, 594–600.

Ayoob, M.M and Zaki, F.A. (2014) Biological control of cabbage butterfly, *Pieris brassicae*, by a locally isolated entomopathogenic nematode, *Heterorhabditis bacteriophora* SKUASTK-EPN-Hr-1 in Kashmir. *SKUAST Journal of Research* 16, 66–70.

Bal, H.K., Michel, A.P. and Grewal, P.S. (2014a) Genetic selection of the ambush foraging entomopathogenic nematode, *Steinernema carpocapsae* for enhanced dispersal and its associated trade-offs. *Evolutionary Ecology* 28, 923–939.

Bal, H.K., Acosta, N., Cheng, Z., Whitehead, H., Grewal, P.S. and Hoy, C.W. (2014b) Effect of soil management on *Heterorhabditis bacteriophora* GPS11 persistence and biological control in a vegetable production system. *Biological Control* 79, 75–83.

Barbercheck, M.E. and Hoy, C.W. (2005) A systems approach to conservation of nematodes. In: Grewal, P.S., Ehlers, R.U. and Shapiro-Ilan, D.I. (eds) *Nematodes as Biocontrol Agents*. CAB International, Wallingford, UK, pp. 331–347.

Baur, M.E., Kaya, H.K. and Thurston, G.S. (1995) Factors affecting entomopathogenic nematode infection of *Plutella xylostella* (L.) (Lepidoptera: Plutellidae) on a leaf surface. *Entomologia Experimentalis et Applicata* 77, 239–250.

Baur, M.E., Kaya H.K., Gaugler, R. and Tabashnik, B.E. (1997a) Effects of adjuvants on entomopathogenic nematode persistence and efficacy against *Plutella xylostella*. *Biocontrol Science and Technology* 7, 513–525.

Baur, M.E., Kaya, H.K. and Tabashnik, B.E. (1997b) Efficacy of a dehydrated steinernematid nematode against black cutworm (Lepidoptera: Noctuidae) and diamondback moth (Lepidoptera: Plutellidae). *Journal of Economic Entomology* 90, 1200–1206.

Baur, M.E., Kaya, H.K., Tabashnik, B.E. and Chilcutt, C.F. (1998) Suppression of diamondback moth (Lepidoptera: Plutellidae) with an entomopathogenic nematode (Rhabditida: Steinernematidae) and *Bacillus thuringiensis* Berliner. *Journal of Economic Entomology* 91, 1089–1095.

Beck, B., Brusselman, E., Nuyttens, D., Moens, M., Temmerman, F., Pollet, S., VanWeyenberga, S. and Spanogheb, P. (2014) Improving the biocontrol potential of entomopathogenic nematodes against *Mamestra brassicae*: effect of spray application technique, adjuvants and an attractant. *Pest Management Science* 70, 103–112.

Bélair, G., Wright, D.J. and Curto, G. (2005) Vegetable and tuber crop applications. In: Grewal, P.S., Ehlers, R.-U. and Shapiro-Ilan, D.I. (eds) *Nematodes as Biocontrol Agents*. CAB International, Wallingford, UK, pp. 255–264.

Bohinc, T., Ban, S.G., Ban, D. and Trdan, S. (2012) Glucosinolates in plant protection strategies: a review. *Archives of Biological Science Belgrade* 64, 821–828.

Bracken, G.K. (1990) Susceptibility of first-instar cabbage maggot *Delia radicum* (L.) (Anthomyiidae: Diptera), to strains of the entomogenous nematodes *Steinernema feltiae* Filipjev, *S. bibionis* (Bovien), *Heterorhabditis bacteriophora* Poinar, and *H. heliothidis* (Khan, Brooks, and Hirschmann). *The Canadian Entomologist* 122, 633–639.

Campos-Herrera, R. (2015) *Nematode Pathogenesis of Insects and Other Pests: Ecology and Applied Technologies for Sustainable Plant and Crop Protection*. Springer International Publishing, Zurich, Switzerland.

Campos-Herrera, R., Barbercheck, M., Hoy, C.W. and Stock, S.P. (2012) Entomopathogenic nematodes as a model system for advancing the frontiers of ecology. *Journal of Nematology* 44, 162–176.

Capinera, J.L., Pelissier, D., Menout, G.S. and Epsky, N.D. (1988) Control of black cutworm, *Agrotis ipsilon* (Lepidoptera: Noctuidae), with entomogenous nematodes (Nematoda: Steinernematidae, Heterorhabditidae). *Journal of Invertebrate Pathology* 52, 427–435.

Cárcamo, H., Olfert, O., Dosdall, L., Herle, C., Beres, B. and Soroka, J. (2012) Resistance to cabbage seedpod weevil among selected Brassicaceae germplasm. *The Canadian Entomologist* 139, 658–669.

Chaplin-Kramer, R., O'Rourke, M.E., Blitzer, E.J. and Kremen, C. (2011) A meta-analysis of crop pest and natural enemy response to landscape complexity. *Ecology Letters* 14, 922–932.

Chen, S. and Moens, M. (2003) Susceptibility of cabbage root maggot, *Delia radicum*, to entomopathogenic nematodes (Steinernematidae and Heterorhabditidae). *Nematologia Mediterranea* 31, 157–162.

Chen, S., Han, X. and Moens, M. (2003) Biological control of *Delia radicum* (Diptera: Anthomyiidae) with entomopathogenic nematodes. *Applied Entomology and Zoology* 38, 441–448.

Dolinski, C., Shapiro-Ilan, D. and Lewis, E.E. (2015) Insect cadaver applications: pros and cons. In: Campos-Herrera, R. (ed.) *Nematode Pathogenesis of Insects and Other Pests: Ecology and Applied Technologies for Sustainable Plant and Crop Protection*. Springer International Publishing, Zürich, Switzerland, pp. 207–230.

Dosdall, L.M., Florence, L.Z. and Conway, P.M. (1998) Tillage regime, row spacing, and seeding rate influence infestations of root maggots (*Delia* spp.) (Diptera: Anthomyiidae) in canola. *Canadian Journal of Plant Science* 78, 671–681.

Dosdall, L.M., Dolinski, M.G., Cowle, N.T. and Conway, P.M. (1999) The effect of tillage regime, row spacing, and seeding rate on feeding damage by flea beetles, *Phyllotreta* spp. (Coleoptera: Chrysomelidae), in canola in central Alberta, Canada. *Crop Protection* 18, 217–224.

Dosdall, L.M., Weiss, R.M., Olfert, O. and Cárcamo, H.A. (2002) Temporal and geographical distribution patterns of cabbage seedpod weevil (Coleoptera: Curculionidae) in canola. *The Canadian Entomologist* 134, 403–418.

Ebssa, L. and Koppenhöfer, A.M. (2011) Efficacy and persistence of entomopathogenic nematodes for black cutworm control in turfgrass. *Biocontrol Science and Technology* 21, 779–796.

Ebssa, L. and Koppenhöfer, A.M. (2012) Entomopathogenic nematodes for the management of *Agrotis ipsilon*: effect of instar, nematode species and nematode production method. *Pest Management Science* 68, 947–957.

Ekbom, B. (1991) The effect of abiotic and biotic factors on flea beetle populations in spring rape in Sweden. *IOBC/WPRS Bulletin* 14, 21–27.

Furlong, M.J., Wright, D.J. and Dosdall, L.M. (2013) Diamondback moth ecology and management: problems, progress, and prospects. *Annual Review of Entomology* 58, 517–541.

Gaugler, R. and Campbell, J.F. (1991) Selection for enhanced host-finding of scarab larvae (Coleoptera, Scarabaeidae) in an entomopathogenic nematode. *Environmental Entomology* 20, 700–706.

Gaugler, R., Campbell, J.F. and McGuire, T.R. (1989) Selection for host-finding in *Steinernema feltiae*. *Journal of Invertebrate Pathology* 54, 363–372.

Gaugler, R., Campbell, J.F. and McGuire, T.R. (1990) Fitness of a genetically improved entomopathogenic nematode. *Journal of Invertebrate Pathology* 56, 106–116.

Gaugler, R., Campbell, J.F. and Gupta, P. (1991) Characterization and basis of enhanced host finding in a genetically improved strain of *Steinernema carpocapsae*. *Journal of Invertebrate Pathology* 57, 234–241.

Gaugler, R., Wilson, M. and Shearer, P. (1997) Field release and environmental fate of a transgenic entomopathogenic nematode. *Biological Control* 9, 75–80.

Georgis, R., Poinar, G.O. Jr and Wilson, A.P. (1983) Practical control of the cabbage root maggot, *Hylemya brassicae* (Diptera; Anthomyiidae), by entomogenous nematodes. *IRCS Journal of Medical Science* 11, 322.

Glazer, I. (2015) Improvement of entomopathogenic nematodes: a genetic approach. In: Campos-Herrera, R. (ed.) *Nematode Pathogenesis of Insects and Other Pests: Ecology and Applied Technologies for Sustainable Plant and Crop Protection*. Springer International Publishing, Zürich, Switzerland, pp. 29–55.

Grewal, P.S. (2012) Entomopathogenic nematodes as tools in integrated pest management. In: Abrol, D.P. and Shankar, U. (eds) *Integrated Pest Management: Principles and Practice*. CAB International, Wallingford, UK, pp. 162–236.

Grewal, P.S., Tomalak, M., Keil, C.B.O. and Gaugler, R. (1993) Evaluation of a genetically selected strain of

Steinernema feltiae against the mushroom sciarid *Lycoriella mali*. *Annals of Applied Biology* 123, 695–702.

Grewal, P.S., Power, K.T., Grewal, S.K., Suggars, A. and Haupricht, S. (2004) Enhanced consistency in biological control of white grubs (Coleoptera: Scarabaeidae) with new strains of entomopathogenic nematodes. *Biological Control* 30, 73–82.

Grewal, P.S., Ehlers, R.U. and Shapiro-Ilan, D.I. (2005) Critical issues and research needs for expanding the use of nematodes in biocontrol. In: Grewal, P.S., Ehlers, R.U. and Shapiro-Ilan, D.I. (eds) *Nematodes as Biocontrol Agents*. CAB International, Wallingford, UK, pp. 479–489.

Henderson, A.E., Hallett, R.H. and Soroka, J. (2004) Prefeeding behavior of the crucifer flea beetle, *Phyllotreta cruciferae*, on host and nonhost crucifers. *Journal of Insect Behavior* 17, 17–39.

Hiltpold, I. (2015) Prospects in the application technology and formulation of entomopathogenic nematodes for biological control of insect pests. In: Campos-Herrera, R. (ed.) *Nematode Pathogenesis of Insects and Other Pests: Ecology and Applied Technologies for Sustainable Plant and Crop Protection*. Springer International Publishing, Zürich, Switzerland, pp. 187–205.

Hiltpold, I., Baroni, M., Toepfer, S., Kuhlmann, U. and Turlings, T.C.J. (2010a) Selection of entomopathogenic nematodes for enhanced responsiveness to a volatile root signal helps to control a major root pest. *Journal of Experimental Biology* 213, 2417–2423.

Hiltpold, I., Baroni, M., Toepfer, S., Kuhlmann, U. and Turlings, T.C.J. (2010b) Selective breeding of entomopathogenic nematodes for enhanced attraction to a root signal did not reduce their establishment or persistence after field release. *Plant Signaling and Behavior* 5, 1450–1452.

Hiltpold, I., Hibbard, B.E., French, D.W. and Turlings, T.C.J. (2012) Capsules containing entomopathogenic nematodes as a Trojan horse approach to control the western corn rootworm. *Plant and Soil* 358, 11–25.

Hoffmann, G.M. and Schmutterer, H. (1983) *Parasitäre Krankheiten and Schädlinge an Land-wirtschaftlichen Kulturpflanzen*. Verlag Eugen Ulmer, Stuttgart, Germany.

Hokkanen, H.M.T. (2008) Biological control methods of pest insects in oilseed rape. *Bulletin OEPP/EPPO Bulletin* 38, 104–109.

Hokkanen, H.M.T., Zec-Vojinovic, M., Büchs, W., Husberg, G-B., Klukowski, Z., Luik, A., Nilsson, C., Ulber, B. and Williams, I.H. (2006) Effectiveness of entomopathogenic nematodes in the control of OSR pests. *Proceedings of the MASTER Final Symposium*. Göttingen, Germany.

Hou, Y.M., Pang, X.F. and Liang, G.W. (2001) On the partial application of *Steinernema carpocapsae* strain A24 against striped flea beetle *Phyllotreta striolata*. *Acta Phytophyl Sinica* 28, 151–156.

Hussaini, S.S. (2003) Progress of research work on entomopathogenic nematodes in India. In: Hussaini, S.S., Rabindra, R.J. and Nagesh, M. (eds) *Current Status of Research on Entomopathogenic Nematodes in India*. Project Directorate on Biological Control (ICAR), Bangalore, India, pp. 27–68.

Jansson, R.K., Lecrone, S.H., Gaugler, R. and Smart, G.C. Jr (1990) Potential of entomopathogenic nematodes as biological control agents of sweetpotato weevil (Coleoptera: Curculionidae). *Journal of Economic Entomology* 83, 1818–1826.

Jaworska, M. (1993) The effect of entomopathogenic nematodes from the families of the Heterorhabditidae and Steinernematidae on *Delia brassicae* Hoffmannsegg (Diptera, Anthomyiidae) and its natural enemies. *Polskie Pismo Entomologiczne* 62, 243–254.

Jindal, V., Bal, H.K., An, R. and Grewal, P.S. (2012) Developments in the genetic improvement of entomopathogenic nematodes. In: Koul, O., Dhaliwal, G.S., Khokhar, S. and Singh, R (eds) *Biopesticides in Environment and Food Security: issues and Strategies*. Scientific Publishers, Jodhpur, India, pp. 115–145.

Kakizaki, M. (2004) Control effect of the striped flea beetle, *Phyllotreta striolata* (Fabricius), by application of the entomopathogenic nematode, *Steinernema carpocapsae*, on a Japanese radish. *Annual Report of the Society of Plant Protection of North Japan* 55, 221–225.

Klein, M.G. and Georgis, R. (1992) Persistence of control of Japanese beetle (Coleoptera, Scarabaeidae) larvae with steinernematid and heterorhabditid nematodes. *Journal of Economic Entomology* 85, 727–730.

Lalramliana and Yadav, A.K. (2010) Laboratory evaluation of the pathogenicity of three entomopathogenic nematodes against larvae of cabbage butterfly, *Pieris brassicae* Linnaeus (Lepidoptera: Pieridae). *Science Vision* 9, 166–173.

Lei, Z., Rutherford, T.A. and Webster, J.M. (1992) Heterorhabditid behavior in the presence of the cabbage maggot, *Delia radicum*, and its host plants. *Journal of Nematology* 24, 9–15.

Lenssen, A.W., Iversen, W.M., Sainju, U.M., Caesar-TonThat, T.C., Blodgett, S.L., Allen, B.L. and Evans, R.G. (2012) Yield, pests, and water use of durum and selected crucifer oilseeds in two-year rotations. *Agronomy Journal* 104, 1295–1304.

Li, X.F. and Wang, G.H. (1990) Preliminary study on the control of *Phyllotreta vittata* larvae by entomopathogenic nematodes. *Acta Phytophylactica Sinica* 17, 229–231.

Lipa, J.J. and Ekbom, B. (2003) Microsporidian, haplosporidian, and eugregarine parasites present in populations of *Meligethes aeneus* in Sweden. *Acta*

Agriculturae Scandinavica: Section B, Soil and Plant Science 53, 87–89.
Mahar, A.N., Jan, N.D., Chachar, Q.I., Markhand, G.S., Munir, M. and Mahar, A.Q. (2005) Production and infectivity of some entomopathogenic nematodes against larvae and pupae of cabbage butterfly, *Pieris brassicae* L. (Lepidoptera: Pieridae). *Journal of Entomology* 2, 86–91.
Mahar, A.N., Jan, N.D., Mahar, G.M., Hullio, M.H., Lanjar, A.G. and Buriro, A.H. (2007) Effectiveness of entomopathogenic nematodes against the larvae of mustard beetle *Phaedon cochleariae* at different temperatures. *International Journal of Agriculture & Biology* 9, 851–856.
Mahar, A.N., Jan, N.D. and Mahar, A.Q. (2012) Comparative effectiveness of entomopathogenic nematodes against the pupae of mustard beetle, *Phaedon cochleariae* F. (Chrysomelidae: Coleoptera). *Pakistan Journal of Zoology* 44, 517–523.
Mahmoud, M.F. (2014) Efficacy of entomopathogenic nematodes to certain insect pests infesting oilseed rape in the laboratory and greenhouse. *Egyptian Journal of Biological Pest Control* 24, 387–391.
Mahmoud, M.F. and Shebl, M. (2014) Insect fauna of canola and phenology of the diamondback moth, *Plutella xylostella* L. (Lepidoptera: Plutellidae) as a key pest. *Redia* XCVII, 125–132.
Mason, J.M. and Wright, D.J. (1997) Potential for the control of *Plutella xylostella* larvae with entomopathogenic nematodes. *Journal of Invertebrate Pathology* 70, 234–242.
Mason, J.M., Matthews, G.A. and Wright, D.J. (1998) Screening and selection of adjuvants for the spray application of entomopathogenic nematodes against foliar pests. *Crop Protection* 17, 463–470.
Mason, J.M., Matthews, G.A. and Wright, D.J. (1999) Evaluation of spinning disc technology for the application of entomopathogenic nematodes against a foliar pest. *Journal of Invertebrate Pathology* 73, 282–288.
Menzler-Hokkanen, I. and Hokkanen, H.M.T. (2003) Developing entomopathogenic nematode delivery systems for biological control of oilseed rape pests. In: *Insect Pathogens and Entomopathogenic Nematodes, Proceedings of 9th European Meeting of the IOBC/WPRS working group*. 23–29 May, 2003, Salzau, Germany, p. 60.
Menzler-Hokkanen, I. and Hokkanen, H.M.T. (2005) Developing entomopathogenic nematode delivery systems for biological control of oilseed rape pests. *IOBC/WPRS Bulletin* 28, 19–22.
Millar, L.C. and Barbercheck, M.E. (2002) Effects of tillage practices on entomopathogenic nematodes in a corn agroecosystem. *Biological Control* 25, 1–11.
Mohan, S., Raman, R. and Gaur, H.S. (2003) Foliar application of *Photorhabdus luminescens*, symbiotic bacteria from entomopathogenic nematode *Heterorhabditis indica*, to kill cabbage butterfly *Pieris brassicae*. *Current Science* 84(11), 1397.
Mordor Intelligence (2016) *Global Pesticides Market – Segmented by Type, Application Area and Geography – Trends and forecasts (2015–2020) – Sustainability, Regulation & Competition*. Available at: www.reportbuyer.com/product/2168901/global-pesticides-market-segmented-by-type-application-area-and-geography-trends-and-forecasts-2015-2020-sustainability-regulation-and-competition.html. (accessed 8 October 2016).
Morison, M.V., Meynard, J.M. and Doré, T. (2007) Effects of crop management and surrounding field environment on insect incidence in organic winter oilseed rape (*Brassica napus* L.). *Crop Protection* 26, 1108–1120.
Morris, O.N. (1985) Susceptibility of 31 species of agricultural insect pests to the entomogenous nematodes *Steinernema feltiae and Heterorhabditis bacteriophora*. *The Canadian Entomologist* 117, 401–407.
Morris, O.N. (1987) Evaluation of the nematode, *Steinenema feltiae* Filipjev, for the control of the crucifer flea beetle, *Phyllotreta cruciferae* (Goeze) (Coleoptera: Chrysomelidae). *The Canadian Entomologist* 119, 95–101.
Narayanan, K. and Gopalakrishnan, C. (2003) Evaluation of entomopathogenic nematode, *Steinernema feltiae* against field population of mustard sawfly, *Athalia lugens proxima* (Klug) on radish. *Indian Journal of Experimental Biology* 41, 376–378.
Nielsen, O. (2003) Susceptibility of *Delia radicum* to steinernematid nematodes. *BioControl* 48, 431–446.
Nielsen, O. and Philipsen, H. (2004) Occurrence of *Steinernema* species in cabbage fields and the effect of inoculated *S. feltiae* on *Delia radicum* and its parasitoids. *Agricultural and Forest Entomology* 6, 25–30.
Nielsen, O. and Philipsen, H. (2005) Susceptibility of *Meligethes* spp. and *Dasyneura brassicae* to entomopathogenic nematodes during pupation in soil. *BioControl* 50, 623–634.
Nilsson, C. (1987) Yield losses in summer rape caused by pollen beetles (*Meligethes* spp.). *Swedish Journal of Agricultural Research* 17, 105–111.
Nyasani, J.O., Kimenju, J.W., Olubayo, F.M. and Wilson, M.J. (2007) Potential of using entomopathogenic nematodes in the management of diamondback moth *Plutella xylostella* (L.) (Lepidoptera: Plutellidae) in Kenya. In: *African Crop Science Conference Proceedings* 8, 1063–1066.
Pedigo, L.P. and Rice, M.E. (2009) *Entomology and Pest Management*, 6th edn. Pearson Education Inc., Upper Saddle River, New Jersey.
Philipsen, H. and Nielsen, O. (2003) Host potential of insects from cruciferous crops to entomopathogenic nematodes and augmentation of nematodes through oil seed rape growing. *IOBC WPRS Bulletin* 26, 141–145.
Pimental, D. (2005) Environmental and economic costs of the application of pesticides primarily in the United

States. *Environment, Development and Sustainability* 7, 229–252.

Poinar, G.O. Jr (1990) Taxonomy and biology of Steinernematidae and Heterorhabditidae. In: Gaugler, R. and Kaya, H.K. (eds) *Entomopathogenic Nematodes in Biological Control*. CRC Press, Boca Raton, Florida, pp. 23–61.

Reddy, G.V.P., Tangtrakulwanich, K., Miller, J.H., Ophus, V.L. and Prewett, J. (2014) Sustainable management tactics for control of *Phyllotreta cruciferae* (Coleoptera: Chrysomelidae) on canola in Montana. *Journal of Economic Entomology* 107, 661–666.

Royer, L., Belair, G., Boivin, G. and Fournier, Y. (1996) Attractiveness of cabbage maggot (Diptera: Anthomyiidae) to entomopathogenic steinernematid nematodes. *Journal of Economic Entomology* 89, 614–620.

Saleh, M.M.E. (1995) Efficiency of the Egyptian entomopathogenic nematode *Heterorhabditis tayserae* (Nematoda: Heterorhabditidae) in controlling the cabbage worm, *Pieris rapae* (L.) Lepidoptera: Pieridae). *Egyptian Journal of Biological Pest Control* 5, 103–105.

Salem, S.A., Abdel-Rahman, H.A., Zebitz, C.P.W., Saleh, M.M.E., Ali, F.I. and El-Kholy, M.Y. (2007) Evaluation of entomopathogenic nematodes in controlling some cabbage pests. *Journal of Applied Sciences Research* 3, 323–328.

Sáringer, G., Fodor, A., Nadasy, M., Lucskai, A. and Georgis, R. (1996) Possibilities of biological control using entomopathogenic nematodes against *Leptinotarsa decemlineata* L. and *Athalia rosae* L. larvae. *Mededelingen Faculteit Landbouwkundige en Toegepaste Biologische Wetenschappen, Universiteit Gent* 61, 961–966.

Sayed, A.M.M. and Teilep, W.M.A. (2013) Role of natural enemies, climatic factors and performance genotypes on regulating pests and establishment of canola in Egypt. *The Journal of Basic and Applied Zoology* 66, 18–26.

Schroeder, P.C., Ferguson, C.S., Shelton, A.M., Wilsey, W.T., Hoffmann, M.P. and Petzoldt, C. (1996) Greenhouse and field evaluations of entomopathogenic nematodes (Nematoda: Heterorhabditidae and Steinernematidae) for control of cabbage maggot (Diptera: Anthomyiidae) on cabbage. *Journal of Economic Entomology* 89, 1109–1115.

Schroer, S. and Ehlers, R.-U. (2005) Foliar application of the entomopathogenic nematode *Steinernema carpocapsae* for biological control of diamondback moth larvae (*Plutella xylostella*). *Biological Control* 33, 81–86.

Schroer, S., Yi, X. and Ehlers, R.-U. (2005a) Evaluation of adjuvants for foliar application of entomopathogenic nematodes against the diamondback moth (*Plutella xylostella*). *Nematology* 7, 37–44.

Schroer, S., Ziermann, D. and Ehlers, R.-U. (2005b) Mode of action of a surfactant–polymer formulation to support entomopathogenic nematodes *Steinernema carpocapsae* for control of diamondback moth larvae (*Plutella xylostella*). *Biocontrol Science and Technology* 15, 601–613.

Selvan, S., Grewal, P., Gaugler, R. and Tomalak, M. (1994) Evaluation of steinernematid nematodes against *Popillia japonica* (Coleoptera: Scarabaeidae) larvae: species, strains, and post-application rinse. *Journal of Economic Entomology* 78, 342–345.

Shapiro-Ilan, D.I. and Dolinski, C. (2015) Entomopathogenic nematode application technology. In: Campos-Herrera, R. (ed.) *Nematode Pathogenesis of Insects and Other Pests: Ecology and Applied Technologies for Sustainable Plant and Crop Protection*. Springer International Publishing, Zürich, Switzerland, pp. 231–254.

Shapiro-Ilan, D.I., Duncan, L.W., Lacey, L.A. and Han, R. (2005) Orchard applications. In: Grewal, P.S., Ehlers, R.U. and Shapiro-Ilan, D.I. (eds) *Nematodes as Biocontrol Agents*. CAB International, Wallingford, UK, pp. 215–229.

Shapiro-Ilan, D.I., Bruck, D.J. and Lacey, L.A. (2012) Principles of epizootiology and microbial control. In: Vega, F.E. and Kaya, H.K (eds) *Insect Pathology*, 2nd edn. Academic Press, Elsevier Inc., San Diego, California, pp. 29–72.

Simser, D. (1992) Field application of entomopathogenic nematodes for control of *Delia radicum* in collards. *Journal of Nematology* 24, 374–378.

Sommer, G. (1981) Biology and parasites of *Phyllotreta* spp. (Coleoptera, Halticinae). CIBC Report, Delemont, Switzerland.

Somvanshi, V.S., Ganguly, S. and Paul, A.V.N. (2006) Field efficacy of the entomopathogenic nematode *Steinernema thermophilum* Ganguly and Singh (Rhabditida: Steinernematidae) against diamondback moth (*Plutella xylostella* L.) infesting cabbage. *Biological Control* 37, 9–15.

Soroka, J. and Grenkow, L. (2013) Susceptibility of brassicaceous plants to feeding by flea beetles, *Phyllotreta* spp. (Coleoptera: Chrysomelidae). *Journal of Economic Entomology* 106, 2557–2567.

Stuart, R.J., Shapiro-Ilan, D.I. James, R.R., Nguyen, K.B. and McCoy, C.W. (2004) Virulence of new and mixed strains of the entomopathogenic nematode *Steinernema riobrave* to larvae of the citrus root weevil *Diaprepes abbreviatus*. *Biological Control* 30, 439–445.

Sulistyanto, D., Peters, A., Hokkanen, H. and Ehlers, R.U. (1994) Evaluation of entomopathogenic nematode strains for control of *Delia radicum*, *Tipula paludosa* and *T. oleracea*. *IOBC/WPRS Bulletin* 17, 140–143.

Susurluk, I.A. (2011) Potential of *Steinernema feltiae* (Rhabditida: Steinernematidae) as a biological control agent against the cabbage maggot *Delia radicum* (Diptera: Anthomyiidae) in oilseed rape. *Turkish Journal of Agriculture and Forestry* 35, 413–419.

Susurluk, A. and Ehlers, R.U. (2008) Field persistence of the entomopathogenic nematode *Heterorhabditis bacteriophora* in different crops. *BioControl* 53, 627–641.

Szczygiel, M. (1980) A study of the possibilities of limiting the cabbage maggot population – *Hylemya brassicae* Bouche (Diptera, Anthomyiidae) – by parasitic nematodes of the Family Steinernematidae. MSc thesis, Agricultural University, Krakow, Poland.

Trdan, S., Vidrih, M., Valic, N. and Laznik, Z. (2008) Impact of entomopathogenic nematodes on adults of *Phyllotreta* spp. (Coleoptera: Chrysomelidae) under laboratory conditions. *Acta Agriculturae Scandinavica, Section B- Soil and Plant Science* 58, 169–175.

Uddin, M.M., Uliruchs, C. and Mewis, I. (2008) *Phaedon cochleariae* (F.) performance on different crucifer varieties with different glucosinolate profiles. *Communications in Agricultural and Biological Sciences* 73, 563–572.

van Tol, R.W.H.M., van der Sommen, A.T.C., Boff, M.I.C., van Bezooijen, J., Sabelis, M.W. and Smits, P.H. (2001) Plants protect their roots by alerting the enemies of grubs. *Ecology Letters* 4, 292–294.

Vänninen, I., Vainio, A. and Hokkanen, H. (1992) Attempts to control *Delia* spp. with entomopathogenic nematodes. *Bulletin OILB SROP* 15, 143–153.

Vänninen, I., Hokkanen, H. and Tyni-Juslin, J. (1999) Screening of field performance of entomopathogenic fungi and nematodes against cabbage root flies (*Delia radicum* L. and *D. floralis* (Fall.); Diptera, Anthomyiidae). *Acta Agriculturae Scandinavica* 49, 167–183.

Vollmann, J. and Eynck, C. (2015) Camelina as a sustainable oilseed crop: contributions of plant breeding and genetic engineering. *Biotechnology Journal* 10, 525–535.

Vyas, R.V., Patel, N.S., Joshi, N.R., Patel, N.B. and Patel, D.J. (2000) Bioefficacy of *S. glaseri* against DBM, *P. xylostella* on cress, recommendation for scientific community. In: *Proceedings of the 35th Plant Protection Subcommittee, AGRESCO*. 3–4 April, Gujarat Agricultural University, Anand, India.

Wei, H.Y. and Wang, G.H. (1993) The control effect of a *Steinernema* nematode against striped flea beetle. *Acta Phytophylacica Sinica* 20, 61–64.

Wei, H.Y., Wang, G.H. and Pang, X.F. (1992) Infectivity of the nematode, *Steinernema feltiae*, to the striped flea beetle, *Phyllotreta striolata*. *Journal of South China Agricultural University* 13, 26–29.

Welch, H.E. and Briand, L.B. (1961) Field experiment on the use of a nematode for the control of vegetable crop insects. *Proceedings of the Entomological Society of Ontario* 91, 197–202.

West, R.J. and Vrain, T.C. (1997) Nematode control of black army cutworm (Lepidoptera: Noctuidae) under laboratory and field conditions. *The Canadian Entomologist* 129, 229–239.

Williams, I.H. (2010) The major pests of oilseed rape in Europe and their management: an overview. In: Williams, I.H. (ed.) *Biocontrol-based Integrated Management of Oilseed Rape Pests*. Springer, Dordrecht, The Netherlands, pp. 1–43.

Wu, H.J. and Chow, Y.S. (1989) Susceptibility of *Pieris rapae* (Lepidoptera: Pieridae) to the imported entomogenous nematode *Steinernema feltiae*. *Bulletin of the Institute of Zoology Academia Sinica* 28, 237–244.

Xu, C., De Clercq, P., Moens, M., Chen, S. and Han, R. (2010) Efficacy of entomopathogenic nematodes (Rhabditida: Steinernematidae and Heterorhabditidae) against the striped flea beetle, *Phyllotreta striolata*. *BioControl* 55, 789–797.

Yadav, A.K. and Lalramliana (2012) Evaluation of the efficacy of three indigenous strains of entomopathogenic nematodes from Meghalaya, India against mustard sawfly, *Athalia lugens proxima* Klug (Hymenoptera: Tenthredinidae). *Journal of Parasitic Diseases* 36, 175–180.

Yan, X., Han, R., Moens, M., Chen, S. and De Clercq, P. (2013) Field evaluation of entomopathogenic nematodes for biological control of striped flea beetle, *Phyllotreta striolata* (Coleoptera: Chrysomelidae). *BioControl* 58, 247–256.

Zec-Vojinovic, M. (2009) *Biological Control of Oilseed Rape Pests with Entomopathogenic Nematodes*. Doctoral thesis in agricultural zoology, University of Helsinki, Helsinki, Finland.

10 The OKANOLA Project: Challenges in Managing Insect Pests of Canola in the Southern Plains

TOM A. ROYER* AND KRISTOPHER L. GILES

Oklahoma State University, Stillwater, Oklahoma, USA

10.1 The OKANOLA Project

The OKANOLA project was conceived in 2003 by Dr Thomas Peeper, Emeritus Professor of Weed Science, and the late Mark Boyles, Extension Oilseed Specialist at Oklahoma State University (OSU), as a joint venture between private industry and two universities (Oklahoma State University and Kansas State University) 'to provide research, education, and demonstration to stimulate the development of winter canola as a major profitable rotational crop with winter wheat'. They wanted to help Oklahoma wheat growers develop a more profitable winter wheat production system by introducing winter-hardy canola as a profitable rotational crop, 'to aid in pest management, improve wheat yields and quality, and facilitate the adoption of no-till crop production practices' (Boyles and Peeper, 2008). Growing conditions in Oklahoma and Kansas are ideal for winter canola varieties and this rotation continues to gain popularity on the 5.9 million hectares of winter wheat in the two states. Land devoted to canola production has increased dramatically from zero in 2003 to > 90,000 ha planted in fall 2014 and Oklahoma is now the number two producer of canola in the USA behind North Dakota (National Agricultural Statistics Service, 2015).

Winter canola is grown almost exclusively as a rotational crop with winter wheat. The rotation typically consists of canola in year 1 and wheat in years 2 and 3 (Bushong *et al.*, 2012). Wheat producers started incorporating canola into their rotations for several reasons. Wheat–canola rotations (both Roundup-Ready and conventional varieties) allow for selective management of troublesome grassy and broadleaf weeds while decreasing disease pressure that is common with continuous wheat production. The rotation stabilizes and/or generates higher profits by improving water and nitrogen use efficiency (Blackshaw *et al.*, 1994, 2001) and reducing pest management inputs (DeVuyst *et al.*, 2009). According to an economic budget developed by OSU researchers, annual net profits for wheat–canola rotations are likely to produce US$49.91/ha greater annual returns than continuous wheat systems (DeVuyst *et al.*, 2009). However, estimated net profits for winter canola and the overall rotation are strictly dependent upon effective management of insect pests, weeds and pathogens. Unmanaged pests cause significant yield losses and canola cannot be grown successfully without managing them.

As producers began to grow canola, many production challenges arose. In 2004, we were invited to join the OKANOLA team to seek solutions for aphid problems that had developed. While working to develop sampling plans and management solutions for canola aphids, we also assessed Oklahoma canola producers to identify additional concerns that they had with canola production issues, especially pest management issues (Franke *et al.*, 2009). Agronomic concerns included optimal planting dates, fertility regimes and harvest options. Pest management challenges included insects, diseases and weeds (Franke *et al.*, 2009). Ninety per cent of producer respondents identified insects as a high (59%) or moderate (31%) level of concern as a production issue (Franke *et al.*, 2009). The most

*Corresponding author. E-mail: tom.royer@okstate.edu

common insects they encountered included aphids (83%), the general term 'cabbageworms' (33%) and *Melanoplus* species grasshoppers (30%). Others listed included 'loopers' (16%), 'cutworms' (13%), 'diamondback moth larvae' (13%), 'false chinch bug' (13%) and 'don't know' (13%) (Franke *et al.*, 2009).

In response, we started assembling pest management information for the high-priority pests identified by our growers. Most of the information available was developed for spring canola production areas, especially in North Dakota (Knodel *et al.*, 2014) and Canada (Goodwin, 2005). We also consulted information generated from Georgia (Buntin *et al.*, 2007). In the end, we discovered that the most important pests of spring canola (e.g. flea beetles *Phyllotreta cruciferae* (Goeze), *Phyllotreta striolata* (Fabricius) and *Psyllodes punctulata* Melsh; cabbage seedpod weevil *Ceutorhynchus obstrictus* (Marsham); and the swede midge *Contarinia nasurtii* Kieffer) were of minor importance compared with the insects of concern in winter canola grown in Oklahoma and Kansas. With that understanding, we had to remain responsive as new pests occurred. Over the course of our involvement with OKANOLA, we have seemingly encountered a new pest outbreak every 2–3 years for which we had to develop a set of management strategies (Royer *et al.*, 2015). The following describes the pests that have presented the greatest challenges for management.

10.2 Challenge 1: Aphids

Since the introduction of winter canola into the South Central USA, producers have battled devastating aphid infestations (Giles and Royer, 2008, 2011; Franke *et al.*, 2009). During the first few years of canola production in the region (2004–2006), aphids were the most important factor limiting production and winter canola could not be successfully grown without their suppression (Southern IPM Center, 2007). Three aphids are found to be important pests of winter canola: the turnip aphid *Lipaphis erysimi* (Kaltenbach), the green peach aphid *Myzus persicae* (Sulzer) and the cabbage aphid *Brevicoryne brassicae* (L.). These aphids can increase rapidly because their life cycle from birth to adult is short (7–21 days) and in most situations they reproduce through a process called parthenogenesis. Without management, the combined feeding effect of three aphid species can cause > 50% yield loss (Royer *et al.*, 2013).

10.2.1 Turnip aphid: *Lipaphis erysimi* (Kaltenbach)

The turnip aphid is present from the time canola plants emerge in the fall and continues to feed and breed at reduced rates throughout the winter. Wild mustards (*Brassica* spp.) and canola are important overwintering host plants. Turnip aphid infestations are favoured by moderate temperatures and dry weather. Heaviest infestations occur in the fall through winter (September to February). Turnip aphid typically forms clusters on the underside of leaves and sap sucking causes infested foliage to curl, wilt, or become distorted. Some infested plants are killed; others demonstrate arrested growth.

10.2.2 Green peach aphid: *Myzus persicae* (Sulzer)

The green peach aphid is present from the time canola emerges in the fall through spring. It passes the winter on canola and wild hosts such as wild *Brassica* and *Rumex* species. Winged forms migrate to other hosts in late spring and migrate to canola in the fall. Infestations in canola are most common during the late winter and spring and decline as temperatures become hot in late spring. The green peach aphid extracts sap from plants and excretes a sweet sticky substance known as honeydew. Black sooty mould grows on honeydew and though not directly harming the plants, it may block out sufficient light to reduce yield. While sporadic, green peach aphids can build into populations of thousands per plant, causing defoliation and poor yields. Insect predators, fungal diseases, high temperatures, damp weather and hard rains help reduce aphid populations.

10.2.3 Cabbage aphid: *Brevicoryne brassicae* (L.)

Cabbage aphid infestations are favoured by moderate temperatures and dry weather. Wild mustards (*Brassica* spp.) and canola are important overwintering host plants. Heaviest infestations occur in the spring (March to May). Cabbage aphids typically form clusters on the flowering panicles of canola during bloom and pod fill. Cibils-Stewart *et al.* (2015) showed that cabbage aphid populations that colonized racemes exhibited a higher reproductive rate compared with those that fed on leaves. Natural control occurs from parasitoids, especially the braconid wasp *Diaeretiella rapae* (M'Intosh), which commonly attack cabbage aphids. Insect predators and environmental factors (such as heavy rainfall) can sometimes reduce cabbage aphid populations.

10.2.4 Management of aphids

Producers became discouraged with aphid issues and acreage began to drop. Because of the crisis caused by aphids, the OKANOLA team focused attention on evaluating effectiveness and profitability of neonicotinoid seed treatments, developing sampling protocols, establishing economic thresholds and clarifying recommendations for use of broad-spectrum insecticides (Giles *et al.*, 2006; Royer *et al.*, 2013).

Cultural and biological control

As stated previously, wild mustards and other winter annual weeds (*Rumex* spp.) serve as overwintering hosts for canola aphids and may speed their colonization into canola; therefore, the destruction of these weeds before planting and maintenance of weed control in fields and along margins were suggested to reduce initial aphid infestations.

Insect natural enemies are often abundant in winter canola fields but their impact on aphid populations is still not well understood. In a 2011–2014 study, yellow sticky traps were deployed in 16 winter canola landscapes to document the activity-density of the most common insect predators (K.L. Giles, unpublished data). The most common predators were the lacewing, *Chrysoperla carnea* (Stephens), and two coccinellids, *Hippodamia convergens* Guérin-Méneville and *Coccinella septempunctata* Linnaeus. They were most abundant just prior to spring insecticide applications. In a separate laboratory feeding study, these three predators were shown to exhibit no preference among canola aphid prey; however, predator development was delayed and adults were smaller when they consumed turnip or cabbage aphids (Jessie *et al.*, 2015). Aphids sequester high levels of indole glucosinolates from canola host plants. Results indicated that brassica-specialist aphids (turnip and cabbage) are partially toxic to aphid predators, which may explain observed lower populations of natural enemies in canola versus neighbouring wheat fields (Jessie *et al.*, 2015).

The braconid wasp *Diaeretiella rapae* (M'Intosh) has been observed parasitizing winter canola aphids during the spring (Elliott *et al.*, 2014a, b) and preliminary studies indicate that these parasitoids may be able to prevent very low aphid populations from increasing during the spring. Further research is clearly needed, particularly on quantifying the impact of natural enemies, and development of novel approaches to conserve their presence in this intensively managed crop, including research on the potential for deployment of beetle banks (Collins *et al.*, 2002; MacLeod *et al.*, 2004; Elliott *et al.*, 2006).

Fall aphid management decisions

Based on 2010 and 2011 surveys of growers at winter canola conferences (K.L. Giles, unpublished data), 100% of winter canola producers justifiably use neonicotinoid-treated seed (either clothianidin, imidacloprid or thiamethoxam) to avoid the devastating effects of fall–winter aphid infestations. Multi-year data combining all varieties tested (Giles and Royer, 2008, 2011) revealed that plots with untreated seed harboured fall–winter aphid populations, which reduced yields by an average of 22%. These losses occurred despite rescue treatments at available action thresholds (see section 'Mid–late spring aphid management decisions', below) in the spring and were estimated to cost producers US$128.44/ ha (at US$0.37/kg). Plots with treated seed had few aphids during the winter and available action thresholds were not reached until February as temperatures increased and residual efficacy of seed treatments declined in plants.

Early spring aphid management decisions

As temperatures warm and plants break dormancy in late January and February, producers should apply a nitrogen application (topdressing) to meet yield goals (DeVuyst *et al.*, 2009; OSU, 2016). Costs associated with travelling over fields and applying nitrogen are significant and producers can potentially save money by tank-mixing herbicides and/or insecticides, essentially removing additional application costs. Currently, we have no information on whether producers tank-mix pesticides in response to documented weed and insect detection/ thresholds or are simply looking to reduce risks associated with potential pest impacts. Although aphid populations are rare during January–February in fields planted with neonicotinoid-treated seed, producers may need to respond to the presence of sporadic pests such as cutworms or diamondback moths (Royer *et al.*, 2013).

Mid–late spring aphid management decisions

Pest management during this period is driven by prevention or suppression of severe insect infestations, primarily aphids. Based on surveys at winter canola conferences (Franke *et al.*, 2009; K.L. Giles, unpublished data, 2010, 2011), 75–90% of producers apply broad-spectrum insecticides at least twice each spring to reduce severe spring aphid infestations (Fig. 10.1). The absence of accurate economic thresholds for commonly grown varieties hinders scouting efforts and thus cost-effective suppression of spring aphid infestations. New efficient sequential sampling plans are being finalized for aphids in winter canola (Alyousuf *et al.*, 2015, 2016), but preliminary thresholds are still based on a 2006–2008 study evaluating the effects of combined turnip and green peach aphid infestations on a little-grown low-yielding glyphosate-resistant variety (Fig. 10.2). This study demonstrated a very

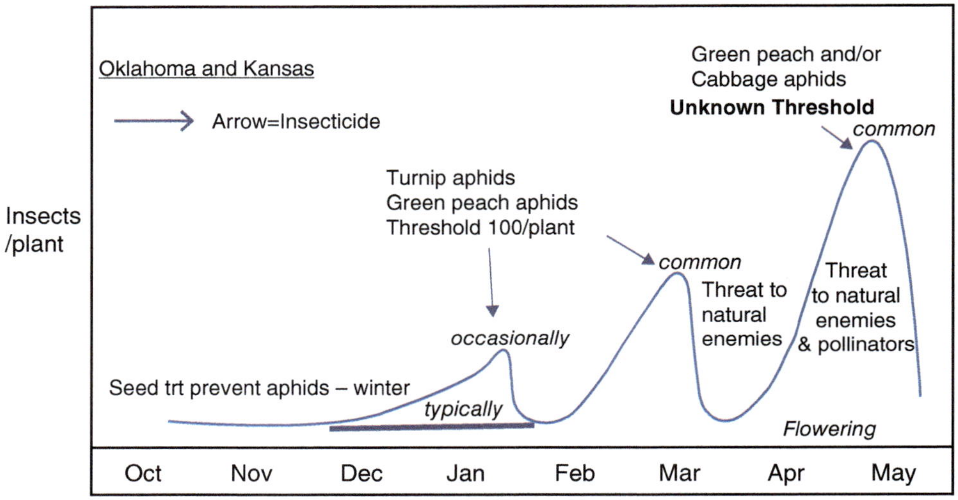

Fig. 10.1. Current pest dynamics with broad-spectrum insecticides in winter canola.

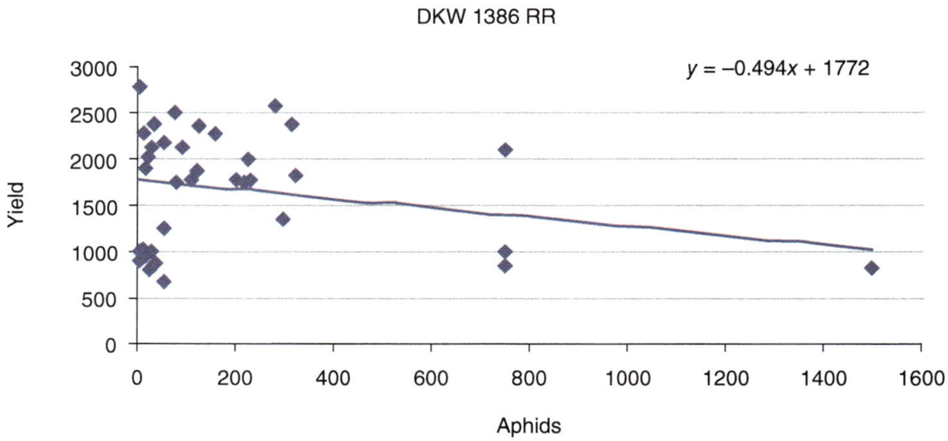

Fig. 10.2. Relationship between aphid intensity during March and seed yield.

weak but significant relationship, that for every aphid/plant, seed yield is lost at the rate of 1.08 kg/ha. Based on this relationship and insecticide costs, a preliminary threshold of ~100–200 aphids per plant was recommended to producers who were willing to scout fields prior to spraying. Fields should be monitored twice per week to obtain representative samples of each area of a field and quickly detect aphid immigrations in a timely manner. Additional recommendations for cabbage aphids infesting racemes were based on recommendations from Australia (Queensland DAF, 2016).

Our management programme relies on the application of broad-spectrum insecticides during the spring, which can result in rapid resurgence of aphids during April–May. A typical spring aphid outbreak in canola is summarized in Table 10.1 and demonstrates: (i) how effective synthetic pyrethroids are in the short term; and (ii) how quickly just one species of aphid can resurge in the absence of natural enemies. In April, these plots were heavily infested with aphids and very few natural enemies, and canola was re-treated to preserve yields (K.L. Giles, unpublished data).

Clearly, beneficial insects (natural enemies and pollinators) appear to be at great risk if synthetic pyrethroid insecticides are used (> 90% of foliar applications are made with synthetic pyrethroids; Franke *et al.*, 2009). The viability of pyrethroid insecticides needs to be carefully preserved. Other than flonicamid, which is registered for control of aphids in canola and has no restrictions on application during bloom, the pyrethroids are the predominant class of insecticides available for use to control aphids. They have restrictions on how they can be used during bloom to protect pollinators. In addition, the green peach aphid has a history of developing resistance to pyrethroid insecticide (Bass *et al.*, 2014). We believe that all stakeholders should be concerned about the long-term viability of the effectiveness of pyrethroid insecticides.

10.3 Challenge 2: Caterpillars

Several caterpillars attack canola at various stages. Army cutworm, *Euxoa auxiliaris* (Grote), lays eggs in newly planted canola fields. Diamondback moth, *Plutella xylostella* (Linnaeus), is resident throughout the growing season. Variegated cutworm, *Peridroma saucia* (Hübner), and corn earworm, *Helicoverpa zea* (Boddie), attack canola in the spring, feeding mainly on seed pods. The biggest challenge is scouting for pod-feeding caterpillars after plants have bolted, as the dense vegetation and narrowly planted rows make it physically difficult to scout.

10.3.1 Army cutworm: *Euxoa auxiliaris* (Grote)

We first encountered army cutworm infestations in 2010 (Royer, 2010a) and had significant infestations in 2012 and 2015. Army cutworm moths migrate to Oklahoma each fall from their summer residence in the Rocky Mountains. They seek bare or sparsely vegetated fields, such as a newly prepared field ready for canola planting, or a field that was planted in dry ground and had not yet emerged because it is awaiting an activating rain event, a practice called 'dusting in'. One army cutworm moth can lay up to 1500 eggs. The eggs hatch soon after being deposited. Egg-laying activity occurs from August through October, which explains why a producer might find different sizes of larvae in a field. Army cutworms feed throughout the winter before they pupate in the soil. Most larvae pupate by mid–late March and moths (also called 'millers') begin to emerge in April and fly north to the Rocky Mountains to spend the summer there. High populations of

Table 10.1. Number of turnip aphids per plant in Perkins, Oklahoma, 2006: comparison of untreated canola versus canola treated with synthetic pyrethroids.

Active ingredient (AI) / formulation[a]	Rate kg (AI)/ha	Aphids/plant pre-treatment (23 Feb)	Aphids/plant 7 days after treatment (2 Mar)	Aphids/plant 22 days after treatment (17 Mar)
Untreated	0	97.4	166.3	257.0
Zeta cypermethrin 0.8EC	0.028	82.3	20.8	83.3
Bifenthrin 2EC	0.046	124.2	1.2	56.3
Gamma cyhalothrin 0.5EC	0.017	105.3	2.4	110.0
Lambda cyhalothrin 2EC	0.034	91.4	0.6	53.5

[a]EC, emulsifiable concentrate

army cutworms can cause severe stand loss in canola if not controlled. Army cutworms feed at the base of canola plants, separating the top of the plant from its roots at the crown base. They can cause direct stand loss or cause canola plants to become stunted and malformed because they are only partially attached to their roots.

10.3.2 Diamondback moth: *Plutella xylostella* (Linnaeus)

Our first significant encounters with diamondback moths also occurred in 2010 (Royer, 2010b); we continued to get reports in 2011 and again in 2015. Diamondback moths are found worldwide and feed on many crops (Knodel and Ganehiarachchi, 2008a). Female moths lay oval, flattened eggs that measure 0.4 mm in groups of one to eight eggs, which will hatch in 5–6 days. One female can lay 150 eggs. They can complete a life cycle in about 32 days, depending on temperature. Typically, all life stages will be found at the same time. When they first hatch, larvae chew small irregular 'windowpane' areas on a leaf. As they get larger, they chew entire leaves, leaving only the veins. Yield loss in winter canola is more likely to be associated with damage to the crown, flowers and seed pods. Overwintering larvae remain in the crown and can damage shoot buds, causing distorted stems as the plants emerge during spring. Infestations of the crown can easily be mistaken for winterkill. When larvae feed on flowers, they cause them to abort. When they feed on seed pods, the pods may fail to produce seed. Feeding associated with flowers and pods can also cause a delay in plant maturity.

Most of the canola grown in Oklahoma is genetically engineered for tolerance to the herbicide glyphosate. Our challenge is to develop recommendations that work with the common grower practice of combining a pyrethroid insecticide with a fall application of glyphosate for army cutworm and diamondback moth, ostensibly to 'save a trip across the field'. Diamondback moth is notorious for developing resistance to pyrethroid insecticides (Liu *et al.*, 1981). Because a majority of the registered insecticides for diamondback moth in winter canola are pyrethroid insecticides, we are working to determine the extent that diamondback moth is present in fall canola and their impact on canola yield potential. This information can be used to support more judicious deployment of a fall tank-mix application of a pyrethroid with a glysophate in order to reduce the possibility of selecting for pyrethroid-resistant diamondback moth populations.

10.3.3 Variegated cutworm: *Peridroma saucia* (Hübner)

Variegated cutworms were found in significant numbers in 2011 (Royer, 2011a). Variegated cutworm populations go through three to four generations per year in Oklahoma and they winter as pupae in the soil. However, they can be active year-around in warmer climates and adults will migrate northward. Larvae are 'climbing cutworms' and literally climb up and down the canola plant to feed. They will strip the green material from the raceme and damage developing pods. We had no established thresholds for managing variegated cutworm but chose to use a threshold of 10.8–21.6/m^2 that had been established for bertha armyworm (*Mamestra configurata* (Walker)) in spring canola (Knodel and Ganehiarachchi, 2008b) because they exhibit similar behaviour.

10.3.4 Corn earworm: *Helicoverpa zea* (Boddie)

Corn earworms became a threat to canola in 2013 (Royer, 2013) and were again found in 2015. Corn earworm is found worldwide and it feeds on many crops. Earworms overwinter as pupae in the soil and emerge as adults in spring. Moths from the farther south will also migrate northwards from Texas. Multiple generations occur in Oklahoma. Larvae feed on pods in a similar manner to variegated cutworms; thus, we use the same threshold (10.8–21.6/m^2) for control.

10.4 Challenge 3: False Chinch Bug: *Nysius raphanus* (Howard)

We encountered significant levels of false chinch bug *Nysius raphanus* (Howard) in 2011 (Royer, 2011b). Adult false chinch bugs are greyish-brown and measure 3.2–4.2 mm long. Eggs are laid and hatch throughout the growing season in the soil surrounding plants and nymphs reach adulthood within 3 weeks. Both nymphs and adults are capable of overwintering in the detritus that remains following harvest and in areas with weedy host plants

(Demirel and Cranshaw, 2005). Multiple generations per growing season are common. False chinch bugs feed predominantly on cruciferous plants (*Brassica* and *Raphanus* species) as well as *Portulaca* species (Demeril and Cranshaw, 2005). False chinch bugs utilize their piercing–sucking mouthparts to feed on sap. Small numbers are typically not injurious. Large numbers on a single plant, sometimes numbering in the thousands, are capable of reducing turgor pressure and causing wilting. Water stress makes canola particularly vulnerable to this type of damage. Seedlings are the most vulnerable stage, though the largest groups are often found on sexually mature flowering plants. They can also damage canola seed and economic thresholds have been developed for false chinch bug on spring canola seed (Demirel, 2003). We found that chemical control is difficult for false chinch bugs due to dry conditions that impede thorough coverage and maximum residual activity. In addition, they often migrate too rapidly and evade pesticide application. Since 2011, we have

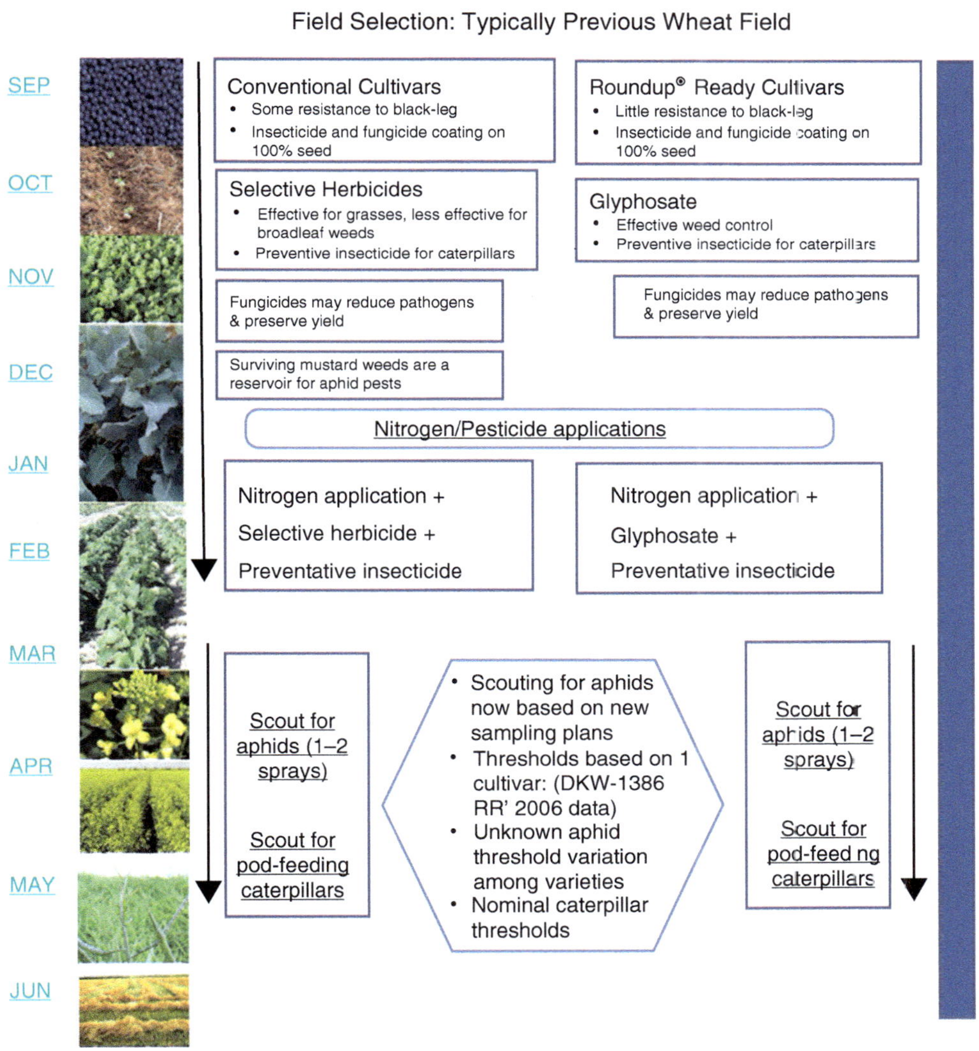

Fig. 10.3. Insect, weed and pathogen management in winter canola.

not had a significant outbreak of false chinch bug in Oklahoma canola.

10.5 Challenge 4: Harlequin Bug: *Murgantia histrionica* (Hahn)

Harlequin bug, *Murgantia histrionica* (Hahn), became numerous in 2015 (Royer, 2015). Wallingford *et al.* (2011) provided a comprehensive overview of its description and life history. This black, shield-shaped stink bug is brightly coloured with orange, red and yellow markings harlequin bugs overwinter as adults among plant debris. Adults emerge early in the spring, usually during late April in Oklahoma. About 2 weeks after resuming activity, females begin to deposit eggs on the undersides of leaves. Eggs are laid in double-row clusters of 10–13 until each female has deposited about 155 eggs. Eggs hatch after 5–20 days, depending on temperature. Nymphs feed for 6–8 weeks before becoming adults. There are three or four generations per year.

Besides canola, the harlequin bug attacks nearly all crucifers, including common weeds of the mustard family such as wild mustard, shepherd's-purse, and peppergrass (Wallingford *et al.*, 2011). Adults and nymphs pierce stalks, leaves and pods with their needlelike mouthparts and extract plant juices. Stems, leaves and pods injured in this manner develop irregular cloudy spots around the puncture wound. Young plants are likely to wilt, turn brown and eventually die, while older plants are only stunted. They will also feed on the pod and can damage developing seeds. Destruction of weeds (especially those in the mustard family) within fields and along fence rows limits overwintering sites. Economic threshold has not been established for this pest but we suggest that insecticide sprays around the perimeter of a field would be appropriate if scouting indicates that the harlequin bugs are numerous (two to three bugs per plant).

10.6 OKANOLA and Pest Management

Pest management in canola is intensive (Fig. 10.3) and starts with a crop rotation which is critical for avoiding pathogen build-up in fields, primarily the blackleg fungus, *Leptosphaeria maculans* (Dominiak-Olson *et al.*, 2010), and mandatory for crop insurance coverage (USDA-RMA, 2016). A number of locally adapted varieties are available and the choice of planting glyphosate-resistant or conventional seed could influence how key pests can be managed, because some pesticide applications can be combined into tank mixes, which reduces the number of 'trips' over the field.

The OKANOLA website (http://canola.okstate.edu/) remains the centralized resource for winter canola production and pest management and contains updated research and extension information as it becomes available. Most of the insect management suggestions (Royer and Giles, 2012) are published. The site provides links to a large number of web pages/files on pest identification and management. Because of the static nature of each web page, management of insect pests, weeds, and fungal pathogens remains largely independent. Optimizing a comprehensive pest management system for winter canola will require continued development of research-based data on the cost-effectiveness of pest management tactics and the utilization of dynamic information delivery of recommendations for management of multiple pests in canola.

References

Alyousuf, A., Giles, K., Payton, M., Elliott, N. and Opit, G. (2015) Development of sequential sampling plans for aphids infesting winter canola in the Southern Plains. Poster abstract P1-3 in: *Proceedings of the 63rd Annual Meeting of the Southwestern Branch of the Entomological Society of America, Tulsa, Oklahoma, 23–26 February 2015*, p. 26.

Alyousuf, A., Giles, K., Opit, G. and Elliott, N. (2016) Defining and efficient sample unit for aphids infesting winter wheat. Poster abstract P1-1 in: *Proceedings of the 64th Annual Meeting of the Southwestern Branch of the Entomological Society of America, Tyler, Texas, 22–25 February 2016*, p. 31.

Bass, C., Puinean, A.M., Zimmer, C.T., Denholm, I., Field, L.M., Foster, S.P., Gutbrod, O., Nauen, R., Slater, R. and Williamson, M.S. (2014) The evolution of insecticide resistance in the peach potato aphid, *Myzus persicae*. *Insect Biochemistry and Molecular Biology* 51, 41–51.

Blackshaw, R.E., Larney, F.O., Lindwall, C.W. and Kozub, K.C. (1994) Crop rotation and tillage effects on weed populations on the semi-arid Canadian prairies. *Weed Technology* 8, 231–237.

Blackshaw, R.E., Larney, F.O., Lindwall, C.W., Watson, P.R. and Dirksen, D.A. (2001) Tillage intensity and crop rotation affect weed community dynamics in a winter wheat cropping system. *Canadian Journal of Plant Science* 81, 805–813.

Boyles, M. and Peeper, T. (2008) Bringing crop rotation to winter wheat producers. Available at: http://canola.okstate.edu/canolaprogram/overview08/overview08.pdf (accessed 26 February, 2016).

Buntin, D., Grey, T., Harris, G.H. Jr, Phillips, D., Buck, J., Prostko, E., Raymer, P., Smith, N., Sumner, P. and Woodruff, J. (2007) B-1331, Canola production in Georgia. University of Georgia Extension, Athens, Georgia.

Bushong, J.A., Griffith, A.P., Peeper, T.F. and Epplin, F.M. (2012) Continuous winter wheat versus a winter canola-winter wheat rotation. *Agronomy Journal* 104, 324–330.

Cibils-Stewart, X., Sandercock, B.K. and McCornack, B.P. (2015) Feeding location affects demographic performance of cabbage aphids on winter canola. *Entomologia Experimentalis et Applicata* 156, 149–156.

Collins, K.L., Boatman, N.D., Wilcox, A., Holland, J.M. and Chaney, K. (2002) Influence of beetle banks on cereal aphid predation in winter wheat. *Agriculture, Ecosystems, and Environment* 93, 337–350.

Demirel, N. (2003) Integrated pest management studies of the insects affecting oilseed brassicas in Colorado. PhD Thesis, Colorado State University, Fort Collins, Colorado.

Demirel, N.W. and Cranshaw, W. (2005) Evaluation of spring canolas and mustards of varying age for relative preference by false chinch bugs, *Nysius raphanus* (Howard). *Pakistan Journal of Biological Sciences* 8, 665–668.

DeVuyst, E.A., Epplin, F., Peeper, T.F. and Boyles, M.C. (2009) *AGEC-257: Oklahoma canola systems vs. continuous wheat budget comparison*. Oklahoma Cooperative Extension Service, Stillwater, Oklahoma.

Dominiak-Olson, J., Boyles, M., Peeper, T. and Sanders, H. (2010) *EPP-7671: Blackleg of canola*. Oklahoma Cooperative Extension Service, Stillwater, Oklahoma.

Elliott, N.C., Tao, F.L., Giles, K.L. and Shufran, K.A. (2006) Ground beetle density in Oklahoma winter wheat fields. *Southwestern Entomologist* 31, 121–128.

Elliott, N.C., Backoulou, G.F., Giles, K.L. and Royer, T.A. (2014a) Aphids and parasitoids in wheat and nearby canola fields in central Oklahoma. *Southwestern Entomologist* 39, 23–28.

Elliott, N.C., Backoulou, G.F., Giles, K.L. and Royer, T.A. (2014b) Parasitism of aphids in canola fields in Central Oklahoma. *Journal of Agricultural and Urban Entomology* 30, 59–64.

Franke, T., Kelsey, K. and Royer, T.A. (2009) *EPP-7085: Pest management needs assessment for Oklahoma canola producers*. Oklahoma Cooperative Extension Service, Stillwater, Oklahoma.

Giles, K. and Royer, T. (2008) Canola insect management. *4th Oklahoma–Kansas Winter Canola Conference, Enid, Oklahoma*. Oklahoma Cooperative Extension Service, Stillwater, Oklahoma.

Giles, K. and Royer, T. (2011) Canola insect management. *7th Oklahoma–Kansas Winter Canola Conference, Enid, Oklahoma*. Oklahoma Cooperative Extension Service, Stillwater, Oklahoma.

Giles, K.L., Peeper, T., Boyles, M., Royer, T., Bushong, J. and Kastl, D. (2006) *Aphids on winter canola in the southern plains*. ASA-CSSA-SSSA International Meetings, US National Canola Association Symposium, Indianapolis, Indiana.

Goodwin, M. (2005) *Crop Profile for Canola in Canada*. Agriculture and Agri-Food Canada, Ottawa, Ontario, Canada.

Jessie, W.P., Giles, K.L., Rebek, E.J., Payton, M.E., Jessie, C.N. and McCornack, B.P. (2015) Preference and performance of *Hippodamia convergens* (Coleoptera: Coccinellidae) and *Chrysoperla carnea* (Neuroptera: Chrosopidae) on *Brevicoryne brassicae*, *Lipaphis erysimi*, and *Myzus persicae* (Hemiptera: Aphididae) from winter-adapted canola. *Environmental Entomology* 44, 880–889.

Knodel, J. and Ganehiarachchi, M. (2008a) *E-1346: Diamondback moth in canola: biology and integrated pest management*. North Dakota State University Extension Service, Fargo, North Dakota.

Knodel, J. and Ganehiarachchi, M. (2008b) *E-1347: Bertha armyworm in canola: biology and integrated pest management*. North Dakota State University Extension Service, Fargo, North Dakota.

Knodel, J.J., Beauzay, P. and Boetel, M. (2014) Canola insects. In: *E-1143: 2015 Field Crop Insect Management Guide*. North Dakota State University Extension Service, Fargo, North Dakota, pp. 24–27.

Liu, M.Y., Tzeng, Y.J. and Sun, C.N. (1981) Diamondback moth resistance to several synthetic pyrethroids. *Journal of Economic Entomology* 74, 393–396.

MacLeod, A., Wratten, S.D., Sotherton, N.W. and Thomas, M.B. (2004) 'Beetle banks' as refuges for beneficial arthropods in farmland: long-term changes in predator communities and habitat. *Agricultural and Forest Entomology* 6, 147–154.

National Agricultural Statistics Service (2015) *Crop acreage, yield, production and value, Oklahoma, 2009–2013*. USDA NASS, Washington DC, pp. 16–17.

OSU (2016) *Fertility*. Department of Plant and Soil Sciences, Oklahoma State University, Stillwater, Oklahoma. Available at: http://canola.okstate.edu/cropproduction/fertility (accessed 27 February 2016)

Queensland DAF (2016) *IPM Guidelines for Grains: Aphids in Canola*. Queensland Department for Agriculture & Fisheries, Brisbane, Australia. Available at: http://ipmguidelinesforgrains.com.au/pests/aphids/aphids-in-canola/ (accessed 12 April 2016).

Royer, T.A. (2010a) Army cutworms reported in canola. *Pest E-alerts* 9(7). Available at: http://entoplp.okstate.edu/pddl/pddl/2010/PA9-7.pdf (accessed 27 February 2016).

Royer, T.A. (2010b) Scouting canola for diamondback moth. *Pest E-alerts* 9(9). Available at: http://entoplp.okstate.edu/pddl/pddl/2010/PA9-9.pdf (accessed 27 February 2016).

Royer, T.A. (2011a) Start looking for variegated cutworms in canola, PRONTO! *Pest E-alerts* 11(21). Available at: http://entoplp.okstate.edu/pddl/pddl/2012/PA11-21.pdf (accessed 27 February 2016).

Royer, T.A. and Giles, K. (2011b) Confusion about false chinch bug thresholds! *Department of Plant and Soil Sciences Extension News*. Available at: http://canola.okstate.edu/cropproduction/insect/insectsNL/Scouting%20Canola%20for%20False%20Chinch%20Bug%202011%20PSS.pdf (accessed 27 February 2016).

Royer, T.A. (2013) Corn earworm search for a new name..... canola podworm? *Pest E-alerts* 12(20). Available at: http://entoplp.okstate.edu/pddl/pddl/2013/PA12-20.pdf (accessed 27 February 2016).

Royer, T.A. (2015) What to do with harlequin bugs in canola? *Pest E-alerts* 14(27). Available at: http://entoplp.okstate.edu/pddl/pddl/2015/PA14-27.pdf (accessed 27 February, 2016).

Royer, T. and Giles, K. (2012) Insect pests. In: Boyles, M., Bushong, J., Sanders, H. and Stamm, M. (eds) *Great Plains Canola Production Handbook*. Kansas State University, Manhattan, Kansas, pp. 20–23.

Royer, T., Giles, K., Zarrabi, A. and Seuhs, K. (2013) *Determine control of foliage-feeding caterpillars and incidence of other sporadic insect pests in winter canola*. Partners in Progress: Oilseed research at OSU 2013, Division of Agricultural Sciences and Natural Resources, Oklahoma Agricultural Experiment Station, Stillwater, Oklahoma.

Royer, T.A., Giles, K.L., Zarrabi, A.A., Damicone, J. and Bushong, J. (2015) Winter canola as a rotational crop for winter wheat in Oklahoma: Pest management challenges and solutions. *8th International IPM Symposium*, Salt Lake City, Utah.

Southern IPM Center (2007) *Crop Profile for Oklahoma Canola*. Available at: http://www.ipmcenters.org/cropprofiles/docs/OKcanola.pdf (accessed 27 February 2016).

USDA Risk Management Agency (2016) *Canola, Oklahoma. A fact sheet for crop insurance, 2016 crop year.* Available at: www.rma.usda.gov/fields/ok_rso/2016/okcanola.pdf (accessed 27 February 2016).

Wallingford, A.K., Kuhar, T.P., Schultz, P.B. and Freeman, J.H. (2011) Harlequin bug biology and pest management in brassicaceous crops. *Journal of Integrated Pest Management* 2(1), H1–H4. doi: http://dx.doi.org/10.1603/IPM10015 (accessed 27 February 2016).

11 Integrated Pest Management in Canola and Other *Brassica* Oilseed Crops: How Far We Have Come and What Is Still Needed

John Gavloski*

Manitoba Agriculture, Carman, Manitoba, Canada

11.1 Defining Integrated Pest Management

The history of integrated pest management (IPM) can be traced back to the late 19th century when ecology was identified as the foundation for scientific plant protection, and there have been many definitions applied to the concept (Kogan, 1998). The concept of IPM was advanced and better defined in California in the 1950s. Early writings on the concept of IPM state that 'chemical and biological control are regarded as two main methods of suppressing insects and spider mites. These two methods are often thought of as alternatives in pest control. This is not necessarily so, for with adequate knowledge they can be made to augment one another' (Stern *et al.*, 1959). Thus according to the original definition of IPM, integrated control sought to identify the best mix of control methods for a given insect pest. Chemical insecticides were a component of integrated pest management but were to be used in the manner least disruptive to biological control.

11.1.1 What does IPM require?

One of the primary purposes of IPM is to encourage the use of crop management strategies, and when needed choosing pesticides, that will enable natural enemies of pests to remain active. Such alternative management strategies can include the use of resistant varieties of crops, the conservation or introduction of biological control agents, cultural controls, and the use of selective insecticides where necessary. The use of economic thresholds, where available, and forecasting methodology are also valuable components of IPM. Thus, one of the requirements of IPM is that where possible and practical, alternatives to non-selective insecticides be used for managing insect populations.

11.2 Main Insects of Concern on Canola in the Northern USA and Canada

There are many insects in canola that potentially can get to levels that cause economic damage. Of those of greatest concern, roughly half are insects or complexes that are crucifer specialists in their feeding habits and are also introduced into North America (Gavloski *et al.*, 2011). These include: the crucifer flea beetle, *Phyllotreta cruciferae* (Goeze) (Coleoptera: Chrysomelidae); the striped flea beetle, *Phyllotreta striolata* (F.) (Coleoptera: Chrysomelidae); the diamondback moth, *Plutella xylostella* (Linnaeus) (Lepidoptera: Plutellidae); the root maggot *Delia radicum* (L.) (Diptera: Anthomyiidae); and the cabbage seedpod weevil, *Ceutorhynchus obstrictus* (Marsham) (Coleoptera: Curculionidae). Several insects or complexes of insects that are more general in their feeding habits, and also native to North America, can also occasionally

*E-mail: john.gavloski@gov.mb.ca

reach economic levels in canola (Gavloski *et al.*, 2011). These include: several species of cutworms; *Lygus* bugs; bertha armyworm, *Mamestra configurata* Walker (Lepidoptera: Noctuidae); and some species of grasshoppers.

This review of IPM progress in canola will focus on the introduced diamondback moth, the introduced species complex of *Phyllotreta* flea beetles (*P. striolata* and *P. cruciferae*), the native defoliator bertha armyworm and a native complex of sap feeders, *Lygus* spp. These are all among the most serious insect concerns for canola in North America.

For each of the four insects that are covered, the following six components of an IPM programme will be evaluated for their current state of implementation for that insect: (i) monitoring and forecasting; (ii) economic thresholds; (iii) cultural control; (iv) resistant varieties; (v) biological control; and (vi) the use of selective insecticides. An overall view of how far we have come in developing IPM programmes in canola and what is still needed will also be presented.

11.3 IPM for Flea Beetles

Flea beetles (*Phyllotreta* spp.) are chronic pests of canola in Canada and the northern USA. Populations can be quite high in many fields in any given year.

11.3.1 Monitoring and forecasting

Techniques for monitoring and forecasting populations of flea beetles in canola have not been well developed. Currently, management decisions are based on estimates of percentage defoliation to seedling plants (Palaniswamy *et al.*, 1992; Tangtrakulwanich *et al.*, 2014). Accurate estimation of percentage defoliation can be difficult. Often our focus is drawn to the damage on the leaf and it is easy to overestimate the damage. Visual guides, showing various levels of percentage defoliation, are helpful when using percentage defoliation as a means of making management decisions.

Sticky traps can be used to determine the most prevalent species of flea beetle in a region. Research is lacking on their usefulness as a means of making management decisions in canola.

Male *P. cruciferae* and *P. striolata* produce aggregation pheromones, which in field trials have been shown to attract both sexes (Soroka *et al.*, 2005; Beran *et al.*, 2011). However, this is currently not something that can be used for monitoring or forecasting purposes.

11.3.2 Economic thresholds

A nominal threshold of 25% of the surface of cotyledons (Fig. 11.1) and first true leaves damaged, with flea beetles still present, is currently used in Canada on canola seedlings. In Montana, where *P. cruciferae* is the only species of flea beetle of concern on canola in the seedling stage, a threshold of 15–20% leaf area damaged has been recommended (Tangtrakulwanich *et al.*, 2014).

Flea beetle feeding on canola in late summer is rarely an economic concern. In terms of late-season economic thresholds, flea beetle feeding that occurs when seeds in lower pods of canola are at the green stage or beyond is unlikely to affect seed yields regardless of the infestation rate of flea beetles. Even when seeds are translucent to green, numbers higher than 100 flea beetles per plant, and for some cultivars higher than 350 per plant, may be necessary to cause significant yield reductions (Soroka and Grenkow, 2012).

Fig. 11.1. Striped (top and middle) and crucifer (bottom) flea beetles on canola cotyledon.

11.3.3 Cultural controls

Seeding rates

Although many factors will be involved in determining an appropriate seeding rate, higher seeding rates may reduce damage from flea beetles. Studies in Alberta found that mean flea beetle damage per plant declined with an increase in seeding rate (Dosdall *et al.*, 1999; Dosdall and Stevenson, 2005). Higher seeding rates and densities per plant may reduce damage to individual plants. Increased cost of the seed and the potential for increased lodging and susceptibility to disease are some of the potential costs of higher seeding rates that need to be considered (Soroka and Elliott, 2011).

Size of seeds

Seedlings from larger seeds are more vigorous and tolerant to flea beetle damage than seedlings from medium or small seeds for both *Brassica rapa* (Elliott *et al.*, 2007a) and *Brassica napus* (Elliott *et al.*, 2008). This is due to a higher initial shoot biomass and higher growth rate when flea beetle damage is high.

Seeding dates

Seeding date can potentially be used to reduce risk of feeding by flea beetles, but the effect of early seeding on injury from flea beetles may vary with location. In Alberta, early seeding resulted in reduced damage by flea beetles in southern Alberta but increased damage in central and northern Alberta (Cárcamo *et al.*, 2008). In North Dakota, early seeding resulted in increased injury by flea beetles (Knodel *et al.*, 2008). Canola producers are, however, urged to seed canola as early as is practical to maximize seed yield (Soroka and Elliott, 2011).

Reduced tillage

Several studies have demonstrated that reduced tillage creates a less favourable environment for flea beetles and can reduce damage to canola from flea beetles (Borstlap and Entz, 1994; Milbrath *et al.*, 1995; Dosdall *et al.*, 1999). The greater structural diversity of no-till plots might interfere with host plant location (Milbrath *et al.*, 1995).

11.3.4 Resistant varieties

There are currently no flea beetle-resistant varieties of canola available, though much research has been done to look for potential sources of flea beetle resistance (Palaniswamy *et al.*, 1992; Gavloski *et al.*, 2000). Antixenosis resistance has been found in *Thlaspi arvense* and several cultivars of *Sinapis alba* (Gavloski *et al.*, 2000). However, introgression of genetic sources of flea beetle resistance from resistant relatives into canola varieties is difficult and has not occurred.

Several studies have evaluated the role of trichomes in deterring flea beetles from feeding on species of *Brassica* and related cruciferous plants. Pods of the mustard *Sinapis alba* (= *Brassica hirta*) 'Gisilba', which have stiff hairs, showed no significant flea beetle damage, while adjacent rapeseed plots showed heavy pod damage (Lamb, 1980). Removal of hairs from the mustard pods caused an increase in feeding damage by the flea beetles. In laboratory tests for resistance to feeding by flea beetles, the wild Mediterranean species *Brassica villosa* Biv. and *B. villosa* Biv. subsp. *drepanensis*, which had a trichome density of > 2172/cm^2, were found to be highly resistant to flea beetle feeding (Palaniswamy and Bodnaryk, 1994). All other species tested had a trichome density of < 30/cm^2 and all suffered significant damage from flea beetle feeding. Behavioural observations indicated that a high density of trichomes on *B. villosa* leaves acted as a physical barrier to flea beetle feeding by preventing the flea beetles from firmly settling on the leaf surface to initiate feeding. Flea beetle feeding was evaluated on canola with genes inserted from *Arabidopsis thaliana* L. for increased trichome production (Soroka *et al.*, 2011). Feeding by flea beetles decreased on canola seedlings with increased trichome density.

11.3.5 Biological controls

Although attempts have been made to release parasitoids of crucifer-feeding flea beetles in the Canadian prairies, biological control has to date not been successful at significantly reducing levels of flea beetles. Little is known of the natural mortality factors for flea beetles. *Townesilitus bicolor* (Hymenoptera: Braconidae), a parasitoid of some crucifer-infesting species of *Phyllotreta* in Europe, was released in Manitoba, Canada, in 1978–1983, in an attempt to reduce populations of *P. cruciferae* and *P. striolata*. Releases totalled 1936 adults of *T. bicolor* and approximately 2639 *Phyllotreta* species collected in Manitoba and parasitized in the laboratory. No evidence of establishment was found at either of the two release sites (Wylie, 1988).

The Nearctic parasitoid *Microctonus vittatae* (Hymenoptera: Braconidae) will parasitize adults of *Phyllotreta* species on the Canadian prairies (Wylie, 1982), but its impact on populations is not known.

Generalist predators, such as *Collops vittatus* (Coleoptera: Melyridae) (Gerber and Osgood, 1975) and the large bigeyed bug, *Geocoris bullatus* (Hemiptera: Geocoridae) (Burgess, 1977), may be incidental predators of adult *Phyllotreta* species. The level of control provided by predators is not known.

Entomopathogenic fungi and the nematode *Steinernema carpocapsae* are potential biopesticides of flea beetles. The fungi *Beauveria bassiana* and *Metarhizium brunneum* reduced feeding by *P. cruciferae* and improved yield of canola in experimental plots (Reddy *et al.*, 2014).

11.3.6 Selective insecticides

Seed treatments containing neonicotinoid insecticides, sometimes combined with a diamide insecticide, are available for providing early-season control of flea beetles. These target specific insects that feed on the canola seedlings. Eventually the effectiveness of the seed treatment will diminish to a level where effective flea beetle control is no longer being provided. Sometimes this happens prior to the canola reaching growth stages where it is able to tolerate feeding by flea beetles effectively, which is generally regarded as being the three- to four-leaf stage (Bracken and Bucher, 1986). In these instances, a foliar application of insecticide may be needed.

In Canada, foliar insecticide options for flea beetles in canola are currently limited to pyrethroid, carbamate and organophosphate insecticides and one insecticide combining a pyrethroid and diamide. All of these are more general in their impact on insect communities and currently more selective foliar insecticide options are not available for flea beetles in canola in Canada.

Spinosad, a somewhat selective insecticide, has potential for use against crucifer flea beetles on canola (Antwi *et al.*, 2007; Elliott *et al.*, 2007b). Spinosad is not harmful to many predatory insects, though it may be harmful to some parasitic Hymenoptera (Williams *et al.*, 2003). Spinosad is registered for the suppression of crucifer flea beetle on *Brassica* leafy vegetables but is not currently registered for flea beetle control in canola in Canada.

11.3.7 Summary of current status of IPM for flea beetles in canola

Monitoring and forecasting for flea beetles in canola needs improvement. Seed treatment decisions need to be made prior to seeding. Yet there is no reliable means of forecasting the risk of flea beetles in canola. Research is needed on this.

The recommendation for making decisions regarding whether foliar insecticides are needed is based on estimating percentage defoliation to seedlings. This is subjective and it is easy to overestimate damage. Improved extension tools that are easy to use in the field, demonstrating examples of various levels of percentage defoliation, may help improve the accuracy of estimating defoliation.

Economic thresholds are nominal for flea beetles in canola. This also needs additional research. Some questions to consider for this research are: is the threshold stage specific (does it change once plants move from the cotyledon to the first and second true leaf stage?) and does it vary for the species of flea beetle (striped vs crucifer)? Temperature and soil moisture would also influence whether canola seedlings can compensate for a given level of feeding by flea beetles, but may be hard to control for in threshold studies and to predict.

Cultural controls have been well researched for flea beetles in canola. However, factors other than insects may often enter into decision making regarding tillage practices, seeding dates and rate of seeding. Although the impacts of some of these cultural practices on flea beetles are known, further research and strategies for managing flea beetles are still needed.

Flea beetle-resistant varieties of canola currently do not exist, despite much research. Advances in developing canola with trichomes to minimize feeding by flea beetles, or the introgression of genetic sources of flea beetle resistance from resistant relatives into canola, may provide future resistant varieties.

Biological control of flea beetles in canola has been attempted in the Canadian prairies, but has not been successful, and natural enemies of flea beetles have been poorly studied. More research is needed in this area.

Seed treatments containing insecticides provide some degree of selectivity and will conserve many natural enemies. However, when seed treatments fail to provide long enough protection and foliar applications of insecticides are needed, all options currently available in Canada are broad-spectrum and harmful to the complex of natural enemies that are present. Foliar insecticides that have at least some degree of selectivity are needed for flea beetles in canola.

The use of seed treatments and an understanding of the impact of some cultural controls (such as effects of seeding date, tillage and seeding rate) have so far been the main contributors towards IPM for flea beetles in canola. More research on economic thresholds, biological controls, resistant varieties and improved methods of monitoring are needed. An impressive amount of research has been done as a base in some of these areas, but more research is needed to provide solutions that can be transferred to the field.

11.4 IPM for Diamondback Moth

Diamondback moths (*Plutella xylostella* (Linnaeus)) are sporadic pests of canola in Canada and the northern USA. They are not able to overwinter well on the Canadian prairies and so damaging populations depend on the moths being moved in on winds from the southern USA or Mexico (Dosdall *et al.*, 2011).

11.4.1 Monitoring and forecasting

Monitoring adult moths

Sex pheromone-baited traps have been developed to monitor levels of adult male diamondback moth (Chisholm *et al.*, 1983), but moth numbers are not directly related to larval density (Miluch *et al.*, 2013). Delta traps (Fig. 11.2) are fitted with removable sticky inserts (Fig. 11.3) and baited with commercially available pheromone for diamondback moth. These traps are used to detect the arrival of migrant diamondback moths into the prairie provinces of Canada and the northern USA. Traps baited with pheromone released from grey rubber septa capture more males than those baited with red septa (Miluch *et al.*, 2014), hence grey rubber septa are currently used to dispense the pheromone. However, commercially available sex pheromone lures attract fewer diamondback moth males than calling virgin female moths (Evenden and Gries, 2010) and research on the development of a more attractive synthetic sex pheromone lure is still needed.

Wind trajectories from potential source regions of diamondback moth can occasionally predict movement of adult diamondback moth between regions successfully (Hopkinson and Soroka, 2010). Used in conjunction with the network of pheromone-baited traps, wind trajectories may help in providing alerts to early northward movement of diamondback moth. Backward trajectories, tracing wind patterns backwards, can be produced from areas where traps are in place; forward trajectories can be produced from potential source regions.

Fig. 11.2. Delta trap for monitoring adults of diamondback moth.

Monitoring larvae

Monitoring techniques for larvae of diamondback moth (Fig. 11.4) have been suggested. Larvae can be monitored by removing the plants in an area measuring 0.1 m^2 (about 1 ft^2), beating them on a clean surface and counting the number of larvae dislodged from the plants. To obtain an accurate count, this procedure should be repeated in at least five locations in the field. This technique has not been researched to test its accuracy or effectiveness.

Fig. 11.3. Diamondback moth on trap insert.

Fig. 11.4. Pupa (left) and larva (right) of diamondback moth.

Sampling with a sweep net can determine the presence and general abundance of diamondback moth in the field, but is currently not a useful means for making management decisions because no studies have been conducted to relate levels caught in a sweep net to levels/ft^2 or yield loss. High levels of diamondback moth caught in sweep sampling can, however, prompt producers to perform counts of larvae per unit area (Dosdall *et al.*, 2011).

Several models have been developed to predict development of diamondback moth based on degree days. Laboratory studies in Saskatchewan suggested that diamondback moth requires 143 days above the lower threshold of 4.23°C to complete their life cycle (Bahar *et al.*, 2014). A study in southern Ontario found that 293 degree days above a threshold temperature of 7.3°C were required to complete one generation (Butts and McEwen, 1981).

11.4.2 Economic threshold

In experiments using simulated defoliation of winter canola, economic injury levels for diamondback moth have been suggested based on the relationship between defoliation and yield loss (Ramachandran *et al.*, 2000). The economic injury levels from this study (32–55 larvae/0.1 m^2 at the rosette stage and 82–140 larvae/0.1 m^2 in the flowering stage) are much higher than the nominal thresholds of 10–15 larvae/0.1 m^2 in immature to flowering plants and 20–30 larvae/0.1 m^2 in plants with flowers and pods (Gavloski, 2016) that are currently used in Canada and the northern USA. However, except in more extreme instances, leaf consumption by diamondback moth larvae is generally not high enough to be an economic threat and it is feeding directly on pods that poses the greatest risk of reducing yield. Feeding on buds and flowers may also be of concern in years of drought conditions (Dosdall *et al.*, 2011), but canola is good at compensating for loss of buds and flowers when soil moisture is adequate. Thresholds currently being used to make management decisions for diamondback moth in canola are nominal and research is needed to develop research-based economic thresholds for use in canola. Such thresholds need to consider potential direct feeding to pods and consider the potential compensatory abilities of canola at younger stages when grown under favourable conditions.

11.4.3 Cultural controls

Cultural controls for diamondback moth have not been developed in canola, though research has been done on cultural controls for diamondback moth in cruciferous vegetable crops. The wild crucifer yellow rocket, *Barbarea vulgaris*, has been identified as a candidate for use as a potential 'dead-end' trap crop for diamondback moth because it is highly attractive for oviposition, but larvae do not survive on it (Badenes-Perez *et al.*, 2004; Lu *et al.*, 2004;

Shelton and Nault, 2004). However, its use in prairie agriculture may not be practical because it is not possible to predict years when diamondback moth will arrive at high levels and there may be reluctance to plant *B. vulgaris* (Dosdall *et al.*, 2011).

11.4.4 Resistant varieties

Lines of transgenic Bt canola, which are effective at completely controlling larvae of diamondback moth, have been developed (Stewart *et al.*, 1996) but are not commercially available. In seed mixtures with non-transgenic lines, diamondback moth larvae would move from transgenic to non-transgenic plants before acquiring lethal doses of toxin, indicating a higher possibility of resistance development (Ramachandran *et al.*, 1998). Thus seed mixtures would not be an effective strategy and, should commercialization be considered, transgenic and non-transgenic plants may need to be grown in separate rows to minimize the rate of resistance development.

11.4.5 Biological controls

Parasitoids can at times result in high levels of regulation of diamondback moth populations. The three main parasitoids of diamondback moth in the Canadian prairies are the larval parasitoids *Diadegma insulare* (Hymenoptera: Ichneumonidae) and *Microplitis plutellae* (Hymenoptera: Braconidae) and the pupal parasitoid *Diadromus subtilicornis* (Hymenoptera: Ichneumonidae) (Braun *et al.*, 2004; Sarfraz *et al.*, 2005). Of these species, *D. insulare* is considered the most important for biological control (Dosdall *et al.*, 2011). Over a 10-year period (1961–1970) in Saskatchewan, 35–81% of first-generation larvae of the diamondback moth were parasitized by *D. insulare* and *M. plutellae,* averaging 68% (Putnam, 1973).

D. insulare is not known to overwinter in the Canadian prairies and is believed to be carried northwards along with its host (Dosdall *et al.*, 2011). Adults spend more time in habitats where flowering plants are abundant so they can feed (Idris and Grafius, 2001) and are attracted to plant tissue damaged by diamondback moth larvae (Mitchell *et al.*, 1999). Having floral resources available throughout the season may help to increase the activity of *D. insulare* and other parasitoids and increase their effectiveness.

Little research has been done on predators of diamondback moth. Green lacewings have been observed feeding on diamondback moth larvae and cocoons (Harcourt, 1960). Some species of ground beetles may also prey on diamondback moth. In Japan the ground beetles *Chlaenius micans* and *Chlaenius posticalis* may be important predators of diamondback moth larvae, as they have high consumption rates and the habit of climbing plants in both immature and adult stages (Suenaga and Hamamura, 1998). More research is needed to determine the role of predators in regulating populations of diamondback moth.

Rainfall may be a major mortality factor of eggs and early instars of diamondback moth (Harcourt, 1963; Kobori and Amano, 2003). The larvae are very susceptible to drowning and may be washed or wriggle to leaf axils or the ground, where they drown in accumulated water (Harcourt, 1963). In cabbage, increased rainfall droplet density and diameter resulted in increased larvae falling from plants (Kobori and Amano, 2003). Mortality rates of up to 74% have been reported from intense rain (Harcourt, 1963). Thus when populations are approaching economic levels, re-sampling is advised after major rain events.

11.4.6 Selective insecticides

Most insecticides registered to manage diamondback moth in canola in Canada are pyrethroid or organophosphate insecticides and are quite general in their effect on insects, also being harmful to populations of beneficial insects. An exception is the diamide insecticide chlorantraniliprole, which has been shown to give good control of diamondback moth in canola (Joshi and Buschman, 2012) and is not harmful to some beneficial insects such as parasitic Hymenoptera (Brugger *et al.*, 2010), but may still cause harm to some generalist predators (Amarasekare and Shearer, 2013).

Selective insecticides derived from the soil bacterium *Bacillus thuringiensis* var. *kurstaki* are registered for the control of diamondback moth in some cruciferous vegetable crops in Canada, but not in canola. XenTari, an insecticide based on *Bacillus thuringiensis* subsp. *aizawai*, is registered for the control of both diamondback moth and bertha armyworm in canola in Canada but is not commercially available.

Spinosad is registered for control of diamondback moth in *Brassica* (cole) leafy vegetables but is not registered for use in canola in Canada. In addition to controlling diamondback moth and being harmless to some predaceous insects, spinosad is toxic to *D. insulare* (Hill and Foster, 2000). There have been, however, instances of diamondback moth resistance to spinosad in some areas of the USA as a consequence of extensive applications in crucifer vegetables in some regions (Zhao *et al.*, 2006).

The insect growth regulator novaluron is also registered in Canada for the control of diamondback moth in *Brassica* vegetable crops but it is not registered in canola in Canada. Novaluron is toxic to some beneficial insects but has no effect on adult insects. In studies on cabbage, *Brassica oleracea* var. *capitata*, in Ethiopia, novaluron effectively controlled diamondback moth and was relatively safe to *Diadegma* sp. (Ayalew, 2011).

11.4.7 Summary of current status of IPM for diamondback moth in canola

Methods for monitoring the arrival of adult populations of diamondback moth are available but need to be used with caution. Pheromone-baited traps can detect the presence of adult moths in a region but models to determine level of risk based on the cumulative number of moths captured are lacking. Combined with wind trajectory information, the traps can determine approximately when adult moths arrive in larger numbers into a region and potentially their source. From this, risk can be determined to some extent based on anticipated numbers of generations that are likely to occur, though whether populations continue to build with each generation is uncertain. No data are available to enable cumulative counts of adult moths in the traps to determine risk levels for a region. Nominal thresholds, based on counts of larvae, are suggested and sampling for larvae is possible but research is needed to develop economic injury levels and economic thresholds. Cultural controls for diamondback moth in canola are lacking and no resistant varieties of canola are available.

Parasitoids of diamondback moth have been well researched and their value in regulating populations of diamondback moth is known. Currently, there are no methods developed to monitor parasitoid populations as additional information to include in attempts to forecast risk of diamondback moth. Such methodology could be a valuable component to a forecasting programme. As identification of parasitoids is often difficult for farmers and agronomists, extension material simplifying parasitoid identification could be of great value. Although parasitoids can have great value in regulating populations of insects such as diamondback moth, their contributions can be underappreciated, as training farmers and agronomists how to identify and quantify levels of crop-feeding insects is much easier than training to identify and quantify parasitoids. Research on predators of diamondback moth is lacking.

There is currently only one selective insecticide option available to canola growers in Canada to control diamondback moth, though more options do exist in cruciferous vegetable crops. The availability of additional selective insecticides that are comparably priced to non-selective options would improve IPM for diamondback moth in canola.

11.5 IPM for Bertha Armyworm

Bertha armyworm (*Mamestra configurata* Walker) is a sporadic pest of canola in Canada and the northern USA, with outbreaks occurring at irregular intervals (Mason *et al.*, 1998).

11.5.1 Monitoring and forecasting

Pheromone-baited traps (Fig. 11.5) are used to monitor adult moths of bertha armyworm. The number of moths captured in traps is influenced by the stage of plant development of canola in which the traps are placed (Turnock, 1984). More moths are captured in fields with plants in the early-bloom stage than in those with plants in earlier or later stages of development. Variability of the relationship between number of moths per trap and density of late-instar larvae is too high to allow accurate prediction of larval density for individual fields (Turnock, 1987). However, the number of moths per trap may be used to predict regionally the risk of some fields having levels of larvae above the economic threshold.

Sampling methods for larvae of bertha armyworm have been developed (Fig. 11.6). These involve using a three-sided frame to outline a 0.25 m^2 sampling area, shaking the plants inside the sampling area, then examining the soil surface for larvae and moving earthen lumps and plant debris to expose hidden larvae (Turnock and Bilodeau, 1985; Wise *et al.*, 2009). Early-maturing fields may contain older larvae than later-maturing fields and so they should be sampled first to determine larval density and the need for insecticide application (Turnock, 1984).

Fig. 11.5. Pheromone-baited trap for bertha armyworm.

Fig. 11.6. Defining area to assess for bertha armyworm larvae.

Bertha armyworm larvae have a moderately clumped distribution in canola (Wise *et al.*, 2009). Females strongly prefer to oviposit on leaves with eggs of a different female than on leaves without eggs (Ulmer *et al.*, 2003).

Degree days for the development of various stages of bertha armyworm have been determined (Bailey, 1976). With a development threshold of 7°C, 82, 356 and 352 accumulated degree days above the threshold are required for development of eggs, larvae and pupae, respectively.

11.5.2 Economic thresholds

A relationship between the larval density of bertha armyworm and yield loss in canola has been determined (Bracken and Bucher, 1977). A loss of 0.058 bushels/acre (3.25 kg/ha) for each larva/m^2 can be expected. This is based on 2 years of data where the average yield loss was 0.325 g/larva. From this, a table of economic injury levels for bertha armyworm, for different values of canola and different control costs, has been developed (Gavloski, 2016). At the current canola value of about US$370.40/t and control costs of US$7.60–12.20/acre, the economic injury level is 16–25 larvae/m^2, with more larvae/m^2 required to achieve the economic injury level as the cost of control increases. These economic injury levels are used to make control decisions, as adjustments to them have not been done to create economic thresholds below those of the economic injury levels as is sometimes done. In a similar test in 1976, where plants were under drought stress, a yield loss of 0.479 g/larva occurred (Bracken and Bucher, 1977). The drought stress on the plants resulted in a very short period of blooming and early dropping of leaves. Lack of leaves may have caused more pod feeding by the larvae and so affected yield more directly, and plants were possibly less able to compensate for tissue loss due to their stressed conditions. Thus for crops under moisture stress, the proposed economic injury levels may need to be lowered.

Feeding on pods by bertha armyworm creates the greatest risk for yield loss. The order of feeding preference on plant parts of canola by sixth-instar larvae of bertha armyworm is leaves, bracts, immature pods and mature pods (Bracken, 1984). Thus canola will be at greatest risk of yield loss when leaves are senescing and larvae (Fig. 11.7) are still actively feeding and are forced to feed on canola pods. Although control measures for bertha armyworm should generally not be undertaken before pod damage is observed, a control decision should be made before many pods are damaged (Bracken, 1987).

11.5.3 Cultural controls

Fall tillage

Depth of snow can affect how successfully bertha armyworm overwinters. Pupae did not survive a Manitoba winter in snow-free field plots but 55% survived in plots with 5 cm or 10 cm of snow (Lamb *et al.*, 1985). Mortality of the overwintering pupae may be lower in untilled soil, because the stubble traps more snow (Turnock and Bilodeau, 1984). Survival of the parasitoids *Banchus flavescens* and *Athrycia cinerea* is less affected by tillage than that of bertha armyworm (Turnock and Bilodeau, 1984). Thus tillage may reduce the frequency and severity of outbreaks of bertha armyworm not only by increasing mortality but also by differentially affecting the survival of its main parasitoids.

11.5.4 Resistant varieties

Although the survival, development and egg-laying preferences of bertha armyworm on different host plants has been studied (Ulmer *et al.*, 2001, 2002; Dosdall and Ulmer, 2004), no canola cultivars resistant to bertha armyworm have been developed. Lines of *Brassica juncea* are less preferred by feeding larvae of bertha armyworm relative to several other crucifer genotypes (Ulmer *et al.*, 2002). Specific foliar glucosinolates, such as sinigrin, which is predominant in *B. juncea*, and sinalbin, which is abundant in *S. alba*, may provide brassicaceous crops with some protection from *M. configurata* (Ulmer *et al.*, 2001). Thus, it may be possible to use foliar glucosinolate levels as predictors of *M. configurata* feeding damage in the development of breeding lines for resistance.

Fig. 11.7. Different colour phases of bertha armyworm larvae.

11.5.5 Biological controls

Parasitoids

The native endoparasitoid *Banchus flavescens* (Hymenoptera: Ichneumonidae) (Fig. 11.8) is the most abundant hymenopteran parasitoid of bertha armyworm. Females attack first-, second- and third-instar larvae (Arthur and Mason, 1985). Parasitism by *B. flavescens* results in significantly decreased food consumption and lower biomass production but does not reduce the time they occur in the crop (Mason *et al.*, 2001). Parasitism of bertha armyworm by *B. flavescens* may exceed 40% (Wylie and Bucher, 1977).

The native parasitoid *Athrycia cineria* (Diptera: Tachinidae) (Fig. 11.9) oviposits on third, fourth, fifth and sixth instars of bertha armyworm (Wylie, 1977). *A. cineria* may kill over 20% of bertha armyworm (Wylie and Bucher, 1977). Larvae of *A. cinerea* are often killed if they compete with larvae of *B. flavescens* (Wylie, 1977).

Adults of *Eurithia consobrina* (Diptera: Tachinidae) were released at three locations in Manitoba in 1986 and 1987 as candidates to help control bertha armyworm (Turnock and Carl, 1995) but probably did not establish (O'Hara, 1999).

Pathogens

A nuclear polyhedrosis virus is often present in populations of bertha armyworm that reach outbreak levels (Turnock, 1988). Infection by nuclear polyhedrosis virus in more than 95% of the bertha armyworm population has been recorded (Erlandson,

Fig. 11.8. *Banchus flavescens.*

Fig. 11.9. *Athrycia cineria.*

1990). Fungal pathogens belonging to one or possibly two species of the genus *Entomophthora* may also cause larval mortality (Wylie and Bucher, 1977). Host larvae that have died from these fungal pathogens may be seen clasping the stems or pods on canola plants in the upper levels of the plant (Wylie and Bucher, 1977).

11.5.6 Selective insecticides

In Canada, most insecticides registered for bertha armyworm in canola are currently pyrethroid, carbamate or organophosphate insecticides, along with one insecticide combining a pyrethroid and diamide. All of these are more general in their impact on insect communities, also being harmful to populations of beneficial insects. An exception is the diamide insecticide chlorantraniliprole, which is registered for bertha armyworm in canola and, as discussed in Section 11.4.6, is not harmful to some beneficial insects such as parasitic Hymenoptera (Brugger *et al.*, 2010).

Many strains of *B. thuringiensis* have been tested for their effectiveness in controlling bertha armyworm and some have a high level of toxicity (Trottier *et al.*, 1988; Morris *et al.*, 1996). The insecticide XenTari, based on *B. thuringiensis* subsp. *aizawai*, is registered for the control of bertha armyworm in canola in Canada but is not commercially available.

11.5.7 Summary of current status of IPM for bertha armyworm in canola

Research has enabled the establishment of good methods of trapping adult moths to predict risk of larval damage, methods of monitoring for larvae, and economic injury levels for making control decisions. Effects of tillage have been researched but, other than this, cultural controls are lacking. No varieties of canola with resistance to bertha armyworm are available.

Some of the key parasitoids of bertha armyworm have been determined and studied. It has been suggested that IPM programmes for *M. configurata* in canola should include evaluating the presence and abundance of parasitoids (Mason *et al.*, 2002). Determining abundance of parasitoids and using a dynamic action threshold as done for soybean aphid (Hallett *et al.*, 2014) would certainly improve our ability to manage bertha armyworm sustainably. However, techniques need to be developed for monitoring key parasitoids; resources are needed to make identification of key parasitoids possible for farmers and agronomists; and research is needed to determine how current thresholds used to make management decisions would be adjusted for various levels of key parasitoids.

There is currently only one selective insecticide option, chlorantraniliprole, available to canola growers in Canada to control bertha armyworm. Insecticides derived from *B. thuringiensis* are additionally available for the control of bertha armyworm in canola in the USA. The availability of additional selective insecticides that are comparably priced to non-selective options would improve IPM for bertha armyworm in canola.

11.6 IPM for Lygus Bugs

Lygus bugs are potential pests of many crops in Canada and the northern USA. They are unpredictable pests of canola because of their ability to produce multiple generations, having up to six species forming a 'complex' and a wide host-plant range (Otani and Cárcamo, 2011).

11.6.1 Monitoring and forecasting

Sequential decision plans have been developed for lygus bugs in canola based on sweep-net sampling (Wise and Lamb, 1998a). Samples taken along the edges of commercial fields and at various distances into the fields all gave similar estimates of plant bug density, justifying the use of edge sampling.

Although currently used to monitor lygus bugs in canola, sweep-net sampling can be difficult when canola gets into the podding stages, especially with the tall dense canopies of some of the modern varieties. Thus, there is still a need to explore other possible means of sampling lygus bugs in canola.

Levels of lygus bugs could potentially decrease if a hard rain event occurs. A study in New Jersey found that years with higher levels of rain in May and June had fewer first generation nymphs of *Lygus lineolaris* in alfalfa (Day, 2006). This reduction was likely because of raindrops knocking the lygus bugs off the plants, rather than biotic, because infections by nematodes and a fungus disease were not significantly increased. Thus, after a hard rainfall, fields should be re-sampled for lygus bug levels rather than using counts made before the rain to make management decisions.

11.6.2 Economic threshold

Economic injury levels were developed in southern Manitoba for lygus bugs in canola based on numbers caught using a 38 cm diameter sweep net (Wise and Lamb, 1998b). Economic injury levels when flowering is complete are calculated based on an assumed loss of 0.1235 bushels/acre (0.007 t/ha) for each lygus bug per ten sweeps. Economic injury levels when seeds in the lower pods are green are calculated based on an assumed loss of 0.0882 bushels/acre (0.005 t/ha) for each lygus bug per ten sweeps. At the current canola value of about CAN$11.00/bushel (US$370.40/t) and control costs of CAN$10.00–16.00/acre (US$7.60–12.20/acre), the economic injury level is 7–11 lygus bugs in ten sweeps when flowering is complete, or 10–16 lygus bugs in ten sweeps when seeds in the lower pods are green (Fig. 11.10). These thresholds, which are currently used, were developed on open-pollinated cultivars that are no longer planted; ongoing studies are validating the thresholds using herbicide-tolerant hybrid cultivars. When precipitation is greater than 100 mm from the onset of bud formation to the end of flowering, the plants may partially compensate for damage by lygus bugs (Wise and Lamb, 1998b). Other areas of the prairies, particularly northern growing regions, may have different growing conditions that may result in different economic injury levels (Otani and Cárcamo, 2011).

Fig. 11.10. Lygus bug on canola pod.

Thus further research is still needed quantifying injury to canola by lygus bugs in different canola-growing regions. Such research should also account for the effect that rainfall and soil moisture can have on the level of injury by lygus bugs.

11.6.3 Cultural controls

There are currently no cultural controls that can be recommended for management of lygus bugs in canola. Alfalfa has been shown to be an effective trap crop for lygus bugs in cotton when interplanted with strips of alfalfa 20 ft (6.1 m) wide at intervals of up to 540 ft (164.6 m) (Sevacherian and Stern, 1974). However, whether this would be effective or practical in canola has not been studied. The effect of seeding date on lygus bugs in canola in Manitoba has been studied (Leferink and Gerber, 1997). The stage of plant development, and not seeding date, is important in attracting colonizing *Lygus* adults to canola.

11.6.4 Resistant varieties

Little work has been done to develop or identify varieties of canola resistant to lygus bugs. Varieties with high or low levels of glucosinolates are both suitable hosts (Butts and Lamb, 1990). Varieties of *S. alba* with high concentrations of the glucosinolate sinalbin in seeds reduced feeding by *L. lineolaris* (Bodnaryk, 1996).

11.6.5 Biological controls

Biological control options for lygus bugs in canola need additional research. Mortality of nymphs of lygus bugs caused by the parasitoid *Peristenus braunae* (Hymenoptera: Braconidae) can reach

70% at peak times in some alfalfa sites but may be virtually absent in canola sites (Braun *et al.*, 2001). *Peristenus digoneutis* Loan, a parasitoid of lygus bugs, was successfully introduced from Europe into North America in the 1980s for controlling native *Lygus* populations. Surveys confirm that *P. digoneutis* populations have become established throughout eastern North America and that the spread of the parasitoid continues, but previous releases of *P. digoneutis* in western Canada were not successful (Haye *et al.*, 2013).

The potential of honey bees, *Apis mellifera* L., to vector *Beauveria bassiana* (Balsamo) to *L. lineolaris* in canola has been studied (Al Mazra'awi *et al.*, 2006). In addition to improving pollination, the bees effectively vectored the inoculum from the hives to the crop, resulting in *Lygus* mortalities as high as 56% in some treatments in caged field trials in Ontario. However, *B. bassiana* is not registered for use in canola in Canada.

Few studies have been done on the predatory complex on lygus bugs, particularly in canola-growing regions. Research in Arizona found that the damsel bug *Nabis alternatus* could be a potentially significant field predator of *L. hesperus* (Perkins and Watson, 1972).

11.6.6 Selective insecticides

Insecticides currently registered to control lygus bugs in canola in Canada are all non-selective, thus of potential harm to natural enemies and pollinators that would be in the field. Therefore, should an insecticide application be economical, the application should be made as late in the day as possible, or early morning, when pollinators and natural enemies would be less active and less likely to come into direct contact with the insecticide.

Flonicamid is registered for the suppression of lygus bugs in some berry crops in Canada. Flonicamid has selective activity against hemipterous and thysanopterous pests (Morita *et al.*, 2014) and is not harmful to beneficial insects and mites.

Novaluron is registered in Canada for the control of nymphs of lygus bugs in strawberries but is not registered for use in canola. In a study to investigate the direct contact toxicity of five insecticides currently used, or with potential for use, in canola IPM on bees that may forage in canola, novaluron was non-toxic to all species of bees tested (Scott-Dupree *et al.*, 2009).

11.6.7 Summary of current status of IPM for lygus bugs in canola

Techniques for monitoring lygus bugs with a sweep net have been developed. As canola advances into the podding stage, vigorously growing fields of canola can be very difficult to walk through to sample insects with a sweep net. Thus research should continue to explore additional means of monitoring lygus bugs in canola. The economic thresholds currently used are under re-evaluation. Thresholds may need to be adjusted for the different growing conditions of some regions and the widespread use of new hybrid canola varieties. Cultural controls are lacking for lygus bugs in canola, as are canola varieties with resistance to feeding by *Lygus* bugs.

More research on natural enemies of lygus bugs in canola is needed. Species of *Peristenus* can cause high levels of mortality of *Lygus* nymphs in some crops but this has not been observed in canola. The use of honey bees to vector pathogens of lygus bugs, and potentially other pests that occur at the flowering stage, deserves further research.

The complete lack of selective insecticides for lygus bugs in canola is a big constraint to the preservation of parasitoids and predaceous insects, and hence to IPM, when *Lygus* populations become high in canola. If flowering is still occurring on some canola plants when insecticide applications for lygus bugs occur, there may additionally be consequences due to potential harm to pollinators and pollination.

11.7 Conclusions

Progress has been made on developing monitoring techniques and economic thresholds for some of the key insect pests of canola but there are insects such as diamondback moth and flea beetles where research on economic thresholds is needed. Research determining thresholds for when multiple defoliators are present on a crop at the same time have been developed in cruciferous vegetable crops (Shelton *et al.*, 1982) and soybeans (Hutchins *et al.*, 1988) and could be of value in canola as well. Grouping insects into injury guilds, based on the plant's physiological response to the injury, forms the theoretical basis for the multiple-species approach (Hutchins *et al.*, 1988). Pod defoliators, such as bertha armyworm and diamondback moth, would be one such combination of simultaneous defoliators.

Cultural controls have been studied for flea beetle and to a lesser extent for bertha armyworm. The effects of reduced tillage on these insects is known, potentially reducing damage from flea beetles but potentially increasing survival of bertha armyworms. Factors other than insect management will likely be the determining factor on choice of tillage practices; however, knowing which risks are reduced or potentially enhanced by a given management choice is of value in crop management planning. As is the case for many of the agronomic practices such as type and level of tillage, and date and rate of seeding, many factors other than insect management are involved in these decisions. Therefore it is a combination of many factors that will determine what seeding and soil management practices are used. Ultimately, growers will implement the best cultural practices to maximize profit. Entomology researchers and IPM practitioners will need to understand the impact of these practices on pest and beneficial arthropods and should develop sustainable management strategies within cropping systems designs to maximize profit.

No insect-resistant varieties are available for any insect pest of canola. This is quite different than for plant pathogens, where resistant varieties of canola are available for several plant pathogens such as black leg (*Leptosphaeria maculans*) (Rimmer, 2006), sclerotinia (*Sclerotinia sclerotiorum*) (Mei *et al.*, 2011) and clubroot (*Plasmodiophora brassicae*) (Diederichsen *et al.*, 2009). Pathogen resistance is often a factor in what variety of canola is seeded, with varieties chosen based on risks of greatest concern. Flea beetles are the largest chronic insect concern in many of the canola-growing areas of Canada; and should flea beetle resistance develop to the point where it is incorporated into high-yielding commercial varieties, it could have great uptake and economic value. Recent research on flea beetle resistance, particularly trichome-based resistance (Soroka *et al.*, 2011), has been encouraging and plant resistance may be the best means of significantly reducing insect damage and insecticide use in canola.

Seeding canola in the fall is practical in some areas but currently not in Canada or the northern USA. Should advances lead to fall-seeded canola becoming viable in these regions, this could potentially dramatically influence the interactions of the crop with insects and possibly reduce the risk of some insects. In one study (Dosdall and Stevenson, 2005), flea beetle damage was lower on canola seeded in the fall than on canola seeded in the spring. Seeding in the fall enabled plants to progress beyond the vulnerable cotyledon stage by the time that most injury from flea beetles occurred. Fall-seeded plants matured 5–21 days earlier than plants seeded in April and about 10–30 days before plants seeded in May. Maturity of canola earlier in the season may also minimize damage by bertha armyworm.

Much work remains to be done on natural enemies of crop-feeding insects in canola. The role of predaceous insects in regulating crop-feeding insects in canola is poorly known. Some parasitoids have been well studied but are usually not factored into management decisions, due to inadequate methods of monitoring, lack of general recognition and lack of means of factoring them into decision making.

Few selective insecticides are available for insect management in canola. The availability of economical selective insecticide options are a core component of many IPM programmes. In canola, flea beetles, diamondback moth, bertha armyworm and *Lygus* bugs can potentially reach pest levels and reduce yield if not controlled. Cultural and biological controls exist to some degree but in outbreak years are often insufficient to prevent substantial yield loss. For biological and chemical controls to augment each other, as suggested in the original definition of IPM, additional selective insecticide options are needed.

For insect pests where selective insecticide options are lacking, it may be worth considering whether currently registered insecticides can be used more efficiently, potentially preserving refuges of parasitoids and predaceous insects. For the management of rangeland grasshoppers, a strategy known as reduced area and agent treatments (RAATs) has been effective. RAATs is a pest management strategy in which the rate of insecticide is reduced from traditional levels and untreated swaths (refuges) are alternated with treated swaths (Lockwood and Schell, 1997; Lockwood *et al.*, 2002). The RAATs strategy appears to depend on movement of grasshoppers from untreated to treated swaths and on the conservation of natural biological control agents. For grasshoppers on rangelands, this technique provides better economic returns than the traditional blanket application (Lockwood *et al.*, 2002). It could be interesting to consider whether such a strategy could provide improved economic returns, while preserving refuges of predaceous and parasitic insects, for more mobile insect pests of canola, such as lygus bugs or possibly flea beetles.

Having areas of wild vegetation or pasture integrated into the farm can be beneficial for enhancing pollination and populations of beneficial arthropods. A study of the value of natural patches in agroecosystems found that the abundance of wild bees was greatest in canola fields that had more uncultivated land within 750 m of field edges and seed set was greater in fields with higher bee abundance (Morandin and Winston, 2006). A cost–benefit model that estimates profit in canola agroecosystems with different proportions of uncultivated land was presented. In a similar study, populations of bumblebees and other wild bees were higher in canola fields that had pasture nearby (Morandin *et al.*, 2007). Having flowering plants available throughout the season may also help to increase the activity and effectiveness of parasitoids, such as *D. insulare*.

IPM as originally conceived requires a shift in mindset, where it is realized that in addition to insects that potentially will be feeding on the crop, there are also many beneficial invertebrates that, if preserved and encouraged, can aid in pest management, pollination and many other vital services on the farm. One of the primary goals of IPM is to learn to manage the crop feeding pests effectively when needed, while doing minimal harm to the assortment of beneficial organisms that will naturally occur on the farm.

References

Al Mazra'awi, M.S., Shipp, J.L., Broadbent, A.B. and Kevan, P.G. (2006) Dissemination of *Beauveria bassiana* by honey bees (Hymenoptera: Apidae) for control of tarnished plant bug (Hemiptera: Miridae) on canola. *Environmental Entomology* 35, 1569–1577.

Amarasekare, K.G. and Shearer, P.W. (2013) Comparing effects of insecticides on two green lacewings species, *Chrysoperla johnsoni* and *Chrysoperla carnea* (Neuroptera: Chrysopidae). *Journal of Economic Entomology* 106, 1126–1133.

Antwi, F.B., Olson, D.L. and Carey, D.R. (2007) Comparisons of ecorational and chemical insecticides against crucifer flea beetle (Coleoptera: Chrysomelidae) on canola. *Journal of Economic Entomology* 100, 1201–1209.

Arthur, A.P. and Mason, P.G. (1985) The life history and immature stages of *Banchus flavescens* (Hymenoptera: Ichneumonidae), a parasitoid of the bertha armyworm, *Mamestra configurata* (Lepidoptera: Noctuidae) in western Canada. *The Canadian Entomologist* 117, 1249–1255.

Ayalew, G. (2011) Effect of the insect growth regulator novaluron on diamondback moth, *Plutella xylostella* L. (Lepidoptera: Plutellidae), and its indigenous parasitoids. *Crop Protection* 30, 1087–1090.

Badenes-Perez, F.R., Shelton, A.M. and Nault, B.A. (2004) Evaluating trap crops for diamondback moth, *Plutella xylostella* (Lepidoptera: Plutellidae). *Journal of Economic Entomology* 97(4), 1365–1372.

Bahar, M.H., Soroka, J.J., Grenkow, L. and Dosdall, L.M. (2014) New threshold temperatures for the development of a North American diamondback moth (Lepidoptera: Plutellidae) population and its larval parasitoid, *Diadegma insulare* (Hymenoptera: Ichneumonidae). *Environmental Entomology* 43, 1443–1452.

Bailey, C.G. (1976) Temperature effects on non-diapause development in *Mamestra configurata* (Lepidoptera: Noctuidae). *The Canadian Entomologist* 108, 1339–1344.

Beran, F., Mewis, I., Srinivasan, R., Svoboda, J., Vial, C., Mosimann, H., Boland, W., Büttner, C., Ulrichs, C., Hansson, B.S. and Reinecke, A. (2011) Male *Phyllotreta striolata* (F.) produce an aggregation pheromone: identification of male-specific compounds and interaction with host plant volatiles. *Journal of Chemical Ecology* 37, 85–97.

Bodnaryk, R.P. (1996) Physical and chemical defences of pods and seeds of white mustard (*Sinapis alba* L.) against tarnished plant bugs, *Lygus lineolaris* (Palisot de Beauvois) (Heteroptera: Miridae). *Canadian Journal of Plant Science* 76, 33–36.

Borstlap, S. and Entz, M.H. (1994) Zero-tillage influence on canola, field pea and wheat in a dry subhumid region: agronomic and physiological responses. *Canadian Journal of Plant Science* 74(3), 411–420.

Bracken, G.K. (1984) Within plant preferences of larvae of *Mamestra configurata* (Lepidoptera: Noctuidae) feeding on oilseed rape. *The Canadian Entomologist* 116, 45–49.

Bracken, G.K. (1987) Relation between pod damage caused by larvae of bertha armyworm, *Mamestra configurata* Walker (Lepidoptera: Noctuidae), and yield loss, shelling, and seed quality in canola. *The Canadian Entomologist* 119, 365–369.

Bracken, G.K. and Bucher, G.E. (1977) An estimate of the relation between density of bertha armyworm and yield loss on rapeseed, based on artificial infestations. *Journal of Economic Entomology* 70, 701–705.

Bracken, G.K. and Bucher, G.E. (1986) Yield losses in canola caused by adult and larval flea beetles, *Phylotreta cruciferae* (Coleoptera: Chrysomelidae). *The Canadian Entomologist* 118, 319–324.

Braun, L., Erlandson, M., Baldwin, D., Soroka, J., Mason, P., Foottit, R. and Hegedus, D. (2001) Seasonal occurrence, species composition, and parasitism of

Lygus spp. in alfalfa, canola, and mustard. *The Canadian Entomologist* 133, 565–577.

Braun, L., Olfert, O., Soroka, J., Mason, P. and Dosdall, L.M. (2004) Diamondback moth biocontrol activities in Canada. In: Kirk, A.A. and Bordat, D. (eds) *Improving Biocontrol of Plutella xylostella*. Proceedings of the International Symposium, 21–24 October 2002. CIRAD, Montpellier, France, pp. 144–146.

Brugger, K.E., Cole, P.G., Newman, I.C., Parker, N., Scholz, B., Suvagia, P., Walker, G. and Hammond, T.G. (2010) Selectivity of chlorantraniliprole to parasitoid wasps. *Pest Management Science* 66, 1075–1081.

Burgess, L. (1977) *Geocoris bullatus*, an occasional predator on flea beetles (Hemiptera: Lygaeidae). *The Canadian Entomologist* 109, 1519–1520.

Butts, R.A. and Lamb, R.J. (1990) Comparison of oilseed *Brassica* crops with high or low levels of glucosinolates and alfalfa as hosts for three species of *Lygus* (Hemiptera: Heteroptera: Miridae). *Journal of Economic Entomology* 83, 2258–2262.

Butts, R.A. and McEwen, F.L. (1981) Seasonal populations of the diamondback moth, *Plutella xylostella* (Lepidoptera: Plutellidae) in relation to day-degree accumulation. *The Canadian Entomologist* 113, 127–131.

Cárcamo, H.A., Otani, J.K., Dosdall, L.M., Blackshaw, R.E., Clayton, G.W., Harker, K.N., O'Donovan, J.T. and Entz, T. (2008) Effects of seeding date and canola species on seedling damage by flea beetles in three ecoregions. *Journal of Applied Entomology* 132, 623–631.

Chisholm, M.D., Steck, W.F., Underhill, E.W. and Palaniswamy, P. (1983) Field trapping of diamondback moth *Plutella xylostella* using an improved four-component sex attractant blend. *Journal of Chemical Ecology* 9, 113–118.

Day, W.H. (2006) The effect of rainfall on the abundance of tarnished plant bug nymphs [*Lygus lineolaris* (Palisot)] in alfalfa fields. *Transactions of the American Entomological Society* 132(3 & 4), 445–450.

Diederichsen, E., Frauen, M., Linders, E.G.A., Hatakeyama, K. and Hirai, M. (2009) Status and perspectives of clubroot resistance breeding in crucifer crops. *Journal of Plant Growth Regulation* 28(3), 265–281.

Dosdall, L.M. and Stevenson, F.C. (2005) Managing flea beetles (*Phyllotreta* spp.) (Coleoptera: Chrysomelidae) in canola with seeding date, plant density, and seed treatment. *Agronomy Journal* 97, 1570–1578.

Dosdall, L.M. and Ulmer, B.J. (2004) Feeding, development, and oviposition of bertha armyworm (Lepidoptera: Noctuidae) on different host plant species. *Environmental Entomology* 33, 756–764.

Dosdall, L.M., Cowle, N.T. and Conway, P.M. (1999) The effect of tillage regime, row spacing, and seeding rate on feeding damage by flea beetles, *Phyllotreta* spp. (Coleoptera: Chrysomelidae), in canola in central Alberta, Canada. *Crop Protection* 18, 217–224.

Dosdall, L.M., Soroka, J. and Olfert, O. (2011) The diamondback moth in canola and mustard: current pest status and future prospects. *Prairie Soils and Crops Journal* 4, 66–76.

Elliott, R.H., Mann, L.W. and Olfert, O.O. (2007a) Effects of seed size and seed weight on seedling establishment, seedling vigour and tolerance of summer turnip rape (*Brassica rapa*) to flea beetles, *Phyllotreta* spp. *Canadian Journal of Plant Science* 87(2), 385–393.

Elliott, R.H., Benjamin, M.C. and Gillott, C. (2007b) Laboratory studies of the toxicity of spinosad and deltamethrin to *Phyllotreta cruciferae* (Coleoptera: Chrysomelidae). *The Canadian Entomologist* 139, 534–544.

Elliott, R.H., Franke, C. and Rakow, G.F.W. (2008) Effects of seed size and seed weight on seedling establishment, vigour and tolerance of Argentine canola (*Brassica napus*) to flea beetles, *Phyllotreta* spp. *Canadian Journal of Plant Science* 88(1), 207–217.

Erlandson, M.A. (1990) Biological and biochemical comparison of *Mamestra configurata* and *Mamestra brassicae* nuclear polyhedrosis virus isolates pathogenic for the bertha armyworm, *Mamestra configurata* (Lepidoptera: Noctuidae). *Journal of Invertebrate Pathology* 56, 47–56.

Evenden, M.L. and Gries, R. (2010) Assessment of commercially available pheromone lures for monitoring diamondback moth (Lepidoptera: Plutellidae) in canola. *Journal of Economic Entomology* 103, 654–661.

Gavloski, J. (2016) *Insect Management in Oilseed Crops in Western Canada*. Western Committee on Crop Pests guide to integrated control of insect pests of crops. Available at: http://www.westernforum.org/Documents/WCCP/WCCP_documents/WCCP_Guidelines/WCCP_16/Oilseeds-WCCP-2016.pdf (accessed 1 June 2016).

Gavloski, J., Ekuere, U., Keddie, A., Dosdall, L., Kott, L. and Good, A.G. (2000) Identification and evaluation of flea beetle (*Phyllotreta cruciferae*) resistance within Brassicaceae. *Canadian Journal of Plant Science* 80, 881–887.

Gavloski, J., Cárcamo, H. and Dosdall, L. (2011) Insects of canola, mustard, and flax in Canadian grasslands. In: Floate, K.D. (ed.) Arthropods of Canadian Grasslands (Volume 2): Inhabitants of a Changing Landscape. Biological Survey of Canada, Ottawa, Canada, pp. 181–214.

Gerber, G.H. and Osgood, C.E. (1975) *Collops vittatus* (Coleoptera: Melyridae): a predator of flea beetle adults in rapeseed. *Manitoba Entomologist* 9, 61.

Hallett, R.H., Bahlai, C.A., Xue, Y. and Schaafsma, A.W. (2014) Incorporating natural enemy units into a dynamic action threshold for the soybean aphid, *Aphis glycines* (Homoptera: Aphididae). *Pest Management Science* 70, 879–888.

Harcourt, D.G. (1960) Biology of the diamondback moth, *Plutella maculipennis* (Curt.) (Lepidoptera: Plutellidae), in eastern Ontario III. Natural enemies. *The Canadian Entomologist* 92, 419–428.

Harcourt, D.G. (1963) Major mortality factors in the population dynamics of the diamondback moth, *Plutella maculipennis* (Curt.) (Lepidoptera: Plutellidae). *Memoirs of the Entomological Society of Canada* 32, 55–66.

Haye, T., Olfert, O., Weiss, R.M., Gariepy, T.D., Broadbent, B. and Kuhlmann, U. (2013) Bioclimatic analyses of distributions of a parasitoid *Peristenus digoneutis* and its host species *Lygus* spp. in Europe and North America. *Agricultural and Forest Entomology* 15, 43–55.

Hill, T.A. and Foster, R.E. (2000) Effect of insecticides on the diamondback moth (Lepidoptera: Plutellidae) and its parasitoid *Diadegma insulare* (Hymenoptera: Ichneumonidae). *Journal of Economic Entomology* 93, 763–768.

Hopkinson, R.F. and Soroka, J.J. (2010) Air trajectory model applied to an in-depth diagnosis of potential diamondback moth infestations on the Canadian prairies. *Agricultural and Forest Meteorology* 150, 1–11.

Hutchins, S.H., Higley, L.G. and Pedigo, L.P. (1988) Injury equivalency as a basis for developing multiple-species economic injury levels. *Journal of Economic Entomology* 81(1), 1–8.

Idris, A.B. and Grafius, E. (2001) Evidence of *Diadegma insulare* (Cresson), a parasitoid of diamondback moth, *Plutella xylostella* (L.), present in various habitats. *Pakistan Journal of Biological Science* 1, 742–743.

Joshi, A. and Buschman, L.L. (2012) Efficacy of foliar insecticide treatments for managing diamondback moth in winter canola, 2010. *Arthropod management tests 2012, Vol.* 37. doi: 10.4182/amt.2012. F9.

Knodel, J.J., Olson, D.L., Hanson, B.K. and Henson, R.A. (2008) Impact of planting dates and insecticide strategies for managing crucifer flea beetles (Coleoptera: Chrysomelidae) in spring-planted canola. *Journal of Economic Entomology* 101, 810–821.

Kobori, Y. and Amano, H. (2003) Effect of rainfall on a population of the diamondback moth, *Plutella xylostella* (Lepidoptera: Plutellidae). *Applied Entomology and Zoology* 38, 249–253.

Kogan, M. (1998) Integrated pest management: historical perspectives and contemporary developments. *Annual Review of Entomology* 43, 243–270.

Lamb, R.J. (1980) Hairs protect pods of mustard (*Brassica Hirta* 'Gisilba') from flea beetle feeding damage. *Canadian Journal of Plant Science* 60, 1439–1440.

Lamb, R.J., Turnock, W.J. and Hayhoe, H.N. (1985) Winter survival and outbreaks of bertha armyworm, *Mamestra configurata* (Lepidoptera: Noctuidae), on canola. *The Canadian Entomologist* 117, 727–736.

Leferink, J.H.M. and Gerber, G.H. (1997) Development of adult and nymphal populations of *Lygus lineolaris* (Palisot de Beauvois), *L. elisus* Van Duzee, and *L. borealis* (Kelton) (Heteroptera: Miridae) in relation to seeding date and stage of plant development on canola (Brassicaceae) in southern Manitoba. *The Canadian Entomologist* 129, 777–787.

Lockwood, J.A. and Schell, S.P. (1997) Decreasing economic and environmental costs through reduced area and agent insecticide treatments (RAATs) for the control of rangeland grasshoppers: empirical results and their implications for pest management. *Journal of Orthoptera Research* 6, 19–32.

Lockwood, J.A., Anderson-Sprecher, R. and Schell, S.P. (2002) When less is more: optimization of reduced agent – area treatments (RAATs) for management of rangeland grasshoppers. *Crop Protection* 21(7), 551–562.

Lu, J., Liu, S. and Shelton, A.M. (2004) Laboratory evaluations of a wild crucifer *Barbarea vulgaris* as a management tool for the diamondback moth *Plutella xylostella* (Lepidoptera: Plutellidae). *Bulletin of Entomological Research* 94, 509–516.

Mason, P.G., Arthur, A.P., Olfert, O.O. and Erlandson, M.A. (1998) The bertha armyworm (*Mamestra configurata*) (Lepidoptera: Noctuidae) in western Canada. *The Canadian Entomologist* 130, 321–336.

Mason, P.G., Erlandson, M.A. and Youngs, B.J. (2001) Effects of parasitism by *Banchus flavescens* (Hymenoptera: Ichneumonidae) and *Microplitis mediator* (Hymenoptera: Braconidae) on the bertha armyworm, *Mamestra configurata* (Lepidoptera: Noctuidae). *Journal of Hymenoptera Research* 10, 81–90.

Mason, P.G., Erlandson, M.A. Elliott, R.H. and Harris, B.J. (2002) Potential impact of spinosad on parasitoids of *Mamestra configurata* (Lepidoptera: Noctuidae). *The Canadian Entomologist* 134, 59–68.

Mei, J., Qian, L., Disi, J.O., Yang, X. Li, Q., Li, J., Frauen, M., Cai, D. and Qian, W. (2011) Identification of resistant sources against *Sclerotinia sclerotiorum* in *Brassica* species with emphasis on *B. oleracea*. *Euphytica* 177(3), 393–399.

Milbrath, L.R., Weiss, M.J. and Schatz, B.G. (1995) Influence of tillage system, planting date, and oilseed crucifers on flea beetle populations (Coleoptera: Chrysomelidae). *The Canadian Entomologist* 127, 289–293.

Miluch, C.E., Dosdall, L.M. and Evenden, M.L. (2013) The potential for pheromone-based monitoring to predict larval populations of diamondback moth, *Plutella xylostella* (L.), in canola (*Brassica napus* L.). *Crop Protection* 45, 89–97.

Miluch, C.E., Dosdall, L.M. and Evenden, M.L. (2014) Factors influencing male *Plutella xylostella* (Lepidoptera: Plutellidae) capture rates in sex pheromone-baited traps on canola in western Canada. *Journal of Economic Entomology* 107, 2067–2076.

Mitchell, E.R, Hu, G.Y., Okine, J.S. and Carpenter, J.E. (1999) Parasitism of diamondback moth (Lepidoptera: Plutellidae) larvae by *Cotesia plutellae* (Hymenoptera: Braconidae) and *Diadegma insulare* (Hymenoptera:

Ichneumonidae) in cabbage fields after inundative releases of *C. plutellae*. *Journal of Entomological Science* 34, 101–112.

Morandin, L.A. and Winston, M.L. (2006) Pollinators provide economic incentive to preserve natural land in agroecosystems. *Agriculture, Ecosystems & Environment* 116, 289–292.

Morandin, L.A., Winston, M.L., Abbott, V.A. and Franklin, M.T. (2007) Can pastureland increase wild bee abundance in agriculturally intense areas? *Basic and Applied Ecology* 8(2), 117–124.

Morita, M., Yoneda, T. and Akiyoshi, N. (2014) Research and development of a novel insecticide, flonicamid. *Journal of Pesticide Science* 39, 179–180.

Morris, O.N., Trotiier, M., Converse, V. and Kanagaratnam, P. (1996) Toxicity of *Bacillus thuringiensis* subsp. *aizawai* for *Mamestra configurata* (Lepidoptera: Noctuidae). *Journal of Economic Entomology* 89, 359–365.

O'Hara, J.E. (1999) Tachinidae (Diptera) parasitoids of bertha armyworm (Lepidoptera: Noctuidae). *The Canadian Entomologist* 131, 11–28.

Otani, J. and Cárcamo, H. (2011) Biology and management of *Lygus* in canola. *Prairie Soils and Crops Journal* 4, 42–53.

Palaniswamy, P. and Bodnaryk, R.P. (1994) A wild *Brassica* from Sicily provides trichome-based resistance against flea beetles, *Phyllotreta cruciferae* (Goeze) (Coleoptera: Chrysomelidae). *The Canadian Entomologist* 126, 1119–1130.

Palaniswamy, P., Lamb, R.J. and McVetty, P.B.E. (1992) Screening for antixenosis resistance to flea beetles, *Phyllotreta cruciferae* (Goeze) (Coleoptera: Chrysomelidae), in rapeseed and related crucifers. *The Canadian Entomologist* 124, 895–906.

Perkins, P.V. and Watson, T.F. (1972) *Nabis alternatus* as a predator of *Lygus hesperus*. *Annals of the Entomological Society of America* 65, 625–629.

Putnam, L.G. (1973) Effects of the larval parasites *Diadegma insularis* and *Microplitis plutellae* on the abundance of the diamondback moth in Saskatchewan rape and mustard crops. *Canadian Journal of Plant Science* 53(4), 911–914.

Ramachandran, S., Buntin, G.D., All, J.N., Raymer, P.L. and Stewart, C.N. Jr (1998) Movement and survival of diamondback moth (Lepidoptera: Plutellidae) larvae in mixtures of nontransgenic and transgenic canola containing a cryIA (c) gene of *Bacillus thuringiensis*. *Environmental Entomology* 27, 649–656.

Ramachandran, S., Buntin, G.D. and All, J.N. (2000) Response of canola to simulated diamondback moth (Lepidoptera: Plutellidae) defoliation at different growth stages. *Canadian Journal of Plant Science* 80, 639–646.

Reddy, G.V.P., Tangtrakulwanich, K., Wu, S., Miller, J.H., Ophus, V.L. and Prewett, J. (2014) Sustainable management tactics for control of *Phyllotreta cruciferae* (Coleoptera: Chrysomelidae) on canola in Montana. *Journal of Economic Entomology* 107(2), 661–666.

Rimmer, R.S. (2006) Resistance genes to *Leptosphaeria maculans* in *Brassica napus*. *Canadian Journal of Plant Pathology* 28, S288–S297.

Sarfraz, M., Keddie, B.A. and Dosdall, L.M. (2005) Biological control of the diamondback moth, *Plutella xylostella* (L.): a review. *Biocontrol Science and Technology* 15, 763–789.

Scott-Dupree, C.D., Conroy, L. and Harris, C.R. (2009) Impact of currently used or potentially useful insecticides for canola agroecosystems on *Bombus impatiens* (Hymenoptera: Apidae), *Megachile rotundata* (Hymentoptera: Megachilidae), and *Osmia lignaria* (Hymenoptera: Megachilidae). *Journal of Economic Entomology* 102, 177–182.

Sevacherian, V. and Stern, V.M. (1974) Host plant preferences of lygus bugs in alfalfa-interplanted cotton fields. *Environmental Entomology* 3, 761–766.

Shelton, A.M. and Nault, B.A. (2004) Dead-end trap cropping: a technique to improve management of the diamondback moth, *Plutella xylostella* (Lepidoptera: Plutellidae). *Crop Protection* 23, 497–503.

Shelton, A.M., Andaloro, J.T. and Barnards, J. (1982) Effects of cabbage looper, imported cabbageworm, and diamondback moth on fresh market and processing cabbage. *Journal of Economic Entomology* 75(4), 742–745.

Soroka, J.J. and Elliott, B. (2011) Innovative methods of managing flea beetles in canola. *Prairie Soils and Crops Journal* 4, 1–7.

Soroka, J.J. and Grenkow, L.F. (2012) When is fall feeding by flea beetles (*Phyllotreta* spp., Coleoptera: Chrysomelidae) on canola (*Brassica napus* L.) a problem? *Canadian Journal of Plant Science* 92, 97–107.

Soroka, J.J., Bartelt, R.J., Zilkowski, B.W. and Cossé, A.A. (2005) Responses of flea beetle *Phyllotreta cruciferae* to synthetic aggregation pheromone components and host plant volatiles in field trials. *Journal of Chemical Ecology* 31(8), 1829–1843.

Soroka, J.J., Holowachuk, J.M., Gruber, M.Y. and Grenkow, L.F. (2011) Feeding by flea beetles (Coleoptera: Chrysomelidae; *Phyllotreta* spp.) is decreased on canola (*Brassica napus*) seedlings with increased trichome density. *Journal of Economic Entomology* 104(1), 125–136.

Stern, V.M., Smith, R.F., van den Bosch, R. and Hagen, K.S. (1959) The integration of chemical and biological control of the spotted alfalfa aphid: the integrated control concept. *Hilgardia* 29(2), 81–101.

Stewart, C.N. Jr, Adang, M.J., All, J.N., Raymer, P.L., Ramachandran, S. and Parrott, W.A. (1996) Insect control and dosage effects in transgenic canola containing a synthetic *Bacillus thuringiensis* cryIAc gene. *Plant Physiology* 112, 115–120.

Suenaga, H. and Hamamura, T. (1998) Laboratory evaluation of carabid beetles (Coleoptera: Carabidae) as predators of diamondback moth (Lepidoptera: Plutellidae) larvae. *Environmental Entomology* 27(3), 767–772.

Tangtrakulwanich, K., Reddy, G.V.P., Wu, S., Miller, J.H., Ophus, V.L. and Prewett, J. (2014) Developing nominal threshold levels for *Phyllotreta cruciferae* (Coleoptera: Chrysomelidae) damage on canola in Montana, USA. *Crop Protection* 66, 8–13.

Trottier, M.R., Morris, O.N. and Dulmage, H.T. (1988) Susceptibility of the bertha armyworm, *Mamestra configurata* (Lepidoptera, Noctuidae), to sixty-one strains from ten varieties of *Bacillus thuringiensis*. *Journal of Invertebrate Pathology* 51, 242–249.

Turnock, W.J. (1984) Effects of the stage of development of canola (*Brassica napus*) on the capture of moths in sex attractant traps and on larval density of *Mamestra configurata* (Lepidoptera: Noctuidae). *The Canadian Entomologist* 116, 579–590.

Turnock, W.J. (1987) Predicting larval abundance of the bertha armyworm, *Mamestra configurata* Wlk., in Manitoba from catches of male moths in sex attractant traps. *The Canadian Entomologist* 119, 167–178.

Turnock, W.J. (1988) Density, parasitism, and disease incidence of larvae of the bertha armyworm, *Mamestra configurata* Walker (Lepidoptera: Noctuidae), in Manitoba, 1973–1986. *The Canadian Entomologist* 120, 401–413.

Turnock, W.J. and Bilodeau, R.J. (1984) Survival of pupae of *Mamestra configurata* (Lepidoptera: Noctuidae) and two of its parasites in untilled and tilled soil. *The Canadian Entomologist* 116, 257–267.

Turnock, W.J. and Bilodeau, R.J. (1985) A comparison of three methods of examining the density of larvae of the bertha armyworm, *Mamestra configurata*, in fields of canola (*Brassica* spp.). *The Canadian Entomologist* 117, 1065–1066.

Turnock, W.J. and Carl, K.P. (1995) Evaluation of the Palaearctic *Eurithia consobrina* (Diptera: Tachinidae) as a potential biocontrol agent for *Mamestra configurata* (Lepidoptera: Noctuidae) in Canada. *Biocontrol Science and Technology* 5, 55–68.

Ulmer, B., Gillott, C. and Erlandson, M. (2001) Feeding preferences, growth, and development of *Mamestra configurata* (Lepidoptera: Noctuidae) on Brassicaceae. *The Canadian Entomologist* 133, 509–519.

Ulmer, B., Gillott, C. and Erlandson, M. (2002) Oviposition preferences of bertha armyworm *Mamestra configurata* Walker (Lepidoptera: Noctuidae) on different crucifer cultivars and growth stages. *Environmental Entomology* 31, 1135–1141.

Ulmer, B., Gillott, C. and Erlandson, M. (2003) Conspecific eggs and bertha armyworm, *Mamestra configurata* (Lepidoptera: Noctuidae), oviposition site selection. *Environmental Entomology* 32(3), 529–534.

Williams, T., Valle, J. and Viñuela, E. (2003) Is the naturally derived insecticide spinosad compatible with insect natural enemies? *Biocontrol Science and Technology* 13, 459–475.

Wise, I.L. and Lamb, R.J. (1998a) Sampling plant bugs, *Lygus* spp. (Heteroptera: Miridae), in canola to make control decisions. *The Canadian Entomologist* 130, 837–851.

Wise, I.L. and Lamb, R.J. (1998b) Economic threshold for plant bugs, *Lygus* spp. (Heteroptera: Miridae), in canola. *The Canadian Entomologist* 130, 825–836.

Wise, I.L., Turnock, W.J. and Gavloski, J. (2009) Sequential decision plan for controlling *Mamestra configurata* in spring canola. *The Canadian Entomologist* 141, 619–626.

Wylie, H.G. (1977) Observations on *Athrycia cinerea* (Diptera: Tachinidae), a parasite of *Mamestra configurata* (Lepidoptera: Noctuidae). *The Canadian Entomologist* 109, 747–754.

Wylie, H.G. (1982) An effect of parasitism by *Microctonus vittatae* (Hymenoptera: Braconidae) on emergence of *Phyllotreta cruciferae* and *Phyllotreta striolata* (Coleoptera: Chrysomelidae) from overwintering sites. *The Canadian Entomologist* 114, 727–732.

Wylie, H.G. (1988) Release in Manitoba, Canada of *Townesilitus bicolor* [*HYM.: Braconidae*], a European parasite of *Phyllotreta* spp. [*Col.: Chrysomelidae*]. *Entomophaga* 33(1), 25–32.

Wylie, H.G. and Bucher, G.E. (1977) The bertha armyworm, *Mamestra configurata* (Lepidoptera: Noctuidae). Mortality of immature stages on the rape crop, 1972–1975. *The Canadian Entomologist* 109, 823–837.

Zhao, J.-Z., Collins, H.L., Li, Y.-X., Mau, R.F.L., Thompson, G.D., Hertlein, M., Andaloro, J.T., Boykin, R. and Shelton, A.M. (2006) Monitoring of diamondback moth (Lepidoptera: Plutellidae) resistance to spinosad, indoxacarb, and emamectin benzoate. *Journal of Economic Entomology* 99, 176–181.

12 Canola Insect Pest Management in the South-eastern USA

G. David Buntin*

Department of Entomology, University of Georgia, Griffin, Georgia, USA

12.1 Introduction

Canola is a major oilseed crop and is grown in many countries for the production of edible vegetable oil, meal for livestock and as a feedstock for biofuel. Canola seed at maturity contains 38–42% oil and meal contains 37–38% crude protein after oil extraction as compared with 44% for soybean meal (Raymer *et al.*, 1990). Canola grown in the South-eastern USA (SE USA) could be used as a source of edible oil, as feedstock for biodiesel production and as meal as feed for livestock, mainly for poultry.

Canola was grown commercially in the SE USA in the states of Georgia, Alabama, Mississippi, North Carolina, South Carolina and northern Florida from the late 1980s to about 2000. Production in this region in the mid-1990s was mostly under contract for a specialty oil-type canola with production peaking at about 10,000 ha (25,000 acres) in 1994 (G.D. Buntin, 1997, unpublished results). When this production ended, canola production declined to low levels after 1998. Production resumed in Georgia and in Alabama, South Carolina and eastern Tennessee when small crush facilities were established. Current annual production in the South-east (Alabama, Georgia, Florida, Mississippi, South Carolina and Tennessee) is about 3000–6000 ha.

Canola production systems are based on autumn planting of winter-type cultivars for the piedmont and mountain areas and spring-type cultivars for the coastal plain region of the South-east (Raymer, 2002). Optimal planting time typically is in October with harvest in May or early June (Buntin *et al.*, 2013). Currently, production in the South-east is with non-GMO (non-genetically modified organism) varieties, which provides a price premium for the oil. Canola typically is grown in a double-crop system with canola being planted following maize, sorghum, tobacco, early maturing soybeans and certain vegetables. Summer crops planted after canola harvest include soybeans, sorghum, sunflower and double-cropped maize. Within double-crop systems, canola is also a useful rotation crop for winter cereal grains, such as winter wheat.

Insects can cause considerable damage and reduce canola yield. Growers must be prepared to scout and prevent damage from insects in canola. Two perennial pests are the cabbage seedpod weevil and several species of aphids. A number of other insects are occasional pests in the South-east.

12.2 Major Insect Pests

12.2.1 Aphids

Of all the insects attacking canola, aphids have been the most damaging to the crop in the South-east (Buntin and Raymer, 1994; Buntin *et al.*, 2013). Three aphids have been found to damage canola: the turnip aphid, *Lipaphis erysimi* (Kaltenbach); cabbage aphid, *Brevicoryne brassicae* (L.); and green peach aphid, *Myzus persicae* (Sulzer). Mild weather conditions in the fall, winter and early spring can allow aphids to reach very large populations, especially in the coastal plain region of the South-east (Buntin and Raymer, 1994). Feeding injury by turnip and cabbage aphids during the seedling and rosette stages can cause curled discoloured leaves, shortened internodes and stunting of plants (Fig. 12.1). Stunted

*E-mail: gbuntin@uga.edu

Fig. 12.1. (a) A mixture of turnip and green peach aphids on a canola leaf; and (b) turnip aphid infestation during pod development.

plants often remain smaller than normal plants throughout the growing season and are more susceptible to cold-temperature mortality. Damage can occur throughout the field, but often is concentrated in injury foci of a few metres where aphids spread from an initial point and plants are dead or severely stunted (Buntin and Raymer, 1994) (Fig. 12.2). Usually *M. persicae* does not cause noticeable plant injury unless very large infestations occur. Aphids, primarily *M. persicae* and *B. brassicae*, have also been reported to transmit several viral diseases to canola, including turnip mosaic, cauliflower mosaic, beet western yellows and broccoli necrotic yellows viruses, in other areas of the world (Lamb, 1989; Blackman and Eastop, 2000). Currently, aphid-transmitted viruses are not considered to be a significant problem in canola in the South-east.

Aphids are most obvious on canola when feeding on flower racemes (Fig. 12.1). Aphids will concentrate in the flower buds and flowering section on the upper stem. Infestations of one or a few adjacent plants of several hundred aphids per plant are common during full bloom. Injury causes flowers to abort and infested areas have deformed or missing pods. Sooty mould may develop on aphid excretions. Aphid populations typically decline and disperse after bloom. A 3-year study in Georgia found that season-long insecticide control of aphids prevented yield losses of 10–35% in the coastal plains region but yield differences in northern Georgia were ≤ 2% in 3 years but 33% in one of four years (Buntin and Raymer, 1994). Infestations during the seedling and rosette stages in the winter caused the greatest yield losses in this study. Infestations in bud stage also cause yield loss, while sprays during full bloom did not affect grain yield (Buntin and Raymer, 1994).

Several natural enemies help to regulate aphid populations. The most important are a parasitic wasp, *Diaeretiella rapae* (M'Intosh) (Boyd and Lentz, 1994a), and predatory coccinellid larvae and adults, primarily *Coccinella septempunctata* L., *Harmonia axyridis* (Pallas) and *Hippodamia convergens* Guérin-Méneville (G.D. Buntin, 1996, unpublished results). Natural enemies, especially predators, are most active in the fall before freezing temperatures occur and in the spring during the bloom period. Coccinellids are usually inactive during the winter months; consequently aphid populations can increase unchecked during warm periods in the winter.

Fig. 12.2. Injury by turnip aphids spreading from an initial point of infestation during the rosette stage, causing severe stunting and plant mortality.

Management options for aphids in canola rely on sampling and control with insecticides. Useful plant resistance for aphids is not currently available in adapted commercial varieties (Buntin *et al.*, 2013). In recent years canola seed has been routinely treated with a neonicotinoid insecticide of imidacloprid, thiamethoxam or clothianidin. These insecticides are systemic in the plant and will control aphids in the seedling and early rosette stages, thereby reducing stand loss and seedling cold-temperature mortality (Rife and Buntin, 2007). Canola in the rosette through flowering stages should be scouted for aphids and treated based on nominal thresholds (Buntin *et al.*, 2013).

12.2.2 Cabbage seedpod weevil

The cabbage seedpod weevil, *Ceutorhynchus obstrictus* (Marsham), formerly *C. assimilis* (Paykull), is the most damaging insect pest of canola in piedmont and mountain regions of the South-east. The insect was not present in the coastal plain and Gulf coast regions of the SE USA until recently, when it was observed causing damage in the upper coastal plain region of Georgia (Buntin and Raymer, 1994; G.D. Buntin, 2012, personal observation). It is not known if there is a climatic barrier preventing the insect from developing and overwintering in the coastal plain region or if the insect simply has not spread into this area. The brassica pod midge, *Dasineura brassicae* Winnertz, which is associated with cabbage seedpod weevil infestation in Europe, does not occur in North America (Dosdall and Moisey, 2004).

Cabbage seedpod weevil has one generation per year, with adults residing in field edges and wooded areas near fields. Adults leave overwintering sites and move to bud-stage and early flowering canola usually in late March and the first week of April in the South-east (Boyd and Lentz, 1994b; Buntin, 1999; Sangireddy *et al.*, 2015). Adults mate and females puncture a hole in the developing canola pod and lay a single egg (Dmoch, 1965; Dosdall and Moisey, 2004). Larvae feed on developing seed inside the seedpods and typically there is one larva per pod. Uncontrolled infestations can reach 70–90% infested pods in the South-east (Buntin, 1998). When infestations exceed 50% of the pods, it is common to find two to three larvae per pod, with one, two and three larvae per pod consuming an average of 3.4, 6.7 and 8.6 seeds per pod, respectively (Buntin, 1999). Mature larvae bore a hole in the canola pod and fall to the ground to pupate. Newly emerging adults occur in May and will feed on stems and pods before dispersing. New adults can puncture and feed on seeds in green pods, which reduces seed size and quality and ability of seed to germinate properly (Buntin *et al.*, 1995).

A total of 13 species of Chalcidoidea were mass-reared from cabbage seedpod weevil larvae in Georgia in 1994–1996, though some species were most likely not primary parasitoids (Gibson *et al.*, 2006). The most prevalent parasitoid was the pteromalid *Lyrcus maculatus* (Gahan), which accounted for 86% of the parasitoid fauna. However, total parasitism rates in larvae are low (1.3–6.6% over a 4-year period) (Buntin *et al.*, 1995). Natural enemies of the adult stage have not been studied in the South-east.

Documented yield losses associated with cabbage seedpod weevil in Georgia range from 3% to 42% (Buntin, 1998, 1999). Buntin (1999) found that yield was not affected until pod infestation exceeded 26%, with yield declining by 1.7% for each 1% increase in the percentage of infested pods. A threshold of two adults per sweep is recommended (Buntin *et al.*, 2013) but usually the insect will exceed this threshold every year after a period of about 5–7 years of production in an area. Cabbage seedpod weevil is managed by use of a single application of an insecticide at about 50–75% bloom, which targets the adult stage before most egg-laying has occurred (Buntin, 1999; Buntin and Ray, 2016). Several pyrethroid insecticides are registered for cabbage seedpod weevil control in canola in the South-east. These products are very effective and inexpensive (Buntin and Ray, 2016) but the compounds are toxic to bees (Zhu *et al.*, 2015) and other pollinators. Concerns about pollinator mortality may limit the use of pyrethroid insecticides during the bloom period in the future, which would make management of cabbage seedpod weevil more challenging.

A trap crop system using early flowering and maturing varieties may be a management option but this approach is not consistently effective in the South-east (Buntin, 1998; Sangireddy *et al.*, 2015). Turnip rape, *Brassica rapa* L., has been successfully used as a trap crop for cabbage seedpod weevil in western Canada (Cárcamo *et al.*, 2007) but *B. rapa* has not been studied as a trap crop in the SE USA. A trap crop system still requires insecticidal control of adults during flowering of the trap crop (Buntin, 1998; Sangireddy *et al.*, 2015). If a pyrethroid insecticide is used, it will be toxic to pollinators but other selective insecticides such as azadirachtin or entomopathogenic fungi should be studied as potential alternatives to pyrethroid insecticides for the control of the cabbage seedpod weevil.

12.3 Minor Insect Pests

12.3.1 Cabbage root aphid

Also known as the poplar petiole gall aphid, *Pemphigus populitransversus* Riley (Hemiptera: Eriosomatidae), this species infests and feeds on roots of *Brassica* plant species such as turnips and canola during the fall, winter and spring. The aphid moves to poplar trees during late spring, where it forms galls on leaf petioles during the summer. When feeding on canola roots, the aphids produce a waxy coating that is easily seen when roots are exposed and may sometimes be seen on the soil surface above infested roots. Feeding by cabbage root aphid can reduce plant growth and vigour, though infestations are often localized in a field and injury is not apparent (Buntin *et al.*, 2013). Yield losses from cabbage root aphids have not been documented and management thresholds are not available but insecticide seed treatments may help to suppress infestations in autumn.

12.3.2 Foliage feeders

Defoliation of canola seedlings by grasshoppers (*Melanoplus* spp.) and crickets (*Gryllus* spp. and *Allenemobius* spp.) has been reported (Buntin and Raymer, 1994). A complex of defoliating lepidopterans also occurs on canola, including the diamondback moth *Plutella xylostella* (L.), cabbage looper *Trichoplusia ni* (Hübner), imported cabbageworm *Pieris rapae* (L.), *Pieris protodice* Boisduval & LeConte, true armyworm *Mythimna unipuncta* (Haworth), cabbage webworm *Hellula rogatalis* (Hulst) and *Spodoptera* spp. (G.D. Buntin, 1994, unpublished results). Of these, diamondback moth has the greatest potential to cause damage and can infest canola throughout the growing season. Extensive defoliation of the rosette stage can occur during warm periods in winter. Diamondback larvae will also chew holes in flowers and green pods. Simulated defoliation studies found that canola is most sensitive to defoliation in the seedling stage and that greater levels of defoliation can be tolerated as crop development progresses (Ramachandran *et al.*, 2000). Defoliation during pod-filling stages did not affect yield. Economic injury levels for defoliation from seedling to bloom by all foliage feeders ranged from 9% to 20%, depending on control costs, canola growth stage and price (Ramachandran *et al.*, 2000).

12.3.3 Flea beetles

Flea beetles (*Phyllotreta* spp.) are the most important pest of spring-planted canola in the northern Great Plains and Canada, where adult flea beetles feed on and destroy seedling stands (Lamb, 1989). In the SE USA, where canola is autumn-planted, flea beetle is a minor pest. Occasionally flea beetles may damage seedling stands in autumn but more often adults are present in the spring, when they can cause small, shot-hole type defoliation during bud and bloom stage. Canola is tolerant of this defoliation if it does not exceed 9–20% of the leaf area (Ramachandran *et al.*, 2000). Flea beetle larvae occur in the soil, where they feed on canola roots. Larval feeding has not been shown to damage the crop (Lamb, 1989).

12.3.4 Pollen beetle

Pollen beetles (*Meligethes* spp.) are common on flowers of canola. The pollen beetle in the SE USA is not the same species as those in Europe, where larvae infest flower buds, causing bud abortion (Lamb, 1989). In the South-east, pollen beetles feed on flower pollen but their effect on canola flower buds, flowers, pod development and productivity is not known.

12.3.5 Lygus and plant bugs

Lygus bugs (primarily *Lygus lineolaris* Palisot de Beauvois) infest canola during bloom and green pod stages. Adults and nymphs occur, with one generation usually completing development in canola before pod maturity. Feeding can deform and blast flower buds and flowers. Feeding will also shrivel seed and cause a feeding blemish on green pods. Studies of yield loss by lygus plant bugs have not been conducted in the South-east but studies in Canada and the western USA suggest an economic threshold of two or more bugs per sweep (Butts and Lamb, 1991). Because canola blooms early in the spring, lygus bug numbers are usually low (0 to 0.6 bugs per sweep) and do not exceed the treatment threshold (Buntin and Raymer, 1994; Boyd and Lentz, 1999). However, canola can serve as an early-season host for lygus bugs, thereby allowing them to build up and disperse to summer crops as canola matures (Boyd and Lentz, 1999). Canola can also serve as an early-season host for various pentatomid stink bugs, including *Murgantia histrionica* (Hahn), *Euschistus servus* (Say) and *Euschistus tristigmus* (Say), *Nezera viridula* (L.) and *Oebalus pugnax* (F.) and coreid leaffooted bugs, including *Leptoglossus phyllopus* (L.) and *Acanthocephala femorata* (F.) (G.D. Buntin, 1994, unpublished results).

12.3.6 False chinch bugs

False chinch bugs, *Nysius raphanus* Howard, typically infest canola during late bloom and pod-fill stages. Populations can become very large immediately before harvest. Nymphs and adults suck sap from the plant. Heavy or prolonged feeding injury can reduce pod set and seed fill and thereby reduce yield (Demirel and Cranshaw, 2006). However, using a nominal threshold of 25 or more bugs per sweep, infestations only occasionally reach damaging levels. Control late in the canola development often is not feasible, because infestations usually occur within the crop harvest interval for registered insecticides. In double-crop and reduced tillage systems, *N. raphanus* may remain in the same field after harvest and cause damage to emerging seedlings of the next summer crop following canola (Buntin *et al.*, 2007).

12.3.7 Thrips

Several species of thrips can occur in large numbers in canola flowers. The main species are flower thrips *Frankliniella tritici* (Fitch), western flower thrips *Frankliniella occidentalis* Pergande, tobacco thrips *Frankliniella fusca* (Hinds) and soybean thrips *Neohydatothrips variabilis* Beach. Their feeding injury can also cause blemishes on pods and cause pods to distort in a curved or circular manner. This damage is not thought to be economically important but has not been studied in the South-east.

12.4 Canola Pollinators

All canola grown commercially in the South-east is *Brassica napus*, which is self-fertile and mostly self-pollinated or wind-pollinated. Nevertheless, the crop is a rich source of pollen and nectar for bees and other insects at the time of spring when flowering canola is one of the few nectar sources available. Insects are helpful supplemental pollinators and often will shorten the time needed for pollination and seed set (Williams, 1978). Older studies indicated that bees are not needed for successful pollination and seed set of *B. napus* canola (Williams, 1978) but a recent large-scale study showed that honey bees may provide a small increase in seed set and yield of open-pollinated cultivars of winter oilseed rape (Lindstrom *et al.*, 2016). Flowering canola is highly attractive to pollinating insects; few other crops promote such intense bee visitation. In Georgia, a survey of pollinators found that bee visitors to canola were honey bees *Apis mellifera* L., the eastern carpenter bee *Xylocopa virginica* L., several bumblebee species *Bombus* spp. and other solitary bees, mostly andrenid and halictid bees, with honey bee being the prevalent species (G. D. Buntin, 1994, unpublished results). However, the number of bees in flowering canola has declined substantially in Georgia in recent years, though these reductions have not been quantified (G.D. Buntin, 2015–2016, personal observation).

12.5 Canola Cropping Sequence and Effects on Insect Pests

In the South-east, canola is grown in the winter and followed by various summer crops in a double-crop system. The crop is also rotated with winter cereal grains such as winter wheat within a double-crop system. One year of canola production can reduce infestations of the Hessian fly, *Mayetiola destructor* (Say), in winter wheat (Buntin *et al.*, 2007). In a cropping system study, the previous summer crop and crop sequence had little effect on stand and yield of canola (Buntin *et al.*, 2007) but spring-planted stands of both soybean and millet were reduced when planted following canola harvest. Seedling injury and stand losses of these crops were associated with seedling injury by the false chinch bug (Buntin *et al.*, 2007). Cotton stands could also be reduced following winter canola but stand loss was due to seedling infection by the fungal pathogen *Rhizoctonia solani* AG-4 (Buntin *et al.*, 2002).

As a winter crop maturing in spring, canola can serve as a spring bridging host for pests of summer crops. Lygus plant bugs, thrips, false chinch bugs, several species of stink bugs (Pentatomidae) and lepidopteran defoliators can all occur at non-damaging levels in canola from flowering to plant maturity. These insects may reproduce and disperse from canola to summer crops such as cotton and vegetables, where they are important pests (Boyd and Lentz, 1999; Buntin *et al.*, 2007). Conversely, canola fields can serve as a refuge and source of beneficial insects, especially those that feed on aphids, which can increase in the crop during the spring and disperse to summer crops. Canola may also serve as a non-insecticidal refuge plant host for insect pests of summer crops that are treated with insecticides and have developed insecticidal resistance, such as diamondback moth, lygus bugs and thrips.

12.6 Future Canola Production and Pest Management in the South-east

Canola has been shown to have good yield potential and economic value in the South-east (Raymer *et al.*, 1990; Raymer, 2002; Buntin *et al.*, 2013). Initial predictions of large acreages in the early 1990s have not developed (Raymer *et al.*, 1990). Canola competes in the South-eastern farmscape with numerous other field and vegetable crops. Canola as a winter crop in a double-crop system does not integrate well with the two largest crops of cotton and peanut, because optimal planting time of these crops occurs before canola harvest. Furthermore, maize is planted 2 months before canola harvest, while cotton and soybeans often mature too late for timely canola planting in the autumn. Additional impediments to wider adoption of canola include a lack of adapted varieties (especially in the coastal plain region), lack of conventional herbicides, lack of herbicide registrations in the South-eastern region for herbicide-resistant traits in canola, susceptibility of current canola varieties to carryover of persistent herbicides from summer crops (especially after cotton and peanuts) and lack of widespread infrastructure and buying points to support canola production (Grey *et al.*, 2006; Buntin *et al.*, 2013).

Currently, insect pest management is not an obstacle to increased area of canola production in the South-east. Judicious use of insecticides within an integrated management approach can prevent losses from insect pests but possible restrictions or loss of key insecticides may limit control options for insect pests of canola in the future.

References

Blackman, R.L. and Eastop, V.F. (2000) *Aphids on the World's Crops*, 2nd edn. J. Wiley & Sons, New York.

Boyd, M.L. and Lentz, G.L. (1994a) Seasonal occurrence of aphids and aphid parasitoid *Diaeratiella rapae* (M'Intosh) (Hymenoptera: Aphidiidae) on rapeseed in Tennessee. *Environmental Entomology* 23, 349–353.

Boyd, M.L. and Lentz, G.L. (1994b) Seasonal incidence of the cabbage seedpod weevil (Coleoptera: Curculionidae) on rapeseed in west Tennessee. *Environmental Entomology* 23, 900–905.

Boyd, M.L. and Lentz, G.L. (1999) Seasonal occurrence and abundance of the tarnished plant bug (Hemiptera: Miridae) and thrips (Thysanoptera: Thripidae) on rapeseed in west Tennessee. *Journal of Agricultural and Urban Entomology* 16, 171–178.

Buntin, G.D. (1998) Cabbage seedpod weevil (*Ceutorhynchus assimilis* Paykull) management by trap cropping and its effect on parasitism by *Trichomalus perfectus* (Walker) in oilseed rape. *Crop Protection* 17, 299–305.

Buntin, G.D. (1999) Damage loss assessment and control of cabbage seedpod weevil (Coleoptera: Curculionidae) in winter canola using insecticides. *Journal of Economic Entomology* 92, 220–227.

Buntin, G.D. and Raymer, P.L. (1994) Pest status of aphids and other insects in winter canola in Georgia. *Journal of Economic Entomology* 87, 1097–1104.

Buntin, G.D. and Ray, L.E. (2016) Cabbage seedpod weevil control in winter canola, 2015. *Arthropod Management Tests* 41, 1, tsw001. doi: http://dx.doi.org/10.1093/amt/tsw001.

Buntin, G.D., McCaffrey, J.P., Raymer, P.L. and Romero, J. (1995) Quality and germination of rapeseed and canola seed damaged by adult cabbage seedpod weevil,

Ceutorhynchus assimilis (Paykull) [Coleoptera: Curculionidae]. *Canadian Journal of Plant Science* 75, 539–541.

Buntin, G.D., Raymer, P.L., Bednarz, C.W., Phillips, D.V. and Baird, R.E. (2002) Winter crop, tillage, and planting date effects in double-crop cotton. *Agronomy Journal* 94, 273–280.

Buntin, G.D., Cunfer, B.M., Phillips, D.V. and Wilson, J.P. (2007) Sequence and rotation effects on pest incidence and grain yield of double-cropped soybean and pearl millet after winter wheat and canola. *Crop Management* 6(1). doi: 10.1094/CM-2007-1023-01-RS.

Buntin, D., Grey, T., Harris, G.H. Jr, Phillips, D., Buck, J., Prostko, E., Raymer, P., Smith, N., Sumner, P. and Woodruff, J. (2013) Canola production in Georgia. University of Georgia Cooperative Extension Service Bulletin 1331. University of Georgia, Athens, Georgia. Available at: http://extension.uga.edu/publications/detail.cfm?number=B1331.

Butts, R.A. and Lamb, R.J. (1991) Pest status of lygus bugs (Hemiptera: Miridae) in oilseed *Brassica* crops. *Journal of Economic Entomology* 84, 1591–1596.

Cárcamo, H.A., Dunn, R., Dosdall, L.M. and Olfert, O. (2007) Managing cabbage seedpod weevil in canola using a trap crop—A commercial field scale study in western Canada. *Crop Protection* 26, 1325–1334.

Demirel, N. and Cranshaw, W. (2006) Plant yield response to artificial infestation of the false chinch bug, *Nysius raphanus*, confined on spring canola. *Phytoparasitica* 34, 477–485.

Dmoch, J. (1965) The dynamics of a population of the cabbage seedpod weevil (*Ceutorhynchus assimilis* Payk.) and the development of winter rape. *Ekologia Polska – Seria A* 13, 249–287.

Dosdall, L.M. and Moisey, D.W.A. (2004) Developmental biology of the cabbage seedpod weevil, *Ceutorhynchus obstrictus* (Coleoptera: Curculionidae), in spring canola, *Brassica napus*, in western Canada. *Annals of the Entomological Society of America* 97, 458–465.

Gibson, G.A.P., Gates, M.W. and Buntin, G.D. (2006) Parasitoids (Hymenoptera: Chalcidoidea) of the cabbage seedpod weevil (Coleoptera: Curculionidae) in Georgia, USA. *Journal of Hymenoptera Research* 15, 187–207.

Grey, T.L., Raymer, P.L. and Bridges, D.C. (2006) Herbicide-resistant canola (*Brassica napus*) response and weed control with postemergence herbicides. *Weed Technology* 20, 551–557.

Lamb, R.J. (1989) Entomology of oilseed *Brassica* crops. *Annual Review of Entomology* 34, 211–229.

Lindstrom, S.A.M., Herbertsson, L., Rundlöf, M., Smith, H.G. and Bommarco, R. (2016) Large-scale pollination experiment demonstrates the importance of insect pollination in winter oilseed rape. *Oecologia* 180, 759–769.

Ramachandran, S., Buntin, G.D. and All, J.N. (2000) Response of canola to simulated diamondback moth (Lepidoptera: Plutellidae) defoliation at different growth stages. *Canadian Journal of Plant Science* 80, 639–646.

Raymer, P.L. (2002) Canola: an emerging oilseed crop. In: Janick, J. and Whipkey, A. (eds) *Trends in New Crops and New Uses*. ASHA Press, Alexandria, Virginia, pp. 122–126.

Raymer, P.L., Auld, D.L. and Mahler, K.A. (1990) Agronomy of canola in the United States. In: Shadahi, P. (ed.) *Canola and Rapeseed: Production, Chemistry, Nutrition and Processing Technology*. Van Nostrand Reinhold, New York, pp. 25–35.

Rife, C. and Buntin, G.D. (2007) Effect of insecticide and fungicide seed treatment on canola production and yield under varying levels of precipitation. *Proceedings of the 12th International Rapeseed Congress.* Wuhan, China, 4, 271–273.

Sangireddy, S.R., Ward, R., Cebert, E. and Ward, K.E. (2015) Temporal distribution of cabbage seedpod weevil [*Ceutorhynchus obstrictus* (Coleoptera: Curculionidae)] on different canola cultivars and potential use of early maturing canola as trap crop. *Agriculture and Biology Journal of North America* 6, 1–7.

Williams, I.H. (1978) The pollination requirements of swede rape (*Brassica napus* L.) and of turnip rape (*Brassica campestris* L.). *Journal of Agricultural Science, Cambridge* 91, 343–348.

Zhu, J.C., Adamczyk, J., Rinderer, T., Yao, J., Danka, R., Luttrell, R. and Gore, J. (2015) Spray toxicity and risk potential of 42 commonly used formulations of row crop pesticides to adult Honey Bees (Hymenoptera: Apidae). *Journal of Economic Entomology*, 108(6), 2640-2647. DOI: 10.1093/jee/tov269.

13 Integrated Management of Insect Pests of Rapeseed (Canola) in China

Zi-Hua Zhao[1]*, Leyun Wang[1] and Gadi V.P. Reddy[2]

[1]*Department of Entomology, College of Plant Protection, China Agricultural University, Beijing, China;* [2]*Western Triangle Agricultural Research Center, Montana State University, Conrad, Montana, USA*

13.1 Introduction

Rapeseed and mustard are widely planted in China (mainly in Anhui, Sichuan, Hubei and Gansu provinces) (Li *et al.*, 2014). In 2014, more than 60 million hectares of rapeseed were grown, according to the Chinese Ministry of Agriculture (Wu, 2014), and total rapeseed production in China exceeds 1 billion seeds per year (Guan, 2011). Rapeseed is grown mainly for oil, but also for food. Mustard is a native vegetable with a long history of cultivation in China (Jiang, 2007). Both rapeseed and mustard host similar arthropod communities, and the dominant pests are the same. Hu (2010a) found 15 pest species in rapeseed fields in Anhui province, while He (2008) reported 38 pest species in rapeseed fields of Gansu province. Pest composition varies among regions due to the effects of different climatic and environmental conditions (Li *et al.*, 2012; Wu, 2014).

Forty-one pest species, covering five orders and 18 families, are pests of rapeseed and mustard in China (Table 13.1). Of these, 12 are major pests. In recent decades, the major pest outbreaks have increased rapidly due to agricultural intensification and increased use of fertilizer (Li *et al.*, 2012), and economic losses caused in China by these pests have reached US$20 million annually (He *et al.*, 1998; He, 2008). To reduce such losses, the species composition, population dynamics of major pests, economic loss and community structure in rapeseed and mustard fields have received increased attention (Zhang *et al.*, 2005; Hu *et al.*, 2010b; Hou *et al.*, 2013). In response, integrated pest management methods have been developed to better manage these pests (Zong *et al.*, 1986; Wang *et al.*, 2001).

13.2 Main Pest Species of Canola (Rapeseed) and Mustard in China

The canola and mustard insect communities in China contain many pest species and natural enemies, among a complicated food web structure that varies among geographical locations and climatic conditions (Hu *et al.*, 2010a). Among the pests, aphids (*Lipaphis erysimi* [Kaltenbach] and *Myzus persicae* [Sulzer]), leaf miners (*Phytomyza horticola* Gourean), diamondback moths (*Plutella xylostella*) and cabbage caterpillars (*Pieris rapae* [L.]) are the most economically important pests. In rapeseed, aphid damage is concentrated in two periods, one in winter and one in spring, the latter causing the most damage. When rapeseed plants are in the seedling stage, radish aphids (*L. erysimi*) and diamondback moths are the dominant pests while green peach aphids (*M. persicae*) and leaf miners are the main pest species during pod formation as crops mature. Most reports in China on this group of pests are taxa descriptions, or descriptions of species' biological traits, pest population dynamics, or insect communities in the crops. The main pest species of these crops are described in the following sections.

*Corresponding author. E-mail: zhzhao@cau.edu.cn

Table 13.1. Pests of rapeseed and mustard and their damage levels in China.

Order	Family	Species	Damage level[a]	References
Coleoptera	Melolonthidae	*Aphodius biformis* Reitter	*	He, 2008; Wu, 2014
		Hoplia aureola (Pallas)	*	He, 2008; Wu, 2014
		Hoplia cincticollis (Faldermann)	*	Zhang *et al*., 2005; He, 2008
		Maladera ovatula (Fairmaire)	*	He, 2008
		Maladera verticalis Fairmaire	*	He, 2008
	Rutelidae	*Popillia quadriguttata* (Fabricius)	*	He, 2008
		Phyllopertha horticola (L.)	*	Li *et al*., 2012
		Phyllopertha diversa Waterhouse	*	Liu *et al*., 2013
	Halticidae	*Phyllotreta striolata* (Fabricius)	**	Fu *et al*., 2005; He *et al*., 2012
		Psylliodes punctifrons Baly	*	He, 2008
	Chrysomelidae	*Collaphellus bowringi* Baly	*	Wu, 2014
		Chrysomela vigintipunctata (Scopoli)	*	He, 2008; Li *et al*., 2012
		Chrysomela populi L.	*	He, 2008
		Entomoscelis suturalis Weise	*	He, 2008
	Cassididae	*Cassida nebulosa* Linnaeus	*	He, 2008
		Strongyllodes variegates Fairmaire	*	He *et al*., 1998
		Meligethes aeneus Fabricius	**	Lei, 2009
	Elateridae	*Agriotes fuscicollis* Miwa	*	He, 2008
		Pleonomus canaliculatus Fald	*	He, 2008
		Melanotus caudex Lewis	*	Wu, 2014
	Meloidae	*Epicauta gorhami* Marseul	*	He, 2008
	Curculionidae	*Epicauta chinensis* Laporte	*	Wu, 2014
		Ceutorhynchus quadridens (Panzer)	*	He, 2008
		Bothynoderus punctiventris Germar	**	He, 2008
Lepidoptera	Pieridae	*Pieris rapae* Linnaeus	***	Huang *et al*., 2015
		Pontia daplidice Linnaeus	*	Wu, 2014
	Noctuidae	*Argyrogramma agnata* Staudinger	*	He, 2008
		Mythimna separata Walker	*	He, 2008
		Mamestra brassicae L.	**	Wu *et al*., 2013
	Plutellidae	*Plutella xylostella* (L.)	****	Hu and Liu, 2003
	Pyralidae	*Evergestis extimalis* Scopoli	*	Wu, 2014
		Hellula undalis Fabricius	**	Huang *et al*., 2014
Hemiptera	Miridae	*Adelphocoris taeniphorus* Reuter	*	He, 2008
	Pentatomidae	*Eurydema gebleri* Kolenati	*	Wu, 2014
	Aphididae	*Lipaphis erysimi* (Kaltenbach)	***	He, 2008
		Myzus persicae (Sulzer)	****	Zhao *et al*., 1995
		Brevicoryne brassicae (Linn.)	*	He, 2008; Li *et al*., 2011
		Lipaphis erysimi (Kaltenbach)	*	He, 2008
	Cicadellidae	*Tettigella viridis* (Linnaeus)	*	He, 2008
Diptera	Agromyzidae	*Phytomyza horticola* Gourean	***	Zhu *et al*., 2006
Hymenoptera	Tenthredinidae	*Athalia rosae japanensis* Rhower	**	Young, 1966

[a]Damage levels: *rare pests; **regular pests; ***serious pests; ****disastrous pests

13.2.1 *Phyllotreta striolata* (Fabricius) (Coleoptera: Chrysomelidae)

Host plants

Cabbage, broccoli, Chinese cabbage, radish, turnip, rapeseed and mustard.

Biology and ecology

Phyllotreta striolata has three to five generations per year in northern China (Jiang, 2007) and seven to eight generations in southern China (Fu *et al*., 2005). Adults of *P. striolata* overwinter in field ditches and under fallen leaves, weeds and soil in North China (Jiang, 2007). Overwintering adults generally begin feeding in late March if the temperature rises to 10°C in North China (Jiang, 2007), with activity strongly correlated with increasing temperature (Fu *et al.*, 2005). Adults begin to breed in early April in Gansu province, and their long life span (adult: 4–5 months) leads to a high degree of overlap among generations, each of which requires about 1 month (He *et al.*, 2012). Beetles of the first and second generations in spring cause more serious damage to most rapeseed and mustard crops in spring than in summer, because the higher temperatures of summer are unfavourable to *P. striolata* development (He *et al.*, 2012).

High humidity facilitates outbreaks of *P. striolata*, especially during egg-laying periods (He *et al.*, 2012). Adults prefer to lay eggs in humid soil, and the rate of egg hatching is very low when humidity is below 90% (He *et al.*, 2012). Temperatures between 21 and 30°C are suitable for *P. striolata* development, with adult activity decreasing significantly out of this temperature range (He *et al.*, 2012). Feeding capacity and reproduction rate also decrease in summer (He *et al.*, 2012). *P. striolata* beetles are oligophagous pests, preferring to feed on brassicas. Large brassica farms thus facilitate outbreaks of *P. striolata.*

13.2.2 *Psylliodes punctifrons* Baly (Coleoptera: Halticidae)

Host plants

Rapeseed and other brassicas.

Biology and ecology

Psylliodes punctifrons has only a single generation per year in Northwest China, where adults overwinter in the soil and under fallen leaves near or in canola fields (Hou and Che, 1998). Adults lay eggs in the soil surrounding rapeseed roots. Hatched larvae begin feeding on rapeseed roots in mid-March (Hou and Che, 1998). In summer, larvae and adults mainly feed on rapeseed leaves. When the plants flower and the leaves die down, larvae transfer their feeding to the roots and stalks. The larval stage lasts about 1 month and the pupal stage lasts about 18 days (Hou and Che, 1998). New adults (of the first new generation) emerge in late May and move to surrounding weeds or soil for aestivation (Hou and Che, 1998). In autumn, when temperatures decrease, adults return to rapeseed. The continuous cropping of rapeseed facilitates outbreaks of *P. punctifrons.*

Adults feed on rapeseed leaves and pods, and deposit faeces at the damage site. The tendency of adult *P. punctifrons* to move upwards, preferring to feed on the tops of rapeseed plants, causes uneven ripening of rapeseed crops (Hou and Che, 1998). In addition, adults are attracted to the colour green and aggregate on immature green rapeseed leaves. If disturbed, they cease movement and fall to the ground. However, they often stay on the same plant and do not transfer to other plants. Larvae mainly damage rapeseed roots and stalks, but they also feed on the leaves and cause plant mortality.

13.2.3 *Meligethes aeneus* Fabricius (Coleoptera: Cassididae)

Host plants

Cabbage, Chinese cabbage, mustard and broccoli in Cruciferae, and other plants including carrot, sunflower and fruit trees

Biology and ecology

Both adults and larvae of *M. aeneus* feed on nectar, stamens, flower stalks and sepals, causing plant mortality (He *et al.*, 1998). Typically, adults cause more serious damage than larvae, due to larger food consumption. *M. aeneus* has only one generation each year in Northwest China. Adults overwinter in the soil or under fallen leaves surrounding the rapeseed fields (He *et al.*, 1998), where adults lay eggs in rapeseed flower buds. The overwintering *M. aeneus* begin feeding on rapeseed in the middle of May. The adults also like eating some weeds (e.g. *Iris ensata* Thunb). In early June, the adults always immigrate into rapeseed from weeds and feed on rapeseed flower buds. Then the adults begin to mate and lay eggs (Lei, 2009). The overwintering adults die at the end of July. The larvae feed on rapeseed from mid-June to mid-July and enter pupation (Lei, 2009). The new adults emerge in early August (Lei, 2009).

13.2.4 *Pieris rapae* L. (Lepidoptera: Pieridae)

Host plants

Vegetables in Cruciferae, Asteraceae, Convolvulaceae and six other families. Cabbage, mustard and broccoli are the most important hosts in China.

Biology and ecology

Pieris rapae undergoes four to eight generations a year, depending on the region in China (Huang *et al.*, 2015). Adults of the first generation emerge at the end of May, feeding on nectar and pollen (Huang *et al.*, 2015). Damage to crops is most serious from August to October. The mature larvae begin to pupate for overwintering at the end of October. Pupae overwinter in woody plants in hedges, crevices, soil, or on the weeds surrounding crops suitable for the species (Huang *et al.*, 2015).

Adults fly during the daytime, especially at noon in warm and sunny weather. Adults prefer to lay eggs on Cruciferae. Eggs are laid singly on host leaves. One female can lay more than 100 eggs (Huang *et al.*, 2015). Newly hatched larvae eat leaves after feeding on the egg shell. The first- and second-instar larvae spin silk thread to attach to foliage. The most suitable temperature and humidity for development are 20–25°C and 76% (He, 2008), respectively. Under the most suitable conditions, the egg, larval, pupal and adult stages last 4–8 days, 11–22 days, 10 days and 5 days, respectively (Huang *et al.*, 2015).

Natural enemies exert a strong control on *P. rapae*. Common natural enemies include the parasitic wasps *Trichogramma evanescens*, *Apanteles rubecula* and *Pteromalus puparum*, which attack eggs, larvae and pupae, respectively (Wu, 2014). The species is also susceptible to artificial applications of the bacterium *Bacillus thuringiensis* (Bt) (Wu, 2014).

13.2.5 *Mamestra brassicae* L. (Lepidoptera: Noctuidae)

Host plants

Cabbage, radish, spinach, carrot, rapeseed and mustard.

Biology and ecology

In northern China, *M. brassicae* undergoes three to four generations each year and overwinters as pupae in the soil (Zhang *et al.*, 2007). Adults emerge from overwintered pupae when temperatures reach 15–16°C (Zhang *et al.*, 2007). In Shandong province, *M. brassicae* has two population peaks, one from mid-June to early July and the second from mid-September to early October (Wu *et al.*, 2013). In the first peak, larvae mainly damage cabbage, spinach and rapeseed. In the second peak, larvae mainly damage radish and cabbage (Zhang *et al.*, 2007). *M. brassicae* has a strong taxis to sweet-sour tastes and is not attracted by light (Wu *et al.*, 2013). Therefore, sweet-sour liquid can be used to trap and monitor *M. brassicae* adults. Adults feed on honeydew and nectar for extra nutrition during the egg-laying period. On average, the number of eggs in an egg mass is 100–200 and a female lays 1000–2000 eggs over her lifetime (Wu *et al.*, 2013). Most eggs are laid on the underside of leaves. The egg stage lasts 4–6 days and the suitable temperature range for egg development is 23–26°C (Zhang *et al.*, 2007)

Larvae undergo six instars. Newly hatched larvae eat the chorions of their eggs and when young they feed in aggregations on the undersides of leaves. Second- and third-instar larvae scatter to feed separately on different plants. Fourth-instar larvae feed at night. At 25°C, larval and pupal development lasts 30–35 days and 10 days, respectively. Aestivation and overwintering of pupae last 50–60 days and 6 months, respectively. The developmental threshold temperature is 15°C (Wu *et al.*, 2013).

M. brassicae populations show an irregular pattern of outbreaks at spatial–temporal scales (He, 2008). Suitable temperature facilitates outbreaks of *M. brassicae* by shortening the emergence period. A temperature range of 18–25°C and a relative humidity range of 70–80% are optimum for *M. brassicae* development (Zhang *et al.*, 2007). High temperature and humidity are unfavourable to *M. brassicae*. In addition, adults need to feed on nectar for extra nutrition (Wu *et al.*, 2013). Abundant available nectar coinciding with the adult stage could therefore trigger outbreaks of *M. brassicae*.

13.2.6 *Plutella xylostella* (L.) (Lepidoptera: Plutellidae)

Host plant

Cabbage, broccoli, leaf mustard, rapeseed, turnip and many other vegetables.

Biology and ecology

Plutella xylostella is an important migratory pest around the world (Hu and Liu, 2003). Larvae mainly feed on leaf parenchyma tissue, forming transparent windows in leaves. Third and fourth instars can feed on the whole leaf tissue, causing simple defoliation, and larvae prefer to feed on newly seeded crops (Hu and Liu, 2003). *P. xylostella* is broadly distributed throughout China, undergoing 4 to 19 generations, depending on region. Pupae overwinter under fallen leaves and weeds in northern China, while all stages (eggs, larvae, pupae and adults) can survive throughout the year in South China. Adults fly at night, displaying peak activity from 19:00–23:00 h. Females mate shortly after eclosion and lay eggs soon thereafter. Even so, females have a long life span (20–30 days) (Hu and Liu, 2003), and the egg-laying period of overwintered adults is longer than the larval developmental period, leading to overlapping generations in most field populations (Hu and Liu, 2003). Females can lay 200–600 eggs per individual and the eggs are dispersed singly over the leaf surface (He, 2008). The optimum temperature for development of *P. xylostella* ranges from 20°C to 30°C. Under suitable conditions, the durations of the egg, larval and pupal stages are in the ranges of 2–11 days, 12–27 days and 8–14 days, respectively (Hu and Liu, 2003).

P. xylostella has many natural enemies, including various parasitic wasps (e.g. *Diadromus collaris* (Gravenhorst) (Hymenoptera: Ichneumonidae), *Cotesia plutellae* (Kurdjumov) (Hymenoptera: Braconidae), *Oomyzus sokolowskii* (Kurdjumov) (Hymenoptera: Eulophidae)), generalist predators (e.g. *Erigonidium graminicolum* (Sundevall) (Araneida: Araneidae), *Oedothorax insecticeps* Boes. et Str. (Lycosidae: Arachnida), *Coleosoma octomaculatum* (Boes. et Str.) (Araneida: Theridiidae)) and specialist predators (e.g. *Chilomenes quadriplagiata* (Swartz) (Coleoptera: Coccinellidae)) (He, 2008).

13.2.7 *Hellula undalis* Fabricius (Lepidoptera: Pyralidae)

Host plant

Cabbage, radish, rapeseed and mustard of Cruciferae.

Biology and ecology

Hellula undalis has three to nine generations per year in China (Huang *et al.*, 2014). Mature larvae can spin and form a spinneret in the soil. The spinneret is the structure the larva uses to spin a cocoon, within which it pupates. In regions where *H. undalis* overwinters, it does so as pupae in diapause (Huang *et al.*, 2014). In South China, *H. undalis* feeds actively on host plants throughout the winter and therefore has no diapause. The population density of *H. undalis* peaks at the end of August or in early September in South China, mainly leading to serious damage in cauliflower. From September to November, it mainly damages late-sown radish. High temperatures and dry conditions facilitate outbreaks of *H. undalis* populations (Huang *et al.*, 2014). In addition, *H. undalis* prefers to feed on crops at the two- to four-leaf stage (Hu *et al.*, 2010b). Adults fly at night and are attracted to lights. Eggs are scattered on leaves (mainly in the centre) and stalks. Newly hatched larvae burrow into the leaf, forming leaf mines. Third-instar larvae spin silk and net leaves together. Fourth- and fifth-instar larvae could burrow into stalks of host plant. Larvae balloon on silk threads to reach other plants for feeding. Mature larvae pupate in the soil surrounding vegetable fields (Huang *et al.*, 2014).

H. undalis undergoes one to three generations in North China. Pupae overwinter in weeds, fallen leaves and soil surrounding the field, and adults emerge from July to September the following year. The first generation damages plants from July to mid-September, the second generation from late August to late September and the third generation from late September to early October (Jiang, 2007). Thus, the three generations are highly overlapping in the field. Adults show limited dispersal ability and lay their eggs on leaf veins, laying an average of 70–100 eggs per female lifetime (Jiang, 2007). Eggs hatch in 3–10 days and larvae feed on plants both day and night. Young larvae feed on leaf mesophyll tissue. Third-instar larvae chew leaves. Larval development lasts 11–26 days (Huang *et al.*, 2014). Mature larvae turn light green and pupate in the soil, or occasionally in fallen leaves or at the base of a petiole in rapeseed field. Pupae enter diapause and overwinter from the end of September or early October in North China (Jiang, 2007).

13.2.8 *Myzus persicae* (Sulzer) (Hemiptera: Aphididae)

Host plants

Cabbage, turnip, rapeseed, broccoli, mustard and many other crops, including vegetables and fruit trees.

Biology and ecology

Myzus persicae is a polyphagous species, with 285 host species (from 74 families) (Zhao *et al.*, 1995). While *M. persicae* needs to transfer between different host plants, its summer herbaceous hosts include pear, peach, plum and cherry. Woody overwintering hosts include cabbage, turnip, pepper, potato, rapeseed and mustard. In addition, *M. persicae* spreads many plant viruses, including *Cucumber mosaic virus* (CMV), *Potato virus Y* (PVY) and *Tobacco etch virus* (TEV) (He, 2008).

M. persicae has a rapid development rate, with ten generations in North China (Zhao *et al.*, 1995) and 20–30 generations in Yangtze River region (Jiang, 2007). In spring, *M. persicae* begins development at 6°C in spring. After two to three generations on the overwintering woody hosts, *M. persicae* migrates to surrounding vegetable fields. A wingless female can produce 60–70 nymphs within 20 days (Zhao *et al.*, 1995). The suitable temperature is 20–25°C for development of *M. persicae*. The nutrition of host plants has a great effect on *M. persicae* populations (Zhao *et al.*, 1995). In late autumn, alate *M. persicae* returns to overwintering hosts (Jiang, 2007).

M. persicae has many natural enemies, including ladybird beetles (*Propylaea japonica* (Thunberg), *Harmonia axyridis* (Pallas) and *Coccinella septempunctata* L. (Coleoptera: Coccinellidae)), hoverflies (*Episyrphus balteatus* De Geer (Diptera: Syrphidae)), lacewings (*Chrysopa sinica* Tjeder (Neuroptera: Chrysopidae)), aphid parasites (*Aphidius avenae* Haliday (Hymenoptera: Aphidiidae)) and fungi in the order Entomophthorales (Jiang, 2007).

13.2.9 *Brevicoryne brassicae* (L.) (Hemiptera: Apididae)

Host plants

Cabbage, radish, rapeseed and Chinese cabbage.

Biology and ecology

Brevicoryne brassicae undergoes eight to ten generations each year, which overlap (He, 2008). In temperate regions the main reproduction mode of *B. brassicae*, apart from the generations on the overwintering hosts, is parthenogenic. Overwintered eggs begin to hatch in April. By the end of May, *B. brassicae* migrates to other host plants (He, 2008). The developmental threshold temperature is 4.5°C, while the effective accumulated temperature for complete development of a wingless aphid is 134.5 degree days. Each wingless female aphid can lay 40–60 eggs in conditions of 15–20°C (He, 2008). The natural enemies of *B. brassicae* are similar to those of *M. persicae* (He, 2008).

13.2.10 *Phytomyza horticola* Gourean (Diptera: Agromyzidae)

Host plant

More than 130 vegetables and crops, including pea, fava bean, celery, cabbage, rapeseed and mustard.

Biology and ecology

Larvae of this leaf miner feed on leaf mesophyll, forming white-coloured mines on leaves. The whole plant turns pale and dies if a large number of larvae are present (ten individuals per plant) (He, 2008). *P. horticola* has 4 to 18 generations in China (Yuan *et al.*, 2013), depending on the region. In northern China, *P. horticola* overwinters as pupae, which hide in fallen leaves (He, 2008). In southern China, all stages, including eggs, larvae and adults, are present in the field year round (Yuan *et al.*, 2013). In North China, *P. horticola* begins to emerge in early spring and peaks at the end of spring (He, 2008). Temperatures above 35°C cause increased mortality. Therefore, *P. horticola* aestivates during summer as pupae in North China (He, 2008). Adults like to move during the day, feed on nectar and show a strong taxis to nectar. Females prefer to lay eggs on tall, dense plants. Eggs are always scattered on leaf margins, especially on the leaf apex (Yuan *et al.*, 2013). Eggs are laid (50–100 eggs/female) on the margins of tender leaves, using the ovipositor to cut into leaf tissue. Newly hatched larvae drill into the leaf immediately and form serpentine mines. In North China, egg development in spring and summer requires 10 days and 4–5 days, respectively. There are three nymphal instars, which collectively last 5–15 days (Yuan *et al.*, 2013). Mature larvae (third instars) pupate at the end of the mine. The pupal stage lasts 10–20 days. The optimal temperature range for development is 16–18°C (Li *et al.*, 2003).

Natural enemies have a good control effect on *P. horticola* populations. For example, in Fujian province in southern China there are four species of parasitic wasps (*Chrysocharis pubicomis* (Zetterstedt)

(Hymenoptera: Pteromalidae), *Diglyphus isaea* (Walker) (Hymenoptera: Eulophidae), *Chrysocharis pentheus* (Walker) (Hymenoptera: Eulophidae) and *Opius pallipes* (Wesmael) (Hymenoptera: Braconidae)) that all use *P. horticola*, and parasitism can reach 80% in late April (Hou and Che, 1998; Zhu *et al.*, 2006; He, 2008).

13.2.11 *Athalia rosae japanensis* (Rhower) (Hymenoptera: Tenthredinidae)

Host plants

Cabbage, radish, broccoli, rapeseed, turnip, broccoli and mustard.

Biology and ecology

Athalia rosae japanensis (turnip sawfly) has five generations in South China, where mature larvae pupate in the soil during the overwintering period (Young, 1966). The lifespan of male and female *A. rosae japanensis* are 2–9 days and 5–12 days, respectively. The egg period is 4–41 days and the hatchability is about 93.3–100%. The larvae have five instars (total 10–36 days). Adults emerge in the middle of May. Adults fly during the day and lay eggs on the underside of leaves (Young, 1966). Newly hatched larvae immediately begin feeding on cole crop foliage. Larvae feed on foliage, flowers and stalks of rapeseed, turnips, mustard and other similar brassicas.

There have been no reports about the natural enemies of *A. rosae japanensis* in China (Wu, 2014).

13.3 Integrated Pest Management in Rapeseed in China

There are many pest species in rapeseed fields in China. For different species, specific management practices have been developed. Chemical control is the most common method used in rapeseed fields. Alternatives like physical control, biological control and plant breeding have not been applied widely in China.

13.3.1 Chemical control

Chemical pesticides are often the most effective strategy in pest control (Liu *et al.*, 2013; Wu *et al.*, 2013). In recent years, the toxicity of many pesticides used on rapeseed pests has been determined. Some pesticides are effective on the major pests of rapeseed (Liu *et al.*, 2013). Imidacloprid, nicotine, chlorpyrifos, nitenpyram and pymetrozine are all widely used against aphids, flea beetles and leaf beetles (Wu *et al.*, 2013). Avermectin is used to control leaf miners, in accordance with regulatory standards governing pesticide use in China (Wu *et al.*, 2013). However, in recent years some pest species have developed resistance to commonly used pesticides, indicating that pesticides should be alternated with other types of control (Hou and Che, 1998; He, 2008).

13.3.2 Physical control

Physical control is also a common strategy in pest management. For example, yellow cards are often used to trap insect pests (aphids and leaf miners) (Li and Feng, 2007). In China, sweet-and-sour liquid is used to attract and kill target pests (Yang *et al.*, 2011). Black light lamps are used to catch some coleopteran pests (e.g. *P. striolata*) and lepidopteran pests (e.g. *P. xylostella*) (Li and Feng, 2007). Mesh enclosures and other barriers are used to exclude insect pests physically. For example, *Liriomyza sativae* Blanchard could be excluded outside the greenhouse by using mesh enclosures (Li and Feng, 2007).

13.3.3 Cultural and biological control

Cultural and biological controls aim to decrease pesticide use and improve ecosystem balance in crop fields, and several are being developed in China for use in canola crops.

Barrier plantings

The castor-oil plant (*Ricinus communis* L.) is a common crop in northern China that can be used to reduce crop colonization by agricultural pests, acting as a repellent barrier (Kong, 2013). The roots, stems and leaves of castor-oil plant contain trace amounts of the poison toxalbumin, which is harmless to mammals and is environmentally safe, but repellent to insect pests (Kong, 2013).

Intercropping and crop rotation

Intercropping is a commonly used planting technology in China to enhance yields and improve biological control of pests by natural enemies. For example, aphids are the most important pests in rapeseed and always reach peak populations in late April to early May. However, natural enemies such as the ladybird beetle *Coccinella septempunctata* only reach peak population by mid-May. Therefore, to mediate the mismatch of aphids and their natural enemies, potato can be planted surrounding rapeseed fields. The strips of potato plants act as reservoirs for natural enemies, supporting ladybird beetles early in the season for later movement into rapeseed when aphids increase (Wang *et al.*, 2008). Rapeseed can also be intercropped with other crops (e.g. wheat, vegetable, soybean) (Wang *et al.*, 2008). Such intercropping can use differences in the phenology of the two crops to sustain natural enemies, facilitating the synchronization of natural enemies (Shen *et al.*, 2007).

Crop rotation is the most important cultural practice used for pest control in China (Bu *et al.*, 2014). For example, a rapeseed–maize rotation can suppress the most common insect pests in northwestern China (Tang and Hong, 2015). A rapeseed–rice rotation can be used to control pests in southern China by disturbing pest population cycles (Bu *et al.*, 2014). In addition, rotation could also enhance fertilizer use efficiency and improve soil health (Tang and Hong, 2015).

Biological pesticides

Application of botanical or microbial pesticides can help to increase the environmental safety of control. The use of both botanical pesticides (azadirachtin, matrine and pyrethrum) and microbial pesticides (validamycin and streptomycin) has been encouraged by the Chinese government (Kong, 2013). Some entomopathogenic fungi, such as *Metarhizium anisopliae* and *Beauveria bassiana*, are effective for control of aphids, leaf miners and diamondback moths (Zhu *et al.*, 2011). When conditions are favourable, the spores of *M. anisopliae* and *B. bassiana* can penetrate insect cuticle and cause mortality. The host ranges of *M. anisopliae* and *B. bassiana* are very broad, including pests of Coleoptera, Lepidoptera and Hemiptera (Lin *et al.*, 2006; Zhang *et al.*, 2006).

Conservation tillage

No-till strategies are often used in sustainable agriculture and are valued for creating a stable environment that conserves natural enemies, especially ground-dwelling predators. No-till production of rapeseed can also help to conserve soil, water and nutrition (Su *et al.*, 2011). However, no-till technology has not yet been applied on large farms in China (Tang *et al.*, 2008). In recent years, placing mulch straw on fields after harvest has been used to increase the abundance of natural enemies and disturb the population cycles of insect pests (Tang and Hong, 2015). Both no-till and straw mulch strategies reduce the harmful effects of agricultural practices on agroecosystems (Zhu *et al.*, 2005). In rapeseed, the application of straw mulch can increase the harbour abundance of natural enemies by providing refuges or alternative prey or hosts (Tang *et al.*, 2008).

13.3.4 Plant breeding and transgenic technology

The screening of insect-resistant varieties is an important field in the breeding of mustard and rapeseed (Liu *et al.*, 2008). However, conventional plant breeding is a time- and labour-intensive task and one that also reveals resistant mechanisms (Chen *et al.*, 2006). In recent years, the Bt gene has been successfully inserted into *Brassica napus* and Bt varieties of the crop provide good control of *Laphygma exigua* (Hübner) and *P. rapae* L. (Li *et al.*, 1999). Many other resistance genes (herbicide tolerance, pathogen resistance, yield increase, quality improvement and drought resistance) have been introduced into crop genomes to enhance pest resistance (Guan, 2002; Du *et al.*, 2008). Plant-breeding technology based on genetic functions is an important field for the future of pest management. Transgenic rapeseed with various profitable traits has recently entered commercial production in China (Guan, 2011). Finally, technology based on mass release of sterile males is a promising strategy for some pest species (Ji *et al.*, 2007).

References

Bu, R., Ren, T., Lu, J., Li, X., Cong, R., Li, Y., Wang, Y. and Lu, J. (2014) Analysis of P fertilizer efficiency under rice–rapeseed rotation system. *Scientia Agricultura Sinica* 47, 1227–1234.

Chen, G.R., Wang, G.H. and Tang, X.Y. (2006) The review of transgenic breeding technology of rapeseed. *Crop Research* 5, 442–447.

Du, J.Z., Sun, Y., Cao, Q.F., Hao, Y.S., Liu, L.L. and He, J. (2008) Performance of agronomic characters of transgenic oilseed and their insect resistant activity. *Biotechnology Bulletin* 3, 147–152.

Fu, J., Cheng, Q., Lin, Z. and You, M. (2005) Ecological disturbance of field *Phyllotreta striolata* population occurrence. *Chinese Journal of Ecology* 24, 917–920.

Guan, C.Y. (2002) The review of transgenic rapeseed. *Engineering Science* 4, 8–10.

Guan, C.Y. (2011) The developed direction of rapeseed in China. *Grain Science and Technology and Economy* 36, 5–6.

He, Y. (2008) Structure and dynamics of insect communities in spring-sowing rape field and resistance of its varieties to rape aphids. Thesis for Master's Degree, Gansu Agricultural University, China.

He, C.G., Wang, G.L., Fan, Y.H., Zou, Y.X. and Deng, H.Y. (1998) Study on a new pest *Strongyllodes variegatus* (Fairmaire, 1891) attacking rape. *Acta Agriculture Boreali-occidentalis Sinica* 7, 18–23.

He, H., Bin, S. and Lin, J. (2012) Review of biological and ecological characteristics of *Phyllotreta striolata* (Fabricius). *Journal of Anhui Agricultural Science* 40, 10683–10686.

Hou, Y.M. and Che, J.Y. (1998) The insect community structure and pest control strategy in rape fields. *Acta Agriculture Boreali-occidentalis Sinica* 7, 51–54.

Hou, S.M., Hu, B.C., Hu, B.J., Li, Q.S., Fei, W.X., Jiang, Y.F., Fan, Z.X. and Rong, S.B. (2013) New pest *Strongyllodes variegatus* on winter oilseed rape in Anhui Province. *Chinese Journal of Oil Crop Sciences* 35, 692–696.

Hu, Q. and Liu, Y. (2003) Progress and summarization of study on diamondback moth (*Plutella xylostella* L.). *Journal of Tianjin Agricultural College* 10, 33–37.

Hu, L.C., Cui, W., Wang, X. and Lou, Y.G. (2010a) Herbivore resistance induced by *Spodoptera litura* (Fabricius) (Lepidoptera: Noctuidae) and its relation to the JA signaling pathway in Chinese cabbage (*Brassica campestris* L.). *Acta Entomologica Sinica* 53, 1001–1008.

Hu, B.J., Zhang, H.S., Li, C.C., Zhang, D.F., Hou, S.M., Shi, L., Gu, J.T. and Zhou, Z.Y. (2010b) Surveys of rape insect pest and screening of aphids controlling agents. *Chinese Bulletin of Entomology* 47, 779–782.

Huang, J., Yu, Y., Yuan, Y., Yang, Y., Huang, S., Li, Y., Li, Y., Li, Hui. (2014) The identification and population dynamics of *Hellula undalis* Fabricius and control technology. *China Vegetables* 8, 65–67.

Huang, R., Zhou, J., Zhang, S., Zeng, X., Hu, L. and Yao, Y. (2015) Research progress in biological control of *Pieris rapae*. *Acta Agriculturae Jiangxi* 27, 46–49.

Ji, Q., Hou, W. and Chen, J. (2007) Development of a genetic sexing strain and the sterile male technique of the Oriental fruit fly, *Bactrocera dorsalis* (Hendel.). *Acta Entomologica Sinica* 50, 1002–1008.

Jiang, W.J. (2007) Studies on structure of arthropod communities in rape-field and the function of natural enemies controlling effects. Thesis for Master's Degree, Anhui Agricultural University, China.

Kong, Z.Q. (2013) The biological control of rapeseed. *China Agriculture Information* 8, 102.

Lei, L. (2009) Happened Law and Control on *Meligethes aeneus* in Datong. *Qinghai Agricultural Technology* 2, 29–30.

Li, J. and Feng, L.F. (2007) Effective physical control way of the pests and its study evolvement. *Journal of Shanxi Agricultural Sciences* 35, 67–70.

Li, X.B., Zheng, S.X., Dong, W.B., Chen, G.R., Mao, H.Z. and Bai, Y.Y. (1999) Insect-resistant transgenic plants of *Brassica napus* and analysis of resistance in the plants. *Acta Genetica Sinica* 26, 262–268.

Li, J., Li, X., Zhang, Y. and Cheng, W. (2003) Progress in study on vegetable and flower leafminers *Liriomyza* spp. (Diptera: Agromyzidae). *Journal of Northwest Science and Technology University of Agriculture and Forest* 31, 162–166.

Li, C., Wu, W.Q., Zhu, L., Zhang, Q.W. and Liu, X.X. (2011) Effects of wheat planted adjacent to rape on the major pests and their natural enemies in wheat field. *Chinese Journal of Applied Ecology* 22, 3371–3376.

Li, J., Chang, J. and Li, H.P. (2012) Species and prevention methods of main pests on North spring-rape. *Inner Mongolia Agricultural Science and Technology* 6, 77–79.

Li, D.D., Liu, J., Zhou, Q.B. and Huang, Q. (2014) Influence of scale variation on precision of rapeseed acreage remote sensing monitoring. *Chinese Journal of Agricultural Resources and Regional Planning* 35, 85–92.

Lin, H., Wang, P., Zhang, L., Yang, X., Li, S. and Zhou, Y. (2006) Growth of *Beauveria brongniartii* and two varieties of *Metarhizium anisopliae*. *Chinese Journal of Biological Control* 22, 123–127.

Liu, D., Liu, G. and Wang, H. (2008) Isolation of resistant genes and their gene engineering research development in rapeseed. *Journal of Agricultural Science and Technology* 10, 6–11.

Liu, W.H., Ma, X.J., Zhang, S.F. and Yan, X.D. (2013) Main pest control effects of spring-sowing rape in the seeding stage. *Biological Disaster Science* 36, 274–278.

Shen, J., Nie, Q., Huang, D., Liu, G. and Tao, L. (2007) Recent advances in controlling plant diseases and insect pests by mixture planting and inter-planting of crops. *Acta Phytophylacica Sinica* 34, 209–216.

Su, W., Lu, J.W., Zhou, G.S., Li, X.K., Han, Z.H. and Lei, H.X. (2011) Effect of no-tillage and direct sowing density on growth, nutrient uptake and yield of rapeseed (*Brassica napus* L.). *Scientia Agricultura Sinica* 44, 1519–1526.

Tang, J. and Hong, S. (2015) The rotation technology of rapeseed, maize, and potato. *Modern Agriculture* 2, 34–36.

Tang, Y., Li, C., Jiang, L., Wu, C. and Lin, S. (2008) Research on different rape cultivation patterns and matching cultivated techniques of direct seeding under no-tillage condition in Chengdu plain. *Southeast China Journal of Agricultural Sciences* 21, 946–949.

Wang, L.Y., Yang, W., Xin, H.P. and Zhong, X.Z. (2001) The test of prevention effect of several pesticides during germination-period of spring-rape. *Heilongjiang Agricultural Science* 1, 20–22.

Wang, L., Liu, Y., Ji, X., Wang, G. and Zhou, H. (2008) Effects of wheat–oilseed rape or wheat–garlic intercropping on the population dynamics of *Sitobion avenae* and its natural enemies. *Chinese Journal of Applied Ecology* 19, 1331–1336.

Wu, L.L. (2014) Arthropod community research on canola. Thesis for Master's Degree, Northwest Agricultural & Forest University of China.

Wu, K., Zhao, X., Yang, J. and Cai, Q. (2013) Advance on the occurrence regulation and control measures of *Mamestra brassicae* in sugar beet. *Journal of Agriculture* 3, 25–28.

Yang, M., Xiong, R., Tuhenimu, A., Zhang, P. and Lu, Z. (2011) Trapping effects of the mixture of sugar and acetic acid on moths and lacewings. *Plant Protection* 37, 170–173.

Young, S. (1966) A preliminary study on the cabbage sawfly *Athalia rosae japanensis* (Rhower). *Acta Entomologica Sinica* 15, 56–62.

Yuan, H., Wang, X. and Yang, Y. (2013) The vertical niche of *Phytomyza horticola* Gourean (Diptera: Agromyzidae) and investigation methods in fields. *China Plant Protection* 33, 50–53.

Zhang, D.F., Xian, W.R., Zhao, X.Y., Wen, H. and Yang, J.L. (2005) Investigation on insect kinds and method of controlling in spring rape field east agri-region of Qinghai. *Qinghai Agriculture Science* 4, 7–10.

Zhang, L.S., Zhang, Z.H., Gao, S. and Nong, X.Q. (2006) Application and development of *Metarhizium* biological pesticide. *Chinese Journal of Biological Control* 22, 141–146.

Zhang, Z., Lian, M., Li, T. and Zhou, Y. (2007) Observation on biological characteristic of *Mamestra brassicae* L. *Journal of Shanxi Agricultural Science* 35, 96–97.

Zhao, H., Yuan, F., Dong, Y. and Zhang, G. (1995) Life table of *Myzus persicae* under different temperature and host plant conditions. *Chinese Journal of Applied Ecology* 6, 83–87.

Zhu, Z.L., Qing, M.F., Zheng, J.A., Jiang, X.L. and Wu, J. (2005) Effect of water utilization efficiency under no tillage and straw mulching to wheat field and rape seed field. *Southeast China Journal of Agricultural Sciences* 15, 565–568.

Zhu, Y., Zhang, Q., Chen, Z., Zhao, L., Bao, M. and He, P. (2006) The research on the time niche of leaf-mining flies and its parasites on pea. *Journal of Yunnan Agricultural University* 21, 725–728.

Zhu, H., Luo, X.M., Song, J.X., Fan, M.Z. and Li, Z.Z. (2011) Effects of *Beauveria bassiana* on *Myzus persicae* and its two predaceous natural enemies. *Chinese Journal of Applied Ecology* 22, 2413–2418.

Zong, B.L., Zhong, C.Z., Lei, C.L. and Zhu, Z.P. (1986) Studies of natural enemies of rape pests in Wuchang. *Chinese Journal of Biological Control* 2, 118–120.

14 Integrated Control of Insect Pests on Canola and Other *Brassica* Oilseed Crops in Pakistan

MUHAMMAD SARWAR*

Pakistan Atomic Energy Commission, National Institute for Biotechnology & Genetic Engineering (NIBGE), Faisalabad, Punjab, Pakistan

14.1 Introduction

Family Brassicaceae, formerly known as Cruciferae, in the order Brassicales, is a mustard family of flowering plants in angiosperm floras distributed throughout the world. The plant's inflorescence is an elongated corymbose raceme, borne terminally on main stem and branches and carrying bright yellow flowers. Brassicaceae species are categorized by the presence of four-petalled cross-shaped flowers which bear two long and two short stamens and yield pod-like fruits recognized as siliques. Though family Brassicaceae comprises nearly 338 genera and more than 3700 species, some of the foremost genera and relevant species are, in genus *Brassica*: cauliflower (*Brassica oleracea* var. *botrytis*), turnip (*Brassica rapa* var. *rapa*), cabbage (*Brassica oleracea* var. *capitata*), napa cabbage (*Brassica rapa* var. *pekinensis*), brown mustard (*Brassica juncea*), broccoli (*Brassica oleracea* var. *italica*), rape (*Brassica napus* var. *napus*), Brussels sprouts (*Brassica oleracea* var. *gemmifera*), bird rape (*Brassica rapa* var. *chinensis*), rutabaga (*Brassica napus* var. *napobrassica*), kohlrabi (*Brassica oleracea* var. *gongylodes*), collard (*Brassica oleracea* var. *acephala*), kale (*Brassica oleracea* var. *acephala*), white mustard (*Brassica alba*), black mustard (*Brassica nigra*); and in other genera: bitter-cress (*Cardamine* spp.); basket-of-gold (*Aurinia saxatilis*); and arugula (*Eruca vesicaria* subsp. *sativa*) and garden rocket (*Eruca sativa*) (Thomas, 2003).

Oilseed *Brassica* crops are grown throughout the world but *Brassica napus* Linnaeus, *Brassica juncea* L. and *Brassica campestris* L. are the most important in Pakistan, where they are more widely grown than some other species. Rapeseed and mustard contain 44–46% good-quality oil and their meal has 38–40% protein. The canola oilseeds are sometimes particular varieties of *B. rapa* (Polish canola) but usually comprise the related species *B. napus* (rapeseed) and *B. juncea* (mustard greens). Among these crops, canola is the most important crop due to its oil of superior quality and varieties of *B. napus* having less than 2% erucic acid in oil and 30 μg glucosinolates/g in free oil meal are termed canola. Canola seeds commonly contain 40% or more oil and produce meals with 35–40% protein. In Pakistan, canola, rapeseed and mustard are primarily grown in Punjab, Sindh and Khyber Pakhtunkhwa provinces. The country's total area grown and production of rapeseed–mustard and canola cultivations during 2014/15 are almost 478,000 acres and 35,000 acres (193,440 ha and 14,164 ha), 181,000 t and 16,000 t seed and 58,000 t and 6000 t oil, respectively, distributed among four provinces (Government of Pakistan, 2015).

Within the tribe Brassiceae, the same species can be utilized for several uses according to different forms categorized into oilseed, forage, condiment and vegetable crops by using their buds, inflorescences, leaves, roots, seeds and stems. Although broadly and comprehensively grown as vegetables wherein young tender leaves are used in salads or older leaves with stems may be eaten fresh or canned, these are being developed more for fodder and their seeds produce an essential oil that is used

*E-mail: drmsarwar64@yahoo.com

for food cooking and as hair oil and lubricants, while seed residue is used as cattle feed and in fertilizers (Cartea *et al.*, 2011; Sarwar *et al.,* 2013).

14.2 Insect Pests of *Brassica* Crops

Damage inflicted due to insects is a main limitation factor for growing of *Brassica* crops in this region, owing to inadequate management of various key pests at different phenological stages of the crop. These crops host a wide range of insect pests, including lepidopterans, sawflies, beetles and aphids. Studies conducted to document the population dynamics of key pests observed on *Brassica* crops revealed that densities are Lepidoptera 46%, followed by Hemiptera 27%, Coleoptera 9%, Orthoptera 9% and Acari 9%. In Pakistan and elsewhere, cabbage aphid (*Brevicoryne brassicae* (L.)), mustard aphid (*Lipaphis erysimi* (Kaltenbach)) and green peach aphid (*Myzus persicae* (Sulzer)) (Hemiptera: Aphididae) are the most destructive pests and major limiting factors for successful production of oilseeds. Both nymphs and adults of aphids cause damage to *Brassica* plants from early vegetative stage to silique formation, resulting in stunted plant growth; flowers wither and pod formation is hindered. Among the other important insect pests of cruciferous crops, attacks by diamondback moth (*Plutella xylostella*), leaf webber (*Crocidolomia binotalis*), cabbage webworm (*Hellula undalis*), cabbage butterfly (*Pieris brassicae*) and tobacco caterpillar (*Spodoptera litura*) result in significant losses in yield and there is a high level of incidence in low-temperature zones. Aphids (*B. brassicae, L. erysimi* and *M. persicae*) are yellow or green winged or wingless insects, whose nymphs and adults suck cell sap and devitalize plants. Populations of *B. brassicae* are more than those of *L. eyrsimi*, and all the *Brassica* varieties evaluated have been found susceptible. Weekly populations of both these species of aphids differ significantly from their appearance till maturity of the crop. The appearance of aphids is not uniform at all locations; however, the highest population is recorded from the last week in February to the second week in March. The incidence of mixed-instar larvae of mustard sawfly, *Athalia lugens proxima* (Klug) (Hymenoptera: Tenthredinidae), is indicated by their feeding on or skeletonizing the foliage, causing severe defoliation in the fields (Sarwar *et al.*, 2004a, b; Aslam and Razaq, 2007). Consequently, it is imperative to have a wide-ranging understanding of these pests and related management control actions so that their damage and spread may be stopped. Therefore, this chapter emphasizes the identification, life cycle, mode of damage and integrated management of major insect pests of *Brassica* oilseed crops.

14.2.1 Aphids (Homoptera: Aphididae)

Aphids are small insects (2–2.5 mm in length) with long slender mouthparts; they have soft pear-shaped bodies with long legs and may be green, yellow, brown, red, blue, grey or black in colour, depending on the species and the plants they feed on. The antennae have six segments and winged forms have two pairs of membranous wings, while a few species appear waxy or woolly due to the secretion of a waxy white or grey substance over the body surface. Most species have a pair of tube-like structures called cornicles projecting backwards out of the hind end of the body. Aphids do not move rapidly when disturbed.

Aphids forage through sucking of sap and enormous colonies can cause the hosts to become deformed and the foliage to become curled, shrivelled and yellowish, while both sides of leaves are infested in case of severe infestations, causing severe stunting of plants and affecting seed yield. Some aphid species inject a toxin into plants, which causes leaves to curl and further distorts growth, whereas a few species cause gall formations and some species are capable of killing plants. In the Punjab, an increase of one aphid beyond 25 aphids/plant can reduce yields of mustard by 1.5 kg/ha. Like other soft-bodied insects, aphids produce a sweet and watery excrement called honeydew, which serves as a medium on which a sooty fungus or sooty mould grows, giving a dirty appearance to crop, and which can decrease the photosynthetic capability of the plant, reducing the crop's market value. Aphids vector many plant diseases, causing greater losses than those caused by the direct feeding injury that is often the greatest impact of an aphid infestation. In the transmission of non-persistent plant viruses, the virus reproduces in the plant and aphids simply aid in dissemination of the pathogen and infection process.

The winged forms of aphids are produced in spring and autumn and they spread from locality to locality and from field to field. The ability to produce winged individuals provides the pest with a way to disperse to other plants when the quality of the food source deteriorates (Sarwar *et al.*, 2011a).

Cabbage aphid, *Brevicoryne brassicae* (Linnaeus)

The cabbage aphid is an oval or pear-shaped insect with piercing/sucking mouthparts; it is 2.0–2.5 mm long and has a pale green body with short cornicles. Wingless forms are somewhat oval in shape. The posterior end of the body tapers greatly and appears greyish-green or greyish-white due to their waxy powder covering (Fig. 14.1). Underneath the wax covering there are eight dark brown or black spots located on the upper abdominal surface. Winged females are somewhat smaller than the wingless forms but not covered with waxy powder (Pal and Singh, 2013). The wings are short and stout with prominent veins. The head and thorax are dark brown to black with dark brown antennae. The winged aphids have a yellow abdomen with two dark spots on the dorsal anterior abdominal segments. These spots merge into a dark band across the top of the last abdominal segments. The aphids reproduce in two ways. In warm climates, aphid colonies consist solely of females and reproduction does not involve mating or egg laying: the females give birth to live female nymphs that remain on the adult female's back until they are large enough to survive on their own. In temperate climates, aphids reproduce as above during the warmer periods of the year but the reproduction method changes in the fall: males are produced in response to a decrease in photoperiod or temperature, mating occurs and females lay eggs. Aphids overwinter as the egg stage near the soil surface in plant debris. Nymphs differ from adults (including wingless adults, known as apterae) in having less developed caudae and siphunculi. The nymphal period varies from 7 to 10 days and nymphs are similar in appearance to adults except for their smaller size. During the crop season there are up to 15 overlapping generations, primarily reliant upon temperature. The total life cycle represents the period from the birth of the nymph until its death as an adult, varying between 16 and 50 days, with a shorter life cycle at higher temperatures (Jahan *et al.*, 2013).

Fig. 14.1. *Brevicoryne brassicae*.

This aphid species feeds on all cultivated and wild plants of the family Brassicaceae and may attack the crop at any stage, causing momentous yield losses. During severe damage to various plants, the colonies of this aphid are found on both lower and upper leaf surfaces, in leaf folds of developing heads, on leaf stalks and near leaf axles. They are occasionally found at the soil level and prefer feeding on young leaves and flowers. Aphids feed by sucking sap from their hosts and infested seedlings may become stunted and distorted. Continued feeding on mature plants causes wilting, yellowing and general curtailing of the foliage. The aphids also produce honeydew on plants, which can eventually lead to leaf death and decay. The cabbage aphid is a vector of 23 virus diseases of Cruciferae and both apterae (wingless) and alate (winged) aphids are able to transmit the virus, but more transmissions are obtained with apterae forms. The use of non-chemical pest controls should be considered and the safer insecticides (including fenitrothion and endosulfan) are very important for control of aphids. When the crop is in its flowering stage, it is advisable to carry out spraying after 5 pm when pollinators like honey bees have stopped visiting the crop (Sarwar, 2014a).

Mustard or turnip aphid, *Lipaphis erysimi* (Kaltenbach)

The apterae female aphids are 1.2–2.4 mm in length and have yellowish green, grey-green or olive-green bodies (Fig. 14.2). The alate female forms are around 1.4–2.2 mm long with a dusky green abdomen having dark lateral stripes separating the body segments and dusky wing veins. Male aphids measure about 1.20–1.35 mm long, noticeably less than the females, and are olive-green to brown in colour. One or two days after emerging from the last moult, female adults initiate a reproductive phase that lasts for 13–20 days, followed by a 2–3-day post-reproductive phase. Winged females produce 31–40 young at an average temperature of 30°C, whereas wingless forms produce 70–87 young in their lifetime of 26–37 days. With sexual reproduction, fertilization of females by males results in the production of eggs in the cooler months. Parthenogenetic birth of live female nymphs takes place without fertilization by males; and when males are very rare throughout the year, then females are almost wholly viviparous, giving birth to live young. Generally, eggs are laid along leaf veins with a total nymphal period of 8–9 days lasting for 1–2, 2, 2 and 3 days for first, second, third and fourth instars, respectively, with slight interval differences between wingless and winged forms. Higher ambient temperatures shorten the lifespan of the turnip aphid and with cooler temperatures there is an upsurge in longevity of the pest. When temperatures during the summer are 85–94°F (29.4–34.4°C), adults can live for 15–18 days. At winter temperatures of 55–68°F (12.8–20°C), the lifespan is considerably longer at 31–61 days. At an average temperature of 86.36°F (30.2°C), 26 young are produced by each female aphid, whereas the number of nymphs produced at an average temperature of 55.4°F (13°C) may be about 132 (Amjad *et al.*, 1999; Sultana *et al.*, 2009).

The turnip aphid induces injuries to the plant right from seedling to maturity stage, by way of the highest populations taking place throughout the flowering and podding formation periods. The aphid is found in great numbers on undersides of leaves, growing points, flowers (inflorescences), young shoots and rhizomes, affecting leaf size, producing chlorosis, rolling, yellowing and patches on leaves, shortening of the internodes of shoots, lesions and distortion on inflorescences and growing points, and dwarfing of whole plants. Feeding damage by *L. erysimi* to oil-bearing crops is responsible for decreases in plant height and terminal shoots, number of branches per plant, leaf and inflorescence size, siliques per plant, grains per silique, oil yield or oil content and seed yield. With regard to the various growth stages and plant parts of crop: pods are the most desirable part for the development and duration of various stages of the aphid, pest fecundity and number of individuals reaching the adult stage, followed by stalk and tender leaves, while hard leaves are the least advantageous portion. Feeding by these aphids on brassica plants also creates honeydew, which helps as a medium for development of sooty moulds that may reduce the market worth of a crop. Insecticidal control of aphids is most effective during the rosette stage, but treatments at first bloom can also enhance yield (Buntin and Raymer, 1994).

Green peach aphid, *Myzus persicae* (Sulzer)

Green peach aphid, *Myzus persicae*, is 1.6–2.2 mm in length. Winged (alate) green peach aphids have a black head and thorax and a yellowish green abdomen with a large dark patch dorsally. Wingless adults vary in colour from green to pale yellow (Fig. 14.3). Winged aphids seemingly attempt to colonize nearly all plants available; they often deposit a few young and then take flight. This highly dispersive nature contributes significantly to their effectiveness as vectors of plant viruses. This aphid does not have a waxy covering and bears long cornicles. It is found throughout the

Fig. 14.2. *Lipaphis erysimi.*

Fig. 14.3. *Myzus persicae.*

country, where it is viewed as a pest principally due to its ability to transmit plant viruses. Nymphs initially are greenish but soon turn yellowish, greatly resembling viviparous (parthenogenetic, nymph-producing) adults (Karimullah *et al.*, 1995). Studies on the developmental biology of viviparous aphids have reported four instars in this aphid, with the duration of each averaging 2.0, 2.1, 2.3 and 2.0 days, respectively. Females give birth to offspring within 6–17 days after birth, with an average age of 10.8 days at first birth. Because females give birth to live female nymphs, the population is solely composed of females and there are no males. The length of reproduction varies considerably but an average of 14.8 days is common. The average length of life is about 23 days, mean total longevity of 41 days, and mean fecundity is 75 offspring. Development can be rapid, often 10–12 days for a complete generation and with more than 20–21 annual generations reported in mild climates. All generations except the autumn generation culminating in egg production are parthenogenetic. Eggs, which are deposited on plants, measure about 0.6 mm long and 0.3 mm wide; they are elliptical in shape and initially are yellow or green but soon turn black (Sarwar *et al.*, 2009a; Sarwar, 2013a).

On *Brassica* crops, green peach aphids are usually present on the underside of the oldest leaves. They are common on seedlings, young plants and lower leaves of older plants. This species is seldom found in the heads and can attain very high densities on young plant tissue, causing water stress, wilting and reduced growth rate of the plant. Contamination of harvestable plant material with aphids, or with aphid honeydew, also causes yield loss (Fig. 14.4). However, green peach aphid does not seem to produce the high volume of honeydew observed with other aphid species. Blemishes to the plant tissue, usually in the form of yellow spots, may result from aphid feeding. Nymphs and adults are equally capable of virus transmission and over 100 viruses transmitted by this species have been listed. For a combined attack on aphid pests, early control is essential to stop the spread of insects. Varieties with some resistance should be used; and surrounding areas should be checked for sources of aphids and these host plants should be removed. Seed treatment with powder insecticides may control aphids for 3–4 weeks after planting. Seedling and rosette-stage plants should be treated if the population exceeds five to ten aphids per leaf or if

Fig. 14.4. Aphid damage.

15% of plants are infested. Bud and early-bloom stage should be treated if infestation exceeds 15% of stalks (racemes) but late-flower and pod stages should not be treated. During bloom, insecticides should be applied early in the morning or late in the day to minimize spray effect on bees. Plant health should be maintained with organic fertilizers and proper watering to allow plants to outgrow and tolerate the pest's damage. Use of a whole-neem extract containing azadirachtin for plant protection demonstrated aphid deterrence with no indication of resistance (Sarwar, 2013b).

14.2.2 Whitefly, *Bemisia tabaci* (Gennadius) (Hemiptera: Aleyrodidae)

Whitefly *Bemisia tabaci* is a minute white-winged insect, 1.5 mm long, that flies up in clouds from the underside of brassica leaves when disturbed. It is visible on its host plants throughout the year and overwinters as an adult. The adults lay ovoid eggs with a peg-like pedicel on the lower leaf surface, from which hatch the scale-like oval nymphs. First-instar nymphs that hatch from the eggs are mobile but the subsequent developmental stages (second, third, fourth and pupal non-feeding fourth instars) are sessile and do not move from the feeding site originally selected by the crawler. Different host plants species can significantly affect whitefly survivorship and reproductive rates. For example, at 25°C, egg to adult development for *B. tabaci* is rapid (lasting for 20 days) or slow (taking 29 days). Both adults and nymphs suck sap and create honeydew, which allows the growth of sooty moulds that can reduce plant height, leaf number, leaf size and dry-weight yields. Declines in the visual quality of plants because of honeydew, black sooty mould contamination and feeding damage are problems associated with high densities of whitefly populations (Sattar *et al.*, 2005). *B. tabaci* is attacked by predatory species representing eight arthropod orders, including members of the families Phytoseiidae (Acari), Coccinellidae (Coleoptera), Syrphidae (Diptera), Anthocoridae, Nabidae and Miridae (all Hemiptera), and Chrysopidae and Coniopterygidae (both Neuroptera) (Sarwar and Saqib, 2010; Sattar *et al.*, 2004; Sarwar, 2014b). The parasites *Encarsia formosa* and some other *Encarsia* species that occur naturally can develop as larvae in whitefly nymphs and are sometimes sufficiently numerous to be effective at keeping whitefly population at a low level. The relatively persistent pesticides deltamethrin or lambda-cyhalothrin can be sprayed on to the lower leaf surfaces and have limited effects on natural enemies but no more than two applications of these insecticides are permitted on brassicas during the growing season. Whitefly is frequently found on brassicas, but not necessarily as a serious problem that requires a serious control (Fekrat and Shishehbor, 2007).

14.2.3 Thrips, *Thrips tabaci* (Linderman) (Thysanoptera: Thripidae)

Thrips are tiny insects < 3 mm in length; they are cream to light brown in colour with long, narrow wings fringed at the margins in fore and hind pairs. Females of most plant-feeding species lay their elongate cylindrical to kidney-shaped eggs on or into leaves or buds. Thrips hatch from an egg and develop through two actively feeding larval stages and two non-feeding stages, the pre-pupa and pupa, even though thrips do not have a true pupal stage before becoming an adult. The pale-coloured pre-pupae and pupae of most thrips species drop to the soil and leaf litter or lodge within plant crevices for development. Thrips have several generations (up to about eight) in a year and when the weather is warm the life cycle from egg to adult may be completed in 2 weeks. Adult and immature thrips can do damage to any *Brassica* plants by sucking their juices and scraping at leaves, flowers and fruits (Syed *et al.*, 2015). Thrips cause most damage to seedlings by rasping tender leaves and terminal buds with their piercing mouth parts. The plant responds by forming scar tissue at the feeding site, giving the leaf a rough, rusty and warty appearance. Thrips overwinter on grasses and they migrate into *Brassica* crop fields as the weedy verges die down and the winter wheat and alfalfa are harvested. Resistance to thrips damage in plants is related to traits like earliness, Brix value and leaf surface wax; and selection for resistance may result in possibly undesired changes in these associated traits (Voorrips *et al.*, 2008).

To control thrips, weeds and grass should be removed from and around field areas, and crop debris should be cleaned up, especially green leaves after harvest, to eliminate alternative hosts. Plants should be inspected in the field for signs of thrips or their damage and any infested plants should be discarded after securely bagging. Commercially available thrips predators may be released, such as minute pirate bugs (which feed on eggs and larvae before they can become adults), lady beetles or lacewings to attack and destroy all stages of this pest. For best results, predator releases are made after first knocking down severe infestations with water sprays or other methods. Blue sticky traps are helpful for monitoring adult populations and sprays of acetamiprid, dimethoate, spinosad or imidacloprid offer better control. Overall, increased emphasis should be placed on breeding brassicas to be resistant to *T. tabaci* as the foundation for its management. Thorough plant coverage is necessary when using natural contact insecticides, especially on the undersides of leaves and where leaves attach to stems, which are favourite places for thrips to congregate (Shelton *et al.*, 2008; Khanzada *et al.*, 2016).

14.2.4 Red-legged earth mite, *Halotydeus destructor* (Tucker), and blue oat mite, *Penthaleus major* (Duges) (Acarina: Penthaleidae)

Adult mites of both these species are eight-legged, oval-shaped and about 1 mm long, and survive over summer as eggs. Red-legged earth mites have somewhat flattened black bodies and pinkish orange legs and mouthparts and they feed gregariously, usually on the upper side of the leaves. Blue oat mites have rounded dark blue to black bodies, bright red or bright pinkish red legs and mouthparts, with a red spot in the centre of the lower back, and they feed either singly or in small groups of five to ten individuals, mostly on the underside of the leaves. Both red-legged earth and blue oat mites develop through two to three generations in a year.

The mites feed on the cotyledons and leaves of seedlings by a rasping and sucking action and the appearance of silvery patches indicates mite damage. Mites attack seedlings as soon as or even before they emerge and cause weakening or even death of them. Sometimes heavily infested plants have mottled and then whitened cotyledons and leaves. Very severely damaged plants die and harshly injured plants usually remain stunted and weak. The mites normally feed from late afternoon until early morning but feeding continues during the day in calm and overcast weather. They are very active and when disturbed on a plant can drop or descend to the ground and quickly hide in the soil or under vegetation.

These mites are similar in appearance in all their life stages and both species prefer light, sandy or loamy well drained soils. They often occur together in crops and pastures, where they may cause crop damage, and canola is a good host for breeding. Three insecticidal control tactics are available for mites: insecticide-treated seed; bare-earth spraying of the soil surface immediately after seed sowing prior to germination; and spraying of the seedling foliage with a systemic insecticide after emergence. For these tactics to be effective the mites must be detected early, with careful monitoring for pests every 2–3 days by estimating mite numbers in a 10 × 10 cm (100 cm^2) ground area, repeated at five to ten sites in the crop, with treatment if ten or more mites are found per 100 cm^2 sample (Baker, 2010; Sarwar, 2013c).

14.2.5 Bud borer, *Helicoverpa armigera* (Hubner) (Lepidoptera: Noctuidae)

Wingspan of the adult moth is 30–45 mm and the forewings are brownish or reddish brown (females) or dull greenish to yellow or light brown (males). Hindwings are pale with a broad, dark outer margin and there is a pale patch near the centre of this dark region. The moths feed on nectar and live for around 10 days. Females lay around 1000 eggs, mainly at night in singles or clusters on growing points, leaves, flower buds, flowers and developing fruits, and sometimes on stems. The oviposition period lasts for 5–24 days and fertile eggs hatch in about 3 days during warm weather (25°C) and 6–10 days in cooler conditions, changing from white to brown to a black-head stage before hatching. The hatching larva (neonate) eats through the egg shell to make an exit hole and is 1–1.5 mm long, with a brown-black head and white or yellowish white, dark-spotted body. Fully grown sixth-instar larvae are 40–50 mm long and there may be considerable variation in colours and markings. Large larvae (longer than 24 mm) are the most damaging stage, since larvae consume about 50% of their overall diet in the fifth and sixth instars. Larvae need to be targeted when they are still small (< 7 mm). Fully grown larvae crawl to the base of the plant, tunnel up to 10 cm into the soil and form a chamber in which they pupate. Pupal duration is determined by temperature, taking around 2 weeks in summer and up to 6 weeks in spring and autumn. However, diapausing (overwintering) pupae take much longer to emerge and they survive in the soil when host plants are scarce. In non-diapausing pupae, the pupal period ranges from about 6 days at 35°C to > 30 days at 15°C, while the diapausing period for pupae may last for several months. In captivity, longevity varies from 1 to 23 days for males and 5 to 28 days for females. Tools used to predict and monitor migrations and diapause emergence include pheromone (chemical lure) traps that attract male moths using the sex pheromone that mimics chemicals emitted by female moths to attract mates (Sarwar *et al.*, 2009b, 2011b).

Bud borer larvae feed on leaves, flower buds and flowers, developing pods, fruits and seeds, and in most *Brassica* crops young larvae can graze on leaves alone. In some *Brassica* crops, larvae infest reproductive structures (flowers) as soon as they hatch. Once established in these concealed feeding locations, larvae are much more difficult to control with insecticides. Accordingly, for implementation of sustainable management of *H. armigera*, more attention should be devoted to some basic information such as monitoring efforts, forecasting activities and economic thresholds. In addition, more studies are needed to evaluate the potential of novel control measures, including selective insecticides and sublethal doses, and genetically modified cultivars and microbial pathogens (especially commercial formulations of *Bacillus thuringiensis* Berliner and nuclear polyhedrosis viruses) for control of this noctuid pest. A number of cultural practices such as time of sowing, spacing, fertilizer application, deep ploughing, inter-culture and flooding have been reported to reduce the survival of and damage by *Helicoverpa* species. Intercropping or strip-cropping with marigold, sunflower, linseed or coriander can minimize the extent of damage to the main crop. Strip-cropping can also increase the efficiency of chemical pest control. Hand-picking of large larvae can reduce *Helicoverpa* damage; however, the adoption of cultural practices depends on the crop husbandry practices in a particular agroecosystem (Fathipour and Sedaratian, 2013; Sarwar, 2013d).

14.2.6 Brown-tail moth, *Euproctis chrysorrhoea* (Linnaeus) (Lepidoptera: Lymantriidae)

The wings of this species are pure white, similar to the body; wingspan is 36–42 mm and there is a tuft of brown hairs at the end of the abdomen. The brown coloration extends along most of the back of the abdomen in the male. In the female, the back of the abdomen is white but the tuft of brown hairs is much bigger. The larva is brown with red and white markings and mainly very hairy. These hairs provide protection for this species throughout its life cycle: the larva incorporates some of these into the cocoon within which it pupates, while the emerging adult female collects some on its tail and uses it to camouflage and protect the eggs as they are laid. The peculiarity of the brown-tail moth's life history is that for 10 months of the year it is in the larval stage, and overwinters as a young larva within a tough silken tent. Pre-diapausing larvae emerge and feed gregariously in August after about 3 weeks of egg incubation. Diapausing larvae, as a response to short photoperiods, build communal winter nests in the fall, within which they overwinter (Frago *et al.*, 2009).

Outbreaks of brown-tail moth are usually for a short duration of a few months at most but it causes major problems, especially to people. They are notorious for urtication hairs on the caterpillars and the scales on the wings of adult moths cause serious skin, eye and inhalation difficulties. The polyphagy of this pest is remarkable and this, together with its tendency to reach outbreak densities, makes this species a major pest of forests that may also attack fruit and ornamental trees. The larvae also feed on leaves of brassicas and breed on oilseed crops.

There are several ways of preventing future outbreaks, including monitoring *Euproctis* populations by use of lights traps. Any significant increases in the adults caught in the traps and the egg masses found on the trees may lead to a decision to administer spot insecticide applications on the preferred host. Checks should be made for any natural enemies already present, as several egg and larval parasitoids have been noticed, with some larval stages dying at a young stage, due to small holes seen on the back of the larvae. In homes, the use of botanical pesticides that have been tested and work well, such as neem, are recommended. Entomopathogenic nematodes *Heterorhabditis bacteriophora* (Poinar) and *Steinernema carpocapsae* (Weiser) can cause significantly greater mortality of third, fourth and fifth instars of *E. chrysorrhoea* (Nikdel *et al.*, 2010).

14.2.7 Green stink bug, *Nezara viridula* (Linnaeus) (Hemiptera: Pentatomidae)

The adult is shield-shaped with an overall dull green colour; small black dots can be found along the sides of the abdomen and the eyes are dark red or black. Males on average are 12.1 mm in length and females 13.15 mm long, and the wings completely cover the abdomen length. Females can lay eggs 3–4 weeks after becoming adults and lay as many as 260 eggs over their lifespan on the undersurface of leaves in the upper portions of canopied crops. The green stink bug can complete its life cycle in 65–70 days and there may be up to four generations per year in warm climates. It overwinters as an adult and hides in leaf litter, under tree bark, or in other locations to obtain protection from the weather. As spring temperatures begin to rise, the bug moves out of its winter cover and initiates feeding and oviposition (Capinera, 2001).

The green stink bug has piercing/sucking mouthparts and is highly polyphagous, despite which it shows preference for leguminous plants and brassicas. The mouth is a long beak-like structure called the rostrum; salivary fluid is pumped down to the salivary duct and liquefied food is pumped up in the food canal. All plant parts are likely to be fed upon by this pest but its preference is for growing shoots and developing fruit, decidedly lowering the market value of the crop. The growth of young plants is retarded so that they often wither and may droop. An accepted economic threshold for green stink bug in oilseed crops is reported as three to four stink bugs per 100 passes with a sweep net, or a population level of 5000 bugs/ha (Follett *et al.*, 2009). Parasites, usually the tachinid fly, *Trichopoda pennipes*, parasitize adults and nymphs and a wasp,

Trissolcus basalis, parasitizes eggs, providing biological control of the green stink bug (Panizzi, 2008).

14.2.8 Mustard leaf miner, *Chromatomyia horticola* (Goureau) (Diptera: Agromyzidae)

Chromatomyia horticola is a regular polyphagous pest of rapeseed and mustard and mostly found infesting brown mustard (*B. campestris*). The adult is a two-winged fly with a greyish black back and yellowish speck near the head. The female's mean lifetime and fecundity are from 27.43 to 32.5 days and 67.57 to 87.15 eggs, respectively, on various canola cultivars. The larvae feed within the leaves of the host plant and feeding can severely reduce yields or kill the plants at high fly densities. In cases of severe infestation, the attacked leaves wither, plant vigour is reduced and damage is often more prominent on the older leaves. This pest remains active from December to May and passes its pupal stage in soil. With respect to infestation, correlation analysis reveals a positive correspondence with temperature and a negative one with relative humidity. The various parasitoids recorded and known to attack *C. horticola* are five eulophids (*Chrysocharis horticola* Mani, *Diglyphus horticola* Khan, *Diglyphus* sp., *Pediobius indicus* Khan and *Euderus agromyzae* Gangrade) and two braconids (*Opius* sp. and *Dacnusa* sp.). The *Diglyphus* spp. are found actively associating with the leaf miner. For sampling, 100 infested leaves are randomly collected from the crop site. The leaf samples are brought to the laboratory and kept in plastic culture containers or rearing jars, covered with muslin cloth, until the emergence of adult flies or their parasitoids. The most common method of controlling leaf miners is to spray general pyrethroid pesticide on the infected plants at the right time (Ahmad *et al.*, 2010).

14.2.9 Pea leaf miner, *Phytomyza horticola* (Goureau) (Diptera: Phytomyzidae)

The pea leaf miner (*Phytomyza horticola*) is an insect of cosmopolitan distribution. The female lays most of its eggs on the lower surface of the leaves. The larvae are legless maggots without a head capsule and never have thoracic or abdominal legs. They do not have chewing mouthparts and form a narrow, whitish, linear mine on the upper or lower leaf surface. The puparia are formed within the hardened last larval skin or puparium and as a result sheaths enclosing head appendages, wings and legs are not visible externally. Pupation is internal at the end of the mine, usually in a puparium chamber on the underside of the leaf with the anterior spiracles projecting through the epidermis. The adult pea leaf miners emerge in early February and the flight period continues until the beginning of June. The average period needed by the insect to complete its life cycle is 23 days and there are four generations per year.

Pea leaf miner is a highly polyphagous insect and considered as a pest of many vegetable crops and ornamental plants, but most commonly occurs on Brassicaceae, Fabaceae and Asteraceae families. Natural and organic control methods work best when fighting leaf miner problems, because they do not harm the naturally occurring beneficial insect populations. Plant leaves should be closely monitored; at the first sign of tunnelling, the leaf should be squeezed at the tunnel between two fingers to crush any larvae. Yellow sticky traps are used to catch egg-laying adults and floating row covers can prevent the fly stage from laying eggs on leaves. Soil under infested plants should be covered with plastic mulches to prevent larvae from reaching the ground and pupating. Neem oil may also have repellent qualities and interferes with egg-laying activities (Bjorksten *et al.*, 2005).

14.2.10 Cabbage butterfly, *Pieris brassicae* (Linnaeus) (Lepidoptera: Pieridae)

The wingspan of the adult butterfly is 55–70 mm, with females being larger than the males. The upper sides of the wings of both sexes are usually gleaming white, with a pronounced black tip on the forewing. This is augmented in the female (which has a larger black tip) by a pair of post-discal black spots, with a black smear along the inner margin below the lower spot. The undersides of both sets of wings are pale yellow dusted with grey, except for the centre and base of the forewings, which are white. In females, the black dots of the forewings also appear on the undersides. The head, thorax and abdomen are black with grey hair-like scales. Eggs are 1.4 mm high, bright yellow, bottle-shaped, ribbed vertically and laid upright in clusters of 20–100, mostly on the underside of leaves. The eggs change to bright orange prior to hatching and hatch in about 1–2 weeks, depending on the temperature. Newly emerged larvae are yellow with shiny black heads and after the first moult the colour changes to yellowish green, with yellow lines running along the length of the body. The larvae pass through five instars, feed gregariously and grow fully in 24 days; fully fed they are 45 mm long. The whole body is covered with fine hair-bearing tubercles, many of which are black. Pupation takes place in solid substrate in about 10–15 days and the second generation starts to emerge (Ying-Zhi *et al.*, 2004; Sarwar, 2014c).

The cabbage butterfly is an oligophagous insect with the larvae feeding specifically on members of family Brassicaceae, owing to the presence of glucosinolates, which have been shown to play a central role in the host selection behaviour of *Pieris* butterflies. Glucosinolates also play a role as feeding stimulants for larvae of cabbage butterfly. Generally, *Pieris* adult females use the glucosinolates or their hydrolysis products as positive signals for the recognition of suitable host plants during oviposition. Young larvae scrape away the lower epidermis of the leaves, whereas the older larvae cause extensive defoliation and often reduce plants to a skeleton of stems and major veins and finally may kill the plant (Sadozai and Khan, 2014).

Control of cabbage butterfly is oriented towards the use of efficient but high-risk insecticides, some of them being endocrine disruptors. Use of a mixture of two products (cypermethrin and *Bacillus thuringiensis* subsp. *kurstaki* (Btk)) is of significance, especially for control of advanced late instars late in the season, when Btk action alone is insufficient. Spinosad is effective in inducing mortality and reducing leaf damage by all larval instars. Feeding rate and mortality are equally important parameters when assessing biopesticide efficacy. This strategy should also reduce the possibility of inducing resistance in pest populations. In addition, it tends to reduce the residues in commodities and is a good solution in production of hygienic and safe food (Klokočar-Šmit *et al.*, 2007).

14.2.11 Diamondback moth, *Plutella xylostella* (Linnaeus) (Lepidoptera: Plutellidae)

The adult is a small slender moth, about 6 mm long, and greyish brown, marked with a broad cream or light brown band along the back and with pronounced antennae. When the wings are folded, the moths have a line of three diamond-shaped markings along the middle of the back, which is the basis for the common name of this insect. When viewed from the side, the tips of the wings can be seen to turn upwards slightly. Adult males and females live for about 12 and 16 days, respectively, and females deposit eggs for about 10 days. The eggs are oval and flattened, yellow or pale green, and measure 0.44 mm long and 0.26 mm wide. Eggs are deposited singly or in small groups of two to eight eggs in depressions on the surface of foliage, or occasionally on other plant parts. Females may deposit 250–300 eggs but average total egg production is probably 150 eggs and development time averages 5.6 days. The larva has four instars wherein average and range of development time are about 4.5 (3–7), 4 (2–7), 4 (2–8) and 5 (2–10) days, respectively. The larval body form tapers at both ends and is colourless in the first instar but thereafter green. The body bears relatively few hairs, which are short and most are marked by the presence of small white patches. If disturbed, larvae often wriggle violently, move backwards and spin down from the plant on a strand of a silken thread to remain suspended temporarily or drop to a lower leaf. When the disturbance has passed, the larva climbs on the silken thread back to the leaf and crawls away, leaving a coil of silk on the leaf. Pupation occurs in a loose silk cocoon, usually formed on the lower or outer leaves. The yellowish pupa is 7–9 mm in length and the duration of the cocoon formation averages about 8.5 days (range 5–15 days) (Syed and Abro, 2003; Ahmad, 2005).

Diamondback moth is highly dispersive and attacks only host plants in the family Cruciferae, including both cultivated and wild plants. Plant damage is caused by larval feeding and initially the feeding habit of first-instar larvae is leaf mining, though they are so small that the mines are difficult to notice. The larvae emerge from their mines at the conclusion of the first instar, moult beneath the leaf and thereafter feed on the lower surface of the leaf. Their chewing results in irregular patches of damage and the upper leaf epidermis is often left intact. Although the larvae are very small, they can be quite numerous, resulting in complete removal of foliar tissue except for the leaf veins, and this is particularly damaging to seedlings. The last-stage larva is a more voracious feeder than the first three instars and all the leaf tissues are consumed except the veins. Average population densities of up to 0.3 larvae per plant are considered to be below the treatment level (Obopile *et al.*, 2008).

Various predators and parasitoids have been recorded feeding upon immature stages of *P. xylostella*. Predators include various species of ants and spiders, coccinellid beetles and *Chrysoperla* spp., while *Cotesia plutellae* and *Oomyzus sokolowskii* are important larval and pupal parasitoids, respectively (Syed *et al.*, 2012; Sarwar *et al.*, 2011c).

14.2.12 Crucifer flea beetle, *Phyllotreta cruciferae* (Goeze) (Coleoptera: Chrysomelidae)

Flea beetles are bluish black, measuring 2–3 mm in length and with enlarged hind femurs. The adult female beetle lays eggs in the soil that hatch in 12–15 days, depending on the temperature. The larvae are whitish and feed on the roots of the plant without causing significant damage to the plant. The larva completes its life cycle in 3–4 weeks and there is a single generation per year.

Adults feed on the first true leaves of seedlings, causing bite holes and resulting in severe damage. The attacked plants emit a decaying odour and economic threshold at seedling stage is 25–50% leaf damage. Currently, most farmers manage flea beetles in canola crops by using insecticide-coated canola seed. Seed treatment with imidacloprid at 5g/kg of seed can control flea beetle at the seedling stage, or the crop may be sprayed with carbaryl at 2 g/l of water. There are several other practices that could reduce dependency on chemical control of this insect. Observation of the levels of flea beetle populations in canola during fall harvesting operations can give a forewarning for potential populations in the following spring. Activities that encourage rapid, even germination of the crop can mitigate flea beetle effects. Sowing large seed with high vigour and germination rates into minimally or zero-tilled fields can effectively reduce flea beetle feeding injury and the impact of any injury that does occur. Thus, by integrating management practices, growers can reduce the need for chemical control of flea beetles while maintaining the productivity of canola (Soroka and Elliott, 2011; Khanobporn *et al.*, 2014).

14.2.13 Green semilooper, *Trichoplusia ni* (Hubner) (Lepidoptera: Noctuidae)

Adults semiloopers are dark brown and grey moths, about 22 mm long and measuring 40 mm across the spread of the wings. There is a distinct figure-of-8-shaped white pattern in the middle of the forewings and the wings are folded over the back at rest. Adult female lays whitish round eggs singly on the leaf surface and larvae emerge from the eggs within 3–7 days, then develop for 16–19 days before pupating. The young larva is initially dusky white but becomes a pale green caterpillar with a white stripe on each side of the body. The pupa turns from initially green to dark brown and the adult emerges 9–10 days later. The entire life cycle lasts for 32–37 days and there are one to two generations in a season.

Semiloopers are leaf feeders and in the first three instars they confine their feeding to the lower leaf surface, leaving the upper surface intact. The fourth and fifth instars chew large holes and cause severe damage by skeletonizing the leaves. Feeding sites are marked by large accumulations of sticky and wet faecal material on the crop (Dornan *et al.*, 1995). Pest management can be by handpicking and destruction of fully grown larvae. Early-maturing varieties are less subjected to attack than late-maturing varieties. Row covers, where economically practicable, are effective at preventing looper moths from depositing eggs on crops. Microbial insecticides currently play a role in looper management and *B. thuringiensis* has long been used for effective suppression of the pest, with the advantage of not disrupting populations of beneficial insects. Mass release of *Trichogramma* spp. has been investigated for looper suppression. Moth catches are monitored effectively by light traps, while black light traps and pheromone traps have been used in an attempt to predict population densities (Khan *et al.*, 2010; Singhamuni *et al.*, 2015).

14.2.14 Cabbage borer, *Hellula undalis* (Fabricius) (Lepidoptera: Pyralidae)

The adult moth of cabbage borer or webworm is slender and pale yellowish brown, with grey wavy lines on the forewings. The caterpillar is yellow with a pinkish tinge and has seven purplish brown longitudinal stripes. The adult female lays eggs on the growing point or on the older leaves and the eggs hatch in 2–3 days. The caterpillars feed in the core of the plant and become fully grown in 7–14 days, after undergoing four moults. The fully grown caterpillar spins a cocoon among leaves touching the ground or even inside the larval burrows. The pupal period is about 7 days and the life cycle is accomplished in 15–25 days.

The caterpillars first mine into the leaves and later they feed on the leaf surface, sheltering within the silken passages. When the attack is heavy, the plants are riddled with worms and outwardly look deformed. It is important to focus on controlling of infestations in younger plants. Crops should be monitored for the presence of larvae and treatment decisions can be made based on the infestation density. There is no established treatment threshold for this pest and so growers should establish their own arbitrary threshold. However, a threshold of 15–25% of plants infested in a random-pattern monitoring programme would be a good starting point. The parasitic wasp *Chelonus blackburni* Cameron oviposits into the eggs, develops within the caterpillar and emerges to spin its own cocoon within the pupal cell (Mewis *et al.*, 2002; Sugie *et al.*, 2003).

14.2.15 Armyworm, *Spodoptera litura* (Fabricius) (Lepidoptera: Noctuidae)

Adult moths measure 15–20 mm in length and have a wingspan of 30–38 mm. Forewings are grey to reddish brown, with a complex pattern of creamy streaks and paler lines along the veins. Hindwings are greyish white with greyish brown margins. Males have a blue-grey band from the upper corner (apex) to the inner margin of each forewing. Larvae have bright yellow stripes along the back and the sides; the main colour varies from pale green to dark green, and then finally brown for the later instars or more mature forms. Brown, mature larvae have three thin yellow longitudinal lines, one on the top or dorsal side and one each arranged on the lateral sides. A row of black dots runs along each lateral side and a row of dark triangles decorates each side of the mid-dorsal line. Field specimens of *S. litura* may be confused with another exotic species of concern, *Spodoptera littoralis* (Boisduval), or other *Spodoptera* species including southern armyworm (*S. eridania*), beet armyworm (*S. exigua*), fall armyworm (*S. frugiperda*), yellow-striped armyworm (*S. ornithogalli*), *S. latifascia*, *S. albula*, *S. androgea*, *S. dolichos*, *S. pulchella* and western yellow striped armyworm (*S. praefica*). Though the markings are variable, a bright yellow stripe along the length of the dorsal surface is characteristic of *S. litura* larvae.

Between 2 and 5 days after emergence, females lay eggs in masses of 200–300. The eggs are approximately 4–7 mm in diameter and cream to golden brown in colour. Egg masses are usually covered with body-hair scales and laid on the underside of the host plant leaf. Eggs usually hatch within 3–4 days. The young larvae or caterpillars are translucent green with a dark thorax. They are smooth-skinned with a pattern of red, yellow and green lines, and with a dark patch on the back of the head (mesothorax). Larvae develop through six instars in 15–23 days at 25–26°C. At lower temperatures the larvae often go through an extra instar and maturation may take up to 3 months. The caterpillar burrows several centimetres deep into the soil and pupates there within a cocoon. While pupating, it produces large amounts of fluid. The pupal stage lasts either a few weeks or several months, depending upon time of year, and takes place in about 11–13 days at 25°C. The average life cycle can be completed in about 5 weeks. As with most moths, *Spodoptera* adults are nocturnal and are not visible during the day. Longevity of adults is about 4–10 days, being reduced by high temperature and low humidity. The moths have a flight range of 1.5 km during a period of 4 h overnight, thus facilitating in dispersion and oviposition on different hosts. Pheromone-based traps are commonly used to detect the presence of various *Spodoptera* species (Meagher *et al.*, 2008).

On most crops, damage arises from extensive feeding by larvae on flowering stage, fruiting stage and vegetative growing stage, leading to complete stripping of the plants. Feeding is initially by skeletonizing, or leaving the outline of the leaf veins on the plant. As growth continues, caterpillars eat entire leaves and even flowers and fruits. Skeletonized leaves provide evidence of the presence of larvae. Early instars (first and second) are likely to be found on the underside of leaves. The older larvae are night feeders and usually found in the soil around the base of plants during the day. They chew large areas of the leaf, can occur at high population densities and may strip the crop of its leaves. In such cases, larvae migrate in large groups from one field to another in search of food (Saljoqi *et al.*, 2015). The use of *B. thuringiensis* may effectively control this pest. Other forms of biological, horticultural and cultural control that have been studied include planting near derris and garlic plants, breeding resistant plants from wild plants, breeding resistant plants using *B. thuringiensis* genes, using a baculovirus, using nematode *Steinernema carpocapsae* (Weiser) and using the fly *Exorista japonica* (Townsend) (Ahmad *et al.*, 2013).

14.2.16 Leaf webber, *Crocidolomia binotalis* (Zeller) (Lepidoptera: Pyralidae)

The moth is small and light brownish in colour; the forewings have distinct wavy lines and wavy spots, while the hindwings have a hyaline pigment. The moth lays eggs on the underside of leaves in masses of 45–100 and the eggs hatch in 5–15 days. The larva is green with red head and longitudinal red stripes on the body and it measures 2 cm in length. In the early stages, larvae feed gregariously on the leaf parenchyma. As they grow, they spread out and start webbing the leaves for feeding on them. The larval stage is completed in 25–30 days in summer and about 50 days in winter. When fully grown, the larva descends to the ground and pupates in the soil after making an earthen cocoon. The adult emerges in 14–40 days and the life cycle is completed in 43–80 days. Usually, more than one generation is completed within the season.

The pest causes substantial damage to rapeseed–mustard crops and other crucifers and newly hatched larvae feed on the chlorophyll content of tender leaves. The leaves are skeletonized by the larvae, which remain on the undersurface of leaves in webs and feed on them. Afterwards, they feed on the upper canopy of the leaves, flower buds and inflorescences, which are webbed together, resulting in stunted plant growth. This pest often assumes serious proportions; severely attacked plants are entirely defoliated and it also attacks flower buds and pods. For prevention and control, the adult moth can be killed through light traps; and dusting of carbaryl (10%) is quite effective, but application of the non-residual insecticides monocrotophos and quinalphos is preferred. The larvae are parasitized by hyperparasites like *Diplazon orientalis* (Cameron), *Apanteles crocidolomia* (Ahmed), *Microbracon melleus* (Cresson) and *Plexorista solennis* (Walker) (Kumaranag *et al.*, 2014).

14.2.17 Mustard sawfly, *Athalia lugens proxima* (Klug) (Hymenoptera: Tenthredinidae)

The adult sawfly is orange-yellow in colour with black head and legs, black markings on the body and smoky wings with black veins. The larvae are yellowish green to dark green with five black stripes on the back. A fully grown larva measures 16–18 mm in length and the body has a wrinkled appearance. The larvae closely resemble lepidopteran caterpillars, with the significant variance that instead of the four prolegs that are common to lepidopteran caterpillars, sawfly larvae have six pairs of prolegs on the abdomen. The mustard sawfly breeds from October to March and undergoes pupal diapause during summer. The adults emerge from these cocoons early in October; they live for 2–8 days and lay 30–35 eggs singly, in slits made with their saw-like ovipositors along the underside of the leaf margins. The egg period is 4–8 days and the larvae feed in groups of three to six, exposed on the leaves during morning and evening. There are five instars in males and six in females, with a larval period of 16–35 days. Pupation takes place in waterproof oval cocoons in the soil and the pupal period is 11–31 days. The life cycle is completed in 31–34 days; there are two to three generations from October to March and adult longevity is 10 days (Shepard *et al.*, 1999).

Sawfly is a winter season pest and adults inflict damage by the act of laying eggs with the help of the saw-like ovipositor. The grubs alone are destructive and rape and mustard are attacked at seedling stage, but crop 3–4 weeks old is most favoured. Several larvae are found feeding on the leaves, with dark scattered excreta. Larvae nibble margins of tender leaves and later on bite holes in the leaves and commonly forage during dawn and dusk. They make uneven holes in leaves, preferring the young growth, and skeletonize the leaves completely. Sometimes, even the epidermis of the shoot is eaten. Although the seedlings succumb, the older plants, when attacked, do not bear seed and in a severe infestation the crop looks as if it has been grazed by animals. The larvae remain hidden during the daytime and, when disturbed, fall to the ground and feign death.

This pest can be controlled effectively by maintaining clean cultivation and by collection and destruction of the grubs in the morning and evening, along with the conservation of *Perilissus cingulator* (Morley) (parasitoids of the grubs) and the bacterium *Serratia marcescens* Bizio, which infects sawfly larvae. The crop may be sprayed with malathion 50 EC at 1000 ml/ha and quinalphos 25 EC at 625 ml/ha in about 600–700 l water/ha. Irrigation at the seedling stage is crucial for sawfly management, because most of the larvae can die due to drowning (Chowdhury, 2009; Jayanthi and Ramesh, 2014).

14.2.18 Black cutworm, *Agrotis ipsilon* (Hufnagel) (Lepidoptera: Noctuidae)

The forewings of adults are long and narrow, darker than the hindwings and marked with black dashes or daggers. Males have plumose (feathered) antennae; female antennae are filiform. Eggs are ribbed, about 0.45 mm high, and newly laid eggs are whitish yellow, becoming darker as hatching approaches. The general body colour of the larva is usually uniform above the spiracles and varies from light grey to black without distinct stripes or markings. The subventral and ventral areas are lighter in colour, with numerous pale flecks. There are six to seven larval instars, with sixth and seventh instars approximately 35–50 mm in length. The diagnostic characteristics of these larvae are that they are heterogeneous, have convex granules and relatively large D2 tubercles. Pupae are brown to dark brown and approximately 17–25 mm in length and 5–6 mm in width. Black cutworm has a very wide host range but seedling crop plants are most seriously damaged. In addition to attacking grasses and weeds, it also feeds on *Brassica* crops. Early-instar larvae can create 'shot holes' while feeding on tender leaves of seedling plants. The third to seventh instars become negatively phototaxic and feed mostly at night. Damage from these instars is usually observed as the cutting of young seedlings, often causing death of the cut seedlings (Gemeno and Haynes, 2000; Talpur *et al.*, 2002). General management tactics are monitoring adults and larvae with pheromone and larval bait traps to predict the pest's attack. If possible, planting crops in fields with a known history of cutworm problems should be avoided. Cutworm predators should be boosted by having conservation strips between fields to encourage their other prey species (Frank and Shrewsbury, 2004).

14.2.19 Field cricket, *Gryllus bimaculatus* (DeGeer) (Orthoptera: Gryllidae)

Adult field crickets are black and can be discriminated from other *Gryllus* species by the two dot-like marks on the base of their wings. Their hindmost legs are much enlarged for powerful and rapid jumping. Females have a tubular organ at the rear end, known as an ovipositor, which is used to lay eggs into the ground. Male crickets produce several distinctive chirps, made by rubbing the two outer wings together. Loud fast-frequency chirps are emitted when males encounter one another and are preparing to fight the rival. A soft clipping sound is made to attract a female to encourage for mating. After mating, the females lay their eggs in moist sand or humid soil, typically in groups of 50. A female can lay up to 400 eggs, which incubate in the soil for 15–25 days and then hatch into nymphs. The adult stage is reached in about 12 weeks but very few individuals actually reach this level. The average lifespan of a field cricket is only 1 week and adult crickets die in late summer, leaving only the nymphs, which reproduce in the next summer. Population explosions in this species typically come after rainfall relieves prolonged drought conditions.

The crickets feed at night and spend most of the daylight hours in warm and dark refugia. Weeds and dense vegetation around the field bunds are other good hiding places. They lay their eggs into soil and the nymphs hatch in about 2 weeks. Large populations can cause significant damage to *Brassica* and various other cultivated crops (Gawałek *et al.*, 2014).

Field crickets are preyed upon by a wide range of predators such as most bird species, including cardinals, turkeys, blackbirds and some hawks, either preferentially or opportunistically. Red foxes, box turtles, toads and many other mammalian, reptilian and amphibian predators also vigorously consume field crickets. There is a virus that causes body paralysis of crickets; fungal infections colonize the intestines; and rickettsia infections, mermithid worms (nematodes) and ectoparasitic mite infestations all affect this pest. There are also species of parasitic wasps that sting and paralyse field crickets and then lay their eggs in the still-living cricket's body. The larvae of these wasps feed upon the cricket as they grow and develop.

It is frequently helpful to spray within, outside and around the field and in ornamental beds. The elimination of piles of bricks, stones, wood or other debris around the fields can help to reduce pest numbers, and nearby trash dumps that provide both food and shelter should be cleaned out (Tyler *et al.*, 2015).

14.2.20 Devastating grasshopper, *Melanoplus devastator* (Scudder) (Orthoptera: Acrididae)

The body of adults is pale grey and tan with dark grey spots; the underside of the abdomen is pale green to yellow. Females are 21–23 mm in length and males 18.5–22 mm. The leathery forewings and the membranous hindwings are held straight back and extend beyond the tip of the abdomen. Antennae are short and thread-like (filiform) with 25–26 segments. Eggs are pale yellow, 3.9–4.4 mm long, and contained within an egg pod approximately 19.05–21.6 mm in length. Newly hatched nymphs are white; however, after several hours of exposure to sunlight, they assume the distinctive colours and markings of adults. The shape of immature nymphs is similar to that of adults, but smaller, and there are six nymphal instars to complete their development. The nymphs begin development when plant forage is green and succulent and they feed on the host's foliage, with a preference for new growth.

The devastating grasshopper prefers semi-arid habitats and also lower areas at the foothills of mountain ranges, ridges, slopes and banks of ravines where soil is rocky. It is a general feeder and has been known to cause damage to brassicas by feeding along leaf edges and occasionally on the epidermis of stems. These grasshoppers are good flyers and they are responsible for enormous damage to other crops. When plants mature and become dry, late nymphs and adults survive on green but less palatable plants. Populations of five medium to large grasshoppers or eight small grasshoppers per 929 cm^2 warrant pest control action. When populations become threatening, chemical recommendations should be adopted (Schell and Schell, 2007; Fontana *et al.*, 2011).

14.2.21 Painted bug, *Bagrada hilaris* (Burmeister) (Hemiptera: Pentatomidae)

The body of painted bug is shield-shaped, black with red and yellow markings running primarily lengthwise, and ranges in size from 3–4 mm wide and 5–7 mm long, the female being somewhat larger than the male. Both dorsal thoracic plates anteriorly are positioned on the body of the adult (pronotum) and a central triangular plate on the dorsal surface of the thorax (scutellum) has a clear longitudinal marking down the centre. Within a season an adult female can lay over 100 eggs in clusters of about ten eggs close together or singly on the undersides of leaves, stems and in the soil near the base of plants. The eggs have an opaque and white to light red shade. During optimum temperature situations (96°F/35.6°C), nymphs emerge after nearly 3–4 days. Nymphs pass through five nymphal instars with an average of 3.35, 7.08, 6.39, 7.33 and 10.25 days, respectively. Freshly emerged first stages are bright red and have somewhat darkened to black coloration on the antennae, head, pronotum and legs, while bodies of late stages are darker but may still have pale to dark red markings. As premature nymphal instars progress, the abdomen remains reddish and develops some white dots and black bands.

Adults may survive in summer and cause severe damage to the germinating crop. They seek green vegetation around the field and are capable of overwintering in areas with cruciferous plants such as wild mustard and rocket. The life cycle lasts for 3–4 weeks and several overlapping generations may occur in a year. During crop growing season, favourable food and environmental circumstances permit the bugs to multiply rapidly; nymphs and adults in hundreds may be found in clustered field distributions (Bundy *et al.*, 2012; Taylor *et al.*, 2015).

The prime host plants of the bug are cruciferous crops, including mustard and canola, while adults may feed on weeds growing around and within fields in the absence of the preferred hosts. The bugs, especially in the early stages of development, gather in masses and suck the sap from plants. Typically, physical feeding damage by the bugs through piercing/sucking-type mouthparts causes small puncture marks visible as white patches starting on the edges of leaves; eventually the leaves wilt and dry. The bugs feed on both upper and lower surfaces of the leaves, and while foraging apparently inject saliva to aid in breaking down the inner leaf tissue. These stink bugs show a high favouritism for feeding on the four-leaf stage of host plants; however, feeding on the cotyledon stage of direct-seeded plants has been observed as the most susceptible stage to feeding damage in the field. During the early stage of plant development, short periods of infestation by this pest result in loss or reduction in chlorophyll content and total leaf area, and this may be part of a feeding-induced stress response. As a result, heavily attacked plants may have a scorched appearance. Research has shown that control measures should start if there is more than one bug/m^2 in the early growing stage. If the crop has completed the early growing stage, a higher threshold level of three bugs/m^2 can be maintained. Economic thresholds depend on many factors (crop stage, crop age and socio-economic and climatic conditions) and cannot be adopted without taking the local conditions into consideration (Huang *et al.*, 2014).

Early detection is crucial, due to the ability of bug populations to increase rapidly. Monitoring should begin before planting, by scouting and inspecting areas around the field intended for production. Sweep netting and careful visual observation of the soil surface, weeds, grasses and other vegetation surrounding the field are essential to determine if the bugs are present. Fields surrounded by weedy areas with an abundance of wild mustards or other host plants may be at a higher risk for bug infestation. Black light trapping is not effective; however, visual observation at night with a fluorescent lamp is effective.

Based on field trials, plant damage and crop losses can be reduced when contact insecticides are applied after populations exceed one adult in 3 ft (1 m) per row of seedlings or transplants. Young seedlings are the most susceptible to feeding damage and should be protected by insecticides if a field has a history of bagrada bugs. Contact foliar insecticide applications are the most effective in the afternoon and early evening during the peak of insect activity on plants; however, because bagrada bugs fly away when disturbed, adults can easily escape before contact with the insecticide and return later on. Additionally, bugs may drop to the soil to avoid contact with insecticides (Reed *et al.*, 2013). Field trials demonstrate that sowing imidacloprid-treated seed provides higher productivity and reduces plant damage. Generalist predators, including spiders, may attack bagrada bugs; parasitoids that attack the eggs of this pest include flies (Sarcophagidae and Tachinidae) and wasps (Scelionidae). Weed suppression and removal of crop residues after harvest in and around fields during the growing season and the overwintering season may help to reduce sources of infestation. Planting clean, non-infested transplants from nursery stock is critical to reduce the spread and potential outbreaks of bagrada bug populations. Screened tunnels, floating row covers and fine-mesh screening material can be used to cover and protect field plantings. Burying the edges of the protective covers can help to prevent bagrada bugs from reaching to the plants (Ahuja *et al.*, 2008).

14.2.22 Seedpod weevil, *Ceutorhynchus assimilis* (Paykull) (Coleoptera: Curculionidae)

The adult seedpod weevil is grey, about 3 mm long and with a pronounced snout. During oviposition, a female makes an opening in the pod wall with its mouthparts and then turns to deposit an egg into the opening. The female then brushes the abdomen over the area, releasing a pheromone that deters other females from ovipositing in the same pod. Larvae develop rapidly in canola and progress through three larval instars in approximately 50 days. Larvae are C-shaped legless grubs with a brown head and are up to 3 mm long. Each larva consumes five to six canola seeds. When mature, the larva chews an opening in the pod wall, drops to the soil, burrows inside, pupates in an earthen cell and the adult emerges about 14 days later. It overwinters as an adult beneath the soil, primarily in wooded areas or grassy vegetation, and there is one generation per year (Cárcamo *et al.*, 2001).

In canola, both larvae and adults can damage the crop. They congregate as adults on flowers, with larvae developing within pods and feeding on seeds and adults feeding through pod pericarp late in the season to further reduce yield and quality. In addition, when the new generation of adults emerges, weevils can feed through the pod walls in late-maturing fields to consume canola seeds, reducing yield and quality. Pods with exit holes shatter before harvest more frequently than non-infested pods and exit holes may serve as entry points for fungal spores that can germinate to reduce yield further.

The development of sustainable control strategies has proceeded in several directions, including chemical, cultural, plant resistance and biological control, and successes in each area have facilitated the development of an integrated management approach for this pest. The chemical compounds and formulations with greatest effectiveness for reducing both adult and larval densities through foliar sprays and seed treatments have been identified. A nominal economic threshold of three to four adult weevils per 180 sweep-net samples when the crop is at 10–20% flower is recommended before applying chemical insecticide. Trap crop technology is tested under prairie field conditions and seeding date recommendations are determined. New canola germplasm has been produced that expresses both antibiotic and antixenotic (non-preference) resistance to the weevil (Dosdall and Cárcamo, 2011). The feasibility of managing the seedpod weevil in winter oilseed rape, *B. napus*, has been studied using an early-flowering trap crop consisting of an autumn-planted winter-type variety with an autumn-planted spring-type variety that flowers 2–3 weeks before the main crop. The trap crop treatment does not adversely affect parasitism of *C. assimilis* by *Trichomalus perfectus* (Walker) but trap crop systems may be effective with lower populations of *C. assimilis* for reliable control of damage (Buntin, 1998).

14.3 Integrated Pest Management (IPM)

Oilseed *Brassica* crops are attacked by several insect pests, of which the aphid is one of the most widespread and important maladies in Pakistan or elsewhere. In an integrated pest-controlling programme of canola and other *Brassica* oilseed crops, several management techniques are combined to provide superior levels of pest control and a dramatic reduction in pesticide usage to have an excellent crop (Figs 14.5, 14.6 and 14.7). The first step is to establish a monitoring programme for accurate assessment of the problem on plants. Pest species are identified, injury levels are established and management

Fig. 14.5. Pollinator on brassica flowers.

Fig. 14.6. Healthy brassica plants.

techniques are applied only if the thresholds are exceeded. Parasites and predators of pest species are located or otherwise imported for release to establish in contributing to the management programme (Sarwar, 2012a, 2013e). Investigations have been conducted on several fronts to manage infestations of these pests, including chemical, cultural, host plant resistance and biological control strategies as given in the following sections.

14.3.1 Monitoring

Monitoring of oilseeds regularly at least twice a week when plants are growing rapidly is meaningful in order to catch pest infestations early, so that growers

Fig. 14.7. Brassica pods.

can knock the pests off or prune out attacked plants. Checks should be made for evidence of natural enemies such as lady beetles, lacewings, syrphid flies and the mummified skins of parasitized aphids, and for disease-killed aphids, which may appear off-colour, bloated, flattened or fuzzy. Substantial numbers of any of these natural control factors can mean that the aphid population may be reduced rapidly without the need for treatment. Aphids tend to be most prevalent along the upwind edge of the field and close to other infested plants of the same species, so a special effort should be made to check these areas.

At least two methods have been used for monitoring aphids or other insects and the most commonly used technique involves visual observation of the plant. On small plants, the entire plant should be examined; and on larger plants the examination should include representative leaves, twigs, stems or other portions of the plant. Where direct observation of aphids is difficult, such as in tall plants, monitoring may be done indirectly by quantifying production of honeydew using water-sensitive spray droplet cards to collect and count aphid honeydew droplets beneath the infested host, and correlating these with abundance of pests in the plant canopy. Normally, observations on the incidence of insect pests are recorded at weekly intervals, starting from initial appearance and continuing to crop maturity, on ten randomly selected plants by counting the numbers of eggs, larvae, nymphs and adults per plant or leaf from various replications (Gullan and Martin, 2003).

As *Brassica* oilseeds suffer heavy losses from aphids, it is impossible to get good grain yields without controlling them. A system of classifying the relative levels of infestation using an aphid population density index on oilseeds is presented in Table 14.1.

14.3.2 Decision-making and thresholds

Data on the number of aphids and other insect pests may be combined with other information, including the injury caused by the pests, the value of the plants being managed and the cost of control activities, to create economic or injury levels and the thresholds. Because of the economic importance of aphids, thresholds and action levels have been established for several aphid–crop systems. However, the work concerning aphids suggests that two aphids per leaf in the spring and five aphids per leaf in the summer can justify intervention. The *B. campestris* crop that is sown early in the Punjab is damaged less, because it matures by the time aphids build up their populations. However, it needs one or two sprays in November and December if necessary. The late-sown *B. campestris* and *B. juncea* in the Punjab suffer a much higher loss owing to severe incidence of the aphid populations. Moreover, excessive delay in sowing would make the wheat sowing late and may also make the crop suffer heavily from aphid invasion. In this way, timely sowing of the crop before 15 October is helpful to escape infestation of aphids. The crop should be threshed as early as possible to avoid further losses and infected plant debris should be disposed of immediately, particularly the green leaves. As the avoidable loss to these crops is very high, it is profitable to spray the crop two to three times, depending upon the level of infestation. The return for every rupee spent on spraying might be five to ten times more profit from the produce (Sarwar *et al.*, 2003).

One of the most important considerations in choosing chemical pesticides is the complex of other pests in the *Brassica* ecosystem. Diamondback moth, cabbage webworm and cabbage looper are of equal or greater importance than the webworm. The assumed impact of insecticides on parasitoids of these and other pests is of prime importance in a pest management programme. While controlling the aphids, care should be taken to use selective pesticides at low dosage. They should be applied at proper intervals, so as to save natural enemies (viz. lady beetles, green lacewings and the syrphid flies), as these predators appear in appreciable numbers in February and March. Where wheat sowing is to follow the oilseed crops, early sowing and the chemical control of aphids would be the right practice. Some of the contact and systematic insecticides, (namely malathion, dimethoate, phosphamidon and menazon) are quite effective but the rate at which aphids build up their populations after the spraying makes it necessary to repeat the application (Sarwar, 2013f). Foliar application of insecticides such as oxydemeton methyl 25 EC or dimethoate 30 EC at 1 ml/l of water is done when 26–28 aphids/plant are observed. If the pest population builds up again, insecticidal spray can be repeated at 15-day intervals, or 5% neem seed kernel extract and 2% neem oil sprays are also effective to manage other mustard insects. In cases of severe infestation the crop is sprayed with endosulfan 35 EC, or quinalphos 25 EC at 1–1.5 ml/l of water, and spraying

Table 14.1. Categorization of relative levels of infestation by aphid populations on canola and other *Brassica* oilseed crops.

Serial no.	Aphid population density index	Relative levels of infestation
1	None (0)	No aphids seen per plant
2	Very light (V)	One to a few aphids per plant and only a few scattered young plants infested, or one to a few aphids per leaf, shoot or other section of larger plant and only a few colonies per large plant with the colonies on young tender leaves or buds are visible
3	Light (L)	About 5–25 aphids per plant and many plants infested, or with many colonies on larger plants, and the colonies are not confined to young shoots
4	Medium (M)	About 26–100 aphids per plant and most plants infested, or with large numbers of aphids on larger plants and not in recognizable colonies, but diffuse and infesting many leaves, stems, etc.
5	Heavy (H)	More than 100 aphids per plant with virtually all plants infested, or with stems, leaves, buds, etc. solidly covered with aphids

should be done in the afternoon for avoidance of toxicity to pollinators. Seed treatment with imidacloprid 70 WS at 5.0 g/kg of seed can be done for management of painted bug and other insects during early crop stages. Infested twigs should be plucked from the rows and destroyed, or dipped in kerosene/ insecticide-treated solution two to three times at 10-day intervals in the crop season to avoid a further increase of insects (Sarwar, 2012b).

14.3.3 Balanced fertilizer

Application of balanced fertilizer is one of the most important components in aphid management. High levels of nitrogen fertilizer favour aphid reproduction, so no greater quantity of nitrogen than necessary should ever be used. Instead, a less soluble form of nitrogen should be applied in small portions throughout the season, rather than all at once. Slow-release fertilizers such as organic fertilizers or urea-based time-release formulations are the best options. Application of balanced dosages of fertilizers (nitrogen, phosphorous and potash at 100:40:40) is required, because use of only nitrogenous fertilizers makes the crop vulnerable to aphids (Shaw-Yhi *et al.*, 2008; Sarwar *et al.*, 2011d).

14.3.4 Non-chemical control

It is very important that fields are ploughed immediately after harvest and this can reduce the spread of pests to other plants. Other useful practices include rotation of plantings with non-host crops, avoiding planting species and cultivars susceptible to aphids, as well as starting to check plants early in the growing season and recording observations on the presence of aphid parasitoids and predators. Crop hygiene, in particular the removal of old crops and destruction of weeds of the family Cruciferae, prevents pest population build-up. Handpicking and destruction of bugs and caterpillars help to reduce plant damage, but handpicking is only practicable in small holdings and is particularly important in the early stages of the crop. Eggs laid in the soil are readily killed by cultivation, so frequent light cultivation (once or twice a week) of field beds can help in controlling the pests. Aluminium foil or silver-coloured reflective mulches have been successfully used to repel invading aphid populations, reduce their numbers on seedlings and small plants and condense transmission of aphid-borne viruses; in addition, yield is usually increased by the greater amount of solar energy reflecting on to leaves. Plastic mulches can inhibit the migration of winged aphids into newly planted areas and these work best with young or small plants up to about 0.3 m tall. For applying a reflective mulch: remove all weeds and cover beds with mulch; bury the edges with soil to hold them down; after the mulch is in place, cut or burn holes (diameter 3–4 inches/75–100 mm) and then plant several seeds or a single transplant into each hole, to repel aphids and some other insects and to enhance crop growth and control weeds (Sarwar, 2015a; Sarwar and Salman, 2015a).

14.3.5 Mechanical and cultural controls

Mechanical and cultural controls should be used where feasible. Before planting, surrounding areas should be checked for sources of aphids and these sources removed. Aphids may be crushed with fingers if infestations are not too extensive. It is important to rid the field and surrounding areas of any alternative host plants such as mustards or other cruciferous weeds. Destruction of plant debris at the end of the season can help to kill overwintering aphid eggs in temperate climates. Planting a nectar plant to attract beneficial insects could also be helpful, e.g. sweet alyssum (*Lobularia maritima* (L.)) as tested in cabbage (Razaq *et al.*, 2011). Replanting on land where an aphid-infested crop has been recently removed should be avoided. Plant spacing can be considered as an effective approach for aphid control in canola (*B. napus*) crops. Where aphid populations are localized on a few curled leaves or new shoots, the best control may be to prune out these areas and dispose of them. In large plants, some aphids thrive in the dense inner canopy; pruning out these areas can make the habitat less suitable. The field should be ploughed in the summer season and clean cultivation followed by weeding, hoeing and burning of debris in and around the field to destroy alternative hosts (Sarwar, 2008).

14.3.6 Biological control

Naturally occurring parasites and predators are important factors in regulating pest population densities. Hundreds of natural enemies have been recorded, principally lady beetles (Coleoptera: Coccinellidae), flower flies (Diptera: Syrphidae), lacewings (Neuroptera: Chrysopidae) and parasitic wasps (Hymenoptera: Braconidae) as the more common types of hunter. Green peach aphid, and many other aphids, may also occasionally be cannibalistic or predatory in habit. Parasitic wasps lay eggs within the aphids and the first sign of parasite activity is the presence of mummified aphids. Syrphid fly maggots and lady beetles are efficient predators of aphids, as are lacewing larvae, called aphid lions and often found among aphid colonies. A big-eyed bug (Hemiptera: Lygaeidae), a predator of the green peach aphid, is important for regulating aphid populations. The parasitic wasp *Diaeretiella rapae* (M'Intosh) (Hymenoptera: Braconidae) lays eggs within half-grown nymphs (preferring second to fourth instars over first instar nymphs and adults) and mummifies them, forming a hard light brown shell around the aphid. Protecting habitat that may foster the populations and survival of natural enemies can help to reduce the need for pesticides. Spot treatment with insecticidal soap, oil, or other approved insecticides when necessary would have less impact on biological control agents than widespread spraying. Integration of chemicals with natural enemies offers promise for enhanced protection from aphid damage. Applying commercially available lady beetles may give some temporary control when properly handled; applications are usually made when aphids first become apparent (Hina *et al.*, 2015; Sarwar and Sattar, 2016).

14.3.7 Biopesticides and botanicals

Biopesticides are more environmentally friendly and safer for operators and consumers but they have low initial toxicity and low efficacy with advanced larval stages and they require a knowledge of pest and host biology. Field studies have been conducted using formulations of neem (extracts of *Azadirachta indica* A. Juss.) against larvae of *Pieris brassicae* and larval mortality remained the highest after 7 days of treatment. The neem formulations are safer for the parasitoid *Cotesia glomerata* L., which parasitizes *P. brassicae* larvae. Aqueous extracts of *Melia azedarach* L. leaves and seeds have also been shown to have significant anti-feedant and deterrent action towards *P. brassicae*. The caterpillars are highly susceptible to purified crystals of strains of *B. thuringiensis* (Bt) and the pests are also susceptible to the fungi *Beauveria bassiana* and *Metarhizium anisopilae*. Klokočar-Šmit *et al.* (2007) investigated the effects of formulations based on *B. thuringiensis* subsp. *kurstaki* (Btk) and spinosad on *P. brassicae* instars, and found spinosad more effective in inducing mortality and reducing of leaf damage by larvae. If an aphid population has exceeded thresholds in the previous season, use of a dormant oil to kill overwintering life stages could be considered. Seed extract of the chinaberry tree (*Melia azedarach* L.), leaf extract of peppermint (*Mentha piperita* L.) and seeds and leaf extract of lantana (*Lantana camara* L.) have shown promising results against cabbage aphid (Sarwar, 2015b, c; 2016a).

14.3.8 Crop irrigation

Another way to reduce aphid or pest populations on robust plants is to knock off the insects with a strong spray of water. Through this practice, most dislodged aphids or pests would not be able to return to the plant and their honeydew can be washed off as well. Using water sprays early in the day allows plants to dry off rapidly in the sun and these can be less susceptible to fungal diseases. Watering and overhead irrigation can disturb the pests and discourage them from feeding on the crop. However, the use of sprinkler irrigation may lead to an increase of diseases such as black rot and downy mildew. It is not surprising that crucifer crops with overhead sprinkle irrigation tend to have fewer diamondback moth larvae than drip- or furrow-irrigated crops and the best results are obtained with daily evening applications. First irrigation should be given 3–4 weeks after sowing as it reduces the pest population significantly. Handpicking of larvae and their destruction especially with irrigation water results in drowning of the pests (Hati *et al.*, 2001).

14.3.9 Mixed cropping

Strong-smelling plants such as garlic, onion or parsley grown near *Brassica* crops are reported to reduce insect infestations. Crop diversity can influence abundance of diamondback moth; and larvae generally are fewer in number and more heavily parasitized when crucifer crops are interplanted with another crop. This does not necessarily lead to reduction in damage; however, surrounding the crops with two or more rows of more preferred hosts such as collard can delay or prevent the dispersal of diamondback moth. Crucifer transplants are often shipped long distances prior to planting and diamondback moth may be included with the transplants. Every effort should be made to assure that transplants are free of insects prior to planting. Oilseed safflower with coriander not only reduces the aphid population on safflower but also increases the predatory population. Crop rotation with non-host crops can be beneficial, because choice of cultivar could also reduce aphid populations and damage. Wheat and barley intercropped with canola (*B. napus*) can reduce the prevalence of aphid populations (Sarwar, 2011).

14.3.10 Alteration of planting time

It has been reported that early-sown *Brassica* crops escape aphid attack but no study has reported that alteration of planting date can reduce or eliminate the application of insecticides. Early sowing can escape the peak pest incidence; for example, crop sown in October and November has maximum infestation, while the September-sown crop experiences less damage, but mid-September is better for sowing to minimize aphid damage. The effect of sowing date for mustard crop (*B. juncea*) on turnip aphid (*L. erysimi*) incidence and population build-up of adults and larvae of its predator, the seven-spotted ladybird beetle (*C. septempunctata*), has been studied under different sowing dates. The aphid incidence has been found to be significantly lowest when sown in October as compared with November; however, a significantly high population of 2.0 beetles/plant was recorded on crop sown in November in comparison with all other sowing dates. The conclusions about changes in sowing time with respect to management of the pest showed that the early-sown crop escaped the aphid damage due to asynchrony in the susceptible crop growth stage and peak aphid population. Consequently, the predator population was also low in the early-sown crop compared with late-sown. Since the crop escaped the pest damage, high predator populations in early-sown crop proved of little importance. Growers, therefore, should sow the crop early in the month of October, preferably by the third week of October to get optimum yield. On the other hand, the aphid population was high in the late-sown crop, which fortunately also harboured high predator populations. Thus, in a late-sown crop, though pest damage is high, there is need for effective decision making by growers for the delayed application of insecticides and measures to conserve the natural enemies (Kular *et al.*, 2012; Sarwar and Sattar, 2013; Khayat, 2015).

Several studies have indicated that weather plays an important role in aphid or other pest appearance, multiplication and disappearance. Summer heat in the valley and desert areas reduces the populations of many species, and aphid activity is also limited during the coldest part of the year. Studies indicate that insect abundance increases with rising temperature and temperature is tightly linked with the density of aphids or other sucking pests. Interestingly, the increase in temperature could also decrease growth of some aphid species, depending on their thermal requirements and host specificity. The appearance of aphids on mustard seems to be largely controlled by temperature and in warm humid locations aphids attain peak population levels earlier compared with relatively cool climates (Whittaker and Tribe, 1998).

14.3.11 Resistant varieties

If resistant varieties are developed, these can provide oil for human consumption that is free from insecticide residues. Most of the work on mechanisms of resistance against aphids has been reported in *Brassica* species other than *B. napus*. Moreover, *B. juncea* is less susceptible to insect pests and disease than other brassicas and there is a difference in aphid population among the varieties tested. Therefore, further research might be undertaken towards morphological or biochemical characters considered responsible for resistance. However, aphid infestation has been proved to be so prolific that none of the germplasm accession is free from it. Extensive screening of *Brassica* germplasm has failed to identify any source of resistance against aphids. It has been argued that, even though some species and varieties among the *Brassica* oilseeds are less susceptible to major insect pests, their resistance is probably not sufficient to prevent significant yield losses due to insect attack. The goal of breeding for resistance has not yet been achieved, as aphids present on crops every year cross the threshold level. Therefore, research on breeding for resistance against aphids in *Brassica* oilseeds is urgently required in Pakistan. Varieties can be graded into highly resistant, resistant, moderately resistant and susceptible, according to their aphid infestation index (Jatoi *et al.*, 2002; Khattak *et al.*, 2002; Sarwar, 2013g, h, i) as given in Table 14.2.

There are many plant characters that are responsible for host plant resistance. The plant structures may have positive as well as negative effects on herbivores and their natural enemies. Morphological characters responsible for resistance against *L. erysimi* in crucifers include smaller and hardy inflorescences with loosely packed buds, darker leaves, non-waxy plants, more branches with a wider

Table 14.2. Grading of oilseed *Brassica* varieties according to the aphid infestation index.

Serial no.	Aphid infestation index	Designation
1	0–1.0	Highly resistant
2	1.1–2.0	Resistant
3	2.1–3.0	Moderately resistant
4	Above 3	Susceptible

angle of orientation and the hairiness of the plant. Among the biochemical characters, presence of higher contents of flavonoids, total sugars and reducing sugars, low quantities of total ash, nitrogen and phosphorus contents in *Brassica* plants induce resistance. Higher concentrations of protein and free amino acids have been reported to be responsible for susceptibility to *L. erysimi* in cruciferous species. Ascorbic acid and glucosinolates contents in *Brassica* species are negatively correlated with aphid populations. Some varieties (canola types Punjab Sarson, Faisal Sarson; or non-canola types Khanpur raya, raya Anmol, hybrid mustard) tested in research trials have shown all-round resistance against aphids; otherwise, insecticides should be applied to manage the pests to avoid economic damage (Anwar and Shafique, 1999; Aslam *et al.*, 2002, 2005).

14.3.12 Chemical control

Chemical insecticides are one of the inevitable components in integrated pest management technology. Despite the numerous options potentially available, many producers are dependent on insecticides for suppression of aphid abundance. Systemic insecticide applications are especially popular at planting time, most of which provide long-lasting protection against aphid population build-up during the critical and susceptible early stages of plant growth and some of which provide protection for 3 months. There is no specific threshold level for aphids on cruciferous crops but insecticides should only be used when aphid populations are high (> 50/ plant) on very young seedlings or transplants. Some aphid populations can be suppressed by conventional insecticides used to control lepidopteran pests. Insecticides such as thiamethoxam at 0.2 g/l or acetamipirid followed by imidacloprid have been found more effective and economical. Application of insecticide at recommended doses (i.e. lambda cyhalothrin 2.5 EC at 300 ml/acre, bifenthrin 10 EC at 250 ml/acre, perfenofos 500 EC at 500 ml/acre) have shown favourable results against aphids. Among several seed dressers, carbosulfan 25 DS at 20 g/kg of seed proved effective as an insecticide (Amer *et al.*, 2009). Because of the waxy nature of the pest and crop, care must be taken that sprays provide good wetting of the crop. Proper rates of surfactants in combination with well adjusted spray equipment are important to achieve insect control with minimum effort. However, when these products are used to control diamondback moth and cabbage webworm (in early season), the beneficial insect complex is maintained and this usually keeps aphid populations under check. A study on insecticide resistance in cabbage aphid carried out in Pakistan reports that aphids have developed resistance to chemicals including methomyl, emamectin benzoate and pyrethroids (cypermethrin, lambda cyhalothrin, bifenthrin and deltamethrin) and to neonicotinoids (imidacloprid, acetamiprid and thiamethoxam). Their resistance level is also found to increase progressively in concurrence with regular use on vegetables (Ahmad and Akhtar, 2013; Sarwar and Salman, 2015b; Sarwar 2016 b, c).

14.3.13 Encouragement of natural enemies

Predator species, such as green lacewing *Chrysoperla carnea* (Stephens) (Neuroptera: Chrysopidae), eleven-spotted ladybird beetle *Coccinella undecimpunctata* (Linn.), seven-spotted ladybird beetle *Coccinella septempunctata* Linn. (Coleoptera: Coccinellidae) and syrphid fly, *Ischiodon scutellaris* (Fabricius) (Diptera: Syrphidae), are recorded when the population of aphids or other pests is sufficiently developed on the canola varieties. Though ladybird beetles, hoverflies and lacewings are frontline natural enemies of *L. erysimi* and *B. brassicae*, their populations are usually insufficient to control aphid populations; however, populations of *Coccinella* beetles are always greater in abundance than those of *Chrysoperla carnea*. Similarly, many other insects feed on aphids, including beetles, flies, earwigs and predaceous bugs, and all are thought to play important roles in reducing aphid populations (Farooq and Tasawar, 2008; Sarwar, 2009; Sarwar and Salman, 2016). One hymenopteran parasitoid, *Diaeretiella rapae* (M'Intosh) (Hymenoptera: Aphidiidae), has been reported as a parasitoid of aphids (*L. erysimi* and *B. brassicae*) on *Brassica* crops; it can be observed parasitizing aphids in some years near maturity of crop, but at this stage the aphids have already damaged the crop and the natural enemy does not perform a noteworthy part in suppression of aphids in *Brassica*. The skin of the parasitized aphid turns crusty and golden brown, which is a form called a mummy and is discoloured and papery in texture. If the adult wasp has emerged, there is a round hole in the mummy where the wasp exited (Sarwar, 2015d).

Another tiny hymenopteran parasitoid, *Trichogramma chilonis* (Ishii) (Hymenoptera: Trichogrammatidae), can be found in the canola crop. The commercial development of this natural enemy and the fact that it attacks so many important caterpillar pests has earned it a place in widespread lists of many pest management parasitoids. Nowadays, *Trichogramma* species are the most extensively used natural enemy of insects in the world, because they are easy to mass rear and can attack many lepidopteran larvae (Sarwar and Salman, 2015c).

Aphids are very susceptible to fungal diseases when it is humid and these pathogens can kill entire colonies of aphids when conditions are appropriate. Signs include dead aphids that have turned reddish or brown and have a fuzzy, shrivelled texture, unlike the shiny, bloated, tan-coloured mummies that are formed when aphids are parasitized. Entomopathogenic fungi *Beauveria bassiana*, *Metarhizium anisopliae* and *Paecilomyces fumosoroseus* are the best examples that attack different types of insects, including for use against the bagrada bug (Khan *et al.*, 1990). Aphids are also susceptible to infection by bacteria, viruses, protozoa and nematodes, but none of these are known to cause high mortality in natural populations.

Growers may encourage natural predators and parasitoids by releasing lacewings, ladybird beetles, syrphid flies, predaceous midges and aphid mummies. Manipulative biological control of oilseed-raiding insect pests can make use of these natural enemies already present in the environment by making conditions as favourable as possible for them and by avoiding spraying chemicals that may interfere with the predators. An augmentative biological control programme recognizes that natural enemies may be present but can be insufficient in numbers to control the pest species and seeks to make up this deficit. Classical biological control introduces a natural enemy species that is not already present in the location with the hope that it may become established and eventually control the pest. Culturing and supplies of these natural enemies are available commercially from insectaries for manipulative, augmentative and classical biological control purposes. If desired, growers can retain these natural enemies to avoid dispersion from crop by providing food to the insects with small amounts of honey or sugar and by the use of small bottle caps or similar items to contain the food. Larvae of predators and aphid mummies may be collected in one area and released in the control area (Sarwar, 2013j). Furthermore, the larvae and adults of these natural enemies may supplement their normal prey in times of scarcity with other types of food. However, when an artificial sugar solution is applied to plant foliage with a spray bottle in an attempt to supplement food further, then predators and parasitoids can be attracted to agroecosystems near crop maturity. To apply ladybird beetles as pest controls, they should be kept refrigerated until just before releasing and allowing them into the field at dusk, as those released in daylight might fly away immediately. The ladybird beetles should be misted with water just before release and the surface of the plant on to which they are being released should also be misted over. The ladybird beetles are placed at the base of infested plants or in the forks of low branches so that the beetles can crawl higher onto the plant in search of aphid prey. It is useful to grow nectar-producing flowering plants that can attract adults of these insects and provide suitable habitat that will encourage predators to remain in the vicinity. Beneficial organisms may also be preserved by using insecticides with short residual activities, such as soap and oil, and avoiding treatments of large numbers of plants in favour of spot treating if it is only individual plants that require intervention (Suhail *et al.*, 1999; Sarwar, 2016d, e).

14.4 Future Development of Integrated Control

In Pakistan, canola, rapeseed and mustard are the third most important source of edible oil after cotton and sunflower. A complex of insect and mite pests inflict damage on oilseed *Brassica* species in this country. Insect pests such as aphids and painted bugs are major pests that cause severe damage at critical crop stages. Cabbage aphid, *B. brassicae*, and turnip aphid, *L. erysimi*, are continuously observed as the most abundant among all insects detected in the field. Only a few plants of *B. napus* are found to be infested by *M. persicae*. Populations of the cabbage aphid are always in greater abundance than those of turnip aphid on *B. napus* and *B. juncea*.

Aphids may infest rapeseed crops at any stage of growth but damage is most common during the flowering and pod formation periods. The painted bug is observed at two distinct crop stages, firstly at seedling stage and secondly at maturity of the plant. Green stink bugs usually occur late in the season when weather is hot and dry; they suck sap from pods and reduce pod setting and seed viability, affecting oil quantity. Diamondback moth and *Helicoverpa* are common at flowering to pod-filling stages and can chew pods and seed; therefore regular monitoring is necessary to check whether these pests have caused damage. Red-legged earth mites can destroy emerging seedlings within days of emergence by sucking sap from plants and typical damage appears as silvering or whitening of host plant. The *Helicoverpa* bud borer, diamondback moth (*P. xylostella*), cabbage aphid (*B. brassicae*), turnip aphid (*L. erysimi*) and green peach aphid (*M. persicae*) are regular and predictable pests at the flowering and pod formation stage of plants. Flea beetles (*P. cruciferae*) damage different oilseed *Brassica* crops in the early season, whereas cabbage aphid, turnip aphid and green peach aphid, diamondback moth and seedpod weevil (*C. assimilis*) are the major late-season insect pests. Incidences of mustard sawfly (*A. lugens proxima*), painted bug (*B. hilaris*) and cabbage butterfly (*P. brassicae*) are found at different growth stages of rapeseed–mustard crops and are essentially noticed from seedling stage to harvest. Mustard sawfly is observed at seedling stage, whereas the incidence of cabbage butterfly is found from the pod-bearing stage of the crop.

Besides the crop stage, climatic conditions and varietal differences play a vital role in the incidence and multiplication of aphids. Cloudy weather conditions along with temperatures in the range of 10–20°C and relative humidity of 82% favour the survival and multiplication of the insects. Among *Brassica* crop species, *B. campestris* is found to be more susceptible to aphids as compared with *B. juncea*, *B. carinata* and *Eruca sativa*. Having a shorter maturity period, *B. campestris* can suffer from greater severity of aphid attack because of early flowering. In contrast, *B. carinata*, with a long maturity period, has lesser attack by mustard aphid which coincides with its late flowering.

From the above discussion it is concluded that aphids are among the important insect pests of *Brassica* crops in Pakistan. Moreover, methods of pest management such as host plant resistance, altered planting time and occurrence of natural enemies cannot significantly reduce losses due to aphids. Therefore, application of sprays is inevitable and insecticides, if applied properly with ecological principles, are not an environmental evil. Timing of insecticides with respect to crop stage, pollinators and economic threshold level needs to be determined and future research should be directed towards these aspects. Control of alternative hosts plants of the pest species in the field vicinity is more desirable and can successfully control aphids. Aluminium foil or white plastic mulches should be used if possible in newly planted areas. The reflection of light from these materials will confuse aphids and prevent them from landing on plants.

Strategically, the key IPM considerations for canola and other brassicas are year-round approaches to pest management which includes off-season operations and planning as well as crop management. There needs to be regular crop monitoring, recording numbers of pests and beneficial enemies and reviewing data for population trends. Canola can compensate for early damage by setting new buds and pods to replace parts damaged by pests; however, excessive early damage may reduce yield. Biopesticides used in vegetative canola prior to flowering can preserve beneficial enemies of insect pests. Syrphid fly maggots, ladybird beetle adults and larvae and lacewing larvae (aphid lions) are common predators of aphids. Nuclear polyhedrosis viruses are effective against *Helicoverpa* larvae < 7 mm in length, and Bt is effective against diamondback moth and *Helicoverpa* (< 7 mm in length). The use of spray oils should be considered where aphid populations are low to moderate and applications should be repeated if required. Where pests invade from adjacent fields, consideration should be given to spraying only field borders and not the whole crop area. The use of aphid-selective products (e.g. pirimicarb) to preserve the beneficial insects potentially reduces the need for follow-up applications. Seed dressings may be the most effective control for some soil insects, as well as the least disruptive to natural enemies. Cultural control methods may be considered but increasing insecticide usage may not solve pest problems, instead selecting for pests that are more difficult to kill. Oilseed brassicas are minor crops in Pakistan and farmers are not well aware about losses due to insect and mite pests. Therefore, awareness should be created through extension workers to reduce the losses due to pest invasions.

References

Ahmad, M. (2005) Diamondback moth, *Plutella xylostella*: a review of its biology, ecology, and control. *Journal of Agricultural Research* 43(4), 361–382.

Ahmad, M. and Akhtar, S. (2013) Development of insecticide resistance in field populations of *Brevicoryne brassicae* (Hemiptera: Aphididae) in Pakistan. *Journal of Economic Entomology* 106, 954–958.

Ahmad, M.J., Yaqoob, M. and Mohi-Uddin, S. (2010) Seasonal occurrence of leaf miner, *Chromatomyia horticola* (Goureau) (Diptera: Agromyzidae) and its parasitoids, *Diglyphus* spp. on brown mustard, *Brassica campestris* in Kashmir valley. *Journal of Biological Control* 24(4), 300–304.

Ahmad, M., Ghaffar, A. and Rafiq, M. (2013) Host plants of leaf worm, *Spodoptera litura* (Fabricius) (Lepidoptera: Noctuidae) in Pakistan. *Asian Journal of Agriculture and Biology* 1(1), 23–28.

Ahuja, R., Singh, S.K., Sundria, M.M. and Dhandapani, A. (2008) Integrated management strategy for painted bug, *Bagrada hilaris* (Burm.), inflicting injury at seedling stage of mustard (*Brassica juncea*) in arid western Rajasthan. *Pesticide Research Journal* 20, 48–51.

Amer, M., Aslam, M., Razaq, M. and Afzal, M. (2009) Lack of plant resistance against aphids, as indicated by their seasonal abundance in canola (*Brassica napus* L.) in southern Punjab, Pakistan. *Pakistan Journal of Botany* 41(3), 1043–1051.

Amjad, M., Islam, N. and Kakakhel, S.A. (1999) Turnip aphid *Lipaphis erysimi* Kalt. (Homoptera: Aphididae) biology, intrinsic rate of increase and development threshold temperature on higher seed *Brassica*. *Pakistan Journal of Biological Science* 2, 599–602.

Anwar, M. and Shafique, M. (1999) Relative development of aphids on different *Brassica* cultivars. *Pakistan Journal of Zoology* 31, 357–359.

Aslam, M. and Razaq, M. (2007) Arthropod fauna of *Brassica napus* and *Brassica juncea* from southern Punjab (Pakistan). *Journal of Agriculture and Urban Entomology* 24(2), 49–50.

Aslam, M., Ahmad, M., Islam, Z. and Anjum, S. (2002) Population of aphid, *Lipaphis erysimi* (Kalt.) on canola (*Brassica napus*). *Science Technology and Development* 21, 41–42.

Aslam, M., Razaq, M., Islam, Z. and Shahzad, A. (2005) Comparison of different canola (*Brassica napus* L.) varieties for resistance against cabbage aphid (*Brevicoryne brassicae* L.). *International Journal of Agriculture and Biology* 7, 781–782.

Baker, G. (2010) Pests of canola and their management. In: *Canola Best Practice Management Guide*. Grains Research and Development Corporation, Canberra, Australia, pp. 47–58.

Bjorksten, T.A., Robinson, M. and La Salle, J. (2005) Species composition and population dynamics of leafmining flies and their parasitoids in Victoria. *Australian Journal of Entomology* 44, 186–191.

Bundy, C.S., Grasswitz, T.R. and Sutherland, C. (2012) First report of the invasive stink bug *Bagrada hilaris* (Burmeister) (Heteroptera: Pentatomidae) from New Mexico, with notes on its biology. *Southwestern Entomologist* 37, 411–414.

Buntin, G.D. (1998) Cabbage seedpod weevil (*Ceutorhynchus assimilis* Paykull) management by trap cropping and its effect on parasitism by *Trichomalus perfectus* (Walker) in oilseed rape. *Crop Protection* 17(4), 299–305.

Buntin, G.D. and Raymer, P.L. (1994) Pest status of aphids and other insects in winter canola in Georgia. *Journal of Economic Entomology* 87, 1097–1104.

Capinera, J.L. (2001) *Handbook of Vegetable Pests*. Academic Press, San Diego, California, 729 p.

Cárcamo, H.A., Dosdall, L.M., Dolinski, M., Olfert, O. and Byers, R. (2001) The cabbage seedpod weevil, *Ceutorhynchus obstrictus* (Coleoptera: Curculionidae) – a review. *Journal of the Entomological Society of British Columbia* 98, 201–209.

Cartea, M.E., Francisco, M., Soengas, P. and Velasco, P. (2011) Phenolic compounds in brassica vegetables. *Molecules* 16, 251–280.

Chowdhury, M. (2009) Incidence of saw fly, *Athalia lugens proxima* Klug. as influenced by level of irrigation and fertilizers on mustard. *The Journal of Plant Protection Sciences* 1(1), 80–82.

Dornan, A.P., Sears, M.K. and Stewart, J.G. (1995) Evaluation of a binomial model for insecticide application to control lepidopterous pests in cabbage. *Journal of Economic Entomology* 88, 302–306.

Dosdall, L.M. and Cárcamo, H.A (2011) Biology and integrated management of the cabbage seedpod weevil in prairie canola crops. *Prairie Soils & Crops Journal* 4, 14–23.

Farooq, A. and Tasawar, Z. (2008) Evaluation of integrated management of aphid pests, *Brevicoryne brassicae* and *Lipaphis erysimi* on canola crop in Southern Punjab, Pakistan. *Pakistan Journal of Zoology* 40(1), 13–17.

Fathipour, Y. and Sedaratian, A. (2013) Integrated management of *Helicoverpa armigera* in soybean cropping systems. In: El-shemy, H. (ed.) *Soybean Pest Resistance*. InTech, Rijeka, Croatia, pp. 232–280.

Fekrat, L. and Shishehbor, P. (2007) Some biological features of cotton whitefly, *Bemisia tabaci* (Homoptera: Aleyrodidae) on various host plants. *Pakistan Journal of Biological Sciences* 10(18), 3180–3184.

Follett, P.A., Wright, M.G. and Golden, M. (2009) *Nezara viridula* (Hemiptera: Pentatomidae) feeding patterns in macadamia nut in Hawaii: nut maturity and cultivar effects. *Environment Entomology* 38(4), 1168–1173.

Fontana, P., Buzzetti, F.M. and Marino-Perez, R. (2011) New Acrididae from Oaxaca state in Mexico

(Orthoptera: Caelifera: Acrididae: Ommatolampinae, Melanoplinae). *Zootaxa* 2862, 39–55.

Frago, E., Selfa, J., Pujade-Villar, J., Guara, M. and Bauce, E. (2009) Age and size thresholds for pupation and developmental polymorphism in the browntail moth, *Euproctis chrysorrhoea* (Lepidoptera: Lymantriidae), under conditions that either emulate diapause or prevent it. *Journal of Insect Physiology* 55, 952–958.

Frank, S.D. and Shrewsbury, P.M. (2004) Effect of conservation strips on the abundance and distribution of natural enemies and predation of *Agrotis ipsilon* (Lepidoptera: Noctuidae) on golf course fairways. *Environmental Entomology* 33(6), 1662–1672.

Gawałek, M., Dudek, K., Ekner-Grzyb, A., Kwieciński, Z. and Sliwowska, J.H. (2014) Ecology of the field cricket (Gryllidae: Orthoptera) in farmland: the importance of livestock grazing. *North-Western Journal of Zoology* 10(2), 325–332.

Gemeno, C. and Haynes, K.F. (2000) Periodical and age related variation in chemical communication system of black cutworm moth, *Agrotis ipsilon*. *Journal of Chemical Ecology* 26(2), 329–342.

Government of Pakistan (2015) *Pakistan Economic Survey, 2014–15*. Oilseed Development Board, Economic Adviser's Wing, Finance Division, Government of Pakistan, Islamabad.

Gullan, P.J. and Martin, J.H. (2003) Sternorrhyncha (jumping plant-lice, whiteflies, aphids and scale insects). In: Resh, V. and Carde, R. (eds) *Encyclopedia of Insects. Academic Press*, Amsterdam, pp. 1079–1089.

Hati, K.M., Mandal, K.G., Mihsra, A.K., Ghosh, P.K. and Acharya, C.L. (2001) Evapotranspiration, water-use efficiency, moisture use and yield of Indian mustard (*Brassica juncea*) under varying levels of irrigation and nutrient management in vertisol. *Indian Journal of Agricultural Sciences* 74, 339–342.

Hina, H.K., Sarwar, M. and Lohar, M.K. (2015) Repellence activity of plant oils against red flour beetle *Tribolium castaneum* (Herbst) (Coleoptera: Tenebrionidae) in wheat. *International Journal of Animal Biology* 1(3), 86–92.

Huang, T.I., Reed, D.A., Perring, T.M. and Palumbo, J.C. (2014) Host selection behavior of *Bagrada hilaris* (Hemiptera: Pentatomidae) on commercial cruciferous host plants. *Crop Protection* 59, 7–13.

Jahan, F., Abbasipour, H., Askarianzade, A., Hasanshahi, G. and Saeedizadeh, A. (2013) Effect of eight cauliflower cultivars on biological parameters of the cabbage aphid, *Brevicoryne brassicae* (L.) (Hem: Aphididae) in laboratory conditions. *Archives of Phytopathology and Plant Protection* 46, 636–642.

Jatoi, M.Y., Javed, H. and Kakakhel, S.A. (2002) Relative resistance among 22 *Brassica napus* cultivars against turnip aphid, *Lipaphis erysimi* Kalt. *Asian Journal of Plant Sciences* 1, 558–559.

Jayanthi, P.D.K. and Ramesh, C.R. (2014) Severe incidence of mustard sawfly, *Athalia lugens proxima* (Klug) on leafy Chinese cabbage, *Brassica rapa* ssp. *chinensis*. *Pest Management in Horticultural Ecosystems* 20(1), 95–96.

Karimullah, Naeem M., Ahmad, S., Paracha, A.M. and Jan, M.T. (1995) Infestation of aphid, *Myzus persicae* (Sulz.) on potato in Peshawar, NWFP Pakistan. *Sarhad Journal of Agriculture* 11, 741–744.

Khan, M., Khalil, S.K. and Karamatullah (1990) Biological control of aphid with an entomopathogenic fungus *Verticillium lecanii*. *Pakistan Journal of Agricultural Research* 11, 174–177.

Khan, M.H., Sarwar, M., Farid, A. and Syed, F. (2010) Compatibility of pyrethroid and different concentrations of neem seed extract on parasitoid *Trichogramma chilonis* (Ishii) (Hymenoptera: Trichogrammatidae) under laboratory conditions. *The Nucleus* 47(4), 327–331.

Khanobporn, T., Reddy, G.V.P., Wu, S., Miller, J.H., Ophus, V.L. and Prewett, J. (2014) Developing nominal threshold levels for *Phyllotreta cruciferae* (Coleoptera: Chrysomelidae) damage on canola in Montana, USA. *Crop Protection* 66, 8–13.

Khanzada, M.S., Syed, T.S., Khanzada, S.R., Abro, G.H., Salman, M., Anwar, S., Sarwar, M., Perzada, A.A., Wang, S. and Abro, A.H. (2016) Occurrence and abundance of thrips, whitefly and their natural enemy, *Geocoris* spp. on cotton crop at various localities of Sindh, Pakistan. *Journal of Entomology and Zoology Studies* 4(1), 509–515.

Khattak, S.U., Hameed, M., Khan, A.U., Zeb, A. and Farid, A. (2002) Pesticidal control of rapeseed aphid, *Brevicoryne brassicae* L. *Pakistan Journal of Zoology* 34, 222–228.

Khayat, M. (2015) Measuring yield and its components of canola genotypes in different planting date. *Indian Journal of Fundamental and Applied Life Sciences* 5(2), 48–55.

Klokočar-Šmit, Z.D., Inđić, D.V., Vuković, S.M., Filipović, M.M. and Červenski, J.F. (2007) Preliminary investigation on the effects of biological and synthetic insecticides on large white butterfly (*Pieris brassicae* L.) larvae. *Proceedings for Natural Sciences, Matica Srpska, Novi Sad* 112, 75–82.

Kular, J.S., Brar, A.S. and Kumar, S. (2012) Population development of turnip aphid *Lipaphis erysimi* (Kaltenbach, 1843) (Hemiptera: Aphididae) and the associated predator *Coccinella septempunctata* Linnaeus, 1758 as affected by changes in sowing dates of oilseed Brassica. *Entomotropica* 27(1), 19–25.

Kumaranag, K.M., Kedar, S.C., Thodsare, N.H. and Bawaskar, D.M. (2014) Insect pests of cruciferous vegetables and their management. *Popular Kheti* 2(1), 80–86.

Meagher, R.L., Brambila, J. and Hung, E. (2008) Monitoring for exotic *Spodoptera* species (Lepidoptera: Noctuidae) in Florida. *Florida Entomologist* 91(4), 517–522.

Mewis, I., Ulrichs, C. and Schnitzler, W.H. (2002) The role of glucosinolates and their hydrolysis products in

oviposition and host-plant finding by the cabbage webworm, *Hellula undalis*. *Entomologia Experimentalis et Applicata* 105, 129–139.
Nikdel, M., Niknam, G. and Dordaei, A.A. (2010) Evaluation of susceptibility of the brown tail moth, *Euproctis chrysorrhoea* (L.), to entomopathogenic nematodes under laboratory conditions. *Nematology Mediterranean* 38, 3–6.
Obopile, M., Munthali, D.C. and Matilo, B. (2008) Farmers' knowledge, perceptions and management of vegetable pests and diseases in Botswana. *Crop Protection* 27(8), 1220–1224.
Pal, M. and Singh, R. (2013) Biology and ecology of the cabbage aphid, *Brevicoryne brassicae* (Linn.) (Homoptera: Aphididae): a review. *Journal of Aphidology* 27, 59–78.
Panizzi, A.R. (2008) Southern green stink bug, *Nezara viridula* (L.) (Hemiptera: Heteroptera: Pentatomidae). *In*: Capinera, J.L. (ed.) *Encyclopedia of Entomology*. Springer, Heidelberg, Germany, pp. 3471–3471.
Razaq, M., Aslam, M., Amer, M. and Shad, S.A. (2011) Insect pest status of aphids on oilseed brassica crops and need for chemical control. *Crop and Environment* 2(2), 60–63.
Reed, D., Palumbo, J.C., Perring, T.M. and May, C. (2013) *Bagrada hilaris* (Hemiptera: Pentatomidae), an invasive stink bug attacking cole crops in the southwestern United States. *Journal of Integrated Pest Management* 4(3), 1–7.
Sadozai, A. and Khan, I.A. (2014) Developmental response of *Pieris brassicae* (L.) (Lepidoptera: Pieridae) on different cauliflower cultivars under laboratory conditions. *American Journal of Plant Sciences* 5, 2611–2616.
Saljoqi, A.U.R., Riaz-ul-Haq, Ehsan-ul-Haq, Khan, J. and Ali, G. (2015) Rearing of *Spodoptera litura* (Fabricius) on different artificial diets and its parasitization with *Trichogramma chilonis* (Ishii). *Pakistan Journal of Zoology* 47(1), 169–175.
Sarwar, M. (2008) Plant spacing – a non polluting tool for aphid (Hemiptera: Aphididae) management in canola, *Brassica napus*. *Journal of Entomological Society of Iran* 27(2), 13–22.
Sarwar, M. (2009) Populations' synchronization of aphids (Homoptera: Aphididae) and ladybird beetles (Coleoptera: Coccinellidae) and exploitation of food attractants for predator. *Biological Diversity and Conservation* 2(2), 85–89.
Sarwar, M. (2011) Effects of wheat and barley intercropping ecosystem on the prevalence of aphid (Hemiptera: Aphididae) population in canola (*Brassica napus* L.) crop. *Biological Diversity and Conservation* 4(1), 11–16.
Sarwar, M. (2012a) Frequency of insect and mite fauna in chilies *Capsicum annum* L., onion *Allium cepa* L. and garlic *Allium sativum* L. cultivated areas, and their integrated management. *International Journal of Agronomy and Plant Production* 3(5), 173–178.
Sarwar, M. (2012b) Competency of natural and synthetic chemicals in controlling gram pod borer, *Helicoverpa armigera* (Hubner) on chickpea crop. *International Journal of Agricultural Sciences* 2(4), 132–135.
Sarwar, M. (2013a) Categorization of selected canola *Brassica napus* L. varieties/lines for tolerance to natural infestation of aphid *Myzus persicae* (Sulzer). *Biological Diversity and Conservation* 6(3), 35–39.
Sarwar, M. (2013b) Comparative suitability of soil and foliar applied insecticides against the aphid *Myzus persicae* (Sulzer) (Aphididae: Hemiptera) in canola *Brassica napus* L. *International Journal of Scientific Research in Environmental Sciences* 1(7), 138–143.
Sarwar, M. (2013c) Comparing abundance of predacious and phytophagous mites (Acarina) in conjunction with resistance identification between Bt and non-Bt cotton cultivars. *African Entomology* 21(1), 108–118.
Sarwar, M. (2013d) Exploration on resource of resistance in chickpea (*Cicer arietinum* L.) genotypes to gram pod borer *Helicoverpa armigera* (Hubner) (Lepidoptera). *African Journal of Agricultural Research* 8(26), 3431–3435.
Sarwar, M. (2013e) Integrated pest management (IPM) – a constructive utensil to manage plant fatalities. *Journal of Agriculture and Allied Sciences* 2(3), 1–4.
Sarwar, M. (2013f) The inhibitory properties of organic pest control agents against aphid (Aphididae: Homoptera) on canola *Brassica napus* L. (Brassicaceae) under field environment. *International Journal of Scientific Research in Environmental Sciences* 1(8), 195–201.
Sarwar, M. (2013g) Relative degree of susceptibility and resistance of different *Brassica campestris* L. genotypes against aphid *Myzus persicae* – a field investigation. *The Nucleus* 50(1), 81–86.
Sarwar, M. (2013h) Study on differences in some new mustard *Brassica campestris* L. genotypes for having resistance and susceptibility feedback infected with aphid *Myzus persicae* (Sulzer) (Homoptera: Aphididae). *Global Journal of Scientific Researches* 1(3), 80–84.
Sarwar, M. (2013i) Assessment of genetic divergence in rapeseeds *Brassica napus* L. and *Brassica campestris* L. crops for exploitation of host plant tolerance to aphid *Myzus persicae* (Sulzer). *Journal of Cereals and Oilseeds* 4(8), 101–105.
Sarwar, M. (2013j) Studies on incidence of insect pests (Aphids) and their natural enemies in canola *Brassica napus* L. (Brassicaceae) crop ecosystem. *International Journal of Scientific Research in Environmental Sciences* 1(5), 78–84.
Sarwar, M. (2014a) The propensity of different larval stages of lacewing *Chrysoperla carnea* (Stephens) (Neuroptera: Chrysopidae) to control aphid *Myzus persicae* (Sulzer) (Homoptera: Aphididae) evaluated on canola *Brassica napus* L. *Songklanakarin Journal of Science and Technology* 36(2), 143–148.
Sarwar, M. (2014b) Knowing about identify and mode of damage by insect pests attacking winter vegetables

and their management. *Journal of Ecology and Environmental Sciences* 2(4), 1–8.
Sarwar, M. (2014c) Some insect pests (Arthropoda: Insecta) of summer vegetables, their identification, occurrence, damage and adoption of management practices. *International Journal of Sustainable Agricultural Research* 1(4), 108–117.
Sarwar, M. (2015a) Usage of biorational pesticides with novel modes of action, mechanism and application in crop protection. *International Journal of Materials Chemistry and Physics* 1(2), 156–162.
Sarwar, M. (2015b) Biopesticides: an effective and environmental friendly insect-pests inhibitor line of action. *International Journal of Engineering and Advanced Research Technology* 1(2), 10–15.
Sarwar, M. (2015c) Information on activities regarding biochemical pesticides: an ecological friendly plant protection against insects. *International Journal of Engineering and Advanced Research Technology* 1(2), 27–31.
Sarwar, M. (2015d) Microbial insecticides – an eco-friendly effective line of attack for insect pests management. *International Journal of Engineering and Advanced Research Technology* 1(2), 4–9.
Sarwar, M. (2016a) A potent folklore of botanical plant materials against insect pests together with their preparations and applications. *Sky Journal of Biochemistry Research* 5(4), 58–62.
Sarwar, M. (2016b) Indoor risks of pesticide uses are significantly linked to hazards of the family members. *Cogent Medicine* 3, 1155373.
Sarwar, M. (2016c) Biological Control to Maintain Natural Densities of Insects and Mites by Field Releases of Lady Beetles (Coleoptera: Coccinellidae). *International Journal of Entomology and Nematology* 2(1), 21–26.
Sarwar, M. (2016d) Food habits or preferences and protecting or encouraging of native ladybugs (Coleoptera: Coccinellidae). *International Journal of Zoology Studies* 1(3), 13–18.
Sarwar, M. (2016e) Recognition of some lady beetles (Coleoptera: Coccinellidae) deadly sighted for insect and mite pests in agroecosystems. *International Journal of Entomology Research* 1(2), 29–34.
Sarwar, M. and Salman, M. (2015a) Success stories of eco-friendly organically acceptable insecticides as natural products discovery. *International Journal of Materials Chemistry and Physics* 1(3), 392–398.
Sarwar, M. and Salman, M. (2015b) Insecticides resistance in insect pests or vectors and development of novel strategies to combat its evolution. *International Journal of Bioinformatics and Biomedical Engineering* 1(3), 344–351.
Sarwar, M. and Salman, M. (2015c) Biological insecticide *Trichogramma* spp. (Hymenoptera: Trichogrammatidae) strikes for caterpillar control. *International Journal of Entomology Research* 1(1), 31–36.
Sarwar, M. and Salman, M. (2016) From production to field application methodology of generalist predator green lacewing, *Chrysoperla carnea* [Stephens] (Neuroptera: Chrysopidae). *International Journal of Zoology Research* 1(1), 35–40.
Sarwar, M. and Saqib, S.M. (2010) Rearing of predatory seven spotted ladybird beetle *Coccinella septempunctata* L. (Coccinellidae) on natural and artificial diets under laboratory conditions. *Pakistan Journal of Zoology* 42(1), 47–51.
Sarwar, M. and Sattar, M. (2013) Varietals variability of winter rapes (*Brassica napus* L.) for their susceptibility to green aphid, *Myzus persicae* (Sulzer) (Homoptera: Aphididae). *Pakistan Journal of Zoology* 45(4), 883–888.
Sarwar, M. and Sattar, M. (2016) An analysis of comparative efficacies of various insecticides on the densities of important insect pests and the natural enemies of cotton, *Gossypium hirsutum* L. *Pakistan Journal of Zoology* 48(1), 131–136.
Sarwar, M., Ahmad, N., Siddiqui, Q.H., Rajput, A.A. and Tofique, M. (2003) Efficiency of different chemicals on canola strain rainbow (*Brassica napus* L.) for aphids control. *Asian Journal of Plant Sciences* 2(11), 831–833.
Sarwar, M., Ahmad, N., Bux, M., Rajput, A.A. and Tofique, M. (2004a) Response of various *Brassica* genotypes against aphids infestation under natural conditions. *Pakistan Journal of Zoology* 36(4), 69–74.
Sarwar, M., Ahmad, N., Siddiqui, Q.H., Ali, A. and Tofique, M. (2004b) Genotypic response in canola (*Brassica* species) against aphid (Aphidae: Homoptera) attack. *The Nucleus* 41(1–4), 87–82.
Sarwar, M., Ahmad, N., Khan, G.Z. and Tofique, M. (2009a) Varietals resistance and susceptibility in mustard (*Brassica campestris* L.) genotypes against aphid *Myzus persicae* (Sulzer) (Homoptera: Aphididae). *The Nucleus* 46(4), 507–512.
Sarwar, M., Ahmad, N. and Tofique, M. (2009b) Host plant resistance relationships in chickpea (*Cicer arietinum* L.) against gram pod borer (*Helicoverpa armigera*) (Hubner). *Pakistan Journal of Botany* 41(6), 3047–3052.
Sarwar, M., Ahmad, N., Bux, M., Nasrullah and Tofique, M. (2011a) Comparative field evaluation of some newer versus conventional insecticides for the control of aphids (Homoptera: Aphididae) on oilseed rape (*Brassica napus* L.). *The Nucleus* 48(2), 163–167.
Sarwar, M., Ahmad, N. and Tofique, M. (2011b) Identification of susceptible and tolerant gram (*Cicer arietinum* L.) genotypes against gram pod borer (*Helicoverpa armigera*) (Hubner). *Pakistan Journal of Botany* 43(2), 1265–1270.
Sarwar, M., Ahmad, N. and Tofique, M. (2011c) Impact of soil potassium on population buildup of aphid (Homoptera: Aphididae) and crop yield in canola (*Brassica napus* L.) field. *Pakistan Journal of Zoology* 43(1), 15–19.
Sarwar, M., Ahmad, N., Tofique, M. and Salam, A. (2011d) Efficacy of some natural hosts on the development

of *Chrysoperla carnea* (Stephens) (Neuroptera: Chrysopidae) – a laboratory investigation. *The Nucleus* 48(2), 169–173.

Sarwar, M.F., Sarwar, M.H., Sarwar, M., Qadri, N.A. and Moghal, S. (2013) The role of oilseeds nutrition in human health: a critical review. *Journal of Cereals and Oilseeds* 4(8), 97–100.

Sattar, M., Sarwar, M., Rahoo, G.M. Rizvi, S.N.H. and Sher, M.U. (2004) Population development of whitefly, *Bemisia tabaci* (Gennadius) (Homoptera: Aleyrodidae) on some cucurbit crops. *Pakistan Journal of Zoology* 36(4), 261–265.

Sattar, M., Sarwar, M. and Hussain, H. (2005) Impact of pH, nitrogen and protein contents in some cucurbits on the population build-up of whitefly, *Bemisia tabaci* (Homoptera: Aleyrodidae). *Pakistan Journal of Zoology* 37(4), 275–279.

Schell, S. and Schell, S. (2007) Devastating grasshopper species fact sheet. (Extract modified for electronic publication from Pfadt, R.E. (1994) *Wyoming Agricultural Experiment Station Bulletin 912: Field Guide to Common Western Grasshoppers*, 2nd edn). Wyoming Agricultural Experiment Station, Laramie, Wyoming.

Shaw-Yhi, H., Cheng-Hsiang, I. and Tse-Chi, S. (2008) Effects of plant nutrient availability and host plant species on the performance of two *Pieris* butterflies (Lepidoptera: Pieridae). *Biochemical Systematics and Ecology* 36(7), 505–513.

Shelton, A.M., Plate, J. and Chen, M. (2008) Advances in control of onion thrips (Thysanoptera: Thripidae) in cabbage. *Journal of Economic Entomology* 101(2), 438–443.

Shepard, B.M., Carner, G.R., Barrion, A.T., Ooi, P.A.C. and Van den Berg, H. (1999) *Insects and Their Natural Enemies Associated with Vegetables and Soybean in Southeast Asia*. Quality Printing Company, Orangeburg, South Carolina.

Singhamuni, S.A.A., Jayasuriya, M.I.U.F., Hemachandra, K.S. and Sirisena, U.G.A.I. (2015) Evaluation of the potential of *Trichogramma chilonis* Ishii (Hymenoptera: Trichogrammatidae) as a bio-control agent for *Trichoplusia ni*, cabbage semi-looper. *Tropical Agricultural Research* 26(2), 223–236.

Soroka, J.J. and Elliott, B. (2011) Innovative methods for managing flea beetles in canola. *Prairie Soils & Crops Journal* 4, 1–7.

Sugie, H., Yase, J., Futai, K. and Shirai, Y. (2003) A sex attractant of the cabbage webworm, *Hellula undalis* Fabricius (Lepidoptera: Pyralidae). *Applied Entomology and Zoology* 38(1), 45–48.

Suhail, A., Sabir A.M., Hussain A. and Saeed, A. (1999) Predatory efficacy of *Coccinella septempunctata* L. on cotton *Brevicoryne brassicae*. *Aphis gossypii* Glov. *Pakistan Journal of Biological Science* 2, 603–605.

Sultana, N.A., Khan, M.A.H., Islam, M.N. and Hasanuzzaman, M. (2009) Integrated management of aphid (*Lipaphis erysimi* Kalt.) in mustard. *World Journal of Zoology* 4(2), 105–108.

Syed, T.S. and Abro, G.H. (2003) Effect of *Brassica* vegetable hosts on biology and life table parameters of *P. xylostella* under laboratory conditions. *Pakistan Journal of Biological Science* 6(22), 1891–1896.

Syed, T.S., Abro, G.H., Awan, M.S. and Sattar, M. (2012) The role of natural enemies to control diamondback moth, *Plutella xylostella* (L.) population in various seasons. *Pakistan Journal of Zoology* 44(6), 1479–1485.

Syed, T.S., Khanzada, M.S., Khanzada, S.R., Abro, G.H., Salman, M., Sarwar, M., Dayo, S.H., Anwar, S. and Su, W. (2015) Population dynamics of thrips, whiteflies and their natural enemies on mustard (*Brassica campestris* L.) crop in different localities of Sindh, Pakistan. *Journal of Entomology and Zoology Studies* 4(1), 7–16.

Talpur, M.A., Qureshi, K.H. and Nizamani, I.A. (2002) Effectiveness of different insect cides against greasy cutworm, *Agrotis ipsilon* (Hufn.) on cauliflower crop. *Pakistan Journal of Applied Sciences* 2(2), 216–218.

Taylor, M.E., Bundy, C.S. and McPherson, J.E. (2015) Life history and laboratory rearing of *Bagrada hilaris* (Hemiptera: Heteroptera: Pentatomidae) with descriptions of immature stages. *Annals of the Entomological Society of America* 108(4), 536–551.

Thomas, P. (2003) *Canola Growers Manual*. Canola Council of Canada, Winnipeg, Manitoba, Canada.

Tyler, F., Fisher, D., D'Ettorre, P., Rodríguez-Muñoz, R. and Tregenza, T. (2015) Chemical cues mediate species recognition in field crickets. *Frontiers in Ecology and Evolution* 3, 48.

Voorrips, R.E., Steenhuis-Broers, G., Tiemens-Hulscher, M. and Lammerts Van Bueren, E.T. (2008) Plant traits associated with resistance to *Thrips tabaci* in cabbage (*Brassica oleracea* var *capitata*). *Euphytica* 163, 409–415.

Whittaker, J.B. and Tribe, N.P. (1998) Predicting numbers of an insect (*Neophilaenus lineatus*: Homoptera) in a changing climate. *Journal of Animal Ecology* 67, 987–991.

Ying-Zhi, C., Li, L., Chih-Wei, W., Chin-Chang, Y. and Shaw-Yhi, H. (2004) Response of two *Pieris* (Lepidoptera: Pieridae) species to fertilization of a host plant. *Zoological Studies* 43(4), 778–786.

15 Cover Crops as a Tool for Insect Pest Management with a Focus on Oilseed Brassicas

R.W.M. Udari M. Wanigasekara[1*] and Barbara J. Sharanowski[2]

[1]*University of Manitoba, Winnipeg, Canada;* [2]*University of Central Florida, Orlando, Florida*

15.1 Introduction

Cover crops have been used for centuries as a part of traditional agricultural practices, as they provide a variety of economic and ecological benefits (Reeves, 1994). Historically, cover crops have been used as off-season, short-term rotation plantings that provide a more continuous cover for soil to reduce erosion, prevent soil nutrient loss due to leaching and protect the ground from extreme freezing (Pieters and McKee, 1938; Reeves, 1994; Tonitto *et al.*, 2006). However, cover crops can also be double cropped or relay cropped (to create a secondary cash crop), or intercropped as a living mulch to provide additional ecological services. While the strategy used is largely dependent on climatic conditions, needs of the ecosystem and length of the growing season, cover crops can provide a whole host of additional ecological benefits, including reduced soil compaction (Rosolem *et al.*, 2002; Williams and Weil, 2004), improved nutrient cycling (DuPont *et al.*, 2009), increased yield (Frye *et al.*, 1985), enhanced soil and water quality (Mendes *et al.*, 1999; Dabney *et al.*, 2001) and suppression of weeds and insect pests (Teasdale, 1996; Kumar *et al.*, 2009; Price and Norsworthy, 2013; Bulan *et al.*, 2015; Mehring *et al.*, 2016).

Cover crops are generally categorized as legumes (e.g. hairy vetch, alfalfa), non-legume broadleaf plants (e.g. brassicas, buckwheat) or grasses (e.g. rye, winter wheat). There are several reviews outlining the ecological services provided by various species (Reeves, 1994; Dabney *et al.*, 2001; Snapp *et al.*, 2005; Clark, 2008). Unfortunately, there is far less information available on using cover crops for insect pest management, especially for large-scale oilseed brassicas. There are a few studies evaluating cover crops for pest control in orchards and vineyards (Bugg *et al.*, 1991; Creamer *et al.*, 1996; Smith *et al.*, 1996; Altieri *et al.*, 2005). For example, cover crops have been used in Californian vineyards to increase populations of natural enemies of grape leafhoppers (Doutt and Nakata, 1973; Daane *et al.*, 1998). Bugg *et al.* (1991) demonstrated that hairy vetch and rye cover crops significantly increased ladybird beetle populations in pecan orchards, thereby controlling pestiferous aphids. However, for adoption in large-scale agriculture, such as oilseed brassicas, the benefits must clearly outweigh the costs and labour associated with planting cover crops, whether integrated into the main crop, double planted, or planted as off-season ground covers (Knowler and Bradshaw, 2007). This chapter discusses the prospect and potential of using cover crops as a tool for sustainable insect pest management, with a focus on brassica crops, including large-scale field crops such as canola and mustard.

*E-mail: udari_madu@yahoo.com

15.2 Cover Crops for Conservation Biocontrol of Arthropod Pests

Predators and parasitoids are efficient natural regulators of insect pest populations. In natural systems, pests rarely reach outbreak population levels as they are controlled by a diverse suite of predators and parasitoids (natural enemies) (DeBach and Rosen, 1991). When those natural enemies are removed from the system, pests can reach uncontrollable levels. This has been observed frequently when foreign species enter new environments without their natural enemies (e.g. emerald ash borer, fire ants), allowing the foreign invader to outcompete native species in enemy-free space (Jeffries and Lawton, 1984). In modern agroecosystems, much of the native flora and fauna has been altered and this creates an unnatural system that decreases the diversity and efficiency of natural enemies and allows pests to flourish (Risch *et al.*, 1983; Matson *et al.*, 1997; Menalled *et al.*, 1999; Bianchi *et al.*, 2006). Decreased landscape diversity, upscaling of field sizes, clearing of hedgerows and high pesticide use have caused cyclical population outbreaks of certain pests as their competitors and natural enemies are eliminated from the system (Landis *et al.*, 2000; Bianchi *et al.*, 2006; Park *et al.*, 2015). However, increasing plant diversity in agroecosystems can re-establish the natural balance of species community dynamics by increasing local abundance and diversity of beneficial insects (Dyer and Landis, 1997; Baggen and Gurr, 1998; Coll, 1998; Bianchi *et al.*, 2006).

Habitat manipulation to enhance and protect predators and parasitoids for pest control is the main strategy of conservation biocontrol (Eilenberg *et al.*, 2001). For agroecosystems, conservation biocontrol includes strategies to protect natural enemies through responsible pesticide use, or to enhance them by providing resources that increase their local abundance, such as additional food sources or overwintering sites (DeBach and Rosen, 1991). Cover crops have an important role in conservation biocontrol by increasing landscape and rotational diversity, improving conditions for beneficial soil fauna and providing requisites for natural enemies, such as shelter or non-host nutritional resources for predators and parasitoids (Jonsson *et al.*, 2008). Cover crops can also interfere with the capacity of pests to colonize hosts by imposing physical barriers, disrupting olfactory and visual cues and creating diversions to non-crop hosts (Altieri and Gliessman, 1983; Trenbath, 1993; Altieri, 1999). If cover crops can enhance the effectiveness of natural enemies, then their use could reduce the need for insecticide applications, further conserving beneficial insects and reducing environmental and food contaminants. Recent research on using cover crops to enhance natural enemies is described below. As most of this research has not been done for large-scale agriculture, prospects for use in these systems are also discussed, with a focus on oilseed brassicas.

15.2.1 Increasing diversity of natural enemies

Increased diversity in the landscape flora is often associated with an increase in the diversity of natural enemies and pollinators in the agroecosystem (Van Emden, 1965; Altieri and Letourneau, 1982; Altieri, 1999; Duffy, 2003). However, assessing the relationship between vegetational diversity and beneficial arthropods is not an easy task, as different species are affected differently by the community composition of plants and animals. Most research has focused on the effects of intercropping, field margin plantings, or nearby native vegetation on one species or specific group of organisms (Dennis and Fry, 1992; Kromp and Steinberger, 1992; Bedford and Usher, 1994; Kajak and Łukasiewicz, 1994; Thomas and Marshall, 1999; Marshall and Moonen, 2002). For example, Hendrickx *et al.* (2007) demonstrated that spider diversity within agroecosystems was highest with increasing abundance and proximity of native natural habitats.

Most studies demonstrate increases in natural enemy diversity but do not specifically address how effective the increased diversity is at controlling specific target pests (Menalled *et al.*, 2003), though there are exceptions (Marino and Landis, 1996). As Landis *et al.* (2000) pointed out, the goals of pest suppression need to be determined to appropriately enhance natural enemies to control pests; increasing diversity for diversity's sake is not an effective strategy. It is critical that specific species interactions be tested, as the individual species' responses (both pest and beneficial) to increased habitat diversity may be less than ideal. For example, Thies *et al.* (2005) found that aphid parasitism was increased in wheat fields with increased overall landscape complexity but aphid colonization also increased, limiting the benefit of the natural enemy response.

For insect pest management, cover crops are more likely to be utilized to conserve and enhance natural enemies (as part of a conservation biocontrol strategy), rather than as a method of increasing overall landscape diversity. Cover crops can increase rotational diversity and thus should also increase complexity within the larger landscape (Reeves, 1994). However, to our knowledge, no studies demonstrating the link between cover crop usage, overall landscape complexity and natural enemy diversity have been performed.

15.2.2 Providing nutritional resources for natural enemies

Cover crops can act as an alternative non-host food source for many adult parasitic wasps that attack agricultural pests by providing honeydew, pollen or nectar to fulfill their nutritional requirements (Jervis *et al.*, 1993; Lee and Heimpel, 2003). For many species, newly emerged adult parasitoids search for floral resources to enhance their longevity, fecundity and energy before searching for their hosts (Takasu and Lewis, 1993; Wäckers, 1994; Kugimiya *et al.*, 2010). Without readily accessible non-host resources, parasitic wasps may spend more time foraging and less time searching for their hosts (Baggen and Gurr, 1998). Thus, providing cover crops for nutrition may increase parasitism in the field. For example, commercial vineyards in New York integrated nectar-producing buckwheat cover crops to control grape leafhoppers by enhancing the effectiveness of *Anagrus* wasps (English-Loeb *et al.*, 2003).

Predatory insects can also get nutritional benefits from cover crops, as some species feed on plant materials to fulfil their nutritional requirements for at least part of their life cycle (Principi and Canard, 1984; Hickman and Wratten, 1996). Flowering non-crop plants, including buckwheat and sunflower, have been used to increase predator diversity in North American vineyards to control pests, such as grape leafhoppers and flower thrips, by increasing the abundance of minute pirate bugs, coccinellids and spiders (Nicholls *et al.*, 2000). Trujillo-Arriaga and Altieri (1990) demonstrated that Mexican maize triculture with faba beans and squash enhanced the abundance of some ladybird beetles (*Hippodamia koebelie* and *H. convergens*) by providing extrafloral nectar. Thus, cover crops can enhance both parasitoids and predator abundance by providing needed nutritional resources that may not be found in more simple agricultural habitats.

15.2.3 Providing overwintering sites for natural enemies

Appropriate microclimatic conditions and shelter are other significant factors that determine the survival of natural enemies residing within agroecosystems (Dyer and Landis, 1996; Coll, 1998; Menalled *et al.*, 1999; English-Loeb *et al.*, 2003). Many natural enemies require overwintering or aestivation sites to complete their life cycles and need refuges for mating or to escape disturbances caused by agricultural practices. The simplification of agricultural landscapes through mowing of hedgerows has minimized available overwintering habitats for beneficial insects (Corbett and Rosenheim, 1996; Pfiffner and Luka, 2000; Marshall, 2004).

Planting cover crops or native vegetation adjacent to agricultural fields has been proposed to provide overwintering sites and refuges, thereby overcoming the effects of landscape simplification (Corbett and Rosenheim, 1996). For example, winter cereals such as rye and winter wheat offer overwintering sites for predators such as carabid beetles (Sotherton, 1984). Field margin plantings of native vegetation (or beetle banks) have been used extensively in European cropping systems to provide shelter for a variety of beneficial arthropods and have been shown to increase the overall abundance of natural enemies (Dennis and Fry, 1992; Dennis *et al.*, 1994; Holland and Fahrig, 2000; Hajek, 2007). Field margin plantings have rarely been studied in Canadian agriculture, particularly for oilseed brassicas. There may be opportunity for winter cover crops to play a role in increasing the survival and abundance of natural enemies but much more research is needed to assess the effectiveness of specific cover crops on increasing the overall abundance of natural enemies that is translated into increased pest control.

15.2.4 Enhancing soil fauna with cover crops

Cover crops have great potential not only to improve soil quality but also to increase the overall survival of multiple beneficial ground-dwelling species (Mathews *et al.*, 2004; Brévault *et al.*, 2007). Cover crops combined with reduced tillage practices can enhance soil health and soil organic matter (Sainju *et al.*, 2002), which is essential for maintaining healthy soil fauna. Hartwig and Ammon (2002) found that a combination of living mulch and no-till cropping techniques enhanced soil-dwelling organisms like earthworms, collembolans, ground beetles and soil microbial biomass, compared with conventional tillage. These soil-dwelling macro- and microorganisms perform a number of vital functions, including nutrient cycling, soil formation and pest management (Coleman *et al.*, 1993; Brévault *et al.*, 2007). Additionally, protecting and enhancing soil faunal abundance and biodiversity acts to regulate pest populations in agroecosystems by balancing the predator–herbivore ratio (Tillman *et al.*, 2004).

Interestingly, Brassicaceae cover crops incorporated into the soil as green manure or mulch (brassica mulch) can suppress pest and disease organisms, as brassicas contain high concentrations of glucosinolates and release biocidal isothiocyanate upon tissue disruption (Kirkegaard and Sarwar, 1998; Garibaldi *et al.*, 2009; Bangarwa *et al.*, 2011). Larkin and Griffin (2007) showed that soil amendments with canola (*Brassica napus*) and Indian mustard (*Brassica juncea*) successfully controlled multiple pathogens in potato fields. Allopathic properties may be an additional benefit of brassica mulches (Wyse, 1994; Teasdale, 1996; Reddy, 2001; Adamavičienė *et al.*, 2009). For example, isothiocyanates released by a turnip–rape mulch have been shown to suppress weed germination (Petersen *et al.*, 2001). Thus, cover crops can be harnessed for their biofumigation properties to improve soil biodiversity and enhance pest suppression, in addition to the many other ecosystem services they offer.

15.3 Cover Crop Selection

Cover crops that have dual roles, such as improving soil quality while contributing to pest suppression, are likely to have the best chance for adoption in *Brassica* agriculture. If cover crops are to be used for enhancing biological control, cover crop selection must be carefully researched and considered before implementation. Careful selection involves consideration of the specific needs and preferences of the biocontrol agent (natural enemy) that is to be enhanced (Knowler and Bradshaw, 2007; Clark, 2008). For example, enhancing the effectiveness of parasitoid wasps by providing additional nutritive resources such as nectar and pollen requires an understanding of their innate preferences, which may be influenced by their ability to locate the plant (visual and olfactory cues) and utilize the resource (e.g. flower shape and size relative to body size of the wasp) (Jervis *et al.*, 1996; Carrié *et al.*, 2012).

Flower colour is an important consideration, as flowers advertise their presence by notable visual stimuli (Horovitz and Cohen, 1972; Chittka *et al.*, 1994; Wäckers, 1994; Giurfa *et al.*, 1996). Many foraging hymenopterans show innate preferences towards yellow, which is the most common colour in natural flowers (Weevers, 1952) as well as the primary signal in the advertisement of pollen (Lunau, 2000). Thus, natural enemies can be attracted by choosing cover crops that have flowers with colours that are preferred by the target (Wäckers, 1994).

Olfaction is another important sensory cue used by natural enemies to locate both their non-host food source and subsequently their hosts. Natural enemies are attracted by the volatiles released by plants in response to herbivory (Whitman and Eller, 1990; Tumlinson *et al.*, 1993; Williams and Cook, 2010), though species have specific preferences that must be tested to assess attractiveness to the biocontrol agent (Wäckers, 2004; Gardiner *et al.*, 2009; James *et al.*, 2014). Parasitoids also use different sets of cues depending on their own hunger level. For example, food-deprived wasps respond to flower odours whereas satiated individuals (sugar-fed) respond to host-associated odours (Wäckers, 1994).

While a flower may have attractive colours and odours, the nectar and pollen may not be accessible to the natural enemy due to floral architecture, such as corolla aperture and depth (Jervis, 1998; Vattala *et al.*, 2006; Bianchi and Wäckers, 2008). For example, alyssum (*Lobularia maritima*) flowers have small gaps between the stamens and petals, which inhibits access to the nectar by small parasitic wasps (Patt *et al.*, 1997). Some species of *Phacelia* have guard hairs that limit nectar access by small parasitoids (Jervis, 1998). Thus, it is critical to identify the target natural enemy to be enhanced beforehand and extend biocontrol research to evaluate the compatibility of plant architecture and floral traits with the preferences of the biocontrol agent (Landis *et al.*, 2000; Vattala *et al.*, 2006; Bianchi and Wäckers, 2008; Carrié *et al.*, 2012).

Cover crops must be chosen with the pest in mind to minimize the risk of high densities of economic pests of cash crops (Risch, 1981; Reeves, 1994; Landis *et al.*, 2000). For example, using brassica cover crops or mulches may create issues by attracting more pests that are specialists on brassicas. There is also potential for pest migration from the cover crop to a primary cash crop if cover crops mature, die, or otherwise become unsuitable to hosts due to normal seasonal changes (Zehnder *et al.*, 2007; Clark, 2008). While cover crops may serve to disrupt host finding in pests (Trenbath, 1993), the strong odour produced from brassica crops may overwhelm any disrupting effect the cover crop may have (Dover, 1985). Alternatively, pests may be preferentially attracted to brassica covers or mulches and reduce overall pest density and damage on the main crop (Altieri and Gliessman, 1983). Overall, it is crucial to consider a farmer's needs, the specific requirements of natural enemies being targeted and specific crop–pest interactions before selecting the appropriate cover crop to maximize benefits and minimize risks.

15.4 Importance of Understanding Native Agroecosystem Insect Diversity

The most critical step for any study involving biocontrol is the identification of the beneficial species involved in the system (Rosen and DeBach, 1973). However, there is limited understanding of native beneficial insects in agroecosystems in general. Part of the reason is that some natural enemies, such as hymenopteran parasitoids, are exceptionally diverse and difficult to identify, either because appropriate taxonomic resources are not available or because the wasps are part of cryptic species complexes that have yet to be teased apart (Rosen, 1978; Heraty, 2009; Sharanowski *et al.*, 2014). As the emphasis of conservation biological control is to enhance the abundance and effectiveness of the natural enemies that are already present (Landis *et al.*, 2000), it is essential to conduct more research on both beneficial and pest taxa to identify their life history characteristics, population dynamics, seasonal abundance and patterns and the differential performance of pests and their natural enemies in various farming systems (Pickett and Bugg, 1998). Appropriate cover crops cannot be selected or even researched if we do not know the specific pests that exist and their suite of natural enemies that we might want to enhance; thus cover crops have been an underutilized tool for agriculture (Altieri, 1995).

Long-term interdisciplinary studies that can integrate taxonomic, ecological and agronomic research on pest and beneficial insects in relation to the other economic ecosystem benefits that cover crops can provide would be ideal but are unlikely to be feasible without large-scale funding (Snapp *et al.*, 2005). A recent study in Europe documented more than 80 parasitoids found on six key pests of oilseed rape (Ulber *et al.*, 2010) through a large collaborative effort across multiple countries. Twelve of these parasitoids were identified as having potential for conservation biocontrol as they were widespread and abundant and had high parasitism levels on the target pests. This type of research, combining taxonomy, biology, ecology and applied agriculture, provides an important model for how research can be funded and conducted to maximize applied benefits. Further, this research provides the starting point to research cover crops that can be utilized to enhance and protect these key parasitoids for long-term approaches to sustainable agriculture. For *Brassica* agriculture at all scales, future studies should focus on cover crops that can diversify rotations while promoting conservation of natural enemies of key pests, particularly specialists, to provide the most immediate economic benefit for farmers.

15.5 Possible Implementations of Cover Crops in *Brassica* Crops

Numerous studies have examined the responses of pest and beneficial arthropods to plant diversification in various agroecosystems (Altieri and Gliessman, 1983; Tonhasca, 1993; Sarrantonio and Gallandt, 2003; Nyoike and Liburd, 2010). Diversified vegetable crops have been shown to reduce pest abundance relative to monocultures in several brassicas, including broccoli (*Brassica oleracea*), cabbage (*B. oleracea*), cauliflower (*B. oleracea*), collards (*B. oleracea*) and Brussels sprouts (*B. oleracea*) (Cromartie, 1981; Altieri and Gliessman, 1983). A legume living mulch between rows of *Brassica* vegetables inhibited pests when compared with fields without cover crops, as it was harder for herbivores to locate the crop in more complex habitats (Costello and Altieri, 1995). Altieri and Gliessman (1983) found that population densities of the crucifer flea beetle, *Phyllotreta cruciferae*, were reduced in diversified California collard crop fields. Theunissen *et al.* (1995) demonstrated that intercropping white cabbage with two species of clover reduced damage caused by the cabbage moth (*Mamestra brassicae*), cabbage aphid (*Brevicoryne brassicae*) and cabbage root fly (*Delia brassicae*).

The benefits of using cover crops in field cropping have long been recognized but the use of cover crops to manage field crop pests, especially pests of brassicas, has been limited (Wyland *et al.*, 1996; Masiunas *et al.*, 1997). Economic factors are frequently considered as main barriers to the adoption of cover crops by farmers (Carlisle, 2016). Since the benefits of cover crops in pest management have not been economically estimated, farmers may consider that the expense and time would outweigh the advantages (Reeves, 1994). Although it has been demonstrated that sustainable practices are as economically viable as conventional practices, farmers might be deterred by costs associated with planting and termination of the cover before planting the cash crop in a large area (Wyland *et al.*, 1996). Another possible reason for limited usage of cover crops in large-scale field agriculture is that unmanaged cover crops can act as weeds by competing with the primary crop for light, moisture, nutrients and space (Bugg, 1992). Cover crops can also deprive primary crops of valuable soil nutrients and water, especially in dry years, and may negatively affect the cash crop yield (Corak *et al.*, 1991). Short growing seasons, such as found in Canada, restrict the ability to relay crop (Carlisle, 2016), which may also prevent adoption (Wandel and Smithers, 2000). However, off-season and double cropping, as well as the use of cover crops as living mulches, holds potential for brassica crops in large-scale field agriculture and vegetable row cropping.

Establishing cover crops for pest management has been limited in Canada but farmers have grown winter wheat, spring wheat, barley, oats and rye to suppress weeds, increase crop yield and enhance soil quality (Martens *et al.*, 2001; Blackshaw *et al.*, 2010). Winter wheat and rye are excellent off-season cover crops because they rapidly produce a ground cover that holds soil in place against the forces of wind and precipitation. Rye can also decrease nitrate leaching into groundwater and it lessens the need for nitrogenous fertilizers (Dabney *et al.*, 2001). Even though pest regulation is not the primary goal, implementing these cover crops does increase rotational diversity and may also create overwintering and oviposition sites for beneficial arthropods (Tiemann *et al.*, 2015).

However, unmanaged cover crops and untested inappropriate species of covers may harbour pest insects, diseases and nematodes that could be harmful to cash crops. For example, rye and winter wheat may help with soil erosion and nitrate leaching but they may also attract army cutworms (Gavloski and Meers, 2011). Thus, cover crop selection and proper maintenance of covers are important to maximize the benefits and prevent unwanted effects. Detailed research is essential to estimate the benefits of specific cover crops for habitat diversification and pest management and to ensure that strategies will be adopted by farmers (Veromann *et al.*, 2008). This is especially true for brassica crops that have a multitude of pests that are attracted to glucosinolates and specialize on Brassicaceae plants.

15.6 Conclusions

Cover crops have numerous ecosystem benefits that can help to create a more sustainable approach to intensified agriculture, including brassicas. They can increase landscape and rotational diversity, improve soils without the need for additional inputs, prevent soil erosion, contribute to soil quality and fertility and control water uptake. However, cover crops also have an underutilized role in integrated pest management by providing resources to protect and enhance natural enemies of target pests. This use of cover crops in conservation biocontrol is attractive, as cover crops can be selected to provide multiple benefits to farmers.

For cover crops to work within a conservation biocontrol strategy, the species of natural enemies that attack key pests must be known. Although key brassica pests are well known in most parts of the world, the suite of natural enemies attacking these pests has only been studied thoroughly in very few regions. Basic taxonomic research to document natural enemy diversity and tease apart cryptic species complexes, if present, is essential for effective biocontrol.

To enhance and protect natural enemies, specific cover crops need to be researched for their utility in providing additional resources for the natural enemies in the system, such as additional nutrition, shelter and overwintering sites. In general, availability of nectar/pollen is the main limiting factor for parasitoids that feed on non-host resources, whereas overwintering sites and refuge zones are the main limitations for predators. As not all resources are created equally, each species of prospective cover crop needs to be tested for its ability to attract natural enemies and increase its efficiency in pest control of target pests. Those cover crop species that already have known ecosystem benefits, such as increasing soil fertility, are ideal choices for initial selection. For parasitic wasps, flower colour, odour and architecture need to be examined to ensure that wasps can find and utilize the cover crop to obtain pollen and nectar. For predators, overwintering, aestivation and reproductive needs must be understood to be able to select species that can provide appropriate requisites. Additionally, community dynamics between a cover crop, the cash crop and the pest must also be understood, so that rotational diversity is created and primary or secondary pests are not attracted to the cover crop. As pests are well known for most brassica crops, they offer great potential for additional research on using cover crops as living mulches, ground covers, or relay crops.

It is critical to understand the factors that limit adoption of cover crops, particularly with respect to large-scale farming of oilseed brassicas. Without information on the specific benefits of cover crops for pest management, farmers are unlikely to adopt strategies that may require the input of time, labour, or money without clear economic returns. Thus, there is a need for detailed research on the use of cover crops in brassica agriculture and how these covers translate into increased pest control in addition to the other ecosystem benefits that they provide. Larger-scale research that incorporates economics and potential incentives for farmers to implement cover crops is highly recommended to ensure that research is translated into economic returns as well as ecosystem benefits.

References

Adamavičienė, A., Romaneckas, K., Šarauskis, E. and Pilipavičius, V. (2009) Non-chemical weed control in sugar beet crop under an intensive and conservation soil tillage pattern. II. Crop productivity. *Agronomy Research* 7, 143–148.

Altieri, M.A. (1995) *Agroecology: The Science of Sustainable Agriculture*, Westview Press, Boulder, Colorado.

Altieri, M.A. (1999) The ecological role of biodiversity in agroecosystems. *Agriculture, Ecosystems & Environment* 74, 19–31.

Altieri, M.A. and Gliessman, S.R. (1983) Effects of plant diversity on the density and herbivory of the flea beetle, *Phyllotreta cruciferae* Goeze, in California collard (*Brassica oleracea*) cropping systems. *Crop Protection* 2, 497–501.

Altieri, M.A. and Letourneau, D.K. (1982) Vegetation management and biological control in agroecosystems. *Crop Protection* 1, 405–430.

Altieri, M.A., Ponti, L. and Nicholls, C.I. (2005) Manipulating vineyard biodiversity for improved insect pest management: case studies from northern California. *The International Journal of Biodiversity Science and Management* 1, 191–203.

Baggen, L.R. and Gurr, G.M. (1998) The influence of food on *Copidosoma koehleri* (Hymenoptera: Encyrtidae), and the use of flowering plants as a habitat management tool to enhance biological control of potato moth, *Phthorimaea operculella* (Lepidoptera: Gelechiidae). *Biological Control* 11, 9–17.

Bangarwa, S.K., Norsworthy, J.K., Mattice, J.D. and Gbur, E.E. (2011) Glucosinolate and isothiocyanate production from Brassicaceae cover crops in a plasticulture production system. *Weed Science* 59, 247–254.

Bedford, S.E. and Usher, M.B. (1994) Distribution of arthropod species across the margins of farm woodlands. *Agriculture, Ecosystems & Environment* 48, 295–305.

Bianchi, F.J. and Wäckers, F.L. (2008) Effects of flower attractiveness and nectar availability in field margins on biological control by parasitoids. *Biological Control* 46, 400–408.

Bianchi, F.J.J.A., Booij, C.J.H. and Tscharntke, T. (2006) Sustainable pest regulation in agricultural landscapes: a review on landscape composition, biodiversity and natural pest control. *Proceedings of the Royal Society B: Biological Sciences* 273, 1715–1727.

Blackshaw, R.E., Molnar, L.J. and Moyer, J.R. (2010) Suitability of legume cover crop-winter wheat intercrops on the semi-arid Canadian prairies. *Canadian Journal of Plant Science* 90, 479–488.

Brévault, T., Bikay, S., Maldes, J. and Naudin, K. (2007) Impact of a no-till with mulch soil management strategy on soil macrofauna communities in a cotton cropping system. *Soil and Tillage Research* 97, 140–149.

Bugg, R.L. (1992) Using cover crops to manage arthropods on truck farms. *HortScience* 27, 741–745.

Bugg, R.L., Dutcher, J.D. and McNeill, P.J. (1991) Cool-season cover crops in the pecan orchard understory: effects on Coccinellidae (Coleoptera) and pecan aphids (Homoptera: Aphididae). *Biological Control* 1, 8–15.

Bulan, M.T.S., Stoltenberg, D.E. and Posner, J.L. (2015) Buckwheat species as summer cover crops for weed suppression in no-tillage vegetable cropping systems. *Weed Science* 63, 690–702.

Carlisle, L. (2016) Factors influencing farmer adoption of soil health practices in the United States: a narrative review. *Agroecology and Sustainable Food Systems* 40, 583–613.

Carrié, R.J.G., George, D.R. and Wäckers, F.L. (2012) Selection of floral resources to optimise conservation of agriculturally-functional insect groups. *Journal of Insect Conservation* 16, 635–640.

Chittka, L., Shmida, A., Troje, N. and Menzel, R. (1994) Ultraviolet as a component of flower reflections, and the colour perception of Hymenoptera. *Vision Research* 34, 1489–1508.

Clark, A. (2008) *Managing Cover Crops Profitably*. Sustainable Agriculture Research and Education Program, DIANE Publishing, Beltsville, Maryland.

Coleman, D.C., Hendrix, P.F., Beare, M.H., Cheng, W.X. and Crossley, D.A. Jr (1993) Microbial and faunal interactions as they affect soil organic matter dynamics in subtropical agroecosystems. In: Paoletti, M.G., Foissner, W. and Coleman, D.C. (eds) *Soil Biota, Nutrient Cycling and Farming Systems*, Lewis Publishers, Boca Raton, Florida.

Coll, M. (1998) Parasitoid activity and plant species composition in intercropped systems. In: Pickett, C.C.H. and Bugg, R.L. (eds) *Enhancing Biological Control: Habitat Management to Promote Natural Enemies of Agricultural Pests.* University of California Press, Berkeley, California, pp. 85–120.

Corak, S., Frye, W. and Smith, M. (1991) Legume mulch and nitrogen fertilizer effects on soil water and corn production. *Soil Science Society of America Journal* 55, 1395–1400.

Corbett, A. and Rosenheim, J.A. (1996) Impact of a natural enemy overwintering refuge and its interaction with the surrounding landscape. *Ecological Entomology* 21, 155–164.

Costello, M.J. and Altieri, M.A. (1995) Abundance, growth rate and parasitism of *Brevicoryne brassicae* and *Myzus persicae* (Homoptera: Aphididae) on broccoli grown in living mulches. *Agriculture, Ecosystems & Environment* 52, 187–196.

Creamer, N.G., Bennett, M.A., Stinner, B.R. and Cardina, J. (1996) A comparison of four processing tomato production systems differing in cover crop and chemical inputs. *Journal of the American Society for Horticultural Science* 121, 559–568.

Cromartie, W.J. Jr (1981) The environmental control of insects using crop diversity. In: Pimentel, D. (ed.) *CRC Handbook of Pest Management in Agriculture.* CRC Press, Boca Raton, Florida, pp. 223–251.

Daane, K.M., Costello, M.J., Yokota, G. and Bentley, W. (1998) Can we manipulate leafhopper densities with management practices? *Grape Grower* 30, 18–36.

Dabney, S.M., Delgado, J.A. and Reeves, D.W. (2001) Using winter cover crops to improve soil and water quality. *Communications in Soil Science and Plant Analysis* 32, 1221–1250.

DeBach, P. and Rosen, D. (1991) *Biological Control by Natural Enemies*. Cambridge University Press, London.

Dennis, P. and Fry, G.L.A. (1992) Field margins: can they enhance natural enemy population densities and general arthropod diversity on farmland? *Agriculture, Ecosystems & Environment* 40, 95–115.

Dennis, P., Thomas, M. and Sotherton, N. (1994) Structural features of field boundaries which influence the overwintering densities of beneficial arthropod predators. *Journal of Applied Ecology* 31, 361–370.

Doutt, R.L. and Nakata, J. (1973) The *Rubus* leafhopper and its egg parasitoid: an endemic biotic system useful in grape-pest management. *Environmental Entomology* 2, 381–386.

Dover, J.W. (1985) The responses of some Lepidoptera to labiate herb and white clover extracts. *Entomologia Experimentalis et Applicata* 39, 177–182.

Duffy, J.E. (2003) Biodiversity loss, trophic skew and ecosystem functioning. *Ecology Letters* 6, 680–687.

DuPont, S.T., Ferris, H. and Van Horn, M. (2009) Effects of cover crop quality and quantity on nematode-based soil food webs and nutrient cycling. *Applied Soil Ecology* 41, 157–167.

Dyer, L.E. and Landis, D.A. (1996) Effects of habitat, temperature, and sugar availability on longevity of *Eriborus terebrans* (Hymenoptera: Ichneumonidae). *Environmental Entomology* 25, 1192–1201.

Dyer, L.E. and Landis, D.A. (1997) Influence of noncrop habitats on the distribution of *Eriborus terebrans* (Hymenoptera: Ichneumonidae) in cornfields. *Environmental Entomology* 26, 924–932.

Eilenberg, J., Hajek, A. and Lomer, C. (2001) Suggestions for unifying the terminology in biological control. *BioControl* 46, 387–400.

English-Loeb, G., Rhainds, M., Martinson, T. and Ugine, T. (2003) Influence of flowering cover crops on *Anagrus* parasitoids (Hymenoptera: Mymaridae) and *Erythroneura* leafhoppers (Homoptera: Cicadellidae) in New York vineyards. *Agricultural and Forest Entomology* 5, 173–181.

Frye, W., Smith, W. and Williams, R. (1985) Economics of winter cover crops as a source of nitrogen for no-till corn. *Journal of Soil and Water Conservation* 40, 246–249.

Gardiner, M.M., Fiedler, A.K., Costamagna, A.C. and Landis, D.A. (2009) Integrating conservation biological control into IPM systems. In: Radcliffe, E.B., Hutchison, W.D. and Cancelado, R.E. (ed.) *Integrated Pest Management Concepts, Tactics, Strategies and Case Studies*. Cambridge University Press, Cambridge, UK, pp. 151–162.

Garibaldi, A., Gilardi, G., Clematis, F., Gullino, M., Lazzeri, L. and Malaguti, L. (2009) Effect of green *Brassica* manure and *Brassica* defatted seed meals in combination with grafting and soil solarization against *Verticillium* wilt of eggplant and *Fusarium* wilt of lettuce and basil. *VII International Symposium on Chemical and Non-Chemical Soil and Substrate Disinfestation 883*, 295–302.

Gavloski, J. and Meers, S. (2011) Arthropods of cereal crops in Canadian grasslands. *Arthropods of Canadian Grasslands* 2, 217–237.

Giurfa, M., Vorobyev, M., Kevan, P. and Menzel, R. (1996) Detection of coloured stimuli by honeybees: minimum visual angles and receptor specific contrasts. *Journal of Comparative Physiology A* 178, 699–709.

Hajek, A.E. (2007) *Natural Enemies: An Introduction to Biological Control*. Cambridge University Press, New York.

Hartwig, N.L. and Ammon, H.U. (2002) Cover crops and living mulches. *Weed Science* 50, 688–699.

Hendrickx, F., Maelfait, J.-P., Van Wingerden, W., Schweiger, O., Speelmans, M., Aviron, S., Augenstein, I., Billeter, R., Bailey, D., Bukacek, R., Burel, F., Diekötter, T.I.M., Dirksen, J., Herzog, F., Liira, J., Roubalova, M., Vandomme, V. and Bugter, R.O.B. (2007) How landscape structure, land-use intensity and habitat diversity affect components of total arthropod diversity in agricultural landscapes. *Journal of Applied Ecology* 44, 340–351.

Heraty, J. (2009) Parasitoid biodiversity and insect pest management. *Insect Biodiversity: Science and Society*, 445–462.

Hickman, J. and Wratten, S. (1996) Use of *Phacelia tanacetifolia* (Hydrophyllaceae) as a pollen source to enhance hoverfly (Diptera: Syrphidae) populations in cereal fields. *Journal of Economic Entomology* 89, 832–840.

Holland, J. and Fahrig, L. (2000) Effect of woody borders on insect density and diversity in crop fields: a landscape-scale analysis. *Agriculture, Ecosystems & Environment* 78, 115–122.

Horovitz, A. and Cohen, Y. (1972) Ultraviolet reflectance characteristics in flowers of crucifers. *American Journal of Botany* 59, 706–713.

James, D.G., Seymour, L., Lauby, G. and Buckley, K. (2014) Beneficial insects attracted to native flowering buckwheats (*Eriogonum Michx*) in central Washington. *Environmental Entomology* 43, 942–948.

Jeffries, M. and Lawton, J. (1984) Enemy free space and the structure of ecological communities. *Biological Journal of the Linnean Society* 23, 269–286.

Jervis, M. (1998) Functional and evolutionary aspects of mouthpart structure in parasitoid wasps. *Biological Journal of the Linnean Society* 63, 461–493.

Jervis, M.A., Kidd, N.A.C., Fitton, M.G., Huddleston, T. and Dawah, H.A. (1993) Flower-visiting by hymenopteran parasitoids. *Journal of Natural History* 27, 67–105.

Jervis, M.A., Kidd, N.A.C. and Heimpel, G.E. (1996) Parasitoid adult feeding behaviour and biological control – a review. *Biocontrol News and Information* 16, 11–26.

Jonsson, M., Wratten, S.D., Landis, D.A. and Gurr, G.M. (2008) Recent advances in conservation biological control of arthropods by arthropods. *Biological Control* 45, 172–175.

Kajak, A. and Łukasiewicz, J. (1994) Do semi-natural patches enrich crop fields with predatory epigean arthropods. *Agriculture, Ecosystems & Environment* 49, 149–161.

Kirkegaard, J. and Sarwar, M. (1998) Biofumigation potential of brassicas. *Plant and Soil* 201, 71–89.

Knowler, D. and Bradshaw, B. (2007) Farmers' adoption of conservation agriculture: a review and synthesis of recent research. *Food Policy* 32, 25–48.

Kromp, B. and Steinberger, K.H. (1992) Grassy field margins and arthropod diversity: a case study on ground beetles and spiders in eastern Austria (Coleoptera: Carabidae; Arachnida: Aranei, Opiliones). *Agriculture, Ecosystems & Environment* 40, 71–93.

Kugimiya, S., Uefune, M., Shimoda, T. and Takabayashi, J. (2010) Orientation of the parasitic wasp, *Cotesia vestalis* (Haliday) (Hymenoptera: Braconidae), to visual and olfactory cues of field mustard flowers, *Brassica rapa* L. (Brassicaceae), to exploit food sources. *Applied Entomology and Zoology* 45, 369–375.

Kumar, V., Brainard, D.C. and Bellinder, R.R. (2009). Suppression of Powell amaranth (*Amaranthus powellii*) by buckwheat residues: role of allelopathy. *Weed Science* 57, 66–73.

Landis, D.A., Wratten, S.D. and Gurr, G.M. (2000) Habitat management to conserve natural enemies of arthropod pests in agriculture. *Annual Review of Entomology* 45, 175–201.

Larkin, R.P. and Griffin, T.S. (2007) Control of soilborne potato diseases using *Brassica* green manures. *Crop Protection* 26, 1067–1077.

Lee, J.C. and Heimpel, G.E. (2003) Nectar availability and parasitoid sugar feeding. In: Van Driesche, R. (ed.) *Proceedings of the 1st International Symposium on Biological Control of Arthropods, 14–18 Jan 2002, Honolulu*. USDA Forest Service, Morgantown, West Virginia, pp. 220–225.

Lunau, K. (2000) The ecology and evolution of visual pollen signals. In: Dafni, A., Hesse, M. and Pacini, E. (eds) *Pollen and Pollination* (special edition of *Plant Systematics and Evolution*, Vol. 222, No. 1–4). Springer, Vienna, Austria, pp. 89–111.

Marino, P.C. and Landis, D.A. (1996) Effect of landscape structure on parasitoid diversity and parasitism in agroecosystems. *Ecological Applications* 6(1), 276–284.

Marshall, E.J.P. (2004) Agricultural landscapes: field margin habitats and their interaction with crop production. *Journal of Crop Improvement* 12, 365–404.

Marshall, E.J.P. and Moonen, A.C. (2002) Field margins in northern Europe: their functions and interactions with agriculture. *Agriculture, Ecosystems & Environment* 89, 5–21.

Martens, J.R.T., Hoeppner, J.W. and Entz, M.H. (2001) Legume cover crops with winter cereals in southern Manitoba. *Agronomy Journal* 93, 1086–1096.

Masiunas, J.B., Eastburn, D.M., Mwaja, V.N. and Eastman, C.E. (1997) The impact of living and cover crop mulch systems on pests and yields of snap

beans and cabbage. *Journal of Sustainable Agriculture* 9, 61–89.
Mathews, C.R., Bottrell, D.G. and Brown, M. (2004) Habitat manipulation of the apple orchard floor to increase ground-dwelling predators and predation of *Cydia pomonella* (L.) (Lepidoptera: Tortricidae). *Biological Control* 30, 265–273.
Matson, P.A., Parton, W.J., Power, A. and Swift, M. (1997). Agricultural intensification and ecosystem properties. *Science* 277, 504–509.
Mehring, G., Stenger, J. and Hatterman-Valenti, H. (2016) Weed control with cover crops in irrigated potatoes. *Agronomy* 6, 3.
Menalled, F.D., Marino, P.C., Gage, S.H. and Landis, D.A. (1999) Does agricultural landscape structure affect parasitism and parasitoid diversity? *Ecological Applications* 9, 634–641.
Menalled, F.D., Costamagna, A.C., Marino, P.C. and Landis, D.A. (2003) Temporal variation in the response of parasitoids to agricultural landscape structure. *Agriculture, Ecosystems & Environment* 96, 29–35.
Mendes, I., Bandick, A., Dick, R. and Bottomley, P. (1999) Microbial biomass and activities in soil aggregates affected by winter cover crops. *Soil Science Society of America Journal* 63, 873–881.
Nicholls, C.I., Parrella, M.P. and Altieri, M.A. (2000) Reducing the abundance of leafhoppers and thrips in a northern California organic vineyard through maintenance of full season floral diversity with summer cover crops. *Agricultural and Forest Entomology* 2, 107–113.
Nyoike, T. and Liburd, O. (2010) Effect of living (buckwheat) and UV reflective mulches with and without imidacloprid on whiteflies, aphids and marketable yields of zucchini squash. *International Journal of Pest Management* 56, 31–39.
Park, M.G., Blitzer, E., Gibbs, J., Losey, J.E. and Danforth, B.N. (2015) Negative effects of pesticides on wild bee communities can be buffered by landscape context. *Proceedings of the Royal Society B: Biological Sciences*, 282, 20150299.
Patt, J.M., Hamilton, G.C. and Lashomb, J.H. (1997) Foraging success of parasitoid wasps on flowers: interplay of insect morphology, floral architecture and searching behavior. *Entomologia Experimentalis et Applicata* 83, 21–30.
Petersen, J., Belz, R., Walker, F. and Hurle, K. (2001) Weed suppression by release of isothiocyanates from turnip-rape mulch. *Agronomy Journal* 93, 37–43.
Pfiffner, L. and Luka, H. (2000) Overwintering of arthropods in soils of arable fields and adjacent semi-natural habitats. *Agriculture, Ecosystems & Environment* 78, 215–222.
Pickett, C.H. and Bugg, R.L. (1998) Enhancing biological control: habitat management to promote natural enemies of agricultural pests. In: Pickett, C.H. and Bugg, R.L. (eds) *Enhancing Biological Control.* University of California Press, Berkeley, California, pp. 1–23.
Pieters, A. and McKee, R. (1938) The use of cover and green-manure crops. In: *Soils and Men.* USDA Yearbook of Agriculture. US Government Printing Office, Washington DC, pp. 431–444.
Price, A.J. and Norsworthy, J.K. (2013) Cover crops for weed management in southern reduced-tillage vegetable cropping systems. *Weed Technology* 27, 212–217.
Principi, M. and Canard, M. (1984) Feeding habits. In: Canard, M., Séméria, Y. and New, T.R. (eds) *Biology of Chrysopidae.* Dr W. Junk, The Hague, pp. 76–92.
Reddy, K.N. (2001) Effects of cereal and legume cover crop residues on weeds, yield, and net return in soybean (*Glycine max*). *Weed Technology* 15, 660–668.
Reeves, D. (1994) Cover crops and rotations. In: Hatfield, J.L. and Stewart, B.A. (eds) *Crop Residue Management*. Lewis Publishers, Boca Raton, Florida, pp. 125–172.
Risch, S.J. (1981) Insect herbivore abundance in tropical monocultures and polycultures: an experimental test of two hypotheses. *Ecology* 62, 1325–1340.
Risch, S.J., Andow, D. and Altieri, M.A. (1983) Agroecosystem diversity and pest control: data, tentative conclusions, and new research directions. *Environmental Entomology* 12, 625–629.
Rosen, D. (1978) The importance of cryptic species and specific identifications as related to biological control. In: Romberger, J. (ed.) *Biosystematics in Agriculture.* Beltsville symposia in agricultural research (2). Allanheld, Osmun & Co. Inc., Montclair, New Jersey, pp. 23–35
Rosen, D. and DeBach, P. (1973) Systematics, morphology and biological control. *Entomophaga* 18, 215–222.
Rosolem, C.A., Foloni, J.S.S. and Tiritan, C.S. (2002) Root growth and nutrient accumulation in cover crops as affected by soil compaction. *Soil and Tillage Research* 65, 109–115.
Sainju, U.M., Singh, B.P. and Yaffa, S. (2002) Soil organic matter and tomato yield following tillage, cover cropping, and nitrogen fertilization. *Agronomy Journal* 94, 594–602.
Sarrantonio, M. and Gallandt, E. (2003) The role of cover crops in North American cropping systems. *Journal of Crop Production* 8, 53–74.
Sharanowski, B.J., Zhang, Y.M. and Wanigasekara, R. (2014) Annotated checklist of braconidae (Hymenoptera) in the Canadian prairies ecozone. In: Giberson, D. and Cárcamo, H. (eds) *Arthropods of Canadian Grasslands: Biodiversity and Systematics Part 2.* Biological Survey of Canada, Ottawa, Canada, pp. 399–425.
Smith, M.W., Arnold, D.C., Eikenbary, R.D., Rice, N.R., Shiferaw, A., Cheary, B.S. and Carroll, B. L. (1996) Influence of ground cover on beneficial arthropods in pecan. *Biological Control* 6, 164–176.

Snapp, S.S., Swinton, S.M., Labarta, R., Mutch, D., Black, J.R., Leep, R., Nyiraneza, J. and O'Neil, K. (2005) Evaluating cover crops for benefits, costs and performance within cropping system niches. *Agronomy Journal* 97, 322–332.

Sotherton, N. (1984) The distribution and abundance of predatory arthropods overwintering on farmland. *Annals of Applied Biology* 105, 423–429.

Takasu, K.E.I.J.I. and Lewis, W.J. (1993) Host-and food-foraging of the parasitoid *Microplitis croceipes*: learning and physiological state effects. *Biological Control* 3, 70–74.

Teasdale, J.R. (1996) Contribution of cover crops to weed management in sustainable agricultural systems. *Journal of Production Agriculture* 9, 475–479.

Theunissen, J., Booij, C. and Lotz, L. (1995) Effects of intercropping white cabbage with clovers on pest infestation and yield. *Entomologia Experimentalis et Applicata* 74, 7–16.

Thies, C., Roschewitz, I. and Tscharntke, T. (2005) The landscape context of cereal aphid – parasitoid interactions. *Proceedings of the Royal Society B: Biological Sciences* 272, 203–210.

Thomas, C.F.G. and Marshall, E.J.P. (1999) Arthropod abundance and diversity in differently vegetated margins of arable fields. *Agriculture, Ecosystems & Environment* 72, 131–144.

Tiemann, L., Grandy, A., Atkinson, E., Marin-Spiotta, E. and McDaniel, M. (2015) Crop rotational diversity enhances belowground communities and functions in an agroecosystem. *Ecology Letters* 18, 761–771.

Tillman, G., Schomberg, H., Phatak, S., Mullinix, B., Lachnicht, S., Timper, P. and Olson, D. (2004) Influence of cover crops on insect pests and predators in conservation tillage cotton. *Journal of Economic Entomology* 97, 1217–1232.

Tonhasca, A. (1993) Effects of agroecosystem diversification on natural enemies of soybean herbivores. *Entomologia Experimentalis et Applicata* 69, 83–90.

Tonitto, C., David, M. and Drinkwater, L. (2006) Replacing bare fallows with cover crops in fertilizer-intensive cropping systems: a meta-analysis of crop yield and N dynamics. *Agriculture, Ecosystems & Environment* 112, 58–72.

Trenbath, B. (1993) Intercropping for the management of pests and diseases. *Field Crops Research* 34, 381–405.

Trujillo-Arriaga, J. and Altieri, M.A. (1990) A comparison of aphidophagous arthropods on maize polycultures and monocultures, in central Mexico. *Agriculture, Ecosystems & Environment* 31, 337–349.

Tumlinson, J.H., Lewis, W.J. and Vet, L.E.M. (1993) How parasitic wasps find their hosts. *Animal Ecology* 266, 100–106.

Ulber, B., Williams, I.H., Klukowski, Z., Luik, A. and Nilsson, C. (2010) Parasitoids of oilseed rape pests in Europe: key species for conservation biocontrol. In: Williams, I.H. (ed.) *Biocontrol-based Integrated Management of Oilseed Rape Pests*. Springer, Dordrecht, The Netherlands.

Van Emden, H.F. (1965) The role of uncultivated land in the biology of crop pests and beneficial insects. *Scientific Horticulture* 17, 121–136.

Vattala, H.D., Wratten, S.D., Phillips, C.B. and Wäckers, F.L. (2006) The influence of flower morphology and nectar quality on the longevity of a parasitoid biological control agent. *Biological Control* 39, 179–185.

Veromann, E., Kevväi, R., Luik, A. and Williams, I.H. (2008) Do cropping system and insecticide use in spring oilseed rape affect the abundance of pollen beetles (*Meligethes aeneus* Fab.) on the crop? *International Journal of Pest Management* 54, 1–4.

Wäckers, F. (1994) The effect of food deprivation on the innate visual and olfactory preferences in the parasitoid *Cotesia rubecula*. *Journal of Insect Physiology* 40, 641–649.

Wäckers, F.L. (2004) Assessing the suitability of flowering herbs as parasitoid food sources: flower attractiveness and nectar accessibility. *Biological Control* 29, 307–314.

Wandel, J. and Smithers, J. (2000) Factors affecting the adoption of conservation tillage on clay soils in Southwestern Ontario, Canada. *American Journal of Alternative Agriculture* 15, 181–188.

Weevers, T. (1952) Flower colours and their frequency. *Acta Botanica Neerlandica* 1, 81–92.

Whitman, D.W. and Eller, F.J. (1990) Parasitic wasps orient to green leaf volatiles. *Chemoecology* 1, 69–76.

Williams, I.H. and Cook, S.M. (2010) Crop location by oilseed rape pests and host location by their parasitoids. In: Williams, I.H. (ed.) *Biocontrol-based Integrated Management of Oilseed Rape Pests*. Springer, Dordrecht, The Netherlands, pp. 1–43.

Williams, S.M. and Weil, R.R. (2004) Crop cover root channels may alleviate soil compaction effects on soybean crop. *Soil Science Society of America Journal* 68, 1403–1409.

Wyland, L., Jackson, L., Chaney, W., Klonsky, K., Koike, S. and Kimple, B. (1996) Winter cover crops in a vegetable cropping system: impacts on nitrate leaching, soil water, crop yield, pests and management costs. *Agriculture, Ecosystems & Environment* 59, 1–17.

Wyse, D.L. (1994) New technologies and approaches for weed management in sustainable agriculture systems. *Weed Technology* 8, 403–407.

Zehnder, G., Gurr, G.M., Kühne, S., Wade, M.R., Wratten, S.D. and Wyss, E. (2007) Arthropod pest management in organic crops. *Annual Review of Entomology* 52, 57–80.

16 Detection, Symptomatology and Management of Aster Yellows Disease in Canola

Chrystel Olivier[1]*, Tim Dumonceaux[2], Edel Pérez-López[3], Tyler J. Wist[2], Bob Elliott[2] and Sally Vail[2]

[1]*Agriculture and Agri-Food Canada, London, Ontario, Canada;* [2]*Agriculture and Agri-Food Canada, Saskatoon, Saskatchewan, Canada;* [3]*Instituto de Biotecnologia y Ecologia Aplicada (INBIOTECA), Universidad Veracruzana, Xalapa, Veracruz, México*

16.1 Introduction

Phytoplasmas are obligate parasites that belong to the class Mollicutes and genus '*Candidatus* Phytoplasma' (McCoy *et al.*, 1989; IRPCM Phytoplasma/Spiroplasma Working Team – Phytoplasma Taxonomy Group, 2004). Phytoplasmas have been associated with diseases affecting over 700 plant species worldwide (Foissac and Wilson, 2010; Bertaccini *et al.*, 2014). Phytoplasmas are wall-less bacteria that are transmitted by phloem-feeding insects, mostly leafhoppers but also planthoppers and psyllids (Weintraub and Beanland, 2006). Phytoplasmas live and reproduce in the phloem of their plant hosts and in most of the organs of their insect vectors. The effect of phytoplasmas on their hosts varies depending on several parameters, such as the phytoplasma strain, host species, vector infectivity and environmental conditions. Most plant species infected with phytoplasmas develop symptoms that are unique to these diseases, such as virescence (greening of flower organs), phyllody (floral organs turning into leaf-like structures), witches' broom (excessive stem and branch production) or dwarfism (McCoy *et al.*, 1989). Symptoms are most severe when leafhoppers infect seedlings with phytoplasmas (Chiykowski, 1981; Starzycki *et al.*, 2003; Olivier *et al.*, 2014). The molecular mechanisms involved in the pathogenicity of the phytoplasma are still unclear. Recently, genes associated with virulence factors or specific type of symptoms have been identified (Sugio *et al.*, 2014; Orlovskis *et al.*, 2015; Ma and Ma, 2016).

Aster yellows (AY) disease is caused by phytoplasmas belonging to the taxon '*Ca.* Phytoplasma asteris' (Lee *et al.*, 2004). Aster yellows disease is the most widespread phytoplasma disease worldwide, mostly because it can be transmitted by more than 20 different species of leafhoppers and can use more than 250 plant species as hosts (Lee *et al.*, 2004; Olivier *et al.*, 2009). In North America, the main AY vector is *Macrosteles quadrilineatus* (Forbes) (Hemiptera: Cicadellidae), a migratory leafhopper carried into the northern USA and Canada during spring by south winds originating from the Gulf coast regions of North America (Hoy *et al.*, 1992; Frost *et al.*, 2013). The origin of the winds arriving on the prairies can be tracked using a model that measures wind trajectories at 50 locations across Canada (Hopkinson and Soroka, 2010), allowing growers and scientists to know if the current winds crossing their areas are originating from the aster leafhopper's breeding and feeding areas. In addition, a small proportion of the aster leafhoppers found in Canada and the northern USA arise from overwintering eggs laid by adults the previous summer (Chiykowski, 1981).

Aster yellows has been known to be a devastating disease since the 1920s, when it nearly wiped out

*Corresponding author. E-mail: drchrystelolivier@gmail.com

the culture of aster in several US states (Kunkel, 1926). In Canada, AY-like symptoms have been described since 2015 on carrot, lettuce and celery (Anon., 1936, 1941). Aster yellows has been prevalent in those crops in the Maritimes (New Brunswick, Nova Scotia and Prince Edward Island), Quebec, Ontario and Manitoba and losses can be very heavy unless crops are sprayed regularly to control the leafhopper vectors (Rankin, 1987; Elliot and Northover, 2007). In less susceptible crops such as canola, AY has been considered of little significance, with disease incidence usually lower than 1%, except for the epidemic in 1957 (Vanterpool, 1963). However, since 2000, three outbreaks of increasing incidence occurred in canola crops grown in the Canadian prairies, with percentages of infected plants per field ranging from 2% to 15% in 2000, 2% to 25% in 2007 and 5% to 84% in 2012 (McLaren and Platford, 2001; Olivier *et al.*, 2006; Pearse *et al.*, 2008). The 2012 AY outbreak was by far the worst and the most damaging AY outbreak for the canola industry. Today, AY is considered to be an emerging disease in crops because of the increase in its frequency, incidence and geographical distribution (Olivier *et al.*, 2009; Foissac and Wilson, 2010).

AY is difficult to control, as there are no chemicals that can kill the pathogen directly and no canola cultivars are known to be resistant. The only viable option to control AY in canola crops is to use insecticidal sprays against the leafhopper vectors. According to the 2016 *Guide to Crop Protection* (Saskatchewan Ministry of Agriculture, 2016), dimethoate (Lagon/Cygon AG) is the only insecticide currently registered for the control of aster leafhoppers in canola in Canada. Facts that are hampering the development of an efficient management strategy of AY in canola crops include: (i) that no economic threshold for aster leafhoppers in *Brassica* crops is available for growers; (ii) that not all aspects of the disease epidemiology, such as the timing of the symptom expression and the exact production losses due to AY, are known; (iii) that the movement of the aster leafhoppers between the crops and the evolution of the aster leafhopper population over the season and subsequent years, as well as the reproductive capability of *M. quadrilineatus* in *Brassica* crops, are not known; and (iv) that the presence of tolerance or resistance to AY among canola germplasms has not been studied.

Recently, several AY strains were identified in various *Brassica* crops, with infection incidence and plant repartition varying between plants. This demonstrates the need for accurate detection and identification of the phytoplasma strains in leafhoppers as well as in plant tissues (Olivier *et al.*, 2011; Chittem and Del Río Mendoza, 2015; Dumonceaux *et al.*, 2015).

Since 2001, leafhopper surveys have been conducted annually in canola and cereal crops grown in Saskatchewan. The species of leafhoppers and the percentage of each species infected with the AY phytoplasma from 2001 until 2013 were partially published and presented in annual meetings along with their abundance in cereal and canola crops (Olivier *et al.*, 2011; Bahar *et al.*, 2014). Similarly, leafhopper population surveys and AY incidence were conducted in nurseries of *Brassica* germplasms and results were discussed in conferences but not yet published (Olivier *et al.*, 2015).

This chapter presents our findings on: (i) the development of molecular tools using *cpn60* as a marker, for accurate detection of AY phytoplasma strains in *Brassica* plants with single or mixed infections; (ii) the evolution/survey of leafhopper populations over the years 2005–2014, with notes on the 2015 growing season; (iii) the effect of soil moisture and leafhopper density on AY symptom expression and seed production in canola, leading to the first estimate of an economic threshold; and (iv) preliminary results regarding the identification of resistance or tolerance to AY in canola germplasms.

16.2 Molecular Tools

Canola (*Brassica napus*) and other *Brassica* species have been associated with '*Ca.* Phytoplasma asteris'-related strains. Most strains infecting *Brassica* have been classified as members of subgroups 16SrI-A and 16SrI-B (Olivier *et al.*, 2009; Salehi *et al.*, 2011; Chittem and Del Río Mendoza, 2015; Dumonceaux *et al.*, 2015). Phytoplasmas are impossible to isolate in axenic media, which is why their taxonomy remains under the criteria followed for unculturable microorganisms (Murray and Stackebrandt, 1995). Based on the similarity of 16S rRNA gene sequences supported by phylogenetic analysis, phytoplasmas are typically classified as '*Candidatus* Phytoplasma' species (IRPCM Phytoplasma/Spiroplasma Working Team – Phytoplasma Taxonomy Group, 2004) based on the sequences of molecular targets located in the 16S rRNA-encoding region, while the restriction fragment length polymorphism (RFLP) of 16S rRNA gene sequences generated with a set of 17 endonucleases forms the basis of the classification of phytoplasmas into 16Sr RFLP groups and subgroups (Lee *et al.*, 1993, 1998). Moreover, the detection and quantification of phytoplasmas in plant and insect samples is reliant on the use of molecular diagnostic techniques such as quantitative PCR (qPCR), because the pathogens are difficult to culture. Many of the molecular diagnostic methods that have been developed target the 16S rRNA-encoding locus, which is normally present in two copies in the phytoplasma genome (Zhao *et al.*, 2014).

However, certain closely related phytoplasma strains are difficult to differentiate using the 16S rRNA gene and sequence heterogeneity of the ribosomal operon is sometimes noted (Duduk and Bertaccini, 2011), which complicates classifications. This situation has led to the use of other genes as part of the scheme of identification and classification of phytoplasmas. The use of *tuf*, *rpl*V (*rpl*22)–*rps*C (*rps*3), *sec*Y, *map*, *uvr*B–*deg*V, *nus*A, *sec*A, *rpo*B and *cpn*60 genes to identify phytoplasmas has increased the accuracy of phytoplasma classification (Marcone *et al.*, 2000; Streten and Gibb, 2005; Lee *et al.*, 2006; Shao *et al.*, 2006; Arnaud *et al.*, 2007; Botti and Bertaccini, 2007; Hodgetts *et al.*, 2008; Mitrović *et al.*, 2011, 2015; Valiunas *et al.*, 2013).

A sequence of approximately 550 bp located within the Cpn60-encoding gene, known as the *cpn*60 universal target (*cpn*60 UT), has been determined to meet the criteria for a barcode marker within the domain Bacteria (Links *et al.*, 2014). As part of our work determining the *cpn*60 UT sequences for phytoplasmas, we detected strains of AY phytoplasma (16SrI) that infect *Brassica* spp. and *Camelina* spp. (Dumonceaux *et al.*, 2015). In particular, we noted two strains, designated AY-SF1 (16SrI-B) and AY-ruta (16SrI-A), that were 97% identical in *cpn60* sequence but shared 99% sequence identity at the 16S F2nR2 locus (Gundersen and Lee, 1996). Moreover, the two strains showed a differential pattern of infection in *Brassica* spp. and *Camelina* spp. wherein strain AY-ruta was less prevalent in *Camelina* spp. compared with strain AY-SF1, which infected both plant species (Dumonceaux *et al.*, 2015). This observation led us to develop two qPCR assays that can target either of these two strains, along with a third assay aimed at quantifying all of the phytoplasma strains with a known *cpn60* sequence. These qPCR assays were highly efficient and selectively targeted the intended phytoplasma (Table 16.1). None of the assays generated a signal with uninfected *Brassica napus* or *Camelina sativa* DNA as the template, suggesting that these assays can be used for the detection and quantification of phytoplasma infections in these plant tissues. For canola producers, a diagnostic test capable of detecting, quantifying and typing phytoplasma infections in asymptomatic plants, and possibly in insects, can improve the management strategy and avoid large economic losses.

16.3 Leafhopper Population

A large-scale AY survey has been implemented in canola and cereal crops grown in Saskatchewan since 2001. Except for 2012, AY incidence in plants

Table 16.1. Primer and probe sequences and amplification conditions for qPCR assays detecting AY-phytoplasma.

Primer/ probe name	Target	Sequence (5′–3′)	PCR product size, bp[a]	PCR efficiency	Correlation coefficient	C_q in non-target template[b]
D0348	AY	GCTTCATTAACAGTTGCA	95	2.06	0.998	ND
D0349	strain WB	CTACCTGATGAAACAGAAG				
AY-ruta	*cpn60*	TTGAATATCTTCTTGGTCGTCTACTT				
D0400	AY	GGAAGAAGTAGTAAAAGC	99	1.85	0.998	ND
D0401	strain SF1	GTTCCTCTTAATTTATTAGC				
AY-SF1	*cpn60*	TAGCTGAAGCTGTGGAAAATG				
D0398	AY	TGGAGTTATTAATGTTGATG	90	1.94	0.999	37.4
D0399	*cpn60*	GGAGAAGCATATCCTTTA				
AY-phyto	(all)	ATCCTTCAACAACTTCTAATTCTG				

[a]Amplification conditions for all assays: 95°C, 3 min (1×); 95°C, 10 sec, 59°C, 10 sec, 72°C, 30 sec (40×)

[b]Non-target template consisted of a mixture of 10^6 plasmid copies of each phytoplasma *cpn60* UT that was not targeted in the assay. Each assay was also tested using DNA from uninfected *B. napus* tissues and no signal was generated

ND, not detected

and leafhoppers was determined at the same locations in canola for 2001–2015 and in cereals for 2005–2015, with partial results from the period 2001–2008 already published (Olivier *et al.*, 2011). Over 20 leafhopper species tested positive for the presence of AY phytoplasma DNA at least once in the course of the study. Overall, the aster leafhopper, *M. quadrilineatus* (Fig. 16.1a), was the most abundant species by far and was found to be six to ten times more abundant in barley and wheat fields than in canola fields. When leafhoppers were infected, most carried the AY phytoplasma strain 16SrI-A, but strains 16SrI-B and -C were also detected. The leafhopper with the highest AY infection rate was the aster leafhopper followed by *Amplicephalus inimicus* and *Psammotettix* sp. (Olivier *et al.*, 2011) The aster leafhopper is implicated as the main vector of AY in North America in all crops (Kunkel, 1926; Hoy *et al.*, 1992) and canola is no exception, hence the focus on the aster leafhopper population in this chapter. A summary of the evolution of leafhopper populations over time from the years with sufficient data to observe trends is presented in Fig. 16.2.

Surveys began when the canola crop was large enough to sweep without damaging the plants and this time point typically corresponded to the last week of May. The canola crop was usually harvested by the end of August, except for a few fields in 2009 that persisted into early September and seemed

Fig. 16.1. (a) Adult and (b) fifth-instar nymph of the aster leafhopper, *Macrosteles quadrilineatus* Forbes, (Hemiptera: Cicadellidae). (c) Development of aster yellows symptom 'bladder-like' pod (***b***) on a canola flower. (d) Aster yellows symptoms of swollen bud (***sb***), virescence (***v***) of typically yellow canola petals, and phyllody (***p***) of sepals, petals and pods. (e) 'Witches' broom' (***wb***) and 'bladder-like' pods (***b***) on raceme of canola. Note the aster yellows 'purpling' symptom on the pods and stem.

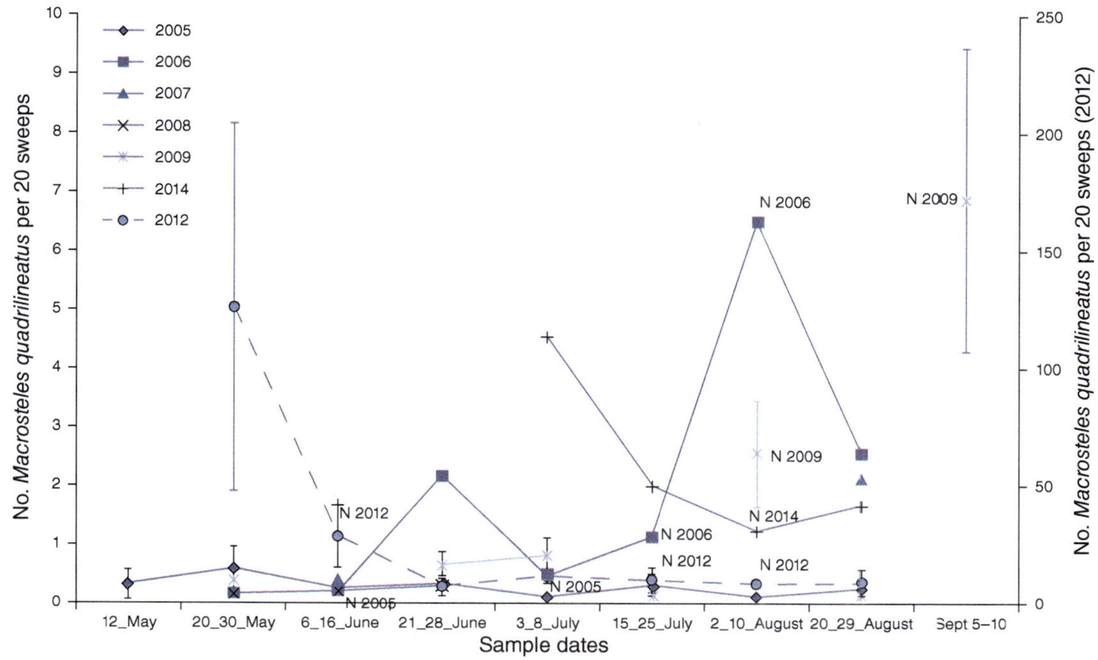

Fig. 16.2. Historical aster leafhopper population evolution in canola crops over the course of the growing season in Saskatchewan and Alberta, Canada. 'N (year)' indicates when nymphs were found in sweep net samples. Note that the leafhopper numbers from 2012 (dashed line) are graphed on the right-hand *x* axis to compensate for the large numbers of leafhoppers per 20 sweeps obtained in this year.

to concentrate the aster leafhoppers (Fig. 16.2). Nymphs of *M. quadrilineatus* (Fig. 16.1b) appeared in small numbers sporadically over the course of the survey and indicate that two generations of leafhoppers can be produced if the initial leafhopper migration arrives early enough in the spring. The appearance of nymphs corresponds to the initial arrival of migrant leafhoppers and the known generation time of *M. quadrilineatus* of approximately 27–34 days (Capinera, 2008). Preliminary experiments, however, suggest that aster leafhoppers do not reproduce successfully on canola (unpublished) and the nymphs that appear in canola sweeps were caught near field edges (data not shown). Knowledge of population numbers and percentage of infection of *M. quadrilineatus* in canola crops at the end of May and the earliest dates in June is critical, because the crop plants are most susceptible to AY infection when infected at the earliest growth stages (cotyledon to fourth true leaf) (Olivier *et al.*, 2014). In most of these years (Fig. 16.2), the *M. quadrilineatus* numbers per 20 sweeps in commercial canola fields on the Canadian prairies were low. The massive AY outbreak in canola during 2012 (Miller *et al.*, 2013) can be attributed to several key differences in the *M. quadrilineatus* populations that year. Firstly, winds from the southern USA that carried migrant leafhoppers arrived on four separate dates (Table 16.2) prior to canola germination. Secondly, temperatures during that period remained above 0°C except for three nights that reached a mean of approximately −1.5°C. Air temperatures slightly below freezing for short periods of time may not be sufficient to kill aster leafhoppers, especially if adults remain in the leaf litter. Thirdly, the number of *M. quadrilineatus* in canola fields at the critical time for crop infection in late May averaged 126 per 20 sweeps, which is extremely high compared with all other years (Fig. 16.2). Lastly, the population of *M. quadrilineatus* in early June 2012 had a high rate of AY infection (9.1%) and the population retained a high infection rate over the growing season (Table 16.3). An additional factor was that the spring climate in 2012 was unusually cold and wet, which slowed plant growth and kept the

Table 16.2. Arrival dates and origin of first and second south winds (when applicable) and post-arrival temperatures with potential to kill aster leafhoppers, *Macrosteles quadrilineatus*.

Year	1st south wind arrival dates	Wind origin	1st south wind post-arrival temperatures	2nd south wind arrival dates from USA SW prior to early June[a]
2005	May 7–9	Texas and Central USA	Remained > 0°C	–
2006	April 1, 2	California, Oregon, Washington	Remained > 0°C	–
2007	April 1	California, Mexico, Texas	6 nights < –5 °C	June 1
2008	April 10–12	Oregon, Washington, Central USA	2 nights < –5 °C	May 29–30
2009	April 11–17	Western Coast and Central USA, Washington, Oregon, California	Remained > 0°C	–
2012	April 2, 5–7	California, Washington, Oregon and Central USA	3 nights at –2°C	April 13, 25

[b]All temperatures > 0°C after 2nd south wind arrival (–, no sampling)

Table 16.3. Percentage of positive samples[a] (number of positive samples / total number of samples) and estimated aster yellows infection rates (mean ± S.E.) in groups of five aster leafhoppers, *Macrosteles quadrilineatus,* sampled during the growing season of 2009 and 2012.

	2009		2012	
Time of season/year	% of positive samples	Estimated infection ratea	% of positive samples	Estimated infection ratea
Early May	–	–	–	–
Mid-May	–	–	–	–
Late May	7.14 (1/14)	1.23 ± 1.2	23.08 (6/26)	4.28 ± 1.7
Early June	–	–	43.6 (17/39)	9.10 ± 2.1
Late June	8.70 (2/23)	1.5 ± 1.1	29.4 (5/17)	5.6 ± 2.4
Early July	11.1 (3/27)	1.9 ± 1.1	37.5 (6/16)	7.5 ± 2.9
Late July	–	–	25.0 (9/36)	4.7 ± 1.5
Early August	26.4 (14/52)	5.0 ± 1.3	34.2 (12/35)	6.8 ± 1.9
Late August	–	–	25.7 (9/35)	4.8 ± 1.6
Early September	43.7 (38/87)	9.1 ± 1.4	53.0 (9/17)	11.8 ± 3.8
Total number of samples tested	*n*=221		*n*=82	

[a]In 2009 and 2012, leafhoppers were tested by groups of 5 and not individually because of the high number of leafhoppers. Therefore, estimated infection rates were calculated using a method described for pooled organisms in epidemiology studies (Bhattacharyya *et al.*, 1979) (–, no sampling)

plants at a susceptible stage for longer, hence exacerbating AY symptoms in canola.

The 2007 growing season was also an AY outbreak year in canola (Pearse *et al.*, 2008) but the leafhopper population dynamic was different. The first wave of migrant leafhoppers would have arrived with the south winds on 1 April, but were most likely killed in the 6-day frost that began the day after their arrival (Table 16.2). However, a second south wind arrived on 1 June, which corresponds to a time when canola is in the highly susceptible cotyledon or two-leaf stage. Based on our PCR tests, the aster leafhoppers in early June were not infected with AY, while 10% were infected by the end of July and 62.5% by the end of August 2007 (Table 16.4). This high infection rate suggests that the leafhoppers that reproduced on cereal crops and/or weedy grass species may have acquired AY infection in or around the canola fields from a weedy disease reservoir. Many plant species located in and around crops have been described as being hosts for AY (Kunkel, 1926; Nagdeve *et al.*, 2015; un Nabi *et al.*, 2015). As well, field observations suggest that weedier fields typically have a higher rate of AY infection (unpublished). We observed that in 2006 and 2008 the leafhoppers appeared to arrive uninfected and then acquired AY infection over the course of the season, probably by feeding on AY-infected weed reservoirs (Table 16.4). In contrast, during the years 2005 and 2014, aster leafhoppers did not arrive infected and did not acquire AY infection over the season (Table 16.4). In part, this may be due to the low numbers of leafhoppers present during the non-AY outbreak years (Fig. 16.2).

Leafhopper diversity results indicate an increase in the number of the silver leafhopper, *Athysanus argentarius* (Hemiptera: Cicadellidae), an alternative AY-vectoring leafhopper species (Chiykowski, 1979) in 2014. This shift in the proportion of leafhopper species in and around canola fields might have consequences for the spread of AY in the Canadian prairies. *Athysanus argentarius* is mostly found in ditches and rarely in the fields and it usually constitutes 0.1% of the overall leafhopper population (Olivier *et al.*, 2011). In 2014, *A. argentarius* constituted 50% (238/476) and 70% (1086/1545) of the leafhopper population in the canola fields and in the ditches, respectively. Preliminary results from 2015 indicated that *A. argentarius* represented 30% (76/253) of the leafhopper population in the fields and 80% of the leafhopper population in the ditches (7017/8772). On several occasions, an unusually high number of *A. argentarius* adults were found in canola fields. For example, 76 *A. argentarius* were found in one site in a canola field in 2015, probably indicating a localized distribution of *A. argentarius*. The high population of *A. argentarius* in ditches bordering crops in 2015 (n = 7017), coupled with its capacity to vector AY, could be one of the key factors in maintaining, increasing and spreading reservoirs of AY infection in perennial weeds that persist across seasons on the prairies. This scenario should be investigated more fully as a potential piece of the AY puzzle on the Canadian prairies.

In addition to population surveys we attempted to find infected overwintering adults in spring, by setting emergence traps every year in early spring throughout Saskatchewan and in five locations in Alberta and five locations in Saskatchewan from 2010 to 2014. Prior to this survey, a single infected adult female had been found in one emergence trap in Saskatoon in 2004. After 4 years of emergence trapping, however, there is not enough evidence to suggest that adults of *M. quadrilineatus* successfully overwinter in high enough numbers to be an issue in canola crops in Alberta and Saskatchewan. However, eggs of *M. quadrilineatus* are reputed to overwinter successfully (Chiykowski, 1981) and some of the increase in numbers of aster leafhoppers over time may be due to local reproducing populations rather than migrants arriving on the winds.

16.4 Symptomatology

In canola, typical symptoms of AY include chlorosis, stunting, phyllody (green petals) and virescence (abnormal development of green pigmentation in plant parts that are not normally green) (Fig. 16.1d) (Lehmann and Skadow, 1971). Plants infected by AY also exhibit bladder-like pods (Fig. 16.1c,e) bearing leaf-like tissues, and the appearance of the silique can range from normal-looking pods bearing normal-looking seeds to small pods bearing malformed and shrivelled seeds (Olivier, 2007; Olivier *et al.*, 2014). Other AY symptoms observed in the field include pod abortion on part or whole branches or on the whole plant, the presence of seeds germinating in pods and purpling coloration of stem, pods and leaves. However, these other symptoms are not specific to AY infection and can be caused by other pathogens or stresses. Purpling can occur when plants are under biotic or abiotic stress, such as a nutrient deficiency, or purpling could be the normal colour of a particular cultivar. Failures in flower bud formation or pod abortion have been associated with phytoplasma infection (Kaminska *et al.*, 2012) but also with heat stress (Young *et al.*, 2004), nutrient deficiencies (Asad *et al.*, 2002) or occasional male sterility in hybrid canola. Heat stress can also cause abnormal development of pods that appear very similar to bladder-like pods and can cause seeds to germinate in malformed and normal pods (Angadi *et al.*, 2000).

Table 16.4. Aster yellows infection rates[a] (mean ± SE) in aster leafhoppers, *Macrosteles quadrilineatus*, sampled[b] during the growing season between 2005–2008 and in 2014.

Time of season/year	2005	2006	2007	2008	2014
Mid-May	0 (0/4)	–	–	–	–
Late May	0 (0/32)	0 (0/1)	–	0 (0/5)	–
Early June	0 (0/13)	0 (0/6)	–	–	0 (0/5)
Late June	0 (0/26)	3.4% (1/29)	0 (0/10)	50% (5/10)	–
Early July	0 (0/6)	20% (1/5)	0 (0/4)	0 (0/3)	0 (0/50)
Late July	0 (0/4)	0 (0/5)	10% (1/10)	0 (0/3)	0 (0/2)
Early August	0 (0/5)	0 (0/35)	–	0 (0/2)	0 (0/5)
Late August	0 (0/8)	0 (0/11)	62.5 ± 10.4 (19/28)	0 (0/3)	0 (0/20)
Early September	–	–	–	–	–
Total number of samples tested	*n*=98	*n*=92	*n*=52	*n*=26	*n*=82

[a]Infection rates were calculated by dividing the number of samples (–, no sampling) that tested positive with the 16Sr PCR test by the total number of tested samples (*n*) and these are shown in parentheses. Leafhoppers were tested individually.

However, in heat-stressed plants, none of the other AY symptoms, such as 'dwarfism' type of stunting (with very short internodes), phyllody and virescence, are expressed. Also, once the heat stress is removed, normal flowers and pods will form on the raceme, differentiating malformations of pods from those caused by AY. In addition, bladder-like pods in AY-infected plants do not bear normal, malformed or sprouted seeds; instead they are filled with leaf-like tissues rather than seeds (Olivier *et al.*, 2014). Abnormal-looking seedlings grown out of seeds harvested from AY-infected plants have been described and phytoplasma DNA was detected in the seeds and any ensuing malformed canola seedlings (Starzycki and Starzycka, 2000; Olivier *et al.*, 2010). Severely malformed plants were obtained from self-pollinated plants grown from AY phytoplasma-positive seedlings but no phytoplasma DNA was detected in the seedlings once they had grown to the second true-leaf stage or in the self-progeny plants (Olivier and Galka, 2008). Although AY phytoplasma DNA has now been identified in seedlings and embryos obtained from seeds harvested on phytoplasma-infected plants belonging to various other species, their transmission via seeds has not been demonstrated (Nipah *et al.*, 2007; Azadvar *et al.*, 2011; Calari *et al.*, 2011; Faghihi *et al.*, 2011; Zwolińska *et al.*, 2012).

16.5 Effect of Soil Moisture and Leafhopper Density on AY Symptoms and Seed Production in Canola

AY symptom expression and estimation of the seed production in canola crops were tested using a bioassay involving canola plants at various growth stages being fed upon by various densities of AY-infected aster leafhoppers under controlled conditions (Olivier *et al.*, 2014). Briefly, hybrid canola plants were planted in plastic cones (one seed/cone) containing an artificial medium and were grown in dry soil (20–30% moisture content) and wet soil (> 70% moisture content) for 10–12 days at 20°C. Plants at the early second true-leaf stage were transferred into cages (*n* = 4 plants/cage) in a growth chamber at 20°C and exposed to eight densities of AY-infected leafhoppers (*n* = 2–16 adults/plant) for 10 hours. The number of leafhoppers on each plant was recorded hourly, in order to obtain the feeding density (expressed in number of leafhoppers/plant), and this number is representative of the feeding length and intensity of leafhoppers on each plant. Plants were then transplanted to pots and grown until maturity at 20°C under high light intensity (> 400 μmol/m^2/s). AY symptoms were rated after 6, 8 and 10 weeks using a five-point scale and seeds were harvested to determine yield and 1000-seed weight (Olivier *et al.*, 2014).

Hourly observations of leafhopper feeding showed that only a portion of the leafhoppers were feeding on the plants and that leafhoppers had a tendency to aggregate on certain plants (Olivier *et al.*, 2014). Leafhopper aggregation on their plant hosts, whether at the imago or nymphal stages, has been observed in several species (Bosco *et al.*, 1997; Bentz and Townsend, 2004; Mizell *et al.*, 2012). Aggregation causes can be non-social and originating from plants or due to the presence of living or dead conspecifics (Chuche *et al.*, 2011; Mizell *et al.*, 2012). Also, a significantly higher percentage of leafhoppers was found to feed on plants grown in wet soil, compared with plants grown in dry soil (Olivier *et al.*, 2014). Indeed, it is well established that leafhoppers prefer to feed on well-watered plants compared with water-limited plants (Krugner and Backus, 2014). Feeding densities had little effect on AY ratings and seed yield when inoculation occurred in dry soil. However, the 1000-seed weight of plants infected in dry soil slightly declined by 25–30% as leafhopper feeding densities increased (Fig. 16.3). In wet soil, AY ratings after 6, 8 and 10 weeks increased as leafhopper feeding densities increased and seed yield and 1000-seed weight declined sharply (Fig. 16.3). Similarly, both the frequency and severity of AY symptoms were higher when infection occurred in wet soil compared with dry soil (Table 16.5). AY ratings were highest when densities were above four leafhoppers per plant in wet soil and above 12 leafhoppers per plant in dry soil. It is important to note that when inoculation occurred in wet soil, with four to six leafhoppers feeding on plants for 10 hours, 67% of the plants produced no seeds, while under dry conditions 20% of the plants reached an AY rating of 3 with more than 12 leafhoppers per plant. This can explain why in some years, such as 2001 with a dry spring in the sampling areas, there was little or no AY incidence despite the presence of a high number of infected leafhoppers (Olivier *et al.*, 2011).

Based on these experiments, plants with AY ratings of 1–3 produced normal-looking and malformed seeds but plants with AY ratings of 3–5 produce no seeds. A first estimation of the economic threshold for aster leafhoppers is fewer than four leafhoppers per plant in wet soil conditions and above 12 leafhoppers per plant in dry soil condition. Also, in dry soil, there is a slight decline in the 1000-seed weight when leafhopper feeding density increases. More experiments are required to assess production losses when infection occurs after the second true-leaf stage as well as with other soil moisture contents and temperatures.

16.6 Aster Yellows Incidence in Canola Nurseries

The possibility of genetic resistance to AY in *Brassica napus* was explored in a collection of diverse spring lines in the 2012 and 2013 growing seasons in Saskatoon, Saskatchewan. In 2012, the

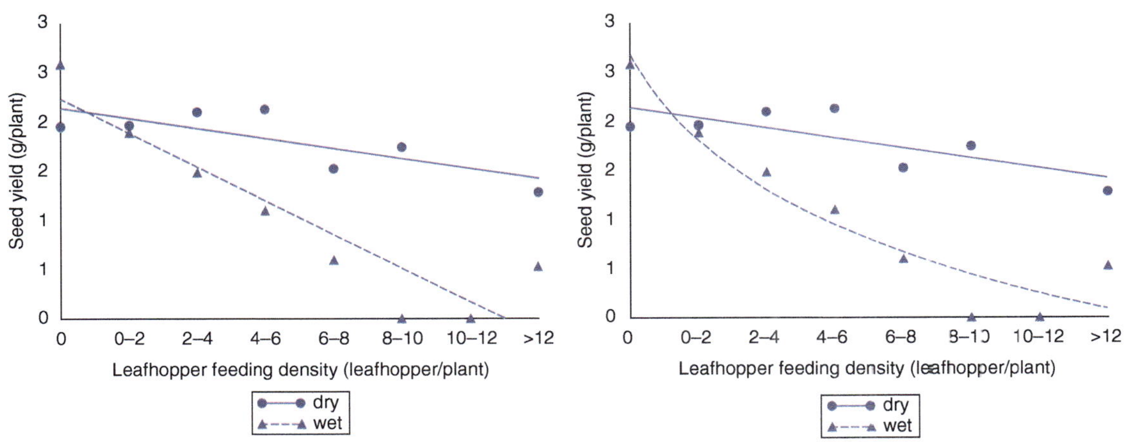

Fig. 16.3. Relationship between (a) seed yield and (b) 1000-seed weight for hybrid canola and feeding density of aster yellows-infected aster leafhoppers, *Macrosteles quadrilineatus*, when plants are grown in dry and wet soil during AY infection.

Table 16.5. Relationship between percentage of hybrid canola plants expressing aster yellows symptoms and aster leafhopper, *Macrosteles quadrilineatus*, feeding density, 10 weeks after infection and with plants grown in wet or dry soil during infection.

Soil moisture	Leafhopper feeding density	No AY symptoms	AY rating 1	AY rating 2	AY rating 3	AY rating 4	AY rating 5
Dry	0	100	0	0	0	0	0
	0.1–2.0	67	28	0	0	0	0
	2.1–4.0	63	37	0	0	0	0
	4.1–6.0	47	53	0	0	0	0
	6.1–8.0	38	50	12	0	0	0
	8.1–10.0	67	33	0	0	0	0
	10.1–12.0	100	0	0	0	0	0
	> 12.0	**40**	**40**	**0**	**20**	**0**	**0**
Wet	0	100	0	0	0	0	0
	0.1–2.0	60	30	0	10	0	0
	2.1–4.0	38	23	0	8	15	15
	4.1–6.0	**0**	**33**	**0**	**0**	**67**	**0**
	6.1–8.0	**0**	**25**	**8**	**0**	**58**	**8**
	8.1–10.0	**0**	**0**	**0**	**0**	**100**	**0**
	10.1–12.0	**0**	**0**	**0**	**0**	**100**	**0**
	> 12.0	**0**	**13**	**0**	**0**	**67**	**20**

visual incidence of AY was determined in a replicated single-row nursery of 218 lines. In 2013, 253 diverse lines of *B. napus* were examined for incidences using the molecular PCR assay in addition to visual counts of visibly infected plants. In both years, significant differences among the lines were found, with approximately 22% of the lines in 2012 having significant AY incidence with up to 60% of plants expressing symptoms in some lines. Infection levels were substantially lower in 2013, with only 2% of the lines having plants expressing symptoms and/or testing positive for the presence of phytoplasma DNA using PCR. In the 2012 and 2013 field trials, a common set of 129 diverse lines of spring *B. napus* were grown. However, AY ranking of those common lines did not correlate between the seasons; and lines showing no AY symptoms in 2012 were not confirmed in 2013 and vice versa. As well, in 2013, visual symptoms were not well correlated with the results of the PCR tests. Although the PCR test was much more sensitive in detecting AY infection than visual surveys, ANOVA-type analysis showed that differences between lines were not a significant source of variation. While visual inspection of infection levels of plots across the fields in 2012 and 2013 showed no distinctive pattern for symptomatic plants, the PCR tests in 2013 showed an edge effect. Indeed, PCR screening showed greater infection on one edge of the field, reaching approximately 12 m into the long edge of the ~120 m long field. Lower PCR incidence was distributed throughout the remainder of the field with random distribution.

Thus, it is postulated that if resistance to AY exists, it is not a highly heritable trait. The possible relationship of AY infection with traits typically evaluated within a canola breeding programme was explored in the diversity collection of spring *B. napus*. However, no correlation between infection and agronomic traits (emergence, pre-bolting biomass, time to flowering and maturity, height, lodging) and harvested seed quality traits (yield, seed size; oil, protein and fibre content; seed colour; contents of fatty acids and glucosinolates within profiles) were found in either season. Interestingly, in 2013, one line consistently showed more symptoms across replications than other entries with the same cultivar name but derived from different seed sources. These results suggest that seed infection initiated in development on the maternal plant could be a source of infection; however, further exploration of this hypothesis is required.

16.7 Conclusion

Three AY outbreaks of increasing incidence were recorded in the Canadian prairies over the past 15 years: in 2000, 2007 and 2012. The last outbreak in 2012 caused several million dollars in production losses to canola producers. Each of these outbreaks was correlated with early arrival of high numbers of aster leafhoppers carried by south winds from the USA. Leafhoppers usually arrived infected with AY but they also presumably picked up the disease from the weed reservoir located in and around the crops. Some weed species are known to be hosts for AY phytoplasma but the vast majority of perennial weeds located in and around canola fields in the Canadian prairies have not yet been tested for the presence of AY. Insecticide sprays are the most common methods to control the leafhopper vectors of AY. Because of the difficulty of timing insecticide sprays and the environmental consequences of insecticide spraying, other disease management strategies are being studied, such as the identification of resistance/tolerance among canola germplasms and the effect of seed treatments on reducing AY infection. Although the identification of resistance/tolerance in canola germplasms was unsuccessful in this study, partly because of the low disease pressure in the past 3 years, the detection of resistance and/or tolerance to AY should continue, as resistant cultivars represent the best long-term sustainable control method. Seed treatment is also an option that has been used successfully to reduce diseases borne by sap feeders (de Oliveira *et al.*, 2007; Zhu *et al.*, 2011). The efficacy of insecticidal seed treatments for leafhopper control and AY suppression in canola is currently being studied.

References

Angadi, S.V., Cutforth, H.W., Miller, P.R., McConkey, B.G., Entz, M.H., Brandt, S.A. and Volkmar, K.M. (2000) Response of three *Brassica* species to high temperature stress during reproductive growth. *Canadian Journal of Plant Science* 80, 693–701.

Anon. (1936) Diseases of vegetable and field crop. *Canadian Plant Disease Survey* 16, 21–47.

Anon. (1941) Diseases of vegetable and field crop. *Canadian Plant Disease Survey* 21, 28–61.

Arnaud, G., Malembic-Maher, S., Salar, P., Bonnet, P., Maixner, M., Marcone, C., Boudon-Padieu, E. and Foissac, X. (2007) Multilocus sequence typing confirms the close genetic interrelatedness of three distinct flavescence doree phytoplasma strain clusters and group 16SrV phytoplasmas infecting grapevine and alder in Europe. *Applied Environmental Microbiology* 73, 4001–4010.

Asad, A., Blamey, F.P.C. and Edwards, D.G. (2002) Dry matter production and boron concentrations of vegetative and reproductive tissues of canola and sunflower plants grown in nutrient solution. *Plant and Soil* 243, 243–252.

Azadvar, M., Baranwal, V.K and Yadava, D.K. (2011) Transmission and detection of toria [*Brassica* rapa L. subsp. *dichotoma* (Roxb.)] phyllody phytoplasma and identification of a potential vector. *Journal of General Plant Pathology* 7, 194–200.

Bahar, M., Olivier, C., Bekkaoui, D., Soroka, J. and Hegedus, D. (2014) Phytoplasma–leafhopper–plant interactions in a changing climate. In: *Joint Annual Meetings of the Entomological Societies of Canada and Saskatchewan: Entomology in a Changing World, 28 Sep–1 Oct 2014, Saskatoon*. Entomological Society of Canada, Winnipeg, Canada, p. 14.

Bentz, J. and Townsend, A.M. (2004) Spatial and temporal patterns of abundance of the potato leafhopper among red maples. *Annals of Applied Biology* 145, 157–164.

Bertaccini, A., Duduk, B., Paltrinieri, S. and Contaldo, N. (2014) Phytoplasmas and phytoplasma diseases: a severe threat to agriculture. *American Journal of Plant Sciences* 5, 1763–1788.

Bhattacharyya, G.K., Karancinos, M.G. and DeFoliart, G.R. (1979) Point estimates and confidence intervals for infection rates using pooled organisms in epidemiologic studies. *American Journal of Epidemiology* 109, 124–131.

Bosco, D., Alma, A. and Arzone, A. (1997) Studies on population dynamics and spatial distribution of leafhoppers in vineyards (Homoptera: Cicadellidae). *Annals of Applied Biology* 130, 1–11.

Botti, S. and Bertaccini, A. (2007) Grapevine yellows in Northern Italy: molecular identification of Flavescence doree phytoplasma strains and of Bois Noir phytoplasmas. *Journal of Applied Microbiology* 103, 2325–2330.

Calari, A., Paltrinieri, S., Contaldo, N., Sakalieva, D., Mori, N., Duduk, B. and Bertaccini, A. (2011) Molecular evidence of phytoplasmas in winter oilseed rape, tomato and corn seedlings. *Bulletin of Insectology* 64, S157–S158.

Capinera, J.L. (2008) Aster leafhopper, *Macrosteles quadrilineatus* Forbes (Hemiptera: Cicadellidae). In: Capinera, J.L. (ed.) *Encyclopedia of Entomology* Springer, Dordrecht, The Netherlands, pp. 320–323.

Chittem, K. and Del Río Mendoza, L.E. (2015) Detection and molecular characterization of '*Candidatus* phytoplasma asteris' related phytoplasmas infecting canola in North Dakota. *Canadian Journal of Plant Pathology* 37, 267–272.

Chiykowski, L. (1979) *Athysanus argentarius*, an introduced European leafhopper, as a vector of aster yellows in North America. *Canadian Journal of Plant Pathology* 1, 37–41.

Chiykowski, L.N. (1981) Epidemiology of diseases caused by leafhopper-borne pathogens. In: Maramorosch, K. and Harris, K.F. (eds) *Plant Diseases and Vectors: Ecology and Epidemiology*. Academic Press, New York, pp. 106–159.

Chuche, J., Boursault, A. and Thiéry, D. (2011) Preliminary study of the aggregative behaviour of *Scaphoideus titanus* larvae. *IOBC/WPRS Bulletin* 67, 239–244.

de Oliveira, C.M., de Oliveira, E., Canuto, M. and Cruz, I. (2007) Chemical control of corn leafhopper and incidence of corn stunting diseases caused by mollicutes. *Pesquisa Agropecuaria Brasileira* 42, 297–303.

Duduk, B. and Bertaccini, A. (2011) Phytoplasma classification: taxonomy based on 16S ribosomal gene, is it enough? *Phytopathogenic Mollicutes* 1, 3–13.

Dumonceaux, T., Green, M., Hammond, C., Perez, E. and Olivier, C. (2015) Molecular diagnostic tools for detection and differentiation of Phytoplasmas based on Chaperonin-60 reveal differences in host plant infection patterns. *PLoS ONE* 9(12), e116039.

Elliot, B. and Northover, P. (2007) Manitoba vegetables insect and diseases situation in review. *Manitoba Weekly Vegetable Report* 15, 17.

Faghihi, M., Bagheri, A., Bahrami, H., Hasanzadeh, H., Rezazadeh, R., Siampour, M., Samavi, S., Salehi, M. and Izadpanah, K. (2011) Witches'-broom disease of lime affects seed germination and seedling growth but is not seed transmissible. *Plant Disease* 95, 419–422.

Foissac, X. and Wilson, M.R. (2010) Current and possible future distributions of phytoplasma diseases and their vectors. In: Weintraub, P.G. and Jone, P. (eds) *Phytoplasma: Genomes, Plant Hosts and Vectors*. CAB International, Wallingford, UK, pp. 309–325.

Frost, K.E., Esker, P.D., Van Haren, R., Kotolski, L. and Groves, R.L. (2013) Seasonal patterns of aster leafhopper (Hemiptera: Cicadellidae) abundance and aster yellows phytoplasma infectivity in Wisconsin carrot fields. *Environmental Entomology* 42, 491–502.

Gundersen, D. and Lee, I.-M. (1996) Ultrasensitive detection of phytoplasmas by nested-PCR assays using two universal primer pairs. *Phytopathologia Mediterranea* 35(3), 144–151.

Hodgetts, J., Boonham, N., Mumford, R., Harrison, N. and Dickinson, M. (2008) Phytoplasma phylogenetics based on analysis of *secA* and 23S rRNA gene sequences for improved resolution of candidate species of '*Candidatus* Phytoplasma'. *International Journal Systematic Evolutionary Microbiology* 58, 1826–1837.

Hopkinson, R.F. and Soroka, J.J. (2010) Air trajectory model applied to an in-depth diagnosis of potential diamondback moth infestations on the Canadian Prairies. *Agricultural and Forest Meteorology* 150, 1–11.

Hoy, C.W., Heady, S.E. and Koch, T.A. (1992) Species composition, phenology, and possible origins of leafhoppers (Cicadellidae) in Ohio vegetable crops. *Journal of Economic Entomology* 85, 2336–2343.

IRPCM Phytoplasma/Spiroplasma Working Team – Phytoplasma Taxonomy Group (2004) '*Candidatus* Phytoplasma', a taxon for the wall-less, non-helical prokaryotes that colonize plant phloem and insects. *International Journal Systematic Evolutionary Microbiology* 54, 1243–1255.

Kaminska, M., Berniak, H. and Kaminski, P. (2012) Failure of flower bud formation in *Brassica* plants associated with phytoplasma infection. *Journal of Agricultural Science* 4, 219.

Krugner, R. and Backus, E.A. (2014) Plant water stress effects on stylet probing behaviors of *Homalodisca vitripennis* (Hemiptera: Cicadellidae) associated with acquisition and inoculation of the bacterium *Xylella fastidiosa*. *Journal of Economic Entomology* 107, 66–74.

Kunkel, L.O. (1926) Studies on aster yellows. *American Journal of Botany* 13, 646–705.

Lee, I.-M., Hammond, R., Davis, R. and Gundersen, D. (1993) Universal amplification and analysis of pathogen 16S rDNA for classification and identification of mycoplasmalike organisms. *Phytopathology* 83, 834–842.

Lee, I.-M., Gundersen-Rindal, D.E., Davis, R.E. and Bartoszyk, I.M. (1998) Revised classification scheme of phytoplasmas based on RFLP analyses of 16S rRNA and ribosomal protein gene sequences. *International Journal of Systematic and Evolutionary Microbiology* 48, 1153–1169.

Lee, I.M., Gundersen-Rindal, D.E., Davis, R.E., Bottner, K.D., Marcone, C. and Seemuller, E. (2004) '*Candidatus* Phytoplasma asteris', a novel phytoplasma taxon associated with aster yellows and related diseases. *International Journal Systematic Evolutionary Microbiology* 54, 1037–1048.

Lee, I.-M., Bottner, K.D., Munyaneza, J.E., Davis, R.E., Crosslin, J.M., du Toit, L.J. and Crosby, T. (2006) Carrot purple leaf: a new spiroplasmal disease associated with carrots in Washington State. *Plant Disease* 90, 989–993.

Lehmann, W. and Skadow, K. (1971) Untersuchungen zur Verbreitung, Ätiologie und Vektorübertragbarkeit der Blütenvergrünung des Rapses. *Archives of Phytopathology & Plant Protection* 7, 323–336.

Links, M.G., Demeke, T., Gräfenhan, T., Hill, J.E., Hemmingsen, S.M. and Dumonceaux, T.J. (2014) Simultaneous profiling of seed-associated bacteria and fungi reveals antagonistic interactions between microorganisms within a shared epiphytic microbiome on *Triticum* and *Brassica* seeds. *New Phytologist* 202, 542–553.

Ma, K.W. and Ma, W. (2016) Phytohormone pathways as targets of pathogens to facilitate infection. *Plant Molecular Biology* 91(6), 713–725.

Marcone, C., Lee, I.M., Davis, R.E., Ragozzino, A. and Seemuller, E. (2000) Classification of aster yellows-group phytoplasmas based on combined analyses of rRNA and tuf gene sequences. *International Journal of Systematic and Evolutionary Microbiology* 50 (5), 1703–1713.

McCoy, R.E., Caudwell, A., Chang, C.J., Chen, T.A., Chiykowski, L.N., Cousin, M.T., Dale, J.L., De Leeuw, G.T.N., Golino, D.A., Hackett, K.J., Kirkpatrick, B.C., Marwitz, R., Petzold, H., Sinha, R.C., Sugiura, M., Whitcomb, R.F., Yang, I.L., Zhu, B.M. and Seemuller, E. (1989) Plant diseases associated with mycoplasma-like organisms. In: Whitcomb, R.F. and Tully, J.G. (eds) *The Mycoplasmas*. Academic Press, New York, pp. 545–560.

McLaren, D.L. and Platford, R.G. (2001) Distribution, prevalence and incidence of canola diseases in Manitoba (2000). *Canadian Plant Disease Survey* 81, 109–110.

Miller, S.G., Anderson, K., Bassendowski, K.A., Britz, L., Buitenhuis, N., Campbell, E., Chant, S., Christopher, J., Cowell, L.E., Cranston, R., Dokken-Bouchard, F.L., Friesen, S., Gugel, R.K., Hicks, J., Ippolito, J., Jurke, C., Kennedy, V., Kirkham, C.L., Martinka, T., Moore, M., Oster, K., Peng, G., Phelps, S.M., Platford, R.G., Senko, S., Stonehouse, K. and Vakulabharanam, V. (2013) Survey of canola diseases in Saskatchewan, 2012. *Canadian Plant Disease Survey* 93, 149–154.

Mitrović, J., Kakizawa, S., Duduk, B., Oshima, K., Namba, S. and Bertaccini, A. (2011) The *groEL* gene as an additional marker for finer differentiation of '*Candidatus* Phytoplasma asteris'-related strains. *Annals of Applied Biology* 159, 41–48.

Mitrović, J., Smiljković, M., Seemüller, E., Reinhardt, R., Hüttel, B., Büttner, C., Bertaccini, A., Kube, M. and Duduk, B. (2015) Differentiation of '*Candidatus* Phytoplasma cynodontis' based on 16S rRNA and *groEL* genes and identification of a new subgroup, 16SrXIV-C. *Plant Disease* 99, 1578–1583.

Mizell III, R.F., Andersen, P.C., Brodbeck, B.V. and Hunter, W.B. (2012) Congener response reduces risks from bottom-up and top-down forces: behavioral parsimony by a xylophage. *American Entomologist* 58, 106–115.

Murray, R.G. and Stackebrandt, E. (1995) Taxonomic note: implementation of the provisional status *Candidatus* for incompletely described procaryotes. *International Journal of Systematic Bacteriology* 45, 186–187.

Nagdeve, D., Bharti, Y. and Kumar, A. (2015) Survey of weeds growing near *Zea mays* fields in western and central Uttar Pradesh. *Agrica* 4, 48–51.

Nipah, J.O., Jones, P., Hodgetts, J. and Dickinson, M. (2007) Detection of phytoplasma DNA in embryos from coconut palms in Ghana, and kernels from maize in Peru. *Bulletin of Insectology* 60, 385.

Olivier, C.Y. (2007) Aster yellows. In: Rimmer, R.S., Shattuck, V. and Buchwaldt, L. (eds) *Compendium of Brassica Diseases*. The American Phytopathological Society Press, St Paul, Minnesota, pp. 62–64.

Olivier, C. and Galka, B. (2008) Consequences of phytoplasma infection on canola crop production in the Canadian prairies. In: *Proceedings of Endure International Conference: Diversifying crop protection, 12–15 October 2008*. Endure, La Grand-Motte, France, O.47.

Olivier, C.Y., Séguin-Swartz, G., Hegedus, D. and Barasubiye, T. (2006) First report of *Candidatus* asteris phytoplasma subgroups AY 16SrI-A (strains WB and CHRY 16S) and 16SrI-B in *Brassica* rapa in Saskatchewan, Canada. *Plant Disease* 90, 832.

Olivier, C.Y., Lowery, D.T. and Stobbs, L.W. (2009) Phytoplasma diseases and their relationships with insect and plant hosts in Canadian horticultural and field crops. *The Canadian Entomologist* 141, 425–462.

Olivier, C.Y., Galka, B. and Séguin-Swartz, G. (2010) Detection of aster yellows phytoplasma DNA in seed and seedlings of canola (*Brassica napus* and *B. rapa*) and AY strain identification. *Canadian Journal of Plant Pathology* 32, 298–305.

Olivier, C.Y., Séguin-Swartz, G., Galka, B. and Olfert, O.O. (2011) Aster yellow in leafhoppers and field crops in Saskatchewan, Canada, 2001–2008. *The Americas Journal of Plant Sciences and Biotechnology* 141, 425–462.

Olivier, C., Elliott, B., Man, L. and Nordin, D. (2014) Development of a rating scale for Aster yellow in canola. *Canadian Plant Disease Survey* 94, 162–176.

Olivier, C., Dumonceaux, T. and Perez, E. (2015) Manejo de artrópodos vectores de patógenos vegetales en cultivos de importancia agrícola en Canadá. In: *Congreso Internacional de Diagnóstico Molecular para el manejo integrado de Plagas y enfermedades en la Producción Agraria*. Universidad Privada Antenor Orreo, Trujillo, Peru, pp. 46–47.

Orlovskis, Z., Canale, M.C., Thole, V., Pecher, P., Lopes, J.R.S. and Hogenhout, S.A. (2015) Insect-borne plant pathogenic bacteria: getting a ride goes beyond physical contact. *Current Opinion in Insect Science* 9, 16–23.

Pearse, P.G., Bassendowski, K., Cross, D.J., Gugel, R.K., Kirkham, C.L., Kutcher, R.H., Morrall, R.A.A. and Yasinowski, J.M. (2008) Survey of canola diseases in Saskatchewan, 2007. *Canadian Plant Disease Survey* 88, 103–104.

Rankin, L. (1987) Lettuce disease survey in New Brunswick, 1986. *Canadian Plant Disease Survey* 67, 25–30.

Salehi, M., Izadpanah, K. and Siampour, M. (2011) Occurrence, molecular characterization and vector transmission of a Phytoplasma associated with rapeseed phyllody in Iran. *Journal of Phytopathology* 159, 100–105.

Saskatchewan Ministry of Agriculture (2016) *Guide to Crop Protection Weeds, Plant Diseases, Insects*. Government of Saskatchewan, Regina, Canada.

Shao, J., Jomantiene, R., Dally, E., Zhao, Y., Lee, I.-M., Nuss, D. and Davis, R. (2006) Phylogeny and characterization of phytoplasma NusA and use of the nusA gene in detection of group 16SrI strains. *Journal of Plant Pathology* 88, 193–201.

Starzycki, M. and Starzycka, E. (2000) Parents and progeny plant deformation of *Brassica napus* L. infected by *Phytoplasma* sp. *Rośliny Oleiste – Oilseed Crops* XXI, 399–408.

Starzycki, M., Starzycka, E. and Pszczola, J. (2003) Research in rapeseed resistance breeding – progress made in recent years. *Rosliny Oleiste – Oilseed Crops* XXIV, 363–372.

Streten, C. and Gibb, K. (2005) Genetic variation in *Candidatus* Phytoplasma australiense. *Plant Pathology* 54, 8–14.

Sugio, A., MacLean, A.M. and Hogenhout, S.A. (2014) The small phytoplasma virulence effector SAP11 contains distinct domains required for nuclear targeting and CIN-TCP binding and destabilization. *New Phytologist* 202, 838–848.

un Nabi, S., Dubey, D.K., Rao, G., Baranwal, V. and Sharma, P. (2015) Molecular characterization of '*Candidatus* Phytoplasma asteris' subgroup IB associated with sesame phyllody disease and identification of its natural vector and weed reservoir in India. *Australasian Plant Pathology* 44, 289–297.

Valiunas, D., Jomantiene, R. and Davis, R.E. (2013) Evaluation of the DNA-dependent RNA polymerase beta-subunit gene (rpoB) for phytoplasma classification and phylogeny. *International Journal Systematic Evolutionary Microbiology* 63, 3904–3914.

Vanterpool, T.C. (1963) Rape diseases in Saskatchewan in 1963. *Canadian Plant Diseases Survey* 43, 212–214.

Weintraub, P.G. and Beanland, L. (2006) Insect vectors of phytoplasmas. *Annual Review Entomology* 51, 91–111.

Young, L.W., Wilen, R.W. and Bonham-Smith, P.C. (2004) High temperature stress of *Brassica* napus during flowering reduces micro-and megagametophyte fertility, induces fruit abortion, and disrupts seed production. *Journal of Experimental Botany* 55, 485–495.

Zhao, Y., Davis, R.E., Wei, W., Shao, J. and Jomantiene, R. (2014) Phytoplasma genomes: evolution through mutually complementary mechanisms, gene loss and horizontal acquisition. In: Gross, D.C., Lichens-Park, A. and Kole, C. (eds) *Genomics of Plant-Associated Bacteria*. Springer, Heidelberg, Germany, pp. 235–271.

Zhu, Y., Loso, M.R., Watson, G.B., Sparks, T.C., Rogers, R.B., Huang, J.X., Gerwick, B.C., Babcock, J.M., Kelley, D., Hegde, V.B., Nugent, B.M., Renga, J.M., Denholm, I., Gorman, K., DeBoer, G.J., Hasler, J., Meade, T. and Thomas, J.D. (2011) Discovery and characterization of sulfoxaflor, a novel insecticide targeting sap-feeding pests. *Journal Agricultural Food Chemistry* 59, 2950–2957.

Zwolińska, A., Krawczyk, K. and Pospieszny, H. (2012) Molecular characterization of stolbur phytoplasma associated with pea plants in Poland. *Journal of Phytopathology* 160, 317–323.

17 Pestiferous Insects of Mustard: Biology and Integrated Management

Dhana Raj Boina* and S. Jesu Rajan

National Institute of Plant Health Management, Hyderabad, India

17.1 Introduction

The crucifers comprising oilseed and vegetable crops are mainly grown during the *rabi* (winter–spring) season all over the world. Of these, mustard along with rapeseed forms an important oilseed crop, the seeds of which are rich in oil (35–45%) (Firake *et al.*, 2013; Anon., 2015a). During 2013/14, mustard and rapeseed together were cultivated in 36.15 million hectares worldwide with a production of 71.09 million tonnes and a productivity of 1970 kg/ha (Anon., 2015a). In decreasing rank order, Canada, China, India and Australia are major players in mustard and rapeseed cultivation (Anon., 2015a). Different mustard species are commercially cultivated in different geographical regions of the world. Indian or oriental or brown mustard, *Brassica juncea* (L.) Ozern, is cultivated commercially in Asia (including India), Canada, Denmark, the UK and the USA (Ziaee, 2012). Black mustard, *Brassica nigra* L., is cultivated in Argentina, Chile, a few European countries and the USA (Ziaee, 2012). White or yellow mustard, *Brassica* (*Sinapis*) *alba* L. and *Brassica hirta* Moench, is cultivated commercially in Canada, the Middle East and Mediterranean Europe and is grown wild in North Africa (Dosdall *et al.*, 2011; Ziaee, 2012). The Ethiopian mustard or Ethiopian kale or Abyssinian mustard, *Brassica carinata* Braun, is cultivated in India, East and North Africa and the USA (Harrison, 2013; Anon., 2015a; AVRDC, 2015). Field mustard or turnip mustard, *Brassica rapa* L. or *Brassica campestris* L. (*B. rapa* is also referred to as canola: Brown *et al.*, 2004; Dosdall *et al.*, 2011) is cultivated commercially in Canada (Dosdall *et al.*, 2011; Ziaee, 2012). Of the several constraints that limit mustard plants in realizing their full genetic potential in terms of yield, biotic stress in the form of pestiferous insects is of prime importance and causes 10–96% or sometimes complete yield loss (Aamir and Khalid, 1961; Singh and Sachan, 1999; Shylesha *et al.*, 2006; Kumar, 2015).

Mustards, due partly to their stored secondary metabolites such as glucosinolates (e.g. sinigrin) and their breakdown products (e.g. isothiocyanates (ITCs)), act as attractants, arrestants and stimulants for *Brassica* specialist feeders (Soroka *et al.*, 2005). The preference by *Brassica* specialists for different mustard species over other *Brassica* species led to ecological engineering of mustards as companion (trap) plants in several cruciferous vegetable crop ecosystems, such as cabbage (*Brassica oleracea* L. var. *capitata* L.), cauliflower (*Brassica oleracea* L. var. *botrytis* L.), broccoli (*Brassica oleracea* L. var. *italica* Plenck) and knol-khol or kohlrabi (*Brassica oleracea* L. var. *gongylodes*) (Srinivasan and Moorthy, 1991, 1992; Silvia-Krott *et al.*, 1995; Luther *et al.*, 1996; Boina, 2000; Charleston and Kfir, 2000; Muniappan *et al.*, 2001; George *et al.*, 2009). This suggests that mustard crop grown for seed oil is highly prone to attack by important insect pests of cruciferous crops. More than 43 insect species, including both generalist and specialist feeders, infest the mustard crop from seedling to harvesting stage and continue even during storage (Aamir and Khalid, 1961; Singh and Sachan, 1999; Shylesha *et al.*, 2006; Singh, 2009; Kumaranag *et al.*, 2014; Anon., 2015b; Kumar, 2015). Of these, 10–12 root feeders, defoliators and sucking pests

*Corresponding author. E-mail: boinaraj@gmail.com

are of economic importance, at times causing 100% yield loss both qualitatively and quantitatively (Yadav and Singh, 2015). However, the destructive pest status of insects is dynamic and varies with season, year, country, plant species, variety, cultivar and cropping system (mono vs multiple cropping) (Ziaee, 2012).

Various plant protection techniques, i.e. cultural, mechanical, physical, biological, genetic and chemical, are available for farmers to deploy for protecting mustard plants by managing pest densities. The majority of farmers, however, rely heavily on a single plant protection method, i.e. application of chemical insecticides. It is well understood and widely established that such heavy reliance on chemical insecticides causes environmental pollution and health hazards to human beings and kills natural enemies and non-target organisms, as well as causing problems of pesticide resistance, resurgence in pests and pesticide residues in mustard oil and cake (Yadav and Singh, 2015). Therefore, the best approach is to adopt integrated pest management (IPM), which combines two or more compatible pest management tactics in a harmonious way. Such a holistic approach not only effectively manages the pests but also reduces the adverse effects of heavy usage of insecticides.

In this chapter, Section 17.2 deals with identification, biology and the nature and symptoms of damage of important root-feeding, leaf-feeding and sap-feeding pestiferous insects of mustard crop. Section 17.3 gives a detailed account of various conventional IPM practices and techniques available for sustainable management of mustard insect pests, including a new pest management decision-making tool (agroecosystem analysis (AESA)) and a revisited and revised cultural pest management tactic based on ecological principles (ecological engineering for pest management). Section 17.4 describes modern and novel techniques that hold potential in future pest management, including genetically engineered (GE) mustard plants along with its merits and demerits. Section 17.5 discusses the influence of changing climate conditions on pest management. Finally, Section 17.6 attempts to put forward an ideal IPM strategy, with concluding remarks.

17.2 Biology of Pestiferous Insects of Mustard

The important insect pests infesting mustard crop grown for vegetable, seed (as condiment) and oil purpose are given in Table 17.1.

17.2.1 Mustard sawfly: *Athalia lugens proxima* Klug

Importance

This pest is mainly distributed in Asian countries. It infests mustard, toria (*Brassica campestris*), rapeseed, cabbage, cauliflower, knol-khol, turnip, radish, etc. It is a sporadic pest, prefers cold weather and infests crops from October to March. It prefers to feed on seedlings and resowing of the crop may be required under heavy infestation.

Identification and biology

Adults are medium-sized wasps. The head, thorax and legs are black and the abdomen is orange in colour (Paul, 2007; Chowdhury, 2009; Kumaranag *et al.*, 2014). Wings are translucent, smoky with black veins. Females possess a saw-like ovipositor, hence the common name, sawfly (Kumaranag *et al.*, 2014). Adults live for 2–8 days. The total life cycle is completed in 30–35 days (Kumaranag *et al.*, 2014). Each female lays 30–35 eggs singly on the lower side of leaf margins in the slits made with the saw-like ovipositor (Kumaranag *et al.*, 2014). Eggs are spherical, light bluish green in colour and measure 0.5 mm in diameter. Eggs hatch in 4–8 days (Kumaranag *et al.*, 2014; Anon., 2015b). Larvae are greenish black with a wrinkled body and possess five dark stripes on lateral sides and eight pairs of abdominal legs (pro-legs) (Kumaranag *et al.*, 2014; Anon., 2015b). Upon hatching, larvae start feeding on leaves in groups of three to six during morning and evening hours while hiding during the daytime. Upon slight disturbance or if touched, they fall to the ground and feign death. A fully grown larva measures 16–18 mm in length (Kumaranag *et al.*, 2014; Anon., 2015b). Larvae go through six to seven instars and complete development in 14–35 days (Kumaranag *et al.*, 2014; Anon., 2015a). The fully grown larvae pupate in the soil in waterproof oval-shaped cocoons (Kumaranag *et al.*, 2014; Anon., 2015b). The cocoons resemble sand particles and the pupal period is completed in 11–31 days (Anon., 2015b). They go into diapause in the pupal stage during summer. Mustard sawfly completes two to three generations in a year.

Nature and symptoms of damage

The early-instar larvae nibble the leaves, resulting in numerous shot holes (Fig. 17.1); later instars

Table 17.1. Details of economically important insect pests infesting mustard crops.

Insect pest				Taxonomic group	
Common name	Scientific name	Pest category	Feeding nature / habitat	Family	Order
Mustard sawfly	*Athalia lugens proxima* Klug	Defoliator (chewing and biting mouthparts)	Feeds on leaves, stems, flower buds, flowers and pods	Tenthredinidae	Hymenoptera
Diamondback moth	*Plutella xylostella* (L.)			Yponomeutidae	Lepidoptera
Leaf webber	*Crocidolomia binotalis* (Zell.)			Pyraustidae	
Head borer / cabbage webworm	*Hellula undalis* (F.)				
Large cabbage butterfly	*Pieris brassicae* (L.)			Pieridae	
Tobacco caterpillar / taro caterpillar / cutworm	*Spodoptera litura* (Fab.)			Noctuidae	
Bihar hairy caterpillar	*Spilarctia obliqua* Walk.			Arctiidae	
Flea beetle	*Phyllotreta cruciferae* (Goeze)		Adult feeds on leaves; grub feeds on roots	Chrysomelidae	Coleoptera
Painted bug	*Bagrada hilaris* (Burmeister)	Sap feeder (piercing and sucking mouthparts)	Feeds from all aerial parts	Pentatomidae	Hemiptera
Green peach aphid	*Myzus persicae* (Sulzer)			Aphididae	
Cabbage aphid	*Brevicoryne brassicae* (L.)				
Mustard aphid	*Lipaphis erysime* (Kalt.)				
Cabbage maggot / root maggot	*Delia radicum* (L.)	Root feeder (hook-like mouthparts)	Grub feeds on roots and shoots	Anthomyiidae	Diptera
Brassica pod midge	*Dasineura brassicae* Winn.	Pod feeder (hook-like mouthparts)	Feeds on pods	Cecidomyiidae	

Fig. 17.1. The mustard sawfly larva and its feeding damage (shot holes) on mustard leaf.

feed from leaf margins towards the midrib, causing complete defoliation (skeletonization) (Fig. 17.1) (Kumaranag *et al.*, 2014; Anon., 2015b). The late-instar larvae also feed on the epidermis of shoots, causing seedling death in the early stages of crop and at times necessitating re-sowing. Later in the season, feeding on shoots leads to poor seed setting (Anon., 2015b). The infestation may result in yield loss of 5–18% (Chowdhury, 2009). The appearance of shot holes in leaves, skeletonized leaves and dead seedlings are symptoms of damage.

17.2.2 Diamondback moth: *Plutella xylostella* (L.)

Importance

With its cosmopolitan distribution, this is one of the most destructive insect pests of cruciferous crops such as cabbage, cauliflower, broccoli, collards, kale, kholrabi, Chinese cabbage, Brussels sprouts, etc. On a global basis, the economic yield loss caused by *P. xylostella* and the cost involved in managing it in various cruciferous crops have been estimated to the tune of US$1 billion (Talekar and Shelton, 1993).

Identification and biology

Adults are greyish brown small slender moths (6 mm in length) with conspicuous antennae (Cordero and Kuhar, 2009). Forewings are narrow with white triangular markings on inner (anal) margins of each wing. When forewings are folded over the back at rest, three diamond-shaped white patches are formed dorsally, deriving the common name 'diamondback'. Hindwings are fringed with long hairs. Mating takes place on the day of emergence. Adult moths live for 7–20 days. Egg to adult emergence (total life cycle) is completed in 15–18 days. It is a multivoltine species, completing several generations in a year (Kumaranag *et al.*, 2014; Anon., 2015b). Adult females lay oval, somewhat flattened, pinhead-sized eggs either singly or in groups of two to 57 on the lower or upper side of leaves (Cordero and Kuhar, 2009; Dosdall *et al.*, 2011; Kumaranag *et al.*, 2014; Anon., 2015b). Eggs are yellow to pale green in colour and measure 0.4 mm in length and 0.2 mm in width. Each female may lay 18–356 eggs in its lifetime. Eggs hatch in 2–9 days (Dosdall *et al.*, 2011; Kumaranag *et al.*, 2014; Anon., 2015b).

The first-instar larvae are colourless to yellow with a dark brown head capsule (Cordero and Kuhar, 2009). They enter (mine) into leaf tissues and feed on the chlorophyll content of mesophyll cells. The subsequent three instars are external feeders and either feed from the lower side of the leaf or nibble leaves, buds, flowers and pods. Larvae are yellowish green with fine erect black hairs on the body. Larvae taper at both the ends. When disturbed, larvae wriggle violently with backwards movement and fall off from the leaf, hanging with a silken thread (Cordero and Kuhar, 2009). The larvae go through four instars. The fully grown larva measures 8–12 mm in length. The larval period lasts for 8–30 days (Cordero and Kuhar, 2009; Dosdall *et al.*, 2011; Kumaranag *et al.*, 2014; Anon., 2015b). The fully grown larva constructs a loosely woven barrel-shaped silken cocoon within which it goes into pupation (Kumaranag *et al.*, 2014; Anon., 2015b). The silken cocoon is open at both ends and is usually attached to the leaf or stem or petiole surface. Pupae are green to yellow and measure 7–9 mm in length (Cordero and Kuhar, 2009). The pupal period lasts for 4–15 days (Cordero and Kuhar, 2009; Dosdall *et al.*, 2011; Kumaranag *et al.*, 2014; Anon., 2015b).

Nature and symptoms of damage

Upon hatching, the first-instar caterpillars start mining through the leaf epidermis and mesophyll, leaving white papery patches on the other side of the leaf surface. The subsequent three instars are leaf-surface feeders and make either papery membranes or holes on leaves. Later they feed on buds, flowers and pods as well (Dosdall *et al.*, 2011; Kumaranag *et al.*, 2014; Anon., 2015b). The typical damage symptoms are white papery patches on the leaves, withered appearance of leaves, holes in the leaves, skeletonization of leaves with frass, bored pods, eaten seeds in pods, etc.

17.2.3 Leaf webber: *Crocidolomia binotalis* Zeller

Importance

Its presence has been reported throughout the world. In addition to mustard, it infests cabbage, radish, turnip, etc. in Cruciferaceae. It is a regular pest of minor importance but at times reaches serious proportions. It is found along with diamondback moth in several *Brassica* crops.

Identification and biology

Adult moths are yellowish brown. Forewings are pale yellow with ferruginous (iron-rust) distinct and indistinct wavy lines and clear black wavy spots (Kumaranag *et al.*, 2014; Anon., 2015b, c; Pujari, 2015). Hindwings are pale white with brown apical margin. Adults possess a black tuft of hairs near the thorax. It completes more than one generation in a season. The adult females lay eggs in masses of 40–100 eggs on the lower side of leaves. The egg period is 5–15 days (Ziaee, 2012; Kumaranag *et al.*, 2014; Pujari, 2015). The neonate larvae feed gregariously on leaves. As they grow, they spread out and web the leaves together, live within and feed on them (Fig. 17.2). They then move up and web flowers and flower buds, feeding on them and boring into pods (Ziaee, 2012; Kumaranag *et al.*, 2014; Pujari, 2015). Mature larvae are green to purple with white to cream dorsal and lateral lines, a red head and black prothoracic shield and they measure 2 cm in length. Larval development is completed in 24–52 days (Ziaee, 2012; Kumaranag *et al.*, 2014; Pujari, 2015). The fully grown larvae descend to the ground and pupate in the soil in an earthen cocoon 2 cm below the soil surface (Ziaee, 2012). They may also pupate in the webbed leaves or flowers. The pupal period is 14–40 days (Kumaranag *et al.*, 2014; Anon., 2015b, c; Pujari, 2015).

Nature and symptoms of damage

Newly hatched larvae feed on the chlorophyll of young leaves. Later on, larvae use silken threads to web older leaves, buds and pods and live within them. Larvae remain on the lower side of leaves or in leaf folds in webbed leaves and skeletonize the leaves (Fig. 17.3). Under severe infestations, flower buds, flowers and pods are completely eaten by the larvae (Kumaranag *et al.*, 2014; Anon., 2015b, c; Pujari, 2015). Webbed leaves, flowers and pods, skeletonized/defoliated plants, bored and eaten seeds in the pods are identification symptoms of attack (Kumaranag *et al.*, 2014; Anon., 2015b, c).

17.2.4 Head borer/cabbage webworm: *Hellula undalis* (F.)

Importance

The pest is distributed throughout the world. It is a sporadic pest but occasionally becomes an important defoliator of cruciferous crops. In addition to mustard it infests broccoli, cabbage, cauliflower, collards, kale, radishes, knol-khol, rutabaga and turnips.

Fig. 17.2. Leaf webber larvae with typical webbing of mustard leaves and feeding damage.

Fig. 17.3. Leaf webber feeding damage (skeletonization and excreta) in mustard.

Identification and biology

The adult is a slender moth with a pale yellowish brown body. The forewings have grey wavy lines, an apical spot and pale-edged dark lunule. Hindwings are pale dusky but darker at the apical area (Kumaranag *et al.*, 2014). Adult females lay oval, flat, grey or yellowish green eggs either singly or in groups on new growth or the underside of older leaves (Reiter *et al.*, 2008; Webb, 2013; Kumaranag *et al.*, 2014; Anon., 2015b). The eggs measure 0.3–0.5 mm in length (Reiter *et al.*, 2008). Eggs are initially pearl-white, turning pink within a day and then to brown before hatching. The egg period is 2–4 days (Anon., 2015b).

The pale whitish brown neonate larvae without stripes on the body start mining through the leaves. Later the larvae feed on leaves and enter into the stem. The entrance holes are plugged with excreta and silk threads. The larvae go through five larval instars. Mature larvae are yellowish grey with five brownish purple longitudinal stripes running from the head to anal region (Reiter *et al.*, 2008; Webb, 2013; Kumaranag *et al.*, 2014; Anon., 2015b); they measure 12–15 cm in length and have a black head, while the body is covered with moderately long yellow or brown hairs. The larvae produce copious amounts of silk with which they web leaves and take refuge inside (Kumaranag *et al.*, 2014; Anon., 2015b). Larval development is completed in 7–17 days (Kumaranag *et al.*, 2014; Anon., 2015b). Fully grown larvae produce a lot of silk and spin their cocoons in the webbed leaves or in the soil (Reiter *et al.*, 2008; Webb, 2013; Kumaranag *et al.*, 2014; Anon., 2015b). Pupae are yellowish brown. The pupal period lasts for 6–7 days (Kumaranag *et al.*, 2014; Anon., 2015b).

Nature and symptoms of damage

The caterpillars initially mine into leaves, causing white papery membranes (Reiter *et al.*, 2008; Webb, 2013; Kumaranag *et al.*, 2014; Anon., 2015b). Later they defoliate the plants, making holes in leaves and boring into the stems.

17.2.5 Large cabbage butterfly: *Pieris brassicae* (L.)

Importance

This pest is present on a global scale. In addition to mustard, it infests cabbage, cauliflower, knol-khol, turnip, radish, sarson, toria and other cruciferous crops. It is one of the most damage-causing (up to 40%) insects in crucifer crops in India (Hasan and Ansari, 2010, 2011).

Identification and biology

Adults are white butterflies with a wingspan of 5.6–6.6 cm. Forewings have a black tinge at the merging corner of costal and apical margins and a black spot on the costal margin of the hindwing. The females have two prominent dark circular spots on the dorsal side of the forewing. Male moths are smaller than female moths and possess black spots on the ventral side of each forewing. Adults are very active flyers during daytime. Adults live for 3–14 days (Kumaranag *et al.*, 2014). They complete two to three generations in a year (Paul, 2007). Each female lays on an average 150 conical, oblong, creamy-yellow eggs in groups of 40–50 on the upper or lower sides of leaves. The incubation period is 3–14 days (Kumaranag *et al.*, 2014). Larvae are velvety bluish green with black dots. The body is covered with white hairs, while yellow stripes occupy the dorsal and lateral sides of the body (Fig. 17.4). Larvae pass through five instars in 13–38 days (Paul, 2007). The fully grown larvae are green and measure 40–50 mm in length (Kumaranag *et al.*, 2014). The mature larvae, which may migrate from one field to another (Kumaranag *et al.*, 2014), pupate either on the plant or away from the host on trees, fences and other materials. The pupal period lasts 8–25 days (Paul, 2007; Kumaranag *et al.*, 2014).

Fig. 17.4. Large cabbage butterfly larva feeding on mustard pod.

Nature and symptoms of damage

The neonates are gregarious and scrape the chlorophyll content of young leaves. As they grow, they disperse and start feeding on leaves by making holes from the margin inwards. Under severe infestation, only veins are left (skeletonization). Larvae feed on young shoots and pods as well.

17.2.6 Tobacco caterpillar/taro caterpillar: *Spodoptera litura* (F.)

Importance

It is distributed in temperate and tropical countries of Asia, Australia and the Pacific Islands (Feakin, 1973). It is a polyphagous pest, attacking more than 120 plant species comprising both annual and perennial crops grown for food and fibre including field, vegetable, fruit, plantation and ornamental crops. In Brassicaceae it infests cabbage, cauliflower, radish, turnip, etc. (CABI, 2015a).

Identification and biology

Adults are stout-bodied light brown moths measuring 22 mm in length. They have a wingspan of 4 cm. The forewings are brown with white criss-cross markings. Hindwings are cream-white with brown margins. The adults live for 7–10 days. The total life cycle is completed in 30–50 days and there are about eight generations in a year (Kumaranag *et al.*, 2014). The female lays 300 cream-coloured eggs in groups on the lower side of the leaf. The egg mass is covered by buff-coloured hairs detached from the female's abdomen tip after oviposition. Eggs hatch into neonates in 3–6 days (Kumaranag *et al.*, 2014). Larvae are velvety and appear in different colours, with dark stripes on the dorsal and lateral sides of the body. Larvae complete six instars in 16–30 days (Kumaranag *et al.*, 2014). Mature larvae enter the soil below the host plant and pupate in an earthen cell. Pupae are brown. The pupal period lasts 7–14 days (Kumaranag *et al.*, 2014).

Nature and symptoms of damage

The larvae are nocturnal, i.e. active during the night. The neonates feed gregariously by scraping the chlorophyll and later instars feed individually by making holes and cause leaf skeletonization (Kumaranag *et al.*, 2014).

17.2.7 Bihar hairy caterpillar: *Spilarctia obliqua* Walker

Importance

It is a sporadic pest infesting mustard, sesame (*Sesamum indicum* L.), mash (*Phaseolus mungo* L.), mung (*Phaseolus aureus* Zuccagni), linseed, sunflower and some vegetables.

Identification and biology

Adult moths have a crimson-red body with black dots. Wings are pinkish with numerous black spots and have a wingspan of 40–50 mm. Adults live for 7 days (Anon., 2015b). Adult moths lay 400–1000 light-green spherical eggs in groups of 50–100 on the lower sides of leaves. The egg period is 8–13 days (Anon., 2015b). The larvae are orange in colour with transverse bands and covered with tufts of yellow hairs of up to 5 mm long, which are darker at both ends. The larvae go through seven instars in 30–56 days (Anon., 2015b). After completing development, the larvae either pupate in plant debris or reach the ground and pupate in the soil. The pupal period is 7–15 days (Anon., 2015b).

Nature and symptoms of damage

The hatched hairy caterpillars feed gregariously by scraping off the chlorophyll on the lower or upper sides of leaves. The larvae spread to different plants as they grow and feed on leaves by making holes in soft portions of stems and shoots. In severe infestations, plants are completely defoliated (Anon., 2015b).

17.2.8 Flea beetle: *Phyllotreta cruciferae* (Goeze)

Importance

It is an important pest of canola and mustard (Brown *et al.*, 2004). In addition, it infests raya, taramira, toria, radish, turnip, cabbage, cauliflower, knol-khol, dahlia, sweet sultan, antirhinium and sweet peas. It is distributed in India, North and South America, Russia, Europe, Australia and Japan. It overwinters as an adult in the soil or beneath hedges (Burgess, 1977; Gavloski *et al.*, 2011).

Identification and biology

Adults are metallic bluish black small beetles measuring 2–3 mm in length. The hind femur is enlarged (Kumaranag *et al.*, 2014). Adults are active during warm and sunny days and overwinter as adults (Gavloski *et al.*, 2011). Adult females lay 50–80 creamy-white eggs singly in the soil around the host plants. The egg period is 5–15 days (Kumaranag *et al.*, 2014). The grubs are dirty white and measure 5 mm in length. They become fully grown in 3–4 weeks (Kumaranag *et al.*, 2014). Mature grubs pupate in the soil. The pupal period is 2–4 days. There are seven to eight generations in a year.

Nature and symptoms of damage

Grubs feed on the roots of seedlings and established plants but do not cause significant damage to the plants. Adults feed initially on the cotyledons, first true leaves and stems of seedlings, leading to decay and death of seedlings. Feeding on leaves by adults ultimately leads to formation of shot holes (Gavloski *et al.*, 2011). Adults also feed on stems, flowers and pods. The weed *Gynandropsis pentaphylla* L. serves as an alternative host (Kumaranag *et al.*, 2014).

17.2.9 Painted bug: *Bagrada hilaris* (Burmeister)

Importance

Its presence is reported in India, Africa (South Africa) and North America (CABI, 2015b). It infests crucifers, including cabbage, cauliflower, broccoli, Chinese cabbage (CABI, 2015b), and non-cruciferous crops such as rice, maize, sugarcane, indigo and coffee.

Identification and biology

The adult is a shield-shaped black bug with orange, red or yellow markings over the body (Fig. 17.5). The adult measures 5–7 mm in length and half of the length at the maximum breadth area. Females are larger than males. The total life cycle is completed in 3–4 weeks and there are several generations in a year (Ziaee, 2012; Reed *et al.*, 2014). Adult females lay up to 150 barrel-shaped pale yellow eggs either singly or in groups (of up to six) on the lower side of leaves or in the soil beneath the host plant. The oviposition period lasts for 2–3 weeks. Eggs appear white initially and turn to orange or red as they age before hatching. The egg period is 4–8 days (Ziaee, 2012; Reed *et al.*, 2014). Nymphs are black with orange, red and yellow markings; they are orange to red immediately after hatching and after every moult but eventually turn black. Nymphs pass through five nymphal instars. The fifth-instar nymphs measure 4 mm in length and 2.66 mm in breadth and develop wing pads on the thorax and white spots on the abdomen before moulting into the adult stage. The nymphal period is completed in 22–34 days (Ziaee, 2012; Reed *et al.*, 2014).

Fig. 17.5. Painted bug adult feeding on mustard pod.

Nature and symptoms of damage

Both nymphs and adults cause plant damage by feeding on aerial parts either singly or in groups. They insert their piercing and sucking mouthparts into plant tissue such as leaf, stem, flower and pod and suck the cell sap. The feeding results in irregular star-shaped lesions on leaves and stems. The affected aerial parts wither, wilt and dry, giving a scorched appearance. The affected plants show stunted growth and seedlings may die, necessitating re-sowing of the crop. The adults excrete a resinous substance, which spoils the appearance of pods. The bugs continue to infest the harvested pods even after reaching the threshing yard. The infestation of mature plants may result in loss of quality and quantity of yield (up to 31%) (Ziaee, 2012; Reed *et al.*, 2014). The appearance of bugs in groups feeding on aerial parts and wilted, dried and scorched leaves are typical symptoms of damage.

17.2.10 Aphids (mustard aphid: *Lipaphis erysimi* (Kalt.); green peach aphid: *Myzus persicae* (Sulzer); cabbage aphid: *Brevicoryne brassicae* (L.))

Importance

Owing to their cosmopolitan distribution coupled with high fecundity and voracious feeding nature, aphids are the most destructive pests of cruciferous crops (Firake *et al.*, 2013) such as mustard, cabbage, cauliflower, knol-khol, toria, sarson, raya, taramara, etc. across the globe. Of the three important aphid species that infest mustard, mustard aphid (*L. erysimi*) is the most dominant and key pest and may cause 9–96% yield loss, 31% seed weight loss and 5–6% oil content loss (Bakhetia and Arora, 1986; Bakhetia and Sekhon, 1989; Singh and Premchand, 1995; Singh and Sharma, 2002; Shylesha *et al.*, 2006).

Identification and biology

LIPAPHIS ERYSIMI. Adults are small (2.0–2.5 mm), soft-bodied, pear-shaped, louse-like insects. Adults possess a pair of cornicles projecting out backwards and upwards from the fifth or sixth abdominal segment. The cornicles release defence secretions (e.g. alarm pheromones). Adults exhibit body colour differentiation. Wingless females are yellowish green, greyish green or olive green with a white waxy bloom covering the body. The abdomen of winged females is dusky green with dark lateral stripes separating the abdominal segments and wings have prominent dusky veins. Males are olive green to brown. When the plant quality deteriorates or a colony is overcrowded, adults develop wings and disperse to new locations. About 45 generations are completed in a year (Ziaee, 2012; Anon., 2015b). Adult females lay white eggs along the veins of leaves (Anon., 2015b). Adults usually reproduce by giving birth to young ones (nymphs) directly (parthenogenesis). Each adult produces 20–133 offspring in its lifespan (Anon., 2015b). Nymphs are greenish yellow and are similar to adults except for the smaller size, which increases in subsequent instars. Nymphs go through four instars in 8–9 days (first, second, third and fourth instars complete their development in 1–2, 2, 2 and 3 days, respectively). Nymphs suck the cell sap from all aerial parts of the plants, i.e. leaf, twig, stem, inflorescence and pods (Anon., 2015b).

MYZUS PERSICAE. The winged green peach aphid adults measure 1.8–2.1 mm in length and are delicate and pear-shaped, with black head and thorax. The abdomen is yellowish green with a large dark patch dorsally. The wingless adults measuring 1.7–2.0 mm in length are yellowish or greenish. They may have medial and lateral green stripes. The cornicles of both forms are moderately long, matching the body colour, and unevenly swollen along their length. The legs and antennae are pale in colour (Capinera, 2005). Each apterous adult produces five to 92 nymphs, while each alate adult produces eight to 49 nymphs. Apterous adults live for 10–25 days and alate adults live for 15–27 days (Anon., 2015b). The adults lay yellow or green eggs, which turn black before hatching. The eggs are elliptical. They usually overwinter in the egg stage (Capinera, 2005). The nymphs are greenish and undergo four to five instars. All the nymphal instars are the same in appearance except that they increase in size. The wingless (apterous) nymphs take 4–7 days and winged (alate) nymphs take 5–8 days to complete the nymphal stage (Anon., 2015b).

BREVICORYNE BRASSICAE. The wingless adult female cabbage aphid measures 2.1–2.6 mm in length and is pale green, oval in shape with the posterior end of the body tapering greatly. The body is covered with white waxy powder and there are eight dark brown or black spots on the dorsal side of the body, which increase in their size towards the posterior end. The winged adult females are smaller than the wingless females and are not covered with waxy powder. The head and thorax are dark brown to black, while the abdomen is yellowish green with two dark spots on the dorsal side of anterior segments. The antennae are dark brown and wings are short with prominent veins (Kessing and Mau, 1991).

Nature and symptoms of damage

Both adults and nymphs possess piercing and sucking mouth parts. They suck the cell sap from all aerial parts of the plant such as leaf, stem, twig, inflorescence and pods (Fig. 17.6). Owing to the cell-sap feeding by both adults and nymphs, leaves start curling up and appear yellow and mottled. The flowers fail to form pods and the developing pods do not set the seeds properly. Under heavy infestation, plant growth is stunted and plants wilt and wither. Heavily affected fields have a sick and blighted appearance. The adults and nymphs excrete copious amounts of honeydew on which sooty mould develops, adversely affecting photosynthesis. The early infestation in the season and infestation of inflorescences lead to poor pod setting/formation, reducing the yield to 20–25% (Anon., 2015b). The cabbage aphid transmits 25 viruses that cause various diseases in brassicaceous plant species, such as *Cauliflower mosaic virus* (Ziaee, 2012). Typical symptoms of aphid damage are curled leaves, yellow

Fig. 17.6. Mustard aphids feeding on the reproductive parts (inflorescence and pods) and shoot of mustard plant.

and mottled leaves and stunted, wilted and withered plants. In addition, the presence of honeydew and ants, the presence of sooty mould growth and white cast skin on the affected plant parts, lower leaves and branches and at the base of the plant indicate aphid damage.

17.2.11 Cabbage maggot/root maggot: *Delia radicum* (L.)

Importance

It infests broccoli, Brussels sprouts, cabbage, kale, turnip, rapeseed, mustard and white mustard crops (Dosdall *et al*., 1994; Klingen *et al*., 2002; Ziaee, 2012). The infestation leads to significant reductions in leaf, stem and root biomass as well as flowering and seed production and ultimately yield (Blossey and Hunt-Joshi, 2003). Of the different *Brassica* species, yellow mustard (*B. rapa*) is most susceptible, oilseed rape (*B. napus*) and Indian mustard (*B. juncea*) are moderately susceptible and white mustard (*B. alba*) is least susceptible (Dosdall *et al*., 1994).

Identification and biology

Adults are dark grey housefly-like dipterans but smaller (5 mm). The abdomen has a dark stripe and is covered with black hairs and bristles (setae). Male and female dimorphism exists. Males have contiguous eyes while females have separated eyes. The anteroventral setae on the hind femur are long in males, while they are uneven in females. Adults feed on nectar. Females lay small white eggs (1 mm in diameter) on cool, moist soil close to the plants. Eggs hatch in 6 days. The white larvae (maggots), bearing a crown of ten small black membranous points at the posterior end, feed on host plants for 3–4 weeks. The mature larvae leave the roots and pupate in soil at 5–20 cm below the soil surface. Pupae are reddish brown. There are one to two generations per year (Ziaee, 2012).

Nature and symptoms of damage

The hatched maggots start feeding on roots and shoots, which causes delayed plant growth, withering of leaves and development of a bluish tinge. At times 300 maggots can be seen feeding inside one plant, causing extensive damage to the main root and disrupting the transportation of water and nutrients to aerial parts, leading to the death of the plant (Ziaee, 2012). Withered plants, bluish-tinged leaves and death of plants individually or in patches are typical symptoms.

17.2.12 Brassica pod midge: *Dasineura brassicae* Winn.

Importance

Distributed in most of the European countries and Morocco (Africa), it infests oilseed rape, turnip rape, rape, Indian mustard, black mustard, cabbage, cauliflower, etc. Of all the crops, it prefers oilseed rape the most (CABI, 2008).

Identification and biology

Adults are small dipteran flies measuring 1–1.5 mm in length. Females are reddish and males are blackish. Females lay eggs in clusters in developing pods through holes left by the pod weevil damage (Ziaee, 2012). The egg period is 4 days. Upon hatching, white maggots start feeding and develop inside the pods. The developed maggots appear whitish yellow and measure 1.5 mm in length; they drop to the soil and pupate inside whitish cocoons 1.4–2.1 mm in length at a depth of 3 cm from the soil surface (Pavela and Kazda, 2007). The maggots go through three instars in 9–15 days. Maggots may also enter into diapause and stay for 1–4 years. There are three to four generations in a year (Jellis, 2003)

Nature and symptoms of damage

The hatched maggots feed on developing seeds inside pods; as a result, pods swell and burst at times. Opened pods or the appearance of fly-emergence holes are typical symptoms.

17.3 Integrated Management

It is estimated that by 2050, on a global scale, the total production of oil crops and products in oil equivalents (including mustard) should increase to 282 million tonnes from the current 178 million tonnes (Alexandratos and Bruinsma, 2012). This projection calls for a significant increase in mustard production and productivity; and effective management of pestiferous insects will act as one of the boosters for achieving this. In order to protect mustard crops from pestiferous insects, several pest management practices have been adopted from time to time with varying degree of success. Until the advent of chemical pesticides a few decades ago, pest management practices were mainly of environmentally friendly cultural, mechanical and biological origin, but these have been gradually replaced with environmentally and economically costly intensive chemical practices for easy, quick and effective pest management. However, heavy reliance on chemical insecticides, together with inappropriate and indiscriminate use, has eclipsed the advantage of this method and generated a multitude of problems, including pesticide resistance and pest resurgence especially in crucifer-specialist herbivores such as *P. xylostella* and *L. erysimi* (Talekar and Shelton, 1993; Yadav and Singh, 2015). *Brassica* specialists (*P. xylostella* and *L. erysimi*) with voracious feeding natures coupled with high fecundity helped them in developing resistance to most of the insecticides used in mustard and other cruciferous crop ecosystems, which led to the failure of pest management programmes dominated by a single component, i.e. insecticides (Facknath, 1997; Cordero and Kuhar, 2009; Dosdall *et al.*, 2011; Yadav and Singh, 2015).

Keeping the above aspects in mind, there is a need to develop, evaluate and adopt sustainable IPM strategies/modules for mustard pests. IPM is an ecological approach for effectively managing the pestiferous insect complex of the mustard ecosystem in the most environmentally friendly and economically sound way by integrating two or more pest management tactics from cultural, mechanical, physical, biological, genetic, quarantine and chemical origin in a harmonious way that ultimately results in high-quality crops produced in a sustainable manner (Bajwa and Kogan, 2002; Guerene, 2006). The information available in the literature on various tactics and practices under each IPM component, along with other novel pest management options that hold potential in future mustard pest management, and the need for development and evaluation of IPM strategies/modules for pestiferous insects of mustard are discussed below.

17.3.1 Agroecosystem analysis-based monitoring and decision making

The decision for timing pest management intervention is usually based on regular monitoring/scouting of the field for different pests and their densities using either economic threshold level (ETL) or an appropriate alternative tool. Agroecosystem analysis (AESA), briefly described here, is a comprehensive on-farm monitoring and evaluation tool, considered a better alternative to ETL-based decision making for management (FAO, 2006, 2011). AESA is a learning tool to help to equip farmers with necessary skills and knowledge about the ecosystem for taking better and informed decisions about pest management (FAO, 2006, 2011). It consists of three components: observation; analysis; and decision making. Unlike ETL, AESA does take into account density-dependent and independent factors in the ecosystem, such as natural enemies and weather factors, that directly influence pest population fluctuations (Boina *et al.*, 2014; Korlapati *et al.*, 2014a, b). Moreover, insect biomass consumption-rate statistics on which ETLs are based may not hold true or may need revision under changing climatic conditions. Under elevated atmospheric CO_2 levels, which are projected to reach 540–970 ppm by 2100, the C:N ratio in plants increases, affecting their secondary metabolism and leading to lower protein concentration in leaves and lower nutritive value to herbivores, which therefore switch to a compensatory feeding mode. For example, insects consume more plant biomass than they would under normal atmospheric CO_2 levels, in order to fulfil their nitrogen requirement (Lincoln *et al.*, 1986; Stiling *et al.*, 1999). Increased oilseed rape (*B. napus*) leaf area consumption by *P. xylostella* larvae under elevated CO_2 and temperature supported the above hypothesis (Zvereva and Kozlov, 2006; Stiling and Cornelissen, 2007).

The AESA is conducted weekly, preferably in morning hours (before 9am), for determining the pest:defender (P:D) ratio as the basis on which informed decisions on pest management interventions can be made. In AESA, 50 plants/ha are randomly selected, leaving 1 m distance from the bund on all sides of the field (Korlapati *et al.*, 2014b). At each selected plant, all the flying insects (both pests and crop defenders/natural enemies) are collected using a sweep net and counted. Similarly, the number of insects (pests and defenders/natural enemies) present on each plant as well as beneath the plant on the ground are counted and recorded. Using this information, the P:D ratio is estimated by dividing the total number of pests by the total number of crop defenders. Since estimated P:D ratios for given pest and defender combinations are not available at this juncture, a universal P:D ratio of 2:1 has been adopted for arriving at a pest management decision. If the estimated P:D ratio is ≤ 2:1 then the ecosystem does not need any external pest management intervention, as defenders are in sufficient numbers (one defender for every two pests) in the ecosystem to deal with pests. On the other hand, if the estimated ratio is > 2:1 then the ecosystem needs external intervention in helping to bring the ratio down to ≤ 2:1 (Boina *et al.*, 2014; Korlapati *et al.*, 2014b). Observations on weather conditions help in predicting favourable weather for pests and responding appropriately to the situation; for example, 'prepone' (i.e. bring forward) the field visit to take appropriate timely intervention, as against regular 7-day intervals (weekly activity).

Currently, AESA is in its infancy and there are several aspects in which it needs improvement and merits extensive research in areas such as estimation of specific P:D ratios for important pest:defender combinations for a given crop ecosystem. Furthermore, in its present format, AESA takes considerable time for farmers to carry out, discouraging its adoption; therefore, it needs to be modified to make it simple and easy for farmers to adopt it effectively and practise AESA with minimal time spent on it.

17.3.2 Cultural control

The role of agronomic and cultural practices in preventing pest infestation and/or reducing pest density on cultivated crops has been well established in several crop ecosystems (Kolte, 1985; Paul, 2007; Firake *et al.*, 2013; Kumaranag *et al.*, 2014; Anon., 2015b). One of the main objectives of these practices is to create unfavourable conditions for pest build-up by having an adverse impact on pest development, survival and reproduction. The commonly recommended cultural practices for pestiferous insects of mustard are: (i) removal and burning of crop residues after harvest (clean cultivation) to reduce the carryover of pests between crops, especially for painted bug (Kolte, 1985; Anon., 2015b); (ii) removal and destruction of weeds in and around crop fields, which serve as alternative hosts for pests (for example, *G. pentaphylla* acts as an alternative host for flea beetle) (Kumaranag *et al.*, 2014); and (iii) deep ploughing the field immediately after harvest and/or in summer to destroy the resting insect life stages such as larvae and pupae of mustard sawfly (*A. lugens proxima*), pupae of leaf webber (*C. binotalis*), pupae of tobacco caterpillar (*S. litura*), eggs of painted bug (*B. hilaris*), etc. These are some of the cultural practices taken before crop sowing and must be a part of any IPM programme.

Planting resistant/tolerant varieties (through conventional breeding) specific to a particular region is an important cultural practice that cannot be ignored. Considerable evidence is documented in the literature for a positive correlation between delay in the sowing time and increased aphid infestation (Kolte, 1985; Firake *et al.*, 2013; Kumaranag *et al.*, 2014). Cold (and cloudy) weather that sets in during November–December is highly favourable for aphid multiplication; therefore, early (by 15 October) or timely (by first week of November) sowing of the crop in some regions escapes the pest attack by reducing the chances for aphid build-up (Paul, 2007; Shekhawat *et al.*, 2012; Firake *et al.*, 2013). In contrast, a recent study based on analysis of weather data and aphid infestations concluded that aphids build up more quickly under warm and humid climates than in cool climates (Rao *et al.*, 2014). As with aphids, early sowing reduces the incidence of mustard sawfly and painted bug infestations (Anon., 2015b). On the other hand, use of early maturing varieties helps to escape the pests that infest the crop late in the season, such as head borer (*H. undalis*) (Reiter *et al.*, 2008).

It seems that the water droplets emanating from sprinklers disrupt diamondback moth's flight, mating and oviposition (Nakahara *et al.*, 1986; Talekar *et al.*, 1986). Hence, irrigating the field through a sprinkler system may decrease pest numbers on mustard plants by dislodging the eggs and larvae (Facknath, 1997) and even aphids from plants. Furthermore, it was reported that providing irrigation to mustard crop through sprinkler systems increased water application efficiency by 10% compared with surface methods (Anon., 2015b). Providing irrigation in seedling stage reduces *A. proxima lugens* larval density (due to drowning effect) (Ziaee, 2012). On the other hand, delaying irrigation up to 3–4 weeks after sowing significantly reduces painted bug populations (Kumaranag *et al.*, 2014; Anon., 2015a). Therefore, timing and method of irrigation can be changed appropriately, depending on the pest type and its severity. It is believed that flying aphids are attracted to the yellow-green colour of sickly plants (less vigour); therefore, maintaining the healthy status of mustard plants in the field may partly deter aphid attack (Day, 2014). Painted bugs are known to infest the pods even after harvesting the crop. By keeping the time gap between harvesting and threshing of the crop as short as possible, the post-harvest damage by painted bug can be kept to a low level (Kumaranag *et al.*, 2014).

17.3.3 Ecological engineering for pest management

Ecological engineering for pest management is the use of cultural techniques to alter the crop habitat in favour of crop defenders (natural enemies) for enhanced biological control, while making it less favourable for insect pests (Gurr *et al.*, 2004a, b; Boina *et al.*, 2014). Ecological engineering for pest management comprises both below- and above-ground cultural practices. Below-ground ecological engineering practices consist of zero or minimal tillage, application of composted manures rich in organic matter, growing nitrogen-fixing green manure crops, practising crop rotation, treating seed with biopesticides, etc. (Boina *et al.*, 2014; Korlapati *et al.*, 2014b). Above-ground ecological engineering practices include raising different types of companion plants such as attractant plants (flowering/non-flowering) to attract natural enemies, repellent plants (flowering/non-flowering) to repel insect pests, trap plants (flowering/non-flowering) to attract and trap insect pests, and barrier/guard plants (tall-growing cereals crops) to prevent the entry of flying soft-bodied insect pests into the ecosystem (Boina *et al.*, 2014; Korlapati *et al.*, 2014b).

Rotating mustard crops with non-host Cucurbitaceae (cucumber, melon), Malvaceae (bhendi) or Solanaceae (tomato, brinjal/egg plant/aubergine) crops (Kumaranag *et al.*, 2014) is believed to interfere with the continuous availability of host plants and thereby the perpetuity of pests. Ecological engineering of mustard with a companion plant such as tomato or garlic or clover cover crops (Kienegger *et al.*, 1996; Theunissen *et al.*, 1996) may reduce the pest density, especially of *P. xylostella* on mustard crop, partly due to the confusing chemical (volatile) and visual cues given off by host and companion plants, resulting in repellence of pests or interference with chemical communication of pests required for mating (Facknath, 1997). For instance, cabbage intercropped with tomato or garlic in alternate rows significantly reduced *P. xylostella* larval/pupal density on cabbage (Buranday and Raros, 1975; Sivapragasam *et al.*, 1982; Talekar *et al.*, 1986; Facknath, 1997). Furthermore, cabbage–tomato intercropping sprayed with neem seed kernel extract (NSKE) gave results comparable to those of a chemical insecticide, cartap hydrochloride, in terms of number of *P. xylostella* larvae/pupae, number of infested plants and the quality of harvested cabbage heads (Facknath, 1996).

Along similar lines, an exposure of broccoli (*B. oleracea* var. *italica*) to a perennial evergreen shrub (*Rhododendron tomentosum* Harmaja) caused significant reductions in oviposition (36–38%) by *P. xylostella* adults on broccoli plants and subsequent larval feeding (Himanen *et al.*, 2015). This is attributed to the passive adsorption and re-release of specific semi-volatiles (palustrol, ledol and ledene) by cabbage leaves that were emitted by *R. tomentosum* plants (associational resistance) and masked or disguised the host location, oviposition and feeding stimulation cues for *P. xylostella* (Himanen *et al.*, 2015). Mixed cropping is another cropping pattern that may be used as an effective pest management cultural practice. In a laboratory study, mustard plants when mixed with related *Brassica* species, i.e. *B. campestris* and *B. oleracea*, reduced *P. xylostella* larval density (numbers/plant) and egg deposition, while increasing the take-off frequency of female moths due to altered behavioural responses (Huang *et al.*, 2014). The simultaneous presence of more or less similar chemical volatile and visual cues from three related and preferred host species led to the confused behaviour in *P. xylostella* in the above study (Huang *et al.*, 2014). Further research is required to test and confirm the above hypotheses under field conditions by inter- or mixed-cropping mustard with other non-host plants, such as tomato, garlic or *R. tomensotum*, for practical utility of the above phenomena in integrated management of *P. xylostella*. In addition, planting tall-growing cereal crops such as maize or sorghum along the border of mustard fields may help in reducing the entry of winged aphids, painted bugs and moths into the field. Planting ovipositional trap crops such as castor at 250 plants/ha attracts the moths of *S. litura* to lay eggs (in mass), which makes it easy to collect and destroy the eggs and gregarious stages of larvae (Korlapati *et al.*, 2014b).

17.3.4 Push–pull technique as a pest management strategy

The push–pull technique originally developed by Pyke *et al.* (1987) in Australia was tremendously improved and adopted for successful management of cereal stem borers in maize/sorghum crop ecosystems in African countries by the combined efforts of the International Centre of Insect Physiology and Ecology in Nairobi, Kenya, and Rothamsted Research in the UK (Cook *et al.*, 2007; Khan and Pickett, 2010; Khan *et al.*, 2014). It mainly involves push and pull components, wherein the push component repels the pest away from the main crop while the pull component attracts the pest on to it (Cook *et al.*, 2007). Taking a cue from this technique and to replicate it for mustard crop, it is a prerequisite to identify attractant and repellent source plants for key pestiferous insects of mustard. Given the fact that mustard itself is a highly attractive crop (rich in glucosinolate content such as sinigrin) for *Brassica* specialists, because of which it is used as a trap crop for several crucifer cash crops (Srinivasan and Moorthy, 1991, 1992; Silvia-Krott *et al.*, 1995; Luther *et al.*, 1996; Boina, 2000; Charleston and Kfir, 2000; Muniappan *et al.*, 2001; Shelton *et al.*, 2008; George *et al.*, 2009), the task of identifying a trap crop for mustard becomes even more challenging. Nevertheless, recent research efforts made in this direction have yielded some success to this effect in terms of finding a *Brassica* plant that possesses glucosinolate content much higher than the mustard.

The biennial crucifer weed species belonging to the genus *Barbarea*, growing wild in the USA and some European countries, have evolved unique traits. The possession of high glucosinolate content in their young leaves lures *Brassica* specialists such as *P. xylostella* for oviposition, while simultaneous possession of high content of saponins as a secondary line of defence, exclusive to *Barbarea* among brassicas, deters and kills the hatched neonates through starvation (Badenez-Pérez *et al.*, 2004, 2006, 2011, 2014a, b; Lu *et al.*, 2004). Cabbage intercropped with *Barbarea* spp. confirmed these findings where *P. xylostella* adults exhibited ovipositional preference and higher oviposition (> 99%) on *Barbarea* plants than on cabbage (Badenez-Pérez *et al.*, 2006). The secondary line of defence (i.e. saponins) probably developed over a considerable time in evolutionary terms as a response to specialist herbivores such as *P. xylostella* having adapted to the first line of defence (i.e. glucosinolates) (Badenez-Pérez *et al.*, 2014a). The combination of these traits gives an additional advantage to *Barbarea* for use as an attractant trap plant, strongly supporting the hypothesis that *Barbarea* perfectly meets the requirements of the 'pull' component in a push–pull technique targeted at effective management of specialists, especially *P. xylostella*, in mustard and other cruciferous crops. None the less, such predictions were based on findings from cabbage–*Barbarea* intercropping; therefore, experimental evidence is needed to corroborate the suggestion that *Barbarea* can be considered as a 'dead-end' trap crop for mustard in a push–pull system. A combined strategy that uses mustard varieties bred to contain lower levels of glucosinolates (less attractive to specialists) as main crop and intercropping it with a non-host repellent crop, e.g. tomato or garlic (Buranday and Raros, 1975; Sivapragasam *et al.*, 1982; Talekar *et al.*, 1986; Facknath, 1997), or neem-based spray (as a 'push' component), surrounded by *Barbarea* plants (as a 'pull' component) in a few rows on the border might turn out to be a good push–pull technique for mustard. Along similar lines, transforming mustard varieties rich in glucosinolate content with the *Bacillus thuringiensis* (Bt) gene and planting on the border might serve as a trap crop for non-Bt cultivars with low glucosinolates content. The border Bt plants attract the herbivorous specialists for oviposition but adversely affect survival of the larvae/offspring by acting as a 'dead-end' trap crop, as proposed by Cao *et al.* (2008) and Shelton *et al.* (2008).

17.3.5 Mechanical control

Collection and destruction (by placing in kerosenized water) of gregarious insects such as cabbage butterfly (*Pieris brassicae*), Bihar hairy caterpiller (*Spilarctia obliqua*), *Crocidolomia binotalis*, *Spodoptera litura*, etc. at early infestation levels reduces pest build-up and further crop damage (Kumaranag *et al.*, 2014). For instance, regular handpicking of cabbage leaves with *C. binotalis* eggs and clusters of early-instar larvae coupled with seven to eight spot applications of Bt per season targeted at *C. binotalis* effectively reduced infestations of both *C. binotalis* and *P. xylostella* by conserving parasitoids of *P. xylostella*, especially *Diadegma semiclausum* Hellen, which was otherwise possible with an average of 26 applications of chemical insecticides (Shepard and Schellhorn, 1996). At very low numbers, painted bugs can be handpicked and destroyed, while large numbers that usually congregate on the leaves and stems can either be jerked or tapped to dislodge them on to a sheet to kill (in kerosenized water) or vacuumed (Kumar, 2015). Different types of traps provide tools for monitoring (keeping track of the onset and progress of pest incidence) and for mass trapping, i.e. pest management. These include: (i) yellow sticky traps at 12–14 traps/ha for aphids; (ii) pheromone traps with species-specific lures at 8–10 traps/ha for *P. xylostella*, *S. litura*, *C. binotalis* and *H. undalis*; and (iii) light traps at 1 trap/ha for attracting and trapping phototactic insects. A specially designed pyramid trap baited with crushed sweet alyssum (inside polypropylene bags) can be placed in fields heavily infested with painted bugs to trap and kill the pests. Erecting bird perches at 50 perches/ha encourages the activity of predatory birds such as the black drongo (*Dicrurus macrocercus* Vieillot), the common mynah (*Acridotheres tristis* (L.)), *Motacilla cospica* (for aphids) and so on in the field (Firake *et al.*, 2013). The birds alight on the perches and catch caterpillars, moths, bugs, aphids, etc. A simple technique of setting up bonfires during evening hours can attract and kill several insects and is helpful in reducing their numbers. Although mating disruption with sex pheromones specific to *P. xylostella* males was found to be an effective technique for reducing the population of *P. xylostella* (up to 95%) in cabbage in Japan (Ohbayashi *et al.*, 1992; Ohno *et al.*, 1992), the cost involved in its application prohibits its wider adoption by farmers.

17.3.6 Biological control

The insects and other animals belonging to the third trophic level in the food web known as natural enemies, mainly parasitoids and predators, play a pivotal regulatory function in the ecosystems. Agroecosystems that are less biodiverse (monocropping) and more dependent on external inputs (chemical fertilizers and pesticides) eliminate natural enemies, leading to pestiferous insect outbreaks. Adoption of conservation biological control practices such as ecological engineering to increase biodiversity and food resources for natural enemies, while restraining from use of broad-spectrum chemicals, creates an environment conducive to the build-up of natural enemies in mustard ecosystems (Table 17.2), thereby enhancing their activity.

In Canada, conservation of parasitoids in mustard ecosystems resulted in 30–45% of *P. xylostella* larvae being parasitized by *Diadegma insulare* (Cresson), while *Microplitis plutellae* (Muesebeck) and *Diadromus subtilicornis* (Gravenhorst) parasitized 15% of larvae and 15% of pupae, respectively (Dosdall *et al.*, 2011). In India, conservation of *Diaeretiella rapae* (M'Intosh), a dominant aphid parasitoid in mustard ecosystems, resulted in 48–62% of aphids being parasitized by it (Firake *et al.*, 2013). This was further supported by the observation of heavy parasitization of *L. erysimi* by *D. rapae* and other parasitoids such as *Aphidius colemani* Viereck on mustard plants ecologically engineered with cabbage and cauliflower in plots that used no chemical insecticide applications (Fig. 17.7) (Boina *et al.*, 2014).

Of the several predators, ladybird beetles and syrphid flies play a dominant role in regulating the density of mustard insect pests, especially the aphid *L. erysimi* (Fig. 17.8). In a laboratory study, adults and grubs of the ladybird beetle (*Coccinella septempunctata* L.) consumed on average 10–15 aphid adults (Firake *et al.*, 2013) and 56 aphid nymphs/day, respectively, while maggots of the syrphid flies *Syrphus confrater* Wied., *S. balteatus* Deg. and *Ischiodon scutellaris* (F.) consumed an average of 49, 42 and 36 aphid nymphs/day, respectively (Singh, 2013).

Among several botanicals tried for managing pestiferous insects of mustard, neem-based pesticides as either freshly prepared/extracted or formulated products seem to have the potential to be included in IPM strategies. The findings of various workers with neem products, either alone or integrated with another IPM tactic, corroborate this statement. Foliar spray with freshly prepared 5% or 10% NSKE (Bakhetia *et al.*, 2002; Firake *et al.*, 2013; Meena *et al.*, 2013; Kumaranag *et al.*, 2014) or azadirachtin 0.03% (300 ppm) neem oil-based water-soluble powder (WSP) at 2.5–5.0 kg/ha is recommended for effective

Table 17.2. Natural enemies of important insect pests of mustard crop.

(a) Defoliators

Insect pest	Parasitoids[a]	Common predators for defoliators
Mustard sawfly (*Athalia lugens proxima*)	*Perlissus cingulator* (Morley) (L)	Ladybird beetles: *Coccinella septempunctata* L. *Menochilus sexmaculatus* (F.) *Micraspis discolor* (F.) Syrphid/hoverflies Green lacewing bugs: *Chrysoperla zastrowi sillemi* (Esben-Petersen) *Chrysoperla carnea* (Stephens) Spiders: *Lycosa pseudoannulata* Boesenberg and Strand *Peucetia viridans* Hentz Dragonfly Robber fly Reduviid bugs: *Rhynocoris marginatus* (F.) *Rhynocoris fuscipes* (F.) Praying mantis Carabid beetles Rove beetles Weaver ants: *Oecophylla smaragdina* F. Predatory birds: Common mynah: *Acridotheres tristis* (L.) Black drongo (King crow): *Dicrurus macrocercus* Vieillot
Diamondback moth (*Plutella xylostella*)	*Trichogramma bactrae* Nagaraja (E) *Trichogramma pretiosum* Riley (E) *Cotesia plutellae* (Kurdj.) (L) *Cotesia sicarius* Marshall (L) *Oomyzus (Tetrastichus) sokolowskii* (Kurdj.) (L) *Diadegma mollipla* (Holmgren) (L) *Diadegma insulare* (Cresson) (L) *Microplitis plutellae* (Muesb.) (L) *Brachymeria excarinata* Gahan (P) *Eriborus* spp. (P) *Diadromus collaris* (Gravenhorst) (L-P) *Diadromus subtilicornis* (Gravenhorst) (pP and P)	
Leaf webber (*Crocidolomia binotalis*)	*Cotesia crocidolomiae* (L) *Microbracon mellus* (L) *Inareolata (Diadegma) argenteopilosa* Cam. (L) *Sturmia inconspicuoides* Bar. (L) *Mesochorus* spp. (L) *Atrometus* spp. (L) *Chelonus tabonus* Sonan (E-L) *Pristomerus* spp. (L) *Eriborus sinicus* Holmgren (L) *Argyropylax* spp. (L)	

Continued

Table 17.2. Continued.

Insect pest	Parasitoids[a]	Common predators for defoliators
Head borer / cabbage webworm (*Hellula undalis*)	*Bracon hebetor* Say (L) *Chelonus blackburni* Cameron (E-L)	
Cabbage butterfly (*Pieris brassicae*)	*Trichogramma* spp. (E) *Cotesia glomerata* (L.) (L) *Cotesia rubecula* (Marshall) (L) *Hyposoter ebeninus* (Gravenhorst) (L) *Diadegma pierisae* (Rao) (L) *Pteromalus puparum* (L.) (P) *Pimpla fuscipes* Brulle (P) *Polistes* spp. (L)	
Tobacco caterpillar / cutworm (*Spodoptera litura*)	*Trichogramma chilonis* (Ishii) (E) *Tetrastichus* spp. (E) *Telenomus* spp. (E) *Carcelia* spp. (L-P) *Campoletis chlorideae* Uchida (L)	
Bihar hairy caterpillar (*Spilarctia obliqua*)	–	
Flea beetle (*Phyllotreta cruciferae*)	*Microctonus vittatae* (Muesb.) (A)	Damsel bug: *Nabis alternatus* Parshley

[a]Parasitoid scientific names are followed by letters in parentheses indicating pest stage being parasitized: (E) egg; (N) nymph; (L) larval; (pP) pre-pupal; (P) pupal; (A) adult; (E-G) egg-larval; (L-P) larval-pupal.

(b) Sap feeders

Insect pest	Parasitoids[a]	Common predators for sap feeders
Painted bug (*Bagrada hilaris*)	*Gryon* spp. (E) *Alophora* spp. (A)	Ladybird beetles: *Coccinella transversalis* F. *Coccinella repanda* Thunberg *Coccinella septempunctata* *Oenopia kirby* Mulsant *Menochilus sexmaculatus* (F.) *Hippodamia variegata* (Goeze) *Hippodamia tredecimpunctata* (L.) *Cheilomenes vicina* (Mulsant) *Micraspis discolor* (F.) *Propylaea japonica* (Thunberg) *Harmonia axyridis* (Pallas) Syrphid/hoverflies: *Allograpta javana* Wiedemann *Sphaerophoria scripta* (L.) *Sphaerophoria* spp. *Eristallis* spp. *Metasyrphus corollae* (F.) *Metasyrphus* spp. *Xanthogramma* spp. *Syrphus serarius* (Wiedmann) *Syrphus balteatus* Deg. *Syrphus alternans* Macquart *Syrphus torvus* Osten Sacken

Continued

Table 17.2. Continued.

Insect pest	Parasitoids[a]	Common predators for sap feeders
		Syrphus confrater Wied. *Syrphus* spp. *Ischiodon scutellaris* (Fab.) Green lacewing bugs: *Chrysoperla zastrowi arabica* Henry *et al.* *Chrysoperla zastrowi sillemi* (Esben-Petersen) *Chrysoperla carnea* (Stephens) *Brinckochrysa scelestes* (Banks) *Mallada boninensis* (Okamato) Spiders: *Marpissa calcuttaensis* Tikader *Phidippus* spp. *Argiope pulchella* Thorell *Oxyopes rubisternum* *Lycosa pseudoannulata* Boesenberg and Strand *Leuchge decorate* *Larinia tabia* *Cryptophora carrisae* Aphid midge: *Aphidoletes aphidimyza* (Rondani) *Aphidoletes urticaria* (Kieffer) Predatory birds: *Motacilla cospica* (Gmelin)
Brassica pod midge (*Dasineura brassicae*)	*Omphale clypealis* (Thomson) (L) *Platygaster subuliformis* (Kieffer) (L)	

[a]Parasitoid scientific names are followed by letters in parentheses indicating pest stage being parasitized: (E) egg; (N) nymph; (L) larval; (pP) pre-pupal; (P) pupal; (A) adult; (E-G) egg-larval; (L-P) larval-pupal.

Fig. 17.7. Parasitization (mummification) of mustard aphid by *Diaeretiella rapae* on mustard plants ecologically engineered with cabbage and cauliflower crops.

management of aphids and several lepidopteran caterpillars, especially *P. xylostella* and *S. litura* (CIBRC, 2015). Similarly, application of azadirachtin (neemix 4.5) at low rates (11.25–22.5 g AI/ha) for aphids and at high rates (22.5–50.0 g AI/ha) for defoliator pests is recommended in mustard (Hollingsworth, 2015). Besides neem, other botanicals that have been found to have some effect on mustard pests are petroleum ether extracts of the parthenium plant (*Parthenium hysterophorus* L.) which, following treatment, reduced lifespan and offspring production in *L. erysimi* (Sohal *et al.*, 2002), and bitter gourd-seed oil emulsion with anti-feedant action against *A. lugens proxima* larvae (Anon., 2015c).

Given the considerable insecticidal potential of ecofriendly neem products on mustard pests, a few efforts have been made to integrate them with other biological control tactics such as parasitoids, predators

Fig. 17.8. Ladybird beetles [adult (a) and grub (b)] predating on aphids on mustard plants ecologically engineered with cabbage and cauliflower crops.

and microbial biopesticides, in an attempt to develop suitable IPM programmes. In this direction, Firake *et al.* (2013), Pandey and Singh (2008) and Yadav and Singh (2015) evaluated two or more pest management tactics applied either simultaneously or in sequence (IPM module) for effective management of destructive aphids in mustard. As per their findings, application of 5% NSKE followed by *Lecanicillium lecanii* (Zimm.) at 10^8 colony-forming spores (CS)/ml or application of 2% neem oil followed by release of green lacewing bug (*Chrysoperla zastrowi sillemi* (Esben-Petersen)) larvae at 50,000/ha effectively managed *L. erysimi* populations in mustard (Firake *et al.*, 2013). Simultaneous release of predatory ladybird beetle (*C. septempunctata*) grubs at 5000/ha and a parasitoid (*D. rapae*) at 5000 mummified aphids/ha coupled with setting up yellow sticky traps at one trap/100 m^2 resulted in an 83% increase in yield compared with the yield of an untreated control (Firake *et al.*, 2013). Similarly, application of 5% NSKE followed by the release of *C. septempunctata* grubs at 5000/ha yielded reduction with *L. erysimi* populations (> 80% control by day 10) comparable to the use of a chemical insecticide (dimethoate at 1 ml/l of water) as a substitute for NSKE with a higher net return to the cost (benefit:cost ratio 6.7:1) (Yadav and Singh, 2015). Furthermore, application of 5% NSKE followed by the release (15 days after NSKE spray) of green lacewing bug larvae (*Chrysoperla carnea* (Stephens)) at 150,000/ha reduced the aphid population to about 17–20% of the numbers in an untreated field; nevertheless, inclusion of a chemical insecticide (endosulfan at 0.07%) into this combination resulted in better aphid management (Pandey and Singh, 2008).

Like the botanicals and natural enemies (parasitoids and predators), microbial biopesticides, i.e. *Bacillus thuringiensis* (Bt) Berliner, nuclear polyhedrosis virus (NPV), entomopathogenic nematodes (EPNs) and entomopathogenic fungi (EPFs), are important biological control components of IPM that hold greater potential for managing pestiferous insects of mustard. As per CIBRC (2015), foliar application of Bt var. *galleriae* or Bt sv. *kurstaki* or Bt sv. *aizawai* at 1 g/l or *Sl*NPV at 100 LE/ha + phagostimulant (e.g. jaggery) + UV protectant protects the crop from *S. litura* when sprayed thrice at intervals of 10–15 days during evening hours as soon as the eggs and neonates are noticed on the crop. Given the promising results obtained in killing mustard sawfly (*A. lugens proxima*) with EPNs, these nematodes may hold a key in managing the populations of mustard sawfly. Based on findings from a laboratory study that evaluated the efficacy of three EPN species after a 48 h incubation period, Yadav and Lalramliana (2012) concluded that *Heterorhabditis indica* Poinar, Karunakar & David was the most potent (LC_{50} = 31 infective juveniles (IJs)/ml) with greater progeny production, followed by *Steinernema thermophilum* Ganguly & Singh (LC_{50} = 37 IJs/ml) and *Steinernema glaseri* Steiner (LC_{50} = 51 IJs/ml) in killing mustard sawfly. Findings from a field study with EPN *Steinernema feltiae* Filipjev on radish crop by Narayanan and Gopalakrishnan (2003) suggested a weekly application of *S. feltiae* at 1.1×10^3–10^5 IJs/ml may effectively manage *A. lugens proxima* in mustard, as it significantly reduced the pest numbers (0.42–0.48 larvae/plant vs 2.95 larvae/plant in untreated control) and increased the radish yield. Similarly, Meena *et al.* (2013) reported that high densities of aphids (*L. erysimi*) can significantly be suppressed for a period of 10 days by foliar application of EPFs (*Lecanicillium lecanii*, *Beauveria bassiana* Balsamo or *Metarhizium anisopliae* (Metch.) Sorokin) at 5 g/l of water.

17.3.7 Genetic control

Use of insects to manage insects of the same species through transfer of damaged genetic material is a novel sterile insect technique (SIT) developed by Knipling (1969). Releasing irradiated substerile parents or their F_1 progeny of *P. xylostella* moth into untreated populations caused sterility in nearly 70% of F_1 progeny of the untreated population. The inclusion of *Diadegma semiclausum* with the above treatment further increased the mortality of *P. xylostella* (Sutrinso, 2005). Similarly, using irradiated males and females of *C. binotalis* alone or together with untreated populations caused a significant reduction in hatching of eggs laid by the parent moths (Sutrinso, 2005). Since there are several limitations in using this technique for effective reductions in field populations of key defoliator pests of mustard, its practical inclusion as a component of IPM strategy for mustard pests has a long way to go.

17.3.8 Chemical control

Owing to the review and revision of regulations governing safety aspects of insecticide manufacture, transportation, storage, sale and use, several conventional insecticides belonging to organophosphate, carbamate and synthetic pyrethroid groups with higher mammalian toxicity and residues have been replaced by the new and more selective insecticide molecules such as indoxacarb, spinosad, spinetoram, chlorantraniliprole, metaflumizone, flubendiamide, pyridalyl, etc. (Reiter *et al.*, 2008; Kumaranag *et al.*, 2014). Undoubtedly, the incorporation of conventional or biorational insecticide molecules in IPM strategies enhances pest management potential of the strategy (Facknath, 1997; Pandey and Singh, 2008; Yadav and Singh, 2015). For instance, a complete control of mustard aphid that resulted in higher net returns (benefit:cost ratio) was made possible only by integrating an insecticide in an IPM strategy with other pest management tactics, while the same treatments excluding insecticide were found to be inferior in suppressing aphids and in net returns (Pandey and Singh, 2008; Yadav and Singh, 2015).

Seed treatment with insecticides offers protection to the mustard seedlings from pests attacking seedlings, i.e. flea beetle and mustard sawfly, and is considered to be the safest method for natural enemies. Therefore, IPM practitioners should give due consideration to this tactic while formulating IPM strategies. For example, seed treatment with imidacloprid 70 WS at 5 g/kg of seed prior to sowing may discourage the early infesting insects from colonizing on seedlings (Kumar, 2015). In spite of the quick action of insecticides upon contact or ingestion, insect pests such as *H. undalis* and *C. binotalis* may escape the exposure to insecticides and resultant death due to their cryptic feeding nature under the silk webbing and folded leaves and inflorescences (Reiter *et al.*, 2008). Therefore, insecticide application should be initiated as soon as the pest appears, or in early-stage instars, and spray should be directed to cover both sides of the leaves (Reiter *et al.*, 2008). It is strongly recommended to read the label carefully and adhere strictly to the instructions specified on the label before application of insecticides. Nevertheless, insecticides incorporated in an IPM programme should be used as a last resort in a safe and judicious way for managing pestiferous insects of mustard, because such careful use is likely to prevent the recurrence of a scenario that would be inevitable with the excessive and indiscriminate use of an insecticide, i.e. insecticide residue, insecticide resistance, insect resurgence and insect replacement.

While developing an IPM strategy, it should be borne in mind that not more than one synthetic chemical insecticide (preferably biorational) should be included. If a second application is necessary, it should be rotated with another insecticide having a different target site, mode of action and possibly with a different chemistry for effective management of pesticide resistance. For instance, the combination of either spinosad and spinetoram or chlorantraniliprole and flubendiamide should not be used, because both the insecticides in the combination share the same target site and mode of action.

17.4 Host Plant Defence-based Mustard Pest Management Tactics

Plants have evolved various physical and chemical mechanisms to defend themselves from continuous stress posed by abiotic and biotic factors, including arthropod insects (Ahuja *et al.*, 2010). Of these, the chemical defence mechanisms could be either constitutive or induced by stress factors such as insect attack (Takabayashi and Dicke, 1996; Karban and Baldwin, 1997; Pare and Tumlinson, 1999; Dicke and Hilker, 2003). Direct defence comprises plant responses, either constitutive or induced by herbivore feeding, that lead to production of volatile and non-volatile secondary metabolites/compounds/chemicals that protect the plants from herbivores via their direct action on them (Dicke, 1999; Mattiacci *et al.*, 2001). Indirect defence comprises plant responses that result in production and emission of volatile blends, either prior or (mostly) in response to herbivore attack, for attracting and recruiting natural enemies (parasitoids and predators) to kill the attacking herbivores (Dicke, 1999; Mattiacci *et al.*, 2001). Exploring and exploiting the vast array of direct and indirect defences of plants could pave the way for inclusion of them as an important tool in mustard IPM strategy.

17.4.1 Direct and indirect defences through natural and induced mechanisms

Brassicaceae family members, including mustard, synthesize and store a specific group of non-volatile sulfur-containing secondary metabolites known as glucosinolates (β-thioglucoside *N*-hydroxysulfates) (Fahey *et al.*, 2001; Bones and Rossiter, 2006; Tripathi and Mishra, 2007), which break down upon herbivore damage, releasing volatile and semi-volatile signal and toxic chemicals (Koristas *et al.*, 1991; Bodnaryk, 1992; Mithen, 1992; Bones and Rossiter, 2006). Additionally, brassicas synthesize and emit a wide array of volatile organic compounds (VOCs), collectively known as semiochemicals, into their microhabitat for intra- and interspecific communication. These non-volatile and volatile and secondary metabolites mediating varied ecological roles in multitrophic interactions (Holopainen, 2004), i.e. attractants, arrestants, stimulants, repellents, deterrents, etc. (Zukalova and Vasak, 2002; Ahuja *et al.*, 2010), can be potential candidates for use in management of both specialist and generalist feeders of brassicas.

For example, glucosinolates (indole glucosinolate, glucobrassicin) and glucosinolate breakdown products (allyl isothiocyanates (AITCs) and isothiocyanates such as iberin and sulforaphane) from brassicas, including mustard (*B. nigra*), were found to be strong stimulators of oviposition in the specialist herbivores, *Pieris rapae* and *P. brassicae* (Traynier and Truscott, 1991; Renwick *et al.*, 1992; van Loon *et al.*, 1992; Huang and Renwick, 1993) and *Plutella xylostella* (Hughes *et al.*, 1997; Renwick *et al.*, 2006). The AITCs also exhibited adverse effects on growth, development and survival of *Pieris rapae* caterpillars (Agrawal and Kurashige, 2003), which is an attract-and-kill strategy (synergistic action of two direct defences) exhibited by *Brassica* plants against cabbage butterflies. On the other hand, cabbage plants sprayed with homogenized cabbage tissue or plant secondary metabolites such as coumarin and rutin deterred oviposition by *P. rapae* (Renwick and Radke, 1985; Tabashnik, 1987). Similarly, spraying of non-host plant (*Chrysanthemum morifolium* Ramat) extracts on Chinese cabbage (*Brassica campestris* L. subsp. *pekinensis*) reduced *Plutella xylostella* ovipositional preference, while increasing attractiveness to the parasitoid *Cotesia plutellae*, leading to reduced host infestation and damage (Liu *et al.*, 2006). Furthermore, spraying with species-specific oviposition-deterring pheromones of *Pieris brassicae* and *P. rapae* and other *Brassica* specialists may deter them from oviposition (Schoonhoven, 1990). Therefore, oviposition stimulators can be explored as potential candidates to attract and trap female moths of *Plutella xylostella*, *Pieris rapae*, *P. brassicae*, etc. in an IPM strategy either by spraying these compounds on trap crops or by placing them as lures in mechanical traps. Similarly, exogenous applications of oviposition-deterrent chemicals of host-plant, non-host-plant or insect origin on mustard plants form an important IPM tactic for mustard specialists. Additionally, breeding efforts can be focused on developing region-specific and key pest-specific mustard varieties that either up-regulate or down-regulate the production of specific glucosinolates to attract (trap crop) or deter (main crop) specialists for reduced infestation.

VOCs (of plant and insect origin) and oviposition-induced plant volatiles (OIPVs) (of plant origin) utilized by insect pests and natural enemies as host and conspecific location cues is another area where efforts need to be concentrated for use in pest management. Flea beetle management is a good example of this. Flea beetles (*Phyllotreta cruciferae*) are attracted to a sesquiterpene aggregation pheromone, (+)-(*6R,7S*)-himachal-9,11-diene, produced by males (Beran *et al.*, 2016) and AITC produced by *Brassica* plants (Soroka *et al.*, 2005). The combined use of these two compounds as lures attracted more flea beetles than either of the lures used separately (Soroka *et al.*, 2005). Therefore, with a suitable trap design and commercial synthesis of AITC and aggregation pheromone, an effective trapping method can be developed for flea beetles. In similar lines, egg deposition by *Pieris* spp. on *Brassica* plants induces changes in the chemical profile of the leaf surface that act as contact cues for the egg parasitoid *Trichogramma evanescens* Westwood, and subsequent parasitization (Fatouros *et al.*, 2005). By application of volatile chemicals involved in *Trichogramma* attraction to host eggs on plants, the egg parasitoid recruitment efforts can be enhanced.

Investigations revealed that exogenous application of phytohormones artificially induced defence pathways in *Brassica* species. Treatment of Brussels sprouts (*B. oleracea*) with jasmonic acid/jasmonate (JA) at a concentration of 1 mM resulted in emission of a volatile blend similar to that induced by *Plutella xylostella* or *Pieris brassicae* larval feeding but at much greater quantities that aided in attraction of three important parasitoids of *Brassica* specialists (*Cotesia glomerata*, *Cotesia rubecula* and *D. semiclausum*) (Bruinsma *et al.*, 2009). Similarly, exogenous application of methyl jasmonate (MeJ) on mustard induced a JA-mediated defence pathway and reduced growth and infestation of aphids (Koramutla *et al.*, 2014). In further exploitation of the use of phytohormones in pest management, methyl salicylate (MeS) emitted by oilseed rape plants upon feeding by larvae of the cabbage moth (*Mamestra brassicae* (L.)) or artificially from dispensers prevented oviposition by other females, i.e. acted as a signal for conspecific females to avoid it as a potential oviposition site (Ulland *et al.*, 2008). Therefore, suffusing crop microhabitat with phytohormones by either (i) exogenous application, (ii) installing dispensers to release, or (iii) transforming mustard plants to emit phytohormones constitutively is a novel non-toxic IPM tactic for a mustard ecosystem.

An alarm pheromone containing a sesquiterpene (*E*)-β-farnesene (EBF) as the principal component (Bowers *et al.*, 1972; Pickett and Griffiths, 1980) is released by several aphid species that infest mustard, including *L. erysimi*, *M. persicae* and *B. brassicae*, to inform other members of an aphid colony of the impending danger of natural enemies (Joachim and Weisser, 2015). This naturally occurring induced pheromone holds immense promise for aphid pest management by eliciting a battery of behavioural and physiological changes in aphids that benefit the plants (Montgomery and Nault, 1977; Calabrese and Sorensen, 1978; Gibson and Pickett, 1983; Ave *et al.*, 1987; Mostafavi *et al.*, 1996; Kunert *et al.*, 2005). Although natural instances of plants (wild potato plant, *Solanum berthaultii* Hawkes) using EBF for aphid management occur (Gibson and Pickett, 1983), there are a few practical issues in the use of EBF emitted naturally by plants. For example: (i) the EBF released by attacked aphids is in pure form and so, ecologically, aphids are adapted to detect even low traces of pure EBF released by conspecifics (Pickett and Griffiths, 1980), whereas plants (e.g. *S. berthaultii*) naturally emit EBF in a mixture of sesquiterpenes (Gibson and Pickett, 1983), which may act as inhibitors of an alarm signal (Dawson *et al.*, 1984; Bruce *et al.*, 2005); and (ii) plants naturally emit EBF at low concentrations while emission at higher concentrations (> 1000 times than that produced naturally by aphids, i.e. < 1 up to 50 ng) is necessary to serve as a kairomone for natural enemies of aphids (Schwartzberg *et al.*, 2008; Cui *et al.*, 2012; Joachim *et al.*, 2013; Joachim and Weisser, 2015). The problems related to ecological functions of artificial EBF can be overcome by either commercial synthesis of EBF and deployment of dispensers to release it, or genetic engineering (GE) of plants (see below) for constitutive emission of pure EBF in large quantities, which appears to be a promising non-toxic and ecofriendly IPM tactic for managing key pests (aphids) in mustard.

Of the several abiotic factors (Gouinguene and Turlings, 2002; Vuorinen *et al.*, 2004a, b; Ibrahim *et al.*, 2008) and biotic factors (Takabayashi *et al.*, 1994; Vuorinen *et al.*, 2004a, b) that influence a plant's inherent and inducible defence systems, those of significant importance include soil health, fertility status and herbivore feeding (Ibrahim *et al.*, 2008). Oilseed rape (*B. napus*) plants grown at higher nutrient levels exhibited higher indirect and direct defence responses with or without *P. xylostella* feeding on them, respectively, compared with at low nutrient level (Ibrahim *et al.*, 2008). This underscores the role of enhanced soil nutritional status, which bolsters the plant's defence system, as an important component to be included in IPM for mustard. Similarly, intensity (quantity and quality) of herbivore feeding (De Moraes *et al.*, 1998; Turlings *et al.*, 1998) determines the *de novo* biosynthesis of herbivore-induced plant volatiles (HIPVs), leading to quantitative and qualitative alterations in the blend of VOCs being emitted by plants (Schular *et al.*, 1999a, b, c; Ibrahim *et al.*, 2008), This finding hints at the possibility of reduced emission of HIPVs from mustard and other *Brassica* plants possessing resistant genes acquired through conventional breeding or genetic engineering.

17.4.2 Direct and indirect defences through genetic breeding and genetic engineering

The development and use of mustard varieties (conventional, genetic breeding or GE) possessing genes responsible for defence traits may be looked at as an effective non-chemical component of IPM for mustard insect pests. However, it is believed that transfer of insect-resistant traits from wild *Brassica* species to cultivated species through conventional breeding is a tedious and difficult task; hence, little progress has been made in this direction (Kanrar *et al.*, 2002; Banga, 2007). None the less, through introgressive breeding, a rare, random and difficult process, Atri *et al.* (2012) developed introgressive lines of mustard using a wild relative, *Brassica fruticulosa* Cirillo, known to possess genes for mustard aphid (*L. erysimi*) resistance. A 2-year screening study revealed consistently greater resistance with respect to number of aphids per plant (average 3 vs 96 aphids in control) and percentage of infested plants (Atri *et al.*, 2012). The advent of biotechnological procedures made the task of transforming plants simple, easy and reliable. For instance, the constitutive production of Bt toxin(s) in transformed plants protects them from being ravaged by a specific target insect pest or pests and precludes or reduces the usage of chemical insecticides (Romeis *et al.*, 2006). Indian mustard (*B. juncea*) plants transformed with a wheat gene coding for wheat germ agglutinin (WGA), a chitin-binding lectin, caused significant aphid mortality and reduced the fecundity in surviving aphids (Kanrar *et al.*, 2002). Similarly, Indian mustard plants transformed with single or two (pyramided) Bt genes coding for *Cry*1Ac, or *Cry*1C, or *Cry*1Ac and *Cry*1C proteins have protected the plants not only from susceptible *P. xylostella* feeding but also from resistant *P. xylostella*, i.e. *Cry*1C plants gave protection from *Cry*1Ac-resistant individuals and vice versa, while *Cry*1Ac and *Cry*1C pyramided plants gave protection from both resistant populations (Cao *et al.*, 2008). Based on the above findings, it was proposed that the significant protection offered by GE mustard varieties against key specialists (*P. xylostella* and *L. erysimi*) makes them good candidates for use as 'dead-end' trap crops for cabbage and other commercially grown *Brassica* vegetable crops (Cao *et al.*, 2008).

As discussed earlier, to realize EBF (alarm pheromone) as a potential aphid management tactic in IPM for mustard and other *Brassica* species, initially a related crucifer plant, *Arabidopsis thaliana* (L.) Heynh. (Beale *et al.*, 2006; de Vos *et al.*, 2010), and subsequently mustard (*B. juncea*) (Verma *et al.*, 2015) were transformed with a sesquiterpene synthase gene to produce EBF. Unlike plants naturally emitting EBF, such as *S. berthaultii* (Gibson and Pickett, 1983), GE plants constitutively emitted high titres of EBF at ~800 ng/h while keeping titres of other sesquiterpenes low (< 70 ng/h) (Beale *et al.*, 2006), which not only successfully repelled green peach aphid (*M. persicae*) on *A. thaliana* (Beale *et al.*, 2006; de Vos *et al.*, 2010) and mustard aphid (*L. erysimi*) on mustard (up to 80%) (Verma *et al.*, 2015) but also acted as a kairomone for the parasitoid *D. rapae* (Beale *et al.*, 2006) and prevented colonization of mustard aphid (*L. erysimi*) on transformed mustard compared with wild-type mustard (Verma *et al.*, 2015). An additional benefit observed with these plants was that aphids (*M. persicae*) exhibited a fitness cost of reduced fecundity in the absence of natural enemies (de Vos *et al.*, 2010).

Although it seemed that constitutive emission of EBF may not serve as a sustainable direct defence, due to the danger of aphids becoming habituated (Kunert *et al.*, 2010) within three generations (de Vos *et al.*, 2010), the finding that this is reversible (no genetic basis) and aphids can revert to becoming EBF responsive within three generations upon migration and feeding on normal plants away from the release source kept the hopes of its sustainable use alive (de Vos *et al.*, 2010). This indicates that non-GE mustard variety fields around GE mustard variety fields may serve as reversion areas for EBF-habituated aphids. Additionally, EBF-habituated aphids are non-responsive to EBF of attacked conspecifics, thereby increasing predation chances and minimizing aphid infestation (de Vos *et al.*, 2010). Furthermore, transformed plants emitting EBF in greater amounts may serve as an indirect defence (a kairomone) for attracting and recruiting natural enemies of aphids into the mustard ecosystem (Micha and Wyss, 1996; Beale *et al.*, 2006; Harmel *et al.*, 2007; Hatano *et al.*, 2008; Kunert *et al.*, 2008; Schwartzberg *et al.*, 2008; Cui *et al.*, 2012; Joachim *et al.*, 2013; Joachim and Weisser, 2015). In order to be an effective and sustainable pest management tactic, EBF should trigger synergistic action by not only repelling aphids but also attracting natural enemies (Beale *et al.*, 2006), because EBF-repelling aphids alone did not cause a decrease in aphid infestation (Calabrese and Sorensen, 1978; Yu *et al.*, 2013). However, it also needs to be determined whether such constitutive emission of EBF at high rates by plants leads to habituation and/or confusion in natural enemies as well (Kunert *et al.*, 2010). None the less, the problem of pest and natural enemies becoming habituated to artificial EBF can be circumvented by simulating the natural release (intermittent or pulsed) of artificial EBF, either by transforming plants to release upon induction (herbivore feeding) or by using programmed dispensers to release EBF at constant or varying intervals.

GE plants offering direct and indirect defences against key pests are a vital and viable component to be included in an IPM strategy for mustard; however, there is an ecological consequence to the use of GE plants in terms of a trade-off between artificial and natural defences. As reported in other crops, whether they are intact (Bt cotton) (Yan *et al.*, 2004) or insect-infested (Bt maize) (Turlings *et al.*, 2005; Dean and De Moreas, 2006), oilseed rape (*B. napus*) plants transformed with Bt gene *Cry*1Ac exhibited reduced emission of both constitutively and *P. xylostella*-induced VOCs (Ibrahim *et al.*, 2008) and consequent reduction in attraction and recruitment of *C. plutellae* (also referred to as *C. vestalis*) compared with conventional non-Bt plants (Schular *et al.*, 1999a, 2003). This may partly be attributed to the reduced feeding damage by *P. xylostella* on Bt oilseed rape (Ibrahim *et al.*, 2008) because the quantity of HIPVs produced are directly related to the intensity (both quantity and quality) of insect feeding damage (De Moraes *et al.*, 1998; Turlings *et al.*, 1998; Schular *et al.*, 1999a, b, c). Additional ecological consequence of Bt plants on tritrophic interactions is increased mortality of *C. plutella* through premature mortality of Bt-susceptible hosts (*P. xylostella*) (Schular *et al.*, 2003, 2004). Under the above situation, maintaining a portion of Bt mustard crop field under non-Bt plants as refuge serves both purposes, i.e. to delay resistance evolution and to maintain an abundance of natural enemy population (Himanen *et al.*, 2009b). Detailed studies are warranted to assess the impact of altered emissions of VOCs and HIPVs from Bt mustard plants manifested through changes in phenotype and physiology on ecological effects with respect to specialized tritrophic interactions and generalized multitrophic interactions in the ecosystem.

Although GE mustard varieties have been developed, they have yet to be released into the market. With this background, the advantages and disadvantages of GE crops in general and GE mustard in particular warrant some discussion. The advantages are: (i) GE mustard plants possess built-in constitutive protection against key pests from seedling to harvesting; (ii) reduction in usage of harmful insecticides; (iii) conservation of natural enemies; (iv) compatibility with any or all components of IPM; (v) pyramiding multiple genes from different sources and modes of action for conferring resistance against multiple key pests or delaying the breakdown of plant resistance, etc.

17.5 Role of Climate Change in Mustard Pest Management

Owing to the spurt in anthropogenic activities on a global scale, there is an enormous increase in the amounts of greenhouse gases such as carbon dioxide (CO_2), ozone (O_3), methane (CH_4) and nitrous oxide (N_2O) released into the atmosphere and this is causing global warming and consequent climate change (Cornelissen, 2011). As coevolving host plants and insect pests are continually exposed to altered environmental conditions under changing climate conditions such as elevated temperature, CO_2 and O_3, pest insects are affected either directly or indirectly (Cornelissen, 2011) leading to changes in pest density and in natural enemies. These impacts on direct and indirect defences of transgenic Bt plants might be different from those of conventional non-Bt plants, due to the inherent differences in resource allocation to different defence mechanisms, especially for constitutive Bt toxin production in Bt plants (Coviella *et al.*, 2002; Chen *et al.*, 2005).

Given the fact that transgenic Bt mustard has not yet been released into the market, available literature on other *Brassica* species where commercialization has taken place is discussed here to predict similar effects with Bt mustard when released. Transgenic oilseed rape plants grown under chronic elevated O_3 exhibited a positive correlation between leaf Bt toxin and O_3 concentrations (Himanen *et al.*, 2009a), giving an impression that Bt oilseed rape plants grown under elevated O_3 atmosphere will have a positive impact on *P. xylostella* management. Unlike O_3, Bt oilseed rape plants grown under elevated temperature and CO_2, alone or in combination, responded with no significant change in leaf Bt toxin concentration (Himanen *et al.*, 2008). However, no study has so far reported the reduction in Bt toxin concentration in brassicas that was documented in a non-*Brassica* plant, Bt cotton, under elevated CO_2 (Coviella *et al.*, 2002; Chen *et al.*, 2005; Wu *et al.*, 2007). Since elevated CO_2 is known to reduce leaf nitrogen content in plants (Lincoln *et al.*, 1986; Stiling *et al.*, 1999; Zvereva and Kozlov, 2006; Stiling and Cornelissen, 2007), which may influence Bt toxin content, conducting more detailed studies by exposing plants to chronic elevated CO_2 and analysing leaf nitrogen and Bt toxin contents is essential for determining the exact role of elevated CO_2 on Bt toxin concentration.

With regard to plant-mediated effects manifested through alterations in defence compounds (mainly glucosinolates) under changing climatic conditions (CO_2 and O_3), the majority view is that the insects cope with the altered plant physiology by reducing larval growth rates and prolonging developmental durations (Reddy *et al.*, 2004; Cornelissen, 2011; Khalig *et al.*, 2015), which enhances biological control by increasing their window of availability and vulnerability to natural enemies. For instance, there were reduced growth rates of specialists *P. xylostella* on *B. napa* subsp. *oleifera* (Reddy *et al.*, 2004), *P. rapae* on collard (*B. oleracea* L. var. *acephala*) (Landosky and Karowe, 2014), *P. brassicae* and *B. brassicae* on *Brassica* plants (*B. oleracea*) (Kleiber *et al.*, 2013) under elevated CO_2 and *P. xylostella* on oilseed rape (Bt and non-Bt) (Himanen *et al.*, 2008) and *P. brassicae* on wild mustard under elevated O_3 (Khalig *et al.*, 2015).

Since emission of HIPVs is directly proportional to the degree of feeding damage (De Moraes *et al.*, 1998; Turlings *et al.*, 1998; Schular *et al.*, 1999a, b, c), minimal feeding by *P. xylostella* on oilseed rape Bt plants under normal (Schular *et al.*, 1999a, 2003) or elevated O_3 (100 ppb) did not lead to emission of the full complement of HIPVs necessary for detection by parasitoids that led to less attraction (Himanen *et al.*, 2008), as was proved in a combination of Bt oilseed rape, *P. xylostella* and *C. vestalis*, which attracted 37% parasitoids as against > 75% in damaged conventional non-Bt plants (Himanen *et al.*, 2009b). On the other hand, increased leaf consumption by *P. xylostella* on oilseed rape under elevated CO_2 and temperature (Zvereva and Kozlov, 2006; Stiling and Cornelissen, 2007) might result in increased emission of HIPVs, benefiting a higher trophic level in a tritrophic interaction. The implied message from these findings is that cultivation of Bt mustard or other *Brassica* varieties resistant to a specialist key pest in normal or O_3-polluted areas may adversely affect the tritrophic signalling, together with supply of hosts that are poor in quantity and quality, resulting in reduced abundance of specialist parasitoids in the ecosystem (Himanen *et al.*, 2009c; Schular *et al.*, 2004).

Under warmer climatic conditions, plants are expected to produce greater amounts of VOCs for a prolonged period of time, saturating the crop habitat (Peñuelas and Staudt, 2010). If these compounds are involved in direct and indirect plant defences against herbivores then the neighbouring plants will be on high alert and help them readily to recruit natural enemies, which might potentially reduce future insect attack (Cornelissen, 2011). Similarly, under increased CO_2 conditions, plants produce higher rates of VOCs, mainly carbon-based secondary metabolites such as terpenes, due to the positive relationship between carbon availability and VOCs production (Lerdau *et al.*, 1994; Yuan *et al.*, 2009). Overall, a literature search on the effect of climate change abiotic factors on plants (both non-Bt and Bt *Brassica* plants) and insects produced a mixed bag of positives and negatives. Furthermore, most of the current findings were based on studies conducted taking one abiotic factor at a time. Hence, detailed investigations need to be planned to simulate natural environmental conditions by chronic exposure of both host plant and pest together to two or more factors in combination that can account for possible interactions among them for drawing meaningful conclusions on effects of changing climatic conditions on specialized tritrophic and generalized multitrophic interactions. With the expanding cultivation of released Bt *Brassica* crops such as oilseed rape (and maybe mustard in future) all over the world, studies to determine the impact of climate change conditions on Bt crops assume significance as they might have the potential to affect the sustainable use of Bt crops in agriculture in future decades (James, 2007).

17.6 Ideal IPM Strategy and Conclusions

Mustard is an important crucifer crop, primarily grown for oil-rich seeds. Under favourable conditions, more than 43 pestiferous insect species exert constant biotic stress on mustard plants and may result in 10–96% or complete yield loss. Therefore, sustainable management of key and destructive pestiferous insects from seedling to harvest stage is vital for realizing the full genetic potential of mustard plants, i.e. maximizing productivity. This assumes significance in the wake of a projected global per capita consumption of oil crops in the form of oil and all uses in 2050 at 16 and 30 kg, respectively, and an increase of about 100 million tonnes of oil to meet this demand by 2050 (Alexandratos and Bruinsma, 2012). However, current management strategies are plagued by the dominance of a single pest management component, i.e. chemical insecticides. The outcomes of such a skewed pest management strategy are short lived and are neither environmentally friendly nor economically sound, warranting a best alternative for this situation in the form of IPM.

Very few efforts have been made to develop effective IPM strategies for mustard pests by integrating two or more tactics. The situation demands development of robust sustainable IPM strategies and calls for more focused research in this direction. Towards realizing this, existing IPM strategies or modules need to be bolstered by taking them as a base and including tactics from new, novel and improved non-chemical, ecofriendly pest monitoring and management methods such as AESA, ecological engineering, inbuilt resistant transgenic plants, insect and plant infochemical-based trapping and management, etc. Typically, the best IPM strategy for sustainable management of mustard pests should integrate and evaluate: (i) a good decision-making tool such as AESA; (ii) best cultural practices such as crop rotation, use of improved resistant varieties (either conventional or GE) for key pests, early planting, ecological engineering with appropriate companion plants; (iii) best mechanical practices such as installing yellow sticky traps, light traps, infochemical-based advanced trapping techniques for key pests; (iv) best natural enemy conservation practices such as provision of nectar- and pollen-rich flowering plants (ecological engineering), mass production and release of parasitoids and predators, application of natural or improved formulation-based botanical (neem) and microbial biopesticide (Bt, NPV, EPN or EPF); and (v) judicious application of environmentally safer and biorational insecticides as a last resort. The successful adoption of an IPM strategy in mustard cultivation, dominated by non-toxic eco-friendly pest management tactics, helps not only to mitigate the adverse impact of sole use of chemical pesticides on the environment, human beings and natural enemies but also to reduce residues in harvested produce as well as resistance and resurgence issues in key pests. Additionally, with a carefully planned and executed IPM approach (from land preparation to harvest), pestiferous insect densities can be kept below the economic damage-causing levels for sustainable increase in productivity.

References

Aamir, K. and Khalid, M. (1961) *Insect Pest Attack on Rapeseed*. Jalal Publisher, Lahore, Pakistan.

Agrawal, A.A. and Kurashige, N.S. (2003) A role for isothiacyanates in plant resistance against the specialist herbivore *Pieris rapae*. *Journal of Chemical Ecology* 29, 1403–1415.

Ahuja, I., Rohloff, J. and Bones, A.M. (2010) Defence mechanisms of Brassicaceae: implications for plant insect interactions and potential for integrated pest management. A review. *Agronomy for Sustainable Development* 30, 311–348.

Alexandratos, N. and Bruinsma, J. (2012) *World Agriculture Towards 2030/2050: the 2012 revision*. ESA working Paper No. 12-03, FAO, Rome. Available at: http://www.fao.org/docrep/016/ap106e/ap106e.pdf (accessed 12 February, 2016).

Anon. (2015a) *Vision 2050*. Directorate of Rapeseed-Mustard Research, Indian Council of Agricultural Research, Sewar, Bharatpur, India, p. 26. Available at: http://www.drmr.res.in/publication/DRMR%20

Rajasthan-%28Vision2050%29-Final.pdf (accessed 1 February, 2016).

Anon. (2015b) *Mustard: insect pests management*. Vikaspedia, Ministry of Electronics and Information Technology, Government of India, Hyderabad, India. Available at: http://vikaspedia.in/agriculture/crop-production/integrated-pest-managment/ipm-for-oilseeds/ipm-strategies-for-mustard-rapeseed/insect-pests-management (accessed 1 August, 2015).

Anon. (2015c) *Important pests of mustard*. Agritech, Tamil Nadu Agricultural University, Coimbatore, India. Available at: http://www.agritech.tnau.ac.in/crop_protection/crop_prot_crop_insect_oil_mustard.html (accessed 1 August, 2015).

Atri, C., Kumar, B., Kumar, H., Kumar, S., Sharma, S. and Banga, S.S. (2012) Development and characterization of *Brassica juncea – fruticulosa* introgression lines exhibiting resistance to mustard aphid (*Lipaphis erysimi* Kalt). *BMC Genomics* 13, 104 (1–9).

Ave, A., Gregory, P. and Tingey, W.M. (1987) Aphid repellent sesquiterpenes in granular trichomes of *Solanum berthaultii* and *S. tuberosum*. *Entomologia Experimentalis et Applicata* 44, 131–138.

AVRDC (2015) Ethiopian kale (*Brassica carinata*). AVRDC, The World Vegetable Center, Shanhua, Taiwan. Available at: http://avrdc.org/ethiopian-kale-brassica-carinata/ (accessed 12 February, 2016).

Badenez-Pérez, F.R., Shelton, A.M. and Nault, B.A. (2004) Evaluating trap crops for diamondback moth, *Plutella xylostella* (Lepidoptera: Plutellidae). *Journal of Economic Entomology* 97, 1365–1372.

Badenez-Pérez, F.R., Shelton, A.M. and Nault, B.A. (2006) Dynamics of diamondback moth oviposition in the presence of a highly preferred non-suitable host. *Entomologia Experimentalis et Applicata* 120, 23–31.

Badenez-Pérez, F.R., Reichelt, M., Gershenzon, J. and Heckel, D.G. (2011) Phylloplane location of glucosinolates in *Barbarea* spp. (Brassicaceae) and misleading assessment of host suitability by a specialist herbivore. *New Physiologist* 189, 549–556.

Badenez-Pérez, F.R., Gershenzon, J. and Heckel, D.G. (2014a) Insect attraction versus plant defense: young leaves high in glucosinolates stimulate oviposition by a specialist herbivore despite poor larval survival due to high saponin content. *PLoS ONE* 9, e95766. doi:10.1371/journal.pone.0095766.

Badenez-Pérez, F.R., Reichelt, M., Gershenzon, J. and Heckel, D.G. (2014b) Using plant chemistry and insect preference to study the potential of *Barbarea* (Brassicaceae) as a dead-end trap crop for diamondback moth (Lepidoptera: Plutellidae). *Phytochemistry* 85, 137–144.

Bajwa, W.I. and Kogan, M. (2002) *Compendium of IPM Definitions (CID) – what is IPM and how is it defined in the worldwide literature?* IPCC Publication no. 998. Integrated Plant Protection Center (IPCC), Oregon State University, Coravallis, Oregon.

Bakhetia, D.R.C. and Arora, R. (1986) Control of insect pests of toria, sarson and rai. *Indian Farming* 36, 41–44.

Bakhetia, D.R.C. and Sekhon, B.S. (1989) Insect pests and their management in rapeseed-mustard. *Journal of Oilseeds Research* 6, 269–273.

Bakhetia, D.R.C., Singh, H. and Chander, H. (2002) IPM for sustainable production of oilseeds. In: Singh, H. and Hegde, D.M. (eds) *Oilseeds and Oils: Research and Development Needs*. Indian Society of Oilseeds Research, Hyderabad, India, pp. 184–218.

Banga, S.S. (2007) Genetic manipulations for oilseeds improvement-conventional. In: Hegde, D.M. (ed.) *Changing Global Vegetable Oils Scenario: Issues and Challenges before India*. Indian Society of Oilseeds Research, Hyderabad, India, pp. 17–34.

Beale, M.H., Birkett, M.A., Bruce, T.J.A., Chamberlain, K., Field, L.M., Huttly, A.K., Martin, J.L., Parker, R., Phillips, A.L., Pickett, J.A., Prosser, I.M., Shewry, P.R., Smart, L.E., Wadhams, L.J., Woodcock, C.M. and Zhang, Y. (2006) Aphid alarm pheromone produced by transgenic plants affects aphid and parasitoid behavior. *Proceedings of the National Academy of Sciences of the USA* 103, 10509–10513.

Beran, F., Rahfeld, P., Luck, K., Nagel, R., Vogel, H., Wielsch, N., Irmisch, S., Ramasamy, S., Gershenzon,, J., Heckel, D.G. and Köllner, T.G. (2016) Novel family of terpene synthases evolved from trans-isoprenyl disphosphate synthases in a flea beetle. *Proceedings of the National Academy of Sciences of the USA* 113, 2922–2927.

Blossey, B. and Hunt-Joshi, H.R. (2003) Belowground by insects: influence on plants and above ground herbivores. *Annual Review of Entomology* 48, 521–547.

Bodnaryk, R.P. (1992) Effects of wounding on glucosinolates in the cotyledons of oilseed rape and mustard. *Phytochemistry* 31, 2671–2677.

Boina, D. (2000) Population dynamics and management of defoliator pests of cabbage. MSc thesis, University of Agricultural Sciences, Dharwad, Karnataka, India.

Boina, D., Jeyakumar, P. and Korlapati, S. (2014) Demonstration of agro-ecosystem analysis based plant health management in conjunction with ecological engineering for pest management in cabbage to reduce reliance on chemical pesticides. *Plant Health News Letter* 4, 2–5.

Bones, A.M. and Rossiter, J.T. (2006) The enzyme and chemically induced decomposition of glucosinolates. *Phytochemistry* 67, 1053–1067.

Bowers, W.S., Webb, R.E., Nault, L.R. and Dutky, S.R. (1972) Aphid alarm pheromone: isolation, identification, synthesis. *Science* 177, 1121–1122.

Brown, J., McCaffrey, J.P., Brown, D.A., Harmon, B.L. and Davis, J.B. (2004) Yield reduction in *Brassica napus*, *B. rapa*, *B. juncea*, and *Sinapis alba* caused by flea beetle (*Phyllotreta cruciferae* (Goeze) (Coleoptera:

Chrysomelidae)) infestation in northern Idaho. *Journal of Economic Entomology* 97, 1642–1647.

Bruce, T.J.A., Birkett, M.A., Blande, J., Hooper, A.M., Martin, J.L., Khambay, B., Prosser, I., Smart, L.E. and Wadhams, L.J. (2005) Response of economically important aphids to components of *Hemizygia petiolata* essential oil. *Pest Management Science* 61, 1115–1121.

Bruinsma, M., Posthumus, M.A., Mumm, R., Mueller, M.J., van Loon, J.J.A. and Dicke, M. (2009) Jasmonic acid-induced volatiles of *Brassica oleracea* attract parasitoids: effects of time and dose, and comparison with induction by herbivores. *Journal of Experimental Biology* 60, 2575–2587.

Buranday, R.P. and Raros, R.S. (1975) Effects of cabbage-tomato intercropping on the incidence and oviposition of the diamondback moth, *Plutella xylostella* (L.). *Philippine Entomology* 2, 369–374.

Burgess, L. (1977) Flea beetles (Coleoptera: Chrysomelidae) attacking rape crops in the Canadian prairie provinces. *The Canadian Entomologist* 109, 21–32.

CABI (2008) *Dasineura brassica* (brassica pod midge) Datasheet. Centre for Agriculture and Biosciences International, Wallingford, UK. Available at: http://www.cabi.org/cpc/datasheet/17969 (accessed 30 March, 2016).

CABI (2015a) *Spodoptera litura* (taro caterpillar) Datasheet. Centre for Agriculture and Biosciences International, Wallingford, UK. Available at: http://www.cabi.org/cpc/datasheet/44520 (accessed 30 March, 2016).

CABI (2015b) *Bagrada hilaris* (painted bug) Datasheet. Centre for Agriculture and Biosciences International, Wallingford, UK. Available at: http://www.cabi.org/cpc/datasheet/8302 (accessed 30 March, 2016).

Calabrese, E.J. and Sorensen, A.J. (1978) Dispersal and recolonization by *Myzus persicae* following aphid alarm pheromone exposure. *Annals of the Entomological Society of America* 71, 181–182.

Cao, J., Shelton, A.M. and Earle, E.D. (2008) Sequential transformation to pyramid two Bt genes in vegetable Indian mustard (*Brassica juncea* L.) and its potential for control of diamond back moth larvae. *Plant Cell Repair* 27, 479–487.

Capinera, J.L. (2005) Common name: green peach aphid; scientific name: *Myzus persicae* (Sulzer) (Insecta: Hemiptera: Aphididae). University of Florida, Gainesville, Florida. Available at: http://entnemdept.ufl.edu/creatures/veg/aphid/green_peach_aphid.htm.

Charleston, D.S. and Kfir, R. (2000) The possibility of using Indian mustard, *Brassica juncea* as a trap crop for the diamondback moth, *Plutella xylostella,* in South Africa. *Crop Protection* 19, 455–460.

Chen, F.J., Wu, G., Ge, F., Parajulee, M.N. and Shrestha, R.B. (2005) Effects of elevated CO_2 and transgenic Bt cotton on plant chemistry, performance, and feeding of an insect herbivore, the cotton bollworm. *Entomologia Experimentalis et Applicata* 115, 341–350.

Chowdhury, M. (2009) Incidence of saw fly, *Athalia lugens proxima* Klug. as influenced by level of irrigation and fertilizers on mustard. *The Journal of Plant Protection Sciences* 1, 80–82.

CIBRC (2015) *Major Uses of Pesticides*. Central Insecticides Board and Registration Committee, Faridabad, India. Available at: http://www.cibrc.nic.in/mup.htm (accessed 15 December, 2015).

Cook, S.M., Khan, Z.R. and Pickett, J.A. (2007) The use of 'push–pull' strategies in integrated pest management. *Annual Review of Entomology* 52, 375–400.

Cordero, R.J. and Kuhar, T. (2009) *Diamondback moth in Virginia*. Virginia Cooperative Extension Publication No. 444-007. Virginia Polytechnic Institute and State University, Blacksburg, Virginia.

Cornelissen, T. (2011) Climate change and its effects on terrestrial insects and herbivory pattern. *Neotropical Entomology* 40, 155–163.

Coviella, C.E., Stipanovic, R.D. and Trumble, J.T. (2002) Plant allocation to defensive compounds: interactions between elevated CO_2 and nitrogen in transgenic cotton plants. *Journal of Experimental Botany* 53, 323–331.

Cui, L.L., Francis, F., Heuskin, S., Lognay, G., Liu, Y.J., Dong, J., Chen, J.L., Song, X.M. and Liu, Y. (2012) The functional significance of (E)-β-farnesene: does it influence the populations of aphid natural enemies in the fields? *Biological Control* 60, 108–112.

Dawson, G.W., Griffiths, D.C., Pickett, J.A., Smith, M.C. and Woodcock, C.M. (1984) Natural inhibition of the aphid alarm pheromone. *Entomologia Experimentalis et Applicata* 36, 197–199.

Day, E. (2014) *Aphids*. Virginia Cooperative Extension Publication No. ENT-82NP. Virginia Polytechnic Institute and State University, Blacksburg, Virginia.

Dean, J.M. and De Moreas, C.M. (2006) Effect of genetic modification on herbivore-induced volatiles from maize. *Journal of Chemical Ecology* 32, 713–724.

De Moraes, C.M., Lewis, W.J., Pare, P.W., Alborn, H.T. and Tumlinson, J.H. (1998) Herbivore-infested plants selectively attract parasitoids. *Nature* 393, 570–573.

de Vos, M., Cheng, W.Y., Summers, H.E., Raguso, R.A. and Jander, G. (2010) Alarm pheromone habituation in *Myzus persicae* has fitness consequences and causes extensive gene expression changes. *Proceedings of the National Academy Sciences of the USA* 107, 14673–14678.

Dicke, M. (1999) Are herbivore-induced plant volatiles reliable indicators of herbivore identity to foraging carnivorous arthropods? *Entomologia Experimentalis et Applicata* 91, 131–142.

Dicke, M. and Hilker, M. (2003) Induced plant defences: from molecular biology to evolutionary ecology. *Basic and Applied Ecology* 4, 3–14.

Dosdall, L.M., Herbut, M.J. and Cowle, N.T. (1994) Susceptibilities of species and cultivars of canola and mustard to infestation by root maggots (*Delia* spp.) (Diptera: Anthomyiidae). *The Canadian Entomologist* 126, 251–260.

Dosdall, L.M., Soroka, J.J. and Olfert, O. (2011) The diamondback moth in canola and mustard: current pest status and future prospects. *Prairie Soils and Crops Journal* 4, 66–76.

Facknath, S. (1996) Application of neem extract and intercropping for the control of some cabbage pests in Mauritius. *Proceedings of International Neem Conference, Queensland, Australia*, Feb, 1996.

Facknath, S. (1997) Integrated pest management of *Plutella xylostella* an important pest of crucifers in Mauritius. In: Lalouette, J.A., Bachraz, D.Y., Sukurdeep, N. and Seebaluck, B.D. (eds) *Proceedings of the Second Annual Meeting of Agricultural Scientists* (AMAS). Food and Agricultural Research Council, Reduit, Mauritius, pp. 103–108.

Fahey, J.W., Zalcmann, A.T. and Talalay, P. (2001) The chemical diversity and distribution of glucosinolates and isothiocyanates among plants. *Phytochemistry* 56, 5–51.

FAO (2006) *Farmer Field School (FFS) Manual*. Guidelines for implementation of a farmer field school (FFS). Special Framework of Assistance (SFA), St Lucia. Food and Agriculture Organization of the United Nations, Rome. Available at: http://www.fao.org/3/a-ap094e.pdf (accessed 1 August, 2015).

FAO (2011) *Farmer Field School Implementation Guide*. Food and Agriculture Organization of the United Nations, Rome. Available at: http://www.fao.org/3/a-i2561e.pdf (accessed 11 February, 2016).

Fatouros, N.E., Bukovinszkine'Kiss, G., Kalkers, L.A., Gamborena, R.S., Dicke, M. and Hilker, M. (2005) Oviposition-induced plant cues: do they arrest *Trichogramma* wasps during host location. *Entomologia Experimentalis et Applicata* 115, 207–215.

Feakin, S.D. (1973) *Pest Control in Groundnuts. PANS Manual No. 2*. Centre for Overseas Pest Research, London.

Firake, D.M., Behere, G.T. and Thakur, N.S.A. (2013) Ecofriendly management of mustard aphid, *Lipaphis erysimi*: a most destructive pest of brassicaceous crops. *Popular Kheti* 1, 31–33.

Gavloski, J., Cárcamo, H. and Dosdall, L. (2011) Insects of canola, mustard, and flax in Canadian grasslands. In: Floate, K.D. (ed.) *Arthropods of Canadian Grasslands: Inhabitants of a Changing Landscape* (Vol. 2). Biological Survey of Canada, Entomological Society of Canada, Ottawa, Canada, pp. 181–214.

George, D.R., Collier, R. and Port, G. (2009) Testing and improving the effectiveness of trap crops for diamondback moth *Plutella xyslotella* (L.): a laboratory-based study. *Pest Management Science* 65, 1219–1227.

Gibson, R.W. and Pickett, J.A. (1983) Wild potato repels aphids by release of aphid alarm pheromone. *Nature* 302, 608–609.

Gouinguene, S.P. and Turlings T.C.J. (2002) The effects of abiotic factors on induced volatile emissions in corn plants. *Plant Physiology* 129, 1296–1307.

Guerene, M. (2006) *Cole Crops and Other Brassicas: Organic Production*. ATTRA Sustainable Agriculture, National Center for Appropriate Technology, NCAT, Butte, Montana. Available at: http://attra.ncat.org.

Gurr, G.M., Wratten, S.D. and Altieri, M.A. (2004a) *Ecological Engineering: Advances in Habitat Manipulation for Arthropods*. CSIRO Publishers, Melbourne, Australia.

Gurr, G.M., Wratten, S.D. and Altieri, M.A. (2004b) Ecological engineering: a new direction for agricultural pest management. *AFBM Journal* 1, 28–35.

Harmel, N., Almohamad, R., Fauconnier, M.L., Du Jardin, P., Verheggen, F., Marlier, M., Haubruge, E. and Francis, F. (2007) Role of terpenes from aphid-infested potato on searching and oviposition behavior of *Episyrphus balteatus*. *Insect Science* 14, 57–63.

Harrison, L. (2013) *Carinata: a crop with potential for Florida's farmers*. University of Florida, IFAS Extension, Gainesville, Florida. Available at: http://nwdistrict.ifas.ufl.edu/phag/2013/ 08/16/carinata-a-crop-with-potential-for-floridas-farmers/ (accessed 12 February 2016).

Hasan, F. and Ansari, M.S. (2010) Effect of different cole crops on the biological parameters of *Pieris brassicae* (L.) (Lepidoptera: Pieridae) under laboratory conditions. *Journal of Crop Science and Biotechnology* 13, 195–202.

Hasan, F. and Ansari, M.S. (2011) Population growth of *Pieris brassicae* (L.) (Lepidoptera: Pieridae) on different cole crops under laboratory conditions. *Journal of Pest Science* 84, 179–186.

Hatano, E., Kunert, G., Michaud, J.P. and Weisser, W.W. (2008) Chemical dues mediating aphid location by natural enemies. *European Journal of Entomology* 105, 797–806.

Himanen, S.J., Nissinen, A., Dong, W-X., Nerg, A.-M., Stewart, C.N. Jr, Poppy, G.M. and Holopainen, J.K. (2008) Interactions of elevated carbon dioxide and temperature with aphid feeding on transgenic oilseed rape: are *Bacillus thuringiensis* (Bt) plants more susceptible to nontarget herbivores in future climate? *Global Change Biology* 14, 1437–1454.

Himanen, S.J., Nerg, A.-M., Nissinen, A., Stewart, C.N. Jr, Poppy, G.M. and Holopainen, J.K. (2009a) Elevated atmospheric ozone increases concentration of insecticidal *Bacillus thuringiensis* (Bt) Cry1Ac protein in Bt *Brassica napus* and reduces feeding of a Bt target herbivore on the non-transgenic parent. *Environmental Pollution* 157, 181–185.

Himanen, S.J., Nerg, A-M., Nissinen, A., Pinto, D.M., Stewart, C.N. Jr, Poppy, G.M. and Holopainen, J.K. (2009b) Effects of elevated carbon dioxide and ozone on volatile terpenoid emissions and multitrophic communication of transgenic insecticidal oilseed rape (*Brassica napus*). *New Phytologist* 181, 174–186.

Himanen, S.J., Nerg, A.-M. and Holopainen, J.K. (2009c) Degree of herbivore feeding damage as an important contributor to multitrophic plant-parasitoid signaling under climate change. *Plant Signaling and Behavior* 4, 249–251.

Himanen, S.J., Bui, T.N.T., Maja, M.M. and Holopainen, J.K. (2015) Utilizing associational resistance for biocontrol: impacted by temperature and aided by indirect defence. *BMC Ecology* 15, 1–12.

Hollingsworth, C. (2015) Pests of cabbage and mustard grown for seed. In: Hollingsworth, C.S. (ed.) *Pacific NorthWest Insect Management Handbook*. Oregon State University, Corvallis, Oregon, p. L1.

Holopainen, J.K. (2004) Multiple functions of inducible plant volatiles. *Trends in Plant Science* 11, 529–533.

Huang, X. and Renwick, J.A.A. (1993) Differential selection of host plants by two *Pieris* species: the role of oviposition stimulants and deterrents. *Entomologia Experimentalis et Applicata* 68, 59–69.

Huang, B., Shi, Z. and Hou, Y. (2014) Host selection behaviour and fecundity of *Plutella xylostella* (Lepidoptera: Plutellidae) on multiple host plants. *Journal of Insect Science* 14, 251. doi: 10.1093/jisesa/ieu113.

Hughes, P.R., Renwick, J.A.A. and Lopez, K.D. (1997) New oviposition stimulants for the diamondback moth in cabbage. *Entomologia Experimentalis et Applicata* 85, 281–283.

Ibrahim, M.A., Stewart-Jones, A., Pulkkinen, J., Poppy, G.M. and Holopainen, J.K. (2008) The influence of different nutrient levels on insect-induced plant volatiles in Bt and conventional oilseed rape plants. *Plant Biology* 10, 97–107.

James, C. (2007) *Executive summary of global status of commercialized biotech/GM crops*. Brief No. 37, ISAAA, Ithaca, NY. Available at: https://www.isaaa.org/resources/publications/briefs/37/executivesummary/default.html (accessed on 28 March, 2016).

Jellis, G. (2003) *Pest Management in Cereals and Oilseed Rape – a Guide*. Home-Grown Cereals Authority (HGCA), London.

Joachim, C. and Weisser, W.W. (2015) Does the aphid alarm pheromone (*E*)-β-farnesene act as a kairomone under field conditions? *Journal of Chemical Ecology* 41, 267–275.

Joachim, C., Hatano, E., David, A., Kunert, M., Linse, C. and Weisser, W.W. (2013) Modulation of aphid alarm pheromone emission of pea aphid prey by predators. *Journal of Chemical Ecology* 39, 773–782.

Kanrar, S., Venkateswari, J., Kirti, P.B. and Chopra, V.L. (2002) Transgenic Indian mustard (*Brassica juncea*) with resistance to the mustard aphid (*Lypaphis erysimi* Kalt.). *Plant Cell Repair* 20, 976–981.

Karban, R. and Baldwin, I.T. (1997) *Induced Responses to Herbivory*. Chicago University Press, Chicago, Illinois.

Kessing, J.L.M. and Mau, R.F.L. (1991) *Brevicoryne brassicae* (Linnaeus). Crop Knowledge Master, Department of Entomology, Honolulu, Hawaii. Available at: http://www.extento.hawaii.edu/kbase/crop/type/brevicor.htm (accessed 12 February, 2015).

Khalig, E., Papazian, S., Poelman, E.H., Holopainen, J.K., Albrectsen, B.R. and Bland, J.D. (2015) Ozone affects growth and development of *Pieris brassicae* on the wild host plant, *Brassica nigra*. *Environmental Pollution* 199, 119–129.

Khan, Z.R. and Pickett, J.A. (2010) Isoschaftoside, a C-glycosylflavonoid from *Desmodium uncinatum* root exudates, is an allelochemical against the development of *Striga*. *Phytochemistry* 71, 904–908.

Khan, Z.R., Midega, C.A.O., Pittchar, J.O., Murage, A.W., Birkett, M.A. and Bruce, T.J.A. (2014) Achieving food security for one million sub-Saharan African poor through push–pull innovation by 2020. *Philosophical Transactions of the Royal Society B: Biological Sciences* 369, 20120284 (1–11).

Kienegger, M., Finch, S., Finch, S. and Brunel, E. (1996) The effect of undersowing with clover on host-plant selection by pest insects of brassica crops. In: *Proceedings of Integrated Control in Field Vegetable Crops*, Guitte, France, 1995. *Bulletin OILB-SROP* 19, 108–114.

Klingen, I., Meadow, R. and Aandal, T. (2002) Morality of *Delia floralis*, *Galleria mellonella*, *Mamestra brassicae* treated with insect pathogenic hypomecetous fungi. *Journal of Applied Entomology* 126, 231–237.

Kleiber, J., Dorn, S. and Nazar-Rodriguez, A.J. (2013) Acclimation to elevated CO_2 increases constitutive glucosinolate levels of *Brassica* plants and affects the performance of specialized herbivores from contrasting feeding guilds. *Journal of Chemical Ecology* 39, 653–665.

Knipling, E.F. (1969) The potential role of sterility for pest control. In: Labrecque, G.C. and Smith, C.N. (eds) *The Principles of Insect Chemosterilization*. Appleton Century Crofts, New York.

Kolte, S.J. (1985) *Diseases of annual edible oilseed crops, Vol. 2: Rapeseed-Mustard and Sesame Diseases*. CRC Press, Boca Raton, Florida.

Koramutla, M.K., Kaur, A., Negi, M., Venkatachalam, P. and Bhattacharya, R. (2014) Elicitation of jasmonate-mediated host defense in *Brassica juncea* (L.) attenuates population growth of mustard aphid *Lipaphis erysimi* (Kalt.). *Planta* 240, 177–194.

Koristas, V.M., Lewis, J.A. and Fenwick, G.R. (1991) Glucosinolate responses of oilseed rape, mustard and kale to mechanical wounding and infestation by cabbage stem flea beetle (*Psylliodes chrysocephala*). *Annals of Applied Biology* 118, 209–221.

Korlapati, S., Jeyakumar, P., Sain, S.K. and Boina, D. (2014a) On-farm production of biocontrol agents and microbial biopesticides to promote AESA based plant health management in conjunction with ecological engineering for pest management. *Plant Health News Letter* 4, 2–7.

Korlapati, S., Sushil, S.N., Jeyakumar, P., Shankar, G., Sharma, O.P., Sain, S.K., Boina, D.R. and Chattopadhyaya, S.D. (2014b) *AESA Based IPM Package No. 11: AESA based IPM –Mustard/Rapeseed.* Department of Agriculture and Cooperation, Ministry of Agriculture, Government of India, New Delhi, India, p. 49. Available at: http://farmer.gov.in/imagedefault/ipm/Mustard.pdf (accessed 2 December, 2015).

Kumar, A. (2015) Integrated pest management in rapeseed-mustard. Available at: http://www.krishisewa.com/cms/articles/disease-management/133-ipm-mustard.html (accessed 1 August, 2015).

Kumaranag, K.M., Kedar, S.C., Thodsare, N.H. and Bawaskar, D.M. (2014) Insect pests of cruciferous vegetables and their management. *Popular Kheti* 2, 80–86.

Kunert, G., Otto, S., Rose, U.S.R., Gershenzon, J. and Weisser, W.W. (2005) Alarm pheromone mediates production of winged dispersal morphs of aphids. *Ecology Letters* 8, 596–603.

Kunert, G., Schmoock-Ortlepp, K., Reissmann, U., Creutzburg, S. and Weisser, W.W. (2008) The influence of natural enemies on wing induction in *Aphis fabae* and *Megoura viciae* (Hemiptera: Aphididae). *Bulletin of Entomological Research* 98, 57–62.

Kunert, G., Reinhold, C. and Gershenzon, J. (2010) Constitutive emission of aphid alarm pheromone, (E)-β-farnese, from plants does not serve as a direct defense against aphids. *BMC Ecology* 10, 1–22.

Landosky, J.M. and Karowe, D.N. (2014) Will chemical defenses become more effective against specialist herbivores under elevated CO_2? *Global Change Biology* 20, 3159–3176.

Lerdau, M.T., Litvak, M. and Monson, R. (1994) Plant chemical defense: monoterpenes and the growth-differentiation balance hypothesis. *Trends in Ecology and Evolution* 9, 58–61.

Lincoln, D.E., Couvet, D. and Sionit, N. (1986) Response of an insect herbivore to host plants grown in carbon dioxide enriched atmospheres. *Oecologia* 69, 556–560.

Liu, S-S., Yue-Hong, L. and Yong-Gen, L. (2006) Non-host plant extract reduce oviposition of *Plutella xylostella* (Lepidoptera: Plutellidae) and enhance parasitism of its parasitoid *Cotesia plutellae* (Hymenoptera: Braconidae). *Bulletin of Entomological Research* 96, 373.

Lu, J.H., Liu, S.S. and Shelton, A.M. (2004) Laboratory evaluation of a wild crucifer *Barbarea vulgaris* as a management tool for the diamondback moth, *Plutella xylostella* (Lepidoptera: Plutellidae). *Bulletin of Entomological Research* 94, 509–516.

Luther, G.C., Valenzuela, H.R. and DeFrank, J. (1996) Impact of cruciferous trap crops on lepidopteran pests of cabbage in Hawaii. *Journal of Economic Entomology* 25, 39–47.

Mattiacci, L., Rudelli, S., Rocca, B.A., Genini, S. and Dorn, S. (2001) Systemically-induced response of cabbage plants against a specialist herbivore, *Pieris brassicae*. *Chemoecology* 11, 167–173.

Meena, H., Singh, S.P. and Nagarm, R. (2013) Evaluatioin of microbial agents and bio-products for the management of mustard aphid, *Lipaphis erysimi* (Kalt.). *The Bioscan* 8, 747–750.

Micha, S.G. and Wyss, U. (1996) Aphid alarm pheromone (E)-beta-farnesene: a host finding kairomone for the aphid primary parasitoid *Aphidius uzbekistanicus* (Hymenoptera: Aphidiinae). *Chemoecology* 7, 132–139.

Mithen, R. (1992) Leaf glucosinolate profiles and their relationship to pest and disease resistance in oilseed rape. *Euphytica* 63, 71–83.

Montgomery, M.E. and Nault, L.R. (1977) Comparative response of aphids to the alarm pheromone, (*E*)-β-farnesene. *Entomologia Experimentalis et Applicata* 22, 236–242.

Mostafavi, R., Henning, J.A., Gardea-Torresday, J. and Ray, I.M. (1996) Variation in aphid alarm pheromone content among glandular and eglandular-haired *Medicago* accessions. *Journal of Chemical Ecology* 22, 1629–1638.

Muniappan, R., Cruz, J. and Bamba, J. (2001) Trap crops for diamondback moth and other crucifer pests in Guam. In: *Proceedings of the Fourth International Workshop, Melbourne, Victoria, Australia, 26–29 November 2001*. The Regional Institute, Gosford, Australia, pp. 141–146. Available at: http://www.regional.org.au/au/esa/2001/06/0603muniappan.htm (accessed 2 August, 2015).

Nakahara, L.M., McHugh, J., Otsuka, C.K., Funusaki, G.Y. and Lai, P.Y. (1986) Integrated control of diamondback moth and other insect pests using an overhead sprinkler system, an insecticide, and biological control agents, on a watercress farm in Hawaii. In: Talekar, N.S. and Griggs, T.D. (eds) *Proceedings of the First International Workshop on Diamondback Moth Management*. Asian Vegetable Research and Development Center (AVRDC), Shanhua, Tainan, Taiwan, AVRDC, pp. 403–413.

Narayanan, K. and Gopalakrishnan, C. (2003) Evaluation of entomopathogenic nematode, *Steinernema feltiae* against field population of mustard saw fly, *Athalia lugens proxima* Klug. on radish. *Indian Journal of Experimental Biology* 41, 376–378.

Ohbayashi, N., Simizu, K. and Nagata K. (1992) Control of diamondback moth by using synthetic sex pheromone. In: Talekar, N.S. (ed.) *Diamondback Moth and Other Crucifer Pests: Proceedings of the Second International Workshop*. Asian Vegetable Research and Development Center (AVRDC), Shanhua, Tainan, Taiwan, pp. 99–104.

Ohno, T., Asayama, T. and Ichikawa, K. (1992) Evaluation of communication disruption method for suppressing diamondback moth infestations at two different areas using synthetic sex pheromone. In: Talekar, N.S. (ed.) *Diamondback Moth and Other Crucifer Pests:*

Proceedings of the Second International Workshop. Asian Vegetable Research and Development Center (AVRDC), Shanhua, Tainan, Taiwan, pp. 109–114.

Pandey, R. and Singh, N.N. (2008) Effectiveness of biocontrol based IPM modules against *Lipaphis erysimi* Kaltenbach (Hemiptera: Aphididae). *Journal of Plant Protection Research* 48, 1–7.

Pare, P.W. and Tumlinson, J.H. (1999) Plant volatiles as a defense against insect herbivores. *Plant Physiology* 121, 325–332.

Paul, A.V.N. (2007) *Insect Pests and their Management*. Indian Agricultural Research Institute, New Delhi, India.

Pavela, R. and Kazda, J. (2007) Influence of application term on effectiveness of some insecticides against brassica pod midge (*Dasineura brassicae* Winn.). *Plant Protection Science-UZPI* 43, 1212–2580.

Peñuelas, J. and Staudt, M. (2010) BVOCs and global change. *Trends in Plant Science* 15, 133–144.

Pickett, J.A. and Griffiths, D.C. (1980) Composition of aphid alarm pheromones. *Journal of Chemical Ecology* 6, 349–360.

Pujari, S. (2015) Leaf webber (*Crocidolomia binotalis*): distribution, life cycle and control. Available at: http://www.yourarticlelibrary.com/zoology/leaf-webber-crocidolomia-binotalis-distribution-life-cycle-and-control/23977/ (accessed 2 August, 2015).

Pyke, B., Rice, M.J., Sabine, B. and Zalucki, M. (1987) The push–pull strategy – behavioral control of *Heliothis*. *Australian Cotton Grower* May–July 1987, 7–9.

Rao, B.B., Rao, V.U.M., Nair, L., Prasad, Y.G., Ramaraj, A.P. and Chattopadhyay, C. (2014) Mustard aphid infestation in India: development of forewarning models. *Journal of Environmental Biology* 35, 683–688.

Reddy, G.V.P., Tossavainen, P., Nerg, A.M. and Holopainen, J.K. (2004) Elevated atmospheric CO_2 affects the chemical quality of Brassica plants and growth rate of the specialist, *Plutella xylostella*, but not the generalist, *Spodoptera littoralis*. *Journal of Agriculture and Food Chemistry* 52, 4185–4191.

Reed, D.A., Perring, T.M., Newman, J.P., Bethke, J.A. and Kabashima, J.N. (2014) *Bagrada bug, integrated pest management for home gardeners and landscape professionals*. Pest notes publication # 74116. UC Statewide Integrated Pest Management Program, University of California, Davis, California.

Reiter, S., Kuhar, T. and Doughty, H. (2008) *Cabbage webworm*. Virginia Cooperative Extension Publication No. 2811-1022. Virginia Polytechnic Institute and State University, Blacksburg, Virginia.

Renwick, J.A.A. and Radke, C.D. (1985) Constituents of host and non-host plants deterring oviposition by the cabbage butterfly, *Pieris rapae*. *Entomologia Experimentalis et Applicata* 39, 21–26.

Renwick, J.A.A., Radke, C.D., Sachdev-Gupta, K. and Stadler, E. (1992) Leaf surface chemicals stimulating oviposition by *Pieris rapae* (Lepidoptera: Pieridae) on cabbage. *Chemoecology* 3, 33–38.

Renwick, J.A.A., Haribal, M., Gouinguene, S. and Stadler, E. (2006) Isothiocyanates stimulating oviposition by the diamondback moth, *Plutella xylostella*. *Journal of Chemical Ecology* 32, 755–766.

Romeis, J., Meissle, M. and Bigler, F. (2006) Transgenic crops expressing *Bacillus thuringiensis* toxins and biological control. *Nature Biotechnology* 24, 63–71.

Schoonhoven, L.M. (1990) Host-marking pheromones in Lepidoptera, with special reference to two *Pieris* spp. *Journal of Chemical Ecology* 16, 3043–3052.

Schular, T.H., Potting, R.P.J., Denholm, I. and Poppy, G.M. (1999a) Parasitoid behavior and Bt plants. *Nature* 400, 825–826.

Schular, T.H., Potting, R.P.J., Denholm, I. and Poppy, G.M. (1999b) Parasitoid behavior and Bt plants. *Nature* 402, 750.

Schular, T.H., Poppy, G.M., Kerry, B.R. and Denholm, I. (1999c) Potential side effects of insect resistant transgenic plants on arthropod natural enemies. *Trends in Biotechnology* 17, 210–216.

Schular, T.H., Potting, R.P., Denholm, I., Clark, S.J., Clark, A.J. *et al.* (2003) Tritrophic choice experiments with the Bt plants, the diamondback moth (*Plutella xylostella*), the parasitoid *Cotesia plutellae*. *Transgenic Research* 12, 351–361.

Schular, T.H., Denholm, I., Clark, S.J., Stewart, C.N. and Poppy, G.M. (2004) Effects of Bt plants on the development and survival of the parasitoid *Cotesia plutellae* (Hymenoptera: Braconidae) in susceptible and Bt-resistant larvae of the diamondback moth, *Plutella xylostella* (Lepidoptera: Plutellidae). *Journal of Insect Physiology* 50, 435–453.

Schwartzberg, E.G., Kunert, G., Stephan, C., Biedermann, A., Rose, U.S.R., Gershenzon, J., Boland, W. and Weisser, W.W. (2008) Real-time analysis of alarm pheromone emission by the pea aphid (*Acyrthosiphon pisum*) under predation. *Journal of Chemical Ecology* 34, 76–81.

Shekhawat, K., Rathore, S.S., Premi, O.P., Kandpal, B.K. and Chauhan, J.S. (2012) Advances in agronomic management of Indian mustard (*Brassica juncea* (L.) Czernj. Cosson): an overview. *International Journal of Agronomy*, 2012. doi: 10.1155/2012/408284.

Shelton, A.M., Hatch, S.L., Zhao, J.Z., Chen, M. Earle, E.D. et al. (2008) Suppression of diamondback moth using Bt-transgenic plants as a trap crop. *Crop Protection* 27, 403–409.

Shepard, B.M. and Schellhorn, N.A. (1996) A *Plutella/Crocidolomia* management program for cabbage in Indonesia. In: *Proceedings of the Third International Symposium on Management of Diamondback Moth and Other Crucifer Pests*, Kuala Lumpur, Malaysia, 29 October – 1 November, 1996, pp. 262–266. Available at: http://

web.entomology.cornell.edu/shelton/diamondback-moth/pdf/1996papers/1996DBM45.pdf (accessed 12 February, 2016).

Shylesha, A.N., Azad, T.N.S., Pathak, K.A., Rao, K.R. Saikia, K., Suroshe, S., Kodandaram, M.H. and Kalaisekhar, A. (2006) *Integrated management of insect pest of crops in north eastern hill region. Technical Bulletin No. 19.* Division of Entomology, ICAR RC for NEH Region, Umiam, Meghalaya, India.

Silvia-Krott, I.U., Singh, P., Lali, T.S. and Muniappan, R. (1995) Development of a trap cropping system for cabbage in Guam. *Pest Management in Horticultural Ecosystems* 1, 28–35.

Singh, S.P. (2009) Insect pest management in oilseed crops. *Indian Farming* 58, 29–33.

Singh, K. (2013) Preying propensity of larvae/grubs of syrphid and coccinellid predators on mustard aphid, *Lipaphis erysimi* (Kalt.). *International Journal of Agriculture and Food Science Technology* 7, 687–694.

Singh, P.K. and Premchand (1995) Yield loss due to mustard aphid, *Lipaphis erysimi* (Kalt.) in Eastern Bihar Plateau. *Journal of Applied Zoological Research* 6, 97–100.

Singh, C.P. and Sachan, G.C. (1999) Ecofriendly management of *Lipaphis erysimi* Kalt. in *Brassica carinata*. *The Proceedings of the 10th international Rapeseed Conference*, Canberra.

Singh, Y.P. and Sharma, K.C. (2002) Integrated approach to manage the mustard aphid, *Lipaphis erysimi* (Kalt.) (Homoptera: Aphididae) in oil seed Brassica crops – a review. *Journal of Aphidology* 16, 77–88.

Sivapragasam, A., Tees, S.P. and Ruwaida, M. (1982) Effects of intercropping cabbage with tomato on the incidence of *Plutella xylostella*. *MAPS News Letter* 6, 6–7.

Sohal, S.K., Rup, P.J., Kaur, H., Kumari, N. and Kaur, J. (2002) Evaluation of the pesticidal potential of the congress grass, *Parthenium hysterophorus* L. on the mustard aphid, *Lipaphis erysimi* (Kalt.). *Journal of Environmental Biology* 23, 15–18.

Soroka, J.J., Bartelt, R.J., Zilkowski, B.W. and Cosse, A.A. (2005) Responses of flea beetle, *Phyllotreta cruciferae*, to synthetic aggregation pheromone components and host plant volatiles in field trials. *Journal of Chemical Ecology* 31, 1829–1843.

Srinivasan, K. and Moorthy, P.N.K. (1991) Indian mustard as a trap crop for management of major lepidopterous pests on cabbage. *Tropical Pest Management* 37, 26–32.

Srinivasan, K. and Moorthy, P.N.K. (1992) Development and adoption of integrated pest management for major pests of cabbage using Indian mustard as a trap crop. In: Talekar, N.S. (ed.) *Diamondback Moth and Other Crucifer Pests: Proceedings of the 2nd International Workshop*, 10–14 December 1990. Asian Vegetable Research and Development Center, Taipei, Taiwan, pp. 511–521.

Stiling, P. and Cornelissen, T. (2007) How does elevated carbon dioxide (CO_2) affect plant–herbivore interactions? A field experiment and meta-analysis of CO_2-mediated changes on plant chemistry and herbivore performance. *Global Change Biology* 13, 1823–1842.

Stiling, P., Rossi, A.M., Hungate, B., Dijkstra, P., Hinkle, C.R., Knott, W.M. and Drake, B. (1999) Decreased leaf minor abundance in elevated CO_2: reduced leaf quality and increased parasitoid attack. *Ecology and Application* 9, 240–244.

Sutrinso, S. (2005) Integrated control of cabbage pests *Plutella xylostella* (L.) and *Crocidolomia binotalis* (Z.) by release of irradiated moths and the parasitoid *Diadegma semiclausum* (H.) under field cage and a small area conditions. *A Scientific Journal of Applications of Isotopes and Radiation* 1, 29–37.

Tabashnik, B.E. (1987) Plant secondary compounds as oviposition deterrents for cabbage butterfly, *Pieris rapae* (Lepidoptera: Pieridae). *Journal of Chemical Ecology* 13, 309–316.

Takabayashi, J. and Dicke, M. (1996) Plant–carnivore mutualism through herbivore-induced carnivore attractants. *Trends in Plant Science* 1, 109–113.

Takabayashi, J., Dicke, M. and Posthumus, M.A. (1994) Volatile herbivore-induced terpenoids in plant-mite interactions: variation caused by biotic and abiotic factors. *Journal of Chemical Ecology* 20, 1329–1354.

Talekar, N.S. and Shelton, A.M. (1993) Biology, ecology, and management of the diamondback moth. *Annual Review of Entomology* 38, 275–301.

Talekar, N.S., Lee, S.T. and Huang, S.W. (1986) Intercropping and modification of irrigation method for the control of diamondback moth. In: Talekar, N.S. and Griggs, T.D. (eds) *Proceedings of the First International Workshop on Diamondback moth Management.* Asian Vegetable Research and Development Center (AVRDC), Shanhua, Tainan, AVRDC, Taiwan, pp. 145–152.

Theunissen, J., Schelling, G., Finch, S. and Brunel, E. (1996) Undersowing crops of white cabbage with strawberry clover and spurrey. *Proceedings of Integrated Control in Field Vegetable Crops*, Guitte, France, 1995. *Bulletin OILB-SROP* 19, 128–135.

Traynier, R.M.M. and Truscott, R.J.W. (1991) Potent natural egg-laying stimulant for cabbage butterfly *Pieris rapae*. *Journal of Chemical Ecology* 17, 1371–1380.

Tripathi, M.K. and Mishra, A.S. (2007) Glucosinolates in animal nutrition. A review. *Animal Feed Science and Technology* 132, 1–27.

Turlings, T.C.J., Lengwiler, U.B., Bernasconi, M.L. and Wechsler, D. (1998) Timing of induced volatile emissions in maize seedlings. *Planta* 207, 146–152.

Turlings, T.C.J., Jeanbourquin, P.M., Held, M. and Degen, T. (2005) Evaluating the induced-odor emission of a Bt maize and its attractiveness to parasitic wasps. *Transgenic Research* 14, 807–816.

Ulland, S., Ian, E., Mozuraitis, R., Borg-Karlson, A.K., Meadow, R. *et al.*, (2008) Methyl salicylate, identified as primary odorant of a specific receptor neuron type, inhibits oviposition by the moth *Mamestra brassicae* L. (Lepidoptera: Noctuidae). *Chemical Senses* 33, 35–46.

van Loon, J.A.A., Blaakmeer, A., Griepink, F.C., Van Beek, T.A., Schoonhoven, L.M. and De Groot, A. (1992) Leaf surface compound from *Brassica oleraceae* (Cruciferae) induces oviposition by *Pieris brassicae* (Lepidoptera: Pieridae). *Chemoecology* 3, 39–44.

Verma, S.S., Sinha, R.K. and Jajoo, A. (2015) (E)-β-farnesene gene reduces *Lipaphis erysimi* colonization in transgenic *Brassica juncea* lines. *Plant Signaling and Behavior* 10, 1–4.

Vuorinen, T., Reddy, G.V.P., Nerg, A-M. and Holopainen, J.K. (2004a) Monoterpene and herbivore-induced emissions from cabbage plants grown at elevated atmospheric CO_2 concentration. *Atmosphere and Environment* 38, 675–682.

Vuorinen, T., Nerg, A-M., Ibrahim, M.A., Reddy, G.V.P. and Holopainen, J.K. (2004b) Emission of *Plutella xylostella*-induced compounds from cabbages grown at elevated CO_2 and orientation behavior of the natural enemies. *Plant Physiology* 135, 1984–1992.

Webb, S.E. (2013) *Insect management for crucifers (cole crops) (broccoli, cabbage, cauliflower, collards, kale, mustard, radishes, turnips).* IFAS Extension Publication No. ENY-464. University of Florida, Gainesville, Florida.

Wu, G., Chen, F-J., Ge, F. and Sun, Y-C. (2007) Effects of elevated carbon dioxide on the growth and foliar chemistry of transgenic Bt cotton. *Journal of Integrative Plant Biology* 49, 1361–1369.

Yadav, A.K. and Lalramliana (2012) Evaluation of the efficacy of three indigenous strains of entomopathogenic nematodes from Meghalaya, India against mustard sawfly, *Athalia lugens proxima* Klug (Hymenoptera: Tenthredinidae). *Journal of Parasites and Diseases* 36, 175–180.

Yadav, S. and Singh, S.P. (2015) Biointensive integrated management strategy for mustard aphid *Lipaphis erysimi* Kalt. (Homoptera: Aphididae). *Journal of Applied and Natural Science* 7, 192–196.

Yan, F., Bengtsson, M., Anderson, P., Ansebo, L., Xu, C. and Witzgall, P. (2004) Antennal response of cotton bollworm (*Helicoverpa armigera*) to volatiles in transgenic Bt cotton. *Journal of Applied Entomology* 128, 354–357.

Yu, X., Zhang, Y., Ma, Y., Xu, Z., Wang, G. and Xia, L. (2013) Expression of an (*E*)-β-farnesene synthase gene from Asian peppermint in tobacco affected aphid infestation. *The Crop Journal* 1, 50–60.

Yuan, J.S., Himanen, S.J., Holopainen, J.K., Chen, F. and Stewart, C.N. (2009) Smelling global climate change: mitigation of function for plant volatile organic compounds. *Trends in Ecology and Evolution* 24, 323–331.

Ziaee, M. (2012) Oilseed pests. In: Akpan, U.G. (ed.) *Oilseeds*. InTech Publications, Rijeka, Croatia, pp. 117–184. Available at: http://www.intechopen.com/books/oilseeds/oilseed-pests (accessed 2 August, 2015).

Zukalova, H. and Vasak, J. (2002) The role and effects of glucosinolates of *Brassica* species – a review. *Rostlinna Vyroba* 48, 175–180.

Zvereva, E.L. and Kozlov, M.V. (2006) Consequences of simultaneous elevation of carbon dioxide and temperature for plant–herbivore interactions: a meta-analysis. *Global Change Biology* 12, 27–41.

18 Volatile Organic Compounds in Integrated Pest Management of *Brassica* Oilseed Crops

Sari J. Himanen[1], Tao Li[2], James D. Blande[3] and Jarmo K. Holopainen[3]*

[1]*Natural Resources Institute Finland (Luke), Mikkeli, Finland;* [2]*University of Copenhagen, Copenhagen, Denmark;* [3]*University of Eastern Finland, Kuopio, Finland*

18.1 Introduction

All plants emit volatile organic compounds (VOCs), which have high vapour pressures that enable their evaporation into the surrounding air and can be perceived by other organisms in the environment, such as insects (Bruce *et al.*, 2005). Understanding the roles of VOCs in mediating insect behaviour potentiates the use and development of VOC-based crop protection strategies. Oilseed brassicas have a specialized secondary chemistry characterized by glucosinolates, which are degraded upon tissue damage into products that include volatile compounds (Halkier and Gershenzon, 2006). Together with terpenoids and other VOCs, these compounds serve as signals to various *Brassica*-feeding arthropods and their natural enemies. Herbivores themselves also emit VOCs for intraspecific communication, such as aggregation, alarm and sex pheromones, and synthetic analogues can be used in pheromone traps (Witzgall *et al.*, 2010). This chapter summarizes the roles of VOCs in herbivore–natural enemy interactions of oilseed brassicas and gives perspectives on the potential for utilizing VOCs to manipulate herbivores and their natural enemies in integrated pest management (IPM) of *Brassica* crops. VOCs have no residual effects in the environment and fit the ideal of using preventive, knowledge-based, environmentally benign and multifunctional strategies in IPM. Ecologically sound means of oilseed pest control are urgently needed to reduce pesticide use and combat pesticide resistance problems.

18.2 Volatile Organic Compounds and Their Ecological Functions in Association with Plants and Herbivores

VOCs are released from green foliage, reproductive structures such as buds, open flowers and pollen, and even the roots of plants (Jönsson *et al.*, 2005; Hiltpold and Turlings, 2012). The most extensively researched plant-originating VOCs, in terms of insect ecology, are the terpenoids, green leaf volatiles (GLVs: C_6 and C_9 fatty acid derivatives), amino acid derivatives such as phenylpropanoids and benzenoids, and certain more specific compound groups such as glucosinolate breakdown products released from Brassicaceae (Pichersky and Gershenzon, 2002).

The composition of plant VOC blends and the magnitude of emissions vary with genotype, phenology, plant part and diurnal rhythm (reviewed by, for example, Dudareva *et al.*, 2013). Plant VOCs are induced and altered by various, often co-occurring, biotic stressors (e.g. oviposition or feeding by herbivores and infection by pathogens) and abiotic stressors (e.g. temperature, drought, ozone) (Yuan *et al.*, 2009; Holopainen and Gershenzon, 2010). In addition, elicitors such as plant hormones and growth promoters can be used to induce VOC defences chemically (Sobhy *et al.*, 2014) (Fig. 18.1).

*Corresponding author. E-mail: jarmo.holopainen@uef.fi

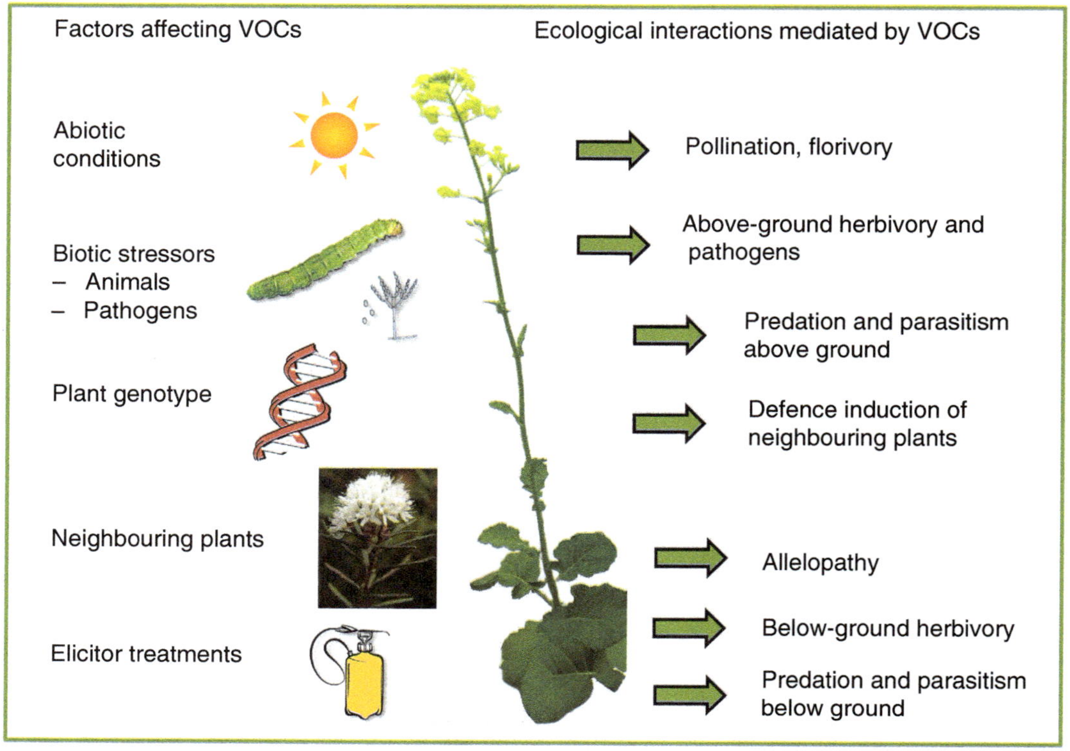

Fig. 18.1. Plant VOC emissions are affected by various biotic and abiotic factors, while the VOCs emitted can serve as infochemicals that structure ecological interactions with associated biota in natural and agroecosystems. The unique feature of VOCs is their ability to influence diverse species community assemblages and connect trophic cascades without direct physical contact.

Plant VOCs act as semiochemicals providing information to other organisms in the environment, which can shape ecological interactions in natural and agroecosystems (Fig. 18.1). VOCs can mediate: (i) direct attraction or repulsion of herbivores (Bruce *et al.*, 2005); (ii) priming and/or induction of defences in conspecific or heterospecific plants (Heil and Karban, 2010); (iii) attraction of the herbivore's natural enemies to herbivore-induced plant volatiles (HIPVs) (Heil and Karban, 2010); (iv) attraction of insect pollinators to ensure successful reproduction (e.g. Blight *et al.*, 1997); and (v) associational resistance whereby a plant coexists with a neighbour that repels herbivores (e.g. Himanen *et al.*, 2015). VOCs can also act in below-ground interactions (Johnson and Nielsen, 2012), though most research has focused on above-ground interactions. Cross-talk between above-ground and below-ground herbivory in inducing VOCs and subsequent responses has also been reported (Soler *et al.*, 2012). Allelopathy, restricting the growth of neighbouring plants such as weeds via allelochemicals, and pathogen defence, while not addressed here, can also be mediated by VOCs (for reviews, see e.g. Kegge and Pierik, 2010; Stenberg *et al.*, 2015).

The volatility together with the high natural variability and responsiveness of plant VOCs demand that insects employing them as cues have sophisticated olfactory detection systems (Bruce and Pickett, 2011). In addition, they can deal with this variability through associative learning (Allison and Hare, 2009) and often combine olfactory sensing with visual or chemotactile cues (e.g. Kühnle and Müller, 2011). Thus, VOCs can impact crop herbivory by altering herbivore attraction to crops directly, through varied predation pressure by their natural enemies using VOC cues or via affecting plant resistance by induced defences.

18.3 Oilseed *Brassica* VOCs

The plant family Brassicaceae contains a range of wild species, with the *Brassica* genus hosting most of the cultivated oilseeds, cabbages and leafy vegetables, with a variety of phenotypes and high economic value. The most common oilseed *Brassica* crops addressed here include oilseed rape (*Brassica napus* L.), turnip rape (*Brassica rapa* L. subsp. *oleifera* (syn. *campestris*)) and mustard rapeseed (*Brassica juncea* L. Czern). Oil is mostly used for food and (bio)industry purposes, and the seed cake meal as a high-protein feed for animals.

A characteristic of all species of the order Brassicales is their glucosinolate–myrosinase-based defence system (reviewed by Halkier and Gershenzon, 2006). Glucosinolates (GSs) are degraded by the myrosinase enzyme upon tissue damage, such as during herbivore feeding, to produce breakdown products, some of which are volatile. The genetically determined GS profile of the plant defines the resulting breakdown products, with the most common being isothiocyanates, thiocyanates, nitriles and epithionitriles. The volatile isothiocyanates (ITCs), toxic to many arthropods, have been extensively studied for their biological activity and they are believed to make brassicaceous plants less suitable for most non-adapted generalist herbivores (Bruce, 2014, and references therein). In turn, specialist pests employ ITCs for host recognition, are able to detoxify them and, in some species (e.g. the cabbage aphid *Brevicoryne brassicae* and the sawfly *Athalia rosae*), can sequester GSs for their own defence (Winde and Wittstock, 2011).

In addition to GS breakdown products, oilseed brassicas emit various mono- and sesquiterpenoids, GLVs and benzenoids, with both species and growth stage specificity (Tollsten and Bergström, 1988; Jönsson *et al.*, 2005). Typical terpenoids emitted from *B. napus* foliage and buds include sabinene, myrcene, limonene and (*E*,*E*)-α-farnesene (Jönsson *et al.*, 2005; Himanen *et al.*, 2009). Benzenoids such as benzaldehyde, methyl benzoate, phenyl acetaldehyde, benzyl cyanide, indole and 2-phenylethanol are emitted at the flowering stage (Jönsson *et al.*, 2005; Cook *et al.*, 2007; Piesik *et al.*, 2013).

Foliar, bud and floral damage has been found to induce emissions of numerous terpenoids, GLVs, ITCs, nitriles, sulfides and cyanides from *B. napus* and *B. rapa* (Tollsten and Bergström, 1988; Himanen *et al.*, 2009; Piesik *et al.*, 2013). However, the transfer from the vegetative to the reproductive stage can partly counteract induced VOC defences. Desurmont *et al.* (2015), for example, found VOCs to be induced more strongly by *Pieris brassicae* caterpillars feeding on vegetative and pre-flowering than on flowering *B. rapa*. In the same study, *Cotesia glomerata* parasitoids were most attracted to host-induced VOCs of plants at the bud stage, followed by the vegetative stage and then flowering plants. However, floral VOC emissions can enhance bee pollination of oilseeds, which benefits yield and can allow vectoring of beneficial entomopathogenic fungi contributing to *Brassica* IPM (Mänd *et al.*, 2010).

18.4 Volatiles as Host Recognition Cues for Important *Brassica* Herbivores

Common economically important *Brassica* herbivores (Fig. 18.2) are also the major pests of oilseed rape (Alford *et al.*, 2003; Williams, 2010). These include early-season *Phyllotreta* spp. leaf beetles (Coleoptera: Chrysomelidae) followed by the rape pollen beetle (*Meligethes aeneus* Fabricius (Coleoptera: Nitidulidae)), the mustard leaf beetle (*Phaedon cochleariae* Fabricius (Coleoptera: Chrysomelidae)) and the turnip sawfly (*Athalia rosae* L. (Hymenoptera: Tenthredinidae)). Several lepidopteran species damage oilseed foliage at their larval stage, including the diamondback moth (*Plutella xylostella* (L.) (Yponomeutidae)), the large cabbage white (*Pieris brassicae* L.) and the small cabbage white (*P. rapae* L.) (Pieridae) and the cabbage moth (*Mamestra brassicae* L. (Noctuidae)). Root-feeding flies (Diptera: Anthomyiidae), including the cabbage maggot (*Delia radicum* L.) and the turnip maggot (*Delia floralis* Fallén), can reduce oilseed yield when extensive larval damage disturbs root function. Oilseeds also host several aphid species (Hemiptera: Aphididae), the most common being the cabbage aphid (*Brevicoryne brassicae* (L.)), the mustard aphid (*Lipaphis erysimi* (Kaltenbach)) and the generalist peach–potato aphid (*Myzus persicae* (Sulzer)). In later season, pod and seed weevils (Curculionidae) and the pod midge *Dasineura brassicae* Winnertz (Diptera: Cecidomyiidae) can damage the developing seeds. The prevalence and damage inflicted by these pest species varies by geographical region. Williams (2010) considered cabbage stem flea beetle (*Psylliodes chrysocephala* L.), pollen beetle, cabbage seed weevil (*Ceutorhynchus assimilis* (Paykull)), cabbage stem weevil (*Ceutorhynchus pallidactylus* Marsh.), rape stem weevil (*Ceutorhynchus napi* Gyll.) and pod midge as the most important pests for oilseed crops in Europe.

Fig. 18.2. Important *Brassica* herbivores that utilize VOCs in host location: (a) flea beetle, *Phyllotreta undulata*; (b) diamondback moth, *Plutella xylostella*; (c) rape pollen beetle, *Meligethes aeneus*; (d) large cabbage white butterfly, *Pieris brassicae*. Photos: *P. brassicae* Sari Himanen; others Jarmo Holopainen.

Most *Brassica* pests are good flyers, limiting the effectiveness of crop rotation in pest management. VOCs enable long-distance host location using upwind anemotaxis and combine with visual and tactile cues that determine host acceptance at shorter distances (Williams and Cook, 2010). Characteristic *Brassica*-emitted VOCs induce olfactory and behavioural responses in many important *Brassica* herbivores (Fig. 18.2). The behavioural responses can, however, range from repellence to attraction, depending on species and growth stage (e.g. Jönsson *et al.*, 2007). Specific ITC-specialized olfactory receptor neurons have been found in antennae of several *Brassica* pests, e.g. the cabbage aphid (*Brevicoryne brassicae* (L.)), but also the black bean aphid (*Aphis fabae*), which is not adapted to feeding on brassicas, thus indicating a role for olfaction in ITC avoidance as well (Nottingham *et al.*, 1991). Typically, blends of several compounds rather than individual VOCs are important for eliciting responses and concentrations need to approximate natural levels of plant emissions (Bruce and Pickett, 2011).

Various flea beetle species, including *Phyllotreta cruciferae* (Goeze), *P. undulata* and *P. atra*, are attracted to allyl ITC-baited traps and the compound also enhances trapping of the cabbage stem flea beetle (*Psylliodes chrysocephala* L.), a pest of winter oilseeds (Tóth *et al.*, 2007). Combining allyl ITC with a recognized aggregation pheromone further increases the attractiveness to *P. cruciferae* (Soroka *et al.*, 2005). In a laboratory study, females of the mustard leaf beetle *Phaedon cochleariae* were attracted to 2-phenylethyl ITC and both sexes were attracted to VOCs of the host plant *Nasturtium officinale* (Kühnle and Müller, 2011).

Yellow colour is an important cue for the pollen beetle *M. aeneus*, but ITCs (Smart and Blight, 2000) and *B. napus* bud emissions are highly attractive to overwintered adults seeking oviposition sites (Jönsson *et al.*, 2007). Baiting yellow traps with ITCs (especially 2-phenylethyl ITC and a mix of four ITCs) enhances catch effectiveness of *M. aeneus* (Blight and Smart, 1999). Many less host-plant specific VOCs, e.g. methyl salicylate (MeSA), certain GLVs and linalool (Smart and Blight, 2000), attract this polyphagous beetle, which also feeds on pollen from various natural plants. Out of 25 tested floral VOCs, *M. aeneus* responded to 20, with 16 proving attractive. The magnitude of emission also mattered, as release rate affected the response.

Plutella xylostella is attracted to GS breakdown products (Pivnick *et al.*, 1994), GLVs (Reddy and Guerrero, 2000) and many terpenoids (Bruce *et al.*, 2005). The cabbage butterflies (*Pieris* spp.) respond to GLVs, benzaldehyde, phenylacetaldehyde and many terpenoids (Bruce *et al.*, 2005) and use GSs in host recognition, although nitriles can also deter *P. rapae* oviposition (de Vos *et al.*, 2008). The generalist herbivore cabbage moth *Mamestra brassicae* is attracted to GLVs and allyl ITC (Rojas, 1999), while MeSA reduces oviposition (Ulland *et al.*, 2008). The sawfly *Athalia rosae* uses GSs to identify host by contact chemoreception; and allyl, benzyl and butyl ITC and iberverin induce electroantennogram responses, with allyl ITC inducing a behavioural attraction, though only in mature females (Barker *et al.*, 2006).

In a field study on broccoli, oviposition of the root fly *D. radicum* was induced by the GLV (*Z*)-3-hexenyl acetate, whereas dimethyl disulfide, typically emitted after *D. radicum* larval root damage, reduced oviposition (Kergunteuil *et al.*, 2012). In the same study, allyl ITC, MeSA and acetophenone had no effect on oviposition. In general, root-emitted ITCs and GLVs attract *D. radicum*, while terpenoids repel it (Johnson and Nielsen, 2012, and references therein). The same review reported that allyl ITC has attracted *D. floralis*, while phenylethyl ITC has repelled it. The aphids *L. erysimi, B. brassicae* and *M. persicae* are attracted to ITCs, whereas the black bean aphid, which cannot feed on brassicas, is repelled by ITCs (Webster, 2012).

Attraction of the cabbage seed weevil *C. assimilis*, which mass-migrates to crops after flowering, varies with time of season: during colonization the highest attraction was found to be a three-ITC mixture (Bartlet *et al.*, 1993). Terpenoids, GS breakdown products, GLVs and MeSA also induced responses (Williams and Cook, 2010, and references therein). The pod midge *D. brassicae* is attracted to allyl ITC in traps, but responses vary by sex and mating status: mated females are most responsive (Murchie *et al.*, 1997).

Thus, many *Brassica* pests respond to VOCs at some points in their life cycle, which potentiates field testing and development of novel VOC-based management strategies. In general, increasing the number of attractive compounds in the VOC blends used in field baits is connected to increased attraction (Szendrei and Rodriguez-Saona, 2010). This improves the chance of having the most behaviourally influential compounds within a sufficient background blend. Understanding species-specific olfaction and behavioural responses as well as testing the effectiveness of different concentrations and blends in release and trap studies is needed to develop practical applications for IPM. It is noteworthy that the dominant plant-emitted compounds are often not the most influential: ITCs often comprise only small proportions of the herbivore-induced VOC emission but, as reviewed above, they are central cues for many *Brassica* pests.

18.5 Plant-emitted VOCs in Attracting Natural Enemies

Brassica herbivores are attacked by numerous generalist predators such as carabids, spiders, ladybird beetles and birds, as well as specialist foragers, often hymenopteran parasitoids (Williams and Cook, 2010). Olfactory responses to VOCs have been widely reported in herbivore natural enemies. Examples range from generalist polyphagous predators such as carabid beetles responding to (*E*)-β-farnesene and wheat (*Triticum* sp.) VOCs (Kielty *et al.*, 1996) to specialized parasitoids that can use plant-emitted VOCs, present in higher quantities, in long-range orientation, and host-induced VOCs in combination with visual, tactile or acoustic signals at shorter distances (Williams and Cook, 2010). A host-induced blend with a relevant background odour is likely to be important for many ichneumonid parasitoids (Braasch *et al.*, 2012).

Herbivore natural enemies can also learn to use host-related cues. For example, acclimation to host-plant odour cues was needed for orientation to host-induced VOCs by both specialist (*Diadegma semiclausum*) and generalist (*D. fenestrale*) parasitoids of *P. xylostella* (Gols *et al.*, 2012), indicating flexible learning by environment. Growth stage-specific oilseed VOCs can also attract different parasitoids. For example, common parasitoids of *M. aeneus* have differential preferences for VOC cues that correspond to their preferred growth stages of the host. *Tersilochus heterocerus* (Thomson) (which mostly parasitizes second-instar larvae of *M. aeneus*) is attracted to floral VOCs, whereas *Phradis interstitialis* (Thomson) (which prefers eggs and first-instar larvae) and *Phradis morionellus* (Holmgren) avoid the floral odour (Jönsson *et al.*, 2005). All three species are attracted to bud VOCs. VOCs might thus enhance the ability of specialized natural enemies to exploit their preferred host stages.

Specificity of host location needs to be good, especially for the highly specialized natural enemies. When testing responses of the *M. aeneus* parasitoid *Phradis* sp. to lavendel odours (potential *M. aeneus* repellents), electrophysiological responses were found but no behavioural responses. *Cotesia vestalis* (Hymenoptera: Braconidae), a parasitoid of *P. xylostella*, also oriented to hosts in the presence of non-host semi-volatiles that were adsorbed to broccoli leaves (Himanen *et al.*, 2015). Furthermore, even host oviposition sufficed to serve as host cues for the *P. brassicae* egg parasitoid *Trichogramma brassicae* and larval parasitoid *Cotesia glomerata* on cabbage (Fatouros *et al.*, 2012). These results indicate both high specificity and low interference of VOC cues for these parasitoids.

GS breakdown products also appear to be important infochemicals for parasitoids. For example, benzyl cyanide and dimethyl trisulfide attract *C. vestalis* in a dose-dependent manner (Kugimiya *et al.*, 2010). The pod midge egg parasitoid *Platygaster subuliformis* (Hymenoptera: Platygastridae) is also attracted to 2-phenylethyl ITC-baited traps and the larval parasitoid *Omphale clypealis* to allyl ITC traps (Murchie *et al.*, 1997). Crop plants also vary in their attractiveness to natural enemies. For example, Kaasik *et al.* (2014) found *Tersilochus heterocerus* to be most abundant and effective on *Brassica nigra*. However, phenology and traits other than VOCs can impact on the preferences.

Parasitoids of below-ground herbivores have also been found to utilize VOC cues. *Aleochara bipustulata* L. and *Aleochara bilineata* (Gyll.) (Coleoptera: Staphylinidae), parasitic beetles of *D. radicum*, responded differently to VOCs of mustard seed meal (where limonene is the major compound): *A. bipustulata* was attracted to the seed meal treated plots, its trap catch in the field was higher and *D. radicum* parasitism increased by 11.7%, whereas *A. bilineata* showed no response (Riley *et al.*, 2007). A larval parasitoid, *Trybliographa rapae* Westwood (Hymenoptera: Figitidae), was attracted to both host-damaged turnip roots and leaves (Neveu *et al.*, 2002), which indicates systemic induction of VOC emission by *D. radicum* feeding.

For effective biological control, increasing natural enemy attraction is not enough. Good foraging efficiency is also needed. The potential for VOCs to enhance parasitism on brassicas has been demonstrated by, for example, Uefune *et al.* (2012). In their study, *C. vestalis* was attracted to a synthetic mixture of (*Z*)-3-hexenyl acetate, *n*-heptanal, α-pinene and sabinene and release of this mixture from dispensers in a semi-field experiment increased *P. xylostella* parasitism on *B. rapa* var. *perviridis* plants. The effectiveness of different natural enemies for biological control also varies with their life history traits, competition encountered and the herbivore stage attacked; the most effective species and strategies to support that effectiveness can also vary with region and be affected by landscape and climate characteristics (e.g. Chaplin-Kramer *et al.*, 2011).

HIPVs released from dispensers can be used to attract numerous natural enemies and augment biological control (Kaplan, 2012). Methyl salicylate (MeSA) is one HIPV that is used commercially in the PredaLure product marketed for attracting lacewings and ladybird beetles. MeSA has also been tested as a potential attractant of natural enemies of *Brassica* pests under field conditions. However, despite appearing to attract the *P. xylostella* parasitoid *D. semiclausum*, field plots with MeSA lures were also significantly more attractive to some non-target arthropods, including the leaf-mining pest *Scaptomyza flava* and the lacewing parasitoid *Anacharis zealandica* (Orre *et al.*, 2010).

MeSA has been used in trials to test its attractiveness to beneficial insects in a strategy known as 'attract and reward', whereby a chemical attractant is utilized to lure beneficial insects into a crop and a food reward is provided to sustain the beneficial insects in the absence of prey (Orre Gordon *et al.*, 2013). The reward in this study was buckwheat, *Fagopyrum esculentum* Moench, which is a flowering plant that provides a floral reward to arthropods. There was no synergistic effect of buckwheat and MeSA, but buckwheat was associated with increases in the populations of five beneficial insects of the third trophic level and one fourth-level parasitoid (Orre Gordon *et al.*, 2013). Thus, attraction of arthropods of the fourth trophic level may lead to an undesirable trophic cascade, leading to a reduced abundance of beneficial natural enemies in the third trophic level. Simpson *et al.* (2011) found Scelionidae parasitoids to occur in greater amounts near MeSA or (*Z*)-3-hexenyl acetate-sprayed broccoli, and the buckwheat reward increased the occurrence of both predators and parasitoids. Therefore, this strategy offers some promise, but efforts should focus on eliminating some of the unwanted side effects. Further, a more suitable reward could be selected to benefit only certain effective natural enemies, since herbivores can also benefit from accessible nectar sources (Winkler *et al.*, 2010). Another general VOC used for natural enemy attraction is 2-phenylethanol (Benallure), targeting mostly the lacewing *Chrysoperla carnea*, syrphid flies and certain ladybird beetles. Interestingly, 2-phenylethanol is also emitted by *Brassica* flowers (Tollsten and Bergström, 1988). Thus, since VOCs can serve as cues for a variety of arthropods ranging from root herbivores to pollinators, their use for pest management purposes carries both opportunities for synergy and risks of unwanted side effects (Kaplan, 2012).

Due to many natural enemies of *Brassica* herbivores from different feeding guilds responding to VOCs, VOC-based attraction strategies could support other components of IPM, including natural enemy-favouring landscapes, perennial vegetation structures that provide habitat, and floral and extrafloral nectar and pollen sources throughout and between growing seasons (Ramsden *et al.*, 2015). VOC-based strategies targeting both generalist predators and specialist parasitoids might be most effective, since parasitoids, despite their high specificity and effectiveness, often reduce pest populations with a delay.

18.6 Pheromones of Important *Brassica* Pest Herbivores

Insect pheromones are typically highly species specific and effective in minute amounts (Witzgall *et al.*, 2010). Due to their specificity, pest pheromones do not harm other species such as natural enemies; some pheromones might even attract them (Vosteen *et al.*, 2016). Sex pheromones allow effective mate finding and reproduction in nature and synthetic mimics can be used for mating disruption. Aggregation and alarm pheromones could be employed for attraction/repulsion. Identification of species-specific pheromones has commercial application potential, if synthesis is economically feasible. Hundreds of pheromones have been identified, most often for lepidopteran species, with commercial pheromones available for an increasing number of horticultural, agricultural and forest pests (Witzgall *et al.*, 2010). Combining pheromones with plant-emitted VOCs, which function as oviposition cues, can be more effective than either VOC source alone (Reddy and Guerrero, 2000). Pheromone lures are most effective if they attract females or both sexes (as males can reproduce with several females). Recognized pheromones, exemplified below, provide potential for using pheromones alone or in combination with plant VOCs in monitoring or attract-and-kill applications for *Brassica* pests.

Extensive work has been done with aphid sex and alarm pheromones. (*E*)-β-farnesene is an alarm pheromone in several aphid species and has been found to require a background of ITCs to repel the mustard aphid *L. erysimi* (Dawson *et al.*, 1987). Verma *et al.* (2015) reported transgenic *B. juncea* emitting (*E*)-β-farnesene to have reduced *L. erysimi* colonization. Aphid predators such as ladybird beetles and lacewings are also attracted by (*E*)-β-farnesene, though the effectiveness seems to require emission quantities resembling the range of a plant-emitted cue rather than a host-emitted one (Vosteen *et al.*, 2016).

Different blends of *P. xylostella* pheromones have been used in field monitoring. Combining the pheromone with (*Z*)-3-hexenyl acetate has increased trap catching of males and females in cabbage fields (Reddy and Guerrero, 2000). However, crop identity may affect this, as no major effect of GLV addition on effectiveness of catching *P. xylostella* males in four-component blend commercial pheromone traps was reported for tests in Canadian canola fields (Miluch *et al.*, 2014). A male-produced aggregation pheromone of *Phyllotreta cruciferae*, consisting of six sesquiterpenes, is attractive to both sexes and increasingly so when combined with allyl ITC in the field (Soroka *et al.*, 2005). Another flea beetle species, *Phyllotreta vittula* Redtenbacher, on the other hand, was attracted by the pheromone combined with 3-butenyl ITC (Tóth *et al.*, 2012). Oviposition-deterring pheromones produced by *Pieris* spp. butterflies have also been found to affect subsequent oviposition and induce plant defences (Blaakmeer *et al.*, 1994).

18.7 Volatiles in Push–pull Strategies for *Brassica* Pests

Modern push–pull or stimulo-deterrent diversion strategies might provide alternatives to the use of synthetic insecticides (Eigenbrode *et al.*, 2016). The traditional push–pull approach is based on guiding or 'pushing' the pests from a valued crop with disruptive, repellent or deterrent VOCs, or by repellent companion crops. The simultaneous 'pull' strategy is used to attract pests to a more preferred trap crop that is economically not (or less) important or by using VOC attractants, arrestants, feeding or oviposition stimulants.

There is a rich history in, for example, China and Western Europe for using fragrant plants and essential oils to repel or control agricultural and horticultural pests (Ibrahim *et al.*, 2001; Nerio *et al.*, 2010). Many weeds rich in essential oils can be important sources of repellent VOCs in low-input agriculture. Intercropping with pest-repelling plants to provide associational resistance has been experimented in brassicas with both success and failure (Hooks and Johnson, 2003). Better understanding of the mechanisms might assist in developing modern companion cropping utilizing VOCs. The essential oil-based botanical insecticide thymol oil is used for *P. xylostella* biocontrol (Khumrungsee *et al.*, 2014) and exemplifies the potential for odorant natural compounds in *Brassica* pest control.

Although natural plants possess a huge arsenal of essential oils with reported repellence and toxicity (Nerio *et al.*, 2010), repellence of herbivores has been less studied than attraction using synthetic VOCs released in the field (Szendrei and Rodriguez-Saona, 2010). However, essential oil of lavender (*Lavandula angustifolia* Miller) effectively repels *M. aeneus* with greater efficacy than the traditional oils of eucalyptus, geranium or peppermint (Mauchline *et al.*, 2005). In field tests, the oil only had an effect on approaching beetles, not those inhabiting the crop (Mauchline *et al.*, 2013), which could be one major constraint of using repellent VOCs more generally. A laboratory study using semi-volatile-emitting *Rhododendron tomentosum* reported reduced *P. xylostella* oviposition on neighbouring broccoli (Himanen *et al.*, 2015). Testing such persistent compounds in the field could be one means to improve the effectiveness of pest control by VOCs.

A push–pull strategy for *M. aeneus* control that combines several mechanisms has been initiated in the UK, involving the aforementioned lavender oil, a less preferred red oilseed rape flower colour (Cook *et al.*, 2013) and an earlier-flowering *B. rapa* as a trap crop (Cook *et al.*, 2007). *B. napus* with reduced ITC emissions has also been tested (Cook *et al.*, 2006). Earlier flowering is likely to be the dominant mechanism of action for the *B. rapa* trap crop but there are also differences in VOCs. Floral emissions of both *B. rapa* and *B. napus* include phenylacetaldehyde and indole (with *M. aeneus* responding to both), which are both lacking from *B. napus* bud emissions (Cook *et al.*, 2007). There is, however, climatic or variety-based specificity, as in Estonian conditions *B. rapa* failed to trap more *M. aeneus* and, instead, *B. nigra*, *B. juncea* and *Sinapis alba* showed potential to protect *B. napus* during the bud stage (Veromann *et al.*, 2012; Kaasik *et al.*, 2014). Several trap crops have been tested for *P. xylostella* control, including *B. juncea*, *Brassica hirta* and a dead-end trap crop, the saponin-producing *Barbarea vulgaris*, which is attractive for oviposition but prevents larval development (Shelton and Nault, 2004). Flea beetles have also been noted to be more abundant on earlier-developing *B. juncea* and *B. nigra* and on late-developing *Raphanus sativa* and *Eruca sativa* compared with *B. napus* (Metspalu *et al.*, 2014).

In general, herbivore characteristics such as colonization pattern, movement speed and sensory mode of host location affect the effectiveness of push–pull strategies for different pests. A meta-analysis on insect pest behavioural manipulation by VOCs found most applications to have targeted lepidopteran, followed by coleopteran and thysanopteran pest species (Szendrei and Rodriguez-Saona, 2010), which might indicate the former being the most responsive to VOCs. For species using visual and olfactory cues, the population regulatory effect is larger, due to high attraction and arrestment (i.e. strength of inhibition of flight by the trap) within the trap crop, compared with species relying more on post-alighting host recognition (Potting *et al.*, 2005). Thus, the biology of the pests is important for designing VOC-based push–pull strategies. Modelling of species biology at the multitrophic level together with differential management and landscape-level factors may also be used to recognize the most influential practices for particular species. For example, Vinatier *et al.* (2012) simulated spatio-temporal dynamics of *M. aeneus* and *T. heterocerus* and reported that trap crop and crop rotation greatly affected pollen beetle densities and parasitism whereas ploughing had only a minor effect. Eigenbrode *et al.* (2016) also proposed that push–pull methods will be better implemented in IPM by producers if there is a better understanding of insect behaviour and how pest insects respond to push and pull factors in each crop system.

18.8 Future Potential of VOC-based Biocontrol Strategies

Advances in our understanding of plant VOC biosynthesis and emission as well as its ecological significance open up exciting new avenues for manipulating plant VOC release in order to improve pest control. Manipulation techniques potentially exploited in future *Brassica* pest biocontrol programmes include supplementing plant VOC blends with synthetic VOCs (Kaplan, 2012) and modifying plant VOC emission via exogenous application of defence elicitors or plant strengtheners (compounds that strengthen the plant's own stress resistance and/or tolerance mechanisms) (Sobhy *et al.*, 2014), conventional breeding or genetic engineering (Birkett and Pickett, 2014), as well as targeted pheromone-VOC trapping (Reddy and Guerrero, 2000).

Manipulating VOC emission of the crop plants themselves can be done by treating plants with chemical elicitors and plant strengtheners, including jasmonic acid, methyl jasmonate, salicylic acid, benzothiadiazole (BTH, a chemical mimic of salicylic acid) and plant volatiles that can either directly induce or prime VOC emission (Sobhy *et al.*, 2014). Compounds that trigger defence priming are particularly promising, as priming is a cost-saving defence mechanism whereby plants are prepared to respond faster or more strongly to future attack. For instance, in several *Brassica* species exogenous application of jasmonates in laboratory conditions increases VOC emission, which enhances attraction of parasitoids (Bruinsma *et al.*, 2009, and references therein). In addition, exposure of *Brassica oleracea* to herbivore-induced VOCs emitted from *Pieris brassicae*-damaged conspecific neighbours has been shown to prime both direct and indirect defences by inhibiting growth of *P. brassicae* and *M. brassicae* caterpillars and augmenting attraction of their parasitoid, *Cotesia glomerata* (Peng *et al.*, 2011). Volatile-induced priming of indirect defence has also been demonstrated in broccoli (Li and Blande, 2015), where primed plants release greater amounts of GLVs than unprimed ones. Despite these extensive laboratory studies, field tests with elicitors have rarely been carried out in *Brassica* crops (but see Loivamäki *et al.*, 2004) and the elicitors can also induce physiological changes other than defence responses, thus lacking specificity. Future efforts are needed to address these aspects.

Transgenic plants that more effectively repel or deter herbivores and/or more effectively recruit herbivores' natural enemies can be generated by genetic engineering of VOC biosynthetic pathways, in particular when key repellents or attractants and genes implicated in the biosynthesis of these compounds have been identified. The prospect for genetic engineering of plant VOCs to improve plant pest resistance has been extensively reviewed (e.g. Birkett and Pickett, 2014). Examples include restoring the emission of (*E*)-β-caryophyllene to maize lines to augment recruitment of entomopathogenic nematodes (Degenhardt *et al.*, 2009) and incorporation of the aphid alarm pheromone (*E*)-β-farnesene into wheat (Bruce *et al.*, 2015). Other emerging techniques for crop protection include RNA interference (RNAi) (e.g. Price and Gatehouse 2008), which involves transgenic plants producing double-stranded RNA (dsRNA) directed against essential insect genes, and genome editing such as CRISPR (Clustered Regularly Interspaced Short Palindromic Repeats)/Cas9 (CRISPR Associated) (e.g. Bortesi and Fischer, 2015), which targets complex genomes in a specific and precise fashion to obtain specific agronomic traits such as pest resistance. These technologies promise to be more efficient in crop improvement and may draw more public support with fewer biosafety issues, as no transgenic protein is expressed in transgenic plants. Classical plant breeding programmes can also be used to incorporate biologically active VOCs into new varieties. A remarkable example comes from *B. oleracea* and *B. napus* (Bradburne and Mithen, 2000) for which enhancing the production of but-3-enyl ITC, while maintaining the total GS levels, has been found to increase the attractiveness of plants to the braconid wasp *Diaeretiella rapae* in a field setting. For all of the aforementioned breeding methods, unintended effects on non-target organisms as well as the 'cry wolf' effects of plants constitutively releasing VOCs should be assessed.

Design of VOC-based applications to reduce pest pressure on oilseeds has the potential to benefit from an improved understanding of pest–natural enemy biology, including olfactory and visual orientation, and should target the early (egg) stages as well as those key to population development. Furthermore, it is important to take into account impacts of VOC manipulation on numerous insects

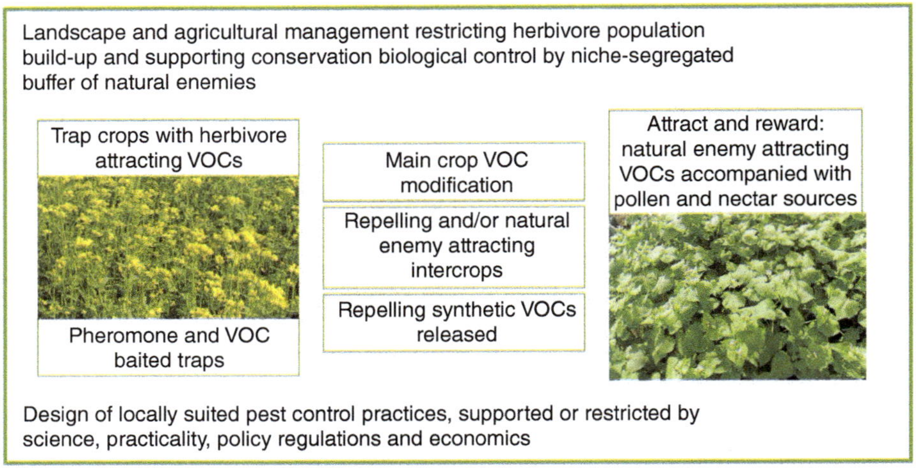

Fig. 18.3. VOC-based biocontrol strategies can work in synergy to buffer *Brassica* agroecosystems from pest problems and help reduce the reliance on pesticides. Photos (turnip rape and buckwheat): Sari Himanen.

of several trophic levels, which could affect food-web dynamics: for example, herbivores and their key predators or parasitoids (Himanen *et al.*, 2015) and potentially also their hyperparasitoids (Orre *et al.*, 2010).

Modifying the field VOC environment as an IPM tactic would be particularly suitable for organic farming, where chemical control, often detrimental to natural enemies, is not used. VOC-based push–pull strategies can include repellent chemicals, selective pheromone and HIPV trapping, attractive trap crop or natural enemy supporting mixtures to assure whole-season effectiveness, and within-field intercrops to reduce the ease of host location by herbivores (Fig. 18.3). Ecological infrastructure such as perennial biodiversity enhancing strips with attractant VOCs and suitable pollen and floral and extrafloral nectar sources (Ramsden *et al.*, 2015), designed to support a niche segregated population of natural enemies, could be added to build a year-by-year increasing buffer against pest attacks.

18.9 Conclusions

Integrated pest management incorporates conservation of natural enemies and biological control, increasing host-plant resistance and cultural control practices, before chemicals. All these can be supported by VOC-based solutions: many natural enemies used in biological control rely on host-induced plant VOCs, crop plant resistance can be increased by manipulating crop VOC emission or by companion plants, and cultural control can involve trap crops, intercrops and living mulches that impact the VOC environment encountered by pests. Plant-emitted VOCs induce responses in many *Brassica* herbivores in the laboratory; and pheromones and VOC-baited trapping has been demonstrated to have good potential in field applications. Practical demonstration of the effectiveness of VOCs in the field could lead to a novel means of pest control with no environmental load or chemical residues. Through combining VOC-based and other IPM strategies, the threshold for using chemical control might be increased to levels allowing populations of natural enemies to remain viable. Cooperation between scientists and agricultural practitioners to target local pest problems utilizing knowledge on pest biology and development of feasible practical management means is important for reducing reliance on pesticides and developing oilseed *Brassica* IPM.

References

Alford, D.V., Nilsson, C. and Ulber, B. (2003) Chapter 2: Insect pests of oilseed rape crops. In: Alford, D.V. (ed.) *Biocontrol of Oilseed Rape Pests*. Blackwell Science, Oxford, UK.

Allison, J.D. and Hare, J.D. (2009) Learned and naïve natural enemy responses and the interpretation of volatile organic compounds as cues or signals. *New Phytologist* 184, 768–782.

Barker, A.M., Molotsane, R., Müller, C., Schaffner, U. and Städler, E. (2006) Chemosensory and behavioural responses of the turnip sawfly, *Athalia rosae*, to glucosinolates and isothiocyanates. *Chemoecology* 16, 209–218.
Bartlet, E., Blight, M.M., Hick, A.J. and Williams, I.H. (1993) The responses of the cabbage seed weevil (*Ceutorhynchus assimilis*) to the odour of oilseed rape (*Brassica napus*) and to some volatile isothiocyanates. *Entomologia Experimentalis et Applicata* 68, 295–302.
Birkett, M.A. and Pickett, J.A. (2014) Prospects of genetic engineering for robust insect resistance. *Current Opinion in Plant Biology* 19, 59–67.
Blaakmeer, A., Hagenbeek, D., Van Beek, T.A., de Groot, A.E., Schoonhoven, L.M. and van Loon, J.J.A. (1994) Plant response to eggs vs host marking pheromone as factors inhibiting oviposition by *Pieris brassicae*. *Journal of Chemical Ecology* 20, 1657–1665.
Blight, M.M. and Smart, L.E. (1999) Influence of visual cues and isothiocyanate lures on capture of the pollen beetle, *Meligethes aeneus* in field traps. *Journal of Chemical Ecology* 25, 1501–1516.
Blight, M.M., le Métayer, M., Pham Delègue, M.-H., Pickett, J.A., Marion-Poll, F. and Wadhams, L.J. (1997) Identification of floral volatiles involved in recognition of oilseed rape flowers, *Brassica napus*, by honeybees, *Apis mellifera*. *Journal of Chemical Ecology* 23, 1715–1727.
Bortesi, L. and Fischer, R. (2015) The CRISPR/Cas9 system for plant genome editing and beyond. *Biotechnology Advances* 33, 41–52.
Braasch, J., Wimp, G.M. and Kaplan, I. (2012) Testing for phytochemical synergism: arthropod community responses to induced volatile blends across crops. *Journal of Chemical Ecology* 28, 1264–1275.
Bradburne, R.P. and Mithen, R. (2000) Glucosinolate genetics and the attraction of the aphid parasitoid *Diaeretiella rapae* to *Brassica*. *Proceedings of the Royal Society B: Biological Sciences* 267, 89–95.
Bruce, T.J.A. (2014) Glucosinolates in oilseed rape: secondary metabolites that influence interactions with herbivores and their natural enemies. *Annals of Applied Biology* 164, 348–353.
Bruce, T.J.A. and Pickett, J.A. (2011) Perception of plant volatile blends by herbivorous insects – finding the right mix. *Phytochemistry* 72, 1605–1611.
Bruce, T.J.A, Wadhams, L.J. and Woodcock, C.M. (2005) Insect host location: a volatile situation. *Trends in Plant Science* 6, 269–274.
Bruce, T.J.A., Aradottir, G.I., Smart, L.E., Martin, J.L., Caulfield, J.C., Doherty, A., Sparks, C.A., Woodcock, C.M., Birkett, M.A., Napier, J.A. Jones, H.D. and Pickett, J.A. (2015) The first crop plant genetically engineered to release an insect pheromone for defence. *Scientific Reports* 5, 11183.
Bruinsma, M., Posthumus, M.A., Mumm, R., Mueller, M.J., van Loon, J.J.A. and Dicke, M. (2009) Jasmonic acid-induced volatiles of *Brassica oleracea* attract parasitoids: effects of time and dose, and comparison with induction by herbivores. *Journal of Experimental Botany* 60, 2575–2587.
Chaplin-Kramer, R., O'Rourke, M.E., Blitzer, E.J. and Cremen, C. (2011) A meta-analysis of crop pest and natural enemy response to landscape complexity. *Ecology Letters* 14, 922–932.
Cook, S.M., Smart, L.E., Martin, J.L., Murray, D.A., Watts, N.P. and Williams, I.H. (2006) Exploitation of host plant preferences in pest management strategies for oilseed rape (*Brassica napus*). *Entomologia Experimentalis et Applicata* 119, 221–229.
Cook, S.M., Rasmussen, H.B., Birkett M.A., Murray, D.M., Pye, P.J., Watts, N.P. and Williams, I.H. (2007) Behavioural and chemical ecology underlying the success of turnip rape (*Brassica rapa*) trap crops in protecting oilseed rape (*Brassica napus*) from the pollen beetle (*Meligethes aeneus*). *Arthropod–Plant Interactions* 1, 57–67.
Cook, S.M., Skellern, M.P., Döring, T. and Pickett, J.A. (2013) Red oilseed rape? The potential for manipulation of petal colour in control strategies for the pollen beetle (*Meligethes aeneus*). *Arthropod–Plant Interactions* 7, 249–258.
Dawson, G.W., Griffiths, D.C., Pickett, J.A., Wadhams, L.J. and Woodcock, C.M. (1987) Plant-derived synergists of alarm pheromone from turnip aphid, *Lipaphis* (Hyadaphis) *erysimi* (Homoptera, Aphididae). *Journal of Chemical Ecology* 13, 1663–1671.
Degenhardt, J., Hiltpold, I., Köllner, T.G., Frey, M., Gierl, A., Gershenzon, J., Hibbard, B.E., Ellersieck, M.R. and Turlings, T.C.J. (2009) Restoring a maize root signal that attracts insect-killing nematodes to control a major pest. *Proceedings of the National Academy of Sciences of the USA* 106 13213–13218.
Desurmont, G.A., Laplanche, D., Schiestl, F.P. and Turlings, T.C.J. (2015) Floral volatiles interfere with plant attraction of parasitoids: ontogeny-dependent infochemical dynamics in *Brassica rapa*. *BMC Ecology* 15, 17.
de Vos, M., Kriksunov, K.L. and Jander, G. (2008) Indole-3-acetonitrile production from indole glucosinolates deters oviposition by *Pieris rapae*. *Plant Physiology* 146, 916–926.
Dudareva, N., Klempien, A., Muhlemann, J.K. and Kaplan, I. (2013) Biosynthesis, function and metabolic engineering of plant volatile organic compounds. *New Phytologist* 198, 16–32.
Eigenbrode, S.D., Birch, A.N.E., Linczey, S., Meadow, R. and Snyder, W.E. (2016) Review: a mechanistic framework to improve understanding and applications of push–pull systems in pest management. *Journal of Applied Ecology* 53, 202–212.
Fatouros, N., Lucas-Barbosa, D., Weldegergis, B.T., Pashalidou, F.G., van Loon, J.J.A., Dicke, M., Harvey, J.A.,

Gols, R. and Huigens, M.E. (2012) Plant volatiles induced by herbivore egg deposition affect insects of different trophic levels. *PLoS ONE* 7, e43607.

Gols, R., Veenemans, C., Potting, R.P.J., Smid, H.M., Dicke, M., Harvey, J.A. and Bukovinszky, T. (2012) Variation in the specificity of plant volatiles and their use by a specialist and a generalist parasitoid. *Animal Behaviour* 83, 1231–1242.

Halkier, B.A. and Gershenzon, J. (2006) Biology and biochemistry of glucosinolates. *Annual Review of Plant Biology* 57, 303–333.

Heil, M. and Karban, R. (2010) Explaining evolution of plant communication by airborne signals. *Trends in Ecology and Evolution* 25, 137–144.

Hiltpold, I. and Turlings, T.C.J. (2012) Manipulation of chemically mediated interactions in agricultural soils to enhance the control of crop pests and to improve crop yield. *Journal of Chemical Ecology* 38, 641–650.

Himanen, S.J., Nerg, A.-M., Nissinen, A., Pinto, D.M., Stewart, C.N. Jr, Poppy, G.M. and Holopainen, J.K. (2009) Effects of elevated carbon dioxide and ozone on volatile terpenoid emissions and multitrophic communication of transgenic insecticidal oilseed rape (*Brassica napus*). *New Phytologist* 181, 174–186.

Himanen, S.J., Bui, T.N., Maja, M. and Holopainen, J.K. (2015) Utilizing associational resistance for biocontrol: impacted by temperature, supported by indirect defence. *BMC Ecology* 15, 16.

Holopainen, J.K. and Gershenzon, J. (2010) Multiple stress factors and the emission of plant VOCs. *Trends in Plant Science* 15, 176–184.

Hooks, C.R.R. and Johnson, M.W. (2003) Impact of agricultural diversification on the insect community of cruciferous crops. *Crop Protection* 22, 223–238.

Ibrahim, M.A., Kainulainen, P., Aflatuni, A., Tiilikkala, K. and Holopainen, J.K. (2001) Insecticidal, repellent, antimicrobial activity and phytotoxicity of essential oils: with special reference to limonene and its suitability for control of insect pests. *Agricultural and Food Science in Finland* 10, 243–259.

Johnson, S.N. and Nielsen, U.N. (2012) Foraging in the dark – chemically mediated host plant location by belowground insect herbivores. *Journal of Chemical Ecology* 38, 604–614.

Jönsson, M., Lindkvist, A. and Anderson, P. (2005) Behavioural responses in three ichneumonid pollen beetle parasitoids to volatiles emitted from different phenological stages of oilseed rape. *Entomologia Experimentalis et Applicata* 115, 363–369.

Jönsson, M., Rosdahl, K. and Anderson, P. (2007) Responses to olfactory and visual cues by over-wintered and summer generations of the pollen beetle, *Meligethes aeneus*. *Physiological Entomology* 32, 188–193.

Kaasik, R., Kovacs, G., Kaart, T., Metspalu, L., Williams, I.H. and Veromann, E. (2014) *Meligethes aeneus* oviposition preferences, larval parasitism rate and species composition of parasitoids on *Brassica nigra*, *Raphanus sativus* and *Eruca sativa* compared with on *Brassica napus*. *Biological Control* 69, 65–71.

Kaplan, I. (2012) Attracting carnivorous arthropods with plant volatiles: the future of biocontrol or playing with fire? *Biological Control* 60, 77–89.

Kegge, W. and Pierik, R. (2010) Biogenic volatile organic compounds and plant competition. *Trends in Plant Science* 15, 126–132.

Kergunteuil, A., Dugravot, S., Mortreuil, A., Le Ralec, A. and Cortesero, A.M. (2012) Selecting volatiles to protect brassicaceous crops against the cabbage root fly, *Delia radicum*. *Entomologia Experimentalis et Applicata* 144, 69–77.

Khumrungsee, N., Pluempanupat, W., Koul, O. and Bullangpoti, V. (2014) Toxicity of essential oil compounds against diamondback moth larvae, *Plutella xylostella*, and their impact on detoxification enzymes. *Journal of Pest Science* 87, 721–729.

Kielty, J.P., Allen-Williams, L.J., Underwood, N. and Easterwood, E.A. (1996) Behavioral responses of three species of ground beetle (Coleoptera: Carabidae) to olfactory cues associated with prey and habitat. *Journal of Insect Behavior* 9, 237–250.

Kugimiya, S., Shimoda, T., Tabata, J. and Takabayashi, J. (2010) Present or past herbivory: a screening of volatiles released from *Brassica rapa* under caterpillar attacks as attractants for the solitary parasitoid, *Cotesia vestalis*. *Journal of Chemical Ecology* 36, 620–628.

Kühnle, A. and Müller, C. (2011) Relevance of visual and olfactory cues for host location in the mustard leaf beetle *Phaedon cochleariae*. *Physiological Entomology* 36, 68–76.

Li, T. and Blande, J.D. (2015) Associational susceptibility in broccoli: mediated by plant volatiles, impeded by ozone. *Global Change Biology* 21, 1993–2004.

Loivamäki, M., Holopainen, J.K. and Nerg, A.M. (2004) Chemical changes induced by methyl jasmonate in oilseed rape grown in the laboratory and in the field. *Journal of Agricultural and Food Chemistry* 52, 7607–7613.

Mänd, M., Williams, I.H., Viik, E. and Karise, R. (2010) Chapter 14: Oilseed rape, bees and integrated pest management. In: Williams, I.H. (ed.) *Biocontrol-based Integrated Management of Oilseed Rape Pests*. Springer, Dordrecht, The Netherlands.

Mauchline, A., Osborne, J., Martin, A., Poppy, G. and Powell, W. (2005) The effects of non-host plant essential oil volatiles on the behaviour of the pollen beetle *Meligethes aeneus*. *Entomologia Experimentalis et Applicata* 114, 181–188.

Mauchline, A.L., Cook, S.M., Powell, W. and Osborne, J.L. (2013) Effects of non-host plant odour on *Meligethes aeneus* during immigration to oilseed rape. *Entomologia Experimentalis et Applicata* 146, 313–320.

Metspalu, L., Kruus, E., Ploomi, A., Williams, I.H., Hiiesaar, K., Jõgar, K., Veromann, E. and Mänd, M.

(2014) Flea beetle (Chrysomelidae: Alticinae) species composition and abundance in different cruciferous oilseed crops and the potential for a trap crop system. *Acta Agriculturae Scandinavica, Section B — Soil & Plant Science* 64, 572–582.

Miluch, C.E., Dosdall, L.M. and Evenden, M.L. (2014) Factors influencing male *Plutella xylostella* (Lepidoptera: Plutellidae) capture rates in sex pheromone-baited traps on canola in western Canada. *Journal of Economic Entomology* 107, 2067–2076.

Murchie, A.L., Smart, L.E. and Williams, I.H. (1997) Responses of *Dasineura brassicae* and its parasitoids *Platygaster subuliformis* and *Omphale clypealis* to field traps baited with organic isothiocyanates. *Journal of Chemical Ecology* 23, 917–926.

Nerio, L.S., Olivero-Verbel, J. and Stashenko, E. (2010) Repellent activity of essential oils: a review. *Bioresource Technology* 101, 372–378.

Neveu, N., Grandgirard J., Nenon, J.P. and Cortesero, A.M. (2002) Systemic release of herbivore-induced plant volatiles by turnips infested by concealed root-feeding larvae *Delia radicum* L. *Journal of Chemical Ecology* 28, 1717–1732.

Nottingham, S.F., Hardie, J., Dawson, G.W., Hick, A.J., Pickett, J.A., Wadhams, L.J. and Woodcock, C.M. (1991) Behavioral and electrophysiological responses of aphids to host and nonhost plant volatiles. *Journal of Chemical Ecology* 17, 1231–1242.

Orre, G.U.S., Wratten, S.D., Jonsson, M. and Hale, R.J. (2010) Effects of an herbivore-induced plant volatile on arthropods from three trophic levels in brassicas. *Biological Control* 53, 62–67.

Orre Gordon, G.U.S., Wratten, S.D., Jonsson, M., Simpson, M. and Hale, R. (2013) 'Attract and reward': combining a herbivore-induced plant volatile with floral resource supplementation – multi-trophic level effects. *Biological Control* 64, 106–115.

Peng, J., van Loon, J.J.A., Zheng, S. and Dicke, M. (2011) Herbivore-induced volatiles of cabbage (*Brassica oleracea*) prime defence responses in neighbouring intact plants. *Plant Biology* 13, 276–284.

Pichersky, E. and Gershenzon, J. (2002) The formation and function of plant volatiles: perfumes for pollinator attraction and defense. *Current Opinion in Plant Biology* 5, 237–243.

Piesik, D., Delaney, K.J., Wenda-Piesik, A., Sendel, S., Tabaka, B. and Buszewski, B. (2013) *Meligethes aeneus* pollen feeding suppresses, and oviposition induces, *Brassica napus* volatiles: beetle attraction/repellence to lilac aldehydes and veratrole. *Chemoecology* 23, 241–250.

Pivnick, K.A., Jarvis, B.J. and Slater, G.P. (1994) Identification of olfactory cues used in host-plant finding by diamondback moth, *Plutella xylostella* (Lepidoptera: Plutellidae). *Journal of Chemical Ecology* 20, 1407–1427.

Potting, R.P.J., Perry, J.N. and Powell, W. (2005) Insect behavioural ecology and other factors affecting the control efficacy of agro-ecosystem diversification strategies. *Ecological Modelling* 182, 199–216.

Price, D.R.G. and Gatehouse, J.A. (2008) RNAi-mediated crop protection against insects. *Trends in Biotechnology* 26, 393–400.

Ramsden, M.W., Menéndez, R., Leather, S.R. and Wäckers, F. (2015) Optimizing field margins for biocontrol services: the relative role of aphid abundance, annual floral resources, and overwinter habitat in enhancing aphid natural enemies. *Agriculture, Ecosystems and Environment* 199, 94–104.

Reddy, G.V.P. and Guerrero, A. (2000) Behavioral responses of the diamondback moth, *Plutella xylostella,* to green leaf volatiles of *Brassica oleracea* subsp. *capitata*. *Journal of Agricultural Food Chemistry* 48, 6025–6029.

Riley, K.J., Kuhlmann, U., Mason, P.G., Whistlecraft, J., Donald, L.J. and Holliday, N.J. (2007) Can mustard seed meal increase attacks by *Aleochara* spp. on *Delia radicum* in oilseed rape? *Biocontrol Science and Technology* 17, 273–284.

Rojas, J.C. (1999) Electrophysiological and behavioral responses of the cabbage moth to plant volatiles. *Journal of Chemical Ecology* 25, 1867–1883.

Shelton, A.M. and Nault, B.A. (2004) Dead-end trap cropping: a technique to improve management of the diamondback moth, *Plutella xylostella* (Lepidoptera: Plutellidae). *Crop Protection* 23, 497–503.

Simpson, M., Gurr, G.M., Simmons, A.T., Wratten, S.D., James, D.G., Leeson, G., Nicol, H. . and Orre-Gordon, G.U.S. (2011) Attract and reward: combining chemical ecology and habitat manipulation to enhance biological control in field crops. *Journal of Applied Ecology* 48, 580–590.

Smart, L.E. and Blight, M.M. (2000) Response of the pollen beetle, *Meligethes aeneus*, to traps baited with volatiles from oilseed rape, *Brassica napus*. *Journal of Chemical Ecology* 26, 1051–1064.

Sobhy, I.S., Erb, M., Lou, Y. and Turlings, T.C.J. (2014) The prospect of applying chemical elicitors and plant strengtheners to enhance the biological control of crop pests. *Philosophical Transactions of the Royal Society B: Biological Sciences* 369, 20120283.

Soler, R., van der Putten, W.H., Harvey, J.A., Vet, L.E.M., Dicke, M. and Bezemer, M. (2012) Root herbivore effects on aboveground multitrophic interactions: patterns, processes and mechanisms. *Journal of Chemical Ecology* 38, 755–767.

Soroka, J.J., Bartelt, R.J., Zilkowski, B.W. and Cossem, A.A. (2005) Responses of flea beetle *Phyllotreta cruciferae* to synthetic aggregation pheromone components and host plant volatiles in field trials. *Journal of Chemical Ecology* 31, 1829–1843.

Stenberg, J.A., Heil, M., Åhman, I. and Björkman, C. (2015) Optimizing crops for biocontrol of pests and disease. *Trends in Plant Science* 20 698–712.

Szendrei, Z. and Rodriguez-Saona, C. (2010) A meta-analysis of insect pest behavioral manipulation with plant volatiles. *Entomologia Experimentalis et Applicata* 134, 201–210.

Tollsten, L. and Bergström, G. (1988) Headspace volatiles of whole plants and macerated plant parts of *Brassica* and *Sinapis*. *Phytochemistry* 27, 4013–4018.

Tóth, M., Csonka, E., Baksca, F., Benedek, P., Szarukán, I., Gomboc, S., Toshova, T., Subchev, M. and Ujváry, I. (2007) Species spectrum of flea beetles (*Phyllotreta* spp., Coleoptera, Chrysomelidae) attracted to allyl isothiocyanate-baited traps. *Zeitschrift für Naturforschung* 62c, 772–778.

Tóth, M., Csonka, E., Bartelt, R.J., Cossé, A.A. and Zilkowski, B.W. (2012) Similarities in pheromonal communication of flea beetles *Phyllotreta cruciferae* Goeze and *Ph. vittula* Redtenbacher (Coleoptera, Chrysomelidae). *Journal of Applied Entomology* 136, 688–697.

Uefune, M., Choh, Y., Abe, J., Shiojiri, K., Sano, K. and Takabayashi, J. (2012) Application of synthetic herbivore-induced plant volatiles causes increased parasitism of herbivores in the field. *Journal of Applied Entomology* 136, 561–567.

Ulland, S., Mozuraitis, I.R., Borg-Karlson, A.K., Meadow, R. and Mustaparta, H. (2008) Methyl salicylate, identified as primary odorant of a specific receptor neuron type, inhibits oviposition by the moth *Mamestra brassicae* L. (Lepidoptera, Noctuidae). *Chemical Senses* 33, 35–46.

Verma, S.S., Sinha, R.K. and Jajoo, A. (2015) (E)-β-farnesene gene reduces *Lipaphis erysimi* colonization in transgenic *Brassica juncea* lines. *Plant Signaling & Behavior* 10, 7.

Veromann, E., Metspalu, L, Williams, I.H., Hiiesaar, K., Mand, M., Kaasik, R., Kovacs, G., Jogar, K., Svilponis, E., Kivimagi, I., Ploomi, A. and Luik, A. (2012) Relative attractiveness of *Brassica napus*, *Brassica nigra*, *Eruca sativa* and *Raphanus sativus* for pollen beetle (*Meligethes aeneus*) and their potential for use in trap cropping. *Arthropod–Plant Interactions* 6, 385–394.

Vinatier, F., Gosme, M. and Valantin-Morison, M. (2012) A tool for testing integrated pest management strategies on a tritrophic system involving pollen beetle, its parasitoid and oilseed rape at the landscape scale. *Landscape Ecology* 27, 1421–1433.

Vosteen, I., Weisser, W.W. and Kunert, G. (2016) Is there any evidence that aphid alarm pheromones work as prey and host finding kairomones for natural enemies? *Ecological Entomology* 41, 1–12.

Webster, B. (2012) The role of olfaction in aphid host location. *Physiological Entomology* 37, 10–18.

Williams, I.H. (2010) Chapter 1: Biocontrol-based integrated management of oilseed rape pests. In: Williams, I.H. (ed.) *Biocontrol-based Integrated Management of Oilseed Rape Pests*. Springer, Dordrecht, The Netherlands.

Williams, I.H. and Cook, S.M. (2010) Chapter 7: Crop location by oilseed rape pests and host location by their parasitoids. In: Williams, I.H. (ed.) *Biocontrol-based Integrated Management of Oilseed Rape Pests*. Springer, Dordrecht, The Netherlands.

Winde, I. and Wittstock, U. (2011) Insect herbivore counteradaptations to the plant glucosinolate-myrosinase system. *Phytochemistry* 72, 1566–1575.

Winkler, K., Wäckers, F.L., Termorshuizen, A.J. and van Lenteren, J.C. (2010) Assessing risks and benefits of floral supplements in conservation biological control. *BioControl* 55, 719–727.

Witzgall, P., Kirsch, P. and Cork, A. (2010) Sex pheromones and their impact on pest management. *Journal of Chemical Ecology* 36, 80–100.

Yuan, J.S, Himanen, S.J., Holopainen, J.K., Chen, F. and Stewart, C.N. Jr (2009) Smelling global climate change: mitigation of function for plant volatile organic compounds. *Trends in Ecology and Evolution* 24, 323–331.

19 Impact of Genetically Modified Herbicide-resistant Oilseed Rape on Non-target Organisms: Natural Enemies of Oilseed Rape Pests

Govinda Shrestha[1*], Prashant Jha[2] and Gadi V.P. Reddy[1]

[1]*Western Triangle Agricultural Research Center, Montana State University, Conrad, Montana, USA;* [2]*Southern Agricultural Research Center, Montana State University, Huntley, Montana, USA*

19.1 Introduction

Genetically modified organisms are the plants, animals, or microorganisms in which the genetic materials have been altered by mating and/or natural recombination (WHO, 2014). The technologies that are used to carry out such genetic alterations are referred to as recombinant-gene technology or genetic engineering (WHO, 2014). In agricultural production systems, the adoption of genetically modified (GM) crops by growers is increasing rapidly in many countries (Phipps and Park, 2002; Qaim and Zilberman, 2003; James, 2014), despite the fact that their impacts on non-target organisms are still fairly unknown (Hails, 2000; Wolfenbarger and Phifer, 2000; Sarewitz, 2004; Snow *et al.*, 2005). GM crops account for more than 181 million hectares under cultivation worldwide, with the most area under cultivation in the USA, followed by Brazil, Argentina, India and Canada (James, 2014) (Fig. 19.1). Based on the novel traits that have been transferred into crops plants, GM crops are categorized into three groups: (i) herbicide resistance (also called herbicide tolerance); (ii) insect/pathogen resistance; and (iii) food improvement parameters such as shelf-life and the amount/composition of fatty acids, protein, vitamins or other nutritional substances (Rogers and Parkes, 1995; Kwon and Kim, 2001). Among these, GM herbicide-resistant crops are the most widely cultivated worldwide, representing 83% of GM crop hectares in 2014, followed by insect-resistant GM crops (Kwon and Kim, 2001; James, 2014).

The GM herbicide-resistant crops are mainly modified to tolerate specific broad-spectrum herbicides that kill weeds, without causing any damage to the GM crop (Schoenenberger and D'Andrea, 2012). Within a little over a decade, several herbicide-resistant crops have been developed, including glyphosate-resistant, glufosinate-resistant, imidazolinones-resistant, bromoxynil-resistant and sethoxydim-resistant crops (Mazur and Falco, 1989; Dekker and Duke, 1995; Dill, 2005; Green, 2012). Among these, glyphosate-resistant and glufosinate-resistant crops are the most widespread successful examples of GM herbicide-resistant crops. Furthermore, glyphosate- and glufosinate-resistant crops are employed much more often than the other herbicide-resistant crops; therefore, the focus of this chapter is on these two GM herbicide-resistant traits.

Glyphosate is a non-selective, broad-spectrum, systemic herbicide (Dill, 2005). It is the only herbicide that kills plants by blocking the enzyme 5-enolpyruvylshikimate-3-phosphate synthase (EPSPS), involved in the biosynthesis of aromatic amino acids and other secondary metabolites in plants (Duke, 2003; Dill, 2005; Tan *et al.*, 2006; Cerdeira and Duke, 2006). The features that make glyphosate a unique and effective

*Corresponding author. E-mail: govinda.shrestha@montana.edu

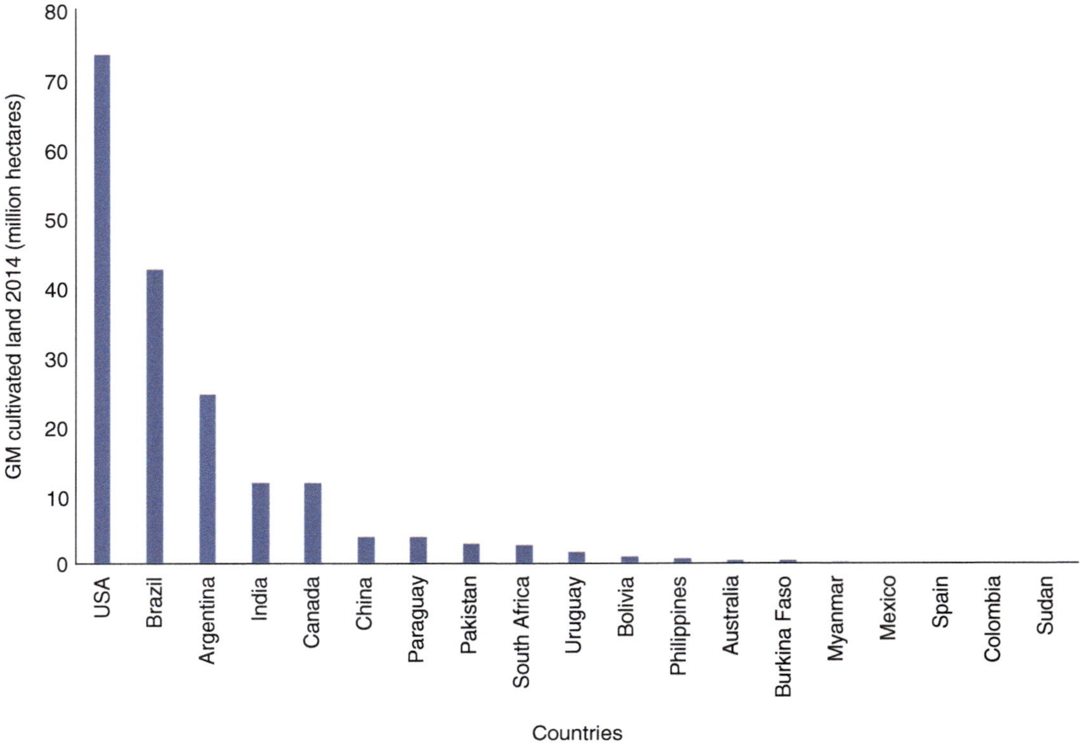

Fig. 19.1. Global status of commercialized genetically modified (GM) crops in 2014 (James, 2014).

post-emergence herbicide are that: (i) most plants metabolically degrade this herbicide slowly or not at all; (ii) it translocates well to metabolically active tissues (meristems); and (iii) the relatively slow mode of action of this herbicide allows movement of the chemical throughout the plant before the symptoms occur (Dill, 2005; Cerdeira and Duke, 2006). Glufosinate is another non-selective, broad-spectrum herbicide and it contains the active ingredient phosphinothricin, which kills plants by blocking the enzyme responsible for nitrogen metabolism and for detoxifying ammonia, a by-product of plant metabolism (Dekker and Duke, 1995; Tan *et al.*, 2006).

Genes from different soil bacteria such as *Agrobacterium* sp., *Ochrobactrum anthropi* Holmes and *Streptomyces viridochromogenes* Krainsky were transferred to create glyphosate- and glufosinate-resistant crops (Dekker and Duke, 1995; Cerdeira and Duke, 2006). The gene encoding a glyphosate-resistant form of EPSPS obtained from *Agrobacterium* sp. has been found to be effective in providing tolerance to glyphosate in glyphosate-resistant crops (Padgette *et al.*, 1991). In addition, a gene from *O. anthropi* that encodes a glyphosate-degrading enzyme (glyphosate oxidase, GOX) has been found to be effective at degrading (metabolizing) glyphosate (Padgette *et al.*, 1991). Glyphosate-resistant GM crops include alfalfa (*Medicago sativa* L.), canola/oilseed rape (*Brassica napus* L. and *Brassica rapa* L.), maize (*Zea mays* L.), soybean (*Glycine max* L.), cotton (*Gossypium hirsutum* L.) and sugar beet (*Beta vulgaris* L.) (Legere, 2005; James, 2014). Glufosinate-resistant crops, meanwhile, contain a gene from the bacterium *S. viridochromogenes* that produces an enzyme that can detoxify phosphinothricin and prevent damage to the crop (Dekker and Duke, 1995). Glufosinate-resistant GM crops include rice (*Oryza sativa* L.) and all those mentioned above, except alfalfa (Legere, 2005; James, 2014).

GM herbicide-resistant oilseed rape (also called canola) is the fourth leading GM crop grown worldwide, after soybean, cotton and maize (James, 2014) (Fig. 19.2). Canola is the second leading oilseed crop in the world, accounting for 13% of global demand

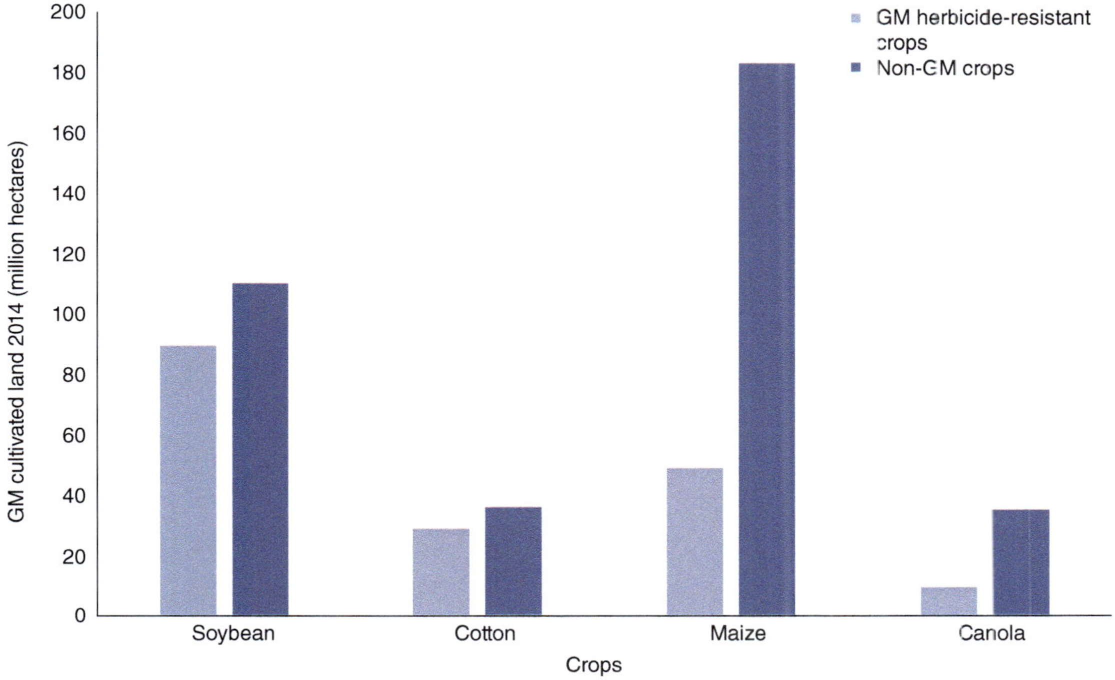

Fig. 19.2. Global adoption of four principal genetically modified (GM) crop cultivars compared with non-GM crop cultivars in 2014 (James, 2014).

for edible oils, and *B. napus* and the *B. rapa* are widely grown species of commercial importance (Raymer, 2002). China, Canada, India, Germany, France, England and Australia are the major oilseed rape producing countries in the world (James, 2014).

In many agricultural systems, broadleaf and grass weeds are often the most common yield-limiting factors of oilseed rape, causing economically significant losses (Harker, 2001; Martin *et al.*, 2001; Gulden *et al.*, 2003). Over the last 10–15 years, world cultivation of oilseed rape has increased, reaching approximately 46 million hectares in 2014 (James, 2014). The adoption of GM herbicide-resistant oilseed rape cultivars by growers has improved weed control and increased crop yield (Graef *et al.*, 2007; James, 2014). The benefits of GM cultivars include: (i) reduced herbicide cost; (ii) more flexible herbicide application timings; (iii) superior control of a broad range of weed spectrum; and (iv) consistent weed control over a range of environments (Graef *et al.*, 2007; Green, 2012; Vencill *et al.*, 2012). In 2014, 25% of the global rapeseed land area was cultivated with GM herbicide-resistant cultivars (James, 2014).

Despite the advantages of GM herbicide-resistant oilseed rape cultivars, their potential impact on non-target organisms cannot be neglected. Non-target organisms in oilseed rape fields include beneficial organisms that protect the crop from pest attack or improve soil health. These include natural enemies of insect pests, such as predators, parasitoids and pathogens (Hokkanen *et al.*, 2003; Williams, 2004; Dosdall and Mason, 2010; Ulber *et al.*, 2010) and pollinators such as honey bees, solitary bees and hoverflies (Morandin and Winston, 2005; Hayter and Cresswell, 2006; Stanley *et al.*, 2013).

The use of GM herbicide-resistant crops, including oilseed rape, may influence beneficial organisms by: (i) directly exposing them to herbicides, thereby reducing populations and communities of predators, parasitoids and insect pathogens (Ahn *et al.*, 2001; Schneider *et al.*, 2009; Menezes *et al.*, 2012); (ii) changing weed density, composition and community and thus affecting the abundance of pollinators and natural enemies (Brooks *et al.*,

2003; Morandin and Winston, 2005); and (iii) being directly toxic to beneficial organisms (Meissle *et al.*, 2005; Romeis *et al.*, 2008; Malone and Burgess, 2009). It should be noted that many other organisms might also be influenced by the use of GM herbicide-resistant crops, as reviewed by Wolfenbarger and Phifer (2000) and Kwon and Kim (2001).

The purpose of this chapter is to provide information not only on the impact of GM herbicide-resistant oilseed crops on natural enemies of pests, but also on possible consequences of spraying those broad-spectrum herbicides (glyphosate and glufosinate) on GM oilseed rape. There are very few studies that have directly compared the impact of GM herbicide-resistant vs non-GM oilseed crops on beneficial organisms, as opposed to numerous studies on the impact of GM insect-resistant crops on non-target organisms (Romeis *et al.*, 2008; Wolfenbarger *et al.*, 2008; Naranjo, 2009). These two broad-spectrum herbicides are frequently and extensively used for weed control on GM herbicide-resistant oilseed rape and so understanding their direct/indirect impact is crucial.

19.2 Effect of GM Herbicide-resistant Oilseed Rape on Predators

Oilseed rape crops are attacked by many pests but also harbour several naturally occurring predators of these pest populations (Büchs and Alford, 2003; Williams, 2004; Ekbom, 2010). Carabid beetles (Carabidae), rove beetles (Staphylinidae) and spiders (Araneae) are the major ground-surface predators, while soldier beetles (Cantharidae), predatory midges (Cecidomyiidae), hoverflies (Syrphidae), big-eyed bugs (Lygaeidae), lacewings (Chrysopidae), long-legged flies (Dolichospodidae) and dance flies (Hybotidae) are the major predators in the crop canopy (Knodel and Olsen, 2002; Williams, 2004; Ekbom, 2010; Williams, 2010). Several studies found the carabid *Clivina fossor* (L.) to be one of the major natural enemies of cabbage seed weevil *Ceutorhynchus assimilis* (Paykull), brassica pod midge *Dasineura brassicae* Winnertz and pollen beetle *Meligethes aeneus* (Fabricius), some of the major pests of oilseed rape in Europe (Alford *et al.*, 2003). Knodel and Olsen (2002) reported that the lacewing *Chrysoperla carnea* (Stephens) and big-eyed bug *Geocoris bullatus* (Say) contribute to the control of crucifer flea beetle *Phyllotreta cruciferae* (Goeze), one of the most serious pests of oilseed rape in North America and Canada.

Long-term field experiments carried out in the UK by Brooks *et al.* (2003) found no significant differences in the total abundance of ground-surface predators (carabid beetles, rove beetles and spiders) between the oilseed rape fields of GM glufosinate-resistant cultivars with the use of glufosinate versus non-GM cultivars with the use of conventional weed control. A similar result was reported by Haughton *et al.* (2003), with no significant difference in the total abundance of epigeal and aerial predators on field margins of the GM herbicide-resistant cultivar versus the non-GM cultivar. However, at the species level, significant differences were observed in the diversity of ground-surface predators (Brooks *et al.*, 2003). This study found carabid beetle species such *Pterostichus niger* Schall, *Pterostichus melanarius* Illiger, *Pterostichus madidus* F., *Bembidion lampros* Herbst, *Loricera pilicornis* F., *Nebria biguttatus* F. and *Nebria brevicollis* F. and spider species such as *Lepthyphantes tenuis* Blackwall, *Erigone atra* Blackwall, *Erigone dentipalpis* Wider and *Pardosa* spp. to be significantly more abundant in non-GM oilseed rape plots with conventional weed management compared with plots of the GM glufosinate-resistant cultivar treated with glufosinate (Brooks *et al.*, 2003). Another study suggested that changes in weed vegetation mediated by multiple applications of glufosinate herbicide in a GM glufosinate-resistant oilseed rape cultivar may reduce the sources of nectar and pollen necessary for the abundance and longevity of many natural enemies, including predators (Raskin *et al.*, 1994).

Glyphosate and glufosinate herbicides are more frequently or heavily used to control weeds in GM herbicide-resistant oilseed rape cultivars as compared with non-GM cultivars. It is therefore, important to determine whether the exposure of predators to these two herbicides in oilseed fields has any negative impact on the predators' biology and population development. To the best of our knowledge, no field studies have examined the impact of these two herbicides on the naturally occurring predators of oilseed rape after direct exposure to the herbicides. However, based on laboratory studies, no harmful effects of glyphosate were found on carabid beetles (*Poecilus chalcites* Say, *Agonum punctiforme* Say, *Amara cupreolata* Putzeys, *Chlaenius laticollis* Say and *Anisodactylus rusticus* Say) (Brust, 1990), on the spider *Lepthyphantes tenuis* Blackwall (Haughton *et al*., 2001), or on the big-eyed bug *Geocoris pallens* Stal. (Yokoyama and Pritchard, 1984) (Table 19.1). Lundgren *et al.* (2009) reported that the lack of toxicity of glyphosate to predators is largely because the amino acid synthesis pathway disrupted by glyphosate is not present in animals, suggesting that this herbicide has little or no toxicity to many predators of oilseed rape pests. Some evidence, however, suggests that glyphosate may be toxic to some predators, including the green lacewing *Chrysopa externa* Hagen (Schneider *et al*., 2009), the predatory mite *Euseius hibisci* Chant (Tanigoshi and Congdon, 1983) and the spider *Alpaida veniliae* Keyserling (Benamú *et al*., 2010). On these species, glyphosate has a negative impact on population development, fecundity and survival (Table 19.1). In contrast, studies have found glufosinate to be highly toxic to several predator species, including the flower bug *Orius strigicollis* Poppius, the multicoloured ladybird beetle *Harmonia axyridis* (Pallas) and the predatory mites *Amblyseius womersleyi* Schicha and *Phytoseiulus persimilis* Athias-Henriot (Table 19.1), indicating that glufosinate may be toxic to the predators of oil seed rape pests as well.

The negative impact of GM herbicide-resistant oilseed rape on predators might be from direct feeding on GM herbicide-resistant plants or from eating prey that have fed on GM herbicide-resistant plants, though there are no studies examining this mechanism. In GM insect-resistant crops, however, various studies have been carried out to determine if feeding by predators directly on GM crops or feeding on prey that have fed on the GM crops has any negative effects on the predators' reproduction, development or survival (Romeis *et al*., 2008; Lövei *et al*., 2009; Naranjo, 2009).

19.3 Effect of GM Herbicide-resistant Oilseed Rape on Parasitoids

Ulber *et al*. (2010) and Dosdall and Mason (2010) reported that parasitoids are one of the key natural enemies of several pest species of oilseed rape crops, mainly because the emergence and seasonal activity periods of parasitoids are usually well synchronized with the phenology of the pre-imaginal life stages of the target host pests. More than 80 hymenopteran species (particularly braconids, chalcids and ichneumonids) are known for their ability to parasitize different pest species of oilseed rape (Dosdall and Mason, 2010; Ulber *et al*., 2010). Among these, 10 to 12 species have been identified as playing key roles in controlling economically significant pests of oilseed rape (Ulber *et al*., 2010). For instance, the parasitoids *Phradis morionellus* Holm, *Diospilus capito* Nees and *Tersilochus heterocerus* Thomson are key natural enemies of the pollen beetle *Trichomalus perfectus* Walker. Other key parasitoids are: (i) *Stenomalina gracilis* Walker for the cabbage seed weevil *Platygaster subuliformis* Kieffer; (ii) *Omphale clypealis* Thomson for the brassica pod midge *Dasineura brassicae* Winnertz; (iii) *Tersilochus fulvipes* Gravenhorst for rape stem weevil *Ceutorhynchus napi* Gyllenhal; (iv) *Tersilochus obscurator* Aubert for cabbage stem weevil *Ceutorhynchus quadridens* Panz.; (v) *Tersilochus microgaster* Szepligeti for cabbage stem flea beetle *Psylliodes chrysocephala* L.; and (vi) *Microctonus vittatae* Muesebeck for some flea beetles (Dosdall and Mason, 2010; Ulber *et al*., 2010).

The impact of GM herbicide-resistant oilseed rape on parasitoids, such as those mentioned above, does not appear to have been investigated extensively, except for the field study by Hawes *et al*. (2003). This study reported significantly lower densities of oilseed rape pest parasitoids in a GM glufosinate-resistant cultivar compared with a non-GM oilseed rape cultivar. Similarly, lower populations of beet (*Beta vulgaris* L.) pest parasitoids have been observed on a GM glyphosate-resistant beet cultivar with the use of glyphosate versus a non-GM cultivar with conventional weed management (Hawes *et al*., 2003). This study in general suggested that the lower abundance of parasitoids on GM glufosinate-resistant spring oilseed rape cultivar is likely to be due to the change in resource availability, such as reduction in weed population or biomass, as these weed resources usually provide a variety of foods for parasitoids. Regarding the impact of exposure of herbicides (glufosinate and glyphosate) on parasitoids of oilseed rape pests, it is fairly unknown whether either of these two herbicides have any toxic effect on parasitoids. However, from other studies, it seems that exposure to glyphosate may have no negative impact on parasitoid biology and population development, while glufosinate may be highly toxic to parasitoids (Bueno *et al*., 2008; Lundgren *et al*., 2009; Menezes *et al*., 2012). For example, Menezes *et al*. (2012) studied the toxicological impacts of several herbicides, including glyphosate and glufosinate, in *Eucalyptus* plantations in Brazil on the parasitoid *Palmistichus elaeisis* Delvare & LaSalle, a

Table 19.1. Impact of exposure to glyphosate or glufosinate on non-target organisms: predators and parasitoids.

Non-target species	Insect taxa	Herbicides	Effects observed	References
(a) Predators				
Chrysopa pallens Rambur	Chrysopidae	Glufosinate	No pupal mortality and slightly harmful to larvae (13–41% mortality)	Ahn *et al*., 2001
C. externa Hagen	Chrysopidae	Glyphosate	Negative effects on development, fecundity and fertility	Schneider *et al*., 2009
Euseius hibisci Chant	Phytoseiidae	Glyphosate	Mortality effect on adults (50–65% mortality)	Tanigoshi and Congdon, 1983
Geocoris pallens Stal	Geocoridae	Glyphosate	No negative effect on female mortality, fecundity and egg viability	Yokoyama and Pritchard, 1984
Orius strigicollis Poppius	Anthocoridae	Glufosinate	High mortality effects on eggs (71%), nymphs (65%) and adults (58%)	Ahn *et al*., 2001
Harmonia axyridis Pallas	Coccinellidae	Glufosinate	No mortality effects on eggs and adults but very harmful to first-instar larvae (100% mortality), harmful to fourth instar (51%) and slightly harmful to pupae (25%)	Ahn *et al*., 2001
Amblyseius womersleyi Schicha	Acari	Glufosinate	No egg mortality but harmful to very harmful for nymphs (46–100% mortality) and very harmful for adults (90–100% mortality)	Ahn *et al*., 2001
Phytoseiulus persimilis Athias-Henriot	Phytoseiidae	Glufosinate	No egg mortality but highly harmful for nymphs (100% mortality) and adults (92% mortality)	Ahn *et al*., 2001
Alpaida veniliae Keyserling	Araneidae	Glyphosate	Negative effects on prey consumption, web building, fecundity, fertility and developmental time of progeny	Benamú *et al*., 2010
Gonatium rubens Blackwall	Linyphiidae	Glyphosate	Negative effects on population development	Haughton *et al*., 2001
Lepthyphantes tenuis Blackwall	Linyphiidae	Glyphosate	No effect on population development	Haughton *et al*., 2001
Poecilus chalcites Say	Carabidae	Glyphosate	No harmful effect on adult longevity and food consumption	Brust, 1990
Agonum punctiforme Say	Carabidae	Glyphosate	No harmful effect on adult longevity and food consumption	Brust, 1990
Amara cupreolata Putzeys	Carabidae	Glyphosate	No harmful effect on adult longevity and food consumption	Brust, 1990
Chlaenius laticollis Say	Carabidae	Glyphosate	No harmful effect on adult longevity and food consumption	Brust, 1990
Anisodactylus rusticus Say	Carabidae	Glyphosate	No harmful effect on adult longevity and food consumption	Brust, 1990
(b) Parasitoids				
Palmistichus elaesis Delvare & LaSalle	Eulophidae	Glufosinate	Negative effects on parasitism and emergence	Menezes *et al*., 2012
P. elaesis	Eulophidae	Glyphosate	No effects on parasitism and adult emergence	Menezes *et al*., 2012
Microctonus hyperodae Loan	Braconidae	Glyphosate	No effect on adult mortality	Addison and Barker, 2006
Telenomus remus Nixon	Scelionidae	Glyphosate	No effects on parasitism and adult emergence	Stecca *et al*., 2016
Trichogramma pretiosum Riley	Trichogrammatidae	Glyphosate	No effects on parasitoid development	Bueno *et al*., 2008

natural enemy of several lepidopteran pests of eucalyptus (such as *Euselasia eucerus* Hewitson and *Thyrinteina arnobia* Stoll). This study indicated that glufosinate reduced the parasitism rate and emergence of the parasitoid *P. elaeisis*, while glyphosate posed little or no risk to this parasitoid. A study by Bueno *et al.* (2008) investigated the effects of several herbicides (including glyphosate) commonly used in GM glyphosate-resistant soybean on the eggs, larvae and pupae of the egg parasitoid *Trichogramma pretiosum*. This parasitoid is an important natural enemy of soybean crop pests (e.g. the soybean looper, *Pseudoplusia includens* Walker, and the soybean webworm, *Omiodes indicata* Fabricius). Similarly, this study suggested that glyphosate is less likely to be toxic to *T. pretiosum* as compared with other herbicides (Table 19.1). Other examples of these effects are shown in Table 19.1.

19.4 Effect of GM Herbicide-resistant Oilseed Rape on Insect Pathogens

Insect pathogens providing natural control of oilseed rape pests include fungi and nematodes (Hokkanen *et al.*, 2003; Williams, 2004). Although the composition and diversity of insect pathogenic fungi in oilseed rape is unclear, the following fungi have been found to infect oilseed rape pests: *Beauveria bassiana* (Balsamo), *Metarhizium anisopliae* (Metschnikoff), *Paecilomyces fumosoroseus* (Wize) and *Entomophthora muscae* (Cohn) (Williams, 2004). For example, *B. bassiana* showed virulence towards *Meligethes aeneus* (Hokkanen, 1993) and *Metarhizium anisopliae* for *Psylliodes chrysocephala, Ceutorhynchus assimilis,* and *Meligethes aeneus* (Husberg and Hokkanen, 2001). Relevant nematodes in canola include species of *Steinernema* and *Heterorhabditis*, which have been shown to be effective against *M. aeneus* and *C. assimilis* (Ehlers and Hokkanen, 1996). Similarly, the microsporidian *Nosema meligethi* (Nageli) is an important source of overwintering mortality of *M. aeneus* (Lipa and Hokkanen, 1992).

Laboratory experiments have shown that the exposure of *B. bassiana, M. anisopliae* and other fungi to glyphosate has deleterious effects on the germination and growth of these fungi (Gardner and Storey, 1985; Morjan *et al.*, 2002; Andaló *et al.*, 2004). In contrast, no significant difference in the infection rate of *B. bassiana* to insect hosts has been found between glyphosate-treated and untreated soils (Harrison and Gardner, 1992).

19.5 Conclusions

GM herbicide-resistant oilseed rape cultivars have undergone a major expansion worldwide over the past decade, with a subsequent increase in the use of two broad-spectrum herbicides, glyphosate and glufosinate. The adoption of GM herbicide-resistant oilseed rape cultivars by growers has provided several agronomic benefits, resulting in improved weed control and increased crop yield. However, the potential impact of GM herbicide-resistant oilseed rape cultivars on non-target organisms, such as natural enemies of oilseed rape pests, cannot be neglected. Currently, very little literature is available addressing these concerns, compared with studies on the impact of GM insect-resistant crops on non-target organisms. In general, glyphosate seems to have less harmful effects on the biology and population development of parasitoids and predators of oilseed rape, while glufosinate may be highly toxic to these natural enemies, especially parasitoids. Further studies are needed to determine the direct/indirect effects of these two common broad-spectrum herbicides on the natural enemy community of oilseed rape pests in the field and the need for this research is even greater, considering the rapid rate of adoption of GM herbicide-resistant crops.

Acknowledgements

We would like to thank the USDA-National Institute of Food and Agriculture Hatch Act (Project #MONB00859; Accession #1009746) for funding. Any opinions, findings, conclusions or recommendations expressed in this publication are those of the authors and do not necessarily reflect the view of the National Institute of Food and Agriculture (NIFA) or the United States Department of Agriculture (USDA).

References

Addison, P.J. and Barker, G.M. (2006) Effect of various pesticides on the non-target species *Microctonus hyperodae*, a biological control agent of *Listronotus bonariensis*. *Entomologia Experimentalis et Applicata* 119(1), 71–79.

Ahn, Y.-J., Kim, Y.J. and Yoo, J.K. (2001) Toxicity of the herbicide glufosinate-ammonium to predatory insects and mites of *Tetranychus urticae* (Acari: Tetranychidae) under laboratory conditions. *Journal of Economic Entomology* 94(1), 157–161.

Alford, D.V., Nilsson, C. and Ulber, B. (2003) Insect pests of oilseed rape crops. In: Alford, D.V. (ed.) *Biocontrol of Oilseed Rape Pests*. Blackwell Science, Hoboken, New Jersey, pp. 9–41.

Andaló, V., Moino, A. Jr, Santa-Cecília, L.V. and Souza, G.C. (2004) Compatibility of *Beauveria bassiana* with chemical pesticides for the control of the coffee root mealybug *Dysmicoccus texensis* Tinsley (Hemiptera: Pseudococcidae). *Neotropical Entomology* 33(4), 463–467.

Benamú, M., Schneider, M. and Sánchez, N. (2010) Effects of the herbicide glyphosate on biological attributes of *Alpaida veniliae* (Araneae, Araneidae), in laboratory. *Chemosphere* 78(7), 871–876.

Brooks, D., Bohan, D., Champion, G., Haughton, A., Hawes, C., Heard, M.S., Clark, S.J., Dewar, A.M., Firbank, L.G., Perry, J.N. *et al.* (2003) Invertebrate responses to the management of genetically modified herbicide-tolerant and conventional spring crops. I. Soil-surface-active invertebrates. *Philosophical Transactions of the Royal Society B: Biological Sciences* 358(1439), 1847–1862.

Brust, G.E. (1990) Direct and indirect effects of four herbicides on the activity of carabid beetles (Coleoptera: Carabidae). *Pesticide Science* 30(3), 309–320.

Büchs, W. and Alford, D.V. (2003) Predators of oilseed rape pests. In: Alford, D.V. (ed.) *Biocontrol of Oilseed Rape Pests*. Blackwell Science, Hoboken, New Jersey, pp. 181–200.

Bueno, A.F., Bueno, R.C.O.F., Parra, J.R.P. and Vieira, S.S. (2008) Effects of pesticides used in soybean crops to the egg parasitoid *Trichogramma pretiosum*. *Ciência Rural* 38(6), 1495–1503.

Cerdeira, A.L. and Duke, S.O. (2006) The current status and environmental impacts of glyphosate-resistant crops. *Journal of Environmental Quality* 35(5), 1633–1658.

Dekker, J.H. and Duke, S.O. (1995) Herbicide-resistant field crops. *Advances in Agronomy* 54, 69.

Dill, G.M. (2005) Glyphosate-resistant crops: history, status and future. *Pest Management Science* 61(3), 219–224.

Dosdall, L.M. and Mason, P.G. (2010) Key pests and parasitoids of oilseed rape or canola in North America and the importance of parasitoids in integrated management. In: Williams, I.H. (ed.) *Biocontrol-based Integrated Management of Oilseed Rape Pests*. Springer, Dordrecht, The Netherlands, pp. 167–213.

Duke, S.O. (2003) Weeding with transgenes. *Trends in Biotechnology* 21(5), 192–195.

Ehlers, R.-U. and Hokkanen, H. (1996) Insect biocontrol with non-endemic entomopathogenic nematodes (*Steinernema* and *Heterorhabditis* spp.): conclusions and recommendations of a combined OECD and COST workshop on scientific and regulatory policy issues. *Biocontrol Science and Technology* 6(3), 295–302.

Ekbom, B. (2010) Pests and their enemies in spring oilseed rape in Europe and challenges to integrated pest management. In: Williams, I.H. (ed.) *Biocontrol-based Integrated Management of Oilseed Rape Pests*. Springer, Dordrecht, The Netherlands, pp. 151–165.

Gardner, W.A. and Storey, G.K. (1985) Sensitivity of *Beauveria bassiana* to selected herbicides. *Journal of Economic Entomology* 78(6), 1275–1279.

Graef, F., Stachow, U., Werner, A. and Schütte, G. (2007) Agricultural practice changes with cultivating genetically modified herbicide-tolerant oilseed rape. *Agricultural Systems* 94(2), 111–118.

Green, J.M. (2012) The benefits of herbicide-resistant crops. *Pest Management Science* 68(10), 1323–1331.

Gulden, R.H., Shirtliffe, S.J. and Thomas, A.G. (2003) Harvest losses of canola (*Brassica napus*) cause large seedbank inputs. *Weed Science* 51(1), 83–86.

Hails, R.S. (2000) Genetically modified plants – the debate continues. *Trends in Ecology and Evolution* 15(1), 14–18.

Harker, K.N. (2001) Survey of yield losses due to weeds in central Alberta. *Canadian Journal of Plant Science* 81(2), 339–342.

Harrison, R.D. and Gardner, W.A. (1992) Fungistasis of *Beauveria bassiana* by selected herbicides in soil. *Journal of Entomological Science* 27, 233–238.

Haughton, A.J., Bell, J.R., Wilcox, A. and Boatman, N.D. (2001) The effect of the herbicide glyphosate on non-target spiders. Part I: Direct effects on *Lepthyphantes tenuis* under laboratory conditions. *Pest Management Science* 57(11), 1033–1036.

Haughton, A.J., Champion, G.T., Hawes, C., Heard, M.S., Brooks, D.R., Bohan, D.A., Clark, S.J., Dewar, A.M., Firbank, L.G., Osborne, J.L. *et al.* (2003) Invertebrate responses to the management of genetically modified herbicide-tolerant and conventional spring crops. II. Within-field epigeal and aerial arthropods. *Philosophical Transactions of the Royal Society B: Biological Sciences* 358(1439), 1863–1877.

Hawes, C., Haughton, A.J., Osborne, J.L., Roy, D.B., Clark, S.J., Perry, J.N., Rothery, P., Bohan, D.A., Brooks, D.R., Champion, G.T. *et al.* (2003) Responses of plants and invertebrate trophic groups to contrasting herbicide regimes in the farm scale evaluations of genetically modified herbicide-tolerant crops. *Philosophical Transactions of the Royal Society B: Biological Sciences* 358(1439), 1899–1913.

Hayter, K.E. and Cresswell, J.E. (2006) The influence of pollinator abundance on the dynamics and efficiency of pollination in agricultural *Brassica napus*: implications for landscape-scale gene dispersal. *Journal of Applied Ecology* 43(6), 1196–1202.

Hokkanen, H. (1993) Overwintering survival and spring emergence in *Meligethes aeneus*: effects of body weight, crowding, and soil treatment with *Beauveria bassiana*. *Entomologia Experimentalis et Applicata* 67(3), 241–246.

Hokkanen, H.M., Menzler-Hokkanen, I. and Butt, T.M. (2003) Pathogens of oilseed rape pests. In: Alford, D.V. (ed.) *Biocontrol of Oilseed Rape Pests*. Blackwell Science Ltd., Hoboken, New Jersey, pp. 299–322.

Husberg, G.-B. and Hokkanen, H.M. (2001) Effects of *Metarhizium anisopliae* on the pollen beetle *Meligethes aeneus* and its parasitoids *Phradis morionellus* and *Diospilus capito*. *BioControl* 46(3), 261–273.

James, C. (2014) *Global status of commercialized biotech/GM crops:* 2013. ISAAA Brief No. 46. International Service for the Acquisition of Agri-Biotech Applications, Ithaca, New York.

Knodel, J. and Olsen, D. (2002) *E-1234: Biology and integrated pest management of the crucifer flea beetle*. North Dakota State University Co-operative Extension Service, Fargo, North Dakota.

Kwon, Y.W. and Kim, D.S. (2001) Herbicide-resistant genetically-modified crop: its risks with an emphasis on gene flow. *Weed Biology and Management* 1(1), 42–52.

Legere, A. (2005) Risks and consequences of gene flow from herbicide-resistant crops: canola (*Brassica napus* L) as a case study. *Pest Management Science* 61(3), 292–300.

Lipa, J.J. and Hokkanen, H.M. (1992) *Nosema meligethi* I. & R.(Microsporida) in populations of *Meligethes* spp. in Europe. *Biocontrol Science and Technology* 2(2), 119–125.

Lövei, G.L., Andow, D.A. and Arpaia, S. (2009) Transgenic insecticidal crops and natural enemies: a detailed review of laboratory studies. *Environmental Entomology* 38(2), 293–306.

Lundgren, J.G., Gassmann, A.J., Bernal, J., Duan, J.J. and Ruberson, J. (2009) Ecological compatibility of GM crops and biological control. *Crop Protection* 28(12), 1017–1030.

Malone, L. and Burgess, E. (2009) Impact of genetically modified crops on pollinators. In: Ferry, N. and Gatehouse, A.M.R. (eds) *Environmental Impact of Genetically Modified Crops*. CAB International, Wallingford, UK, pp. 199–222.

Martin, S.G., Van Acker, R.C. and Friesen, L.F. (2001) Critical period of weed control in spring canola. *Weed Science* 49(3), 326–333.

Mazur, B.J. and Falco, S.C. (1989) The development of herbicide resistant crops. *Annual Review of Plant Biology* 40(1), 441–470.

Meissle, M., Vojtech, E. and Poppy, G.M. (2005) Effects of Bt maize-fed prey on the generalist predator *Poecilus cupreus* L. (Coleoptera: Carabidae) *Transgenic Research* 14(2), 123–132.

Menezes, C., Soares, M., Santos, J., Assis Júnior, S., Fonseca, A. and Zanuncio, J. (2012) Reproductive and toxicological impacts of herbicides used in *Eucalyptus* culture in Brazil on the parasitoid *Palmistichus elaeisis* (Hymenoptera: Eulophidae). *Weed Research* 52(6), 520–525.

Morandin, L.A. and Winston, M.L. (2005) Wild bee abundance and seed production in conventional, organic, and genetically modified canola. *Ecological Applications* 15(3), 871–881.

Morjan, W., Pedigo, L. and Lewis, L. (2002) Fungicidal effects of glyphosate and glyphosate formulations on four species of entomopathogenic fungi. *Environmental Entomology* 31(6), 1206–1212.

Naranjo, S.E. (2009) Impacts of Bt crops on non-target invertebrates and insecticide use patterns. *CAB Reviews: Perspectives in Agriculture, Veterinary Science, Nutrition and Natural Resources* 4(11), 1–11.

Padgette, S., Re, D.B., Gasser, C., Eichholtz, D.A., Frazier, R., Hironaka, C.M., Kishore, G.M. *et al.* (1991) Site-directed mutagenesis of a conserved region of the 5-enolpyruvylshikimate-3-phosphate synthase active site. *Journal of Biological Chemistry* 266(33), 22364–22369.

Phipps, R. and Park, J. (2002) Environmental benefits of genetically modified crops: global and European perspectives on their ability to reduce pesticide use. *Journal of Animal and Feed Sciences* 11(1), 1–18.

Qaim, M. and Zilberman, D. (2003) Yield effects of genetically modified crops in developing countries. *Science* 299(5608), 900–902.

Raskin, R., Gluck, E. and Pflug, W. (1994) The development of fauna and flora on herbicide free agricultural fields. *Natur und Landschaft* 67, 7–14.

Raymer, P.L. (2002) Canola: an emerging oilseed crop. *Trends in New Crops and New Uses* 1, 122–126.

Rogers, H. and Parkes, H. (1995) Transgenic plants and the environment. *Journal of Experimental Botany* 46(5), 467–488.

Romeis, J., Bartsch, D., Bigler, F., Candolfi, M.P., Gielkens, M.M., Hartley, S.E., Hellmich, R.L., Huesing, J.E., Jepson, P.C., Layton, R. *et al.* (2008) Assessment of risk of insect-resistant transgenic crops to nontarget arthropods. *Nature Biotechnology* 26(2), 203–208.

Sarewitz, D. (2004) How science makes environmental controversies worse. *Environmental Science and Policy* 7(5), 385–403.

Schneider, M., Sanchez, N., Pineda, S., Chi, H. and Ronco, A. (2009) Impact of glyphosate on the development, fertility and demography of *Chrysoperla externa* (Neuroptera: Chrysopidae): ecological approach. *Chemosphere* 76(10), 1451–1455.

Schoenenberger, N. and D'Andrea, L. (2012) Surveying the occurrence of subspontaneous glyphosate-tolerant genetically engineered *Brassica napus* L. (Brassicaceae) along Swiss railways. *Environmental Sciences Europe* 24(1), 1–8.

Snow, A.A., Andow, D.A., Gepts, P., Hallerman, E.M., Power, A., Tiedje, J.M. and Wolfenbarger, L. (2005) Genetically engineered organisms and the environment: current status and recommendations. *Ecological Applications* 15(2), 377–404.

Stanley, D.A., Gunning, D. and Stout, J.C. (2013) Pollinators and pollination of oilseed rape crops (*Brassica napus* L.) in Ireland: ecological and economic incentives for pollinator conservation. *Journal of Insect Conservation* 17(6), 1181–1189.

Stecca, C., Bueno, A., Pasini, A., Silva, D. and Andrade, K. (2016) Side-effects of glyphosate to the parasitoid *Telenomus remus* Nixon (Hymenoptera: Platygastridae). *Neotropical Entomology* 45(2), 192–200.

Tan, S., Evans, R. and Singh, B. (2006) Herbicidal inhibitors of amino acid biosynthesis and herbicide-tolerant crops. *Amino Acids* 30(2), 195–204.

Tanigoshi, L. and Congdon, B. (1983) Laboratory toxicity of commonly-used pesticides in California citriculture to *Euseius hibisci* (Chant) (Acarina: Phytoseiidae). *Journal of Economic Entomology* 76(2), 247–250.

Ulber, B., Williams, I.H., Klukowski, Z., Luik, A. and Nilsson, C. (2010) Parasitoids of oilseed rape pests in Europe: key species for conservation biocontrol. In: Williams, I.H. (ed.) *Biocontrol-based Integrated Management of Oilseed Rape Pests*. Springer, Dordrecht, The Netherlands, pp. 45–76.

Vencill, W.K., Nichols, R.L., Webster, T.M., Soteres, J.K., Mallory-Smith, C., Burgos, N.R., Johnson, W.G. and McClelland, M.R. (2012) Herbicide resistance: toward an understanding of resistance development and the impact of herbicide-resistant crops. *Weed Science* 60(sp1), 2–30.

WHO (2014) Food safety. World Health Organization, Geneva. Available at: http://www.who.int/foodsafety/areas_work/food-technology/faq-genetically-modified-food/en/.

Williams, I. (2004) Advances in insect pest management of oilseed rape in Europe. In: Horowitz, A.R. and Ishaaya, I. (eds) *Insect Pest Management*. Springer, Berlin Heidelberg, pp. 181–208.

Williams, I.H. (2010) The major insect pests of oilseed rape in Europe and their management: an overview. In: Williams, I.H. (ed.) *Biocontrol-based Integrated Management of Oilseed Rape Pests*. Springer, Dordrecht, The Netherlands, pp. 1–43.

Wolfenbarger, L.L. and Phifer, P.R. (2000) The ecological risks and benefits of genetically engineered plants. *Science* 290(5499), 2088–2093.

Wolfenbarger, L.L., Naranjo, S.E., Lundgren, J.G., Bitzer, R.J. and Watrud, L.S. (2008) Bt crop effects on functional guilds of non-target arthropods: a meta-analysis. *PLoS ONE* 3(5), 2118.

Yokoyama, V.Y. and Pritchard, J. (1984) Effect of pesticides on mortality, fecundity, and egg viability of *Geocoris pallens* (Hemiptera: Lygaeidae). *Journal of Economic Entomology* 77(4), 876–879.

20 Insect-transmitted Viruses in Canola

RAJAGOPALBABU SRINIVASAN[1]* AND DAVID BUNTIN[2]

[1]*University of Georgia, Tifton, Georgia, USA;* [2]*University of Georgia, Griffin, Georgia, USA*

20.1 Introduction

Canola production could be severely affected by aphids and flea beetles and by viruses transmitted by them in many regions in the USA and elsewhere. Direct feeding damage by aphids and flea beetles often by itself could be a yield-limiting factor. Besides feeding, aphids and flea beetles also transmit a suite of canola-infecting viruses via multiple transmission modes. This chapter discusses aphid and flea beetle species infesting canola, their ecology, their feeding mechanisms, virus transmission modes, virus species and symptoms on canola, and management options to mitigate losses.

20.2 Aphids on Canola

Three species of aphids (Hemiptera: Aphididae) are commonly found on canola. They include the cabbage aphid *Brevicoryne brassicae* (Linn.), green peach aphid *Myzus persicae* (Sulzer) and turnip aphid *Lipaphis erysmi* (Kaltenbach) (Blackman and Eastop, 2006). They are usually found on the underside of leaves as well as on the flowering stalk (Buntin and Raymer, 1994). Various colour morphs are commonly associated with aphids and they vary based on geography, ecological conditions and host availability (Blackman and Eastop, 2006). Grey morphs are usually associated with cabbage and turnip aphids, while yellow or greenish yellow is commonly associated with green peach aphid (Blackman and Eastop, 2006) (Fig. 20.1).

Aphids have piercing and sucking mouthparts and are prolific phloem feeders. When infested at the seedling stage, plants are stunted with shortened internodes and leaves are often yellow and curled (Fig. 20.2). Infested plants are also less cold tolerant; this is critical, as canola is typically grown during the cool seasons (Buntin, 2013). Later-stage infestations, particularly at the flowering stage, could lead to wilting of flowers, ill-filled or deformed pods and reduced pod yields. Severe infestations could also lead to excess honeydew production and sooty mould infections, which could in turn reduce yields. In some instances, canola foliage turns purplish due to induced stress. Aphids are cosmopolitan pests and are uniquely adapted to exploit plant hosts in multiple canola production systems.

20.3 Aphid Ecology and Adaptations

The aphid life cycle has evolved uniquely to exploit multiple ecological zones adeptly. Depending upon the climatic zone, aphids adopt holocyclic or anholocylic reproductive modes. Under temperate conditions, aphids typically alternate between sexual and asexual reproduction (holocyclic), as well as between primary and secondary hosts. They typically overwinter in their least expensive metabolic stage – as eggs on woody perennials. Under subtropical and tropical climatic zones, aphids typically persist year round through asexual (anholocyclic) reproduction (Robert and Lemaire, 1999). Parthenogenetic reproductive abilities allow aphids to increase populations quickly, and effectively exploit ephemeral habitats constituted by agricultural crops or secondary hosts such as canola. Also, many aphids species are highly polyphagous, one example being the green peach aphid, which is capable of using secondary hosts in over 50 plant families (Blackman and Eastop, 2006). Other canola-infecting aphids do not have such a wide host range; turnip and cabbage aphids typically feed on multiple crop and weed hosts within the Brassicaceae. As diverse as their host range is,

*Corresponding author. E-mail: babusri@uga.edu

 Integrated Management of Insect Pests on Canola and Other Brassica Oilseed Crops (ed. G.V.P. Reddy)

Fig. 20.1. (a) Green peach aphids on the underside of a canola leaf; (b) turnip aphids on the underside of a canola leaf; and (c) cabbage aphids on canola stalk. Photographs (a) and (c) by Svetlana Micic, copyright Western Australian Agricultural Authority (WAAA); photograph (b) by David Buntin, UGA.

Fig. 20.2. Aphid infestation and stunting in canola. Photograph by David Buntin, UGA.

they are also capable of adapting and evolving into host-specialized biotypes or races; the evolutionary outcomes of such adaptations have been advantageous under some circumstances (Jaenike, 1990; Srinivasan and Alvarez, 2011) but it can also be construed that such specializations could serve as bottlenecks to function as generalists (Moran, 1988; Losey and Eubanks, 2000).

Besides their ecological adaptations, aphids have a unique feeding mechanism that allows them to extract plant sap while causing relatively less damage than insects with chewing mouthparts. Aphids typically probe through intracellular spaces before they reach the sieve elements. In the process they secrete gelling saliva, which later hardens and seals the lumen once occupied by the stylets. Once the aphids reach the sieve elements they secrete watery saliva with enzymes to digest plant materials; it is during this salivation process that aphids could introduce viruses into the plant. Because of the minimal damage inflicted to cell integrity during this process, the cells survive and permit virus replication thereafter. Such a refined feeding mechanism allows aphids to be more efficient vectors of plant viruses than any other category of insects (de Zoeten, 1968; Dixon, 1973; Pollard, 1977; McLean and Kinsey, 1984). Canola aphids are no exception.

Aphids transmit several viruses to canola in both non-persistent and persistent modes. Non-persistent viruses are often referred to as 'stylet-borne', as the viruses are retained at the very tip of the aphid stylet – the acrostyle (Uzest *et al.*, 2007). These viruses are typically acquired through probing and the acquisition period ranges from a few seconds to minutes. The retention period also ranges only to a few minutes, with no latent period necessary for transmission, which means that the virus can be acquired and inoculated quickly. At the same time, the virus can also be lost quickly with probing and has to be reacquired to inoculate a non-infected plant again.

The other modes of transmission are persistent and persistent propagative transmission. As the term indicates, persistent viruses remain in insects for a longer period (even up to an entire lifetime) upon acquisition. Persistent viruses that multiply within their vectors are called propagative. Contrary to non-persistent viruses, persistent or persistent propagative viruses have an acquisition period lasting several hours to days and the retention period ranges from days to weeks. These viruses also have a latent/refractory period lasting from hours to days. Latent period is the time between acquisition and the ability of an insect to inoculate a plant. This is essentially the time that is required for the virus to reach the salivary glands following acquisition. Only upon reaching the salivary glands could such viruses be inoculated into non-infected plants following salivation that accompanies feeding (reviewed by Watson and Roberts, 1939; Hull, 1994; Nault, 1997; Gray and Banerjee, 1999; Blanc *et al.*, 2001).

20.4 Canola-infecting Viruses

The American Phytopathology Society lists at least seven viruses that are known to infect canola, five of which are transmitted by aphids and two by flea beetles (Kharbanda *et al.*, 2001). Aphid-transmitted viruses will be discussed first. Even though five viruses are listed, not all of them are known to infect canola at once and their distribution varies geographically. Each virus is reviewed below.

20.4.1 Aphid-transmitted viruses

Cauliflower mosaic virus (CaMV)

Cauliflower mosaic virus is the type member of the genus *Caulimovirus* in the family *Caulimoviridae*. This is the only DNA virus that infects canola and it possesses a double-stranded circular genome. It was the first plant virus identified as a DNA virus (Shepherd *et al.*, 1970) and has been extensively studied (Scholthof *et al.*, 2011). Prior to that, all plant viruses were assumed to be RNA viruses. Earlier studies suggested that this virus was transmitted in a semi-persistent manner (Blanc *et al.*, 2001). Semi-persistent viruses are retained in the foregut and are referred to as 'foregut-borne'. Insect transcription factor (ITF) from the viral genome binds with the aphid to facilitate transmission. However, a relatively recent study identified a viral protein 'P2' binding to the insect receptors – a non-glycosylated protein embedded in the chitin matrix located at the very tip of the maxillary stylets where the salivary and food canals converge (Uzest *et al.*, 2007). These results suggested that the virus could be 'stylet-borne', which indicates that the virus could be acquired and inoculated in short intervals (seconds to minutes) in a non-persistent manner. Colonizing as well as non-colonizing aphids could typically transmit such viruses (Kennedy *et al.*, 1962). Colonizing aphids are described above; non-colonizing aphids are those that do not colonize canola. Non-colonizing species could vary with region. Some examples on canola include cowpea aphid *Aphis craccivora* Koch, English grain aphid *Sitobion avenae* (Fabricius), potato aphid *Macrosiphum euphorbiae* (Thomas) and bird cherry-oat aphid *Rhopalosiphum padi* (L.).

CaMV is distributed worldwide and affects members of Brassicaceae, including canola, and also some Solanaceae members under experimental conditions (Lung and Pirone, 1972). CaMV infection in canola is often characterized by yellow ring spots and mottling. CaMV infection can also cause stunting, especially when the plant is infected at the seedling stage. Though not severely yield limiting, heavy yield losses could occur when conditions are optimal. The virus is also known to persist in wild brassicas such as wild mustard, *Brassica arvensis* (Linnaeus), and such hosts often serve as inoculum sources.

Cucumber mosaic virus (CMV)

Cucumber mosaic virus is a single-stranded RNA virus with a tripartite genome in the genus *Cucumovirus* in the family *Bromoviridae*. The virus has a very wide host range extending over 1200 plant species in more than 100 plant families (Palukaitis *et al.*, 1992; Palukaitis and García-Arenal, 2003). The virus is transmitted by about 80 species of aphids in a non-persistent manner (Palukaitis and García-Arenal, 2003); therefore, canola colonizing and non-colonizing aphids could transmit this virus in a non-persistent manner. The virus could be acquired and/or inoculated in very short intervals ranging from a few seconds to minutes. The virus is also seed-transmitted, with up to 50% infection in some hosts (Palukaitis and García-Arenal, 2003), but seed transmission of CMV in canola is not extensively studied.

Even though canola is listed as a host of CMV in the disease catalogue developed by the American Phytopathology Society (Kharbanda *et al.*, 2001), its impact on canola production seems to be minimal overall, as none of the major canola-producing regions include it as a major limiting factor. It has been noticed in canola in Hungary and Iran (Kharbanda *et al.*, 2001; Shahraeen *et al.*, 2003). The symptoms of CMV on canola are also not that clear; the most common symptoms that CMV causes include mosaic and stunting in many plant families. However, even non-symptomatic or less severe infections of CMV could lead to synergistic interactions with other co-occurring viruses and in turn enhance susceptibility to other viruses. The CMV 2b protein 'silencing suppressor' is known to suppress plant defences and enhance susceptibility to other infecting viruses (Palukaitis *et al.*, 1992; Palukaitis and García-Arenal, 2003). Such synergistic interactions have not been well characterized in canola. Overall, this virus by itself, as of now, does not seem to be all that economically relevant to canola production.

Turnip mosaic virus (TuMV)

Turnip mosaic virus is a single-stranded or monopartite RNA virus in the genus *Potyvirus* in the family *Potyviridae*. As in the previous two viruses, TuMV has a worldwide distribution. This virus is not seedborne and is exclusively transmitted by aphids. The multifunctional helper component protein (HC-Pro) encoded by potyviruses is known to facilitate aphid transmission. HC-Pro is known to form a bridge between the aphid stylet and virus capsid protein and this interaction presumably assists in virus retention by the aphid (Raccah *et al.*, 2001; López-Moya and García, 2008). The virus is transmitted by tens of aphid species (Kennedy *et al.*, 1962) in a non-persistent manner, as described for CaMV and CMV above. Similarly, both canola-colonizing and non-colonizing hosts can transmit TuMV.

TuMV has a narrow host range when compared with CMV but is still known to infect several members of Brassicaceae, Chenopodiaceae and Solanaceae. Several such hosts are weeds in canola production fields and could serve as inoculum sources for TuMV and aphid reservoirs. TuMV incidences are quite common in canola regions of Australia, Canada and the USA (Stobbs and Shattuck, 1989; Buntin, 2013). Typical TuMV symptoms on canola include mosaic and stunting; in some cases necrotic spots and vein necrosis could be noticed as well (Fig. 20.3). Plants can be severely stunted when infected at the seedling stage. This virus has the potential to cause serious losses; its severity of incidence seems to be localized though the virus has a wide geographical distribution (van Leur *et al.*, 2014).

Beet western yellows virus (BWYV)

Among others, a synonym for *Beet western yellows virus* is *Turnip yellows virus* (TuYV). BWYV is in the genus *Luteovirus* and family *Luteoviridae* (Duffus, 1960, 1961). This virus also has a wide host range: it is known to infect over 100 host plants in 20 dicotyledonous plant families (Duffus, 1960). BWYV is a single-stranded positive-sense RNA virus. Typically *Luteovirus* members cause yellowing of infected plants (*luteo* in Latin translates as yellow). *Luteovirus* members are phloem restricted, meaning that the virus is only found in the phloem, and insects feeding in the phloem alone can transmit this virus (Smith and Barker, 1999; Domier and D'Arcy, 2008). Therefore, only canola-colonizing aphids can be associated with transmission of this virus. Typically, aphids transmit *Luteoviridae* members in a persistent circulative manner (Miller, 1999; Smith and Barker, 1999). Persistent viruses require longer acquisition and

Fig. 20.3. *Turnip mosaic virus* symptoms on canola. Photographs by Joop van Leur, NSW Department of Primary Industries, Australia.

inoculation access periods stretching from hours to days with a latent period (Nault, 1997). This virus is not seed-transmitted; the rate of BWYV spread is directly proportional to the degree of infestation of colonizing aphids.

Due to its wide host range, BWYV is commonly found infecting alternative weed hosts in the farmscapes and could persist in the area. Unlike the typical 'yellows' symptom associated with luteoviruses in many host plants, the symptoms vary in canola. Infected canola plants often display purpling or reddening of lower leaves (Fig. 20.4). Stunting can occur in plants when infected at the seedling stage and can severely affect yield and oil quality. In Australia, among the viruses infecting canola, BWYV causes more losses than the others. BWYV infection and severe yield losses are becoming more common in Canada and the USA as well (van Leur *et al.*, 2014).

Broccoli necrotic yellows virus (BNYV)

Broccoli necrotic yellows virus is unique among others described above. It is an enveloped bullet-shaped RNA virus, in the genus *Cytorhabdovirus* in the family *Rhabdoviridae* (Hills and Campbell, 1968; Campbell and Lin, 1972; Tomlinson *et al.*, 1972; Sylvester and Richardson, 1992). Even though BNYV has been identified as a virus infecting *Brassica* crops in Australia, the UK and the USA (Lin and Campbell, 1972; Garrett and O'Loughlin, 1977) and listed by the American Phytopathology Society as a canola-infecting virus (Kharbanda *et al.*, 2001), it is not all that prevalent in canola. Some BNYV relatives are animal- and plant-infecting viruses but it is not clear in which direction evolution occurred. However, it is hypothesized that plant-infecting rhabdoviruses such as BNYV replicate in their insect vector and also are horizontally transmitted; it could have had its origin in the animal kingdom (Garrett and O'Loughlin, 1977; Sylvester and Richardson, 1992).

Fig. 20.4. BWYV-infected canola plant with thickened and reddening leaves. Photograph by Brenda Coutts, copyright Western Australian Agricultural Authority (WAAA).

BNYV is not seed-transmitted and is exclusively transmitted by aphids in a persistent and propagative manner (Sylvester and Richardson, 1992). The acquisition and inoculation access periods of cytorhabdoviruses could be long-lasting up to a few days. The latent period is also quite long (several days) (Duffus, 1963). Cytorhabdoviruses are also transmitted transovarially, meaning through the offspring, as parthenogenetic aphids typically produce nymphs (Prentice and Woollcombe, 1951; Duffus, 1963; Sylvester and McClain, 1978). The symptoms of BNYV on canola are not all that clear, or have not been studied extensively, but symptoms such as mild vein clearing, leaf rolling and reduced growth have been observed on *Brassica* hosts (Tomlinson *et al.*, 1972; Garrett and Martindale, 1973).

20.4.2 Viruses transmitted by flea beetles

Flea beetles (*Phyllotreta* spp. in the family Chrysomelidae and order Coleoptera) are important pests in some canola-growing areas. They are major pests in spring-planted canola in the USA but not in fall-planted canola in southern USA. They are also considered important pests in Australia and Canada. In contrast to aphids, these insects can cause substantial defoliation, as they possess biting and chewing mouthparts. Their feeding is characterized by shot holes. Severe defoliation (beyond 25%) during seedling stage and/or budding stage could cause serious yield losses. The economic threshold level is 25% of defoliation with flea beetles continuing to feed (Canola Encyclopedia, 2014). Besides causing rapid defoliation, flea beetles can also transmit viruses in the process. Below are a few examples.

Radish mosaic virus (RaMV)

Radish mosaic virus is a single-stranded RNA virus in the genus *Comovirus* in the family *Secoviridae*

(Tompkins, 1939; Campbell, 1964). As the name indicates, this virus is known to cause mosaic. In addition, symptoms such as ring spots, leaf crinkling, enations and systemic necrosis have been observed on *Brassica* hosts (Campbell, 1964). Despite being listed as a virus infecting canola by the American Phytopathology Society catalogue (Kharbanda *et al.*, 2001), not much is known about this virus in canola. It does not appear to be causing major yield reduction in any major canola-producing part of the world but the virus has been observed in *Brassica* hosts in the USA, Asia and Eastern Europe (Campbell, 1964; Stefanac and Mamula, 1971).

This virus is not believed to be seed-transmitted. The only known vectors are beetles such as *Phyllotreta* spp. The mode of transmission of this virus is not all that clear but studies conducted with other beetle-transmitted viruses, including comoviruses, suggest that the virus is externally borne, that it can be acquired with short feeding access and that there seems to be no latent period for inoculating the acquired virus (Fulton *et al.*, 1987). The regurtitant (a combination of food materials and saliva) is believed to contain RNases (RNA digesting enzymes) that operate with some degree of specificity, i.e. the RNases in the regurgitant are ineffective against the viruses that the beetles transmit but could prevent establishment of other viruses by binding to the virus or degrading it. Perhaps this is the reason why beetles do not transmit other mechanically transmissible viruses such as *Tobacco mosaic virus* (TMV) (Gergerich *et al.*, 1983; Gergerich and Scott, 1988).

Turnip crinkle virus (TCV)

Turnip crinkle virus is a single-stranded positive-sense RNA virus in the genus *Carmovirus* in the family *Tombusviridae* (Broadbent and Heathcote, 1958). The virus is found in the UK as well as in Eastern Europe. Though it is described as a canola-infecting virus (Kharbanda *et al.*, 2001), losses induced by this virus in canola seem to be minimal. The virus has a wide host range of up to or more than 20 families. Symptoms on *Brassica* hosts include crinkling, mottling and rosetting of leaves. Leaf distortion and stunting have also been reported (Broadbent, 1957).

No information exists to assess if the virus is seed transmissible. In general, it is mechanically transmissible and is also transmitted by larvae and adults of flea beetles in the genera *Phyllotreta* and *Psylliodes* (Martini, 1958; Qu and Morris, 2008). At least ten species of flea beetles in these two genera could potentially serve as vectors of TCV. The virus, as in the previous case, can be acquired within a few minutes of acquisition access period and the beetles typically remain viruliferous for less than 24 h (Martini, 1958). This suggests that this virus is also borne externally (non-persistent), as in the case of RaMV. Flea beetles probably transmit TCV by following the same mechanism as described for RMV.

Other beetle-borne viruses

Though these viruses are not considered very important or are not observed in canola production, they have been identified in other *Brassica* spp. One such is a single-stranded RNA virus, *Turnip yellow mosaic virus* (TYMV). TYMV is in the genus *Tymovirus* and causes mosaic in various *Brassica* hosts. It is not clear if it would impact canola or what its symptoms are on canola. The other single-stranded positive-sense RNA virus is *Turnip rosette virus* (TRoV) in the genus *Sobemovirus*. This virus is known to cause leaf twisting, vein banding, rosetting, petiole necrosis and stunting (Broadbent and Heathcote, 1958). Flea beetles in the family Chrysomelidae presumably transmit these viruses. It is also speculated that members of the genus *Tymovirus* could be transmitted by beetles/weevils in the family Curculionidae (Gibbs, 1999; Haenni and Dreher, 2008). The cabbage seedpod weevil (*Ceutorhynchus obstrictus* (Marsham)) is commonly found in regions such as the southern USA. No studies have been conducted so far to assess the virus transmission capabilities of this insect in canola production.

20.5 Management of Vectors and Viruses in Canola

The thresholds applicable for management of insects such as aphids and flea beetles are further reduced when they have the capability to transmit viruses that infect canola. As in other row-cropping systems, a suite of management options could be adopted when available. The most common management options include: (i) resistant cultivars; (ii) insecticides to suppress vector populations; and (iii) cultural tactics, including weed management.

20.5.1 Host-plant resistance

Host resistance is probably the most important tool for vector and virus management, especially in a low-input production system such as that for canola. Currently, in canola, host resistance against any of the aphids or viruses described above is minimal, at the most. Resistant cultivars are not commercially available in most (if not all) canola-producing regions of the world. However, spikes in virus incidences in Australia and in Canada have coerced breeders and plant pathologists to screen for resistance. Researchers identified a major quantitative trait locus that confers resistance to BWYV (also known as *Turnip yellows virus*, TuYV) in a rapeseed cross that involved the lines R54 and Express in Europe (Dreyer *et al.*, 2001). They found that the locus explained 50% of the phenotypic variation and suggested a single mode of inheritance.

Genetic maps for resistance and molecular markers to identify resistance have been made available (Piquemal *et al.*, 2005; Sun *et al.*, 2007). Juergens *et al.* (2010) characterized such resistance to BWYV as a monogenic trait. However, the resistance seems to be incomplete and the virus load seems to be influenced by temperature × genotype interaction. Though simple, such traits could be influenced by interactions and they complicate introgression into desirable cultivars. Due to recent spikes of viruses such as BWYV and TuMV, there are efforts to select and/or breed for resistance in Australia. Most breeding lines evaluated were susceptible to TuMV; some canola breeding lines seem to exhibit tolerance/resistance to BWYV (van Leur et al., 2014). It is possible that such efforts could result in release of virus-resistant cultivars in the near future.

Resistance against vectors is another approach that has been undertaken, at least with reference to flea beetles. None of the commonly grown canola cultivars are immune to flea beetles but screening has revealed varying degrees of tolerance to flea beetles. The level of trichome density seemed to affect flea beetle feeding in canola germplasm entries that were screened in Canada. Increased trichome density was correlated with decreased flea beetle feeding (Gruber *et al.*, 2006; Soroka *et al.*, 2011). Research is currently under way to examine if trichome characteristics that thwart flea beetle feeding substantially could be bred to desirable cultivars and be used as a management tool, either alone or in combination with other cultural and chemical management options.

20.5.2 Chemical management of vectors

Frequent insecticide applications on canola are not cost effective. Both aphids and flea beetles cause serious yield losses when infestation levels are high at the seedling stage as well as at the early rosette stage. In southern USA, in Georgia, two aphids per leaf at the seedling stage and five aphids per leaf at the early rosette stage are considered as thresholds for insecticide application. The threshold at flowering is 15% of infested stems. It is advisable not to treat canola at the late blooming stage, to prevent toxicity to bees (Buntin, 2013). Flea beetles are not so important in some places such as southern USA where canola is planted in the fall but they could become serious pests where canola is planted in the spring, in places such as the northern Great Plains in the USA and Canada. Unlike aphids, economic damage by flea beetles is based on percentage defoliation in the seedling stage; in Canada, 25% seedling defoliation is considered as the economic threshold level (Canola Encyclopedia, 2014). These thresholds are helpful when virus incidences are not so common or severe but should be followed with caution when severe virus incidences are the norm. In such situations it may pay to be proactive and plant treated seeds from the start.

Seed treatment with a systemic neonicotinoid such as imidacloprid (Gaucho®) is a common practice in many places, including Australia, Canada and the USA. This typically protects the plants against both aphids and flea beetles. A carbamate insecticide, pirimicarb (Pirimor®), is also being used against aphids in some places. Besides these, insecticides such as flonicamid (Beleaf®), sulfoxaflor (Transform®) and pyrethroids such as bifenthrin and cyhalothrin are registered for use against aphids on canola. For flea beetles, also, a number of pyrethroids are registered for management on canola (Horton, 2015). In all these cases, care should be taken to prevent resistance development by judiciously following recommended rates and rotating insecticide chemistries whenever possible. For instance, in Australia there is already green peach aphid resistance to pirimicarb (Umina and Edwards, 2015); in such situations, care should be taken to use appropriate insecticides and prevent spread of future spray failures.

20.5.3 Cultural practices/tactics

Planting date

Planting date could be critical to mitigate losses: planting after the peak aphid flight could prevent exposure of the crop at the most vulnerable stage. In order to avoid such exposure, it is imperative to have an extensive knowledge of pest population dynamics, cropping patterns in farmscapes and aphid dispersal patterns from adjoining crops. In some canola-producing regions, as in Australia, planting date seems to have an effect on virus incidence. For instance, planting in mid to late May rather than in mid to late April resulted in decreased virus incidence; these effects were more pronounced when optimal planting was combined with the usage of an aphicide at planting (Hertel *et al.*, 2004). Similarly, planting date also affects flea beetle induced damage on canola (Cárcamo *et al.*, 2008).

Tillage and row spacing

Tillage and row spacing could be other factors that affect vectors and virus incidences. Direct seeding or planting with stubble residues has been known to reduce aphid populations and it is presumed that reducing the area of bare ground prevents aphids from locating hosts and reduces landing on canola. Direct seeding in stubbles is also known to reduce flea beetle populations, because soil temperature increases with deep tillage and flea beetles typically perform poorly when the temperatures are cooler. Reducing row spacing in canola to 10 cm from 20–30 cm is known to interfere with host location of flea beetles for the same reason as for aphids (Dosdall *et al.*, 1999).

Weed management

Weed management is another critical aspect for vector and virus management. Several of the viruses that infect canola have broad host ranges within Brassicaceae as well as others families. These weed species can serve as vector reservoirs and virus inoculum sources and thus influence virus epidemics in canola. Therefore, managing weeds in the farmscape could reduce vectors and virus incidences in canola. Good examples include weedy brassicas such as wild mustard, which could serve not only as a virus host but also as a vector reservoir. Care should be taken to avoid weed removal just prior to planting, especially to prevent vectors dispersing to canola. Doing so could increase virus incidence in canola. To reduce infection in canola, removing weeds a few weeks before scheduled planting/seeding could be useful.

Biological control

Biological control using natural enemies in canola is only based on conservation. A number of natural enemies, including parasitoids belonging to genera such as *Aphelinus*, *Diaeretiella* and *Lysiphlebus*, are often found in canola fields (Boyd and Lentz, 1994; Elliott *et al.*, 2014). Coccinellid beetle larvae of various species have been commonly observed in canola fields (Buntin, 2013). Natural enemies, in general, are good at suppressing aphid populations on canola. However, naturally occurring populations are limited by weather patterns and by flowering in canola. For instance, in Georgia, USA, coccinellids are seen later in the spring at flowering but not in winter months. During their absence, aphid populations tend to increase exponentially without check (Buntin, 2013). When viruses are brought into the management equation with biological control as one of the parameters, the complexity increases. It is also possible that natural enemies can potentially increase virus incidence by enhancing vector movement. Natural enemies are known to agitate aphids, resulting in increased movement as well as increased probing (Tamaki *et al.*, 1970; Irwin and Thresh, 1990). The resultant increased probing could serve as a mechanism for increased virus transmission, especially of non-persistent or non-circulative viruses (Weber *et al.*, 1996). On the other hand, natural enemies could be effective in lowering the incidence of viruses that are persistently transmitted, such as BWYV (Smyrnioudis *et al.*, 2001). In the latter study, the authors found that even the incidence of a persistent virus increased in the presence of a coccinellid beetle, presumably due to increased vector mobility. Research in this area in general is limited and more so in the case of canola.

20.6 Conclusions

In summary, based on the information presented above, even though a number of management options are available, none of them is capable of serving as a 'silver bullet'. In order to maximize control, a suite of the management options described above has to be integrated such that the combinations produce additive/synergistic effects in managing vectors and viruses. Such a combination should be tailored to each location based on prevailing conditions. Until cultivars with appreciable levels of resistance against vectors and/or viruses in canola are available commercially, an integrated management strategy in canola production is warranted to mitigate losses to vectors and viruses.

Acknowledgements

We gratefully acknowledge Ms Svetlana Micic and Dr Brenda Coutts from the Department of Agriculture and Food, WA, and Dr Joop van Leur from NSW, Department of Primary Industries, Australia, for their permission to use images of aphids and virus symptoms on canola. We also acknowledge the Western Australian Agriculture Authority (WAAA) for providing copyright clearance for use of images.

References

Blackman, R.L. and Eastop, V.F. (2006) *Aphids on the World's Herbaceous Plants and Shrubs*. John Wiley & Sons, Chichester, UK.

Blanc, S., Hebrard, E., Drucker, M. and Froissart, R. (2001) Molecular basis of vector transmission: Caulimovirus. In: Harris, K., Duffus, J.E. and Smith, O.P. (eds) *Virus-Insect-Plant Interactions*. Academic Press, San Diego, pp. 143–166.

Boyd, M.L. and Lentz, G.L. (1994) Seasonal occurrence of aphids and aphid parasitoid *Diaeratiella rapae* (M'Intosh) (Hymenoptera: Aphidiidae) on rapeseed in Tennessee. *Environmental Entomology* 23, 349–353.

Broadbent, L. (1957) *Investigation of Virus Diseases of Brassica Crops,* Agricultural Research Council Report No. 14. Cambridge University Press, London.

Broadbent, L. and Heathcote, G.D. (1958) Properties and host range of turnip crinkle, rosette and yellow mosaic viruses. *Annals of Applied Biology* 46, 585–592.

Buntin, D. (ed.) (2013) *Canola Production in Georgia*. UGA Extension Publication B1331. University of Georgia Cooperative Extension, Athens, Georgia.

Buntin, G.D. and Raymer, P.L. (1994) Pest status of aphids and other insects in winter canola in Georgia. *Journal of Economic Entomology* 87, 1097–1104.

Campbell, R.N. (1964) Radish mosaic virus, a crucifer virus serologically related to strains of Bean pod mottle virus and to Squash mosaic virus. *Phytopathology* 54, 1418–1424.

Campbell, R.N. and Lin, M.T. (1972) *Broccoli Necrotic Yellows Virus*. Commonwealth Mycological Institute/ Association of Applied Biologists Descriptions of Plant Viruses, No. 85. Commonwealth Mycological Institute, Kew, UK.

Canola Encyclopedia (2014) *Canola Encyclopedia*. Canola Council of Canada, Winnipeg, Manitoba, Canada. Available at: http://www.canolacouncil.org/canola-encyclopedia/insects/flea-beetles/#resistance-breeding

Cárcamo, H.A., Otani, J.K., Dosdall L.M., Blackshaw, R.E., Clayton, G.W., Harker, K.N., O'Donovan, J.T. and Entz, T. (2008) Effects of seeding date and canola species on seedling damage by flea beetles in three ecoregions. *Journal of Applied Entomology* 132, 623–631.

de Zoeten, G.A. (1968) Application of scanning electron microscopy in the study of virus transmission by aphids. *Journal of Virology* 2, 745–751.

Dixon, A.F.G. (1973) *The Biology of Aphids*. Edward and Arnold, London.

Domier, L.L. and D'Arcy, C.J. (2008) Luteoviruses. In: Mahy, B.W.J. and Van Regenmortel, M.H.V. (eds) *Encyclopedia of Virology*, 3rd edn, Vol. 3. Academic Press, San Diego, pp. 231–238.

Dosdall, L.M., Dolinski, M.G., Cowle, N.T. and Conway, P.M. (1999) The effect of tillage, row spacing, and seeding rate on feeding injury by flea beetles, *Phyllotreta* spp. (Coleoptera: Chrysomelidae) in canola in central Alberta, Canada. *Crop Protection* 18, 217–224.

Dreyer, F., Graichen, K. and Jung, C. (2001) A major quantitative trait locus for resistance to *Turnip yellows virus* (TuYV, syn. Beet western yellows virus, BWYV) in rapeseed. *Plant Breeding* 120, 457–462.

Duffus, J.E. (1960) Radish yellows, a disease of radish, sugar beet, and other crops. *Phytopathology* 50, 389–394.

Duffus, J.E. (1961) Economic significance of beet western yellows (radish yellows) on sugar beet. *Phytopathology* 51, 605–607.

Duffus, J.E. (1963) Possible multiplication in the aphid vector of sowthistle yellow vein virus, a virus with an extremely long insect latent period. *Virology* 21, 194–202.

Elliott, N.C., Backoulou, G.F., Giles, K.L and Royer, T.A. (2014) Aphids and parasitoids in wheat and nearby canola fields in central Oklahoma. *Southwestern Entomologist* 39, 23–28.

Fulton, J.P., Gergerich, R.C. and Scott, H.A. (1987) Beetle transmission of plant viruses. *Annual Review of Phytopathology* 25, 111–123.

Garrett, R.G. and Martindale, H.A. (1973) The occurrence of broccoli necrotic yellows virus in Victoria. *Australian Plant Pathology Society Newsletter* 2, 17–18.

Garrett, R.G. and O'Loughlin, G.T (1977) Broccoli necrotic yellows virus in cauliflower and in the aphid *Brevicoryne brassicae* L. *Virology* 76, 653–663.

Gergerich, R.C. and Scott, H.A. (1988) The enzymatic function of ribonuclease determines plant virus transmission by leaf feeding beetles. *Phytopathology* 78, 270–272.

Gergerich, R.C., Scott, H.A. and Fulton, J.P. (1983) Regurgitant as a determinant of specificity in the transmission of plant viruses by beetles. *Phytopathology* 73, 936–938.

Gibbs, A. (1999) Tymoviruses. In: Granoff, A. and Webster, R.G. (eds) *Encyclopedia of Virology*, 2nd edn. Academic Press, San Diego, pp. 1859–1853.

Gray, S. and Banerjee, N. (1999) Mechanisms of arthropod transmission of plant viruses. *Microbiology and Molecular Biology Reviews* 63, 128–148.

Gruber, M.Y., Wang, S., Ethier, S., Holowachuk, J., Bonham-Smith, P.C. Soroka, J.J. and Lloyd, A. (2006) "Hairy Canola" – Arabidopsis *GL3* induces a dense coating of trichomes on *Brassica napus* seedlings. *Plant Molecular Biology* 60, 679–698.

Haenni, A.-L. and Dreher, A. (2008) Tymoviruses. In: Mahy, B.W.J. and Van Regenmortel, M.H.V. (eds) *Encyclopedia of Virology*, 3rd edn, Vol. 5. Academic Press, San Diego, pp. 199–207.

Hertel, K., Schwinghamer, M. and Bambach, R. (2004) *Virus Diseases in Canola and Mustard*. Agnote DPI 495. New South Wales Department of Primary Industries, Orange, NSW, Australia.

Hills, G.J. and Campbell, R.N. (1968) Morphology of broccoli necrotic yellows virus. *Journal of Ultrastructure Research* 24, 134–144.

Horton, D. (ed.) (2015) *Georgia Pest Management Handbook*. UGA Extension publication. University of Georgia Cooperative Extension, Athens, Georgia.

Hull, R. (1994) Molecular biology of plant virus–vector interactions. *Advances in Disease Vector Research* 10, 361–386.

Irwin, M.E. and Thresh, J.M. (1990) Epidemiology of barley yellow dwarf virus: a study in ecological complexity. *Annual Review of Phytopathology* 28, 393–424.

Jaenike, J. (1990) Host specialization in phytophagous insects. *Annual Review of Ecology and Systematics* 21, 243–273.

Juergens, M., Paetsch, J.M., Krämer, I., Zahn, M. Rabenstein, F., Schondelmaier, J., Schliephake, E., Snowdon, R., Friedt, W. and Ordon, F. (2010) Genetic analyses of the host-pathogen system *Turnip yellows virus* (TuYV)–rapeseed (*Brassica napus* L.) and development of molecular markers for TuYV-resistance. *Theoretical and Applied Genetics* 120, 735–744.

Kennedy, D., Day, M.F. and Eastop, V.F. (1962) *A Conspectus of Aphids as Vectors of Plant Viruses*. Commonwealth Institute of Entomology, London.

Kharbanda, P.D., Fitt, B.D.L., Lange, R.M., West, J.S., Lamey, A.H. and Phillips, D.V. (2001) Diseases of Rapeseed = Canola (*B. napus* L. and *Brassica rapa* L. (= *B. campestris* L.)). American Phytopathology Society publications: common names of diseases. Available at: http://www.apsnet.org/publications/commonnames/Pages/Rapeseed.aspx.

Lin, M.T. and Campbell, R.N. (1972) Characterisation of broccoli necrotic yellows virus. *Virology* 48, 30–40.

López-Moya, J.J. and García, J.A. (2008) Potyviruses. In: Mahy, B.W.J. and Van Regenmortel, M.H.V. (eds) *Encyclopedia of Virology*, 3rd edn, Vol. 4. Academic Press, San Diego, pp. 313–322.

Losey, J.E. and Eubanks, M.D. (2000) Implications of a pea aphid specialization for the potential colonization of vegetables following post-harvest emigration from forage crops. *Environmental Entomology* 29, 1283–1288.

Lung, M.C.Y. and Pirone, T.P. (1972) *Datura stramonium*, a local lesion host for certain isolates of cauliflower mosaic virus. *Phytopathology* 62, 1473.

Martini, C. (1958) The transmission of turnip viruses by biting insects and aphids. In: Quak, F., Dijkstra, J., Beemster, A.B.R. and van der Want, J.P.H. (eds) *Proceedings 3rd Conference Potato Virus Diseases, Lisse-Wageningen* 1957, pp. 106–113.

McLean, D.L. and Kinsey, M.G. (1984) The precibarial valve and its role in the feeding behavior of the pea aphid, *Acyrthosiphon pisum*. *Bulletin of the Entomological Society of America* 30, 26–31.

Miller, W.A. (1999) Luteovirus (Luteoviridae). In: Granoff, A. and Webster, R.G. (eds) *Encyclopedia of Virology*, 2nd edn. Academic Press, San Diego, pp. 901–908.

Moran, N.A. (1988) The evolution of host plant alternation in aphids: evidence for specialization as a dead end. *The American Naturalist* 132, 681–706.

Nault, L.R. (1997) Arthropod transmission of plant viruses: a new synthesis. *Annals of the Entomological Society of America* 90, 521–541.

Palukaitis, P. and García-Arenal, F. (2003) Cucumoviruses. *Advances in Virus Research* 62, 241–323.

Palukaitis, P., Roossinck, M.J., Dietzgen, R.G. and Francki, R.I.B. (1992) Cucumber mosaic virus. *Advances in Virus Research* 41, 281–348.

Piquemal, J., Cinquin, E., Couton, F., Rondeau, C., Seignoret, E., Doucet, I., Perret, D., Villenger, M.-J., Vincourt, P. and Blanchard, P. (2005) Construction of an oilseed rape (*Brassica napus* L.) genetic map with SSR markers. *Theoretical and Applied Genetics* 111, 1514–1523.

Pollard, D.G. (1977) Aphid penetration of plant tissues. In: Harris, K.F. and Maramorosch, K. (eds) *Aphids as Virus Vectors*. Academic Press, New York, pp. 105–118.

Prentice, I.W. and Woollcombe, T.M. (1951) Resolution of strawberry virus complexes IV. The latent period of virus 3 in the vector. *Annals of Applied Biology* 38, 389–394.

Qu, F. and Morris, T.J. (2008) Tymoviruses. In: Mahy, B.W.J. and Van Regenmortel, M.H.V. (eds) *Encyclopedia of Virology*, 3rd edn, Vol. 1. Academic Press, San Diego, pp. 453–457.

Raccah, B., Blanc, S. and Huet, H. (2001) Molecular basis of vector transmission: Potyvirus. In: Harris, K., Duffus, J.E. and Smith, O.P. (eds) *Virus-Insect-Plant Interactions*. Academic Press, San Diego, pp. 181–206.

Robert, Y. and Lemaire, O. (1999) Epidemiology and control strategies. In: Smith, H.G. and Barker, H. (eds) *The Luteoviridae*. CAB International, Wallingford, UK, pp. 211–279.

Scholthof, K.G., Adkins, S., Czosnek, H., Palukaitis, P., Jacquot, E., Hohn, T., Hohn, B., Saunders, K., Candresse, T., Ahlquist, P., Hemenway, C. and Foster, G.D. (2011) Top 10 plant viruses in molecular plant pathology. *Molecular Plant Pathology* 12, 938–954.

Shahraeen, N., Farzadfar, S. and Lesemann, D.E. (2003) Incidence of viruses infecting winter oilseed rape (*Brassica napus* ssp. *olifera*) in Iran. *Journal of Phytopathology* 151, 614–616.

Shepherd, R.J., Bruening, G.E. and Wakeman, R.J. (1970) Double-stranded DNA from *Cauliflower mosaic virus*. *Virology* 41, 339–347.

Smith, H.G. and Barker, R.J. (eds) (1999) *The Luteoviridae*. CAB International, Wallingford, UK.

Smyrnioudis, I.N., Harrington, R., Clark, S.J. and Katis, N. (2001) The effect of natural enemies on the spread of *Barley yellow dwarf virus* (BYDV) by *Rhopalosiphum padi* (Hemiptera: Aphididae). *Bulletin of Entomological Research* 91, 301–306.

Soroka, J.J., Holowachuk, J.M., Gruber, M.Y. and Grenkow, L.F. (2011) Feeding by flea beetles (*Phyllotreta* spp.) (Coleoptera: Chrysomelidae) is decreased on canola *(Brassica napus)* seedlings with increased trichome density. *Journal of Economic Entomology* 104, 125–136.

Srinivasan, R. and Alvarez, J.M. (2011) Specialized host utilization of *Macrosiphum euphorbiae* on a nonnative weed host, *Solanum sarrachoides*, and competition with *Myzus persicae*. *Environmental Entomology* 40(2), 350–356.

Stefanac, Z. and Mamula, D. (1971) A strain of Radish mosaic virus occurring in turnip in Yugoslavia. *Annals of Applied Biology* 69, 229–234.

Stobbs, L.W. and Shattuck, V.I. (1989) Turnip mosaic strains in southern Ontario. *Plant Pathology* 73, 208–212.

Sun, Z., Wang, Z., Tu, J., Zhang, J., Yu, F., McVetty, P.B.E. and Li, G. (2007) An ultradense genetic recombination map for *Brassica napus*, consisting of 13551 SRAP markers. *Theoretical and Applied Genetics* 114, 1305–1317.

Sylvester, E.S. and McClain, E. (1978) Rate of transovarial passage of sowthistle yellow vein virus in selected subclones of the aphid *Hyperomyzus lactucae*. *Journal of Economic Entomology* 71, 17–20.

Sylvester, E.S. and Richardson, J. (1992) Aphid-borne rhabdoviruses relationships with their vectors. *Advances in Disease Vector Research* 9, 313–341.

Tamaki, G., Halfhill, J.E. and Hathaway, D.D. (1970) Dispersal and reduction of *Acyrthosiphon pisum* (Harris) by *Aphidius smithii*. *Annals of the Entomological Society of America* 63, 973–980.

Tomlinson, J.A., Webb, M.J.W. and Faithfull, E.M. (1972) Studies on broccoli necrotic yellows virus. *Annals of Applied Biology* 71, 127–134.

Tompkins, C.M. (1939) A mosaic disease of radish in California. *Journal of Agriculture Research* 58, 119–130.

Umina, P. and Edwards, O. (2015) *Resistance management strategy for the green peach aphid in Australian grains*. GRDC – Grain Research and Development Corporation, Canberra, Australia.

Uzest, M., Gargani, D., Drucker, M., Hebrard, E., Garzo, E., Candresse, T., Fereres, A. and Blanc, S. (2007) A protein key to plant virus transmission at the tip of the insect vector stylet. *Proceedings of the National Academy of Sciences of the USA* 13, 17959–17964.

van Leur, J., Aftab, M., Sharman, M., and Lindbeck, K. (2004) *Viral diseases in canola and winter pulses*. Available at: https://grdc.com.au/Research-and-Development/GRDC-Update-Papers/2014/02/Viral-diseases-in-canola-and-winter-pulses (accessed 13 February 2017).

Watson, M.A. and Roberts, F.M. (1939) A comparative study of the transmission of *Hyoscyamus* virus 3, Potato virus Y, and Cucumber Mosaic virus by the vector *Myzus persicae* (Sulzer), *M. circumflexus* (Buckton), and *Macrosiphum gei* (Koch). *Proceedings of the Royal Society B: Biological Sciences* 127, 543–576.

Weber, C.A., Godfrey, L.D. and Mauk, P.A. (1996) Effects of parasitism by *Lysiphlebus testaceipes* (Hymenoptera: Aphidiidae) on transmission of beet yellows closterovirus by bean aphid (Homoptera: Aphididae). *Journal of Economic Entomology* 89, 1431–1437.

21 Present and Potential Impacts of Insects on Camelina and Crambe

Juliana J. Soroka*, Chrystel Olivier, Tyler J. Wist and Larry Grenkow

Agriculture and Agri-Food Canada, Saskatoon Research and Development Centre, Saskatoon, Saskatchewan, Canada

21.1 Introduction

This chapter examines the insect fauna reported on camelina, *Camelina sativa* (Linnaeus) Cranz., and crambe, *Crambe abyssinica* Hochst ex R.E. Fries, two Old World crops that have seen renewed interest for development both as sources of oil for consumption and for industrial use of oil and meal. A cursory examination of the literature reveals few insect pests of camelina and crambe. As a result, the expectation may be that camelina and likely crambe will not support high populations of crucifer-feeding insect pests. However, 60 years ago in Canada, oilseed rape *Brassica napus* Linnaeus and *Brassica rapa* Linnaeus (now called canola in North America with the reduction of glucosinolate levels in the seed) likewise had few insect pests with minor economic impact, because the crop was in early development and production area was limited. Therefore, while the present economic impact of insects on camelina and crambe may be minimal, host selection and the economic impact of insects may change should production increase and should more insects come into contact with camelina and crambe. This chapter examines the insects that are pests of canola and the insects that are found on camelina and crambe and tries to determine the reasons for their feeding preferences. In this manner it may be possible to predict the pest status of such insects on camelina and crambe more accurately in the future.

21.2 The Crops

Both camelina, *C. sativa*, and crambe, *Cr. abyssinica*, are flowering herbaceous plants in the family Brassicaceae. Currently, while production of oilseed rape/canola *B. napus* ranks in the millions of hectares worldwide, combined commercial production of camelina and crambe ranks in the thousands of hectares. However, both camelina and crambe are receiving interest as alternatives to traditional cruciferous oilseed crops because of their relatively high oilseed content, unique uses, distinct seed morphology, low environmental burden and self-pollinating characteristics with little potential for outcrossing to wild relatives. Thus, increase in production area of both crops is a real possibility.

21.2.1 Camelina

Camelina sativa was domesticated in Europe and Central Asia in Neolithic times as a vegetable oil for human consumption (Knörzer, 1978; Francis and Warwick, 2009) and became a source of animal fodder, fibre and lamp oil (Porcher, 1869; Sturtevant, 1919; Francis and Warwick, 2009). This utility resulted in common names such as gold-of-pleasure, false flax, German sesame, Siberian oilseed and linseed dodder (Porcher, 1869; Putnam *et al.*, 1993; Francis and Warwick, 2009). The ongoing small-scale cultivation of camelina in Eurasia was virtually eliminated following the widespread introduction of electricity in the 19th century and the development of hydrogenated vegetable oils after World War II (Knörzer, 1978). Interest in developing camelina as a competitive crop resurfaced with the discovery that camelina oil is a rich source of essential fatty acids, especially omega-3 fatty acids (Zubr,

*Corresponding author. E-mail: julie.soroka@agr.gc.ca

  Integrated Management of Insect Pests on Canola and Other Brassica Oilseed Crops (ed. G.V.P. Reddy)

1997), prompting its development as a salad and cooking oil. Camelina meal is high in protein (Korsrud *et al.*, 1978) and its components have a variety of other uses, such as additives to nutraceuticals, cosmetics, animal feeds, jet fuels and biolubricants (Pilgeram *et al.*, 2007).

Camelina was thought to have reached North America with European settlers as a weedy contaminant in seeds of flax or other crops (Putnam *et al.*, 1993; Francis and Warwick, 2009). It has been considered periodically as a potential oilseed crop in North America in the past (Plessers *et al.*, 1962; Robinson, 1987), but the drive for energy self-sufficiency in the last decade (e.g. United States Congress, 2007) prompted research in the use of camelina as a source of biofuel for the continent (Pilgeram *et al.*, 2007). As a result, significant commercial cultivation began in the late 1990s in the northern Great Plains.

Camelina initially grows as a rosette and then bolts to reach a height of 30–90 cm at maturity, with its branched stems becoming woody as it matures (Fleenor, 2011). The lanceolate leaves of camelina are 5–8 cm long with entire margins and are covered with simple trichomes mixed with two- and three-armed trichomes (Ančev and Goranova, 2006), with trichome density up to 45 times higher on upper leaf surfaces than on lower ones (Deng *et al.*, 2004a). The small, pale or greenish yellow flowers have four petals arranged in a typical cruciate formation. The seed pods or siliques, which somewhat resemble flax bolls, are 6–14 mm long (Putnam *et al.*, 1993) and contain an average of 15 seeds per pod, with a 1000-seed weight of 0.8–1.8 g (Zubr, 1997). Seeds are one-quarter to one-half the size of canola seeds, pale yellow-brown, oblong and rough.

Camelina is most commonly cultivated as a spring annual (Gugel and Falk, 2006; Urbaniak *et al.*, 2008), though winter annual biotypes have been grown in central and northern Europe (Zubr, 1997) and in milder climates of North America (Hunter and Roth, 2010). When sown in the spring camelina is a short-season crop, with seeding to harvest times in the range of 85–100 days (Putnam *et al.*, 1993; Gugel and Falk, 2006; Ehrensing and Guy, 2008). The crop is well suited to cold, semi-arid climate zones, as the seed is able to germinate at low temperatures (Plessers *et al.*, 1962; Robinson, 1987; Putnam *et al.*, 1993; Ehrensing and Guy, 2008) and seedlings have been observed to survive frosts as low as −11°C without damage (Plessers *et al.*, 1962; Ehrensing and Guy, 2008). Likewise, camelina can tolerate drought (Zubr, 1997; Gugel and Falk, 2006; Francis and Warwick, 2009); in some conditions the crop has minimal water requirements that are much less than typical grain and vegetable crops (Hunsaker *et al.*, 2011). It is able to grow in most soil types except for heavy clay and organic soils (Andersson and Olsson, 1950; Zubr, 1997; Gugel and Falk, 2006) and its nitrogen requirements are lower than for most other oilseed crops (Plessers *et al.*, 1962; Robinson, 1987). Characteristics such as frost and drought tolerance (Putnam *et al.*, 1993) and low fertility requirements make camelina suitable for production with few input costs under marginal conditions.

Camelina has a seed oil content of 290–430 g/kg (Putnam *et al.*, 1993; Vollmann *et al.*, 1996; Zubr, 1997), with a typical oil composition of the fatty acids oleic (C18:1), linoleic (C18:2), alpha linolenic (C18:3) and eicosatrienoic (C20:1). The oil has a wide range of applications, including human consumption in baking, cooking and frying, and uses for industrial soaps and varnishes, cosmetic oils and creams and biofuels (Zubr, 1997; review in Waraich *et al.*, 2013). As a biofuel, camelina methyl-ester had similar properties and expected yields as canola methyl-ester (Fröhlich and Rice, 2005). The meal by-product from oil extraction has potential for use in animal feed. Feed augmented with camelina meal can increase omega-3 fatty acid content in meat and eggs (Rokka *et al.*, 2002; Peiretti and Meineri, 2007). Consumption of dietary omega-3 fatty acid, through enriched foods or directly, can lead to positive health effects in humans and animals (Zubr, 1997). This broad range of end uses and flexibility of production factors has led to camelina being promoted as a new cropping opportunity for cereal cropping systems that will generate competitive returns at low risk to producers.

21.2.2 Crambe

Crambe is an Old World monophyletic genus of the Brassicaceae family with approximately 40 species in three major lineages or clades (Francisco-Ortega *et al.*, 1999). The Mediterranean clade includes three annual species: *Cr. abyssinica* (n=45), *Cr. hispanica* Linnaeus (n=30) and *Cr. glabrata* DC (n=15). In a comparison of the genetic relationships of this complex, Warwick and Gugel (2003) determined that *Cr. glabrata* is genetically distinct from

Cr. abyssinica and *Cr. hispanica* and merits separate species status, while the genetic differences between the two others are less clear.

Cr. abyssinica is native to highlands of north and central Africa, principally Ethiopia, while *Cr. hispanica* is native to the broader Mediterranean and Middle Eastern regions and *Cr. glabrata* is endemic to Spain and Portugal (Francisco-Ortega *et al.*, 1999; Warwick and Gugel, 2003; Oyen, 2007). Most of the effort in domestication of crambe has been on *Cr. abyssinica*, hereafter designated crambe.

Crambe has been planted in Africa, Asia, Europe, the USA, Mexico and South America (Oplinger *et al.*, 1991). In Ethiopia, crambe is traditionally grown locally as a medicinal plant and as a minor oil crop. It was first examined as an oil crop in the former USSR in the 1930s. Sporadic investigations into crambe as an alternative oilseed crop occurred there and in other parts of Europe (Oyen, 2007) but did not elicit much initial interest, as there was no developed market for it. The crop was introduced into North America in the 1940s, with sporadic research efforts until the 1980s, when the United States Department of Agriculture (USDA) organized the High Erucic Acid Development Consortium to examine crambe's potential for commercialization. Production of crambe on a commercial scale began in North Dakota in 1990 (Endres and Schatz, 2013). Crambe production has been stimulated by the recent increased interest in sources of biodiesel and more environmentally friendly lubricants. In Europe, an 18-member European Union consortium recently began a 4-year project with the goal of turning both crambe and camelina into profitable, sustainable, multipurpose, non-genetically modified oil crops, thereby reducing the dependence of Europe's oleochemical industry on imported plant oils (COSMOS, 2015).

Cr. abyssinica is an erect, multi-branched annual herb with the common names of crambe, Abyssinian mustard and Abyssinian kale (Erickson and Bassin, 1990). The plant has large, pinnately lobed, asymmetric glabrous to pubescent leaves with trichomes on the petioles and grows 60–100 cm tall. It is a rapidly growing relatively drought-tolerant plant that is an efficient water user (Merrill *et al.*, 2002), flowering about 45 days after seeding, with seed maturing in 90–100 days after seeding (Oplinger *et al.*, 1991). The small yellowish-white flowers are borne singly on long racemes and bear single round seeds, each of which is enclosed in a hull or pod that usually remains on the seed after harvest. The related species *Cr. glabrata* and *Cr. hispanica* have a more indeterminate flowering pattern, with seed set continuing late in the season, though early-formed pods usually remain on the plant until later ones mature (Lessman, 1990; Warwick and Gugel, 2003; Endres and Schatz, 2013). *Cr. hispanica* has trichome levels and types similar to those found in *Cr. abyssinica*, while hairs on *Cr. glabrata* are sparse and coarse (Warwick and Gugel, 2003).

The renewed interest in crambe utilization is due primarily to the fact that 55–60% of the glycosides in crambe oil, which comprises about 35% of the seed, are composed of erucic acid (Mikolajczak *et al.*, 1961). Erucic acid is a 22-carbon unsaturated fatty acid with a variety of proven and potential uses. Many of these uses are similar to those of camelina: as high-temperature lubricants, constituents in plastics and nylon, paints and coatings, waxes and rubbers, corrosion inhibitors, surfactants and slip agents, adhesives, and others (Nieschlag and Wolff, 1971; van Dyne *et al.*,1990; Bart *et al.*, 2013). Further, because crambe is a fast-growing, high-biomass crop that accumulates significantly higher levels of arsenic and chromium than other Brassicaceae, it holds promise as a crop suitable for phytoremediation of heavy-metal contaminated soils and sediments (Artus, 2006; Paulose *et al.*, 2010; Zulfiquar *et al.*, 2011). While high levels of glucosinolates render crambe oil unsuitable for human consumption, the seed meal of crambe contains 45–60% protein with well-balanced amino acid content. Suitable treatment of defatted crambe can eliminate anti-nutritive factors while increasing palatability, leaving a high-value ruminant feedstock (Korsrud *et al.*, 1978; Lessman, 1990). The defatted seed meal of *Cr. abyssinica* has been found to be toxic to a variety of insects (Tsao *et al.*, 1996, 2002; Peterson *et al.*, 2000). Innovatively, one of the goals of the COSMOS (2015) project is to investigate arthropod fauna that is unaffected by the toxicity of *Cr. abyssinica* meal, with the view of turning residues of the crop into a source of feed for insects to produce high-value protein for human consumption. Production of crambe has fluctuated as industry partners have changed, and large-scale commercial production of crambe remains to be achieved. The crop's future may depend on the future of bio-renewable resources and initiation of research to develop additional innovative markets for it.

At present, both camelina and crambe are under-utilized oilseed crops that can be turned into multi-use, non-genetically modified oilseed crops for the production of oleochemicals and multiple other products (COSMOS, 2015). One factor in their potential development as principal food and industrial crops is their exploitation as food sources by arthropods.

21.3 Arthropod Fauna Associated with Camelina and Crambe: Pollinators

In view of the perceived decline in numbers of managed and native pollinators worldwide, scientific and public interest has turned to determining means of reclaiming, restoring or enhancing land for pollinator use (e.g. Vilsack and McCarthy, 2015). Methods of achieving this goal include evaluating the pollinator forage value of crop plants such as canola and new/alternative crops like camelina and crambe and encouraging the planting of suitable crops or pollinator-friendly seed mixtures for maximum pollinator health as well as agronomic benefit.

Both camelina and crambe have potential as providers of food for pollinators. Although camelina is primarily self-pollinating (Plessers *et al.*, 1962; Zubr, 1997; Groeneveld and Klein, 2014), cross-pollination by insects, especially honey bees and solitary bees, benefits the crop as well as the insects (Groeneveld and Klein, 2014). A Swedish study found that camelina was visited by worker bumblebees (*Bombus* spp.) at a frequency similar to that on flax (*Linum usitatissimum* Linnaeus) and rapeseed (*Brassica napus* var. *oleifera*) (Fridén, 1972). Camelina may be a valuable source of pollinator forage. In a comparison of *Thlaspi arvense* Linnaeus (pennycress), camelina and *B. napus* (winter canola), a recent South Dakota study reported that winter camelina provided the highest combined agroecosystem value through pollinator energy sources early in the spring, green cover and crop seed yields (Eberle *et al.*, 2015). In the South Dakota study, the insect communities visiting camelina and canola flowers were similar to one another, with small bees, flies and other insects making up over 90% of all the visits with approximately one-third in each category (Eberle *et al.*, 2015).

Like camelina, crambe is mainly self-pollinated, but about 15% cross-pollination occurs (Oyen, 2007). When Forcella *et al.* (2014) examined flowering times and pollinator visitations of nine spring-sown crops, including calendula *Calendula officinalis* Linnaeus, camelina, canola, crambe and flax, they found numerous pollinators on all crops except crambe and flax, which attracted few insects during anthesis. Although all oilseed species that were evaluated attracted a wide diversity of insects, honey bees (*Apis mellifera* Linnaeus) often represented half of the individuals observed in the study. While crambe had many fewer pollinators than camelina and other species in the bioassays, the study found that native pollinators tended to be more common, proportionally, on calendula, camelina and crambe than on other crops, attesting to the value and potential value of these plants for maintenance of the health of native pollinators (Forcella *et al.*, 2014).

21.4 Arthropod Fauna Associated with Camelina and Crambe: Pests

There are few major cosmopolitan arthropod pests of camelina or crambe. Indeed, few insect infestations of any sort have been reported on the crops. Reports of insect damage to either crop have been infrequent, sporadic and anecdotal in nature. Insect control has rarely been undertaken on camelina (Porcher, 1869; Robinson, 1987; Putnam *et al.*, 1993; Zubr, 1997; Gugel and Falk, 2006; Ehrensing and Guy, 2008). Robinson (1987) reported that damage to camelina from insects had been insufficient to warrant control measures in 30 years of production on research plots; to date the crop has required little to no chemical protection against insects. Crambe, likewise, has had few reports of insect damage in the field. To determine the potential of insects to become pests of these two new alternative crops, we examined the arthropod fauna associated with other cruciferous crops, especially canola (oilseed rape), and how established hosts compare biochemically and agronomically with camelina and crambe.

21.4.1 Specialist insects

Many of the insects that are pests of crucifers are specialist feeders, monophagous or oligophagous consumers of plants in the family Brassicaceae (Feeny *et al.*, 1970), which contains many secondary plant compounds that are detrimental to generalist feeders (Städler, 1992). Glucosinolates are a specific group of secondary anionic thioglucoside compounds with variable side chains; they and their metabolites are common in Brassicaceae and can act as semiochemicals, regulating insect behaviour in a positive or negative manner (Chew, 1988; Hopkins *et al.*, 1997, 2009; Soroka and Grenkow, 2013). Insects that are specialist crucifer feeders often possess adaptations to the defence chemistry of their hosts, using secondary compounds and their breakdown products as aggregation, oviposition or feeding stimulants (Renwick and Radke, 1990; Hopkins *et al.*, 2009; Hegedus and Erlandson, 2012). These same secondary compounds often act as allomones, detrimental to the biology of more polyphagous, generalist herbivorous insects that are not adapted to this plant chemistry (Hegedus and Erlandson, 2012).

Flea beetles (Phyllotreta spp.) (Coleoptera: Chrysomelidae)

Flea beetles, principally *Phyllotreta cruciferae* (Goeze) and *Phyllotreta striolata* (Fabricius) (Coleoptera: Chrysomelidae; Alticinae), are the most economically devastating chronic pests of canola in North America (Knodel and Olson, 2002), with average annual losses and control costs in the hundreds of millions of dollars (Lamb and Turnock, 1982; Knodel and Olson, 2002; Canola Council of Canada, 2014a). More insecticide is applied annually to control these insects on canola in North America than any other insect pest in the crop (Lamb and Turnock, 1982; Madder and Stemeroff, 1988). *P. cruciferae* is also the main pest of oilseed crops in India and, along with *Phyllotreta undulata* Kutsch., is the main flea beetle pest of crucifers in eastern Europe (Crop Protection Compendium, 2015a), while the cabbage stem flea beetle *Phyllotreta chrysocephala* Linnaeus, the large striped flea beetle *Phyllotreta nemorum* Linnaeus and *Phyllotreta nigripes* Fabricius are pests of crucifers in western Europe, including Great Britain (Anon., 2015). *P. striolata* is a pest of oilseed and vegetable cruciferous crops around the world (Crop Protection Compendium, 2012a).

In North America flea beetles overwinter as adults in plant debris near and in fields, emerge when temperatures rise in spring and feed on winter annual crucifers or volunteer canola before moving into canola fields as the seedlings germinate. After mating, female flea beetles lay up to 25 eggs at the base of host plants. Larvae are subterranean, feeding on root hairs and small roots, before making small earthen cells and pupating. Adults of the next generation emerge in July or August, feeding on whatever crucifers are available until cold weather induces them to seek suitable overwintering sites. The greatest damage to field crops is caused in spring by overwintered adult flea beetles feeding when the crop is at its most vulnerable seedling stage. *P. cruciferae* male adults emit an aggregation pheromone upon finding a suitable host plant (Peng *et al.*, 1992; Soroka *et al.*, 2005), and although each flea beetle feeding pit is tiny in area, the sheer volume of pits created when large populations feed can cause cotyledons and young leaves to have a 'shot hole' appearance. Stem feeding and/or the loss of leaf photosynthetic area can result in seedling mortality, delayed crop maturity or decreased growth and vigour, leading ultimately to decreased seed yield (Lamb, 1984). Because immigration into fields and the realization of economic thresholds can be very rapid, most canola fields in North America are seeded with canola seed coated with an insecticide to combat flea beetles (Knodel and Olson, 2002; Soroka *et al.*, 2008). Recently, neonicotinoid seed treatments have been found to be less efficacious against *P. striolata* than *P. cruciferae* (Tansey *et al.*, 2008, 2009), which may be a factor in the recent increase in prevalence of *P. striolata* in areas where it was once uncommon (J.J. Soroka, unpublished results).

Although flea beetles are one of the most devastating insect pests on oilseed rape in North America, they appear to have little impact on camelina. *P. cruciferae* were observed alighting on camelina in Minnesota but in small plot trials over many years feeding by the beetle never warranted control (Robinson, 1987). In a 4-year study in Montana, Lenssen *et al.* (2012) found flea beetles present on *Brassica juncea* Linnaeus (Czern.) and *Cr. abyssinica*, but not camelina. In host-plant studies in the laboratory, Pachagounder *et al.* (1998) reported little feeding (0–10% consumption) by *P. cruciferae* on 51 camelina accessions compared with those on seven other Brassicaceae (59–100% consumption). These researchers observed flea beetles sitting but not feeding on camelina in the field near Winnipeg, Manitoba (Pachagounder *et al.*, 1998), but did not undertake field trials. In multiple laboratory trials comparing flea beetle feeding damage on camelina and crambe with other crucifers, Soroka and Grenkow (2013) found that camelina entries consistently suffered the lowest feeding levels of any species tested (Table 21.1) (Soroka and Grenkow, 2013; Soroka *et al.*, 2015; J.J. Soroka, unpublished

Table 21.1. Summary of 29 laboratory and eight field bioassays testing feeding levels of flea beetle *Phyllotreta cruciferae* on various accessions of *Camelina sativa* and three *Crambe* spp., along with the susceptible control *Brassica napus* (cv. Westar or AC Excel) and moderately resistant *Sinapis alba* (cv. Ochre or Pennant).[a]

	Tissue area eaten (%) – laboratory trials				Tissue area eaten (%) – field trials			
Plant species[b]	No. trials	No. accessions or cultivars tested	Average over all trials	Range over all trials	No. trials	No. accessions or cultivars tested	Average over all trials	Range over all trials
Brassica napus	29	2	80	23–100	8	2	33	19–60
Sinapis alba	28	1	32	5.0–72	5	2	14	6.4–25
Crambe glabrata	8	7	45	11–96	4	6	13	9.1–26
Crambe abyssinica	15	26	14	0.23–59	6	7	5.6	0.92–13
Crambe hispanica	14	15	4.0	0–22	4	5	2.9	0.31–13
Camelina sativa	14	58	0.49	0–10	8	29	6.0	1.1–20

[a]Data from Soroka and Grenkow, 2013; Soroka *et al.*, 2015; and J.J. Soroka (unpublished results)
[b]Not all accessions or species were present in every trial

results). In only one trial did leaf area eaten exceed 10% of young camelina foliage; this was in a bioassay in which one camelina entry had a maximum average defoliation of 10.2%, while the preferred *B. napus* control was completely defoliated and the semi-resistant entry of *Sinapis alba* Linnaeus averaged 72% defoliation (Soroka *et al.*, 2015). In eight field trials at Saskatoon, Saskatchewan, flea beetle feeding levels on camelina were much lower than on *B. napus* or *S. alba* (Table 21.1). The small amount of flea beetle feeding that was evident on camelina typically occurred shortly after emergence of the plant; by 3 weeks after emergence, flea beetle feeding levels on camelina were negligible (Soroka *et al.*, 2015). While the Winnipeg and Saskatoon laboratory experiments tested only *P. cruciferae*, the field trials at Saskatoon were exposed to mixtures of *P. cruciferae* and *P. striolata* and it is likely that both flea beetle species are averse to feeding on camelina. No detectable damage to camelina accessions from striped flea beetle *P. striolata* was found in field trials in Norway (Henriksen *et al.*, 2009). Similarly, camelina was the least acceptable host of ten brassicaceous lines for the flea beetles *P. cruciferae* and *P. nemorum* in a greenhouse trial in Switzerland (Nielsen *et al.*, 2001). In field trials in Estonia, camelina was the least attractive of eight brassicaceous oilseed crops to a naturally occurring mixture of *P. undulata*, *P. nigripes*, *P. nemorum*, *P. striolata* and *Chaetocnema concinna* Marsh (Metspalu *et al.*, 2014).

The reasons for limited feeding by flea beetles on camelina are not clear. In the field, agronomic characters did not appear to influence flea beetle choice of canola over camelina (Soroka *et al.*, 2015). In an analysis of pre-feeding behaviour of *P. cruciferae* on *B. napus* and the less preferred hosts *S. alba*, *Cr. abyssinica* and camelina, Henderson *et al.* (2004) found that while the initial pre-feeding pattern of flea beetles on cotyledons of camelina followed a similar sequence to that observed for *B. napus*, only 50% of the flea beetles that were tested proceeded from the acclimation or settling phase to the stimulation or host assessment phase with antennal contact of the cotyledon surface. Further, only one of 12 flea beetles tested initiated feeding on camelina, with a feeding duration of 9 seconds, while the equivalent average feeding period for flea beetles on *B. napus* was over 2 minutes (Henderson *et al.*, 2004). These and the results of Pachagounder *et al.* (1998), who found that some beetles eventually fed on cotyledons and leaves and, once feeding was initiated, it tended to continue, suggest that resistance in *C. sativa* may be principally antixenotic rather than antibiotic in nature, resulting from the absence of stimulatory or the presence of inhibitory volatile phytochemicals. Camelina contains three relatively unique sulfinyl glucosinolates (Berhow *et al.*, 2013) but very low levels of other glucosinolates typically found in *Brassica* field crops (Finch,

1978; Schuster and Friedt, 1998), including indole and hydroxybutyl glucosinolates that are common in *B. napus* (Soroka and Grenkow, 2013). Flea beetles may not recognize and be attracted by the novel glucosinolates, or may not encounter sufficient positive stimuli to initiate and maintain feeding on camelina (Soroka and Grenkow, 2013). Alternatively, Onyilagha *et al.* (2014) found that leaves and cotyledons of *C. sativa* contained large amounts of quercetin glycosides, while only trace amounts of these compounds were found in *B. napus* leaves and cotyledons. Onyilagha *et al.* (2014) tested extracts of quercetin glycosides from 3-week-old leaves of *C. sativa* for their effects on crucifer flea beetle feeding and found that extract fractions moderately deterred flea beetle feeding in mixtures containing rutin but not when the glycosides were presented as individual compounds. These results suggest that quercetin glycosides may contribute to flea beetles' non-preference of camelina. While the precise cause or causes of resistance to flea beetles in camelina are not yet known, they appear to be effective in maintaining resistance to the insect pests.

Although early reports indicated little infestation of insects, including flea beetles, on *Cr. abyssinica* in North America and worldwide (e.g. White and Higgins, 1966; Oyen, 2007), a study by Anderson *et al.* (1992) was one of the first to compare specifically the feeding preferences of *P. cruciferae* on *Cr. abyssinica* and traditional oilseed crops. In laboratory and field feeding bioassays in North Dakota, flea beetles fed much less on *Cr. abyssinica* seedlings than they did on *B. napus*, *B. rapa* or *S. alba* (Anderson *et al.*, 1992). Although flea beetle presence was not inhibited on crambe, feeding pits were usually smaller in diameter and less deep, suggesting the presence of a gustatory deterrent in crambe. In Montana, Milbrath *et al.* (1995) observed flea beetles on crambe in numbers lower than on *B. napus* or *B. rapa*, but did not comment on actual flea beetle damage to the crop. A further investigation (Peng *et al.*, 1992) found that flea beetles moved more frequently and spent less time on crambe than on *B. napus* plants and those that fed on *B. napus* lived significantly longer than those that fed on crambe, suggesting both antixenotic and antibiotic modes of resistance in the plant. In their assessment of pre-feeding behaviours of *P. cruciferae* on various Brassicaceae, Henderson *et al.* (2004) found flea beetle responses to *Cr. abyssinica* varying from movement solely restricted to climbing on and off the cotyledon all the way to the same suite of pre-feeding behaviours, including prolonged feeding, seen when feeding on *B. napus*. These actions suggest both a lack of feeding stimuli and the presence of feeding-deterrent phytochemicals in crambe seedlings, as well as the possibility of genetic variability in crambe resistance to flea beetles.

As with camelina, plant secondary compounds may cause the lack of flea beetle feeding seen on *Cr. abyssinica*. In laboratory trials under high feeding pressure, despite morphological similarity among crambe species (Warwick and Gugel, 2003), the order of flea beetle feeding damage consistently was *Cr. glabrata* > *Cr. abyssinica* > *Cr. hispanica* (Table 21.1) (Soroka and Grenkow, 2013). This order was maintained in field trials, with *Cr. glabrata* ranking with the moderately resistant *S. alba*, but considerably more fed upon than either *Cr. abyssinica* or *Cr. hispanica* (Table 21.1) (Soroka and Grenkow, 2013). The primary glucosinolate present in seed of all three *Crambe* species is 2-hydroxy-3-butenyl (HOBU) glucosinolate (Tsao *et al.*, 1996; Warwick and Gugel, 2003). *Cr. abyssinica* and *Cr. hispanica* have similar high amounts of HOBU glucosinolate in the seed, but *Cr. glabrata* has about half the levels of HOBU as the other two species; on average, HOBU makes up 93%, 94% and 64% of the total glucosinolates present in *Cr. abyssinica*, *Cr. hispanica* and *Cr. glabrata* seed, respectively (Warwick and Gugel, 2003). Further, *Cr. glabrata* has greater quantities of 2-propenyl glucosinolate (24% of the total amount) in its seed than the other species, while the remaining glucosinolates apart from these two are present in low or trace amounts in all three species (Warwick and Gugel, 2003). These lower levels of glucosinolates, if expressed in seedling leaf tissue in similar ratios as in seed, may explain why *Cr. glabrata* is fed upon more than the two other species. A thorough analysis of the secondary plant compounds, including glucosinolate composition and content in seedling tissue of the three *Crambe* species, is necessary to fully explain flea beetle feeding levels on them.

Diamondback moth, *Plutella xylostella* (Linnaeus) (Lepidoptera: Plutellidae)

Diamondback moth, *Plutella xylostella* (Linnaeus) (Lepidoptera: Plutellidae), is one of two principal lepidopteran pests of canola in North America and one of the most damaging insect pests of brassicaceous crops around the world (Dosdall *et al.*, 2011; Furlong *et al.*, 2013; Canola Council of Canada, 2014b), in good measure because of its rapid life cycle and propensity to develop resistance to several classes of insecticides. Diamondback moth larvae develop through four instars by feeding first within leaf mesophyll tissue and then on the surfaces of exposed leaves, buds, flowers or pods (Harcourt, 1957). Although overwintering of the pest on the northern Great Plains of North America is a possibility (Dosdall, 1994), the moth is more typically a migrant on jet-stream winds from source areas in the southern regions of the continent (Smith and Sears, 1982; Hopkinson and Soroka, 2009). The moth is multivoltine, with the number of generations occurring in a summer season dependent on the time of arrival of the first migrant moths, the suitability of host plants and local conditions, and to some extent the efficacy of natural control agents (Philip and Mengersen, 1989; Bahar *et al.*, 2013; Philip, 2015).

Host location, oviposition stimulation and feeding initiation by diamondback moth are governed principally by chemical cues, including host glucosinolate composition, level and breakdown products (Renwick and Radke, 1990; Pivnick *et al.*, 1994; Sun *et al.*, 2009; Muller *et al.*, 2010). The presence of specific indole and aliphatic glucosinolates from damaged plants is necessary for initiation of oviposition by diamondback moth females (Sun *et al.*, 2009). In laboratory studies, Deng *et al.* (2004b) found that diamondback moth adults had the same orientation response to both cabbage *Brassica oleracea* Linnaeus and camelina. However, in other trials diamondback moths laid fewer eggs on *C. sativa* than on *B. napus* plants at the bolting to early flowering growth stages (Soroka *et al.*, 2015), suggesting that actual oviposition may be mediated by factors other than volatile odours alone.

In field trials in Montana, camelina had lower densities of diamondback moth larvae than *Cr. abyssinica* or *B. juncea* (Lenssen *et al.*, 2012) and the authors suggested that *C. sativa* is not preferred by the moths. In laboratory studies larvae of diamondback moth fed less, had lower survival, lower pupation rates and weights and shorter longevity of subsequent adult moths when fed on *C. sativa* compared with cabbage *B. oleracea* L. (Deng *et al.*, 2004b). Diamondback moth larvae moved off camelina plants much more frequently, consumed much less *C. sativa* leaf tissue and had a longer larval developmental period and lower pupation rates on *C. sativa* than on *B. napus* at the rosette stage (Soroka *et al.*, 2015). Unlike the results of the study of Deng *et al.* (2004b), Soroka *et al.* (2015) found no differences in diamondback moth pupal weight when larvae were fed *B. napus* or camelina. In a comparison of nine brassicaceous species, Syed *et al.* (2002) found the rate of larval pupation to be lowest in diamondback moths reared on *B. camelina* (= *C. sativa*), while differences in length of pupation on different hosts were not significant.

Both indole and aliphatic glucosinolates promote feeding by diamondback moth larvae (Muller *et al.*, 2010). Similar to the situation with flea beetles (see above), *C. sativa*, with its low level of glucosinolates typical of Brassicaceae, may not be recognized as a potential host by ovipositing moths and/or may be toxic to diamondback moth larvae.

While plant chemistry plays a key role in resistance by camelina to diamondback moth, trichomes may be a morphological factor in the plant's resistance to the pest. Deng *et al.* (2004a) reported trichome densities on upper leaf surfaces of camelina that were about 70 times higher than those found on a typical cultivar of *B. napus* (Gruber *et al.*, 2006). Increased (Talekar *et al.*, 1994), decreased (Handley *et al.*, 2005), or no consistent difference in (Alahakoon *et al.*, 2016) oviposition by diamondback moth females have been noted in hosts with increased trichome number on leaves. The increased number of diamondback moth eggs on *B. napus* than on *C. sativa* plants reported by Soroka *et al.* (2015) may have been caused at least in part by greater trichome density on the less preferred plants. Both antixenosis and antibiosis resistance modalities were observed when diamondback moth larvae interacted with lines of *B. napus* canola possessing elevated densities of trichomes (Alahakoon *et al.*, 2016) and camelina trichomes may elicit the same responses in diamondback moth larvae.

In more than 10 years of field investigation in North Dakota, Weiss (cited in Kmec *et al.*, 1998) observed that diamondback moth regularly occurred on *Cr. abyssinica* and that the crop could be infested at any time from its emergence through maturity. The first peak of diamondback moth adults occurred in patches of pennycress (*T. arvense*) weeds near fields, while subsequent peaks, of which there were two to four depending on year, occurred in crambe at approximately 300 degree-day (base 7.5°C) intervals (Kmec and Weiss, 1997). From a comparison of phenology of the crop host and *P. xylostella*, these authors suggested that the first-generation larvae may be injurious if the crop is planted early and no weeds are available near the crop for ovipositing females; likewise, larvae present in the second half of July may potentially injure the ripening crop (Kmec and Weiss, 1997). Lenssen *et al.* (2012) found that while populations of diamondback moth on camelina remained low across development and maturation of the crop, populations on crambe increased over the same period. However, in a further investigation of the impact of defoliation at anthesis on crambe seed yield using infestation with diamondback moth eggs and artificial defoliation, Kmec *et al.* (1998) observed no reduction in seed yield or thousand-seed weight in any infestation or defoliation experiment. This result suggests that crambe can withstand substantial injury after the start of anthesis (25–50% of leaf area lost) without significant yield loss, due in part to its indeterminate growth pattern which allows compensation for flower injury, albeit with delayed seed set. Research on the effects of diamondback moth larval feeding on crambe seedlings is lacking. Heavy feeding by diamondback moth larvae on canola buds and flowers, especially when plants are under abiotic stress such as drought, can cause delay in plant maturity, uneven crop development, and significant seed yield reduction (Canola Council of Canada, 2014b). However, diamondback moth larval feeding on canola leaves is considered to have only a minor effect on yield (Canola Council of Canada, 2014b), possibly because heavy larval feeding on canola at the most vulnerable seedling stage is extremely rare on the northern Great Plains at present. Should overwintering of the moths become more common at northern latitudes in the future, feeding on canola and crambe seedlings by diamondback moth may become an economic problem.

Crucifer-feeding root maggots (Delia spp.) (Diptera: Anthomyiidae)

Although few reliable figures for crop loss are available, several species of *Delia* flies (Diptera: Anthomyiidae) are serious pests of cruciferous vegetable and field crops throughout the world (Crop Protection Compendium, 2012b, 2015b). A five-species complex of root maggots (*Delia* spp.) can be a serious threat to canola (*Brassica rapa* L. and *B. napus*) production in parts of the Canadian prairies (Griffiths, 1986a; Soroka *et al.*, 2004), with reported yield losses caused by root maggot feeding of up to 50% in *B. rapa* stands and 20% in *B. napus* stands (Griffiths, 1991a). Two of the principal pests of the root maggot complex occurring in North America, cabbage root fly or cabbage maggot (*Delia radicum* Linnaeus) and turnip maggot (*Delia floralis* (Fallén)), are Holarctic in distribution and restricted to cruciferous hosts (Griffiths, 1991b), while a third common member of the complex, bean seed fly or seedcorn maggot (*Delia platura* (Meigen)), is cosmopolitan in distribution and polyphagous in feeding habit (Griffiths, 1993). *D. platura* and a less common congener, *Delia florilega* Zetterstedt, occurring alone or together, are thought to be mainly secondary invaders of cruciferous roots following primary invasion by larvae of *D. radicum* or *D. floralis* (Griffiths, 1993). However, primary infestation by *D. platura* and *D. florilega* has been recorded on radish (*Raphanus sativus* Linnaeus) roots (Nair and McEwen, 1973) and high numbers of *D. platura* adults have been collected in canola fields in Saskatchewan, with few associated *D. radicum* or *D. floralis* adults (J.J. Soroka, unpublished observation). The fifth member of the root maggot complex in canola in Canada is *Delia planipalpis* (Stein), a primary invader of crucifer roots but occurring in fewer numbers than *D. radicum* or *D. floralis*. *D. floralis* is prevalent in northern areas of Alberta and Saskatchewan, with *D. radicum* more southerly in distribution (Griffiths, 1986a, 1991a, b, 1993; Turnock *et al.*, 1992; Soroka and Dosdall, 2011). Mixtures of *D. radicum* and *D. floralis* occur in canola in Canada, depending on location and year (Soroka and Dosdall, 2011), though co-occurrence is rare in Scandinavia (Darvas and Szappanos, 2003).

Larvae of *D. radicum*, *D. floralis* and *D. planipalpis* feed on a wide range of cultivated crucifers, with *D. radicum* preferring cabbage as a principal host and *D. floralis* preferring swede turnips, though both species readily attack canola and other brassicaceous field crops as well as a variety of vegetable crucifers (Crop Protection Compendium, 2012b, 2015b) and cruciferous weeds (Griffiths, 1991b, 1993). *D. platura* and *D. florilega* often occur together, laying their eggs near rotting vegetative material (Crop Protection Compendium, 2015c). Both species have been reported as primary pests of seedlings of many crops, including legumes, cereals, cucurbits, lettuce and carrot, but not canola seedlings (review in Soroka and Dosdall, 2011).

Delia larvae consume root hairs and lateral roots and then bore into the tap root and sometimes the base of the stem of a host plant. Attacked plants wilt, the leaves discolour and the plant usually dies or remains stunted (Hill, 1987). In North America, *Delia* spp. overwinter as puparia, with first-generation adults emerging with warm spring weather. Female flies lay about 100 eggs in cracks in soil near the host stem or sometimes on the plant. First-generation eggs are laid during late April to early May and take 3–7 days to hatch. Larval development takes approximately 3 weeks, followed by pupariation in the soil for 15–35 days. Second-generation flies emerge in late June or July, pupariate in August and either overwinter in diapause or emerge in late August to September, forming a partial third generation that overwinters as larvae and pupariates in the spring. In the USA and southern Canada there are three to four generations of root maggots a year, with emergence from overwintered puparia peaking in May, whereas in northern Alberta and Prince Edward Island, there is just one generation a year with emergence peaking in July or August (Griffiths, 1986b, 1991b, 1993).

While many genera of Brassicaceae have been recorded as hosts of crucifer-feeding *Delia* spp., camelina is not a preferred host of this pest complex. Finch (1978) found that female cabbage maggot flies, *D. radicum*, do not oviposit on *C. sativa* plants, possibly because of an absence of volatile glucosinolates emanating from the plants. Root maggot flies require a complex set of chemical and physical recognition cues in order to select an oviposition host (Hopkins *et al.*, 1997) and camelina plants either may not have positive cues or else may have deterrent factors that prevent *Delia* oviposition and/or larval development on them. Whatever the resistance mechanism, *Delia* maggots inflict little damage to camelina roots. In field trials testing camelina along with *B. napus* controls, 23 of 26 accessions of camelina were completely free of maggot damage, while the *B. napus* controls were moderately fed upon (Soroka *et al.*, 2015). Field trials at two Saskatchewan sites assessing root maggot damage to entries of nine brassicaceous species showed that, under high infestation rates in small plots surrounded by preferred hosts, infestation of *Delia* spp. on camelina can occur (Soroka *et al.*, 2015). However, while infestation rates in these bioassays were high and damage rates were moderate for most *Brassica*, *Sinapis* and *Crambe* entries tested, camelina entries had by far the lowest levels of infestation and damage of any entries at both locations (Soroka *et al.*, 2015), indicating that oviposition and feeding by *Delia* root maggots on camelina is unlikely to be of economic importance.

White and Higgins (1966) reported cabbage maggots occurring on *Cr. abyssinica* in Oregon and Washington, with damage, most serious during the first month of growth, sufficient to induce stand reduction. However, no appreciable seed yield loss occurred unless stand reduction was severe. Except for such anecdotal observations, little research has been conducted on the susceptibility of *Crambe* species to *Delia* root maggots. In field trials at Vegreville, Alberta, Dosdall *et al.* (2000) found *Cr. abyssinica* to have low damage ratings from root maggots among 12 brassicaceous species tested, with the order of susceptibility to root maggot injury being *Raphanus sativus* Linnaeus > *Brassica oxyrrhina* (Coss.) Willk. (= *Brassica barrelieri* subsp. *oxyrrhina* (Coss.) (P.W. Ball & Heywood) > *B. carinata* A. Braun > *B. rapa* > *B. oleracea* Linnaeus > *B. juncea* > *B. tournefortii* Gouan. > *B. napus* > *Cr. abyssinica* > *B. nigra* (Linnaeus) Koch > *Eruca sativa* Mill. > *Sinapis alba*. In a field trial of 30 lines of nine brassicaceous crop species at two sites in Saskatchewan, Soroka *et al.* (2105) found that *Cr. abyssinica* had root maggot damage ratings similar to those on *B. rapa*, *B. carinata*, *Cr. glabrata*, *B. napus* and *B. juncea*. Damage to *S. alba* roots were the lowest of the commonly grown brassicas and were similar to damage on *Cr. hispanica*, while damage to *C. sativa* entries was minimal (Soroka *et al.*, 2015). Although little work has been done to identify factors in crambe species that affect *Delia* biology, Ellis *et al.* (1980) found that preference of *R. sativus* lines for oviposition by *Delia radicum* was correlated with the levels of two volatile glucosinolate hydrolysis products, 4-methylthio-3-butenyl isothiocyanate and 1-cyano-4-methylthio-3-butene, in the ether extracts of macerated radishes.

Cabbage seedpod weevil *Ceutorhynchus obstrictus* (Marsham) and other *Ceutorhynchus* spp. (Coleoptera: Curculionidae)

Native to Europe, cabbage seedpod weevil *Ceutorhynchus obstrictus* (Marsham) has spread to North America, where it has recently become a major pest of canola production in some areas (Dosdall and Cárcamo, 2011). Cabbage seedpod weevils overwinter as adults in the soil, emerging from their overwintering sites in spring, with peak emergence occurring when soil temperatures reach approximately 15°C. Adults then fly to patches of early-flowering brassicaceous weeds and invade canola crops in the bud to early flowering stages in June, feeding on canola pollen, nectar, buds and racemes (Dosdall and Cárcamo, 2011). Mating occurs on canola plants, with ovariole development requiring that females feed on raceme tissue. Most eggs are laid soon after canola flowering. Infestation of pods by larvae is higher when flowering coincides with an increase in cabbage seedpod weevil populations (Dmoch, 1968). Larvae develop rapidly in spring canola, feeding on developing canola seeds and progressing through three larval instars in about 7 weeks. When mature, the larvae chew openings in the pod wall, drop to the soil and pupate in earthen cells, with adults emerging about 14 days later and feeding until cold temperatures induce them to seek overwintering sites (Dosdall and Cárcamo, 2011).

In tests evaluating cabbage seed pod weevil injury to brassicas near Lethbridge, Alberta, results on camelina and crambe entries were inconclusive. Although, in general, few or no exit holes were found in these entries, in most cases the plants grew poorly and pod production occurred too late for weevil infestation (Cárcamo *et al.*, 2007). *C. obstrictus* oviposition had fair synchrony with camelina flowering, however, and the potential exists for *C. obstrictus* exploitation of the crop (Cárcamo *et al.*, 2007). Regarding other weevil species, Ellis (2014) listed *Ceutorhynchus assimilis* (Paykull), *Ceutorhynchus chalybaeus* Germar and *Ceutorhynchus minutus* (Reich) as occurring on *C. sativa* in Europe, but Andersson and Olsson (1950) found no evidence of infestation of camelina by the weevil *C. assimilis* in Sweden. *C. minutus* is listed as occurring on camelina in Great Britain (Pitkin *et al.*, 2016), although its economic impact is unknown. *Ceutorhynchus syrites* Germar is reported to be the most important pest of camelina in Poland (Dmoch, 1968), where larvae damage 25–50% of the seed pods or more. This insect has also been reported on camelina in Eastern Europe (Filipiev, 1929; Rakhmaninov and Vuirzhikovskaya, 1930), England and southern Germany (Madel, 1950). Madel (1950) mentioned that *C. syrites* might attack young camelina plants without causing serious damage, but in southern Germany insecticide application gave satisfactory results when control was warranted. In Sweden, camelina was also attacked by the weevils *Ceutorhynchus rapae* Gyllenhal, *Ceutorhynchus quadridens* (Panzer) and possibly *Ceutorhynchus erysimi* (Fabricius), though usually not seriously (Wahlin, 1951; Borg, 1952). The cabbage stem weevil *Ceutorhynchus pallidactylus* (Marsham) was found to feed significantly less on the related species *Camelina alyssum* (Mill.) than on standard oilseed rape cultivars (Eickermann and Ulber, 2010).

Unlike camelina, in the study of Cárcamo *et al.* (2007) *Cr. abyssinica*, *Cr. glabrata* and *Cr. hispanica* all had flowering times that were asynchronous with peak abundance of *C. obstrictus*, which may have resulted in the very low levels of damage seen on these entries. It is uncertain whether this temporal escape of pod infestation holds across environments and locations. In Montana, Lenssen *et al.* (2012) found cabbage seedpod weevil adults in sweep samples collected from *B. juncea* but never from camelina or crambe. Genetic variability in crambe may influence resistance to cabbage seedpod weevil. For unknown reasons, although most plants of all three crambe species were not or were only slightly damaged by *C. obstrictus*, in the study of Cárcamo *et al.* (2007) one *Cr. abyssinica* plant had 19 larval exit holes. Ellis (2014) listed the weevil *C. chalybaeus* as occurring on *Cr. abyssinica*, with congeners *Ceutorhynchus griseus* Brisout and *C. pallidactylus* occurring on *Crambe* (no species listed) in Europe, but did not mention severity of damage by the pests.

Yellow mustard, *S. alba*, is relatively resistant to *C. obstricus* (Doucette, 1947; Dosdall and Kott, 2006). Decreased levels of 2-phenylethyl glucosinolate and elevated levels of methoxy-3-indolemethyl glucosinolate in resistant germplasm contribute to antixenosis and antibiosis resistance to the weevil (Tansey *et al.*, 2010a). Whether these glucosinolate levels influence the degree of feeding typical on camelina and crambe remains to be seen. Cabbage seedpod weevil females are also influenced by light reflectance properties of the flowers and leaves of their hosts, responding especially to amounts of yellow and ultraviolet light reflected by flowers (Tansey *et al.*, 2010b). The small whitish to pale yellow flowers of both camelina and *Cr. abyssinica* may reduce their level of attractiveness to *C. obstrictus*.

Swede midge Contarinia nasturtii (Kiefer) (Diptera: Cecidomyiidae)

Of relatively minor economic importance to oilseeds in its native Eurasia, swede midge has become a major impediment to canola production in central Canada in a very short time (Phillips, 2015) and is now commonly found in crucifer vegetable crops in central Canada and north-eastern USA (Chen *et al.*, 2011).

In a review of the host plant range of *Contarinia nasturtii*, Ellis (2014) listed *Camelina sativa* but not *Crambe* spp. as a host of the swede midge in Europe, noting that it is a minor pest of Brassicaceae there. A congener, *Camelina microcarpa* Andrz. ex DC., is listed as a host of swede midge in Hallett's (2007) compilation of hosts of the pest but this weedy species exhibited no damage from swede midge when tested by Hallett (2007). In field trials currently under way in Saskatchewan, midge injury to *C. sativa* is much lower than levels seen on *B. rapa*, *B. carinata*, *B. juncea* or *B. napus* and somewhat lower than seen on *S. alba* (L.D. Andreassen and J.J. Soroka, unpublished results). Historically, another cecidomyiid pod midge, possibly *Dasyneura brassicae* Winn., was reported infesting about 10% of the pods of camelina in a district of the USSR (Rakhmaninov and Vuirzhikovskaya, 1930). Andersson and Olsson (1950), however, found no evidence of camelina infestation by *D. brassicae* in Sweden, though Ellis (2014) listed *Delia napi* (Loew) (= *D. brassicae* (Winn.)) as a pest of camelina in Europe. No records have been found of *Contarinia* spp. or any other gall midges on crambe species.

Other specialist insects

Several other insects have been recorded on camelina and/or crambe but their economic impact is negligible, unclear or unknown. There are few records of the brassica aphid complex of *Brevicoryne brassicae* (Linnaeus) and *Lypaphis erysimi* (Kaltenbach) (Homoptera: Aphididae) causing economic damage to camelina (Porcher, 1869; Ehrensing and Guy, 2008), though the cabbage aphid, *B. brassicae*, can reproduce on the plant (Chesnais *et al.*, 2015). Jarvis (1982) found that *B. brassicae* infested *Cr. abyssinica* in greenhouse trials but at lower survival levels than in more commonly grown brassicaceous crops. Singh *et al.* (2014), when testing single entries of 22 different cultivated and wild species of Brassicaceae, found both *C. sativa* and *Cr. abyssinica* to be resistant to turnip aphid, *L. erysimi*. However, White and Higgins (1966) and Oyen (2007) stated that crambe is attacked by aphids (though neither report listed aphid species or circumstance); and 19 of 20 accessions of *Cr. abyssinica* and one of *Cr. hispanica* were found to suffer severe damage from *L. erysimi* (= *Hydaphis erysimi* Davis) in greenhouse trials (Jarvis, 1969), affirming the value of testing multiple entries of a species before making generalizations on host resistance.

The pollen beetle *Brassicogethes aeneus* (Fabricius) (Coleoptera: Nitidulidae), a serious European pest of crucifer oilseed crops, was found infesting camelina in Sweden but not significantly (Wahlin, 1951). Andersson and Olsson (1950) also reported that camelina was less attacked by *B. aeneus* than were other crucifers, and Henriksen *et al.* (2009) reported no detectable damage by *B. aeneus* on camelina in Norwegian field trials. Similarly, in oviposition experiments Ekbom and Borg (1996) found plants of *Cr. abyssinica* had fewer *B. aeneus* eggs than did *Brassica* species; likewise, feeding damage by pollen beetle larvae was similar or lower than on *Brassica* species. No information is known on the reasons for the resistance of the two plant species to these pests. *Brassicogethes viridescens* (Fabricius) has recently become established in eastern Canada (Mason *et al.*, 2003) but to date there have been no reports of *Brassicogethes* spp. feeding on camelina or crambe in North America.

A cryptic note in the 1973 annual report of the North Central Regional Project NC-7, coordinated by the United States Department of Agriculture, stated that crambe was found to be susceptible to cabbage looper *Trichoplusia ni* (Hübner) (Lepidoptera: Noctuidae) and imported cabbageworm *Pieris rapae* (Linnaeus) (Lepidoptera: Pieridae) (Anon., 1973). In a review of the entomofauna of potential oilseed crops of Romania, Palagesiu (2003) listed the seed chalcid, *Eurytoma crambae* Zerova (Hymenoptera: Eurytomidae), to be the characteristic insect species for *Cr. abyssinica* in that country. In an investigation of the seed chalcids of the Palaearctic, *E. crambae* and its relative *E. crambicola* Zerova (Hymenoptera: Eurytomidae) were reported to infest *Crambe tataria* Sebeok and *Crambe kotschyana* Boiss., respectively, in eastern Europe (Zerova and Seregina, 1994). No impact of these wasps on their hosts is known and no Chalcidoidea are known to infest camelina. In a growth chamber trial assessing development of the mustard sawfly *Athalia proxima* Klug (Hymenoptera: Tenthredinidae) on 16 different brassicaceous species, Singh and Sachan (1997) found that larvae of the sawfly did not feed or develop on *C. sativa*. Crambe species were not tested in the investigation and no information is available on their host susceptibility to *A. proxima*.

21.4.2 Generalist insects

Of the common generalist insect pests on the northern Great Plains of North America, several will feed on camelina but their economic impact has not been determined.

Leafhoppers (Homoptera: Cicadellidae)

Leafhoppers (Homoptera: Cicadellidae) can cause injury to plant hosts through direct extraction of photosynthate and through vectoring of plant diseases. The aster leafhopper, *Macrosteles quadrilineatus* Forbes, is a generalist feeder and the main vector of aster yellows (AY) disease in many crops in North America, including canola on the Canadian prairies (Olivier *et al.*, 2009). *M. quadrilineatus* is a migratory leafhopper that arrives in Canada in the spring via winds originating from the southern USA (Valk and Stevenson, 1994). Aster yellows disease can cause severe yield losses in canola crops (Olivier *et al.*, 2009). Leafhopper and AY appearance may be increasing in frequency and extent in northern Great Plains canola (Olivier *et al.*, 2009; Miller *et al.*, 2013).

Symptoms typical of AY disease have been observed in *C. sativa* in North America since the late 1980s (Robinson, 1987). Characteristic symptoms of AY in *C. sativa* include stunting, red coloration of stems and pods, and flattened sterile pods (Séguin-Swartz *et al.*, 2009). Recently, bladder-like pods, 'witches' broom' of branch ends and virescence of petals were observed 28 days after AY-infected *M. quadrilineatus* were caged with *C. sativa* seedlings at the four-leaf stage for 1 week (T.J. Wist, unpublished results). The bladder-like pods often contained masses of undifferentiated tissues, while seeds were set in others.

Nurseries of *C. sativa* grown at the Agriculture and Agri-Food Canada farm at Saskatoon, Saskatchewan, have been monitored for the incidence and severity of AY since 2007. AY phytoplasma belonging to the subgroups 16SrI-A and -B has been detected in *C. sativa* and in *M. quadrilineatus* and other leafhopper species (Olivier *et al.*, 2011; Dumonceaux *et al.*, 2014). In 2009 and 2010, *M. quadrilineatus* represented 79% and 95% of the leafhopper population captured in *C. sativa*, respectively; the next most numerous leafhoppers were *Euscelis maculipennis* Severin and *Endria inimicus* (Say), with the number of adults captured on yellow sticky cards far lower than the number of *M. quadrilineatus* (Séguin-Swartz *et al.*, 2009; Soroka *et al.*, 2015). *C. sativa* is considered susceptible to AY, as shown by the incidence of AY symptoms ranging from 13% to 39% in 2007 and 25–68% in 2012 in various accessions, with both years considered to be AY outbreak years on the Canadian prairies (C. Olivier, unpublished results). However, many *C. sativa* accessions have relatively low incidences of AY symptoms compared with others, demonstrating the presence of genetic biodiversity that can be used in breeding programmes to incorporate AY resistance and/or tolerance into *C. sativa* (Séguin-Swartz *et al.*, 2009).

In 2013 and 2014 the species diversity, population development and AY infectivity of field-collected leafhoppers at Saskatoon were analysed using yellow sticky cards placed in breeding nurseries comprising 28 accessions of *C. sativa* (C. Olivier and T.J. Wist, unpublished results). Sticky cards (18 × 14 cm) were placed in plots from mid to late spring and changed weekly until August. In both years, AY incidence was estimated visually and by using PCR on 20 plant samples per accession per replicate (Dumonceaux *et al.*, 2014; Soroka *et al.*, 2015). All trapped leafhoppers were tested for the presence of AY phytoplasma using 16S conventional PCR (Olivier *et al.*, 2011).

In both years, *M. quadrilineatus* was the most prevalent species, constituting 95% and 98% of the leafhopper population in 2013 and 2014, respectively (Table 21.2). In both years the next most abundant leafhopper species were *Balclutha* spp., followed closely by *Erythroneura comes* (Say) (Table 21.2). In both 2013 and 2014, nymphs of *M. quadrilineatus* were observed on sticky cards (Figs 21.1, 21.2 and 21.3), confirming previous results (Soroka *et al.*, 2015) which suggested that *M. quadrilineatus* feeds and also reproduces on *C. sativa*. In 2013, high numbers of aster leafhopper adults were collected on sticky traps from 11 June to 28 June, corresponding with the arrival of a series of winds from the southern USA on 23 May (Canola Council of Canada, 2013) (Fig. 21.1). A second peak of adults occurred in mid-August. Aster leafhopper nymphs were found on the traps at the beginning of July and again in mid-August (Fig. 21.1), strongly suggesting that completion of two full generations on camelina is possible when

Table 21.2. Leafhopper (Cicadellidae) species (numbers per year and percentage of total number per year) trapped on yellow sticky cards in *Camelina sativa* (Brassicaceae) plots at the Saskatoon Research and Development Centre Farm, Saskatoon, Saskatchewan, Canada, over the summer 2013 and 2014.

Leafhoppers in *Camelina sativa*	2013 (72 cards)	% of total	2014 (711 cards)	% of total
Aceratagallia humilis Oman	0	0.00	4	0.03
Aphrodes spp. Curt.	0	0.00	1	0.01
Athysanus argentarius Metcalf	3	0.16	12	0.08
Balclutha spp. Kirk.	21	1.13	39	0.25
Balclutha punctata (Fabricius)	1	0.05	1	0.01
Ballana spp. DeLong	1	0.05	0	0.00
Chlorotettix spp. Van D.	13	0.70	2	0.01
Colladonus spp. Ball	0	0.00	15	0.10
Deltocephalus grex Oman	1	0.05	0	0.00
Dicraneura mali (Provancher)	1	0.05	4	0.03
Diplocolenus configuratus (Uhler)	0	0.00	7	0.04
Doratura stylata (Boh.)	0	0.00	1	0.01
Draeculacephala angulifera (Walk.)	1	0.05	2	0.01
Empoasca fabae (Harr.)	4	0.21	17	0.11
Empoasca sp.Walsh	5	0.27	7	0.04
Endria inimica (Say)	20	1.07	33	0.21
Erythroneura comes (Say)	0	0.00	13	0.08
Erythroneura sp. Fitch	0	0.00	10	0.06
Euscelidius shenki (Kirsch.)	1	0.05	0	0.00
Eucelis sp. Brulle	0	0.00	2	0.01
Hebecephalus sp. DeLong	1	0.05	30	0.19
Latalus personatus DeLong & Sleesman	0	0.00	5	0.03
Limottetix sp. Sahlb.	2	0.11	0	0.00
Macropsis sp. Lew.	0	0.00	5	0.03
Macrosteles laevis (Ribault)	1	0.05	0	0.00
Neokolla hieroglyphica (Say)	1	0.05	2	0.01
Macrosteles quadrilineatus Forbes	1772	95.01	15,465	98.37
Palus sp. DeLong & Sleesman	0	0.00	3	0.02
Psammotettix sp. Haupt	9	0.48	0	0.00
Psammotetix lividellus (Zett.)	0	0.00	12	0.08
Scaphytopius sp. Ball	0	0.00	13	0.08
Sorhoanus sp. Ribault	7	0.38	15	0.10
Stragania rufoscutellata (Bak.)	0	0.00	0	0.00
Verdanus evansi (Ashmead)	0	0.00	2	0.01
Total leafhoppers	1865		15,722	

migrant leafhoppers arrive early in spring. In 2014, sustained winds from the south occurred relatively late in the season (Prairie Pest Monitoring Network, 2014), with a corresponding late increase in the numbers of aster leafhoppers from 21 July to 11 August (Fig. 21.2). Aster leafhopper nymphs were collected in mid-August only, suggesting that only one generation occurred in 2014 (Fig. 21.2).

In 2013, the incidence of AY in *C. sativa* plants in the field nursery ranged from 0% to 50% with an average of 9% based on visual observations, and from 0% to 80% with an average of 13.3% based on PCR tests. In 2014, only two plants (< 1%) expressing AY symptoms were observed in the nursery; PCR tests showed an AY incidence averaging 1.35% and ranging from 0% to 15%.

In 2013, 10.2% of the *M. quadrilineatus* samples tested positive for the presence of phytoplasma, while 6.25% of the *E. inimica* and *Balclutha* sp. samples were PCR-positive for presence of AY phytoplasma. Two of ten samples of *Chlorotettix* spp. were positive for the phytoplasma. In 2014, few leafhoppers trapped in the *C. sativa* nurseries tested positive for the presence of AY phytoplasma. None of the leafhoppers in the initial population increase through July were infected with the AY

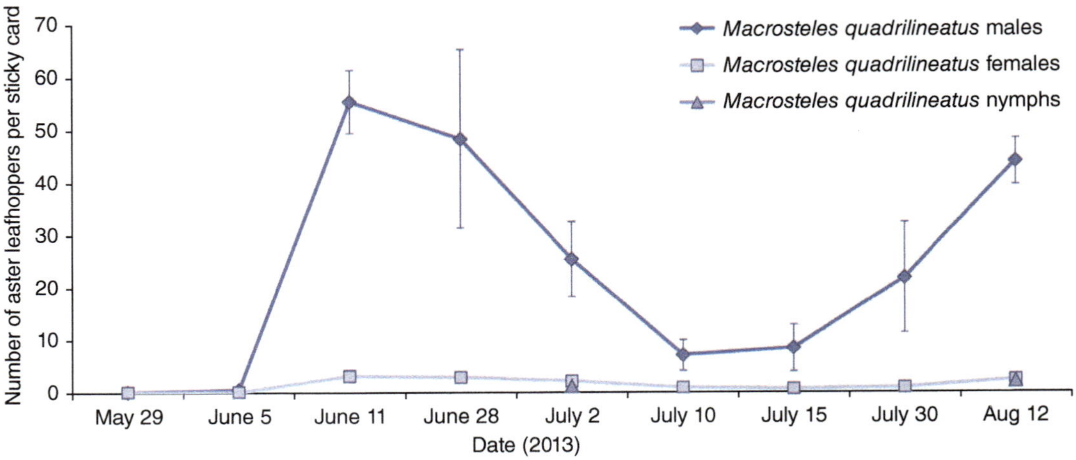

Fig. 21.1. Mean (±SE) number of *Macrosteles quadrilineatus* per yellow sticky card in *Camelina sativa* at the Saskatoon Research and Development Centre Farm, Saskatoon, Canada (2013).

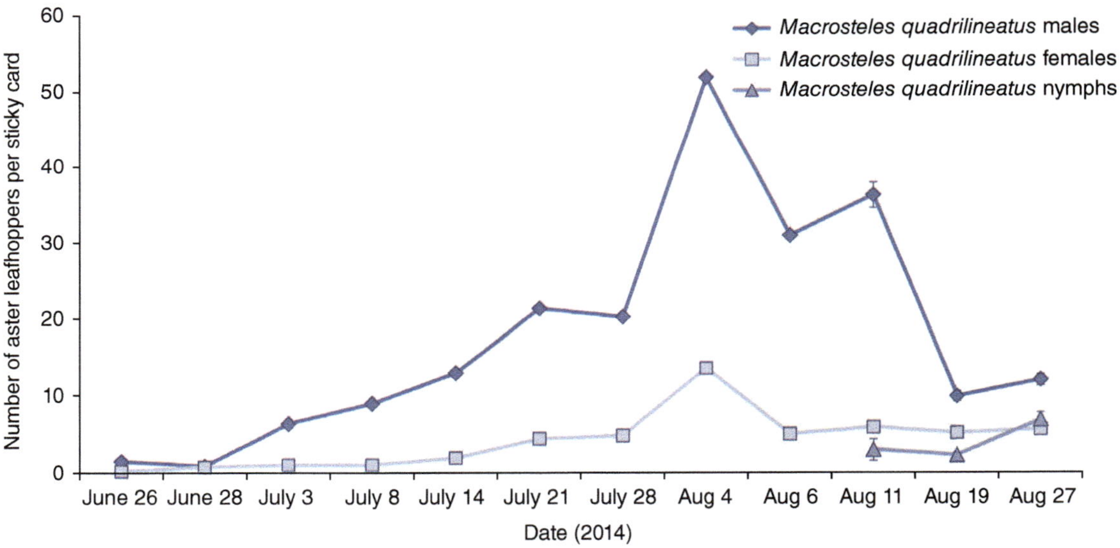

Fig. 21.2. Mean (±SE) number of *Macrosteles quadrilineatus* per sticky card in *Camelina sativa* at the Saskatoon Research Centre and Development Centre Farm, Saskatoon, Canada (2014).

phytoplasma, resulting in very low AY symptoms seen in camelina plants later in the year. The greater amount of AY symptoms in *C. sativa* plants in 2013 than in 2014 demonstrates the magnitude of the effect on *C. sativa* crops of an early arrival of *M. quadrilineatus* in the growing season coupled with an above-average percentage of infected adults.

Incidence of infestation by leafhoppers (Cicadellidae, species not given) has been reported on crambe in the field (White and Higgins, 1966), but damage to the crop was minor.

Lygus species (Heteroptera: Miridae)

Several *Lygus* species (Heteroptera: Miridae), both native and introduced to North America, can be serious pests of canola and many other crops (Otani and Cárcamo, 2011). Many lygus bugs

Fig. 21.3. Aster leafhopper *Macrosteles quadrilineatus* (a) nymphs and (b) adults on *Camelina sativa* plants. Photos: T. Wist.

have broad host ranges, with plant growth stage – especially the presence of reproductive tissue, a preferred feeding site – often as important as plant species in determining the amount of damage inflicted on hosts. The cosmopolitan tarnished plant bug, *Lygus lineolaris* (Palisot de Beauvois), for example, has a host range of over 300 plant species (Young, 1986). In early lygus descriptions neither Kelton (1975) nor Young (1986) listed camelina as a host of the tarnished plant bug or other *Lygus* species in North America but this may reflect a lack of contact with the crops by the insect. The western tarnished plant bug, *Lygus hesperus* Knight, will feed on camelina but in laboratory trials its level of probing and feeding on camelina was lower than on other hosts tested (Naranjo and Stefanek, 2012). The European plant bug, *Lygus rugulipennis* Poppius, is listed as the characteristic insect species of camelina grown as a prospective oilseed crop in Romania (Palagesiu, 2003) but the economic impact of the bug is not clear.

White and Higgins (1966) reported lygus bugs (species not listed) on *Cr. abyssinica* in the USA, with little damage to the crop. In a comparison of insect populations on three oilseed crops, camelina, *Cr. abyssinica* and canola-quality *B. juncea*, in the field in Montana, Lenssen *et al.* (2012) found that later-planted crambe had higher populations of lygus (possibly *Lygus elisus* Van Duzee, the pale legume bug, which is the most commonly occurring lygus species in the area; Ritter *et al.*, 2010) than either camelina or *B. juncea*. This study found that stage of crop maturity had a significant effect on lygus numbers, with populations increasing as crops matured. It is not known if *Lygus* spp. vary in their potential to damage either camelina or crambe.

Bertha armyworm *Mamestra configurata* Walker and other cutworms (Lepidoptera: Noctuidae)

Native to North America, the bertha armyworm, *Mamestra configurata* Walker, is one of the two main lepidopteran pests of canola in Canada (Mason *et al.*, 1998; Canola Council of Canada, 2014c). It has a broad host range, and if natural population regulators fail, bertha armyworm numbers can rise rapidly and cause widespread damage to a variety of dicotyledonous crops (Canola Council of Canada, 2014c). In choice tests in the laboratory, defoliation levels on *C. sativa* by bertha armyworm were similar to those on *B. napus* but in no-choice tests increase in armyworm biomass was less, pupae were lighter and developmental time from second instar to pupation was longer when larvae fed on camelina than on *B. napus* (Soroka *et al.*, 2015), suggesting antibiosis factors resistant to bertha armyworm may exist in the plant. Neither camelina nor crambe was mentioned as a host for bertha armyworm in a review of bertha armyworm in western Canada by Mason *et al.* (1998) but this absence may be a function of lack of contact with the crops by invading Noctuidae rather than host plant resistance. There are no reports of *M. configurata* occurring on crambe but Lenssen *et al.* (2012) noted that in field experiments in Montana *Cr. abyssinica* required replanting in the fourth week of May 2007 because of damage by the cutworm *Euxoa messoria* (Harris) (Lepidoptera: Noctuidae); it may be that bertha armyworm and cutworms in general are potential pests of *Cr. abyssinica*.

Aphids (Heteroptera: Aphididae)

While *Brassica*-feeding aphids are not typically an economic problem on camelina (see above), the generalist-feeding green peach aphid, *Myzus persicae* (Sulzer) (Hemiptera: Aphididae), was observed feeding and reproducing on camelina in the laboratory, with no apparent reduction in colony growth (Soroka *et al.*, 2015), though there was no observation of this pest feeding on camelina in the field in that investigation. In an effort to determine the potential for transmission of phytoviruses to the crop, Chesnais *et al.* (2015) conducted laboratory investigations to determine the interactions between camelina and the Brassicaceae specialist aphid *B. brassicae*, the polyphagous species *Aphis fabae* (Scopoli) and *M. persicae*, and the cereal specialist *Rhopalosiphum padi* (Linnaeus). *A. fabae* and *M. persicae* were able to land, feed and reproduce on camelina, performing better than on *B. brassicae*. Further, *C. sativa* could also be a suitable host for the cereal specialist *R. padi*. Thus, camelina could serve as a suitable host for aphid species and their associated pathogens from surrounding crops and in turn could serve as a reservoir for disease transmission to neighbouring crops. Likewise, Jarvis (1969) reported that both *Cr. abyssinica* and *Cr. hispanica* were satisfactory hosts for *M. persicae* in greenhouse trials and this susceptibility may hold in the field.

Others

Various other generalist insects have been observed settling or feeding on camelina and/or crambe, with little apparent effect. Most reports of grasshopper infestation in camelina and crambe are anecdotal in nature. Grasshoppers (Orthoptera: Acrididae, no species given) were observed causing extensive defoliation of camelina plants in a rangeland ecosystem in Montana in 2009 (Davis, 2010). False chinch bugs, *Nysius ericae* (Schilling), have been reported on *Cr. abyssinica* in Iowa, with little damage to the crop (White and Higgins, 1966). The leafminers *Scaptomyza flava* (Fallen) and *Liriomyza xanthocera* (Czerny) (Diptera: Agromyzidae) have been recorded mining leaves of camelina in Europe, while *L. xanthocera* and *Liriomyza strigata* (Meigen) as well as *Chromatomyia horticola* (Goureau) are listed as infesting *Crambe* (no species given) (Nellis, 2009), though the economic impact of such mining is not known. Knights (2006) proposed that insects that attack canola in Australia will also attack *Cr. abyssinica* and suggested but did not offer evidence that such pests could include redlegged earthmite *Halotydeus destructor* Tucker (Acarina: Penthaleidae), blue oatmite *Penthaleus* spp. (Acarina: Penthaleidae), lucerne flea *Sminthurus viridis* Linnaeus (Collembola: Sminthuridae), native budworm *Helicoverpa punctigera* Wallengren (Lepidoptera: Noctuidae), aphids (Homoptera: Aphididae) and Rutherglen bug *Nysius vinitor* Bergroth (Heteroptera: Lygaeidae).

Alternatively, active components of the defatted seed meal of *Cr. abyssinica* were toxic to western corn rootworm *Diabrotica virgifera virgifera* LeConte larvae (Tsao *et al.*, 1996), acutely toxic to larvae of housefly *Musca domestica* Linnaeus (Tsao *et al.*, 1996; Peterson *et al.*, 2000) and mosquito *Aedes aegypti* (Linnaeus) and chronically toxic to the red flour beetle *Tribolium castaneum* (Herbst) and the sawtoothed grain beetle *Oryzaephilus surinamensis* (Linnaeus), but had no effect on German cockroach *Blattella germanica* (Linnaeus) (Tsao *et al.*,1996). Thus crambe glucosinolates have potential as possible control agents for certain agricultural and public health insect pests.

21.5 Insect Pest Management in Camelina and Crambe

Because there are so few well documented reports of insects inflicting economic injury to camelina or crambe, it is difficult to formulate an insect pest management strategy for the crops, or even to know if one is needed. Some assumptions may be valid but others may not. Just as the assumption that insects that attack oilseed rape will also attack camelina or crambe may not be accurate, so is the assumption that strategies that work for management of insect pests in oilseed rape will also work in camelina and crambe. For example, specific agronomic practices for canola production such as site selection, cultivation and fertilization choices may not be suitable for camelina or crambe production, and the different choices in these alternative crops may influence arthropod community development differently than in canola. Similarly, while information is available on natural enemies of many brassicaceous insect pests, little information has been generated on tri-trophic interactions of camelina/crambe, insect pests and their biological control agents. It should be noted that many natural enemies find their hosts by cueing on to the hosts' host plants and many brassicaceous pests and their natural enemies may not recognize camelina/crambe as hosts.

21.6 Conclusion

This review considers the interactions of *C. sativa* and/or *Cr. abyssinica* with over 60 arthropod species, genera and complexes. Few insects appear to be pests of camelina or crambe. More is known about the insect pests of camelina than crambe and, overall, crambe may be somewhat more susceptible to insect feeding than camelina. However, both of these alternative crops rank low in terms of insect susceptibility compared with the brassicaceous field crops *B. carinata, B. juncea, B. napus* and *B. rapa*. Of the insects that have been reported to inhabit camelina or crambe, few are specialist crucifer feeders and the unique glucosinolate profiles of the two crops may provide deterrents, a lack of attractants, or antibiosis factors to most brassicaceous specialist feeders. The very different feeding levels of insects such as *Phyllotreta* flea beetles or *Delia* root maggots on *Cr. abyssinica, Cr. glabrata* and *Cr. hispanica*, three species that are very similar in morphology and phenology, suggest that plant biochemistry is a main determinant of host selection of brassicaceous specialist insects in the crops. Determination of glucosinolate composition and amount at all growth stages of camelina and crambe species will help to elucidate factors that influence host choice.

Most of the insects that have fed extensively on camelina and crambe are polyphagous in host range. It is possible that insects such as leafhopper, aphid and lygus species may circumvent deleterious secondary compounds in camelina and crambe by means of their piercing–sucking feeding mechanisms.

One other factor to consider is that favourable past experience does not guarantee that insects will not be a problem in camelina or crambe in the future. If production of camelina and/or crambe increases so that the crops become major components of the agricultural landscape over a wide region, generalist feeders that currently are not thought of as crucifer pests may become economically important. While canola is not considered a host of grasshoppers, in certain conditions and especially if preferred hosts are absent, extensive grasshopper injury to canola (Olfert and Weiss, 2002; Begna and Fielding, 2003), and possibly camelina and crambe, can occur. Likewise, many species of cutworms (Coleoptera: Noctuidae) feed on canola and may do the same on camelina or crambe if circumstances warrant. On the other hand, an increase in plant species diversity may facilitate natural pest control in annual cropping systems and a change in crop rotation pattern or duration through planting of camelina or crambe may decrease the incidence of pest infestation for a variety of reasons. Inclusion of camelina or crambe into cropping systems in the southern Great Plains of North America that are currently cereal based may decrease the build-up of cereal pests in such systems. Alternation of canola with crambe or camelina in more northerly areas may decrease the incidence of specialist brassicaceous feeders that prefer canola to either of these alternatives. If there is an increase in the scale of production of camelina and crambe worldwide, such potential changes in the insect fauna of annual cropping systems are possible and perhaps even probable; however, significant increase in production of the two crops is a scenario that may or may not become reality.

References

Alahakoon, U., Adamson, J., Grenkow, L., Soroka, J., Bonham-Smith, P. and Gruber, M. (2016) Field growth traits and insect-host plant interactions of two transgenic canola lines with elevated trichome numbers. *The Canadian Entomologist* 148(5), 603–615.

Ančev, M. and Goranova, V. (2006) Trichome morphology of eleven genera of the tribe Alysseae (Brassicaceae) occurring in Bulgaria. *Willdenowia* 36(1), 193–204.

Anderson, M.D., Peng, C. and Weiss, M.J. (1992) Crambe, *Crambe abyssinica* Hochst., as a flea beetle resistant crop (Coleoptera: Chrysomelidae). *Journal of Economic Entomology* 85, 594–600.

Andersson, G. and Olsson, G. (1950) Feldversuche mit Leindotter—*Camelina sativa* Crantz. *Sveriges Utsaedesfoerengen Tidskrift* 60, 455–458.

Anon. (1973) *Annual report of co-operative regional projects. North Central Regional Project NC-7*. Agricultural Research Service, US Department of Agriculture, Washington, DC. Available at: www.ars.usda.gov/SP2UserFiles/Place/36251200/Annual_Reports/1973_Annual_Report.pdf (accessed 22 February 2016).

Anon. (2015) *Cabbage stem flea beetle*. HGCA Information sheet 24. Agriculture and Horticulture Development Board, Kenilworth, UK. Available at: https://www.nfuonline.com/assets/44446 (accessed 22 February 2016).

Artus, N.N. (2006) Arsenic and cadmium phytoextraction potential of crambe compared with Indian mustard. *Journal of Plant Nutrition* 29, 667–679.

Bahar, M.H., Soroka, J.J., Dosdall, L.M and Olfert, O.O. (2013) Occurrence of diamondback moth, *Plutella xylostella* (Lepidoptera: Plutellidae), and its larval parasitoids across Saskatchewan, Canada. *Biocontrol Science and Technology* 23, 724–729.

Bart, J.C.J., Gucciardi, E. and Cavallaro, S. (2013) *Biolubricants: Science and Technology.* Woodhead Publishing Series in Technology No. 46. Woodhead, Cambridge, UK.

Begna, S.H. and Fielding, D.J. (2003) Damage potential of grasshoppers (Orthoptera: Acrididae) on early growth stages of small-grains and canola under subarctic conditions. *Journal of Economic Entomology* 96, 1193–1200.

Berhow, M.A., Polat, U., Glinski, J.A., Glensk, M., Vaughn, S.F., Isbell, T., Ayala-Diaz, I., Marek, L. and Gardner, C. (2013) Optimized analysis and quantification of glucosinolates from *Camelina sativa* seeds by reverse-phase liquid chromatography. *Industrial Crops and Products* 43, 119–125.

Borg, A. (1952) Further parasite attacks on camelina. *Vaxtskyddsnotiser* 2, 20–23.

Canola Council of Canada (2013) *South winds arrive*. CanolaWatch Issue 8, 23 May 2013. Canola Council of Canada, Winnipeg, Manitoba, Canada. Available at: http://www.canolawatch.org/2013/05/23/south-winds-arrive/ (accessed 22 February 2016).

Canola Council of Canada (2014a) *Flea beetles*. Canola Encyclopedia, Canola Council of Canada, Winnipeg, Manitoba, Canada. Available at: http://www.canolacouncil.org/canola-encyclopedia/insects/flea-beetles/ (accessed 22 February 2016).

Canola Council of Canada (2014b) *Diamondback moth*. Canola Encyclopedia, Canola Council of Canada, Winnipeg, Manitoba, Canada. Available at: http://www.canolacouncil.org/canola-encyclopedia/insects/diamondback-moth/ (accessed 22 February 2016).

Canola Council of Canada. (2014c) *Bertha armyworm*. Canola Encyclopedia, Canola Council of Canada, Winnipeg, Manitoba, Canada. Available at http://www.canolacouncil.org/canola-encyclopedia/insects/bertha-armyworm/# (accessed 22 February 2016).

Cárcamo, H.A., Olfert, O.O., Dosdall, L.M., Herle, C.E., Beres, B.L. and Soroka, J.J. (2007) Resistance to cabbage seedpod weevil among selected Brassicaceae germplasm. *The Canadian Entomologist* 139, 658–669.

Chen, M., Shelton, A., Hallett, R., Hoepting, C.A., Kikkert, J. and Wang, P. (2011) Swede midge (Diptera: Cecidomyiidae), ten years of invasion of crucifer crops in North America. *Journal of Economic Entomology* 104, 709–716.

Chesnais, Q., Verzeaux, J., Couty, A., Le Roux, V. and Ameline, A. (2015) Is the oil seed crop *Camelina sativa* a potential host for aphid pests? *Bioenergy Research* 8, 91–99.

Chew, F.S. (1988) Biological effects of glucosinolates. In: H.G. Cutler (ed.) *Biologically Active Natural Products*. American Chemical Society, Washington, DC, pp. 155–181.

COSMOS (2015) H2020 – COSMOS – Camelina & Crambe Oil Crops as Sources for Medium-chain Oils for Specialty Oleochemicals. Research. Wageningen UR Food & Bio-based Research, Wageningen, The Netherlands. Available at: http://cosmos-h2020.eu/research/ (accessed 22 February 2016).

Crop Protection Compendium (2012a) *Phyllotreta striolata*. CABI, Wallingford, UK. Available at: http://www.cabi.org/cpc/datasheet/40784 (accessed 22 February 2016).

Crop Protection Compendium (2012b) *Delia radicum* cabbage root fly. CABI, Wallingford, UK. Available at: http://www.cabi.org/cpc/datasheet/28164 (accessed 22 February 2016).

Crop Protection Compendium (2015a) *Phyllotreta cruciferae*. CABI, Wallingford, UK. Available at: http://www.cabi.org/cpc/datasheet/40780 (accessed 22 February 2016).

Crop Protection Compendium (2015b) *Delia floralis* turnip maggot. CABI, Wallingford, UK. Available at: http://www.cabi.org/cpc/datasheet/28167 (accessed 22 February 2106).

Crop Protection Compendium (2015c) *Delia platura* bean seed fly. CABI, Wallingford, UK. Available at: http://www.cabi.org/cpc/datasheet/28168 (accessed 22 February 2016).

Darvas, B. and Szappanos, A. (2003) Male and female morphology of some central European *Delia* (Anthomyiidae) pests. *Acta Zoologica Hungarica* 49, 88–101.

Davis, P.B. (2010) The invasion potential and competitive ability of *Camelina sativa* (L.) Crantz (Camelina) in Rangeland Ecosystems. MSc Thesis, Montana State University, Bozeman, Montana. Available at: scholarworks.montana.edu/xmlui/bitstream/handle/1/1148/DavisP0510.pdf (accessed 22 February 2016).

Deng, S.D., Yun, G.L., Zhang, Q.W., Xu, H.L. and Cai, Q.N. (2004a) Studies on morphological resistance of false flax to *Plutella xylostella*. *Plant Protection* 30, 39–42.

Deng, S.D., Yun, G.L., Zhang, Q.W., Xu, H.L. and Cai, Q.N. (2004b) Effect of false flax (*Camelina sativa*) on larval feeding and adult behavioural response of the diamondback moth (*Plutella xylostella*). *Acta Entomologica Sinica* 47, 474–478.

Dmoch, J. (1968) Observations on the population dynamics and injuriousness of the false-flax curculionid (*Ceutorhynchus syrites*). *Rocznik Nauk Rolniczych* 93 (A) pt. 4, 633–654.

Dosdall, L.M. (1994) Evidence for successful overwintering of diamondback moth, *Plutella xylostella* (L.) (Lepidoptera: Plutellidae), in Alberta. *The Canadian Entomologist* 126, 183–185.

Dosdall, L.M. and Cárcamo, H.A. (2011) Biology and integrated management of the cabbage seedpod weevil in Prairie canola crops. *Prairie Soils and Crops* 4, 14–23. Available at: http://www.prairiesoilsandcrops.ca/volume4.php (accessed 22 February 2016).

Dosdall, L.M. and Kott, L.S. (2006) Introgression of resistance to cabbage seedpod weevil to canola from yellow mustard. *Crop Science* 46, 2437–2445.

Dosdall, L.M., Good, A., Keddie, B.A., Ekuere, U. and Stringham, G. (2000) Identification and evaluation of root maggot (*Delia* spp.) (Diptera: Anthomyiidae) resistance within Brassicaceae. *Crop Protection* 19, 247–253.

Dosdall, L.M., Soroka, J.J. and Olfert, O. (2011) The diamondback moth in canola and mustard: current pest status and future prospects. *Prairie Soils and Crops* 4, 66–76. Available at: http://www.prairiesoilsandcrops.ca/volume4.php (accessed 22 February 2016).

Doucette, C.F. (1947) Host plants of the cabbage seedpod weevil. *Journal of Economic Entomology* 40 (6), 838–840. doi: http://dx.doi.org/10.1093/jee/40.6.838.

Dumonceaux, T., Green, M., Hammond, C., Perez, E. and Olivier, C. (2014) Novel molecular diagnostic tools for detection and differentiation of phytoplasmas based on chaperonin-60 reveal differences in host plant infection patterns. *PLoS ONE* 9(12): e116039. doi:10.1371/journal.pone.0116039.

Eberle, C.A., Thom, M.D., Nemec, K.T., Forcella, F., Lundgren, J.G., Gesch, R.W., Riedell, W.E., Papiernik, S.K., Wagner, A., Peterson, D.H. and Eklund, J.J. (2015) Using pennycress, camelina, and canola cash cover crops to provision pollinators. *Industrial Crops and Products Part B* 75, 20–25.

Ehrensing, D.T. and Guy, S.O. (2008) *Oilseed crops – Camelina*. Bulletin EM 8953-E. Oregon State University Extension Service, Corvallis, Oregon.

Eickermann, M. and Ulber, B. (2010) Screening of oilseed rape and other brassicaceous genotypes for susceptibility to *Ceutorhynchus pallidactylus* (Mrsh.). *Journal of Applied Entomology* 134, 542–550.

Ekbom, B. and Borg, A. (1996) Pollen beetle (*Meligethes aeneus*) oviposition and feeding preference on different host plant species. *Entomologia Experimentalis et Applicata* 78, 291–299.

Ellis, P.R., Cole, R.A., Crips, P. and Hardman, J.A. (1980) The relationship between cabbage root fly egg laying and volatile hydrolysis products of radish. *Annals of Applied Biology* 95, 207–214. doi: 10.1111/j.1744-7348.1980.tb04749.x.

Ellis, W.N. (2014) Leafminers and plant galls of Europe. Available at: http://www.bladmineerders.nl/index.htm (accessed 22 February 2016).

Endres, G. and Schatz, B. (2013) *Crambe production.* Bulletin A1010 (revised). North Dakota State University Extension Service, Fargo, North Dakota.

Erickson, D.B. and Bassin, P. (1990) Rapeseed and crambe. Alternative crops with potential industrial uses. *Bulletin of the Kansas Agricultural Experimental Station* 656, 1–33. Available at: www.k-state.edu/historicpublications/pubs/SB656.pdf (accessed 22 February 2016).

Feeny, P., Paauwe, K.L. and Demong, N.J. (1970) Flea beetles and mustard oils: host plant specificity of *Phyllotreta cruciferae* and *P. striolata* adults (Coleoptera: Chrysomelidae). *Annals of the Entomological Society of America* 63, 832–841.

Filipiev, I.N. (1929) Annual Report of the Division of Applied Entomology. *Annals of the Institute of Experimental Agronomy* 7 (1), 94–106.

Finch, S. (1978) Volatile plant chemicals and their effect on host plant finding by the cabbage root fly (*Delia brassicae*). *Entomologia Experimentalis et Applicata* 24, 150–159.

Fleenor, R. (2011) Plant guide for camelina (*Camelina sativa*). USDA Natural Resources Conservation Service, Spokane, Washington. Available at: http://plants.usda.gov/core/profile?symbol=casa2 (accessed 22 February 2016).

Forcella, F., Eberle, C.A., Thom, M.D and Gesch, R.W. (2014) Good crop/bad crop: interrogating oilseeds for floral resources. Abstract in: Miller, T., Alexopoulou, E. and Berti, M.T. (eds) *International Conference in Industrial Crops and 26th Annual Meeting of the Association for the Advancement of Industrial Crops* (AAIC). Program and Abstracts. September 13–19, 2014, Athens, Greece.

Francis, A. and Warwick, S.I. (2009) The biology of Canadian weeds. 142. *Camelina alyssum* (Mill.) Thell.; *C. microcarpa* Andrz. ex DC.; *C. sativa* (L.) Crantz. *Canadian Journal of Plant Science* 89, 791–810.

Francisco-Ortega, J., Fuentes-Aguilar, J., Gómez-Campo, C., Santos-Guerra, A. and Jansen, R.K. (1999) Internal transcribed spacer sequence phylogeny of *Crambe* L. (Brassicaceae): molecular data reveal two Old World disjunctions. *Molecular Phylogenetics and Evolution* 11, 361–380.

Fridén, F. (1972) Humlor och jordbruksväxter. *Svensk Frötidning* 4, 77–81.

Fröhlich, A. and Rice, B. (2005) Evaluation of camelina oils as feedstock for biodiesel production. *Industrial Crops and Products* 21, 25–31.

Furlong, M.J., Wright, D.J. and Dosdall, L.M. (2013) Diamondback moth ecology and management: problems, progress, and prospects. *Annual Review of Entomology* 58, 517–541.

Griffiths, G.C.D. (1986a) Relative abundance of *Delia floralis* (Fallén) and *D. radicum* (L.) (Diptera: Anthomyiidae) infesting canola in Alberta. *Quaestiones Entomologicae* 22, 23–26.

Griffiths, G.C.D. (1986b) Phenology and dispersion of *Delia radicum* (L.) (Diptera: Anthomyiidae) in canola fields at Morinville, Alberta. *Quaestiones Entomologicae* 22, 29–50.

Griffiths, G.C.D. (1991a) Economic assessment of cabbage maggot damage in canola in Alberta. *Proceedings of the GCIRC Eighth International Rapeseed Congress* 2, 528–535 (Saskatoon, July 9–11, 1991).

Griffiths, G.C.D. (1991b) Cyclorrhapha II (Schizophora: Calyptratae) Anthomyiidae. *Flies of the Nearctic Region.* Volume VIII, Part 2, Number 7. E. Schweizerbart'sche Verlagsbuchhandlung (Naegele u. Obermiller), Stuttgart, Germany.

Griffiths, G.C.D. (1993) Cyclorrhapha II (Schizophora: Calyptratae) Anthomyiidae. *Flies of the Nearctic Region.* Volume VIII, Part 2, Number 10. E. Schweizerbart'sche Verlagsbuchhandlung (Naegele u. Obermiller), Stuttgart, Germany.

Groeneveld, J.H. and Klein, A-M. (2014) Pollination of two oil-producing plant species: camelina (*Camelina sativa* L. Crantz) and pennycress (*Thlaspi arvense* L.) double-cropping in Germany. *GCB Bioenergy* 6, 232–251. doi: 10.1111/gcbb.12122 Available at: http://onlinelibrary.wiley.com/doi/10.1111/gcbb.12122/suppinfo (accessed 22 February 2016).

Gruber, M.Y., Wang, S., Ethier, S., Holowachuk, J., Bonham-Smith, P.C., Soroka, J.J. and Lloyd, A. (2006) 'HAIRY CANOLA' – Arabidopsis *GL3* induces a dense coating of trichomes on *Brassica napus* seedlings. *Plant Molecular Biology* 60, 679–698.

Gugel, R.K. and Falk, K.C. (2006) Agronomic and seed quality evaluation of *Camelina sativa* in western Canada. *Canadian Journal of Plant Science* 86, 1047–1058.

Hallett, R.H. (2007) Host plant susceptibility to the swede midge (Diptera: Cecidomyiidae). *Journal of Economic Entomology* 100, 1335–1343.

Handley, R., Ekbom, B. and Ågren, J. (2005) Variation in trichome density and resistance against a specialist insect herbivore in natural populations of *Arabidopsis thaliana*. *Ecological Entomology* 30, 284–292.

Harcourt, D.G. (1957) Biology of the diamondback moth, *Plutella maculipennis* (Curt.) (Lepidoptera: Plutellidae), in eastern Ontario. II. Life-history, behaviour, and host relationship. *The Canadian Entomologist* 12, 554–564.

Hegedus, D.D. and Erlandson, M.E. (2012) Genetics and genomics of insect resistance in Brassicaceae crops. In: Edwards, D., Batley, J., Parkin, I. and Kole, C. (eds) *Genetics, Genomics and Breeding of Oilseed Brassicas*. Science Publishers, Enfield, New Hampshire, pp. 319–372.

Henderson, A.E., Hallett, R.H. and Soroka, J.J. (2004) Pre-feeding behavior of the crucifer flea beetle, *Phyllotreta cruciferae*, on host and non-host crucifers. *Journal of Insect Behavior* 17, 17–39.

Henriksen, B.I.F., Lundon, A.R., Prestløkken, E., Abrahamsen, U. and Eltun, R. (2009) Nutrient supply for organic oilseed crops, and quality of potential organic protein feed for ruminants and poultry. *Agronomy Research* 7, 592–598.

Hill, D.S. (1987) *Agricultural Insect Pests of Temperate Regions and Their Control*. Cambridge University Press, Cambridge, UK.

Hopkins, R., Birch, A.N.E., Griffiths, D.W., Baur, R., Stadler, E. and McKinley, R.G. (1997) Leaf surface compounds and oviposition preference of the turnip root fly *Delia floralis*: the role of glucosinolate and nonglucosinolate compounds. *Journal of Chemical Ecology* 23, 629–643.

Hopkins, R.J., van Dam, N.M. and van Loon, J.J.A. (2009) Role of glucosinolates in insect-plant relationships and multitrophic interactions. *Annual Review of Entomology* 54, 57–83.

Hopkinson, R.F. and Soroka, J.J. (2009) Air trajectory model applied to an in-depth diagnosis of potential diamondback moth infestations on the Canadian Prairies. *Agricultural and Forest Meteorology* 150, 1–11.

Hunsaker, D.J., French, A.N. and Thorp, K.R. (2011) Camelina water use and seed yield response to irrigation scheduling in an arid environment. *Irrigation Science* 2, 27–43.

Hunter, J. and Roth, G. (2010) *Camelina Production and Potential in Pennsylvania. Agronomy Facts 72*. College of Agricultural Sciences, Crop and Soil Sciences, Pennsylvania State University. Available at: http://extension.psu.edu/plants/crops/grains/small/production/camelina-production-and-potential-in-pennsylvania (accessed 22 February 2016).

Jarvis, J. (1969) Differential reaction of introduction of crambe to the turnip aphid and the green peach aphid. *Journal of Economic Entomology* 62, 69–698.

Jarvis, J. (1982) Susceptibility of some Brassicaceae oilseed plant introductions to the cabbage aphid (Homoptera: Aphididae). *Journal of the Kansas Entomological Society* 55, 283–289.

Kelton, L.A. (1975) The lygus bugs (genus *Lygus* Hahn) (Heteroptera: Miridae). *Memoirs of the Entomological Society of Canada* No. 95. Entomological Society of Canada, Winnipeg, Manitoba, Canada.

Kmec, P. and Weiss, M.J. (1997) Seasonal abundance of diamondback moth (Lepidoptera: Yponomeutidae) on *Crambe abyssinica*. *Environmental Entomology* 26, 483–488.

Kmec, P., Weiss, M.J., Milbrath, L.R., Schatz, B.G., Hanzel, J., Hanson, B.K. and Eriksmoen, E.D. (1998) Growth analysis of crambe. *Crop Science* 38, 108–112.

Knights, S. (2006) Crambe – a new crop with potential in southern Australia. In: *BCG Crop and Pasture Production Manual*. Birchip Cropping Group, Birchip, Victoria, Australia, pp. 275–276.

Knodel, J.J. and Olson, D.L. (2002) *Crucifer flea beetle: biology and integrated pest management in canola*. North Dakota State University Co-operative Extension Service Publication E1234. North Dakota State University, Fargo, North Dakota.

Knörzer, K.H. (1978) Evolution and spread of gold of pleasure (*Camelina sativa* S.L.). *Bererichte der Deutschen Botanischen Gesellschaft* 91, 187–195.

Korsrud, G.O., Keith, M.O. and Bell, J.M. (1978) A comparison of the nutritive value of crambe and camelina seed meals with egg and casein. *Canadian Journal of Animal Science* 58, 493–499.

Lamb, R.J. (1984) Effects of flea beetles, *Phyllotreta* spp. (Coleoptera: Chrysomelidae), on the survival, growth, seed yield and quality of canola, rape and yellow mustard. *The Canadian Entomologist* 116, 269–280.

Lamb, R.J. and Turnock, W.J. (1982) Economics of insecticidal control of flea beetles (Coleoptera: Chrysomelidae) attacking rape in Canada. *The Canadian Entomologist* 114, 827–840.

Lenssen, A.W., Iversen, W.M., Sainju, U.M., Caesar-Tonthat, T.C., Blodgett, S.L., Allen, B.L. and Evans, R.G. (2012) Yield, pests, and water use of durum and selected crucifer oilseeds in two-year rotations. *Agronomy Journal* 104, 1295–1304.

Lessman, K.J. (1990) Crambe: a new industrial crop in limbo. In: Janick, J. and Simon, J.E. (eds) *Advances in New Crops.* Timber Press, Portland, Oregon. pp. 217–222. Available at: https://hort.purdue.edu/newcrop/proceedings1990/V1-217.html (accessed 22 February 2016).

Madder, D.J. and Stemeroff, M. (1988) The economics of insect control on wheat, corn, and canola in Canada, 1980–1985. *Bulletin of the Entomological Society of Canada* 20 (Supplement), 1–2.

Madel, W. (1950) Observations on the occurrence of 'camelina weevil'. *Nachrichtenblatt des Deutschen Pflanzenschutzdienstes* 2, 90–92.

Mason, P.G., Arthur, A.P., Olfert, O. and Erlandson, M.A. (1998) The bertha armyworm (*Mamestra configurata*) (Lepidoptera: Noctuidae) in western Canada. *The Canadian Entomologist* 130, 321–336.

Mason, P.G., Olfert, O., Sluchinski, L., Weiss, R.M., Boudreault, C., Grossrieder, M. and Kuhlmann, U. (2003) Actual and potential distribution of an invasive canola pest, *Meligethes viridescens* (Coleoptera: Nitidulidae), in Canada. *The Canadian Entomologist* 135, 405–413.

Merrill, S.D., Tanaka, D.L., Krupinsky, J.M., Liebig, M.A., Hendrickson, J.R., Hanson, J.D., Fehmi, J.S. and Ries, R.E. (2002) Soil water use and soil residue coverage by sunflower compared to other crops. *Proceedings of the 24th Sunflower Research Workshop*. National Sunflower Research Forum, Bismarck, North Dakota, pp. 88–96. Available at: http://www.ars.usda.gov/research/publications/publications.htm?SEQ_NO_115=132622 (accessed 22 February 2016).

Metspalu, L., Kruus, E., Ploomi, A., Williams, I.H., Hiiesaar, K., Jõgar, K., Veromann, E. and Mänd, M. (2014) Flea beetle (Chrysomelidae: Alticinae) species composition and abundance in different cruciferous oilseed crops and the potential for a trap crop system. *Acta Agriculturae Scandinavica, Section B – Soil & Plant Science* 64, 572–582. doi: 10.1080/09064710.2014.933871.

Mikolajczak, K.L., Miwa, T.A., Earle, F.R., Wolff, I.A. and Jones, Q. (1961) Search for new industrial oils. V. Oils of Cruciferae. *Journal of the American Oil Chemists' Society* 38, 678–681.

Milbrath, L.R., Weiss, M.J. and Schatz, B.G. (1995) Influence of tillage, planting date and oilseed crucifers on flea beetle populations (Coleoptera: Chrysomelidae). *The Canadian Entomologist* 127, 289–293.

Miller, S.G., Anderson, K., Bassendowski, K.A., Britz, L., Buitenhuis, N., Campbell, E., Chant, S., Christopher, J., Cowell, L.E. and Cranston, R. (2013) Survey of canola diseases in Saskatchewan, 2012. *Canadian Plant Disease Survey* 93, 149–153.

Muller, R., de Vos, M., Sun, J.Y., Sonderby, I.E. and Halkier, B.A. (2010) Differential effects of indole and aliphatic glucosinolates on lepidopteran herbivores. *Journal of Chemical Ecology* 36, 905–913.

Nair, K.S.S. and McEwen, F.L. (1973) The seed maggot complex, *Hylemya* (*Delia*) *platura* and *H.* (*D.*) *litura* (Diptera: Anthomyiidae), as primary pests of radish. *The Canadian Entomologist* 105, 445–447.

Naranjo, S.E. and Stefanek, M.A. (2012) Feeding behavior of a potential insect pest, *Lygus hesperus*, on four new industrial crops for the arid southwestern USA. *Industrial Crops and Products* 37, 358–361.

Nellis, W.M. (2009) Leafminers and plant galls of Europe. Available at: http://www.bladmineerders.nl/index.htm (accessed 22 February 2016).

Nielsen, J.A., Hansen, M.L., Agerbirk, N., Petersen, B.L. and Halkier, B.A. (2001) Responses of the flea beetles *Phyllotreta nemorum* and *P. cruciferae* to metabolically engineered *Arabidopsis thaliana* with an altered glucosinolate profile. *Chemoecology* 11, 75–83.

Nieschlag, H.J. and Wolff, I.A. (1971) Industrial uses of high erucic oils. *Journal of the American Oil Chemists' Society* 48, 723–727.

Olfert, O.O. and Weiss, R.M. (2002) Impact of grasshopper feeding on selected cultivars of cruciferous oilseed crops. *Journal of Orthoptera Research* 11, 83–86. Available at: http://www.jstor.org/stable/3503665 (accessed 22 February 2016).

Olivier, C.Y., Lowery, D.T. and Stobbs, L.W. (2009) Phytoplasma diseases and their relationships with insect and plant hosts in Canadian horticultural and field crops. *The Canadian Entomologist* 141, 425–462.

Olivier, C., Séguin-Swartz, G., Galka, B. and Olfert, O. (2011) Aster yellows in leafhoppers and field crops in Saskatchewan, Canada, 2001–2008. In: Al-Mughrabi, K. (ed.) Plant Science and Biotechnology in North America: Focus on Canada III. *The Americas Journal of Plant Science and Biotechnology* 5 (Special Issue 2), 88–94.

Onyilagha, J.C., Gruber, M.Y., Hallett, R.H., Holowachuk, J. and Soroka, J.J. (2014) Constitutive flavonoids deter flea beetle insect feeding in *Camelina sativa* L. *Biochemical Systematics and Ecology* 42, 128–133.

Oplinger, E., Oelke, E., Kaminski, A., Putnam, D., Teynor, D., Doll, J., Kelling, K., Durgan, B. and Noetzel, D. (1991) *Crambe. Alternative Field Crops Manual*. University of Wisconsin, Madison, Wisconsin. Available at: https://hort.purdue.edu/newcrop/afcm/crambe.html (accessed 22 February 2016).

Otani, J. and Cárcamo, H. (2011) Biology and management of *Lygus* in canola. *Prairie Soils and Crops* 4, 42–53. Available at: http://www.prairiesoilsandcrops.ca/volume4.php (accessed 22 February 2016).

Oyen, L.P.A. (2007) *Crambe hispanica* L. In: van der Vossen, H.A.M. and Mkamilo, G.S. (eds) *Plant Resources of Tropical Africa Volume* 14. *Vegetable Oils*. PROTA Foundation. Backhuys Publishers, Wageningen, The Netherlands, pp. 65–68. Available at: http://uses.plantnet-project.org/en/Crambe_hispanica_(PROTA) (accessed 22 February 2016).

Pachagounder, P., Lamb, R.J. and Bodnaryk, R.P. (1998) Resistance to the flea beetle *Phyllotreta cruciferae* (Coleoptera: Chrysomelidae) in false flax, *Camelina sativa* (Brassicaceae). *The Canadian Entomologist* 130, 235–240.

Palagesiu, I. (2003) Contributions to the knowledge of the entomofauna of some prospective oil plants in Romania. Stiintifice – Agricultura, Universitatea de Stiinte Agricole si Medicina Veterinara a Banatului Timisoara 2000, 32, 983–992.

Paulose, B., Kandasamy, S. and Dhankher, O.P. (2010) Expression profiling of *Crambe abyssinica* under arsenate stress identifies genes and gene networks involved in arsenic metabolism and detoxification. *BMC Plant Biology* 10, 108. doi: 10.1186/1471-2229-10-108.

Peiretti, P.G. and Meineri, G. (2007) Fatty acids, chemical composition and organic matter digestibility of seeds and vegetative parts of false flax (*Camelina sativa* L.) after different lengths of growth. *Animal Feed Science and Technology* 133, 341–350.

Peng, C., Weiss, M. and Anderson, M.D. (1992) Flea beetle (Coleoptera: Chrysomelidae) response, feeding, and longevity on oilseed rape and crambe. *Environmental Entomology* 21, 604–609.

Peterson, C.J., Cossé, A. and Coats, J.R. (2000) Insecticidal components in the meal of *Crambe abyssinica*. *Journal of Agricultural and Urban Entomology* 17, 27–36.

Philip, H. (2015) *Field crop and forage pests and their natural enemies in Western Canada: Identification and management field guide*. Agriculture and Agri-Food Canada, Saskatoon, Saskatchewan, Canada. Available at: http://publications.gc.ca/collections/collection_2015/aac-aafc/A59-23-2015-eng.pdf (accessed 22 February 2016).

Philip, H. and Mengersen, E. (1989) *Insect Pests of the Prairies*. University of Alberta, Edmonton, Alberta, Canada.

Phillips, T. (2015) President's Message. In: *Ontario Canola Growers Association January 2015 Newsletter*. Ontario Canola Growers Association, Markdale, Ontario, Canada, p.1.

Pilgeram, A.L., Sands, D.C., Boss, D., Dale, N., Wichman, D., Lamb, P., Lu, C., Barrows, R., Kirkpatrick, M., Thompson, B. and Johnson, D.L. (2007) *Camelina sativa*, a Montana omega-3 and fuel crop. In: Whipkey, A. (ed.) *Issues in New Crops and New Uses*. ASHS Press, Alexandria, Virginia, pp. 129–131. Available at: https://hort.purdue.edu/newcrop/nexus/Camelina_sativa_nex.html (accessed 22 February 2016).

Pitkin, B., Ellis, W., Plant, C. and Edmunds, R. (2016) The leaf and stem mines of British flies and other insects. Available at: http://www.ukflymines.co.uk/index.php (accessed 22 February 2016).

Pivnick, K.A., Jarvis, B.J. and Slater, G.P. (1994) Identification of olfactory cues used in host plant finding by diamondback moth, *Plutella xylostella* (Lepidoptera: Plutellidae). *Journal of Chemical Ecology* 20, 1407–1427.

Plessers, A.C., McGregor, W.G., Carson, R.B. and Nakoneshny, W. (1962) Species trials with oilseed plants. II. Camelina. *Canadian Journal of Plant Science* 42, 452–459.

Porcher, F.E. (1869) *Resources of the Southern Fields and Forests, Medical, Economical, and Agricultural. Being also a Medical Botany of the Confederate States; with Practical Information on the Useful Properties of the Trees, Plants, and Shrubs* (revised edn). Walker, Evans & Cogswell, Charleston, North Carolina. Available at: https://books.google.ca/books?id=pjfTuqmKfk4C&pg=PA73&lpg=PA73&dq=Porcher+camelina&source=bl&ots=tTuRUXaymF&sig=DIlTxt_EghjNHER_kZlVGOS-t3k&hl=en&sa=X&ved=0ahUKEwij9YOEpLvJAhVHdR4KHXkmC3UQ6AEIOzAG#v=onepage&q=Porcher%20camelina&f=false (accessed 22 February 2016).

Prairie Pest Monitoring Network (2014) Weekly pest and crop updates. 10 June 2014. Available at: http://www.westernforum.org/IPMNWeeklyUpdates.html (accessed 22 February 2016).

Putnam, D.H., Budin, J.T., Field, L.A. and Breene, W.M. (1993) Camelina: a promising low-input oilseed. In: Janick, J. and Simon, J.E. (eds) *New Crops*. Wiley, New York. pp. 314–322.

Rakhmaninov, A.N. and Vuirzhikovskaya, A.V. (1930) On *Ceuthorhynchus syrites* Germ., a pest of *Camelina sativa*. *Report on Applied Entomology* 4(2), 345–350. (English abstract).

Renwick, J.A.A. and Radke, C.D. (1990) Plant constituents mediating oviposition by the diamondback moth, *Plutella xylostella* (L.). *Phytophaga* 3, 37–46.

Ritter, R.A., Lenssen, A., Blodgett, S. and Taper, M.A. (2010) Regional assemblages of *Lygus* (Heteroptera: Miridae) in Montana canola fields. *Journal of the Kansas Entomological Society* 83, 297–305. doi: 10.2317/jkes0912.23.1.

Robinson, R.G. (1987) Camelina: a useful research crop and a potential oilseed crop. *Minnesota Agricultural Experimental Station Bulletin* 579, 1–12.

Rokka, T., Alen, K., Valaja, J. and Ryhanen, E.L. (2002) The effect of a *Camelina sativa* enriched diet on the composition and sensory quality of hen eggs. *Food Research International* 35, 253–256.

Schuster, A. and Friedt, W. (1998) Glucosinolate content and composition as parameters of quality of camelina seed. *Industrial Crops and Products* 7, 297–302.

Séguin-Swartz, G., Eynck, C., Gugel, R.K., Strelkov, S.E., Olivier, C.Y., Li, J.L., Klein-Gebbinck, H., Borhan, H., Caldwell, C. and Falk, K.C. (2009) Diseases of *Camelina sativa* (L.) Crantz (false flax). *Canadian Journal of Plant Pathology* 31, 375–386.

Singh, S.P. and Sachan, G.C. (1997) Effect of different temperatures and host plants on the developmental behaviour of mustard sawfly, *Athalia proxima*. *Indian Journal of Entomology* 59, 34–40.

Singh, S.P., Kumar, S., Singh, Y.P. and Singh, R. (2014) Reliable screening technique for evaluation of wild crucifers against mustard aphid *Lipahis erysimi* (Kalt.). *Indian Journal of Experimental Biology* 52, 1201–1205.

Smith, D.B. and Sears, M.K. (1982) Evidence for dispersal of diamondback moth, *Plutella xylostella* (Lepidoptera: Plutellidae), into southern Ontario. *Proceedings of the Entomological Society of Ontario* 113, 21–27.

Soroka, J.J. and Dosdall, L.M. (2011) Coping with root maggots in Prairie canola crops. *Prairie Soils and Crops* 4, 24–29. Available at: prairiesoilsandcrops.ca/articles/volume-4-4-screen.pdf (accessed 22 February 2016).

Soroka, J. and Grenkow, L. (2013) Susceptibility of brassicaceous plants to feeding by flea beetles, *Phyllotreta* spp. (Coleoptera: Chrysomelidae). *Journal of Economic Entomology* 106, 2557–2567.

Soroka, J.J., Dosdall, L.M., Olfert, O.O. and Seidle, E. (2004) Root maggots (*Delia* spp., Diptera: Anthomyiidae) in prairie canola (*Brassica napus* L. and *B. rapa* L.): spatial and temporal surveys of root damage and prediction of damage levels. *Canadian Journal of Plant Science* 84, 1171–1182.

Soroka, J.J., Bartelt, R.J., Zilkowski, B.W. and Cossé, A.A. (2005) Responses of flea beetle *Phyllotreta cruciferae* in the field to synthetic aggregation pheromone components and plant host volatiles. *Journal of Chemical Ecology* 31, 1829–1843.

Soroka, J.J., Grenkow, L.F. and Irvine, R.B. (2008) Impact of decreasing ratios of insecticide-treated seed on flea beetle (Coleoptera: Chrysomelidae, *Phyllotreta* spp.) feeding levels and canola seed yields. *Journal of Economic Entomology* 101, 1811–1820.

Soroka, J., Olivier, C., Grenkow, L. and Séguin-Swarz, G. (2015) Interactions between *Camelina sativa* (Brassicaceae) and insect pests of canola. *The Canadian Entomologist* 147, 193–214.

Städler, E. (1992) Behavioural responses of insects to plant secondary compounds. In: Rosenthal, G. and Berenbaum, M. (eds) *Herbivores: Their Interactions with Plant Secondary Metabolites; Evolutionary and Ecological Processes*. Academic Press. San Diego, California, pp 44–88.

Sturtevant, E.L. (1919) Sturtevant's notes on edible plants. In: Hedrick, U.P. (ed.) *State of New York Department of Agriculture Twenty-seventh Annual Report, Vol. 2 Part II*. Available at: https://archive.org/stream/sturtevantsnotes00sturuoft/sturtevantsnotes00sturuoft_djvu.txt (accessed 22 February 2016).

Sun, J.Y., Sonderby, I.E., Halkier, B.A., Jander, G. and de Vos, M. (2009) Non-volatile intact indole glucosinolates are host recognition cues for ovipositing *Plutella xylostella*. *Journal of Chemical Ecology* 35, 1427–1436.

Syed, T.S., Abro, G.H., Lu, Y.Y. and Liang, G.W. (2002) Effect of host plant on biological parameters of *Plutella xylostella* under laboratory condition. *Journal of South China Agricultural University* 23, 18–22.

Talekar, N.S., Liu, S.-H., Chen, C.-L. and Yiin, Y.-F. (1994) Characteristics of oviposition of diamondback moth (Lepidoptera: Yponomeutidae) on cabbage. *Zoological Studies* 33, 72–77.

Tansey, J.A., Dosdall, L.M., Keddie, B.A. and Sarfraz, R.M. (2008) Differences in *Phyllotreta cruciferae* and *Phyllotreta striolata* (Coleoptera: Chrysomelidae) responses to neonicotinoid seed treatments. *Journal of Economic Entomology* 101, 159–167.

Tansey, J.A., Dosdall, L.M. and Keddie, B.A. (2009) *Phyllotreta cruciferae* and *Phyllotreta striolata* responses to insecticidal seed treatments with different modes of action. *Journal of Applied Entomology* 133, 201–209.

Tansey, J.A., Dosdall, L.M., Keddie, B.A., Fletcher, R.S. and Kott, L.S. (2010a) Responses of *Ceutorhynchus obstrictus* (Marsham) (Coleoptera: Curculionidae) to olfactory cues associated with novel genotypes developed by *Sinapis alba* L. x *Brassica napus* L. *Arthropod–Plant Interactions* 4, 95–106.

Tansey, J.A., Dosdall, L.M., Keddie, B.A. and Noble, S. (2010b) Contributions of visual cues to cabbage seedpod weevil, *Ceutorhynchus obstrictus* (Marsham) (Coleoptera: Curculionidae), resistance in novel host genotypes. *Crop Protection* 29, 476–481.

Tsao, R., Reuber, M., Johnson, L. and Coats, J.R. (1996) Insecticidal toxicities of glucosinolate-containing extracts from crambe seeds. *Journal of Agricultural Entomology* 13, 109–120.

Tsao, R., Peterson, C.J. and Coats, J.R. (2002) Glucosinolate breakdown products as insect fumigants and their effect on carbon dioxide emission of insects. *BMC Ecology* 2, 5. doi: 10.1186/1472-6785-2-5.

Turnock, W.J., Timlick, B., Galka, B.E. and Palaniswamy, P. (1992) Root maggot damage to canola and the distribution of *Delia* spp. (Diptera: Anthomyiidae) in Manitoba. *The Canadian Entomologist* 124, 49–58.

United States Congress (2007) *Energy Independence and Security Act of 2007. Act to Move the United States Toward Greater Energy Independence and Security, to Increase the Production of Clean Renewable Fuels, to Protect Consumers, to Increase the Efficiency of Products, Buildings, and Vehicles, to Promote Research On and Deploy Greenhouse Gas Capture and Storage Options, and to Improve the Energy Performance of the Federal Government, and for Other Purposes*. United States Government Publication. Public Law 110-140. 10 December 2007. SuDoc. No. AE 2.110: 110-140. Available at: www.gpo.gov/fdsys/pkg/PLAW-110publ140/content-detail.html (accessed 12 May 2016).

Urbaniak, S.D., Caldwell, C.D., Zheljazkov, V.D., Lada, R. and Luan, L. (2008) The effect of seeding rate, seeding date and seeder type on the performance of *Camelina sativa* L. in the Maritime Provinces of Canada. *Canadian Journal of Plant Science* 88, 501–508.

Valk, M. and Stevenson, A.B. (1994) Aster leafhoppers. In: Howard, R.J., Garland, J.A. and Seaman, W.L. (eds) *Diseases and Pests of Vegetables Crops in Canada*. The Canadian Phytopathological Society and the Entomological Society of Canada Press, Ottawa, Canada, pp. 158–159.

Van Dyne, D.L., Blase, M.G. and Carlson, K.D. (1990) *Industrial Feedstocks and Products from High Erucic Acid Oil: Crambe and Industrial Rapeseed*. University of Missouri, Columbia, Missouri.

Vilsack, T. and McCarthy, G. (2015) *Pollinator Research Action Plan*. Report of the Pollinator Health Task Force, The White House, Washington, DC. May 19, 2015. Available at: https://www.whitehouse.gov/sites/default/files/microsites/ostp/Pollinator%20Health%20Strategy%2015.pdf (accessed 22 February 2016).

Vollmann, J., Damboeck, A., Eckl, A., Schrems, H. and Ruckenbauer, P. (1996) Improvement of *Camelina sativa*, an underexploited seed. In Janick, J. (ed.) *Progress in New Crops*. American Society of Horticulture Science Press, Alexandria, Virginia, pp. 357–362.

Wahlin, B. (1951) A few parasites attacking *Camalina sativa*. *Vaxtskyddsnotiser* 5–6, 90–93.

Waraich, E.A., Ahmed, Z., Ahmad, R., Ashraf, M.Y., Saifullah, Naeem, M.S. and Rengel, Z. (2013) *Camelina sativa*, a climate proof crop, has high nutritive value and multiple uses: a review. *Australian Journal of Crop Science* 7, 1551–1559.

Warwick, S.I. and Gugel, R. (2003) Genetic variation in the *Crambe abyssinica – C. hispanica – C. glabrata* complex. *Genetic Resources and Crop Evolution* 50, 291–305.

White, G.A. and Higgins, J.J. (1966) *Culture of crambe ... a new industrial oilseed crop*. United States Department of Agriculture Production Research Report No. 95. USDA, Washington, DC.

Young, O.P. (1986) Host plants of the tarnished plant bug, *Lygus lineolaris* (Heteroptera: Miridae). *Annals of the Entomological Society of America* 79, 747–762.

Zerova, M.D. and Seregina, L.Y. (1994) *The seed-eating Chalcidoidea of Palearctics*. National Academy of Sciences in the Ukraine, Kiev, Ukraine.

Zubr, J. (1997) Oil-seed crop: *Camelina sativa*. *Industrial Crops Production* 6, 113–119.

Zulfiquar, A., Paulose, B., Chhikara, S. and Dhankher, O.P. (2011) Identifying genes and gene networks involved in chromium metabolism and detoxification in *Crambe abyssinica*. *Environmental Pollution* 159, 3123–3128.

22 Integrated Pest Management and Pollination Services in *Brassica* Oilseed Crops

Francesc R. Badenes-Pérez,[1]* Tulsi Bhardwaj[2] and Raj K. Thakur[3]

[1]*Institute of Agricultural Sciences (CSIC), Madrid, Spain;* [2]*Indian Agricultural Research Institute, New Delhi, India;* [3]*Indian Council of Agricultural Research, New Delhi, India*

22.1 Introduction

Approximately 75% of crop species benefit from insect pollination, representing about 9.5% of the value of the world agriculture production devoted to human food (Klein *et al.*, 2007; Gallai *et al.*, 2009). In Brazil, the economic contribution of pollinators totals US$12 billion (Giannini *et al.*, 2015). In the USA, the estimated values of honey bees (*Apis mellifera* L.) and non-*Apis* pollinators to agriculture are, approximately, US$11.7 billion and US$3.4 billion, respectively (Calderone, 2012). Honey bees are considered the most valuable pollinators of crops worldwide, though solitary bees and non-bee insects are also important pollinators (Winfree *et al.*, 2007; Rader *et al.*, 2009, 2016; Garibaldi *et al.*, 2013). Canola, *Brassica napus* L. ssp. *oleifera*, is considered to have a modest dependence on pollinators for seed set (Klein *et al.*, 2007), but in open-pollinated *Brassica* oilseed crops insect pollination usually results in increased yield and quality (Stanley *et al.*, 2013; Bartomeus *et al.*, 2014, 2015; Lindström *et al.*, 2016). Pollinators such as *A. mellifera* can also disseminate microbial biocontrol agents targeting insect pests of *Brassica* oilseed crops (Carreck *et al.*, 2007). Because of their nectar content, *Brassica* oilseed crops can be a good bee pasture for honey production (Nedić *et al.*, 2013; Thom *et al.*, 2016) (Fig. 22.1). However, because of the loss of habitats that can provide sufficient abundance and diversity of flowering plants, and because of exposure to pesticides and parasites such as *Varroa destructor*, bee populations have been on the decline (Roulston and Goodell, 2011; Goulson *et al.*, 2015). The main integrated pest management (IPM) practices that can affect pollinator abundance are the use of pesticides and the provision of flowering plants that feed pollinators. *Brassica* oilseed crops are among the crops most studied with regard to bee exposure to toxic insecticides (Lundin *et al.*, 2015; Rundlöf *et al.*, 2015; David *et al.*, 2016). This chapter reviews the importance of pollination services in *Brassica* oilseed crops and how IPM practices can be altered to avoid harm to highly valuable pollinating insects.

22.2 Importance of Insect Pollination in *Brassica* Oilseed Crops

Brassica oilseed crops are mass-flowering species producing nectar with relatively high concentrations of sugars and large quantities of pollen, which makes them attractive to a wide variety of insect pollinators (Kevan *et al.*, 1991; Davis *et al.*, 1998; Thom *et al.*, 2016; Eberle *et al.*, 2015). *Brassica napus* and other *Brassica* oilseed crops are self-fertile, being even pollinated by wind when grown in dense stands, but the high efficiency of insect pollination makes these plants mainly entomophilous

*Corresponding author. E-mail: frbadenes@ica.csic.es

Fig. 22.1. Apiaries of *Apis mellifera* next to fields of *Brassica* oilseed crops in (a) India and (b) Nepal.

(Hayter and Cresswell, 2006; Hoyle and Cresswell, 2007, 2009; Abrol and Shankar, 2012). Studies with both winter and spring cultivars of *Brassica* oilseed crops have shown that insect pollination can result in yield increases ranging from 11% to 30% (yield measured as pod number, seed number and/or seed weight) (Adegas and Nogueira Couto, 1992; Mussury and Fernandes, 2000; Atmowidi *et al.*, 2007; Rosa *et al.*, 2011; Bommarco *et al.*, 2012; Shakeel and Inayatullah, 2013; Stanley *et al.*, 2013; Bartomeus *et al.*, 2014; Chambó *et al.*, 2014; Lindström *et al.*, 2016). In one case, bee pollination increased seed production in *B. campestris* by 80% (Singh *et al.*, 2004). However, the increase in yield as a result of pollination is mainly observed in open-pollinated but not on hybrid cultivars (Lindström *et al.*, 2016). In the case of *B. rapa*, which is an obligate outcrossing species, insect pollination is always necessary (Morandin and Winston, 2005). Insect pollination also ensures uniform ripening and earlier pod setting and it increases seed germination and seed oil content in *Brassica* oilseed crops (Kevan and Eisikowitch, 1990; Abrol, 2007; Atmowidi *et al.*, 2007; NRC, 2007; Abrol and Shankar, 2012; Bartomeus *et al.*, 2014).

22.3 Pollinators Attracted to *Brassica* Oilseed Crops

In insect-pollinated plants, the overall richness of pollinators is very important because of complementary pollination and facilitation among pollinator species (Greenleaf and Kremen, 2006; Winfree *et al.*, 2007; Hoehn *et al.*, 2008; Potts *et al.*, 2010; Blüthgen and Klein, 2011; Carvalheiro *et al.*, 2011; Bartomeus *et al.*, 2015). *Apis mellifera* and other *Apis* spp., such as *A. cerana*, *A. dorsata* and *A. florea*, are considered the most efficient pollinators of *Brassica* oilseed crops, but this is in part because they are often the most abundant pollinators (Rader *et al.*, 2009, 2016; Ali *et al.*, 2011; Woodcock *et al.*, 2013; Kunjwal *et al.*, 2014; Abrol and Thakur, 2016) (Fig. 22.2). Compared with wild bees, *A. mellifera* seems to be less efficient in transferring pollen (Jauker *et al.*, 2012b; Garibaldi *et al.*, 2013; Woodcock *et al.*, 2013) (Table 22.1).

Wild bees, such as bumblebees (*Bombus* spp.) and solitary bees, tend to have greater rates of stigma contact than *A. mellifera* (Woodcock *et al.*, 2013). Among wild bees, *Bombus* spp. are some of the most important pollinators of *Brassica* oilseed crops in temperate ecosystems (Corbet *et al.*, 1991; Cresswell *et al.*, 1996; Cresswell, 1999; Hayter and Cresswell, 2006; Rader *et al.*, 2009; Bartomeus *et al.*, 2015). Other important wild bees pollinating *Brassica* oilseed crops include mining bees (Andrenidae), plasterer bees (Colletidae), mason bees and leafcutter bees (Megachilidae), sweat bees (Halictidae), carpenter bees (Xylocopidae) and other solitary bees (Prasad *et al.*, 1991; Mahindru *et al.*, 1998; Rader *et al.*, 2009; Goswami *et al.*, 2014; Kunjwal *et al.*, 2014; Kumari *et al.*, 2015; Abrol and Thakur, 2016). In tropical and neotropical countries, stingless bees (tribe Meliponini) can be important pollinators of *Brassica* oilseed crops (Kunjwal *et al.*, 2014; Witter *et al.*, 2015).

Non-bee insects are also important pollinators. Depending on the location, the relative contribution of non-bee insects compared with bee insects as pollinators of *Brassica* oilseed crops can be very variable, from almost insignificant to approximately 80%

(Rader *et al.*, 2009). Among the non-bee insects pollinating *Brassica* oilseed crops are hoverflies (Fig. 22.3), such as *Episyrphus balteatus* and *Eristalis tenax* (Diptera: Syrphidae), non-syrphid Diptera, Coleoptera, Hemiptera, Neuroptera and Lepidoptera (Jauker and Wolters, 2008; Rader *et al.*, 2009; Bijaya *et al.*, 2011; Jauker *et al.*, 2012a; Kunjwal *et al.*, 2014; Bartomeus *et al.*, 2015; Kumari *et al.*, 2015; Orford *et al.*, 2015).

Fig. 22.2. *Apis dorsata* bees foraging on a *Brassica* oilseed crop in India.

22.4 Best IPM Practices for *Brassica* Oilseed Crops in the Context of Pollinator Conservation

Both pest management and pollinator management are necessary components of modern agriculture. Unfortunately, pesticide use is often negatively correlated with the abundance of pollinators (Cunningham *et al.*, 2002; Godfray *et al.*, 2014; Goulson *et al.*, 2015). Thus, IPM should provide solutions for minimizing pesticide use and managing pests without harming pollinators. Pollinator-friendly pest management includes crop-specific

Table 22.1. Some of the most common non-*Apis* insects that are pollinators of *Brassica* oilseed crops species: rate of visitation and/or effect on pollination.

Pollinator	Rate of visitation and/or pollination	Reference
Bumblebees (*Bombus* spp.)	Visited flowers at a higher rate than other pollinators	Rader *et al.*, 2009
	150 pollen grains delivered to the stigma in a single flower visit	Cresswell, 1999
	Similar rate of flower visitation as *A. mellifera*	Pierre, 2001
	Can be highly constant to the crop	Kreyer *et al.*, 2004
	Foraging twice as fast as *A. mellifera* and providing the maximum rate of pollination	Hayter and Cresswell, 2006
	Mostly deposited pollen on immediate neighbouring plants but pollen dispersal can extend over 20–40 intervening plants from the originating plant	Cresswell *et al.*, 1996
Plasterer bees (*Leioproctus* spp.)	Similar rate of flower visitation as *A. mellifera*	Rader *et al.*, 2009
Other solitary bees (Andrenidae, Colletidae, Halictidae, Megachilidae and Xylocopidae)	Often seen as pollinators	Abrol and Thakur, 2016; Rahman, 1940; Prasad *et al.*, 1991; Mahindru *et al.*, 1998; Kunjwal *et al.*, 2014; Kumari *et al.*, 2015
Mason bees (*Osmia rufa*)	Equally effective as *A. mellifera* and more effective than hoverflies as a pollinator. At much lower densities than *A. mellifera*. Lower fruit set compensated by higher seed weight per pod	Jauker *et al.*, 2012a, b
Hoverflies:	Similar rate of flower visitation as *A. mellifera*	Rader *et al.*, 2009
Eristalis tenax	Enhanced 10% annual seed production	Jauker and Wolters, 2008
Episyrphus balteatus	Approximately five-fold densities of hoverflies required to reach a similar fruit set and yield as with *O. rufa* as a pollinator	Jauker *et al.*, 2012a

mitigation strategies that are targeted at conserving pollinators in order to reduce pollination deficits and meet the demands of crop production (Garratt *et al.*, 2014). Pollinators may have a role in IPM strategies because of their capacity to vector entomopathogenic fungal spores for management of different pests (Carreck *et al.*, 2007; Mänd *et al.*, 2010; Hokkanen *et al.*, 2015), though pollinators are themselves susceptible to the attack of some entomopathogenic fungi, such as *Beauveria bassiana* (Hokkanen *et al.*, 2003). Focusing on biological and cultural control and not using insecticides is the safest IPM option to avoid harming bees.

Fig. 22.3. Hoverflies, like the drone fly *Eristalis tenax* shown on the photograph, can be important pollinators of *Brassica* oilseed crops.

Among the insecticides used to control some of the main insect pests in *Brassica* oilseed crops, neem-based insecticides can be used against aphids and lepidopteran pests and are safe to bees (Sontakke and Dash, 1996). *Bacillus thuringiensis* is also considered safe for bees and natural enemies and provides good control of lepidopteran pests (Furlong *et al.*, 2008). Other insecticides and repellents considered safe to bees are garlic extract, potassium salt of fatty acids, and petroleum oil (Abrol and Thakur, 2016).

Insecticides that are harmful to pollinators (Table 22.2) should be avoided when *Brassica* oilseed crops are at the flowering stage. Insecticides should also not be applied when pollinators are most active, which depends on the pollinator, but is usually during the day for most pollinators in *Brassica* oilseed crops (Willmer and Stone, 2004; Polatto *et al.*, 2014). Insecticide applications should therefore be preferably conducted in the early or late hours of the day, when pollinators are not foraging. The type of application and formulation of pesticides also affects their toxicity to pollinators. Because of drift and inadvertent exposure to non-target organisms, dust formulations are more likely to be toxic to pollinators than emulsions and granular formulations applied in the soil (Johansen, 1977; Ingram *et al.*, 1996). Microencapsulated pesticides can be very toxic to bees, because the polymeric microcapsules are about the same size as pollen grains and adhere

Table 22.2. Insecticides that are applied in *Brassica* oilseed crops against particular target pests, but which are harmful to pollinators (based on Abrol and Thakur, 2016).

Chemical group	Active ingredient	Target pest
Oxadiazines	Indoxacarb	Diamondback moth
Thiourea derivatives	Diafenthiuron	Diamondback moth
Avermectins	Emamectin benzoate	Diamondback moth
Spinosyns[a]	Spinosad	Diamondback moth
Diamides	Chlorantraniliprole	Diamondback moth
Benzoylureas	Flufenoxuron, lufenuron	Diamondback moth
Pyridine azomethine derivatives	Pymetrozine	Aphids, whiteflies, hoppers
Phenylpyrazoles	Fipronil	Aphids, thrips, whiteflies, bollworms
Neonicotinoids	Clothianidin, imidacloprid, thiamethoxam,	Aphids, thrips, whiteflies, hoppers
Organophosphate	Profenofos	Aphids, thrips, whiteflies

[a]Less toxic when dry

readily to bees (Johansen, 1977). Kaolin particles also adhere to bumblebees, causing cuticular water loss and decreasing their longevity (Karise *et al.*, 2016). In pesticides applied by aircraft, as much as 75% of the application sprayed can miss its target and exposure of pollinators and natural enemies is much more likely (Pimentel, 2005; Barnett *et al.*, 2007). Several cases of bee poisoning after aircraft application of insecticides have been documented (Barnett *et al.*, 2007).

22.5 Improvement of Pollination in *Brassica* Oilseed Crops

Mass-flowering *Brassica* oilseed crops are considered highly beneficial to attract and feed wild bees and other pollinators (Holzschuh *et al.*, 2013; Eberle *et al.*, 2015; Riedinger *et al.*, 2015; Thom *et al.*, 2016). However, because of the population decline reported in both domesticated and wild pollinators (Potts *et al.*, 2010), pollination deficit resulting in yield reduction can occur in *Brassica* oilseed crops. In these cases, increasing the abundance of domesticated pollinators with the help of apiaries (Fig. 22.1) can improve crop pollination (Manning and Wallis, 2005).

Not using pesticides that are lethal or harmful to pollinators is the first step towards improving pollination of *Brassica* oilseed crops. Among the most toxic pesticides to honey bees are neonicotinoid, pyrethroid, organophosphate and carbamate insecticides (Frazier *et al.*, 2015; Kessler *et al.*, 2015; Lundin *et al.*, 2015; Rundlöf *et al.*, 2015; Zhu *et al.*, 2015; Hladik *et al.*, 2016). Besides insecticides, acaricides, fungicides and herbicides, as well the solvents and surfactants used in their formulations, can be harmful to pollinators (Simon-Delso *et al.*, 2014; Artz and Pitts-Singer, 2015; Frazier *et al.*, 2015; Mullin *et al.*, 2015; Zhu *et al.*, 2015; Hladik *et al.*, 2016). That is why pollinator abundance was found to be greatest in organic fields, followed by conventional fields, and lowest in herbicide-resistant genetically modified canola (Morandin and Winston, 2005). Consequently, genetically modified and conventional canola plants often suffer from pollination deficit (Morandin and Winston, 2005). Among insecticides, not only the ones that are systemic or are applied to flowers are damaging to bees. Studies have shown that even insecticides applied via seed coating can have negative effects on bees (Krupke and Long, 2015; Rundlöf *et al.*, 2015).

Another way of improving pollination is to plant, next to the crop, a variety of flowering plants that provide nectar and pollen to attract and feed pollinators (Decourtye *et al.*, 2010; Nicholls and Altieri, 2013). Both the quality and the diversity of the nectar and pollen provided by flowering plants are important for pollinators (Di Pasquale *et al.*, 2013). Weeds can also be part of the habitat used in pollinator conservation (Bretagnolle and Gaba, 2015). However, wildflowers can also be contaminated with pesticide drift and pesticide residues from past crop treatments (Botías *et al.*, 2015). Thus, when setting wildflower patches as habitats for pollinators, it is important to establish them in areas where neonicotinoids and other persistent pesticides that are toxic to bees have not been used (Botías *et al.*, 2015). Wildflower patches should also not be set adjacent to crops treated with these pesticides, as those locations may also be contaminated because of pesticide drift (Botías *et al.*, 2015). Providing a flower-rich habitat for pollinators also enhances other ecosystem services, such as biological control (Wratten *et al.*, 2012), which in turn also reduces the need to use insecticides. In addition to flower-rich habitats, solitary bees require nesting sites, which, depending on the species, would be located either on the ground or in vegetation (Klein *et al.*, 2002; Woodcock *et al.*, 2013). The efficiency of wild bees as pollinators compensates the effort required to attract and provide a habitat for them (Woodcock *et al.*, 2013).

22.6 Conclusions and Recommendations

Given the importance of pollinators in agriculture in general and in open-pollinated *Brassica* oilseed crops in particular, all IPM methods used to control pests need to be respectful of pollinators, especially when used during the flowering period. In the case of systemic and persistent insecticides such as neonicotinoids, their use should be avoided even in seed treatments. Not using pesticides that are harmful to pollinators is the first step towards improving pollination of *Brassica* oilseed crops. Alternatives to insecticides, such as biological and cultural control, should be preferred for management of insect pests. Including a variety of flowering plant species that provides high-quality nectar and pollen can attract pollinators in the field where *Brassica* oilseed crops are planted, ensuring that pollinators have sufficient food to live on.

References

Abrol, D.P. (2007) Honeybees and rapeseed: a pollinator–plant interaction. *Advances in Botanical Research* 45, 337–367.

Abrol, D.P. and Shankar, U. (2012) Pollination in oil crops: recent advances and future strategies. In: Gupta, S.K. (ed.) *Technological Innovations in Major World Oil Crops*. Springer, New York, pp. 221–267.

Abrol, D.P. and Thakur, R.K. (2016) *Recent Advances in Pollination Biology*. Kalyani Publishers, New Delhi, India.

Adegas, J.E.B. and Nogueira Couto, R.H. (1992) Entomophilous pollination in rape (*Brassica napus* L. var. *oleifera*) in Brazil. *Apidologie* 23, 203–209.

Ali, M., Saeed, S., Sajjad, A. and Whittington, A. (2011) In search of the best pollinators for canola (*Brassica napus* L.) production in Pakistan. *Applied Entomology and Zoology* 46, 353–361.

Artz, D.R. and Pitts-Singer, T.L. (2015) Effects of fungicide and adjuvant sprays on nesting behavior in two managed solitary bees, *Osmia lignaria* and *Megachile rotundata*. *PLoS ONE* 10(8), e0135688. doi: 10.1371/journal.pone.0135688

Atmowidi, T., Buchori, D., Manuwoto, S., Suryobroto, B. and Hidayat, P. (2007) Diversity of pollinator insects in relation to seed set of mustard (*Brassica rapa* L.: Cruciferae). *HAYATI Journal of Biosciences* 14 155–161.

Barnett, E.A., Charlton, A.J. and Fletcher, M.R. (2007) Incidents of bee poisoning with pesticides in the United Kingdom, 1994–2003. *Pest Management Science* 63, 1051–1057.

Bartomeus, I., Potts, S.G., Steffan-Dewenter, I., Vaissière, B.E., Woyciechowski, M., Krewenka, K.M., Tscheulin, T., Roberts, S.P.M., Szentgyörgyi, H., Westphal, C. and Bommarco, R. (2014) Contribution of insect pollinators to crop yield and quality varies with agricultural intensification. *PeerJ,* 2014, Mar 27:2, e.328. doi: 10.7717/peerj.328

Bartomeus, I., Gagic, V. and Bommarco, R. (2015) Pollinators, pests and soil properties interactively shape oilseed rape yield. *Basic and Applied Ecology* 16, 737–745.

Bijaya, P., Sharmila, M. and Singh, T.K. (2011) Biodiversity and abundance of syrphid fauna on major cruciferous crops in Manipur. *Journal of Advanced Zoology* 32, 12–18.

Blüthgen, N. and Klein, A.M. (2011) Functional complementarity and specialisation: the role of biodiversity in plant–pollinator interactions. *Basic and Applied Ecology* 12, 282–291.

Bommarco, R., Marini, L. and Vaissière, B.E. (2012) Insect pollination enhances seed yield, quality, and market value in oilseed rape. *Oecologia* 169, 1025–1032.

Botías, C., David, A., Horwood, J., Abdul-Sada, A., Nicholls, E., Hill, E. and Goulson, D. (2015) Neonicotinoid residues in wildflowers, a potential route of chronic exposure for bees. *Environmental Science and Technology* 49, 12731–12740.

Bretagnolle, V. and Gaba, S. (2015) Weeds for bees? A review. *Agronomy for Sustainable Development* 35, 891–909.

Calderone, N.W. (2012) Insect pollinated crops, insect pollinators and US agriculture: trend analysis of aggregate data for the period 1992–2009. *PLoS ONE* 7(5), e0037235.

Carreck, N.L., Butt, T.M., Clark, S.J., Ibrahim, L., Isger, E.A., Pell, J.K. and Williams, I.H. (2007) Honey bees can disseminate a microbial control agent to more than one inflorescence pest of oilseed rape. *Biocontrol Science and Technology* 17, 179–191.

Carvalheiro, L.G., Veldtman, R., Shenkute, A.G., Tesfay, G.B., Pirk, C.W.W., Donaldson, J.S. and Nicolson, S.W. (2011) Natural and within-farmland biodiversity enhances crop productivity. *Ecology Letters* 14, 251–259.

Chambó, E.D., De Oliveira, N.T.E., Garcia, R.C., Duarte-Júnior, J.B., Ruvolo-Takasusuki, M.C.C. and Toledo, V.A. (2014) Pollination of rapeseed (*Brassica napus*) by Africanized honeybees (Hymenoptera: Apidae) on two sowing dates. *Anais da Academia Brasileira de Ciencias* 86, 2087–2100.

Corbet, S.A., Williams, I.H. and Osborne, J.L. (1991) Bees and the pollination of crops and wild flowers in the European Community. *Bee World* 72, 47–59.

Cresswell, J.E. (1999) The influence of nectar and pollen availability on pollen transfer by individual flowers of oil-seed rape (*Brassica napus*) when pollinated by bumblebees (*Bombus lapidarius*). *Journal of Ecology* 87, 670–677.

Cresswell, J.E., Bassom, A.P., Bell, S.A., Collins, S.J. and Kelly, T.B. (1996) Predicted pollen dispersal by honey-bees and three species of bumble-bees foraging on oil-seed rape: a comparison of three models. *Functional Ecology* 9, 829–841.

Cunningham, S.A., FitzGibbon, F. and Heard, T.A. (2002) The future of pollinators for Australian agriculture. *Australian Journal of Agricultural Research* 53, 893–900.

David, A., Botías, C., Abdul-Sada, A., Nicholls, E., Rotheray, E.L., Hill, E.M. and Goulson, D. (2016) Widespread contamination of wildflower and bee-collected pollen with complex mixtures of neonicotinoids and fungicides commonly applied to crops. *Environment International* 88, 169–178.

Davis, A.R., Pylatuik, J.D., Paradis, J.C. and Low, N.H. (1998) Nectar-carbohydrate production and composition vary in relation to nectary anatomy and location within individual flowers of several species of Brassicaceae. *Planta* 205, 305–318.

Decourtye, A., Mader, E. and Desneux, N. (2010) Landscape enhancement of floral resources for honey bees in agro-ecosystems. *Apidologie* 41, 264–277.

Di Pasquale, G., Salignon, M., Le Conte, Y., Belzunces, L.P., Decourtye, A., Kretzschmar, A., Suchail, S., Brunet, J.-L. and Alaux, C. (2013) Influence of pollen nutrition on honey bee health: do pollen quality and diversity matter? *PLoS ONE* 8, e72016.

Eberle, C.A., Thom, M.D., Nemec, K.T., Forcella, F., Lundgren, J.G., Gesch, R.W., Riedell, W.E., Papiernik, S.K., Wagner, A., Peterson, D.H. and Eklund, J.J. (2015) Using pennycress, camelina, and canola cash cover crops to provision pollinators. *Industrial Crops and Products* 75, 20–25.

Frazier, M.T., Mullin, C.A., Frazier, J.L., Ashcraft, S.A., Leslie, T.W., Mussen, E.C. and Drummond, F.A. (2015) Assessing honey bee (Hymenoptera: Apidae) foraging populations and the potential impact of pesticides on eight U.S. crops. *Journal of Economic Entomology* 108, 2141–2152.

Furlong, M.J., Ju, K.H., Su, P.W., Chol, J.K., Chang Il, R. and Zalucki, M.P. (2008) Integration of endemic natural enemies and *Bacillus thuringiensis* to manage insect pests of *Brassica* crops in North Korea. *Agriculture, Ecosystems & Environment* 125, 223–238.

Gallai, N., Salles, J.-M., Settele, J. and Vaissière, B.E. (2009) Economic valuation of the vulnerability of world agriculture confronted with pollinator decline. *Ecological Economics* 68, 810–821.

Garibaldi, L.A., Steffan-Dewenter, I., Winfree, R., Aizen, M.A., Bommarco, R. *et al.*, (2013) Wild pollinators enhance fruit set of crops regardless of honey bee abundance. *Science* 339, 1608–1611.

Garratt, M.P.D., Coston, D.J., Truslove, C.L., Lappage, M.G., Polce, C., Dean, R., Biesmeijer, J.C. and Potts, S.G. (2014) The identity of crop pollinators helps target conservation for improved ecosystem services. *Biological Conservation* 169, 128–135.

Giannini, T.C., Cordeiro, G.D., Freitas, B.M., Saraiva, A.M. and Imperatriz-Fonseca, V.L. (2015) The dependence of crops for pollinators and the economic value of pollination in Brazil. *Journal of Economic Entomology* 108, 849–857.

Godfray, H.C.J., Blacquière, T., Field, L.M., Hails, R.S., Petrokofsky, G., Potts, S.G., Raine, N.E., Vanbergen, A.J. and McLean, A.R. (2014) A restatement of the natural science evidence base concerning neonicotinoid insecticides and insect pollinators. *Proceedings of the Royal Society B: Biological Sciences* 281, 20140588

Goswami, V., Khan, M.S. and Srivastava, P. (2014) Association of different insect pollinators and their relative abundance on blossoms of mustard (*Brassica juncea* L.). *Environment and Ecology (Kalyani)* 32, 368–371.

Goulson, D., Nicholls, E., Botías, C. and Rotheray, E.L. (2015) Bee declines driven by combined stress from parasites, pesticides, and lack of flowers. *Science* 347.

Greenleaf, S.S. and Kremen, C. (2006) Wild bees enhance honey bees' pollination of hybrid sunflower. *Proceedings of the National Academy of Sciences of the USA* 103, 13890–13895.

Hayter, K.E. and Cresswell, J.E. (2006) The influence of pollinator abundance on the dynamics and efficiency of pollination in agricultural *Brassica napus*: implications for landscape-scale gene dispersal. *Journal of Applied Ecology* 43, 1196–1202.

Hladik, M.L., Vandever, M. and Smalling, K.L. (2016) Exposure of native bees foraging in an agricultural landscape to current-use pesticides. *Science of the Total Environment* 542, 469–477.

Hoehn, P., Tscharntke, T., Tylianakis J.M. and Steffan-Dewenter, I. (2008) Functional group diversity of bee pollinators increases crop yield. *Proceedings of the Royal Society B: Biological Sciences* 275, 2283–2291.

Hokkanen, H.M.T., Zeng, Q.Q. and Menzler-Hokkanen, I. (2003) *Assessing the impacts of Metarhizium and Beauveria on bumblebees*. In: Hokkanen, H.M.T. and Hajak, A.E. (eds) *Environmental Impacts of Microbial Insecticides*. Kluwer, Dordrecht, The Netherlands, pp. 63–72.

Hokkanen, H.M.T., Menzler-Hokkanen, I. and Lahdenperä, M.-L. (2015) Managing bees for delivering biological control agents and improved pollination in berry and fruit cultivation. *Sustainable Agriculture Research* 4.

Holzschuh, A., Dormann, C.F., Tscharntke, T. and Steffan-Dewenter, I. (2013) Mass-flowering crops enhance wild bee abundance. *Oecologia* 172, 477–484.

Hoyle, M. and Cresswell, J.E. (2007) The effect of wind direction on cross-pollination in wind-pollinated GM crops. *Ecological Applications* 17, 1234–1243.

Hoyle, M. and Cresswell, J.E. (2009) Maximum feasible distance of windborne cross-pollination in *Brassica napus*: a 'mass budget' model. *Ecological Modelling* 220, 1090–1097.

Ingram, M., Nabhan, G. and Buchmann, S. (1996) Our forgotten pollinators: protecting the birds and bees. *Global Pesticide Campaigner* 6(4), 18.

Jauker, F. and Wolters, V. (2008) Hover flies are efficient pollinators of oilseed rape. *Oecologia* 156, 819–823.

Jauker, F., Bondarenko, B., Becker, H.C. and Steffan-Dewenter, I. (2012a) Pollination efficiency of wild bees and hoverflies provided to oilseed rape. *Agricultural and Forest Entomology* 14, 81–87.

Jauker, F., Peter, F., Wolters, V. and Diekötter, T. (2012b) Early reproductive benefits of mass-flowering crops to the solitary bee *Osmia rufa* outbalance post-flowering disadvantages. *Basic and Applied Ecology* 13, 268–276.

Johansen, C.A. (1977) Pesticides and pollinators. *Annual Review of Entomology* 22, 177–192.

Karise, R., Muljar, R., Smagghe, G., Kaart, T., Kuusik, A., Dreyersdorff, G., Williams, I.H. and Mänd, M. (2016) Sublethal effects of kaolin and the biopesticides Prestop-Mix and BotaniGard on metabolic rate, water loss and longevity in bumble bees (*Bombus terrestris*). *Journal of Pest Science* 89, 171–178.

Kessler, S.C., Tiedeken, E.J., Simcock, K.L., Derveau, S., Mitchell, J., Softley, S., Stout, J.C. and Wright, G.A. (2015) Bees prefer foods containing neonicotinoid pesticides. *Nature* 521, 74–76.

Kevan, P.G. and Eisikowitch, D. (1990) The effects of insect pollination on canola (*Brassica napus* L. cv. O.A.C. Triton) seed germination. *Euphytica* 45, 39–41.

Kevan, P.G., Lee, H. and Shuel, R.W. (1991) Sugar ratios in nectars of varieties of canola (*Brassica napus*). *Journal of Apicultural Research* 30, 99–102.

Klein, A.-M., Steffan-Dewenter, I., Buchori, D. and Tscharntke, T. (2002) Effects of land-use intensity in tropical agroforestry systems on coffee flower-visiting and trap-nesting bees and wasps. *Conservation Biology* 16, 1003–1014.

Klein, A.M., Vaissière, B.E., Cane, J.H., Steffan-Dewenter, I., Cunningham, S.A., Kremen, C. and Tscharntke, T. (2007) Importance of pollinators in changing landscapes for world crops. *Proceedings of the Royal Society B: Biological Sciences* 274 (1608), 303–313.

Kreyer, D., Oed, A., Walther-Hellwig, K. and Frankl, R. (2004) Are forests potential landscape barriers for foraging bumblebees? Landscape scale experiments with *Bombus terrestris* agg. and *Bombus pascuorum* (Hymenoptera, Apidae). *Biological Conservation* 116, 111–118.

Krupke, C.H. and Long, E.Y. (2015) Intersections between neonicotinoid seed treatments and honey bees. *Current Opinion in Insect Science* 10, 8–13.

Kumari, S., Chhuneja, P.K., Singh, J. and Choudhary, A. (2015) Relative abundance and diversity of insects on *Brassica juncea* L. Czern under north-western plains of India. *Journal of Experimental Zoology India* 18, 165–171.

Kunjwal, N., Kumar, Y. and Khan, M.S. (2014) Flower-visiting insect pollinators of brown mustard, *Brassica juncea* (L.) Czern and Coss and their foraging behaviour under caged and open pollination. *African Journal of Agricultural Research* 9, 1278–1286.

Lindström, S.A.M., Herbertsson, L., Rundlöf, M., Smith, H.G. and Bommarco, R. (2016) Large-scale pollination experiment demonstrates the importance of insect pollination in winter oilseed rape. *Oecologia* 180, 759–769.

Lundin, O., Rundlöf, M., Smith, H.G., Fries, I. and Bommarco, R. (2015) Neonicotinoid insecticides and their impacts on bees: a systematic review of research approaches and identification of knowledge gaps. *PLoS ONE* 10, e0136928.

Mahindru, N., Singh, G. and Grewal, G.S. (1998) Comparative abundance and foraging behaviour of insect pollinators of raya, *Brassica juncea* L. and role of *Apis mellifera* L. in crop pollination. *Journal of Insect Science* 11, 34–37.

Mänd, M., Williams, I.H., Viik, E. and Karise, R. (2010) Oilseed rape, bees and integrated pest management. In: Williams, I.H. (ed.) *Biocontrol-based Integrated Management of Oilseed Rape Pests*. Springer, Dordrecht, The Netherlands, pp. 357–379.

Manning, R. and Wallis, I.R. (2005) Seed yields in canola (*Brassica napus* cv. Karoo) depend on the distance of plants from honeybee apiaries. *Australian Journal of Experimental Agriculture* 45, 1307–1313.

Morandin, L.A. and Winston, M.L. (2005) Wild bee abundance and seed production in conventional, organic, and genetically modified canola. *Ecological Applications* 15, 871–881.

Mullin, C.A., Chen, J., Fine, J.D., Frazier, M.T. and Frazier, J.L. (2015) The formulation makes the honey bee poison. *Pesticide Biochemistry and Physiology* 120, 27–35.

Mussury, R.M. and Fernandes, W.D. (2000) Studies of the floral biology and reproductive system of *Brassica napus* L. (Cruciferae). *Brazilian Archives of Biology and Technology* 43, 111–117.

Nedić, N., Mačukanović-Jocić, M., Rančić, D., Rørslett, B., Šoštarić, I., Stevanović, Z.D. and Mladenović, M. (2013) Melliferous potential of *Brassica napus* L. subsp. *napus* (Cruciferae). *Arthropod–Plant Interactions* 7, 323–333.

Nicholls, C.I. and Altieri, M.A. (2013) Plant biodiversity enhances bees and other insect pollinators in agroecosystems. A review. *Agronomy for Sustainable Development* 33, 257–274.

NRC (2007) *Status of Pollinators in North America*. Committee on the Status of Pollinators in North America, National Research Council. The National Academies Press, Washington, DC.

Orford, K.A., Vaughan, I.P. and Memmott, J. (2015) The forgotten flies: the importance of non-syrphid Diptera as pollinators. *Proceedings of the Royal Society B: Biological Sciences* 282, 20142934

Pierre, J. (2001) The role of honeybees (*Apis mellifera*) and other insect pollinators in gene flow between oilseed rape (*Brassica napus*) and wild radish (*Raphanus raphanistrum*). *Acta Horticulturae* 561, 47–51.

Pimentel, D. (2005) Environmental and economic costs of the application of pesticides primarily in the United States. *Environment, Development and Sustainability* 7, 229–252.

Polatto, L.P., Chaud-Netto, J. and Alves-Junior, V.V. (2014) Influence of abiotic factors and floral resource availability on daily foraging activity of bees: influence of abiotic and biotic factors on bees. *Journal of Insect Behavior* 27, 593–612.

Potts, S.G., Biesmeijer, J.C., Kremen, C., Neumann, P., Schweiger, O. and Kunin, W.E. (2010) Global pollinator declines: trends, impacts and drivers. *Trends in Ecology and Evolution* 25, 345–353.

Prasad, D., Hameed, S.P., Singh, R. and Singh, B. (1991) Foraging behaviour of insect pollinators on brown mustard, *Brassica juncea* in Bihar, India. *Indian Bee Journal* 51, 131–133.

Rader, R., Howlett, B.G., Cunningham, S.A., Westcott, D.A., Newstrom-Lloyd, L.E., Walker, M.K., Teulon, D.A.J. and Edwards, W. (2009) Alternative pollinator taxa are equally efficient but not as effective as the honeybee in a mass flowering crop. *Journal of Applied Ecology* 46, 1080–1087.

Rader, R., Bartomeus, I., Garibaldi, L.A., Garratt, M.P.D., Howlett, B.G. *et al.* (2016) Non-bee insects are important contributors to global crop pollination. *Proceedings of the National Academy of Sciences of the USA* 113, 146–151.

Rahman, K.A. (1940) Insect pollinators of toria (*Brassica napus* Linn. var. *dichotoma* Prain) and sarson (*Brassica campestris* Linn. var. *sarson* Prain) at Lyallpur. *Indian Journal of Agricultural Science* 10, 422–444.

Riedinger, V., Mitesser, O., Hovestadt, T., Steffan-Dewenter, I., Holzschuh, A. and Rosenheim, J.A. (2015) Annual dynamics of wild bee densities: attractiveness and productivity effects of oilseed rape. *Ecology* 96, 1351–1360.

Rosa, A.D., Blochtein, B. and Lima, D.K. (2011) Honey bee contribution to canola pollination in Southern Brazil. *Scientia Agricola* 68, 255–259.

Roulston, T.A.H. and Goodell, K. (2011) The role of resources and risks in regulating wild bee populations. *Annual Review of Entomology* 56, 293–312.

Rundlöf, M., Andersson, G.K.S., Bommarco, R., Fries, I., Hederström, V., Herbertsson, L., Jonsson, O., Klatt, B.K., Pedersen, T.R., Yourstone, J. and Smith, H.G. (2015) Seed coating with a neonicotinoid insecticide negatively affects wild bees. *Nature* 521, 77–80.

Shakeel, M. and Inayatullah, M. (2013) Impact of insect pollinators on the yield of canola (*Brassica napus*) in Peshawar, Pakistan. *Journal of Agricultural and Urban Entomology* 29, 1–5.

Simon-Delso, N., Martin, G.S., Bruneau, E., Minsart, L.A., Mouret, C. and Hautier, L. (2014) Honeybee colony disorder in crop areas: the role of pesticides and viruses. *PLoS ONE* 9, e0103073.

Singh, B., Kumar, M., Sharma, A.K. and Yadav, L.P. (2004) Effect of bee pollination on yield attributes and seed yield of tori (*Brassica campestris* var. *toria*). *Environment and Ecology (Kalyani)* 22, 571–573.

Sontakke, B.K. and Dash, A.N. (1996) Studies on the effects of insecticides and neem formulation against major pests of mustard and their safety to honey bees. *Indian Journal of Environment and Toxicology* 6, 87–88.

Stanley, D., Gunning, D. and Stout, J. (2013) Pollinators and pollination of oilseed rape crops (*Brassica napus* L.) in Ireland: ecological and economic incentives for pollinator conservation. *Journal of Insect Conservation* 17, 1181–1189.

Thom, M.D., Eberle, C.A., Forcella, F., Gesch, R., Weyers, S. and Lundgren, J.G. (2016) Nectar production in oilseeds: food for pollinators in an agricultural landscape. *Crop Science* 56, 727–739.

Willmer, P.G. and Stone, G.N. (2004) Behavioral, ecological, and physiological determinants of the activity patterns of bees. *Advances in the Study of Behavior* 34, 347–466.

Winfree, R., Williams, N.M., Dushoff, J. and Kremen, C. (2007) Native bees provide insurance against ongoing honey bee losses. *Ecology Letters* 10, 1105–1113.

Witter, S., Nunes-Silva, P., Lisboa, B.B. Tirelli, F.P., Sattler, A., Hilgert-Moreira, S.B. and Blochtein, B. (2015) Stingless bees as alternative pollinators of canola. *Journal of Economic Entomology* 108, 880–886.

Woodcock, B.A., Edwards, M., Redhead, J., Meek, W.R., Nuttall, P., Falk, S., Nowakowski, M. and Pywell, R.F. (2013) Crop flower visitation by honeybees, bumblebees and solitary bees: behavioural differences and diversity responses to landscape. *Agriculture, Ecosystems & Environment* 171, 1–8.

Wratten, S.D., Gillespie, M., Decourtye, A., Mader, E. and Desneux, N. (2012) Pollinator habitat enhancement: benefits to other ecosystem services. *Agriculture, Ecosystems & Environment* 159, 112–122.

Zhu, Y.C., Adamczyk, J., Rinderer, T., Yao, J., Danka, R., Luttrell, R. and Gore, J. (2015) Spray toxicity and risk potential of 42 commonly used formulations of row crop pesticides to adult honey bees (Hymenoptera: Apidae). *Journal of Economic Entomology* 108, 2640–2647.

23 Role of Glucosinolates in Resistance and Attraction to Insects: Applications in Trap Cropping and Pest Management in *Brassica* Oilseed Crops

Francesc R. Badenes-Pérez*

Institute of Agricultural Sciences (CSIC), Madrid, Spain

23.1 Introduction

Glucosinolates are plant secondary metabolites used for defence and are present mainly in the order Brassicales (Halkier and Gershenzon, 2006). The main defence mechanism of glucosinolates occurs when they are hydrolysed by myrosinases upon plant damage, producing isothiocyanates and other compounds that can be toxic to insects (Bones and Rossiter, 1996). Studies have shown that glucosinolates are often detrimental for generalists, while specialist insects in general seem not to be negatively affected (Hopkins *et al.*, 2009). Specialist insects have evolved mechanisms to avoid the toxicity of glucosinolates and can also use them in host plant recognition (Hopkins *et al.*, 2009). For example, larvae of the diamondback moth, *Plutella xylostella* (L.) (Lepidoptera: Plutellidae), have a sulfatase that desulfates glucosinolates; additionally, glucosinolates and their hydrolysis products act as feeding and oviposition stimulants for this insect (Ratzka *et al.*, 2002; van Loon *et al.*, 2002; Renwick *et al.*, 2006, Badenes-Pérez *et al.*, 2011). Thus, glucosinolates in *Brassica* oilseed crops can act as a 'double-edged sword', providing resistance against generalist insects but also increasing susceptibility against specialists (Björkman *et al.*, 2011; Bruce, 2014). On the other hand, this enhanced attractiveness to some specialist insects could be used to increase the effectiveness of trap crops (Badenes-Pérez *et al.*, 2010). This chapter reviews how glucosinolates and trap cropping can affect some of the main insect pests of *Brassica* oilseed crops.

23.2 Glucosinolates in *Brassica* Oilseed Crops

There is intraspecific variation in glucosinolate content in *Brassica* oilseed crops, depending on the subspecies and cultivar, and also depending on environmental conditions (Liersch *et al.*, 2013; Martínez-Ballesta *et al.*, 2013; Tong *et al.*, 2014). However, the dominant glucosinolate in *Brassica carinata* and *Brassica nigra* is allylglucosinolate (sinigrin); the dominant glucosinolates in *Brassica juncea* are allylglucosinolate and 3-butenylglucosinolate (gluconapin); the dominant glucosinolates in *Brassica napus* are indol-3-ylmethylglucosinolate (glucobrassicin), (*R*)-2-hydroxy-3-butenylglucosinolate (progoitrin), and 4-pentenylglucosinolate (glucobrassicanapin); and the dominant glucosinolate in *Brassica rapa* is 3-butenylglucosinolate (Giamoustaris and Mithen, 1995; Bellostas *et al.*, 2007; Badenes-Pérez *et al.*, 2011). Other glucosinolates found in small amounts in *Brassica* oilseed crops include 2-phenylethylglucosinolate (gluconasturtiin), 4-methylsulfonylbutylglucosinolate (glucoerysolin), and hydroxy and

*E-mail: frbadenes@ica.csic.es

  Integrated Management of Insect Pests on Canola and Other Brassica Oilseed Crops (ed. G.V.P. Reddy)

methoxy derivatives of indol-3-ylmethylglucosinolate (Giamoustaris and Mithen, 1995; Bellostas *et al.*, 2007; Badenes-Pérez *et al.*, 2011). Because of their glucosinolate content, *Brassica* oilseed crops can also be used in biofumigation against soil-dwelling insects and other pests (Kirkegaard and Sarwar, 1999; Sukovata *et al.*, 2015; Fourie *et al.*, 2016). Furthermore, given the health benefits of glucosinolates, new *Brassica* oilseed crop varieties with a high content of specific glucosinolates have been developed (Augustine and Bisht, 2015). Besides glucosinolates, there are other plant secondary metabolites, such as leaf-surface waxes, which are important in conferring resistance or enhanced attraction to insects in *Brassica* oilseed crops (Björkman *et al.*, 2011).

23.3 Trap Cropping in *Brassica* Oilseed Crops

Trap crops are plant stands deployed next to a crop in order to protect it from target pests (Hokkanen, 1991; Shelton and Badenes-Pérez, 2006) (Fig. 23.1). Trap cropping has been proposed as an environmentally friendly pest management technique, especially appropriate with insects that are likely to develop resistance to insecticides (Hokkanen, 1991; Shelton and Badenes-Pérez, 2006). Among the economic pests attacking *Brassica* oilseed crops, *Plutella xylostella* and *Meligethes aeneus* are the ones with the highest number of insecticide resistance reports (Furlong *et al.*, 2008, 2013; Riggi *et al.*, 2015). For the management of these and other insect pests attacking *Brassica* oilseed crops, different applications of trap cropping are described below. For insect pests that are specialists of crucifers, glucosinolates and their hydrolysis products may act as attractants and oviposition stimulants; thus, glucosinolate content might be used to increase the effectiveness of trap crops (Badenes-Pérez *et al.*, 2010) (Fig. 23.2).

23.4 Most Important Insect Pests Attacking *Brassica* Oilseed Crops

Brassica oilseed crops can be attacked by a wide variety of insect pests. Among them, some of the most

Fig. 23.1. Effect of plant glucosinolate content on some of the main insect pests attacking *Brassica* oilseed crops.

Fig. 23.2. A larva of *Coccinella septempuntacta* feeding on *Brevicoryne brassicae* on a plant of *Barbarea vulgaris*. With the reduction of insecticides, trap cropping can allow natural enemies to thrive.

important are: the lepidopterans diamondback moth *Plutella xylostella* (L.) (Plutellidae), cabbage looper *Trichoplusia ni* Hübner (Noctuidae), cabbage moth *Mamestra brassicae* (L.) (Noctuidae) and bertha armyworm *Mamestra configurata* Walker (Noctuidae); the coleopterans cabbage weevil *Ceutorhynchus* spp. (Curculionidae), flea beetle *Phyllotreta* spp. and *Psylliodes* spp. (Chrysomelidae), and pollen beetle *Meligethes aeneus* F. (Nitidulidae); the dipterans cabbage maggot *Delia* spp. (Anthomyiidae) and swede midge *Contarinia nasturtii* Kieffer (Cecidomyiidae); and hemipterans such as the aphids *Brevicoryne brassicae* (L.) (Fig. 23.1), *Lipaphis erysimi* (Kaltenbach), and *Myzus persicae* (Sulzer) (Aphididae), and *Lygus* spp. (Miridae) (Hokkanen, 2000; Soroka *et al.*, 2004; Cárcamo *et al.*, 2008; Furlong *et al.*, 2008, 2013; Dosdall and Mason, 2010; Ekbom, 2010; Williams, 2010; Dosdall *et al.*, 2011; Cárcamo, 2012; Zalucki *et al.*, 2012; Soroka and Grenkow, 2013; Tangtrakulwanich *et al.*, 2014; Kumar and Singh, 2015).

The effect of glucosinolates (Fig. 23.2) and the potential of trap cropping in *Brassica* oilseed crops for each of these orders are described in the following sections.

23.4.1 Lepidopteran pests

Among the four main lepidopterans attacking *Brassica* oilseed crops, *Plutella xylostella* is a specialist while *Trichoplusia ni*, *Mamestra brassicae* and *M. configurata* are considered generalists. For *P. xylostella*, glucosinolates are considered attractants as well as feeding and oviposition stimulants (van Loon *et al.*, 2002; Badenes-Pérez *et al.*, 2011). Experiments with *B. napus* have shown that increases in plant glucosinolate content can be associated with oviposition preference and increased damage by *P. xylostella* (Marazzi and Städler, 2004a; Marazzi *et al.*, 2004b; Nikooei *et al.*, 2015). However, feeding by *P. xylostella* on *B. napus* did not seem to be affected by glucosinolate content when comparing a standard line with two lines with low glucosinolate content (Bodnaryk, 1997). Experiments using the crucifer *Barbarea vulgaris* R. Br. (Fig. 23.1), which can also be used as an oilseed crop and has been proposed as a trap crop for *P. xylostella*, have shown that increases in plant glucosinolate content can be used to enhance the attractiveness of the trap crop to ovipositing *P. xylostella* (Andersson *et al.*, 1999; Badenes-Pérez *et al.*, 2010). Besides *Barbarea* spp., among the trap crops proposed for management of *P. xylostella* are the *Brassica* oilseed crops *B. juncea* and *B. rapa* (Shelton and Badenes-Pérez, 2006; Satpathy *et al.*, 2010; Badenes-Pérez *et al.*, 2014). The use of glossy varieties of *Brassica* oilseed crops, preferred by ovipositing *P. xylostella* compared with waxy varieties, could also have applications in trap cropping (Justus *et al.*, 2000; Ulmer *et al.*, 2002). In *Brassica* oilseed crops, glucosinolates are considered a source of host-plant resistance against *M. brassicae* and *M. configurata* (Bodnaryk, 1991; McCloskey and Isman, 1993; Bodnaryk, 1997; Ulmer *et al.*, 2001; Santolamazza-Carbone *et al.*, 2014; Ahuja *et al.*, 2015). No studies have been conducted with regard to the potential of trap cropping in *Mamestra* spp. Although no studies have been conducted with *Brassica* oilseed crops with regard to glucosinolate content and host-plant resistance to *T. ni*, experiments with pure glucosinolates and with other Brassicaceae have shown that glucosinolates can be a source of host-plant resistance against this pest (Shields and Mitchell, 1995; Kliebenstein *et al.*, 2002). A study conducted using *B. juncea* as a trap crop showed that it had some potential for management of *T. ni* in cabbage, but its practical effectiveness has not been demonstrated (Luther *et al.*, 1996).

23.4.2 Coleopteran pests

For *Meligethes aeneus*, a specialist of crucifers, experiments have shown an increased attraction to plants and traps containing glucosinolate hydrolysis products (Free and Williams, 1978; Blight and Smart, 1999; Smart and Blight, 2000; Cook *et al.*, 2006). Lines of *B. napus* with low levels of alkenyl glucosinolates (releasing low levels of isothiocyanates) also seemed to suffer lower levels of colonization by *M. aeneus* compared with conventional *B. napus* cultivars with higher levels of alkenyl glucosinolates (Cook *et al.*, 2006). However, presence and distribution of adults, biting duration and clutch size in *M. aeneus* does not seem to be affected by glucosinolate content in flower buds (Milford *et al.*, 1989; Hervé *et al.*, 2015). *Meligethes aeneus* shows distinct host-plant preference patterns; and different trap crops, including *B. napus*, *B. nigra*, *B. rapa*, *Eruca sativa* and *Raphanus sativus* show potential for management of *M. aeneus* in *Brassica* oilseed crops (Hokkanen *et al.*, 1986; Hokkanen, 1989; Ekbom and Borg, 1996; Cook *et al.*, 2006; Kaasik *et al.*, 2014). These potential trap crop species also differ in their capacity to influence parasitism rates by different parasitoids attacking *M. aeneus* larvae (Jönsson and Anderson, 2007; Kaasik *et al.*, 2014).

Ceutorhynchus obstrictus (Marsham) (= *C. assimilis* (Paykull)) (Coleoptera: Curculionidae) is known to be attracted to the breakdown products of glucosinolates (Free and Williams, 1978). *Brassica napus* lines with low levels of alkenyl glucosinolates (thus releasing low levels of isothiocyanates) were less attractive and seemed to suffer lower levels of colonization by *C. obstrictus* compared with conventional *B. napus* cultivars with higher levels of alkenyl glucosinolates (Cook *et al.*, 2006). However, glucosinolate profile was not correlated with damage by *C. obstrictus* in different lines of several *Brassica* oilseed crops (Cárcamo *et al.*, 2007a). In *C. obstrictus* and *C. pallidactylus*, the effect of glucosinolates in *B. napus* on feeding by larvae has been found to depend on specific individual glucosinolates (Tansey *et al.*, 2010; Eickermann and Ulber, 2011). Oviposition preference by *C. obstrictus* is influenced by plant fertilization (Blake *et al.*, 2011). This effect of plant fertilization could be used in trap cropping, especially considering that *C. obstrictus* has been successfully managed using *B. rapa* as a trap crop in *B. napus* in Canada (Cárcamo *et al.*, 2007b). However, although *B. rapa* was tested as a trap crop, it did not successfully reduce populations of *C. pallidactylus* in *B. napus* in the UK (Barari *et al.*, 2005).

Phyllotreta spp. and *Psylliodes* spp. are also specialist insects. In *Phyllotreta* spp., glucosinolate content seems to be necessary for attraction to the host plant (Nielsen, 1989; Nielsen *et al.*, 2001), though feeding by *Phyllotreta cruciferae* (Goeze) on *B. napus* did not seem to be affected by glucosinolate content when comparing a standard line with two lines with low glucosinolate content (Bodnaryk, 1997). In *Psylliodes chrysocephala* L., pure glucosinolates acted as feeding stimulants in artificial diet and *B. napus* lines with high glucosinolate content suffered more damage by this insect than lines with low glucosinolate content (Bartlet *et al.*, 1994; Giamoustaris and Mithen, 1995). Although *Phyllotreta* spp. show distinct preferences for host plants that indicate that it could be used as a trap crop in *Brassica* oilseed crops (Metspalu *et al.*, 2014), the use of trap cropping to manage *Phyllotreta* spp. has yielded mixed results in crops of *Brassica* spp. (Altieri and Gliessman, 1983; Altieri and Schmidt, 1986; Kloen and Altieri, 1990; Hamid *et al.*, 2006; Badenes-Pérez *et al.*, in press). A trap crop of *B. rapa* has been shown to effectively reduce the populations of *P. chrysocephala* in *B. napus* (Barari *et al.*, 2005). As in the case of *P. xylostella*, damage by *P. cruciferae* and *P. chrysocephala* is inversely correlated with the presence of leaf wax in the plant (Bodnaryk, 1992; Lambdon *et al.*, 1998) and this preference could be used in trap cropping.

23.4.3 Dipteran pests

The two most damaging dipteran pests in *Brassica* oilseed crops are the crucifer specialists *Delia* spp. and *Contarinia nasturtii*. For *D. radicum* and *D. floralis*, glucosinolates and non-glucosinolate compounds have been identified as attractants and oviposition stimulants (Hopkins *et al.*, 1997; Städler *et al.*, 2002; Marazzi *et al.*, 2004a; Marazzi and Städler, 2004b). The oviposition preference of *D. radicum* for certain *Brassica* oilseed crops indicates their susceptibility to this insect pest, but it also indicates that trap cropping has the potential to be used against this insect (Rousse *et al.*, 2003; Kergunteuil et al., 2015a, b). Although *C. nasturtii* is a crucifer specialist, nothing is known about its interaction with glucosinolate content in its host plants. The potential of trap cropping in the management of *C. nasturtii* has so far not been tested.

23.4.4 Hemipteran pests

In *Lygus lineolaris* Palisot de Beauvois, when comparing cultivars of different *Brassica* oilseed crops with different glucosinolate content, oviposition preference and presence of adult insects seemed to be unaffected by glucosinolate content (Butts and Lamb, 1990; Gerber, 1997). However, feeding by adult *L. lineolaris* seemed to be higher on *Sinapis alba* cultivars with low glucosinolate content (Bodnaryk, 1996). In a peach crop, *B. napus* was successfully used as a trap crop for *L. lineolaris* (Foshee *et al.*, 2003). Among *Brassica* oilseed crops, *B. carinata* and *B. napus* were preferred over *B. juncea* by ovipositing *L. lineolaris* (Gerber, 1997), indicating the potential of these *Brassica* spp. as trap crops for this insect.

For the generalist aphid *Myzus persicae*, glucosinolates provide some degree of host-plant resistance (Kim and Jander, 2007) and, as a result of the lack of toxic glucosinolate hydrolysis products, *M. persicae* prefers transgenic *B. napus* without myrosinase compared with conventional *B. napus* (Borgen *et al.*, 2012). *Myzus persicae* also seems to prefer plants with lower glucosinolate content (Staley *et al.*, 2010). However, a positive correlation between the performance of *M. persicae* and host-plant glucosinolate content has been found in some *Brassica* oilseed crops and other *Brassica* spp. (Cole, 1997). The specialist aphids *Brevicoryne brassicae* and *Lipaphis erysimi* can sequester glucosinolates and use them in self-defence (Bridges *et al.*, 2002). *Brevicoryne brassicae* even prefers conventional *B. napus* lines compared to transgenic *B. napus* without myrosinase (Borgen *et al.*, 2012). There is a stronger association between insect performance and host-plant glucosinolate content in *B. brassicae* than in *M. persicae* (Cole, 1997; Staley *et al.*, 2010). Experiments with *B. rapa* have shown, however, that glucosinolate content in this plant can reduce infestation by *L. erysimi* (Kumar and Sangha, 2013). In general, conventional trap cropping has not been very successful in aphid management (Shelton and Badenes-Pérez, 2006), but companion plants and nectar provisioning can increase parasitism and predation of aphids in *Brassica* oilseed crops (Bruce, 2014; Jamont *et al.*, 2014; Banks and Gagic, 2016) (Fig. 23.1).

23.5 Conclusions

Brassica oilseed crops are rich in indolic and aliphatic glucosinolates. Glucosinolates act as a 'double-edged sword' in *Brassica* oilseed crops, providing resistance against some generalist insects, while making plants more susceptible to specialist insects (Fig. 23.2). This enhanced attractiveness to some specialist insects could be used to increase the effectiveness of trap crops in *Brassica* oilseed crops. In *Brassica* oilseed crops, different trap crops show high potential and/or have been successfully tested to manage *Ceutorhynchus obstrictus*, *Delia radicum*, *Lygus lineolaris*, *Meligethes aeneus*, *Plutella xylostella* and *Psylliodes chrysocephala*. Management of these insects with trap cropping is particularly important in those cases, such as *M. aeneus* and *P. xylostella*, for which the risk of insecticide resistance is high. In some cases, some *Brassica* oilseed crops have been proposed as trap crops in other *Brassica* oilseed crops. Management of other insect pests with trap cropping has been either unsuccessful or very inconsistent, such as in the case of *Phyllotreta* spp.; and in other cases, such as *Contarinia nasturtii*, no research has been conducted to find potential trap crops. Further research is needed, especially in cases where trap cropping has not yet been tested and in cases where inconsistency in the effectiveness of trap crops has been shown. Inconsistency in the effectiveness of trap crops also indicates that there is a need for better understanding of insect preference and insect movement dynamics in relation to the phenology of the different crop and trap crop plants.

References

Ahuja, I., Van Dam, N.M., Winge, P., Trælnes, M., Heydarova, A., Rohloff, J., Langaas, M. and Bones, A.M. (2015) Plant defence responses in oilseed rape MINELESS plants after attack by the cabbage moth *Mamestra brassicae*. *Journal of Experimental Botany* 66, 579–592.

Altieri, M.A. and Gliessman, S.R. (1983) Effects of plant diversity on the density and herbivory of the flea beetle, *Phyllotreta cruciferae* Goeze, in California collard (*Brassica oleracea*) cropping systems. *Crop Protection* 2, 497–501.

Altieri, M.A. and Schmidt, L.L. (1986) Population trends and feeding preferences of flea beetles (*Phyllotreta cruciferae* Goeze) in collard-wild mustard mixtures. *Crop Protection* 5, 170–175.

Andersson, A.A.M., Merker, A., Nilsson, P., Sørensen, H. and Åand, P. (1999) Chemical composition of the potential new oilseed crops *Barbarea vulgaris*, *Barbarea verna* and *Lepidium campestre*. *Journal of the Science of Food and Agriculture* 79, 179–186.

Augustine, R. and Bisht, N.C. (2015) Biofortification of oilseed *Brassica juncea* with the anti-cancer compound glucoraphanin by suppressing *GSL-ALK* gene family. *Scientific Reports* 5, 18005.

Badenes-Pérez, F.R., Reichelt, M. and Heckel, D.G. (2010) Can sulfur fertilisation increase the effectiveness of trap crops for diamondback moth, *Plutella xylostella* (L.) (Lepidoptera: Plutellidae)? *Pest Management Science* 66, 832–838.

Badenes-Pérez, F.R., Reichelt, M., Gershenzon, J. and Heckel, D.G. (2011) Phylloplane location of glucosinolates in *Barbarea* spp. (Brassicaceae) and misleading assessment of host suitability by a specialist herbivore. *New Phytologist* 189, 549–556.

Badenes-Pérez, F.R., Reichelt, M., Gershenzon, J. and Heckel, D.G. (2014) Using plant chemistry and insect preference to study the potential of *Barbarea* (Brassicaceae) as a dead-end trap crop for diamondback moth (Lepidoptera: Plutellidae). *Phytochemistry* 98, 137–144.

Badenes-Pérez, F.R., Parrado Márquez, B. and Petitpierre, E. (2017) Can flowering Barbarea spp. (Brassicaceae) be used simultaneously as a trap crop and in conservation biological control? *Journal of Pest Science* 90(2), 623–633.

Banks, J.E. and Gagic, V. (2016) Aphid parasitoids respond to vegetation heterogeneity but not to fragmentation scale: an experimental field study. *Basic and Applied Ecology* 17(5), 438–446.

Barari, H., Cook, S.M., Clark, S.J. and Williams, I.H. (2005) Effect of a turnip rape (*Brassica rapa*) trap crop on stem-mining pests and their parasitoids in winter oilseed rape (*Brassica napus*). *BioControl* 50, 69–86.

Bartlet, E., Parsons, D., Williams, I.H. and Clark, S.J. (1994) The influence of glucosinolates and sugars on feeding by the cabbage stem flea beetle, *Psylliodes chrysocephala*. *Entomologia Experimentalis et Applicata* 73, 77–83.

Bellostas, N., Sørensen, J.C. and Sørensen, H. (2007) Profiling glucosinolates in vegetative and reproductive tissues of four *Brassica* species of the U-triangle for their biofumigation potential. *Journal of the Science of Food and Agriculture* 87, 1586–1594.

Björkman, M., Klingen, I., Birch, A.N.E., Bones, A.M., Bruce, T.J.A., Johansen, T.J., Meadow, R., Mølman, J., Seljåsen, R., Smart, L.E. and Stewart, D. (2011) Phytochemicals of Brassicaceae in plant protection and human health – influence of climate, environment and agronomic practice. *Phytochemistry* 72, 538–556.

Blake, A.J., Dosdall, L.M. and Keddie, B.A. (2011) Bottom-up effects of *Brassica napus* nutrition on the oviposition preference and larval performance of *Ceutorhynchus obstrictus* (Coleoptera: Curculionidae). *Arthropod–Plant Interactions* 5, 39–48.

Blight, M.M. and Smart, L.E. (1999) Influence of visual cues and isothiocyanate lures on capture of the pollen beetle, *Meligethes aeneus* in field traps. *Journal of Chemical Ecology* 25, 1501–1516.

Bodnaryk, R.P. (1991) Developmental profile of sinalbin (p-hydroxybenzyl glucosinolate) in mustard seedlings, *Sinapis alba* L., and its relationship to insect resistance. *Journal of Chemical Ecology* 17, 1543–1556.

Bodnaryk, R.P. (1992) Leaf epicuticular wax, an antixenotic factor in Brassicaceae that affects the rate and pattern of feeding of flea beetles, *Phyllotreta cruciferae* (Goeze). *Canadian Journal of Plant Science* 72, 1295–1303.

Bodnaryk, R.P. (1996) Physical and chemical defences of pods and seeds of white mustard (*Sinapis alba* L.) against tarnished plant bugs, *Lygus lineolaris* (Palisot de Beauvois) (Heteroptera: Miridae). *Canadian Journal of Plant Science* 76, 33–36.

Bodnaryk, R.P. (1997) Will low-glucosinolate cultivars of the mustards *Brassica juncea* and *Sinapis alba* be vulnerable to insect pests? *Canadian Journal of Plant Science* 77, 283–287.

Bones, A. and Rossiter, J. (1996) The glucosinolate-myrosinase system, its organisation and biochemistry. *Physiologia Plantarum* 97, 194–208.

Borgen, B.H., Ahuja, I., Thangstad, O.P., Honne, B.I., Rohloff, J., Rossiter, J.T. and Bones, A.M. (2012) 'Myrosin cells' are not a prerequisite for aphid feeding on oilseed rape (*Brassica napus*) but affect host plant preferences. *Plant Biology* 14, 894–904.

Bridges, M., Jones, A.M.E., Bones, A.M., Hodgson, C., Cole, R., Bartlet, E., Wallsgrove, R., Karapapa, V.K., Watts, N. and Rossiter, J.T. (2002) Spatial organization of the glucosinolate-myrosinase system in brassica specialist aphids is similar to that of the host plant. *Proceedings of the Royal Society B: Biological Sciences* 269, 187–191.

Bruce, T.J.A. (2014) Glucosinolates in oilseed rape: secondary metabolites that influence interactions with herbivores and their natural enemies. *Annals of Applied Biology* 164, 348–353.

Butts, R.A. and Lamb, R.J. (1990) Comparison of oilseed *Brassica* crops with high or low levels of glucosinolates and alfalfa as hosts for three species of *Lygus* (Hemiptera: Heteroptera: Miridae). *Journal of Economic Entomology* 83, 2258–2262.

Cárcamo, H.A. (2012) Effect of insect herbivory on the architecture and seed production of canola, *Brassica napus* L. *Journal of Entomological Science* 47, 44–55.

Cárcamo, H., Olfert, O., Dosdall, L., Herle, C., Beres, B. and Soroka, J. (2007a) Resistance to cabbage seedpod weevil among selected Brassicaceae germplasm. *The Canadian Entomologist* 139, 658–669.

Cárcamo, H.A., Dunn, R., Dosdall, L.M. and Olfert, O. (2007b) Managing cabbage seedpod weevil in canola using a trap crop–A commercial field scale study in western Canada. *Crop Protection* 26, 1325–1334.

Cárcamo, H.A., Otani, J.K., Dosdall, L.M., Blackshaw, R.E., Clayton, G.W., Harker, K.N., O'Donovan, J.T. and Entz, T. (2008) Effects of seeding date and canola species on seedling damage by flea beetles in three ecoregions. *Journal of Applied Entomology* 132, 623–631.

Cole, R.A. (1997) The relative importance of glucosinolates and amino acids to the development of two aphid pests *Brevicoryne brassicae* and *Myzus persicae* on wild and cultivated *Brassica* species. *Entomologia Experimentalis et Applicata* 85, 121–133.

Cook, S.M., Smart, L.E., Martin, J.L., Murray, D.A., Watts, N.P. and Williams, I.H. (2006) Exploitation of host plant preferences in pest management strategies for oilseed rape (*Brassica napus*). *Entomologia Experimentalis et Applicata* 119, 221–229.

Dosdall, L.M. and Mason, P.G. (2010) Key pests and parasitoids of oilseed rape or canola in North America and the importance of parasitoids in integrated management. In: Williams, I.H. (ed.) *Biocontrol-based Integrated Management of Oilseed Rape Pests*. Springer Science, Dordrecht, The Netherlands, pp. 167–214.

Dosdall, L.M., Soroka, J.J. and Olfert, O. (2011) The diamondback moth in canola and mustard: current pest status and future prospects. *Prairie Soils and Crops* 4, 66–76.

Eickermann, M. and Ulber, B. (2011) Resynthesized lines and cultivars of *Brassica napus* L. provide sources of resistance to the cabbage stem weevil (*Ceutorhynchus pallidactylus* (Mrsh.)). *Bulletin of Entomological Research* 101, 287–294.

Ekbom, B. (2010) Pests and their enemies in spring oilseed rape in Europe and challenges to integrated pest management. In: Williams, I.H. (ed.) *Biocontrol-based Integrated Management of Oilseed Rape Pests*. Springer Science, Dordrecht, The Netherlands, pp. 151–165.

Ekbom, B. and Borg, A. (1996) Pollen beetle (*Meligethes aeneus*) oviposition and feeding preference on different host plant species. *Entomologia Experimentalis et Applicata* 78, 291–299.

Foshee, W., Boozer, R. and Horton, D. (2003) New insecticides and trap crop techniques for peach pest management. *Hortscience* 38, 1291.

Fourie, H., Ahuja, P., Lammers, J. and Daneel, M. (2016) Brassicacea-based management strategies as an alternative to combat nematode pests: a synopsis. *Crop Protection* 80, 21–41.

Free, J.B. and Williams, I.H. (1978) The responses of the pollen beetle, *Meligethes aeneus*, and the seed weevil, *Ceuthorhynchus assimilis*, to oil-seed rape, *Brassica napus*, and other plants. *Journal of Applied Ecology* 15, 761–774.

Furlong, M.J., Spafford, H., Ridland, P.M., Endersby, N.M., Edwards, O.R., Baker, G.J., Keller, M.A. and Paull, C.A. (2008) Ecology of diamondback moth in Australian canola: landscape perspectives and the implications for management. *Australian Journal of Experimental Agriculture* 48, 1494–1505.

Furlong, M.J., Wright, D.J. and Dosdall, L.M. (2013) Diamondback moth ecology and management: problems, progress, and prospects. *Annual Review of Entomology* 58, 517–541.

Gerber, G.H. (1997) Oviposition preferences of *Lygus lineolaris* (Palisot de Beauvois) (Heteroptera: Miridae) on four *Brassica* and two *Sinapis* species (Brassicaceae) in field cages. *The Canadian Entomologist* 129, 855–858.

Giamoustaris, A. and Mithen, R. (1995) The effect of modifying the glucosinolate content of leaves of oilseed rape (*Brassica napus* ssp. *oleifera*) on its interaction with specialist and generalist pests. *Annals of Applied Biology* 126, 347–363.

Halkier, B.A. and Gershenzon, J. (2006) Biology and biochemistry of glucosinolates. *Annual Review of Plant Biology* 57, 303–333.

Hamid, H.A., Dalla Monta, L. and Battisti, A. (2006) Undersowing cruciferous vegetables with clover: the effect of sowing time on flea bettles and diamondback moth. *Bulletin of Insectology* 59, 121–127.

Hervé, M.R., Garcia, N., Trabalon, M., Ralec, A., Delourme, R. and Cortesero, A.M. (2015) Oviposition behavior of the pollen beetle (*Meligethes aeneus*): a functional study. *Journal of Insect Behavior* 28, 107–119.

Hokkanen, H., Granlund, H., Husberg, G.B. and Markkula, M. (1986) Trap crops used successfully to control *Meligethes aeneus* (Col., Nitidulidae), the rape blossom beetle. *Annales Entomologici Fennici* 52, 115–120.

Hokkanen, H.M.T. (1989) Biological and agrotechnical control of the rape blossom beetle *Meligethes aeneus* (Coleoptera: Nitidulidae). *Acta Entomologica Fennica* 53, 25–30.

Hokkanen, H.M.T. (1991) Trap cropping in pest management. *Annual Review of Entomology* 36, 119–138.

Hokkanen, H.M.T. (2000) The making of a pest: recruitment of *Meligethes aeneus* onto oilseed Brassicas. *Entomologia Experimentalis et Applicata* 95, 141–149.

Hopkins, R.J., Birch, A.N.E., Griffiths, D.W., Baur, R., Städler, E. and Mckinlay, R.G. (1997) Leaf surface compounds and oviposition preference of turnip root fly *Delia floralis*: the role of glucosinolate and nonglucosinolate compounds. *Journal of Chemical Ecology* 23, 629–643.

Hopkins, R.J., Van Dam, N.M. and van Loon, J.J.A. (2009) Role of glucosinolates in insect-plant relationships and multitrophic interactions. *Annual Review of Entomology* 54, 57–83.

Jamont, M., Dubois-Pot, C. and Jaloux, B. (2014) Nectar provisioning close to host patches increases parasitoid recruitment, retention and host parasitism. *Basic and Applied Ecology* 15, 151–160.

Jönsson, M. and Anderson, P. (2007) Emission of oilseed rape volatiles after pollen beetle infestation; behavioural and electrophysiological responses in the parasitoid *Phradis morionellus*. *Chemoecology* 17, 201–207.

Justus, K.A., Dosdall, L.M. and Mitchell, B.K. (2000) Oviposition by *Plutella xylostella* (Lepidoptera: Plutellidae) and effects of phylloplane waxiness. *Journal of Economic Entomology* 93, 1152–1159.

Kaasik, R., Kovács, G., Kaart, T., Metspalu, L., Williams, I.H. and Veromann, E. (2014) *Meligethes aeneus* oviposition preferences, larval parasitism rate and species composition of parasitoids on *Brassica nigra*, *Raphanus sativus* and *Eruca sativa* compared with on *Brassica napus*. *Biological Control* 69, 65–71.

Kergunteuil, A., Cortesero, A.M., Chaminade, V., Dourlot, S., Paty, C., Le Ralec, A. and Dugravot, S. (2015a) Field and laboratory selection of brassicaceous plants that differentially affect infestation levels by *Delia radicum*. *Journal of Applied Entomology* 139, 487–495.

Kergunteuil, A., Dugravot, S., Danner, H., Van Dam, N.M. and Cortesero, A.M. (2015b) Characterizing volatiles and attractiveness of five brassicaceous plants with potential for a 'push–pull' strategy toward the cabbage root fly, *Delia radicum*. *Journal of Chemical Ecology* 41, 330–339.

Kim, J.H. and Jander, G. (2007) *Myzus persicae* (green peach aphid) feeding on *Arabidopsis* induces the formation of a deterrent indole glucosinolate. *The Plant Journal* 49, 1008–1019.

Kirkegaard, J.A. and Sarwar, M. (1999) Glucosinolate profiles of Australian canola (*Brassica napus annua* L.) and Indian mustard (*Brassica juncea* L.) cultivars: implications for biofumigation. *Australian Journal of Agricultural Research* 50, 315–324.

Kliebenstein, D., Pedersen, D., Barker, B. and Mitchell-Olds, T. (2002) Comparative analysis of quantitative trait loci controlling glucosinolates, myrosinase and insect resistance in *Arabidopsis thaliana*. *Genetics* 161, 325–332.

Kloen, H. and Altieri, M.A. (1990) Effect of mustard (*Brassica hirta*) as a non-crop plant on competition and insect pests in broccoli (*Brassica oleracea*). *Crop Protection* 9, 90–96.

Kumar, S. and Sangha, M.K. (2013) Biochemical mechanism of resistance in some *Brassica* genotypes against *Lipaphis erysimi* (Kaltenbach) (Homoptera: Aphidiae). *Vegetos* 26, 387–395.

Kumar, S. and Singh, Y.P. (2015) Insect pests. In: Kumar, A., Banga, S.S., Meena, P.D. and Kumar, P.R. (eds) *Brassica Oilseeds: Breeding and Management*. CABI, Wallingford, UK, pp. 193–232.

Lambdon, P.W., Hassall, M. and Mithen, R. (1998) Feeding preferences of woodpigeons and flea-beetles for oilseed rape and turnip rape. *Annals of Applied Biology* 133, 313–328.

Liersch, A., Bocianowski, J. and Bartkowiak-Broda, I. (2013) Fatty acid and glucosinolate level in seeds of different types of winter oilseed rape cultivars (*Brassica napus* L.). *Communications in Biometry and Crop Science* 8, 39–47.

Luther, G.C., Valenzuela, H.R. and DeFrank, J. (1996) Impact of cruciferous trap crops on lepidopteran pests of cabbage in Hawaii. *Journal of Economic Entomology* 25, 39–47.

Marazzi, C. and Städler, E. (2004a) Influence of plant sulphur nutrition on oviposition and larval performance of the diamondback moth. *Entomologia Experimentalis et Applicata* 111, 225–232.

Marazzi, C. and Städler, E. (2004b) Influence of sulphur nutrition on oviposition and larval performance of the cabbage root fly. *Agricultural and Forest Entomology* 7, 277–282.

Marazzi, C., Patrian, B. and Städler, E. (2004a) Secondary metabolites of the leaf surface affected by sulphur fertilisation and perceived by the cabbage root fly. *Chemoecology* 14, 87–94.

Marazzi, C., Patrian, B. and Städler, E. (2004b) Secondary metabolites of the leaf surface affected by sulphur fertilisation and perceived by the diamondback moth. *Chemoecology* 14, 81–86.

Martínez-Ballesta, M.D.C., Moreno, D.A. and Carvajal, M. (2013) The physiological importance of glucosinolates on plant response to abiotic stress in *Brassica*. *International Journal of Molecular Sciences* 14, 11607–11625.

McCloskey, C. and Isman, M.B. (1993) Influence of foliar glucosinolates in oilseed rape and mustard on feeding and growth of the bertha armyworm, *Mamestra configurata* Walker. *Journal of Chemical Ecology* 19, 249–266.

Metspalu, L., Kruus, E., Ploomi, A., Williams, I.H., Hiiesaar, K., Jõgar, K., Veromann, E. and Mänd, M. (2014) Flea beetle (Chrysomelidae: Alticinae) species composition and abundance in different cruciferous oilseed crops and the potential for a trap crop system. *Acta Agriculturae Scandinavica, Section B — Soil & Plant Science* 64, 572–582.

Milford, G.F.J., Porter, A.J.R., Fieldsend, J.K., Miller, C.A., Leach, J.E. and Williams, I.H. (1989) Glucosinolates in oildseed rape (*Brassica napus*) and the incidence of pollen beetles (*Meligethes aeneus*). *Annals of Applied Biology* 115, 375–380.

Nielsen, J.K. (1989) The effect of glucosinolates on responses of young *Phyllotreta nemorum* larvae to non-host plants. *Entomologia Experimentalis et Applicata* 51, 249–259.

Nielsen, J.K., Hansen, M.L., Agerbirk, N., Petersen, B.L. and Halkier, B.A. (2001) Responses of the flea beetles *Phyllotreta nemorum* and *P. cruciferae* to metabolically engineered *Arabidopsis thaliana* with an altered glucosinolate profile. *Chemoecology* 11, 75–83.

Nikooei, M., Fathipour, Y., Javaran, M.J. and Soufbaf, M. (2015) How different genetically manipulated *Brassica* genotypes affect life table parameters of *Plutella xylostella* (Lepidoptera: Plutellidae). *Journal of Economic Entomology* 108, 515–524.

Ratzka, A., Vogel, H., Kliebenstein, D.J., Mitchell-Olds, T. and Kroymann, J. (2002) Disarming the mustard oil bomb. *Proceedings of the National Academy of Sciences of the USA* 99, 11223–11228.

Renwick, J.A.A., Haribal, M., Gouinguené, S. and Stadler, E. (2006) Isothiocyanates stimulating oviposition by the diamondback moth, *Plutella xylostella*. *Journal of Chemical Ecology* 32, 755–766.

Riggi, L.G., Gagic, V., Bommarco, R. and Ekbom, B. (2015) Insecticide resistance in pollen beetles over

7 years – a landscape approach. *Pest Management Science* 72(4), 780–786.

Rousse, P., Fournet, S., Porteneuve, C. and Brunel, E. (2003) Trap cropping to control *Delia radicum* populations in cruciferous crops: first results and future applications. *Entomologia Experimentalis et Applicata* 109, 133–138.

Santolamazza-Carbone, S., Velasco, P., Soengas, P. and Cartea, M.E. (2014) Bottom-up and top-down herbivore regulation mediated by glucosinolates in *Brassica oleracea* var. *acephala*. *Oecologia* 174, 893–907.

Satpathy, S., Shivalingaswamy, T., Kumar, A., Rai, A. and Rai, M. (2010) Potentiality of Chinese cabbage (*Brassica rapa* subsp. *pekinensis*) as a trap crop for diamondback moth (*Plutella xylostella*) management in cabbage. *Indian Journal of Agricultural Sciences* 80, 238–241.

Shelton, A.M. and Badenes-Pérez, F.R. (2006) Concepts and applications of trap cropping in pest management. *Annual Review of Entomology* 51, 285–308.

Shields, V.D.C. and Mitchell, B.K. (1995) Sinigrin as a feeding deterrent in two crucifer-feeding, polyphagous lepidopterous species and the effects of feeding stimulant mixtures on deterrency. *Philosophical Transactions of the Royal Society B: Biological Sciences,* 347, 439–446.

Smart, L.E. and Blight, M.M. (2000) Response of the pollen beetle, *Meligethes aeneus*, to traps baited with volatiles from oilseed rape, *Brassica napus*. *Journal of Chemical Ecology* 26, 1051–1064.

Soroka, J. and Grenkow, L. (2013) Susceptibility of brassicaceous plants to feeding by flea beetles, *Phyllotreta* spp. (Coleoptera: Chrysomelidae). *Journal of Economic Entomology* 106, 2557–2567.

Soroka, J.J., Dosdall, L.M., Olfert, O.O. and Seidle, E. (2004) Root maggots (*Delia* spp., Diptera: Anthomyiidae) in prairie canola (*Brassica napus* L. and *B. rapa* L.): spatial and temporal surveys of root damage and prediction of damage levels. *Canadian Journal of Plant Science* 84, 1171–1182.

Städler, E., Baur, R. and De Jong, R. (2002) Sensory basis of host-plant selection: in search of the 'fingerprints' related to oviposition of the cabbage root fly. *Acta Zoologica Academiae Scientiarum Hungaricae* 48 (Suppl. 1), 265–280.

Staley, J.T., Stewart-Jones, A., Pope, T.W., Wright, D.J., Leather, S.R., Hadley, P., Rossiter, J.T., Van Emden, H.F. and Poppy, G.M. (2010) Varying responses of insect herbivores to altered plant chemistry under organic and conventional treatments. *Proceedings of the Royal Society B: Biological Sciences* 277, 779–786.

Sukovata, L., Jaworski, T. and Kolk, A. (2015) Efficacy of *Brassica juncea* granulated seed meal against *Melolontha* grubs. *Industrial Crops and Products* 70, 260–265.

Tangtrakulwanich, K., Reddy, G.V.P., Wu, S., Miller, J.H., Ophus, V.L. and Prewett, J. (2014) Developing nominal threshold levels for *Phyllotreta cruciferae* (Coleoptera: Chrysomelidae) damage on canola in Montana, USA. *Crop Protection* 66, 8–13.

Tansey, J.A., Dosdall, L.M., Keddie, A., Fletcher, R.S. and Kott, L.S. (2010) Antixenosis and antibiosis resistance to *Ceutorhynchus obstrictus* in novel germplasm derived from *Sinapis alba* x *Brassica napus*. *The Canadian Entomologist* 142, 212–221.

Tong, Y., Gabriel-Neumann, E., Ngwene, B., Krumbein, A., George, E., Platz, S., Rohn, S. and Schreiner, M. (2014) Topsoil drying combined with increased sulfur supply leads to enhanced aliphatic glucosinolates in *Brassica juncea* leaves and roots. *Food Chemistry* 152, 190–196.

Ulmer, B., Gillott, C. and Erlandson, M. (2001) Feeding preferences, growth, and development of *Mamestra configurata* (Lepidoptera: Noctuidae) on Brassicaceae. *The Canadian Entomologist* 133, 509–519.

Ulmer, B., Gillot, C., Woods, D. and Erlandson, M. (2002) Diamondback moth, *Plutella xylostella* (L.), feeding and oviposition preferences on glossy and waxy *Brassica rapa* (L.) lines. *Crop Protection* 21, 327–331.

van Loon, J.J.A., Wang, C.Z., Nielsen, J.K., Gols, R. and Qiu, Y.T. (2002) Flavonoids from cabbage are feeding stimulants for diamondback moth larvae additional to glucosinolates: chemoreception and behaviour. *Entomologia Experimentalis et Applicata* 104, 27–34.

Williams, I.H. (2010) The major insect pests of oilseed rape in Europe and their management: an overview. In: Williams, I.H. (ed.) *Biocontrol-Based Integrated Management of Oilseed Rape Pests*. Springer Science, Dordrecht, The Netherlands, pp. 1–43.

Zalucki, M.P., Shabbir, A., Silva, R., Adamson, D., Shu-Sheng, L. and Furlong, M.J. (2012) Estimating the economic cost of one of the world's major insect pests, *Plutella xylostella* (Lepidoptera: Plutellidae): just how long is a piece of string? *Journal of Economic Entomology* 105, 1115–1129.

24 Arthropod Pests of Australian Canola During Crop Emergence: IPM and Future Directions

Sarina Macfadyen* and Matthew P. Hill

CSIRO, Canberra, Australia

24.1 Introduction

Canola (also known as oilseed rape, mostly spring-type *Brassica napus*) is a common winter oilseed crop grown in a range of soil types across Australia, typically as part of a cereal crop rotation. When *B. napus* was originally introduced and grown in Australia in the 1960s (using Canadian varieties), farmers suffered significant problems with the fungal disease 'blackleg' throughout the 1970s (Van de Wouw *et al.*, 2016). It was not until new varieties were developed for Australian conditions that the crop became attractive to farmers. Although *B. napus* was used as a break crop in the traditional cereal rotation on many farms across southern and western Australia (Australian Government, 2008; Kirkegaard *et al.*, 2016), it is now an important cash crop in its own right. *B. napus* is grown in Mediterranean and temperate climatic regions and, depending on rainfall, is planted in autumn (late April to June) each year. The area of land under *B. napus* in Australia in 2014 was 2.72 million hectares, which represents about 12% of all the area of broad-acre cropping, and this trend has been increased in the past 7–10 years, though productivity has not been improved compared with the mid-1990s (Kirkegaard *et al.*, 2016) (Fig. 24.1).

B. napus is especially susceptible to damage from invertebrate pests from emergence to the five-leaf stage (Miles and McDonald, 1999; Gu *et al.*, 2007); consequently pesticide applications in seed dressings and pesticide sprays at the early stages of development are frequent. Today, widespread losses of *B. napus* crops due to pest damage at crop establishment is a factor that limits productivity and is a continual risk for growers. Furthermore, the increasingly widespread practices of reduced tillage and stubble retention for soil conservation and water-use efficiency may have altered the risks associated with certain pests in recent years (Hoffmann *et al.*, 2008) (Fig. 24.2). There is a limited understanding of the biology and ecology of some of these pest species that attack the plant during the early stages (hereafter referred to as emergence pest species). Due to lack of established economic thresholds, decision making for pest management is not reliable. Additionally, some pest species show a high tolerance (and in some cases resistance) to conventional pesticides. Examples include *Balaustium medicagoense* (Acarina: Erythraeidae) and *Bryobia* sp. (Acarina: Tetranychidae).

There are over 28 invertebrate species that are considered pests of *B. napus* at the crop emergence period (Micic *et al.*, 2008). These include mites (e.g. *B. medicagoense* and *Halotydeus destructor* (Acarina: Penthaleidae)), springtails (lucerne flea *Sminthurus viridis* (Collembola: Sminthuridae)), true and false wireworms (e.g. various species in Coleoptera: Tenebrionidae and Coleoptera: Elateridae families), weevils (e.g. *Listroderes difficilis* (Coleoptera: Curculionidae)) and noctuid caterpillars (e.g. *Agrotis infusa* (Lepidoptera: Noctuidae)). Later in the season, when the crop is flowering and at pod-filling stage, aphids and migratory lepidopterans (e.g. *Plutella xylostella* (Lepidoptera: Plutellidae))

*E-mail: sarina.macfadyen@csiro.au

 Integrated Management of Insect Pests on Canola and Other Brassica Oilseed Crops (ed. G.V.P. Reddy)

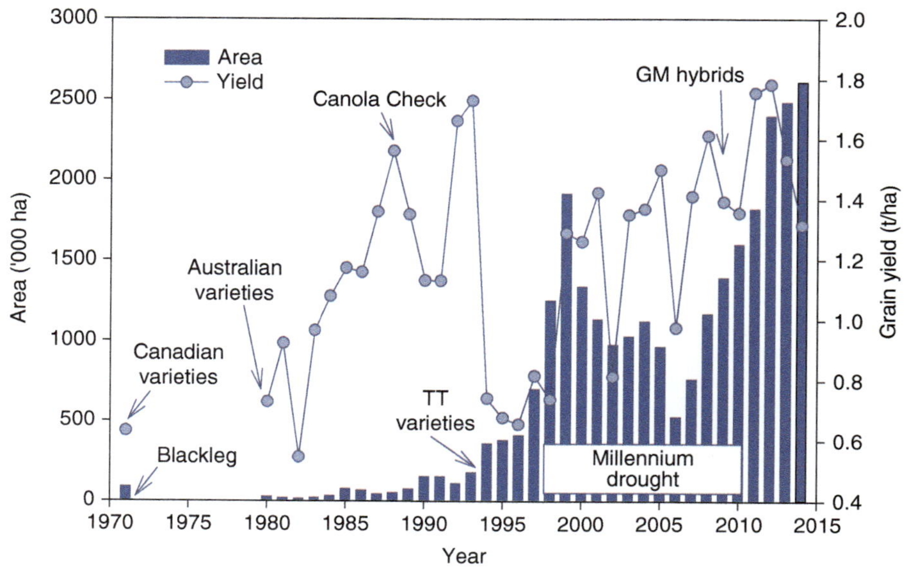

Fig. 24.1. Area and average national grain yield for *Brassica napus* (canola) in Australia highlighting some of the significant events influencing the observed trends. Data compiled from ABARES estimates and Australian Oilseed Federation estimates. Reproduced with permission from Kirkegaard *et al.*, 2016.

can cause sporadic damage (Gu *et al.*, 2007). These pest species are different to the community found attacking *B. napus* (oilseed rape) throughout Europe (Williams, 2010) and North America (Dosdall *et al.*, 2012). The Australian pest species are mostly non-native but many have been present in Australian grain production landscapes for many years.

24.2 Emergence Arthropod Pest Communities in Australian Agroecosystems

This section focuses on pest species that are economically important during the crop emergence period (Table 24.1): lucerne flea (*Sminthurus viridis*); and the Acari species redlegged earth mite (*H. destructor*), blue oat mites (*Penthaleus* spp.), clover mites (*Bryobia* spp.) and *Balaustium* mite (*B. medicagoense*) (Micic *et al.*, 2008). Whilst *Bryobia* mites are mainly active in warmer conditions closer to crop flowering, they can also be pests of *B. napus* at emergence and are included here (as elsewhere, e.g. Micic *et al.*, 2008). The following sections summarise what is known about these species in Australian cropping systems and what control strategies can be used as a part of integrated pest management (IPM) approaches and discuss new monitoring and management tools that are being developed for more effective management of these pests.

24.2.1 *Sminthurus viridis* (L.) (sub-class Collembola: Sminthuridae)

The lucerne flea, *S. viridis*, is a 1–3 mm wingless collembolan that lives in the soil and feeds on living plants, causing skeletonization of the leaf. It is a commonly occurring pest of pastures and broadacre grain crops in autumn (March–May) (Wallace and Mahon, 1971b; Bailey, 2007) (Table 24.1). It feeds on a range of plant types (pasture legumes, grain legumes and *B. napus*) but is mostly restricted to heavy (loamy) soils or fine-textured soils that are slightly acidic (Micic *et al.*, 2008). Lucerne flea is regarded as a resident pest that is relatively non-mobile (Table 24.1), though distribution within a field can be highly dynamic and patchy (Wallace, 1967). It is winter active (June–August) and avoids dry summer conditions by producing diapausing eggs that are resistant to desiccation (Wallace, 1968; Roberts *et al.*, 2011a). The genetic data showed significant population structure between regions, suggesting that there is limited gene flow between populations (Roberts and Weeks, 2011). Although human-facilitated long-distance movement events

Fig. 24.2. *Brassica napus* (canola) planted into a retained stubble field: (a) at emergence (plants just visible); (b) post-emergence. Quarter-quadrat is shown for scale. Photo credit: CSIRO.

occur to some degree, this has implications for how this resident pest is managed within and between seasons. Currently, pesticides are the main control option used by growers; however, this species is known to display high tolerance to a range of pesticides, e.g. α-cypermethrin, bifenthrin, omethoate, methidathion and phosmetin (Roberts *et al*., 2009).

24.2.2 *Halotydeus destructor* (Acarina: Penthaleidae)

The redlegged earth mite, *Halotydeus destructor*, is approximately 1 mm in length with a distinctive globular black velvet body and red-orange legs. This species is active in autumn and winter (March–August) and considered a widespread pest in both pastures and emerging field crops, especially *B. napus* (Ridsdill-Smith *et al*., 2008) (Table 24.1). Females produce diapausing eggs in spring that are able to withstand the hot and dry summer conditions prevalent across cropping and pastoral regions of southern Australia (Ridsdill-Smith *et al*., 2008). *H. destructor* is highly polyphagous and can also exploit alternative broadleaf hosts, such as *Arctotheca calendula* (capeweed), that are common along roadsides, in native vegetation and other areas near crop fields (Fig. 24.3a). On *B. napus*, mite feeding results in a 'silvering' pattern (Fig. 24.3b) on the seedlings and can lead to death or slow growth; however, the plant does show some tolerance to feeding at later growth stages (Arthur *et al*., 2013a).

This species was introduced into Western Australia from South Africa in around 1908 (Halliday, 1991) and has since spread widely across all southern regions. Its distributional limits were previously linked to upper temperatures and a rainfall isohyet (Wallace and Mahon, 1971a), though the species has recently expanded its range in drier and hotter inland regions (Hill *et al*., 2012a). Genetic analysis suggests that the invasive population that arrived in Australia (likely from soil ballast used on ships) was from a lineage similar to that found near Cape Town today. Populations in Australia remain genetically similar across western and eastern Australia, suggesting ongoing gene flow between these large areas (Hill *et al*., 2016). This could occur via the movement of diapausing eggs in soil adhering to farm machinery and livestock and through the transportation of plant material. Laboratory studies on the thermal limits of Australian populations and of populations from the native range in South Africa have shown that Australian populations have an increased upper thermal threshold for movement and recover from cold stress more rapidly (Hill *et al*., 2013). This may explain why this species can persist in drier inland regions of Australia, which experience frosts frequently.

H. destructor is active in autumn and winter during the cool and wet months. The adults begin to lay diapausing eggs in spring in response to increased photoperiod (Ridsdill-Smith *et al*., 2005) and this allows farmers to time control measures (Timerite) (AWI, 2016) that target the adult mites prior to egg laying in the spring (Ridsdill-Smith and Pavri, 2015). However, there continues to be damage to a range of crops caused by this pest each year (Gower *et al*., 2008). The diapause eggs sit within the cadavers of the dead females over the summer (December–February) (Ridsdill-Smith *et al*.,

Table 24.1. Summary of the biological and ecological characteristics of the economically important emergence pest species of *Brassica napus* (canola) in Australia.

Characteristics	*Sminthurus viridis* (L.) (Collembola: Sminthuridae)	*Halotydeus destructor* Tucker (Acarina: Penthaleidae)	*Penthaleus* spp. (Acarina: Penthaleidae)[a]	*Bryobia* spp. (Acarina: Tetranychidae)[b]	*Balaustium medicagoense* Meyer and Ryke (Acarina: Erythraeidae)
Common name	Lucerne flea	Redlegged earth mite	Blue oat mite	Clover mite	*Balaustium* mite
Sexual or asexual reproduction	Sexual	Sexual	Asexual	Asexual (but not confirmed for all species)	Asexual
Diapause	Yes, over-summer as eggs, laid on soil surface	Yes, over-summer as eggs, retained in cadavers of female	Yes, over-summer as eggs, laid on leaves, stems, roots	No, over-wintering eggs laid early–mid winter	Yes, when conditions are unfavourable
Number of generations per year	2–3, active autumn/winter and spring	Multiple, active autumn/winter	2 (on average), active May–November	Multiple, active in warm conditions, spring/autumn	2 per season
Resistance to pesticides[c]	Not confirmed	Confirmed	Not confirmed	Not confirmed	Not confirmed
Tolerance to pesticides[c]	Documented	Not documented	Documented	Documented	Documented
Likely native range	Europe	South Africa	Not sure	Not sure	South Africa
Crops attacked	*B. napus*, pulses, lucerne, clovers and sometimes annual winter cereals	*B. napus*, pulses, lucerne, clovers, sometimes annual winter cereals, vegetables	*B. napus*, cereals, lucerne, pastures	*B. napus*, lupins, wheat, lucerne, vetch and clovers	*B. napus*, lupins, cereals, lucerne, pasture legumes, grasses
Potential loss in *B. napus* if not controlled (AUS$ million)[d]	29.7	96.6	28.8	19.3	

[a]Three species recognized as pests in Australia (*P. major, P. falcatus, P. tectus*)
[b]Four species have been recorded in Australia but their taxonomy is complicated
[c]Recorded in Australia
[d]Murray *et al.* (2013a) report (see Table 7.9)

2005). A significant rainfall event and period of cooling temperatures will trigger egg hatch (McDonald *et al.*, 2015). Pesticides are currently the predominant method for controlling this pest but high levels of resistance to synthetic pyrethroids have been documented (Umina *et al.*, 2012).

24.2.3 *Penthaleus* spp. (Acarina: Penthaleidae)

The *Penthaleus* species group (blue oat mites) consists of a morphologically cryptic complex of three pest species, each approximately 1 mm long: *Penthaleus major*, *Penthaleus falcatus* and *Penthaleus tectus* (Table 24.1) (Qin and Halliday, 1995). These mites are winter-active, with three generations per season. They reproduce parthenogenetically, with populations composed of clones, and produce diapause eggs soon after adult emergence in autumn. Early-sown crops, in years when summer rainfall has supported weed growth, are most at risk from damage. Blue oat mites have been the focus of less prolonged research effort than *H. destructor* but important differences in the

Fig. 24.3. (a) Example of a roadside verge that contains grasses and weeds that support emergence pests of *Brassica napus* (canola). (b) Mites and mite damage to *B. napus* (canola), as seen as the silvering on the leaf. Photo credit: CSIRO.

biology and ecology of the species have been reported, with implications for management of the individual species.

Species distribution, host plant preferences, competitive ability on different host plants and pesticide tolerances differ between the *Penthaleus* species (Gu *et al.*, 2007). Historical sampling has shown that *P. major* can tolerate a drier climate than *H. destructor* (Wallace and Mahon, 1971a). More recent species distribution models show that the potential distributions of the three *Penthaleus* species across southern Australia are best described with different sets of climatic variables (Hill *et al.*, 2012b), suggesting differences in physiological tolerances of the species. The most common *Penthaleus* species on *B. napus* in field surveys was *P. falcatus* but it has a limited range of host plants compared with *H. destructor* (Umina and Hoffmann, 2004). In a shade-house study, *B. napus* was a suitable host for *P. falcatus* but *P. major* could not breed successfully on *B. napus* (Umina and Hoffmann, 2004). However, studies have shown that *P. major* can feed on *B. napus*, though it may not be an ideal host. McDonald *et al.* (1995) found that *P. major* caused about 50% less damage than *H. destructor* in a canola variety, Oscar. In the same shade-house study, feeding on *B. napus* by *P. falcatus* led to delayed flowering time in *B. napus* (Umina and Hoffmann, 2004).

24.2.4 *Bryobia* spp. (Acarina: Tetranychidae)

Four species of the genus *Bryobia* (clover mites) have been recorded in Australia but the taxonomy of these species remains unclear (Table 24.1). Adult mites are about 0.75 mm long with pale red or orange legs (Bailey, 2007). *Bryobia praetiosa* has been identified as being involved in outbreaks but other species are also likely to be causing problems (Micic *et al.*, 2008). These species are active from autumn to spring (March–November) and are likely to reproduce parthenogenetically, though there have been no studies in Australia (Micic *et al.*, 2008). Unlike the other mite pests, they produce overwintering eggs that hatch as conditions get warmer in spring, so damage to crops can occur at crop emergence but also later in the season. *Bryobia* species have been found to damage *B. napus*, pasture and lupins and could reproduce successfully on *B. napus* in a shade-house study (Arthur *et al.*, 2010). They are tolerant of a range of commonly used pesticides, e.g. bifenthrin, methidathion and α-cypermethrin (Arthur *et al.*, 2008). Given the general similarities in appearance with other pest mites, and hence the confusion with identification in the field, the economic impact and pest status of *Bryobia* spp. is largely unknown (Arthur *et al.*, 2010).

24.2.5 *Balaustium medicagoense* (Acarina: Erythraeidae)

Balaustium medicagoense is the only species of *Balaustium* that has been recorded in Australia to date and it is thought this species was introduced from South Africa through the same pathway as *H. destructor*. It typically undergoes two generations from March until December before entering a summer diapause (Table 24.1). It is a relatively large mite and at 3 mm long is over twice the size of *H. destructor*. It is more commonly found on cereals and grasses but also attacks broadleaf crops such as *B. napus*, lucerne and lupins (Arthur *et al.*, 2010). However, outbreak reports from farmers suggest that *B. napus* is the crop most commonly attacked or damaged by *B. medicagoense* (Arthur *et al.*, 2010). Many common pesticides are ineffective at controlling *B. medicagoense* in a field situation (Arthur *et al.*, 2013b). Furthermore, since the 1980s there has been an increase in the frequency of outbreaks of both *B. medicagoense* and *Penthaleus* spp. reported by growers in grain production regions (Hoffmann *et al.*, 2008).

24.3 Impact of Emergence Pests on *B. napus*

The high incidence of *S. viridis* and pest mites, combined with the severity of damage they cause, results in them being the most economically costly pests of *B. napus* (Fig. 24.4). Aphids are the only other pest group that causes large economic losses but not often at crop emergence (Fig. 24.4). The direct economic cost of these pests in *B. napus* is difficult to estimate as other factors, such as frost and competition with weeds, can cause more widespread crop losses within a season than the activities of emergence pests. Seedling death and poor germination due to pests may represent 100% yield loss at the plant level; however, small-scale plant loss at emergence may not reduce yields significantly at the field level (Gu *et al.*, 2007; Umina *et al.*, 2015). Pest damage to *B. napus* has been estimated to cause greater than 10% of crop establishment failures (Miles and McDonald, 1999). The potential loss (if a pest is not controlled) has been

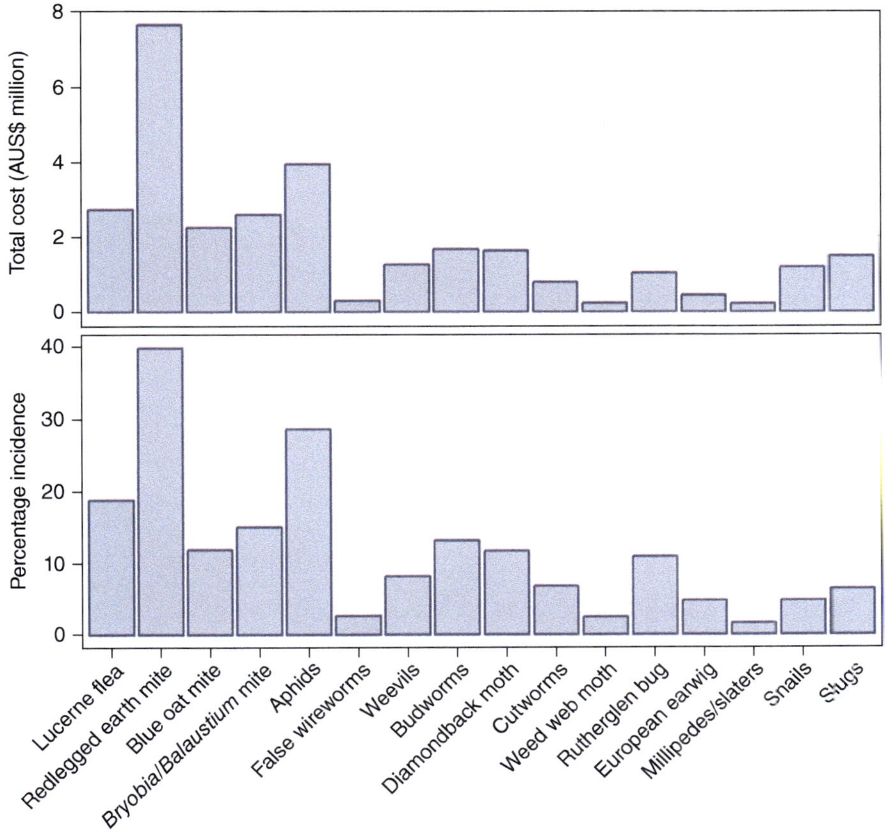

Fig. 24.4. Invertebrate pest treatment costs in *Brassica napus* (canola) across Australia (not limited to emergence pests). Top graph: total cost of pesticide treatments for each pest group; bottom graph: percentage incidence of damage from each pest group (= incidence years × incidence area). Taken from Table 7.8 in Murray *et al.* (2013a).

estimated at AUS$96.6 million for *H. destructor*, AUS$28.8 million for *Penthaleus* species, AUS$29.7 million for *S. viridis* and AUS$19.3 million for *Bryobia* and *Balaustium* species combined (Table 24.1) (Murray *et al.*, 2013a, b).

The total average cost of treatment of invertebrate pests in *B. napus* is AUS$29.5 million (AUS$20.18 per ha) (Murray *et al.*, 2013b), with the use of pesticides for the control of emergence pests contributing a significant portion of this total amount. However, most of the insecticide products are relatively cheap to purchase and apply. In small-plot trials of *B. napus* across Australia the cost of insecticide inputs (including application costs) was an average of AUS$12.67/ha (Macfadyen *et al.*, 2014). At one trial site the economic cost of using pesticides was substantially greater (almost ten times) due to the late-season application of an expensive selective insecticide against aphid pests (Macfadyen *et al.*, 2014).

Emergence pests and their management present a range of indirect costs to growers, such as increased risk of secondary pest outbreaks and the development of resistance to pesticides. There are anecdotal reports that the application of broad-spectrum pesticides early in the season (as bare-earth sprays) to address *H. destructor* risk can lead to secondary pest outbreaks later in the season from the relatively more tolerant *S. viridis*. There is no quantitative evidence to show how often, and under what circumstances, the risk of secondary pest outbreaks in *B. napus* is increased. In cotton, spider mite *Tetranychus urticae* populations reached higher densities in plots treated with pesticides compared with untreated cotton

(Wilson *et al.*, 1998). A significant negative relationship was found between early-season abundance of predators and the mid-season abundance of *T. urticae*, suggesting that the pesticides reduced predator numbers and released *T. urticae* from control (Wilson *et al.*, 1998). Evidence of resistance in *H. destructor* to synthetic pyrethroids has been documented in Western Australia (Umina *et al.*, 2012) prior to 2010 and is likely to have become more widespread since. For other pests, such as the green peach aphid *Myzus persicae*, widespread and high levels of resistance to carbamates and synthetic pyrethroids have been documented (Umina *et al.*, 2014). The economic impact of resistance is difficult to quantify, especially for emergence pests that may attack a range of crops, including *B. napus*, but is likely to lead to costly and long-term management issues.

24.4 Control Options for Emergence Pests as Part of an Integrated Pest Management Approach

To date, pest management in *B. napus* has been heavily dependent on the use of broad-spectrum pesticides, with the use of some cultural control practices, despite evidence that many pests are highly tolerant to the commonly used products (Arthur *et al.*, 2008, 2013b; Roberts *et al.*, 2009). However, increasing levels of pesticide resistance and scrutiny over the use of certain pesticide groups suggest that farmers must start considering alternative approaches. IPM involves the use of a diversity of biological and cultural control practices, combined with the judicious use of pesticides only after a pest has reached threshold (Dent, 1991; Horne and Page, 2008). Generally, IPM is practised across multiple crops on a farm and across a rotation sequence, and so determining how it can be used specifically for *B. napus* pest management can be challenging. However, a study across Europe demonstrated that an integrated strategy in winter oilseed rape that was both energy efficient and used less pesticide resulted in similar yields compared with a conventional approach (Nilsson *et al.*, 2015). The following sections summarize how components of an IPM approach are currently being used against emergence pests in *B. napus*.

24.4.1 Cultural controls

Cultural control techniques involve practices such as crop rotation, planting density, weed control and tillage and stubble management (Horne and Page, 2008). These practices can be effective at controlling emergence pests in *B. napus* by making the environment unsuitable for the growth and development of the pest. Often these practices impact more than one pest species at the same time. For example, *B. napus* crops that follow pastures with grasses and clover are most at risk from pest mite damage at emergence. Growers tend to avoid planting crops that are susceptible to damage into pasture fields, such as *B. napus* (canola), *Lupinus* spp. (lupins), *Vicia sativa* (vetch) and *Medicago sativa* (lucerne). Conversely, when *Triticum aestivum* (wheat) is sown into the rotation, with cultivation, there has been a demonstrated significant decline in *H. destructor* numbers (Merton *et al.*, 1995). The early control of summer and autumn weeds within and around fields, especially broadleaf weeds such as *Arctotis calendula* (capeweed) and *Trifolium* spp. (clovers), can help to reduce the risk of mite outbreaks (Gu *et al.*, 2007). A greater understanding of the host-plant (both crop and weed species) preferences of emergence pests may lead to better crop rotation sequences (Umina and Hoffmann, 2004; Arthur *et al.*, 2010).

Increased planting density through increased seeding density is another technique for managing pest mite species. Increased planting density reduces the impact of the loss of seedlings to mites at emergence and reduces the abundance of weeds that support pest populations. These techniques have not been well studied in Australia. However, one study demonstrated that, despite higher *H. destructor* densities early in the season (seeding rate 10 kg/ha), the increased plant density meant that the crop was able to survive mite damage and significantly reduce weed biomass (Burnett *et al.*, 2003). This led to improved economic yields.

24.4.2 Natural enemies and other biological control

Natural enemies include both generalist predators and parasitoids that can exert some level of suppression of pest abundance (Dent, 1991; Macfadyen *et al.*, 2015). There is little information available about the pathogens of the emergence pests considered here, despite examples of successful control of mite pests in other agroecosystems using bacterial, viral or fungal pathogens (Christian *et al.*, 1991; Rath, 1991). Whether or not natural enemies have any impact on emergence pest species, and therefore lowering associated crop damage, is difficult to quantify. In the cooler part of the year, natural enemy populations are usually low in number and may appear to be present in low numbers in a relatively bare crop field. However, sampling studies have shown the presence and activity of a range of generalist predators (e.g. ants, spiders, predatory beetles and predatory mites) that have the potential to reduce pest mite and collembolan abundance. Twenty natural enemies (19 arthropod predators and one microbial pathogen) of *H. destructor* and *P. major* were recorded in pastures in southern New South Wales and it was concluded that natural enemies could play an important role in reducing earth mite populations in unsprayed pastures (James, 1995).

Most research has focused on predatory mite species but their impact on pest reduction and yield protection has not yet been quantified in *B. napus* (but see Tsitsilas *et al.*, 2011, who showed a direct connection in pastures). The predatory mite *Anystis* sp. (later called *Anystis salicinus* (Acarina: Anystidae); Michael *et al.*, 1991) was introduced from France in the 1960s to control *H. destructor* in Western Australia. It was found to be capable of killing large numbers of *H. destructor* under laboratory conditions but was less effective when alternative foods were available (Otto and Halliday, 1991). In later studies, *Anystis wallacei* (Acarina: Anystidae) was found to feed on *H. destructor*; and another introduced predatory mite, *Neomolgus capillatus* (Acarina: Bdellidae), was found to feed on *S. viridis* (Michael, 1995). When these predators were introduced into pasture plots, they reduced pest numbers by two-thirds. *N. capillatus* was investigated in pastures in Tasmania and reduced *S. viridis* density to about 93% (Ireson *et al.*, 2002). *Bdellodes lapidaria* (Acarina: Bdellidae), the predatory snout mite, was found to be tolerant to several of the pesticides used to control *S. viridis* but did not have any impact on *S. viridis* abundance in a semi-field study. In pasture plots, predatory mites were added at different densities (400, 100 or 50) and the numbers of *S. viridis* were compared with plots without predator addition 9 weeks later. At the end of the study there were no differences in *S. viridis* abundance in samples taken from treatment and control plots (Roberts *et al.*, 2011b). *B. lapidaria* is a relatively widespread species, whereas *A. salicinus* and *N. capillatus* have a more restricted distribution (Michael *et al.*, 1991). There have been no natural enemies identified in broad-acre crops (e.g. *B. napus*, cereals, lupins, faba beans, soybean, cotton) in Australia that are effective in controlling species of *Bryobia*. However, there are a number of predator species known to attack other earth mites such as *H. destructor*, which may potentially prey upon species of *Bryobia*. These include the aforementioned predatory mites (*N. capillatus*, *A. wallacei, A. salicinus, B. lapidaria*) as well as small beetles (e.g. *Diomus notescens* (Coleoptera: Coccinellidae) and other Coccinellidae) and spiders (e.g. Linyphiidae).

A number of pesticide trials that have included natural enemies as well as pests have shown that natural enemies are relatively more tolerant than the pests. For example, this has been demonstrated for omethoate and dimethoate against *A. wallacei* (Michael, 1995). More generally, *B. napus* field trials using a range of pesticides have shown inconsistent treatment effects on non-target arthropods, including natural enemies (Jenkins *et al.*, 2013). This bodes well for the complementary use of pesticides alongside the activities of natural enemies in an IPM management strategy. However, strategies to increase the abundance of predators in *B. napus* at emergence are few. Besides using more selective pesticides, manipulation of ground cover can have an impact on predator abundances. Tsitsilas *et al.* (2011) found that increasing the height and complexity of ground cover in a windbreak led to greater numbers of natural enemies and the suppression of pests in the adjacent pasture.

24.4.3 Pesticides

Selective pesticides can be an effective means of minimizing yield loss when used based on economic thresholds. However, for many of the emergence pests in *B. napus* there are limited economic thresholds (Arthur *et al.*, 2015), few easy monitoring techniques (Arthur *et al.*, 2014) and high levels of tolerance and resistance to commonly used pesticides (Arthur *et al.*, 2008, 2013b; Roberts *et al.*, 2009; Umina *et al.*, 2012). Furthermore, *B. napus* is planted just prior to the wheat crop on most farms, meaning that it is a critical time for staff and a difficult period to spend time on monitoring emerging *B. napus* fields. This is why seed dressings using neonicotinoid (e.g. imidacloprid) pesticides have become popular on *B. napus*. These seed dressings have both a repellent and anti-feeding effect on *H. destructor* and can provide residual protection for up to 28 days (Seidel, 1999).

However, their role in IPM is debatable as they are applied prior to a pest issue arising. Furthermore, imidacloprid (on the seed or in the soil) can translocate into the pollen and nectar of flowers, as well as move through the soil, potentially impacting non-target species and natural enemies (Krischik *et al.*, 2015). None the less, seed dressings have proved useful for controlling low pest densities and preventing yield loss in *B. napus* (Macfadyen *et al.*, 2014; Umina *et al.*, 2015).

In Australia there are very few selective pesticides registered against crop establishment pests (Umina *et al.*, 2015). Jenkins *et al.* (2013) compared the impact of selective pesticides (e.g. paraffinic oil, spinosad, diafenthiuron) and broad-spectrum pesticides on arthropods in field trials. They found that the selective pesticides had fewer negative impacts on non-target arthropods but a few products did reduce numbers of certain groups. Given the difference in costs associated with some selective products (in comparison with conventional broad-spectrum products), it is not surprising that demand for selective products is currently relatively low amongst growers.

24.5 Future Prospects: New Tools for Management

The use of pesticides as the main option for control of emergence pest species (both in seed dressings and as bare-earth or foliar sprays) comes with an increasing risk of resistance. Despite the existence of a diversity of tools and strategies for controlling pests, which have been successful in other agroecosystems around the world, the suite of tools currently used by Australian growers is relatively limited. This is partially due to a lack of research and development into new tools well suited to Australian agroecosystems and partly to complacency caused by cheap and readily available pesticides. The following sections identify some of these new approaches, which, if developed further, may provide a more sustainable approach to the management of emergence pests in *B. napus*.

24.5.1 Predictive models

Given the constraints already identified with managing emergence pests, it is critical that growers are able to understand the risk of an outbreak some time before sowing commences. This gives growers time to alter their cultivation practices, manage weeds, alter planting density and consider alternative crops with lower susceptibility if the risk is great. To date, a simple predictive model has been developed to identify when the median emergence for *H. destructor* will occur in autumn (McDonald *et al.*, 2015). Rainfall and temperature thresholds were evaluated across multiple models and their performance varied between regions. In Western Australia, 5 mm of rain followed by mean day temperatures of below 20.5°C for 10 days were identified as key factors; in south-eastern Australia, a lower temperature threshold of 16°C was employed (McDonald *et al.*, 2015). These models are currently used solely for research purposes but they have the potential to provide useful information to growers if validated further.

24.5.2 Monitoring protocols and thresholds for decision making

Economic thresholds are an important component of any IPM strategy. When a pest population reaches this threshold, beyond which the cost of control is less than the cost of the potential yield loss, intervention is warranted (Dent, 1991). For most of the emergence pest species, economic thresholds are not available for *B. napus* and monitoring guidelines to estimate density at the scale of commercial operations are rare. Arthur *et al.* (2015) attempted to develop an economic injury level for *H. destructor*. At the first true leaf stage a threshold of ten mites per plant was proposed but at other crop development stages a relationship between mite numbers and yield could not be established (Arthur *et al.*, 2015). Compounding this problem is the challenge associated with estimating densities of these emergence pest species. Visual assessments are often the easiest for growers and consultants to implement in the field but this technique is heavily influenced by observer bias, sampling date and time of day (Arthur *et al.*, 2014). Further, the method of sampling most of these emergence pests is by blower-vac, which has the potential to get blocked with dirt, especially in wet conditions. Plant densities of below 30–40 plants/m^2 could be used as a proxy for mite damage when densities could not be estimated properly (Arthur *et al.*, 2015). Quantifying the relationship between plant density, plant damage and yield and pest densities is required in order to develop thresholds for the other emergence pest species. There are some nominal thresholds available for certain regions, which have not been empirically derived but provide some guidelines for growers. For example, in Victoria a threshold of ten holes per leaf has been suggested for *S. viridis* (http://ipmguidelinesforgrains.com.au/pests/lucerne-flea-in-winter-seedling-crops/).

24.5.3 Impact of natural enemies and other biological control agents

The impact of natural enemies at the critical emergence period in *B. napus* is still unclear. A large diversity of natural enemy species is present in or near *B. napus* fields at this time of the year but their impact on emergence pest populations and the associated reduction in pest outbreak risk has not been quantified (Macfadyen *et al.*, 2015; Zalucki *et al.*, 2015). A significant amount of early research was conducted on the potential of predatory mites to control these pests (Michael *et al.*, 1991; Otto and Halliday, 1991; Michael, 1995) and more recently some work has been conducted on pesticide impacts on predatory mites (Roberts *et al.*, 2011b). However, there are many other predatory groups, including lacewings and ladybird beetles (James, 1995), that might also prove useful for controlling emergence pest populations early in the season. Observation of direct feeding interactions between pests and predators can be difficult but there is a range of new tools for detection of prey DNA in the guts of predators that could be used. For example, Layman and Lundgren (2015) used molecular techniques and gut content analysis to document the complex interactions between pest and non-pest herbivores and shared predators in *B. napus* fields. Furthermore, the use of pathogens for the biological control of emergence pests has not progressed since it was initially identified in the early 1990s (Christian *et al.*, 1991; Rath, 1991). An ecological approach considering the assemblage of natural enemies that are part of the *B. napus* ecological community, and provide a significant reduction in pest numbers, is required. This will allow for determining if highly efficient individual species or a range of different species with different functional roles is required for pest suppression. This could be aided by knowledge generated from co-occurrence studies and predictive models.

24.5.4 Host plant resistance

Brassica crops display differences in susceptibilities to damage by insect pests and this could perhaps be exploited to develop varieties that are more tolerant of emergence pests (Gu *et al.*, 2008; Shaw *et al.*, 2009; Karimi *et al.*, 2012). There is a large number of potential *B. napus* varieties that could be used for screening in Australia but to date little research has been conducted in this area (Gu *et al.*, 2008). McDonald *et al.* (1995) screened some northern hemisphere *B. napus* varieties and found that they were more prone to cotyledon damage from *H. destructor*, in comparison with local varieties. Furthermore, other pests, such as cotton bollworm, *Helicoverpa armigera* (Karimi *et al.*, 2012), and diamondback moth, *Plutella xylostella* (Fathi *et al.*, 2010), display differences in development time and fecundity across different *B. napus* varieties documented in overseas studies. It is known that competitive interactions between earth mite species occur and vary by the host plant. Umina and Hoffmann (2005) showed that, on pasture, the competitive advantage swayed between *P. major*, *H. destructor* and *P. falcatus* but that *P. falcatus* and *H. destructor* were consistently superior competitors on *B. napus*. A better understanding of the impact of host-plant and varietal differences on population dynamics of emergence pests could underpin management strategies for novel ways of controlling these species.

24.6 Conclusions

Whilst the focus of this chapter has been on *B. napus* as an isolated crop, *B. napus* often sits in a farming landscape that contains many crop types, pastures and semi-natural vegetation patches. The spatial and temporal spread of host plant resources for pest mite and collembolan species varies across the year. Bottom-up processes related to host plant preferences can be used for better management of these pests. Likewise, top-down processes associated with natural enemies and other biological control agents hold much promise for more effective and sustainable pest control. Importantly, thinking about how pesticides can be combined with cultural and biological control options is the key to better management of these pests in the future.

There has been some research conducted on the basic biology and ecology of these pests but there is still a lack of key information that is useful for developing more sustainable control practices. We have highlighted that our understanding of the impact of natural enemies on emergence pests in *B. napus* is unclear. Both the tolerance of natural enemy species to commonly used pesticides and their role in the frequency or likelihood of secondary pest outbreaks need to be further quantified. We have shown that new tools such as predictive models, simple monitoring protocols and better use of host plant specificities could be used to improve management in an IPM context. Currently, only a very small suite of pesticides is being used to manage these pests and their efficacy is being challenged by the development of resistance.

Acknowledgements

The authors' research is funded by the Grains Research and Development Corporation, Canberra, Australia.

References

Arthur, A.L., Hoffmann, A.A., Umina, P.A. and Weeks, A.R. (2008) Emerging pest mites of grains (*Balaustium medicagoense* and *Bryobia* sp.) show high levels of tolerance to currently registered pesticides. *Australian Journal of Experimental Agriculture* 48, 1126–1132.

Arthur, A.L., Weeks, A.R., Umina, P.A. and Hoffmann, A.A. (2010) Survival and reproduction of the pest mites *Balaustium medicagoense* and *Bryobia* spp. on winter grain crops. *Experimental and Applied Acarology* 52, 141–53.

Arthur, A.L., Hoffmann, A.A. and Umina, P.A. (2013a) Impact of *Halotydeus destructor* on crop seedlings at different plant developmental stages and levels of moisture stress. *Environmental Entomology* 42, 998–1012.

Arthur, A.L., McColl, S.A. and Umina, P.A. (2013b) Synthetic pesticides show little efficacy against *Balaustium medicagoense* (Prostigmata: Erythraeidae) in the field compared with other pest mites. *Australian Journal of Entomology* 52, 259–267.

Arthur, A.L., Hoffmann, A.A. and Umina, P.A. (2014) Estimating densities of the pest *Halotydeus destructor* (Acari: Penthaleidae) in canola. *Journal of Economic Entomology* 107, 2204–2212.

Arthur, A.L., Hoffmann, A.A. and Umina, P.A. (2015) Challenges in devising economic spray thresholds for a major pest of Australian canola, the redlegged earth mite (*Halotydeus destructor*). *Pest Management Science* 71, 1462–1470.

Australian Government (2008) *The Biology of Brassica napus L. (Canola)*. Office of the Gene Technology Regulator, Canberra, Australia.

AWI (2016) TIMERITE. Australian Wool Innovations Ltd, Sydney, Australia. Available at: http://www.wool.com/en/woolgrower-tools/timerite

Bailey, P.T. (2007) *Pests of Field Crops and Pastures: Identification and Control*. CSIRO Publishing, Collingwood, Australia.

Burnett, V.F., Schneider, J.H., Dempsey, F.W. and Rohitha, H.B. (2003) Effect of seeding rate on organic canola emergence, red legged earthmite (*Halyotydeus destructor*) density, weed biomass and grain yield in Northeast Victoria. *Proceedings of the 11th Australian Agronomy Conference, Solutions for a Better* Environment. Australian Society of Agronomy, Geelong, Victoria.

Christian, P.D., Dall, D.J. and Oakeshott, J.G. (1991) The potential of entomopathogens for the control of mites and springtails in Australian pastures. *Plant Protection Quarterly* 6, 175–177.

Dent, D. (1991) *Insect Pest Management*. CABI, Wallingford, UK.

Dosdall, L.M., Zalucki, M.P., Tansey, J.A. and Furlong, M.J. (2012) Developmental responses of the diamondback moth parasitoid *Diadegma semiclausum* (Hellen) (Hymenoptera: Ichneumonidae) to temperature and host plant species. *Bulletin of Entomological Research* 102, 373–384.

Fathi, S.A.A., Bozorg-Amirkalaee, M. and Sarfaraz, R.M. (2010) Preference and performance of *Plutella xylostella* (L.) (Lepidoptera: Plutellidae) on canola cultivars. *Journal of Pest Science* 84, 41–47.

Gower, J.M.C., Hoffmann, A.A. and Weeks, A.R. (2008) Effectiveness of spring spraying targeting diapause egg production for controlling redlegged earth mites and other pests in pasture. *Australian Journal of Experimental Agriculture* 48, 1118–1125.

Gu, H., Fitt, G.P. and Baker, G.H. (2007) Invertebrate pests of canola and their management in Australia: a review. *Australian Journal of Entomology* 46, 231–243.

Gu, H., Edwards, O.R., Hardy, A.T. and Fitt, G.P. (2008) Host plant resistance in grain crops and prospects for invertebrate pest management in Australia: an overview. *Australian Journal of Experimental Agriculture* 48, 1543–1548.

Halliday, R.B. (1991) Taxonomic background of the redlegged earth mite *Halotydeus destructor* Tucker Acarina: Penthalidae. *Plant Protection Quarterly* 6, 162–165.

Hill, M.P., Hoffmann, A.A., Macfadyen, S., Umina, P.A. and Elith, J. (2012a) Understanding niche shifts: using current and historical data to model the invasive redlegged earth mite, *Halotydeus destructor*. *Diversity and Distributions* 18, 191–203.

Hill, M.P., Hoffmann, A.A., McColl, S.A. and Umina, P.A. (2012b) Distribution of cryptic blue oat mite species in Australia: current and future climate conditions. *Agricultural and Forest Entomology* 14, 127–137.

Hill, M.P., Chown, S.L. and Hoffmann, A.A. (2013) A predicted niche shift corresponds with increased thermal resistance in an invasive mite, *Halotydeus destructor*. *Global Ecology and Biogeography* 22, 942–951.

Hill, M.P., Hoffmann, A.A., Umina, P.A., Cheng, X., Miller, A.D. and Austin, J. (2016) Genetic analysis along an invasion pathway reveals endemic cryptic taxa, but a single species with little population structure in the introduced range. *Diversity and Distributions* 22, 57–72.

Hoffmann, A.A., Weeks, A.R., Nash, M.A., Mangano, G.P. and Umina, P.A. (2008) The changing status of invertebrate pests and the future of pest management in the Australian grains industry. *Australian Journal of Experimental Agriculture* 48, 1481–1493.

Horne, P. and Page, J. (2008) *Integrated Pest Management for Crops and Pastures*. Landlinks Press, Melbourne, Australia.

Ireson, J.E., Holloway, R.J., Chatterton, W.S. and Mccorkell, B.E. (2002) Further investigations into the efficacy of *Neomolgus capillatus* (Kramer) (Acarina: Bdellidae) as a predator of *Sminthurus viridis* (L.) (Collembola: Sminthuridae) in Tasmania. *Australian Journal of Entomology* 41, 88–93.

James, D.G. (1995) Biological control of earth mites in pasture using endemic natural enemies. *Plant Protection Quarterly* 10, 58–59.

Jenkins, S., Hoffmann, A.A., McColl, S., Tsitsilas, A. and Umina, P.A. (2013) Synthetic pesticides in agro-ecosystems: are they as detrimental to nontarget invertebrate fauna as we suspect? *Journal of Economic Entomology* 106, 756–775.

Karimi, S., Fathipour, Y., Talebi, A.A. and Naseri, B. (2012) Evaluation of canola cultivars for resistance to *Helicoverpa armigera* (Lepidoptera: Noctuidae) using demographic parameters. *Journal of Economic Entomology* 105, 2172–2179.

Kirkegaard, J.A., Lilley, J.M. and Morrison, M.J. (2016) Drivers of trends in Australian canola productivity and future prospects. *Crop and Pasture Science* 67(3), i–ix.

Krischik, V., Rogers, M., Gupta, G. and Varshney, A. (2015) Soil-applied imidacloprid translocates to ornamental flowers and reduces survival of adult *Coleomegilla maculata*, *Harmonia axyridis*, and *Hippodamia convergens* lady beetles, and larval *Danaus plexippus* and *Vanessa cardui* butterflies. *PLoS ONE* 10, e0119133.

Layman, M.L. and Lundgren, J.G. (2015) Mutualistic and antagonistic trophic interactions in canola: The role of aphids in shaping pest and predator populations. *Biological Control* 91, 62–70.

Macfadyen, S., Hardie, D.C., Fagan, L., Stefanova, K., Perry, K.D., Degraaf, H.E., Holloway, J., Spafford, H. and Umina, P.A. (2014) Reducing insecticide use in broad-acre grains production: an Australian study. *PLoS ONE* 9, e89119.

Macfadyen, S., Davies, A.P. and Zalucki, M.P. (2015) Assessing the impact of arthropod natural enemies on crop pests at the field scale. *Insect Science* 22, 20–34.

McDonald, G., Moritz, K., Merton, E. and Hoffmann, A.A. (1995) The biology and behaviour of redlegged earth mite and blue oat mite on crop plants. *Plant Protection Quarterly* 10, 52–54.

McDonald, G., Umina, P.A., Macfadyen, S., Mangano, P. and Hoffmann, A.A. (2015) Predicting the timing of first generation egg hatch for the pest redlegged earth mite *Halotydeus destructor* (Acari: Penthaleidae). *Experimental and Applied Acarology* 65, 259–276.

Merton, E., McDonald, G. and Hoffmann, A. (1995) Cultural control of redlegged earth mite, blue oat mite and lucerne flea in canola. *Plant Protection Quarterly* 10, 65–66.

Michael, P. (1995) Biological control of redlegged earth mite and lucerne flea by the predators *Anystis wallacei* and *Neomologus capillatus* *Plant Protection Quarterly* 10, 55–57.

Michael, P.J., Dutch, M.E. and Pekin, C.J. (1991) A review of the predators of redlegged earth mite, blue oat mite and lucerne flea. *Plant Protection Quarterly* 6, 178–180.

Micic, S., Hoffmann, A.A., Strickland, G., Weeks, A.R., Bellati, J., Henry, K., Nash, M.A. and Umina, P.A. (2008) Pests of germinating grain crops in southern Australia: an overview of their biology and management options. *Australian Journal of Experimental Agriculture* 48, 1560–1573.

Miles, M. and McDonald, G. (1999) Insect pests of canola. In: Salisbury, P.A., Potter, T.D., McDonald, G. and Green, A.G. (eds) *Canola in Australia: The First Thirty Years*. The Canola Association of Australia Inc., Young, NSW, Australia, pp. 53–58.

Murray, D.A.H., Clarke, M.B. and Ronning, D.A. (2013a) *The Current and Potential Costs of Invertebrate Pests in Grain Crops*. Grains Research and Development Corporation, Canberra, Australia.

Murray, D.A.H., Clarke, M.B. and Ronning, D.A. (2013b) Estimating invertebrate pest losses in six major Australian grain crops. *Australian Journal of Entomology* 52, 227–241.

Nilsson, C., Büchs, W., Klukowski, Z., Luik, A., Ulber, B. and Williams, I.H. (2015) Integrated crop and pest management of winter oilseed rape (*Brassica napus* L.). *Zemdirbyste-Agriculture* 102, 325–334.

Otto, J.C. and Halliday, R.B. (1991) Systematics and biology of a predatory mite (*Anystis* sp.) introduced into Australia for biological control of redlegged earth mite. *Plant Protection Quarterly* 6, 181–185.

Qin, T.K. and Halliday, R.B. (1995) Systematic studies of redlegged earth mite *Halotydeus destructor* (Tucker) and related species (Acarina: Penthaleidae). *Plant Protection Quarterly* 10, 50–52.

Rath, A.C. (1991) Pathogens for the biological control of mite and collembolan pests. *Plant Protection Quarterly* 6, 172–174.

Ridsdill-Smith, T.J. and Pavri, C.C. (2015) Controlling redlegged earth mite, *Halotydeus destructor* (Acari: Penthaleidae), with a spring spray in legume pastures. *Crop and Pasture Science* 66, 938–946.

Ridsdill-Smith, J., Pavri, C., De Boer, E. and Kriticos, D. (2005) Predictions of summer diapause in the redlegged earth mite, *Halotydeus destructor* (Acari:

Penthaleidae), in Australia. *Journal of Insect Physiology* 51, 717–726.

Ridsdill-Smith, T.J., Hoffmann, A.A., Mangano, G.P., Gower, J.M., Pavri, C.C. and Umina, P.A. (2008) Strategies for control of the redlegged earth mite in Australia. *Australian Journal of Experimental Agriculture* 48, 1506–1513.

Roberts, J.M. and Weeks, A.R. (2011) Genetic structure and long-distance dispersal in populations of the wingless pest springtail, *Sminthurus viridis* (Collembola: Sminthuridae). *Genetics Research (Camb.)* 93, 1–12.

Roberts, J.M.K., Umina, P.A., Hoffmann, A.A. and Weeks, A.R. (2009) The tolerance of the lucerne flea, *Sminthurus viridis* (Collembola: Sminthuridae), to currently registered pesticides in Australia. *Australian Journal of Entomology* 48, 241–246.

Roberts, J.M.K., Umina, P.A., Hoffmann, A.A. and Weeks, A.R. (2011a) Population dynamics and diapause response of the springtail pest *Sminthurus viridis* (Collembola: Sminthuridae) in southeastern Australia. *Journal of Economic Entomology* 104, 465–473.

Roberts, J.M.K., Weeks, A.R., Hoffmann, A.A. and Umina, P.A. (2011b) Does *Bdellodes lapidaria* (Acari: Bdellidae) have a role in biological control of the springtail pest, *Sminthurus viridis* (Collembola: Sminthuridae) in south-eastern Australia? *Biological Control* 58, 222–229.

Seidel, J.E. (1999) A new approach to the control of earth mites using Gaucho seed dressing. In: Matthiessen, J.N. (ed.) *Proceedings of the 7th Australasian Conference on Grassland Invertebrate Ecology*. CSIRO, Wembley, Australia, pp. 132–136.

Shaw, E.J., Fletcher, R.S., Dosdall, L.L. and Kott, L.S. (2009) Biochemical markers for cabbage seedpod weevil (*Ceutorhynchus obstrictus* (Marsham)) resistance in canola (*Brassica napus* L.). *Euphytica* 170, 297–308.

Tsitsilas, A., Hoffmann, A.A., Weeks, A.R. and Umina, P.A. (2011) Impact of groundcover manipulations within windbreaks on mite pests and their natural enemies. *Australian Journal of Entomology* 50, 37–47.

Umina, P.A. and Hoffmann, A.A. (2004) Plant host associations of *Penthaleus* species and *Halotydeus destructor* (Acari: Penthaleidae) and implications for integrated pest management. *Experimental and Applied Acarology* 33(1-2), 1–20.

Umina, P.A. and Hoffmann, A.A. (2005) Competitive interactions among four pest species of earth mites (Acari: Penthaleidae). *Journal of Economic Entomology* 98, 307–316.

Umina, P.A., Weeks, A.R., Roberts, J., Jenkins, S., Mangano, G.P., Lord, A. and Micic, S. (2012) The current status of pesticide resistance in Australian populations of the redlegged earth mite (*Halotydeus destructor*). *Pest Management Science* 68, 889–896.

Umina, P.A., Edwards, O., Carson, P., Van Rooyen, A. and Anderson, A. (2014) High levels of resistance to carbamate and pyrethroid chemicals widespread in Australian *Myzus persicae* (Hemiptera: Aphididae) populations. *Journal of Economic Entomology* 107, 1626–1638.

Umina, P.A., Jenkins, S., McColl, S., Arthur, A. and Hoffmann, A.A. (2015) A framework for identifying selective chemical applications for IPM in dryland agriculture. *Insects* 6, 988–1012.

Van de Wouw, A.P., Marcroft, S.J. and Howlett, B.J. (2016) Blackleg disease of canola in Australia. *Crop and Pasture Science* 67(4), 273–283.

Wallace, M.M.H. (1967) The ecology of *Sminthurus viridis* (L.) (Collembola). I. Processes influencing numbers in pastures in Western Australia. *Australian Journal of Zoology* 15, 1173–1206.

Wallace, M.M.H. (1968) The ecology of *Sminthurus viridis* (L.) (Collembola). II. Diapause in the aestivating egg. *Australian Journal of Zoology* 16, 871–883.

Wallace, M.M.H. and Mahon, J.A. (1971a) The distribution of *Halotydeus destructor* and *Penthaleus major* (Acari: Eupodidae) in Australia in relation to climate and land use. *Australian Journal of Zoology* 19, 65–76.

Wallace, M.M.H. and Mahon, J.A. (1971b) The ecology of *Sminthurus viridis* (Collembola). III. The influence of climate and land use on its distribution and that of an important predator, *Bdellodes lapidaria* (Acari: Bdellidae). *Australian Journal of Zoology* 19, 177–188.

Williams, I.H. (2010) The major insect pests of oilseed rape in Europe and their management: an overview. In: Williams, I.H. (ed.) *Biocontrol-Based Integrated Management of Oilseed Rape Pests*. Springer Science, Dordrecht, The Netherlands, pp. 1–43.

Wilson, L.J., Bauer, L.R. and Lally, D.A. (1998) Effect of early season insecticide use on predators and outbreaks of spider mites (Acari: Tetranychidae) in cotton. *Bulletin of Entomological Research* 88, 477–488.

Zalucki, M.P., Furlong, M.J., Schellhorn, N.A., Macfadyen, S. and Davies, A.P. (2015) Assessing the impact of natural enemies in agroecosystems: toward “real” IPM or in quest of the Holy Grail? *Insect Science* 22, 1–5.

25 Use of Entomopathogenic Fungi in the Insect Pest Management of *Brassica* Oilseed Crops

Heikki M.T. Hokkanen* and Ingeborg Menzler-Hokkanen

University of Helsinki, Finland

25.1 Introduction

Insect pathogenic fungi offer excellent opportunities for effective pest management in oilseed *Brassica* crops. Before this can happen, several ecological, technical and socio-economic conditions must be made conducive for the pathogens to exert their full potential. This chapter focuses on the potential and constraints of employing entomopathogenic fungi (EPF) for pest management in oilseed *Brassica* crops (OSB). The taxonomy of many important groups of EPF is under revision, using molecular tools (e.g. Kepler *et al.*, 2014) and resulting in renaming of many taxa known in the literature under different name(s). As the purpose of this review is not to sort out the nomenclature, no attempt will be made to identify the current names for taxa mentioned in older literature; instead, we use the names as given in the source publications.

The pest insect fauna on OSB crops is quite different in different continents. For example, six of the eight key insect pest species on OSB in Europe are coleopterans and the remaining two are dipterans (Menzler-Hokkanen *et al.*, 2006). In contrast, the pest complex in North America consists of coleopteran flea beetles (similar to those in Europe) and several lepidopteran, hemipteran and dipteran species, which usually are not a problem in Europe. Two pests, the *Phyllotreta* flea beetles and the diamondback moth (DBM) *Plutella xylostella*, are common to both continents.

25.2 Entomopathogenic Fungi for Key Oilseed *Brassica* Pests

Known pathogens of key OSB pests are mainly generalist entomopathogens. Among these, only entomopathogenic nematodes (EPN) and EPF are widely distributed (Hokkanen *et al.*, 2003a).

The EPF attacking coleopterans are usually facultative pathogens but those attacking dipterans and the DBM include also obligate insect pathogens. All of these are widely distributed in soils throughout the world and occur throughout the OSB crops. Several studies have been published concerning their distribution in America, Australia and Europe (e.g. St Leger *et al.*, 1992; Roberts and St Leger, 2004), often to find general relationships of these pathogens with the soil type, preferences of habitat type and other abiotic and biotic factors affecting their abundance and role as insect antagonists (Vänninen, 1996). However, their reported natural abundances in agricultural fields have always been low (e.g. Vänninen *et al.*, 1989).

The most thoroughly studied EPF for potential control of OSB pests include *Beauveria bassiana* (Balsamo) Vuillemin and *Metarhizium anisopliae* (Metschnikoff) Sorokin (both Hypocreales: Clavicipitaceae). Both taxa have repeatedly been shown to kill effectively the adults and larvae of OSB pests (e.g. Butt *et al.*, 1994; Husberg and Hokkanen, 2001; H.M.T. Hokkanen *et al.*, unpublished). For example, spray treatment in the field with *M. anisopliae* caused 75% mortality in pollen

*Corresponding author. E-mail: heikki.hokkanen@helsinki.fi

beetle larvae (H.M.T. Hokkanen *et al.*, unpublished) (Fig. 25.1). Despite the apparent potential, field treatments have not reduced the new-generation pollen beetle numbers, either via soil treatment or via fungal sprays (Hokkanen, 1993), but soil treatments with EPF have caused dramatic indirect effects via increased overwintering mortality. Soil treatment with *B. bassiana* reduced overwintering survival of beetles developed under unlimited food resources by 50% (from 14% to 7%), compared with the untreated control. The weight (fat reserves) of the beetles is a crucial factor explaining overwintering mortality of pollen beetle *Meligethes aeneus* (Hokkanen, 1993). Soil treatment with *B. bassiana* lowered the weight of pollen beetles developed under unlimited food resources by 16%, which alone is likely to be enough to result in the measured reduction in survival by 50%. However, beetles developing under normal conditions (collected from the wild) and under shortage of autumn food sources (pollen plants) were 46% lighter than the experimental reference beetles before overwintering. This lighter weight resulted in only 3% survival over the winter (Hokkanen, 1993). If it is assumed that all OSB fields would contain effective levels of EPF either naturally or via augmentation, the pollen beetle populations would be expected to crash completely during overwintering (survival would be reduced from 3% to maybe around 1%).

Meligethes spp. also have a specific EPF, *Nosema meligethi* I. &R. (Dissociodihaplophasida: Nosematidae), previously classified as Protozoa. *N. meligethi* is an intracellular obligate parasite

Fig. 25.1. Pollen beetle *Meligethes aeneus* adult, killed by the entomopathogenic fungus *Metarhizium anisopliae*. Photo by H. Hokkanen.

specific to the genus *Meligethes*. It is an 'ideal' insect pathogen from a population dynamics point of view: it causes chronic disease with lowered fecundity and lifespan, and high overwintering mortality, and is transmitted both horizontally and vertically (Hokkanen and Lipa, 1995). Lipa and Hokkanen (1992) studied the occurrence of *N. meligethi* across Europe. Samples of *Meligethes* spp. from 12 countries in Eastern and Western Europe were inspected for the occurrence of infection. A total of 13,910 individual beetles were checked, of which 561 were found to be infected (4.03%). Infection was not detectable in samples from the UK, Germany, Sweden or Switzerland. One infected beetle was found among 444 beetles inspected from Austria and one from Denmark (1/85). In contrast, the disease was found rather regularly in samples from Finland and from Eastern European countries. No obvious reason for this pattern of infection is known but pesticide usage at or close to the sampling sites may play a role, because most samples from which *Nosema* was detected originated from areas where few pesticides are used.

Nosema infection was also shown to lower pollen beetle weights on the average by 13% – similar to the effect of soil treatment with *B. bassiana* – which is enough to explain the reported increases in the overwintering mortality of the beetles due to *Nosema* infection (Hokkanen and Lipa, 1995).

Many other coleopteran pests of OSB have also been reported to be susceptible to EPF. Reddy *et al.* (2014) reported EPF induced mortality in the flea beetle *Phyllotreta cruciferae* (Goeze), caused by *B. bassiana* and *Metarhizium brunneum*; also other species of *Phyllotreta* on OSB in Europe are infected by these generalist EPF taxa (H.M.T. Hokkanen *et al.*, unpublished). Carreck *et al.* (2007) showed that adults of the cabbage seed weevil *Ceutorhynchus assimilis* are susceptible to *M. anisopliae* under controlled field conditions. Butt *et al.* (1994) had already reported on the susceptibility of the cabbage stem flea beetle *Psylliodes chrysocephala* and three other OSB pests (*Phaedon cochleariae*, *Myzus persicae* and *Lipaphis erysimi*) to several isolates of *B. bassiana* and *M. anisopliae*. In that study, flea beetle mortality rates increased with dose; the estimated LC_{50} of the most virulent strains at 14 days post-inoculation were 10^6 conidia/ml. The respective estimated LT_{50} values of the best two strains for *Psylliodes chrysocephala* at 1×10^7 conidia/ml were 10.0 and 9.3 days, and at 1×10^{10} conidia/ml were 3.8 and 4.2 days. Both strains were pathogenic to *Phaedon cochleariae*, *M. persicae* and *L. erysimi*. *M. persicae* and *L. erysimi* died within 4 days of inoculation.

The mustard aphid *L. erysimi* is reported to be a key production limiting pest on OSB in Pakistan (Ujjan and Shahzad, 2012). The effect of topical treatment of the aphids in laboratory and screen-house bioassays was studied using four species and several strains of the entomopathogenic fungi *Lecanicillium lecanii*, *Paecilomyces lilacinus*, *B. bassiana* and *M. anisopliae*, at various spore concentrations. There was a high variability between the efficacy of the different strains but all species contained strains causing 70–100% mortality of the aphids under the test conditions. The authors conclude that these strains have potential for use as biopesticides against the mustard aphid under field conditions (Ujjan and Shahzad, 2012).

The diamondback moth *Plutella xylostella* has been a target for many studies with EPF. Duarte *et al.* (2016) tested the pathogenicity of isolates of *B. bassiana*, *Metarhizium rileyi*, *Isaria fumosorosea*, *Isaria sinclairii* and *Lecanicillium muscarium* standardized at a concentration of 10^7 conidia/ml, towards second-instar larvae of *P. xylostella*. Several isolates of *B. bassiana* and single isolates of *M. rileyi* and of *I. sinclairii* caused mortality between 80% and 100%, with LC_{50} and LT_{50} from 2.504 to 6.775×10^4 conidia/ml and 52.22 to 112.13 h, respectively. The authors suggested that the use of these isolates could be an important alternative in the management of *P. xylostella*.

For the control of the DBM, the naturally occurring fungal entomopathogen *Zoophthora radicans* (Zygomycetes: Entomophthorales) has been shown to be a potential biological control agent (Pell *et al.*, 1993; Furlong *et al.*, 1995). *Z. radicans* kills infected *P. xylostella* larvae and adults within 3–4 days of infection. Furlong *et al.* (1997) showed that the pre-lethal effects of *Z. radicans* infection in *P. xylostella*, although not immediately apparent, can be considerable, with overall feeding and oviposition rates being reduced by as much as 44% and 51%, respectively. Furthermore, Reddy *et al.* (1998) showed that *Z. radicans* infection inhibits the response to, and production of, the sex pheromone in the DBM. The authors concluded that the ability of *Z. radicans* to induce such reductions in larval voracity and adult fecundity and physiology must have an important role to play in developing pest management systems, into which *Z. radicans* can be integrated.

25.3 The Bioinsecticide Approach

In the laboratory, topical application of EPF by spraying or by pipetting on the target insect is a usual method of testing the efficacy of the fungal preparations. Spraying EPF as mycoinsecticides for the control of OSB pests in the open field has rarely been tried, probably due to poor availability of suitable products, expected lack of efficacy, or the logistics of the farm operations. Butt and Copping (2000) discussed some of the constraints. Several pest species (e.g. *Psylliodes chrysocephala*, *Delia radicum*, *Ceutorhynchus* spp.) are difficult to reach by spraying and treatment operations against other pests during full bloom are technically challenging. However, good reduction of the target pest has been obtained when EPF sprays have been tried. For example, spray treatment in the field with *M. anisopliae* caused 75% mortality in pollen beetle larvae (H.M.T. Hokkanen *et al.*, unpublished). Reduction of pollen beetle larvae on the inflorescences may not be interesting from a practical crop protection point of view, because the main damage to the crop has already been done by the adults and even a high reduction in the size of the larval population is not likely to lower the population of the new generation emerging during the summer (Hokkanen, 2000). *Metarhizium* sprays could possibly be used to kill adult OSB pest beetles in the spring via fungal sprays at the bud stage early in the season; the best option might be to grow and treat an early-flowering trap crop (e.g. turnip rape) with the EPF (H.M.T. Hokkanen and S.M. Cook, unpublished).

Recently, encouraging results have been obtained in the USA in protecting OSB crop against flea beetle injury by using EPF. Reddy *et al.* (2014) evaluated the effectiveness of several biopesticide spray treatments for the control of *P. cruciferae* under field conditions and compared their efficacy with two insecticides (deltamethrin and bifenthrin), commonly used as foliar sprays, as well as seed treatment with an imidacloprid insecticide. The biopesticides included two EPF (*B. bassiana* and *M. brunneum*), as well as the biorational product neem, and petroleum spray oils. The control agents were delivered in combination, or alone in a single or repeated applications at different times. Combined use of *B. bassiana* and *M. brunneum* in two repeated applications, and bifenthrin in five applications, were most effective in reducing feeding injuries and improving yield levels at both trial locations. Reddy *et al.* (2014) considered that EPF are effective against *P. cruciferae* and may serve as alternatives to conventional insecticides or seed treatments in managing this pest.

25.4 Autodissemination of Fungal Insect Pathogens

Autodissemination involves attracting pest insects into artificial devices, where they are contaminated with infective conidia of EPF. When the insects return to the crop, they will disseminate the pathogen amongst their own population (Vega *et al.*, 2000; Vickers *et al.*, 2004). Autodissemination has been studied for many insect and fungal species, including DBM and the *Delia* flies attacking OSB (Meadow *et al.*, 2000; Vickers *et al.*, 2004). According to Baverstock *et al.* (2010), the method has several advantages over the spray application of EPF, including: (i) it is highly efficient, especially when incorporating a target specific pheromone; (ii) it can be species specific; (iii) dissemination devices are simple to construct and maintain; (iv) it is cost effective, as the ratio of fungal inoculum to hosts is low; and (v) contaminated insects will return to their habitats, therefore dispersing the pathogen (Vega *et al.*, 2000). Vickers *et al.* (2004) listed an additional benefit: (vi) inside the inoculum chamber the fungus can be protected from the damaging effects of ultraviolet radiation and provided with an abiotic environment that favours sporulation and infection.

Of the OSB pests, research on autodissemination has mostly targeted the DBM. A laboratory study examined the ecological attributes of two EPF, *Z. radicans* and *B. bassiana*, and evaluated their relative suitability for inclusion in autodissemination strategies against *P. xylostella* (Furlong and Pell, 2001). It was shown that transmission of *Z. radicans* infection from contaminated male moths to adults and larvae was much lower than the transmission of *B. bassiana*. However, *B. bassiana* was not able to initiate epizootics in *P. xylostella*, while *Z. radicans* showed significant potential due to a high level of secondary transmission from moths that had previously died of *Z. radicans* infection (Furlong and Pell, 2001).

The potential for using artificial sex pheromones, compared with pheromones released from virgin females, to lure adult male DBM into dissemination traps to be contaminated with *Z. radicans*, was assessed by Pell *et al.* (1993) and Furlong *et al.* (1995). While males entered traps containing virgin female moths only between dusk and dawn (times when the sex pheromone is naturally released), males entered traps containing synthetic pheromone throughout the day. Males spent a mean of 88 seconds within the traps before leaving and were contaminated in that time with a lethal dose of *Z. radicans*, which they subsequently were able to transmit to the DBM population. Proof for this concept of autodissemination was obtained in another study, where *P. xylostella* adults were inoculated with *Z. radicans* and released into field cages containing plants infested with *P. xylostella* larvae (Vickers *et al.*, 2004). After 6 days, 79% of the larvae in the cage were found to be infected with *Z. radicans*.

Although the technique proved successful in a field cage environment, field-level tests or applications of this technique have apparently not been reported. Vickers *et al.* (2004) considered that further studies would be required concerning the inoculation chamber design and placement, choice of pathogen species and isolate, replacement times (persistence) for the pathogen in the inoculator and the possible effects on other species. They further considered that for autodissemination to be a useful tool in IPM, a practical and commercially viable means of producing and storing inoculum would have to be developed. The technique should also be effective when population densities are low, such as at the beginning of the growing season when small numbers of adult males are available for infection and therefore can only provide few sources of conidia to infect DBM larvae.

25.5 Entomovectoring for Targeted Delivery of Biopesticides

Targeted precision delivery of microbiological control agents such as EPF has been developed in the past two or three decades to improve the practical and economic efficiency of the use of biocontrol agents. Usually, managed pollinating insects such as honey bees and bumblebees are employed as vectors, hence the term 'entomovector technology' (first used by Hokkanen and Menzler-Hokkanen, 2007). The vectoring insects are loaded with the powdery biocontrol preparate at the exit from their hive and while visiting the flowers of the target crop they deliver the biocontrol agent on to the crop inflorescences (Mommaerts and Smagghe, 2011; Hokkanen *et al.*, 2015). As many OSB pests reside in or at least visit the open flowers, they are good targets for entomovectoring, because OSB flowers are highly attractive also to pollinators. These opportunities have been explored during the past 20 years.

Butt *et al.* (1998) reported that honey bees foraging from a hive fitted with an inoculum dispenser at the entrance effectively delivered dry conidia of *M. anisopliae* to the flowers of oilseed rape in caged field plots. In both winter- and spring-sown rape experiments, a greater mortality of pollen beetles occurred in treated plots than in control plots. The mortality (61% on winter rape, 100% on spring rape) was greatest during peak flowering, when the feeding activity of both bees and beetles from the flowers was maximal, providing optimal conditions for inoculum dissemination and infection. Conidial sporulation occurred on a significant proportion of the dead pollen beetles.

Carreck *et al.* (2007) show in a later study that *M. anisopliae*, disseminated to the flowering canopy of oilseed rape by the honey bee, infects seed weevil adults and pollen beetle adults and larvae in the flowers. Subsequent mortality of adults of both species and the larvae of pollen beetle, following incubation in the laboratory, was high (always exceeding 70%) and a high proportion of cadavers showed external conidiation, confirming infection by *M. anisopliae* (Fig. 25.1).

Experiences so far show that bees are more efficient than conventional sprayers in delivering the EPF inoculum to the pest-infested flowers (Butt and Copping, 2000). Entomovectoring improves targeting and provides cost-effective control of pests (Hokkanen *et al.*, 2015) and can also be adopted for improving pest management in OSB.

Entomovectoring may be most appropriate in OSB for dissemination to part of the crop into which the pests are concentrated. In particular, it can be utilized in connection with an early-flowering trap crop (Hokkanen, 1991). Field experiments have indicated excellent potential for employing *Brassica rapa* as a trap crop for *Brassica napus*, as it clearly flowers earlier and is highly attractive to OSB pests as well as to bees (S.M. Cook and H.M.T. Hokkanen, unpublished). Additionally, to encourage pests into the trap crop, lures such as chemical attractants in the trap crop (sex pheromones, gustatory stimulants) (Butt and Copping, 2000), or repellents in the main crop (Mauchline *et al.*, 2013), could be used. As yet, very few inexpensive but effective lures or deterrents have been developed for commercial use.

Entomovectoring has consistently been reported to be safe for the vectoring bee colonies (Butt *et al.*, 1994; Hokkanen *et al.*, 2015; Menzler-Hokkanen, 2016), while individual bees, in particular bumblebees, may become infected if applied to vector entomopathogenic fungi (Hokkanen *et al.*, 2003b).

25.6 Endophytic Entomopathogenic Fungi

Vidal and Jaber (2015) recently reviewed the data on the ability of EPF to colonize and grow endophytically in plants and presented results of their study with 14 isolates or strains of *B. bassiana* on oilseed rape and faba bean.

Published data show that most, if not all, entomopathogenic fungi are able to colonize tissues of at least some plant species as symptomless endophytes. Most data so far refer to *B. bassiana* as an endophyte, but also *M. anisopliae* and *L. lecanii* have been shown to colonize plant tissues. Depending on the plant species and the isolate of the EPF, these interactions could be beneficial to both the plant and the fungus, neutral or even antagonistic (Vidal and Jaber, 2015). Data published in this respect are highly variable and not consistent with regard to the underlying mechanisms that might explain the effects.

Vidal and Jaber (2015) evaluated the potential of *B. bassiana* isolates to grow endophytically in oilseed rape plants, using 14 different isolates originating from different herbivore insects in different regions of the world. Of the 14 screened isolates, 12 were able to colonize inoculated leaves of oilseed rape. Colonization significantly varied among the screened isolates, so that colonization of *B. napus* by *B. bassiana* was significantly higher with three of the strains (including the commercial formulation Naturalis®) in contrast to four other isolates. Two isolates colonized *B. napus* plants more consistently than *Vicia faba* plants, while two other isolates colonized *V. faba* plants better than *B. napus* plants.

The first report demonstrating the endophytic colonization of oilseed rape *B. napus* plants by *M. anisopliae* was by Batta (2013), who also demonstrated the endophytic effect of the fungus against *P. xylostella* larvae feeding on the *B. napus* plants. He also compared the dynamics of DBM larval mortality caused by the fungus either by topical spraying on the leaves, or via endophytic activity. The results indicated that the direct effect of the fungus, inoculated by spraying, initially caused larval mortality of 87%, which then decreased to zero 3 weeks after the spraying. In contrast, the endophytic effect of the fungus caused no mortality immediately after inoculation but the DBM larval mortality increased to 63% after 4 weeks post-inoculation. No information was given on the cause of larval mortality, e.g. whether mycosis was involved or not. The results of Batta (2013), however, are encouraging as they clearly show that the strain of *M. anisopliae* used in the experiment was effective against *P. xylostella* larvae and successfully colonized the tissues of *B. napus* plants. It thus appears that at least some strains of EPF can be used as an effective tool in OSB pest management via the endophytic method of administering the fungus.

Although several other studies have reported adverse effects of endophytic EPF on herbivorous insects, most have either not tested, or not observed, direct mycosis on the herbivores feeding on the plant tissues (Vidal and Jaber, 2015). Therefore, the mode of action of endophytic EPF in most of the studies remains obscure.

Vidal and Jaber (2015) considered that probably the most effective option for utilizing the ability of EPF to colonize plants endophytically for crop protection would be to inoculate host plants at the start of the germination of seeds, either by producing seeds already containing these fungi, or by coating the seeds with fungal conidia (Quesada-Moraga *et al.*, 2006; Biswas *et al.*, 2013).

25.7 Augmentation and Ecosystem Management

Agricultural soils throughout Europe have been shown to be almost deserts with respect to the occurrence of insect pathogens capable of infecting the most important OSB pests (Zec-Vojinovic *et al.*, 2006). In contrast, soils in natural ecosystems such as forests typically harbour these pathogens in quantities several orders of magnitude higher than those in the agricultural fields (Fig. 25.2) (Vänninen *et al.*, 1989). As EPF easily survive natural abiotic conditions in the field, it appears that their absence in field soils is caused by some current agricultural practices that are not compatible with insect pathogens. Soil-dwelling entomopathogens probably suffer from multiple stresses, such as frequent exposure to strong UV light when fields are without vegetation cover, dilution in soil via inversion tillage, lack

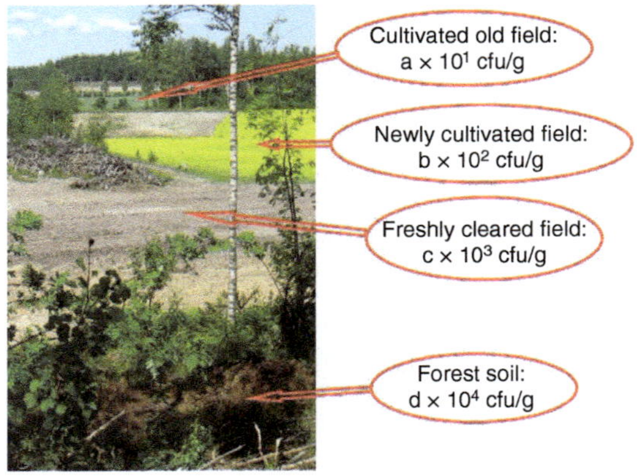

Fig. 25.2. Illustration of the decline in the abundance of entomopathogenic fungi (per gram of soil) in the continuum from natural habitat (forest), to newly cleared forest, to a new field having been cultivated for a few years only, to an old agricultural field. At each step, approximately one order of magnitude in the abundance of EPF colony-forming units (cfu) is lost. Location: Lohja, Southern Finland. Photo and illustration by H. Hokkanen.

of suitable host insects over long periods of time during the rotation and the use of suppressive agrochemicals (e.g. fungicides, herbicides, mineral fertilizers). A specific case is *N. meligethi*, the intracellular obligate parasite of the pollen beetle *M. aeneus*, which occurs at endemic levels outside the OSB growing areas but is virtually absent from areas where the crop is grown (Lipa and Hokkanen, 1992; Hokkanen and Lipa, 1995). This incompatibility pattern is a likely result of frequent insecticide treatments, which probably kill diseased (weakened) individuals more effectively than healthy ones, thus practically curing the pollen beetle populations from the *Nosema* disease.

In order to restore the natural capacity of ecosystems to buffer against pest outbreaks, agricultural soils need to be managed so that the key population-regulating factors, including EPF, have a possibility to inhabit the fields so that they will be suppressive to the main pests. To achieve that, it is necessary to reintroduce (inoculate) the core components of the EPF guild into the fields and to conserve the selected pathogens via insightful management. The conservation biological control approach, following inoculative releases of EPF, involves designing optimum crop rotations that favour the activity of EPF, such as via the use of white clover as undercrop (Fig. 25.3) in rapeseed and in the following crops to support alternative hosts such as *Sitona* weevils for the EPF. The fields can easily be (re-)colonized with these fungi in order to add a significant new natural mortality factor for many key pests of OSB. The technology to do this exists and is used, for example, in Switzerland to colonize meadows with *Beauveria brongniartii* to control *Melolontha melolontha* (Enkerli *et al.*, 2004). EPF-colonized barley kernels are drilled into the soil, where the fungi still grow and sporulate as saprophytes on the barley, ready to attack suitable insects as they pupate into the soil. Such inoculations could be done every time OSB is sown, to ensure the suppression of these pests on an area-wide basis in a sustainable manner. After reintroduction, conservation biological control methods should be used to enhance the possibility of insect pathogens remaining effective in the system for a long time. Such methods include the avoidance of inversion tillage, keeping continuous vegetation cover in the field, growing continuously at least some plants that harbour alternative hosts for soil-dwelling EPF and EPN (e.g. clover undersown in OSB and in following crops) and minimizing (or completely abolishing) pesticide use.

A major obstacle to these developments in Europe is that currently no suitable EPF are commercially available for augmentative use in OSB

Fig. 25.3. White clover undersown in spring turnip rape helps to support insect pathogens by providing alternative hosts also after rapeseed harvest; in addition it helps to control weeds and to fix nitrogen, and can be grown in the following year as a cover crop (or honey crop), serving also for conservation of parasitoids of oilseed *Brassica* crop pests (no tillage). Photo: H. Hokkanen.

and the EPF have to go through a stringent registration procedure before such use. This approach might be more easily utilized on other continents, for example in North America, where suitable EPF are available.

25.8 Entomopathogenic Fungi in Integrated Pest Management

When considering the possible role of EPF in integrated pest management, at least two questions need to be addressed: (i) compatibility of EPF with the use of chemical pesticides (and agricultural practices in general); and (ii) the safety of EPF to non-target organisms.

The standardized testing of the side effects of pesticides on natural enemies has included the effects of chemical pesticides on *B. bassiana* and *M. anisopliae* since 1992 (Hokkanen and Kotiluoto, 1992) and results have been published by the IOBC-WPRS (International Organisation for Biological and Integrated Control – West Palaearctic Regional Section) working group in subsequent articles (e.g. Sterk *et al.*, 1999). Vänninen and Hokkanen (1988) had already tested the effects of 13 different pesticides commonly used in OSB growing and in the following crops in rotation in Finland, on four species of EPF (*B. bassiana, M. anisopliae, Paecilomyces farinosus, Paecilomyces fumoso-roseus*) *in vitro*. Insecticides in general had only a slight negative impact, while (as could be expected) practically all fungicides were detrimental to EPF growth and sporulation. Surprisingly, however, many herbicides also strongly inhibited the growth and sporulation of all EPF. In the context of OSB growing, particular attention has to be paid to the most commonly used pre-emergence (soil incorporated) herbicide in OSB, trifluralin, which strongly inhibited EPF growth. Its effect was third strongest among the 13 tested pesticides (stronger than that of several fungicides), which could almost alone partly explain the observed disappearance of EPF in the field soil of agricultural land.

Some compatibility of EPF with chemical pesticides in OSB pest management was reported by Duarte *et al.* (2016), who suggested that the active ingredients thiamethoxam and azadirachtin were compatible with the EPF in their study in the management of *P. xylostella*. This finding is in line with those of Vänninen and Hokkanen (1988), as insecticides in general are the least harmful pesticides to EPF.

Extensive studies concerning possible non-target effects of using EPF in the OSB ecosystem were conducted in the European Union project ERBIC ('Evaluating Environmental Risks of Biological Control Introductions into Europe') and did not reveal any significant impacts on natural enemies such as carabid and staphylinid beetles, coccinellids, hymenopterous parasitoids or non-target Diptera (Husberg and Hokkanen, 2001; H.M.T. Hokkanen *et al*., unpublished); therefore, these agents are expected to be fully compatible with other biocontrol agents.

Many non-selective control methods, however, severely suppress the action of EPF (e.g. solarization, pesticide use, thermal treatments) and need to be taken into account when designing OSB pest management systems incorporating the use of entomopathogenic fungi.

25.9 Outlook for Using Entomopathogenic Fungi in Oilseed *Brassica* Pest Management

Several effective, reliable ways of utilizing entomopathogenic fungi in OSB pest management have been developed over the past decades and new exciting approaches are currently being studied (e.g. endophytic colonization of crop plants). To our knowledge, however, none of these methods have been taken up by growers, who continue to rely on chemical pesticides for pest control. The lack of uptake is likely due to simple socio-economic factors, such as convenience, training, tradition or lack of trust. The most important factor might be that there is no market push by commercial companies for pest control products based on EPF, so farmers actually have no access to the methods reviewed in this chapter. Agrochemical companies have no interest in promoting biocontrol agents such as EPF and indeed sometimes actively block the market entry of such products. For example, a highly efficient *Metarhizium anisopliae*-based product was developed by a research institute in Europe in the 1980s but the rights to it were bought by an agrochemical company, which never brought it to the market (H.M.T. Hokkanen, personal information).

Probably the easiest and most effective way of incorporating EPF into OSB pest management would be via recolonization of agricultural fields with selected EPF (augmentation), followed by cropping techniques that maintain naturally high and effective inoculum levels for pest management. It would be worth researching whether EPF in such soils would also colonize the crop plants endophytically on their own, or whether additional seed treatment would be needed to achieve an increased resistance to pests. Another, compatible approach deserving further development would be using trap cropping associated with entomovectoring of EPF, to knock out pests before they reach the main crop.

Acknowledgements

Our research on EPF has been supported in part by the EU projects ERBIC ('Evaluating Environmental Risks of Biological Control Introductions into Europe', FAIR5-CT3489), MASTER ('Integrated pest management strategies incorporating bio-control for European oilseed rape pests', QLK5-CT-2001-01447) and REBECA ('Regulation of Biological Control Agents', FP6-SSPE-CT-2005-022709); NordForsk project 70066 EVETECH ('Entomovector technology for sustainable plant protection') and NordForsk project 45941 BICOPOLL-NET ('Nordic Research Network for Biocontrol and Pollination'); and EU-ERA-NET CORE-ORGANIC 2 project 155 BICOPOLL ('Targeted precision biocontrol and pollination enhancement in organic cropping systems').

References

Batta, Y.A. (2013) Efficacy of endophytic and applied *Metarhizium anisopliae* (Metch.) Sorokin (Ascomycota: Hypocreales) against larvae of *Plutella xylostella* L. (Yponomeutidae: Lepidoptera) infesting *Brassica napus* plants. *Crop Protection* 44, 128–134. doi: 10.1016/j.cropro.2012.11.001.

Baverstock, J., Roy, H.E. and Pell, J.K. (2010) Entomopathogenic fungi and insect behaviour: from unsuspecting hosts to targeted vectors. *BioControl* 55, 89–102. doi: 10.1007/s10526-009-9238-5.

Biswas, C., Dey, P., Satpathy, S., Satya, P. and Mahapatra, B.S. (2013) Endophytic colonization of white jute (*Corchorus capsularis*) plants by different *Beauveria bassiana* strains for managing stem weevil (*Apion corchori*). *Phytoparasitica* 41, 17–21.

Butt, T.M. and Copping, L.G. (2000) Fungal biological control agents. *Pesticide Outlook* 11, 186–191. doi: 10.1039/B008009H.

Butt, T.M., Ibrahim, L., Ball, B.V. and Clark, S.J. (1994) Pathogenicity of the entomogenous fungi *Metarhizium anisopliae* and *Beauveria bassiana* against crucifer pests and the honey bee. *Biocontrol Science & Technology* 4, 207–214.

Butt, T.M., Carreck, N.L., Ibrahim, L. and Williams, I.H. (1998) Honey-bee-mediated infection of pollen beetle (*Meligethes aeneus* Fab.) by the insect-pathogenic fungus, *Metarhizium anisopliae*. *Biocontrol Science and Technology* 8, 533–538. doi: 10.1080/09583159830045.

Carreck, N.L., Butt, T.M., Clark, S.J., Ibrahim, L., Isger, E.A., Pell, J.K. and Williams, I.H. (2007) Honey bees can disseminate a microbial control agent to more than one inflorescence pest of oilseed rape. *Biocontrol Science and Technology* 17, 179–191. doi: 10.1080/09583150600937485.

Duarte, R.T., Gonçalves, K.C., Espinosa, D.J.L., Moreira, L.F., De Bortoli, S.A., Humber, R.A. and Polanczyk, R.A. (2016) Potential of entomopathogenic fungi as biological control agents of diamondback moth (Lepidoptera: Plutellidae) and compatibility with chemical insecticides. *Journal of Economic Entomology* 109, 594–601. doi: 10.1093/jee/tow008.

Enkerli, J., Widmer, F. and Keller, S. (2004) Long-term field persistence of *Beauveria brongniartii* strains applied as biocontrol agents against European cockchafer larvae in Switzerland. *Biological Control* 29, 115–123.

Furlong, M.J. and Pell, J.K. (2001) Horizontal transmission of entomopathogenic fungi by the diamondback moth. *Biological Control* 22, 288–299.

Furlong, M.J., Pell, J.K., Ong, P.C. and Syed, A.R. (1995) Field and laboratory evaluation of a sex pheromone trap for the autodissemination of the fungal entomopathogen *Zoophthora radicans* by the diamondback moth, *Plutella xylostella*. *Bulletin of Entomological Research* 85, 331–337.

Furlong, M.J., Pell, J.K. and Reddy, G.V.P. (1997) Premortality effects of *Zoophthora radicans* infection in *Plutella xylostella*. *Journal of Invertebrate Pathology* 70, 214–220.

Hokkanen, H.M.T. (1991) Trap cropping in pest management. *Annual Review of Entomology* 36, 119–138. doi: 10.1146/annurev.en.36.010191.001003.

Hokkanen, H.M.T. (1993) Overwintering survival and spring emergence in *Meligethes aeneus*: effects of body weight, crowding, and soil treatment with *Beauveria bassiana*. *Entomologia Experimentalis et Applicata* 67, 241–246. doi: 10.1111/j.1570-7458.1993.tb01674.x.

Hokkanen, H.M.T. (2000) The making of a pest: recruitment of *Meligethes aeneus* onto oilseed brassicas. *Entomologia Experimentalis et Applicata* 95, 141–149. doi: 10.1046/j.1570-7458.2000.00652.x.

Hokkanen, H.M.T. and Kotiluoto, R. (1992) Bioassay of the side-effects of pesticides on *Beauveria bassiana* and *Metarhizium anisopliae*: standardized sequential testing procedure. *IOBC/WPRS Bulletin* XV(3), 148–151.

Hokkanen, H.M.T. and Lipa, J.J. (1995) Occurrence and dynamics of *Nosema meligethi* (Microsporida) in populations of *Meligethes aeneus* (Coleopera, Nitidulidae) in Finland. *Entomologica Fennica* 6, 11–18.

Hokkanen, H.M.T. and Menzler-Hokkanen, I. (2007) Use of honeybees in the biological control of plant diseases. *Entomological Research* 37(Suppl. 1), A62–A63. doi: 10.1111/j.1748-5967.2007.00082.x.

Hokkanen, H.M.T., Menzler-Hokkanen, I. and Butt, T.M. (2003a) Pathogens of oilseed rape pests. In: Alford, D.V. (ed.) *Biocontrol of Oilseed Rape Pests*. Blackwell Science, Oxford, UK, pp. 299–322.

Hokkanen, H.M.T., Zeng, Q.-Q. and Menzler-Hokkanen, I. (2003b) Assessing the impacts of *Metarhizium* and *Beauveria* on bumblebees. In: Hokkanen, H.M.T. and Hajek, A.E. (eds) *Environmental Impacts of Microbial Insecticides*. Kluwer Academic Publishers, Dordrecht, The Netherlands, pp. 63–72.

Hokkanen, H.M.T., Menzler-Hokkanen, I. and Lahdenperä, M.-L. (2015) Managing bees for delivering biological control agents and improved pollination in berry and fruit cultivation. *Sustainable Agriculture Research* 4, 89–102. doi: 10.5539/sar.v4n3p89.

Husberg, G.-B. and Hokkanen, H.M.T. (2001) Efects of *Metarhizium anisopliae* on the pollen beetle *Meligethes aeneus* and its parasitoids *Phradis morionellus* and *Diospilus capito*. *BioControl* 46, 261–273. doi: 10.1023/A:1011479616787.

Kepler, R.M., Humber, R.A., Bischoff, J.F. and Rehner, S.A. (2014) Clarification of generic and species boundaries for *Metarhizium* and related fungi through multigene phylogenetics. *Mycologia* 106, 811–829. doi: 10.3852/13-319.

Lipa, J.J. and Hokkanen, H.M.T. (1992) *Nosema meligethi* I. & R. (Microsporida) in populations of *Meligethes* spp. in Europe. *Biocontrol Science and Technology* 2, 119–125.

Mauchline, A.L., Cook, S.M., Powell, W. and Osborne, J.L. (2013) Effects of non-host plant odour on *Meligethes aeneus* during immigration to oilseed rape. *Entomologia Experimentalis et Applicata* 146, 313–320. doi: 10.1111/eea.12030.

Meadow, R., Vandberg, J.D. and Shelton, A.M. (2000) Exchange of inoculum of *Beauveria bassiana* (Bals.) Vuill. (Hyphomycetes) between adult flies of the cabbage maggot *Delia radicum* L. (Diptera: Anthomyiidae). *Biocontrol Science and Technology* 10, 479–485. doi: 10.1080/09583150050115061.

Menzler-Hokkanen, I. (2016) Safety of Prestop Mix to pollinators: a critical review of its properties and use in entomovectoring. *Journal of Pest Science* 89, 179–182. doi: 10.1007/s10340-015-0703-x.

Menzler-Hokkanen, I., Hokkanen, H., Büchs, W., Klukowski, Z., Luik, A., Nilsson, C., Ulber, B. and Williams, I.H. (2006) Insect problems in European oilseed rape cultivation, and how to deal with them: the OSB farmers' perspective. *IOBC/WPRS Bulletin* 29(7), 91–94.

Mommaerts, V. and Smagghe, G. (2011) Entomovectoring in plant protection. *Arthropod–Plant Interactions* 5, 81–95. doi: 10.1007/s11829-011-9123-x.

Pell, J.K.,Wilding, N., Player, A.L. and Clark, S.J. (1993) Selection of an isolate of *Zoophthora radicans* (Zygomycetes: Entomophthorales) for biocontrol of the diamondback moth *Plutella xylostella* L (Lepidoptera: Yponomeutidae). *Journal of Invertebrate Pathology* 61, 75–80.

Quesada-Moraga, E., Landa, B.B., Muñoz-Ledesma, J., Jiménez-Diáz, R.M. and Santiago-Álvarez, C. (2006) Endophytic colonization of opium poppy, *Papaver somniferum*, by an entomopathogenic *Beauveria bassiana* strain. *Mycopathologia* 161, 323–329.

Reddy, G.V.P., Furlong, M.J., Pell, J.K. and Poppy, G.M. (1998) *Zoophthora radicans* infection inhibits the response to and production of sex pheromone in the diamondback moth. *Journal of Invertebrate Pathology* 72, 167–169.

Reddy, G.V., Tangtrakulwanich, K., Miller, J.H., Ophus, V.L. and Prewett, J. (2014) Sustainable management tactics for control of *Phyllotreta cruciferae* (Coleoptera: Chrysomelidae) on canola in Montana. *Journal of Economic Entomology* 107(2), 661–666. doi: 10.1603/EC13503.

Roberts, D.W. and St Leger, R.J. (2004) *Metarhizium* spp., cosmopolitan insect–pathogenic fungi: mycological aspects. *Advances in Applied Microbiology* 54, 1–70. doi: 10.1016/S0065-2164(04)54001-7.

Sterk, G., Hassan, S.A., Baillod, M. and 29 others (1999) Results of the seventh joint pesticide testing programme carried out by the IOBC/WPRS-Working group 'Pesticides and Beneficial Organisms'. *BioControl* 44, 99–117. doi: 10.1023/A:1009959009802.

St Leger, R.J., Allee, L.L., May, B., Staples, R.C. and Roberts, D.W. (1992) World-wide distribution of genetic variation among isolates of *Beauveria* spp. *Mycological Research* 96, 1007–1015. doi: 10.1016/S0953-7562(09)80108-1.

Ujjan, A.A. and Shahzad, S. (2012) Use of entomopathogenic fungi for the control of mustard aphid (*Lipaphis erysimi*) on canola (*Brassica napus* L.). *Pakistan Journal of Botany* 44, 2081–2086.

Vänninen, I. (1996) Distribution and occurrence of four entomopathogenic fungi in Finland: effect of geographical location, habitat type and soil type. *Mycological Research* 100, 93–101.

Vänninen, I. and Hokkanen, H. (1988) Effect of pesticides on four species of entomopathogenic fungi in vitro. *Annales Agriculturae Fenniae* 27, 345–353.

Vänninen, I., Husberg, G.-B. and Hokkanen, H. (1989) Occurrence of entomopathogenic fungi and entomophilic nematodes in cultivated soils in Finland. *Acta Entomologica Fennica* 53, 65–71.

Vega, F.E., Dowd, P.F., Lacey, L.A., Pell, J.K., Jackson, D.M. and Klein, M.G. (2000) Dissemination of beneficial microbial agents by insects. In: Lacey, L.A. and Kaya, H.K. (eds) *Field Manual of Techniques in Invertebrate Pathology*. Kluwer, London, pp. 153–177.

Vickers, R.A., Furlong, M.J., White, A. and Pell, J.K. (2004) Initiation of fungal epizootics in diamondback moth populations within a large field cage: proof of concept for auto-dissemination. *Entomologia Experimentalis et Applicata* 111, 7–17.

Vidal, S. and Jaber, L.R. (2015) Entomopathogenic fungi as endophytes: plant–endophyte–herbivore interactions and prospects for use in biological control. *Current Science* 109, 46–54.

Zec-Vojinovic, M., Hokkanen, H.M.T., Büchs, W., Klukowski, K., Luik, A., Nilsson, C., Ulber, B. and Williams, I.H. (2006) Natural occurrence of pathogens of oilseed rape pests in agricultural fields in Europe. In: Alford, D.V. (ed.) *Proceedings of the International Symposium 'Integrated Pest Management of Oilseed Rape Pests'*, Paulinerkirche, Göttingen, Germany, 3–5 April 2006.

Index

Page numbers in **bold** type refer to figures and tables.